P. Debye

THE COLLECTED PAPERS OF
PETER J. W. DEBYE

OX BOW PRESS
Woodbridge, Connecticut

1988 reprint published by:

OX BOW PRESS
P.O. Box 4045
Woodbridge, Connecticut 06525

Library of Congress Cataloging-in-Publication Data

Debye, Peter J. W. (Peter Josef William), 1884-1966
[Prose works. Selections]
The collected papers of Peter J. W. Debye.
p. cm.
Reprint. Originally published: New York : Interscience Publishers, 1954.

ISBN 0-918024-58-7

1. Physics -- Collected works. I. Title II. Title: Collected papers.

QC3. D42 1988 87-36895
530--dc19 CIP

Printed in the United States of America

CONTENTS

CONTENTS

DIPOLE MOMENTS

ELECTROLYTES

LIGHT SCATTERING

CONTENTS

MISCELLANEOUS

CONTENTS

PETER J. W. DEBYE

The Debye saga begins on March 24, 1884, at Maastricht in the Netherlands with the birth to William and Maria Reumkens Debije of their son Peter Josef William. Debye's education began in the Hoogere Burger School of Maastricht, where he received his first formal training in mathematics and science. He entered the University of Aachen at the age of seventeen, matriculating in the field of electrical engineering. He received his diploma in 1905; in the meantime his interest in chemistry and physics began to develop. The latter tendency was encouraged both along experimental lines by Max Wien, who was then at Aachen, as well as into the theoretical realm by Sommerfeld, who was also there. This unusual combination of practical engineering with experimental and theoretical physics from two of the world's masters was undoubtedly decisive in determining the career the young genius was to follow.

His first original work was a theoretical treatment of Foucault currents in a rectangular conductor (Zeitschrift fur Mathematik und Physik, *54*, 418-437 (1907)); a mathematically elegant solution was obtained by means of Green's theorem. Other early work at Aachen included a study of diffraction of light by cylindrical and spherical bodies. Debye was invited to accompany Sommerfeld to Munich when the latter was called to the chair of theoretical physics in 1906. He presented his doctoral thesis there on July 1, 1908; it dealt with the pressure of radiation on spheres of arbitrary electrical properties. It contains an excellent example of Debye's mathematical skill in the extended treatment of Hankel functions in which the variable and the index may increase to infinity. Mathematical tools were not available for the problem, so Debye created them. He became Privat-Dozent in 1910 at Munich. This year also saw the publication with Hondros (Annalen der Physik, *32*, 465-467 (1910)) of a paper which as far as content goes belongs in today's literature of radar and waveguides.

In 1911, Einstein was invited to Prague. To fill his place, the University of Zurich offered the twenty-seven-year-old Debye a full professorship in theoretical physics. He accepted and stayed a year, during which association with Kleiner and Werner freshened his interest in experimental physics and in chemistry. During this year, two of Debye's most fundamental theories were developed: the theory of polar molecules, and the theory of specific heats of solids. Debye then returned to his homeland, as Professor of Theoretical Physics at

Utrecht. He especially wanted to see some experimental work on dielectric constants started, in order to test his ideas about permanent dipoles, and in 1913 moved to Göttingen where experimental facilities were more abundant than at Utrecht. Here Debye settled down for seven years. Despite the research difficulties which were the inevitable consequence of World War I, this period produced the now indispensable powder method for solids which was worked out in collaboration with Scherrer. Debye's title at Göttingen was Professor of Theoretical and Experimental Physics. It proved to be prophetic in associating his name with both theory and experiment because, in 1920, he returned to Zurich as Professor of Experimental Physics and Director of the physics laboratory at Eidgenössische Technische Hochschule. Scherrer also went to Zurich at the same time, so the x-ray school they founded continued to flourish. By now, students from all over the world were crowding Debye's laboratory. Like his own teacher Sommerfeld, Debye was embarked on a career of making scientists as well as science. The period at Zurich is marked by two more Debye triumphs: the interionic attraction theory and the concept of magnetic cooling.

The University of Leipzig was the next to benefit from a Debye sojourn. Debye went there in 1927 as Director of the Physical Institute. Here work on determination of interatomic distance in molecules by x-ray scattering in gases, of dipole moments, and of properties of electrolytes was actively continued. Leipzig now became the Mecca of the world's physicists and especially of the physical chemists, who were eager to learn directly from the man who had solved so many of their fundamental problems. An invitation in 1934 from Berlin appeared to offer greater opportunities for research. In that year, Debye left Leipzig for Berlin, to accept the joint posts of Professor at the University of Berlin and Director at the Kaiser Wilhelm Institute in Berlin-Dahlem. The latter laboratory, built with funds made available by the Rockefeller Foundation, was later called the Max Planck Institute. It was one of the most magnificently equipped laboratories the world has ever known. It included high voltage (three megavolts) apparatus, a cryogenic laboratory with facilities for experimenting with magnetic cooling, a complete chemical laboratory, an independent library, and all the shops one could desire. During the Berlin period, Debye's outstanding work received formal international recognition through the award to him in 1936 of the Nobel Prize in Chemistry. The versatility of his genius is fittingly acknowledged by this award in chemistry to a man whose official title was in physics; indeed, of all contemporary scientists, Debye is the closest to the natural philosopher of classical tradition one for whom the artificial boundaries of various -ologies have no significance.

The 1930's also saw the sinister rise of dictatorship and the police state in Germany. Politics began to interfere with Debye's research at the very beginning of the Berlin period. A short time after the laboratory had started to function, the state wanted Debye to accept German citizenship with the Berlin appointments. Debye rebelled: aside from his natural patriotism to the Netherlands, the whole philosophy of the Nazi regime was abhorrent to him. The story is best told in his own words:

"At the time I accepted to go to Berlin I was still a Dutch citizen. According to the Dutch law a Dutch citizen can accept state positions in another state, without losing his citizenship, only with permission of the queen. I applied for this permission. It was granted. At the same time the German government conceded in a letter signed by the Minister of Education, Dr. Rust, that in accepting the positions offered, I did not become and would not be asked to become a German citizen. The positions, as usual, were lifetime positions. During the time the laboratory was still under construction, I received an offer from Harvard University, which I declined because I did not feel free to quit before having finished what I had undertaken to do. . . The war broke out and one day without previous warning I was informed by Dr. Telschow from the Kaiser-Wilhelm Gesellschaft that I could no longer enter the laboratory except by becoming a German citizen. I refused. I was advised by the Ministry of Education to stay at home and occupy myself by writing a book. Instead, I was able to overcome the difficulties put in my way by different German authorities and to leave for the U.S.A. by way of Italy in order to give the Baker Lectures at Cornell University, to which I had been invited."

Europe thus lost one of her foremost scientists. Cornell University and the United States became Debye's home. Cornell appointed him Professor of Chemistry in 1940 and Head of the Department of Chemistry, and named him Todd Professor of Chemistry in 1948. In 1946, Debye became a citizen of the United States. At Cornell, he quickly became the central figure in a lively research group. Again he entered a new field, and as usual, brought to it new ideas and techniques. This time it was macromolecules and colloid chemistry which caught his attention. During the war years, Debye was active in the synthetic rubber program and rendered invaluable service to his adopted country as a consultant. He proposed the now standard absolute method of determining molecular weights of polymers by their light scattering ability and, together with his son Peter P. Debye, developed the method to a practically usable stage. His micelle theory was also worked out during this time, as well as the technique of thermal diffusion. In 1950, he became Professor Emeritus. This promotion simply meant that he was relieved of the chores of administration and was free to devote all his time and efforts to research. The latest Debye contribution is an electrical method of determining sedimentation constants of macromolecules.

Learned societies and universities the world over have acknowledged the debt science owes Debye by awarding to him medals and honorary degrees and membership. The list includes the Rumford Medal of the Royal Society of London (1930), the Lorentz Medal of the Royal Dutch Academy at Amsterdam (1935), the Nobel Prize in Chemistry (1936), the Franklin Medal of the Franklin Institute (1937), the Willard Gibbs Medal of the American Chemical Society (1949), the Max Planck Medal of the West German Physical Society (1950). He has received the doctor's degree, *honoris causa*, from the University of Liège, the University of Brussels, Oxford, the University of Sofia, Harvard, the Polytechnic Institute of Brooklyn, and St. Lawrence University. The following academies have elected him to membership: American Academy (Boston), Franklin Institute (Philadelphia), National Academy (Washington, D. C.), New York Academy (New York), Royal Dutch Academy (Amsterdam), Royal Society (London), Royal Institution of Great Britain (London), Royal Danish Academy (Copenhagen), Academies of Berlin, Göttingen, and Munich (Germany), Academies of Brussels and Liège (Belgium), Royal Irish Academy (Dublin), Papal Academy (Rome), Indian Academy (Bangalore), National Institute of Science (India), and the Real Sociedad Española de Física y Química (Madrid).

Among the many honors which Debye has received, the one which gave him the greatest pleasure was probably the ceremony in the town of his birth, where in 1939 Maastricht unveiled his statue in the City Hall.

Debye is known to his colleagues through his published work, but better by his active participation in scientific meetings, and by the lectures he delivers on various occasions. When it is known that Debye will be at a meeting, it is invariably the cue for everyone who can to get there because we have learned that worthwhile things will be said. And to hear a Debye lecture is a real treat: he has an uncanny skill in presenting seemingly complicated subjects in a fashion which gets to the nub of the problem with a penetrating clarity. Best of all, though, Debye is known as one to whom we can go for research advice. As described in the citation from Harvard when he was made an honorary alumnus, Debye is "a large-hearted physicist who gladly lends to the chemist a helping hand." To some, science is a cold business of logic and symbols and experiment; to others, it is a creative art. Debye is the Leonardo of today's science.

RAYMOND M. FUOSS

THE COLLECTED PAPERS: A SURVEY

This selection of the classical papers of Professor Debye is being presented to him by his pupils, friends, and the publishers on the occasion of his **seventieth** birthday on March 24, 1954.

The selection was made by Professor Debye himself. The papers originally published in German and Dutch have been translated into English. Papers published originally in the English language have been reproduced directly from the original journals.

We wish to take this opportunity to thank Professors I. Fankuchen, R. M. Fuoss, H. Mark, C. P. Smyth, and H. S. Sack for editorial advice and assistance, and the holders of the copyrights for the original papers for permission to reproduce from their publications.

The papers presented here have been of fundamental importance for many different branches of physics and physical chemistry and constitute important sources for any detailed studies in these particular fields. They are grouped in the following main sections: X-Ray Scattering, Electrolytes, Dipole Moments, Light Scattering, and Miscellaneous.

To place the papers in the proper context, and to indicate some of the later developments of the lines of research they initiated, we present in the following pages a brief survey of the papers from the first four sections.

Papers on X-Ray Scattering

Debye's contributions to the physics of x-rays are numerous and eminent. Very soon (1913) after the discovery of x-ray diffraction on crystal lattices by von Laue and W. H. and W. L. Bragg (1912), he studied the influence of temperature on this phenomenon and in a straightforward and convincing manner computed a factor which expresses the intensity decrease of the individual diffraction spots as a function of wavelength, diffraction angle, and absolute temperature; it is known and has been used since as the "Debye factor."

A few years later (1916), Debye and Scherrer started their famous investigations on the scattering of x-rays from randomly distributed small crystalline particles and paved the way for a practical and rational study of crystal powders, polycrystalline metals, and colloidal systems. The "Debye-Scherrer method" became, in the hands of many investigators, the most powerful tool for establishing the structure of crystals with high symmetry.

After only two years (1918), Debye and Scherrer tackled the problem of elucidating the electron distribution *inside* the individual atom by a quantitative analysis of the diffracted intensities, and introduced the "atomic form factor" in the evaluation of x-ray patterns. The fundamental idea and the mathematical approach of this study proved to be extremely fruitful and was resumed by Debye again (1929-1930) when he applied, with several co-workers, x-ray diffraction to the exploration of the structure of isolated molecules in gases and of randomly packed atoms and molecules in liquids. His investigations stimulated somewhat later the application of electron diffraction to the structural analysis of single molecules in the gaseous state, a method which yielded a wealth of important information on interatomic distances in the hands of Pauling and his co-workers.

During the period of interest focused on the classical wave theory of scattering of radiation by matter (1913-1930), Debye did not neglect the light quantum approach and gave (1923), independently and only a little later than A. H. Compton himself, a quantitative theory of the Compton effect which was one of the key experiments of that period and elucidated in striking way the dualism of the wave and particle theory of light.

HERMAN MARK

Papers on Molecular Dipole Moment and Polarizability

The first of this group of papers laid the foundation for the modern molecular theory of dielectrics by proposing that dielectrics contained not only elastically bound electrons, as previously assumed, but also permanent dipoles of constant electric moment. This explained the temperature dependence of the dielectric constant, which had not been previously understood, and made possible the calculation of molecular dipole moment values of a reasonable magnitude. In the following year, the concept of a permanent electric dipole in the molecule was employed to explain the anomalous dispersion of long electric waves in terms of the molecular radius and internal friction of the material. As in the first paper, the calculations gave results of a reasonable magnitude. The equations for anomalous dielectric dispersion were developed into a relation between dielectric constant and loss, identical in form with one developed without reference to molecular mechanism. However, the molecular mechanism embodied in the Debye relation has been widely employed in the discussion and interpretation of loss measurements, increasing in importance as technical developments have increased the frequency range in which loss is observed.

Two papers published shortly after the First World War treated intermolecular attractive forces in general as of purely electrical origin and considered molecular polarizability, dipole moment, and quadrupole moment as their cause. With molecules regarded as electric quadrupoles characterized by electric moments of inertia, reasonable values were calculated for the mean electric moment of inertia from the van der Waals constant *a* and related to the surface tension and the Sutherland constant in the expression for the viscosity of a gas. It was concluded that the mean electric moment of inertia could be obtained from the empirical reduced equation of state of a gas better than from the van der Waals *a*.

A few measurements indicated the formal correctness of the relation between dielectric constant and molecular dipole moment, but applications of the old quantum theory to the dielectric problem for the limited case of "dumb-bell" molecules gave equations similar in form but with coefficients differing from those obtained by Debye by means of classical mechanics. From the point of view of quantum theory, Debye examined the Zeeman and Stark effects in relation to the Curie law for the temperature dependence of magnetic susceptibility and the analogous relation for dielectric susceptibility which he had derived. At about the same time, other investigators applied the new quantum mechanics to the "dumb-bell" molecules and obtained essentially the same equation as that derived by Debye in terms of classical theory. Debye and Manneback applied the wave mechanics to the rigid polyatomic molecule having two equal moments of inertia to obtain an expression for the dielectric constant of a perfect gas consisting of such molecules. At high temperatures, this was identical with the classical expression and it was concluded that, at usual temperatures, the departure from this law could not exceed a few per cent. In a short time, the quantum mechanical treatment was completely generalized to give an equation identical with the classical Debye equation except for a small term containing the molecular moments of inertia, which was negligible for all but very small molecules at moderate or low temperatures.

Even before the removal of all doubt as to the value of the factor in the dipole moment term in the relation between dielectric constant and molecular dipole moment, dielectric constant mesurements were employed to determine dipole moments for use in molecular structure studies. The dipole moment method was soon recognized as a powerful tool in structure determination and employed in hundreds of investigations. The continued development of the method is shown by the fact that, thirty-nine years after the first dipole moment paper, Debye and Bueche examined the relation between the dipole moments of polymer molecules and hindered rotation around the bonds in the molecules, the moment of a molecular chain depending on the preponderance of positions of parallelism or of antiparallelism of the elementary electric moments.

CHARLES P. SMYTH

Papers on Electrolytes

The section on electrolytes opens, quite naturally, with the now classical 1923 paper "Zur Theorie der Elektrolyte" by Debye and Hückel. The historical introduction to this paper summarizes the bewildering contradictions which haunted the physical chemist of a generation ago; re-reading it today, when complete dissociation of strong electrolytes in solvents of high dielectric constant is taken for granted, we realize how completely the physical chemistry of electrolytes is founded on the ideas presented in this single paper. A few pioneers, notably Sutherland (1907), Bjerrum (1909), and Milner (1912), had had the intuition to see that in the long range of the Coulomb field lay the explanation of all of the properties of electrolytic solutions which the Ostwald school of necessity was forced to classify as anomalous, but they were unable to formulate their ideas in simple mathematical language. In one brilliant step, the combination of the Poisson and the Boltzmann equations, Debye solved the problem of handling slowly convergent sums of interionic potential energies by eliminating them and replacing the point set by a continuum, thereby reducing the problem to the solution of a differential equation. This equation is nonlinear, but for many cases of practical interest (low concentrations in solvents of high dielectric constant) reduces to a linear equation in the coefficient of which the now familiar square root limiting laws are already implicit. The paper contains the development of the fundamental thermodynamics of electrolytic solutions; furthermore, the key to the solution of the conductance problem is announced.

Much of the earlier confusion in the field of conductance had arisen from the existence of two distinct (limiting) types of electrolytes: the strong electrolyte, as exemplified by sodium chloride in water, and the weak electrolyte, for which acetic acid in water is characteristic. The latter class satisfied to a fair degree of approximation the classical law of mass action, and the change of conductance with concentration could be accounted for on the basis of the Arrhenius hypothesis that the relative number of ions was a function of concentration: the conductance increases on dilution because dissociation increases. But all attempts to apply these ideas to strong electrolytes led at best to quibbles about the true degree of dissociation. This question was, of course, irrelevant: conductance is, as Debye and Hückel emphasized, an irreversible process, and hydrodynamics had to be put in order before possible superimposed dissociation equilibria could be considered. The second paper (also 1923) presents the details of the conductance theory, basing the discussion on the two long-range interionic effects of electrophoretic resistance and relaxation braking. Implicit in the existence of a relaxation time for the ionic atmosphere is the existence of a dispersion of conductance at a suitable range of frequencies, as well as a dependence on field strength.

The other papers in this section are essentially elaborations and further applications of the basic ideas contained in the 1923 papers. The work with McAulay (1925) on salting-out again demonstrates the significance of ionic fields in determining the thermodynamic properties of multicomponent systems: the previously mysterious separation of organic solutes from water on the addition of salt is shown to be a consequence of the inhomogeneous field produced by the localized charge on the ions. In collaboration with Falkenhagen, the detailed theory of the dependence of conductance on frequency is presented in the two 1928 papers; these papers predict an observable effect which was later verified in Debye's and Wien's laboratories. The dependence of conductance on field strength had already been observed by Wien; Debye and Falkenhagen correlated this result with the model used in their theoretical treatment.

Since this volume is a collection of Deybe's papers, it is obvious that other work on electrolytes is not included. But it is impossible in these introductory comments to ignore the work of others, for the simple reason that much of it was inspired by Debye's original clarification of the fundamental concepts and all of it is based on the model which he proposed. The modern school of electrolyte chemists can in all propriety be described as disciples of Debye.

RAYMOND M. FUOSS

Papers on Light Scattering

Debye's contributions to polymer chemistry began with the application of Einstein's light-scattering theory to macromolecular solutions and with the demonstration that such systems can be sufficiently purified to make their results of diffraction experiments reproducible and significant. His own old work on the atomic form factor in x-ray scattering led shortly afterwards in a straightforward manner to the scattering functions of randomly coiled chain molecules. Thus his light-scattering method permits not only the absolute determination of weight-average molecular weights but also the direct measurement of the spatial extension of macromolecules in dilute solutions. As a consequence, it can be considered today as the most powerful tool for the study of the state of polymer molecules in solution and is widely used by many contributors. Once Debye's interest in the behavior of high polymers was aroused, he investigated other problems, such as those of intrinsic viscosity, sedimentation, and diffusion, and succeeded very rapidly in advancing them into a domain of better and clearer understanding. Quite recently Debye inaugurated a new method for determining the polymolecularity of macromolecular solutions, and it can only be hoped in the interest of the progress of polymer science that he will be as successful with this endeavor as with many others before.

HERMAN MARK

"The development, after the introduction of any new subject, is never presented as an example of strictly logical deduction from experimental evidence, carefully arranged post factum. Nor is it derived as a logical consequence of a mathematical formulation, appearing seemingly from nowhere at the very beginning of the argument.

"The sequence is described as it really happened, with historical truthfulness and the idea is developed from its inception concurrently with its mathematical formulation from its starting point to the end. This, it seems to me, is the way to convey to the uninitiated reader that our science is essentially an art which could not live without the occasional flash of genius in the mind of some sensitive man, who, alive to the smallest of indications, knows the truth before he has the proof."

From a preface of Professor
Debye to a recent textbook
on physical chemistry.

X-RAY SCATTERING

X-RAY INTERFERENCE AND THERMAL MOVEMENT
(Interferenz von Röntgenstrahlen und Wärmebewegung*)

P. Debye

Translated from
Annalen der Physik, Vol. 43, 1914, pages 49-95

In a few notes I have tried to present evidence[1] proving that the thermal movement of the crystal atoms has a considerable effect on the x-ray interferences discovered by Friedrich-Knipping-Laue and treated theoretically by Laue[2]. As was the case in the original theory of specific heats by Einstein, all atoms were considered independent of one another for these computations, and each was introduced in the calculation as a simple oscillating structure. The further restriction was introduced that the temperature of observation was so high that no indication of a deviation from the equipartition energy was noticeable. The summation of the effects on the incident beam due to the atomic structure, no longer rigid as assumed by Laue, resulted in the absence of an effect on the sharpness of the interference maxima and in the existence of an intensity effect. An explanation was also found for the appearance of noticeable intensities mostly in those directions that deviate only by small angles from the direction of the incident beam, and at the same time for the fact that the so-called reflection of x-rays is observed under ordinary circumstances only if the incidence is not too steep. The computation further shows that, as a consequence of the thermal movement, the interference intensity must always be accompanied by a scattered radiation which has its maximum where the interference intensity is weakest. Moreover the formulas show how the small amplitude of the thermal movement of the atoms in diamond, essentially due to its small compressibility, is the reason for the special position this substance again assumes.

*Received October 10, 1913.

Because of the limitations indicated before, the theory can be considered as only a first step toward a complete theoretical description. Until now only the anomalous behavior of diamond has been observed; even the question as to whether the appearance of the interference phenomena is affected by variations in temperature has not yet been approached by experiments. Under these circumstances it may appear premature to proceed beyond the results attained by approximate methods. Nevertheless we have done so in the following in view of the consideration that it should be easier for experimental testing methods to decide on the correctness of our concepts if the theory has been completed to a certain extent.*

We have extended the theory in the two respects mentioned before.

First we have dropped the assumption of the mutual independence of the atoms. In this note their movement is composed of superposed elastic waves whose wave numbers assume all values in the elastic spectrum of the body, a method that has proved successful in the theory of specific heats.

Second we thus created the opportunity to apply the quantum hypothesis in a definite manner to our case. We have not decided for or against the existence of zero-point energy. The only justified point of view is at present to test both hypotheses thoroughly as to their consequences, though different new articles by A. Einstein and O. Stern, H. Kamerling Onnes, W. H. Keesom, E. Oosterhuis present weighty reasons for the assumption of a zero-point energy. This was the more indicated in our case, since it appeared that the question can be decided by a method involving observation of x-ray interferences, which is probably not difficult to apply. In fact it is clear from the very beginning that, if the fundamental concepts of the theory are correct, the mean square of the amplitude of the atomic movements itself and not its differential coefficient with respect to temperature (as in the case of specific heats) will be decisive. In view of this decision, to be made on the basis of experiments, we have developed the theory to a point where a formula is available which can be used numerically, and have appended numerical and graphical discussions. However, the latter object could be secured only by transforming the general formula to a more readily usable form by an approximate evaluation of the elastic spectrum, but this approximate method has proved successful once before in the computation of the specific heats, and presumably does not lead to essentially erroneous results in this instance, at least if we confine ourselves to monatomic bodies or bodies that may be treated as monatomic, as for instance sylvite:

A summary at the end presents the different laws originating from the arguments presented here.

*Compare footnote at bottom of page 14.

I. Mathematical Formulation of the Principal Problem

Let the primary wave be plane and monochromatic and have the frequency ω in 2π sec. Let it be referred to a rectangular x, y, z coordinate system propagating in a direction specified by the directional cosines α_0, β_0, γ_0. Let us introduce the abbreviation:

$$\varkappa = \frac{2\pi}{\lambda} = \frac{2\pi c}{\omega}$$

where λ designates the wave length and c the velocity of light. The amplitude of the wave will then be proportional to:

$$e^{-i\varkappa(\alpha_0 x + \beta_0 y + \gamma_0 z)}$$

except for a factor depending on time. If x, y, z denote the coordinates of an atom, this function also measures the phase of the secondary radiation proceeding from this atom.

We now put for an arbitrary atom:

$$x = x_0 + u, \quad y = y_0 + v, \quad z = z_0 + w$$

where x_0, y_0, z_0 designate the coordinates of this atom in the absence of thermal movement, if no zero energy exists, *i.e.*, the coordinates at absolute zero temperature. The quantitites u, v, w are the components of the displacement which varies irregularly with time.

Let us observe the body from a far distance, designate by R the distance from the arbitrarily selected atom to the point of observation, and by r the distance to the origin of the coordinate system; then we can write with sufficient accuracy:

$$R = r - [\alpha(x_0 + u) + \beta(y_0 + v) + \gamma(z_0 + w)]$$

where α, β, γ indicate the directional cosines of the direction from the origin to the point of observation, provided the radiating crystal is positioned in the immediate vicinity of the origin of the coordinate system.

At an arbitrary instant, in view of the formula given before, we find, for the amplitude of the secondary radiation emitted by the complete crystal the value:

$$A\frac{e^{-i\varkappa r}}{r}\sum e^{i\varkappa[(\alpha-\alpha_0)(x_0+u)+(\beta-\beta_0)(y_0+v)+(\gamma-\gamma_0)(z_0+w)]} \tag{1}$$

The summation has to be taken over all atoms of the body; the factor A can be a function of $\alpha, \beta, \gamma, \lambda_0, \beta_0, \gamma_0$, and will assume a form which is known and readily found if, for instance, the source of the secondary radiation behaves similarly to an oscillating electron. This relation will not be discussed in the following; we will, for the sake of simplicity, give all the following results as if A were constant. In one respect the above formula for the amplitude is already specialized. Provided the crystal molecule consists of different atoms, the different chemical entities will in general radiate with different intensity for equal excitation. To obtain complete generality, A would have to be included under the summation sign, and the differences in the coefficients A would have to be considered in the summation. We eliminate this complication from the beginning, and treat the crystal as if all its atoms were identical.

The fluctuations of u, v, w caused by the thermal movement are very slow compared with the oscillations of the electrical amplitude characteristic for the wave. During a short time interval, which we may consider long compared to the duration of one oscillation of the x-ray, but as short compared to the time interval required for a noticeable variation of the arrangement of the atoms, we find the intensity J of the secondary radiation by multiplying equation (1) with its conjugated complex expression, while maintaining u, v, w constant. Thus we obtain:

$$\begin{cases} J = \frac{A^2}{r^2} \sum\sum e^{i\varkappa[(\alpha - \alpha_0)(x_0 - x_0') + (\beta - \beta_0)(y_0 - y_0') + (\gamma - \gamma_0)(z_0 - z_0')]} \\ \qquad e^{i\varkappa[(\alpha - \alpha_0)(u - u') + (\beta - \beta_0)(v - v') + (\gamma - \gamma_0)(w - w')]} \end{cases} \tag{2}$$

the same quantitites which are indicated by $x_0 \ldots w$ for the first summation are provided with an apostrophe for the second summation.

The observed mean intensity refers to a time interval which is long compared with the duration of the fluctuations of the atomic arrangement. To find it, we therefore must establish the average of equation (2) with respect to the displacements u, v, w which appear only in the last factor of formula (2). If the probability of an arbitrary arrangement is known, the mean value of this factor can be computed. Let this mean value be represented by:

$$e^M$$

then the final expression for J_m is given by:

$$J_m = \frac{A^2}{r^2} \sum\sum e^M e^{i\varkappa[(\alpha - \alpha_0)(x_0 - x_0') + (\beta - \beta_0)(y_0 - y_0') + (\gamma - \gamma_0)(z_0 - z_0')]} \tag{3}$$

In the following sections we will concern ourselves first with the computation of the above-mentioned probability and the expression e^M derived therefrom. As demonstrated by equation (3), the determination of this mean value constitutes the principal problem which has to be solved.

II. Introducing Normal Coordinates

Just as in the case of the calculation of specific heats, the introduction of normal coordinates is also of great value for the representation of the atomic movement. For this purpose we might proceed from the conventional elastic differential equations for the condition of a continuously occupied space, and would thus be in a position to develop an approximate theory, as has been done for the computation of the mean heat energy.[3] Instead we will here introduce the atomic structure of the body from the beginning, and therefore define our new coordinates following the exposition by Born and v. Kármán.[4] At the present extent of our knowledge of the atomic forces, not much is gained thereby, at least temporarily, since, as is known, the exact calculation of atomic movements taking account of the shortest effective elastic waves requires the knowledge of a much larger number of "elastic constants" than the conventional measurements based on the elastic differential equations can supply. Even at low temperatures, where only the long elastic waves have to be considered, the exact computation of the beginning of the elastic spectrum, which in this case is the only thing required, leads to calculations which are so complicated that they have not been completed as yet for the simplest case of a cubic crystal. We will attempt to exploit the advantages of both methods as much as possible: on the one hand we shall follow Born and v. Kármán in introducing normal coordinates so that a formula can be developed for the desired mean value which is completely determined mathematically and may claim general validity; on the other hand we shall subsequently introduce different simplifications which are unavoidable if a definite, practically usable result is to be secured, and which essentially amounts to a return to a method based on the elastic differential equations.

In accordance with this arrangement, we list several formulas in this section, which will have to be used for the introduction of the normal coordinates. They are given only for convenience, and do not contain anything new beyond the treatment of Born and v. Kármán.

If instead of the finite crystal a crystal of infinite extension is studied, any standing wave may be considered as eigenoscillation. Let the elementary region of the crystal be a rectangular parallelepiped with edges a, b, c; let the rest position of an arbitrary atom be defined by means of three integers l, m, n by the coordinates la, mb, nc. If the standing wave is defined by planes of constant phase:

$$\Omega = l\varphi + m\psi + n\chi = \text{const} \tag{4}$$

introducing three phase constants φ, ψ, χ, the displacements can be represented in the form:

$$\begin{cases} u = \mathfrak{A} \cos \Omega \, e^{i\omega t} \\ v = \mathfrak{B} \cos \Omega \, e^{i\omega t} \\ w = \mathfrak{C} \cos \Omega \, e^{i\omega t} \end{cases} \tag{5}$$

or:

$$\begin{cases} u = \mathfrak{A} \sin \Omega \, e^{i\omega t} \\ v = \mathfrak{B} \sin \Omega \, e^{i\omega t} \\ w = \mathfrak{C} \sin \Omega \, e^{i\omega t} \end{cases} \tag{5'}$$

if care is taken that the relations between the constants $\mathfrak{A}$, $\mathfrak{B}$, $\mathfrak{C}$ and the angular frequency ω in 2π sec. satisfy the symmetrical equations:

$$\begin{cases} A\mathfrak{A} + F\mathfrak{B} + E\mathfrak{C} = \mu\omega^2\mathfrak{A} \\ F\mathfrak{A} + B\mathfrak{B} + D\mathfrak{C} = \mu\omega^2\mathfrak{B} \\ E\mathfrak{A} + D\mathfrak{B} + C\mathfrak{C} = \mu\omega^2\mathfrak{C} \end{cases} \tag{6}$$

Here μ designates the mass of the crystal atom, A, B, C, D, E, F are in general periodic functions of φ, ψ, χ. The coefficients of this representation are not all known in practice; they would be available only if we had complete information regarding the mutual forces between the atoms. However, for small values of the phase constants φ, ψ, χ, *i.e.*, for waves whose wave length is long compared with the distances between atoms, the coefficients $A \ldots C$ can be expressed by the conventional elastic constants. Using the Voigt relations, we find by comparison with the elastic equations for the rhombohedral system:

$$\left\{\begin{aligned} A &= \frac{bc}{a} c_{11} \varphi^2 + \frac{ca}{b} c_{66} \psi^2 + \frac{ab}{c} c_{55} \chi^2 \\ B &= \frac{ca}{b} c_{22} \psi^2 + \frac{ab}{c} c_{44} \chi^2 + \frac{bc}{a} c_{66} \varphi^2 \\ C &= \frac{ab}{c} c_{33} \chi^2 + \frac{bc}{a} c_{55} \varphi^2 + \frac{ca}{b} c_{44} \psi^2 \\ D &= a(c_{23} + c_{44}) \psi \chi \\ E &= b(c_{31} + c_{55}) \chi \varphi \\ F &= c(c_{12} + c_{66}) \varphi \psi \end{aligned}\right. \tag{7}$$

if c_{11}, c_{22}, c_{33}, c_{44}, c_{55}, c_{66}, c_{23}, c_{31}, c_{12} designate the nine elastic constants belonging to this crystal system.

For the cubic system $a = b = c$, we have:

$$c_{11} = c_{22} = c_{33}, \qquad c_{44} = c_{55} = c_{66}, \qquad c_{23} = c_{31} = c_{12}$$

and for the same limiting case as before we obtain:

$$\left\{\begin{aligned} A &= a[c_{11} \varphi^2 + c_{44}(\psi^2 + \chi^2)] \\ B &= a[c_{11} \psi^2 + c_{44}(\chi^2 + \varphi^2)] \\ C &= a[c_{11} \chi^2 + c_{44}(\varphi^2 + \psi^2)] \\ D &= a(c_{12} + c_{44}) \psi \chi \\ E &= a(c_{12} + c_{44}) \chi \varphi \\ F &= a(_{12} + c_{44}) \varphi \psi \end{aligned}\right. \tag{7'}$$

For an isotropic body, equation (7') might be further simplified, since we know that in this case:

$$2 c_{44} = c_{11} - c_{12} \tag{8}$$

and two elastic constants suffice to define the body in the conventional manner.

Assuming $A \ldots F$ to be known, it follows that equation (6) can only be satisfied if ω^2 satisfies an equation of the third degree which has three positive and real roots $\omega_1^2, \omega_2^2, \omega_3^2$. With each of these roots are associated definite ratios $\mathfrak{A}:\mathfrak{B}:\mathfrak{C}$ so that a constant multiplying factor can be assigned arbitrarily. This constant factor we choose such that the sum of the squares of the functions of position preceding the factor $e^{i\omega t}$ in equations (5) and (5') calculated for a finite crystal with a number of atoms equal to N assumes the value 1. Then:

$$\left\{\begin{aligned} \mathfrak{A}_1^2 + \mathfrak{B}_1^2 + \mathfrak{C}_1^2 &= \frac{1}{N} \\ \mathfrak{A}_2^2 + \mathfrak{B}_2^2 + \mathfrak{C}_2^2 &= \frac{1}{N} \\ \mathfrak{A}_3^2 + \mathfrak{B}_3^2 + \mathfrak{C}_3^2 &= \frac{1}{N} \end{aligned}\right. \tag{9}$$

is valid at any frequency. If after this normalization the functions of position:

$$\begin{cases} U' = \mathfrak{A}\cos\Omega, & U'' = \mathfrak{A}\sin\Omega \\ V' = \mathfrak{B}\cos\Omega, & V'' = \mathfrak{B}\sin\Omega \\ W' = \mathfrak{C}\cos\Omega, & W'' = \mathfrak{C}\sin\Omega \end{cases} \tag{10}$$

are introduced as eigenfunctions, where the functions associated with the three different roots of the equation for ω^2 are indicated by the indices $k = 1, 2, 3$, respectively, then Born and v. Kármán showed that the displacements u, v, w can be represented by means of these eigenfunctions in the form:

$$\begin{cases} u = \dfrac{N}{(2\pi)^3}\iiint d\varphi\, d\psi\, d\chi \sum_k (Q_k'\, U_k' + Q_k''\, U_k'') \\ v = \dfrac{N}{(2\pi)^3}\iiint d\varphi\, d\psi\, d\chi \sum_k (Q_k'\, V_k' + Q_k''\, V_k'') \\ w = \dfrac{N}{(2\pi)^3}\iiint d\varphi\, d\psi\, d\chi \sum_k (Q_k'\, W_k' + Q_k''\, W_k'') \end{cases} \tag{11}$$

The quantitites Q_k' and Q_k'', which are arbitrary functions of φ, ψ, χ, may be considered as the desired normal coordinates; the integration is to be extended over the different possible phase angles, *i.e.*, for each of the quantitites φ, ψ, χ, for instance, from $-\pi$ to $+\pi$. The expressions (11) constitute a representation which is in fact reached only for the limit $N = \infty$. For future use we write them in a slightly different form, which can be considered identical with them in view of the large value of N, but which brings out more clearly the existence of $3N$ degrees of freedom. For this purpose the phase cube in a φ, ψ, χ coordinate system is constructed, which corresponds to the region of integration in (11), and this cube is subdivided into N equal elementary cubes by planes parallel to the coordinate planes. Each triplet:

$$\sum_k (Q_k'\, U_k' + Q_k''\, U_k'')$$

etc. can then be imagined as belonging to the center of such an elementary cube. The integration in equation (11) is now accordingly replaced by a summation over the N points of the phase cube which, in view of the fact that the space element $d\varphi d\psi d\chi$ can be replaced with $(2\pi)^3/N$, and introducing the summation sign $\mathfrak{S}$, results in the following expressions for the displacements u, v, w:

$$\begin{cases} u = \mathfrak{S}\sum_k (Q_k'\, U_k' + Q_k''\, U_k'') \\ v = \mathfrak{S}\sum_k (Q_k'\, V_k' + Q_k''\, V_k'') \\ w = \mathfrak{S}\sum_k (Q_k'\, W_k' + Q_k''\, W_k'') \end{cases} \tag{12}$$

It can be shown in the same manner that the potential energy Φ and the kinetic energy T can be expressed in terms of the normal coordinates Q in the form:*

*The dot in eq. (13) indicates in conventional manner differentiation with respect to time.

$$\begin{cases} \Phi = \frac{\mu}{2} \mathfrak{S} \sum_k \omega_k^2 (Q_k'^2 + Q_k''^2) \\ T = \frac{\mu}{2} \mathfrak{S} \sum_k (\dot{Q}'^2 + \dot{Q}_k''^2) \end{cases} \tag{13}$$

We also note that besides equations (9) which express normalization:

$$\begin{cases} \mathfrak{A}_1^2 + \mathfrak{B}_1^2 + \mathfrak{C}_1^2 = \frac{1}{N} \\ \mathfrak{A}_2^2 + \mathfrak{B}_2^2 + \mathfrak{C}_2^2 = \frac{1}{N} \\ \mathfrak{A}_3^2 + \mathfrak{B}_3^2 + \mathfrak{C}_3^2 = \frac{1}{N} \end{cases} \tag{14}$$

three further relations exist between the nine coefficients which can be readily obtained from equation (6), namely:

$$\begin{cases} \mathfrak{A}_1 \mathfrak{A}_2 + \mathfrak{B}_1 \mathfrak{B}_2 + \mathfrak{C}_1 \mathfrak{C}_2 = 0 \\ \mathfrak{A}_2 \mathfrak{A}_3 + \mathfrak{B}_2 \mathfrak{B}_3 + \mathfrak{C}_2 \mathfrak{C}_3 = 0 \\ \mathfrak{A}_3 \mathfrak{A}_1 + \mathfrak{B}_3 \mathfrak{B}_1 + \mathfrak{C}_3 \mathfrak{C}_1 = 0 \end{cases} \tag{15}$$

The two equation triplets (14) and (15) create further equation triplets of the form:

$$\begin{cases} \mathfrak{A}_1^2 + \mathfrak{A}_2^2 + \mathfrak{A}_3^2 = \frac{1}{N} \\ \mathfrak{B}_1^2 + \mathfrak{B}_2^2 + \mathfrak{B}_3^2 = \frac{1}{N} \\ \mathfrak{C}_1^2 + \mathfrak{C}_2^2 + \mathfrak{C}_3^2 = \frac{1}{N} \end{cases} \tag{14'}$$

and:

$$\begin{cases} \mathfrak{A}_1 \mathfrak{B}_1 + \mathfrak{A}_2 \mathfrak{B}_2 + \mathfrak{A}_3 \mathfrak{B}_3 = 0 \\ \mathfrak{B}_1 \mathfrak{C}_1 + \mathfrak{B}_2 \mathfrak{C}_2 + \mathfrak{B}_3 \mathfrak{C}_3 = 0 \\ \mathfrak{C}_1 \mathfrak{A}_1 + \mathfrak{C}_2 \mathfrak{A}_2 + \mathfrak{C}_3 \mathfrak{A}_3 = 0 \end{cases} \tag{15'}$$

as is shown in the transformation of two coordinate systems.

III. Computation of the Desired Mean Value for the Case of Vanishing Zero-Point Energy

Each eigenoscillation of the crystal was, in the preceding section, combined from one sine and one cosine wave, which could occur with arbitrary amplitudes characterized by $Q'$and Q''. We designate the phase, which is constant in space, of the wave combined from these two waves by δ, and put accordingly:

$$Q' = Q \cos \delta, \quad Q'' = Q \sin \delta \tag{16}$$

The phase δ in the following is considered as constant with regard to time so that the ratio $Q':Q''$ is fixed. The formulas for the potential and kinetic energy, respectively:

$$\Phi = \frac{\mu}{2} \mathrm{S} \sum_k \omega_k^2 Q_k^2$$
$$T = \frac{\mu}{2} \mathrm{S} \sum_k \dot{Q}_k^2$$

then follow from equation (13), or by introducing the momentum coordinate P associated with Q:

$$\begin{cases} \Phi = \frac{\mu}{2} \mathrm{S} \sum_k \omega_k^2 Q_k^2 \\ T = \frac{1}{2\mu} \mathrm{S} \sum_k P_k^2 \end{cases} \tag{17}$$

We now consider a volume element in the $6N$-dimensional Q-P space:

$$dR = dQ_1 \, dP_1 \ldots . \, dQ_{3N} \, dP_{3N}$$

The probability that our system of atoms has coordinates and impulses corresponding to this element will be equal to:

$$\mathfrak{w} \, dR$$

where $\mathfrak{w}$ designates a function which depends on the hypothesis made regarding the thermal movement. In any case, the introduction of the normal coordinates offers the possibility of writing $\mathfrak{w}$ as a product of separate factors $\mathfrak{w}_1 \ldots . . 3N$, each of which depends only on two associated impulse and position coordinates. The same is true of the function mentioned at the end of section III, whose average is to be determined, since the displacements are expressed as linear functions of P and Q. From equation (12) we have in view of equation (16):

$$u = \mathrm{S} \sum_k Q_k \mathfrak{A}_k \cos(\Omega - \delta)$$
$$v = \mathrm{S} \sum_k Q_k \mathfrak{B}_k \cos(\Omega - \delta)$$
$$w = \mathrm{S} \sum_k Q_k \mathfrak{C}_k \cos(\Omega - \delta)$$

taking the meaning of $U', V', W', U'', V'', W''$ according to equation (10) into account. The exponent of the exponential function, whose average is to be determined, can then be reduced to the following form, in accordance with section I:

$$\begin{cases} i\varkappa[(\alpha-\alpha_0)(u-u') + (\beta-\beta_0)(v-v') + (\gamma-\gamma_0)(w-w')] = \\ i\varkappa \mathrm{S} \sum_k Q_k \{\cos(\Omega-\delta) - \cos(\Omega'-\delta)\}[(\alpha-\alpha_0)\mathfrak{A}_k + (\beta-\beta_0)\mathfrak{B}_k \\ \qquad + (\gamma-\gamma_0)\mathfrak{C}_k] \end{cases} \tag{18}$$

provided Ω is, as previously, an appreviation for the expression:

$$\Omega = l\varphi + m\psi + n\chi$$

while Ω' belongs to another atom characterized in its rest position by the numbers $l', m' n'$, and is defined by the relation:

$$\Omega' = l'\varphi + m'\psi + n'\chi$$

If an arbitrary term of the sum (18) be designated by the index s and multiplied by the elementary probability corresponding to the coordinates Q_s and P_s, then, in keeping with the separation recognized as possible before, merely the value:

$$\left\{\begin{array}{l} K_s = \\ \iint e^{i\varkappa Q_s\{\cos(\Omega-\delta)-\cos(\Omega'-\delta)\}[(\alpha-\alpha_0)\mathfrak{A}_s+(\beta-\beta_0)\mathfrak{B}_s+(\gamma-\gamma_0)\mathfrak{C}_s]}\, \mathfrak{w}_s\, dQ_s\, dP_s \end{array}\right. \tag{19}$$

must be determined. The product of these double integrals taken over all possible values of s (from 1 to $3N$) results in the mean value designated by e^M in section I. Thus we obtain:

$$e^M = \prod_s K_s \tag{20}$$

or upon taking the logarithm:

$$M = \sum_s \log K_s \tag{20'}$$

In the computation for K_s we introduce the temporary abbreviation:

$$\left\{\begin{array}{l} \varrho_s = \varkappa\{\cos(\Omega-\delta)-\cos(\Omega'-\delta)\}[(\alpha-\alpha_0)\mathfrak{A}_s \\ \qquad + (\beta-\beta_0)\mathfrak{B}_s + (\gamma-\gamma_0)\mathfrak{C}_s] \end{array}\right. \tag{21}$$

then:

$$K_s = \iint e^{i\varrho_s Q_s}\, \mathfrak{w}_s\, dQ_s\, dP_s \tag{22}$$

If the curves:

$$\frac{\mu}{2}\omega_s^2 Q_s^2 + \frac{P_s^2}{2\mu} = \text{const.} = E \tag{23}$$

are plotted in a Q_s-P_s plane, on which curves the partial energy E associated with the coordinates Q_s, P_s is constant according to equation (17), then Planck's quantum theory in its original form states that only those curves whose area is an integral multiple of the quantum h determine the probability of an elementary state defined by Q_s, P_s. For the constant in equation (23) we obtain, as is well known, the value:

$$E = \frac{z h \omega_s}{2\pi} \tag{24}$$

where z is the integer with which h has to be multiplied to determine the area belonging to the respective curve.

We first calculate the mean value of $e^{i\rho_s Q_s}$ belonging to one of these curves. For this purpose we introduce instead of the coordinates Q_s, P_s the phase φ of the movement taking place on the curve E = const., by putting:

$$Q_s = \sqrt{\frac{2E}{\mu\omega_s^2}}\cos\varphi, \quad P_s = \sqrt{2\mu E}\sin\varphi$$

each point on the ellipse (23) is then characterized by a value of φ between 0 and 2π, and all values of φ are equally probable. The mean value of $e^{i\rho_s Q_s}$ associated with a curve E = const. is given by:

$$\frac{1}{2\pi}\int_0^{2\pi} e^{i\varrho_s Q_s}\, d\varphi = \frac{1}{2\pi}\int_0^{2\pi} e^{i\varrho_s\sqrt{\frac{2E}{\mu\omega_s^2}}\cos\varphi}\, d\varphi$$

In general, evaluation of this integral leads to Bessel functions. However, it will be readily seen that in our case an expansion in powers of the factor of cos φ in the exponent is permissible* which may be cut short at the second power of this factor. In fact the normalization equation (9) shows that such an expansion proceeds according to negative powers of $N^{\frac{1}{2}}$ so that, in view of the very large values of N, the expansion can be confined, without introducing any noticeable error, to those terms which upon completion of all steps in the calculation result in expressions of the order N^0. This approximation may also be understood from a practical point of view, since, firstly, in accordance with our assumption, the total movement of an atom is composed of $3N$ elementary movements, whereas, secondly, only one of these necessarily extremely small movements is involved in the computation of K.

If the expansion is carried out, we find:

$$\frac{1}{2\pi}\int_0^{2\pi} e^{i\varrho_s Q_s}\,d\varphi = \frac{1}{2\pi}\int_0^{2\pi} d\varphi\left[1 + i\varrho_s\sqrt{\frac{2E}{\mu\,\omega_s^2}}\cos\varphi - \varrho_s^2\frac{E}{\mu\,\omega_s^2}\cos^2\varphi\ldots\right]$$

$$= 1 - \frac{\varrho_s^2}{2\mu\,\omega_s^2}E$$

or after insertion of the value (24) for E:

$$\frac{1}{2\pi}\int_0^{2\pi} e^{i\varrho_s Q_s}\,d\varphi = 1 - z\frac{h\,\varrho_s^2}{4\pi\,\mu\,\omega_s} \tag{25}$$

of the ellipse with the area zh is proportional to:

$$e^{-\frac{zh\,\omega_s}{2\pi\,kT}}$$

The probability that the point in question is located at any ellipse in the Q_s, P_s space is equal to unity; thus the required probability for a definite ellipse with the index z is numerically equal to:

$$\frac{e^{-z\frac{h\omega_s}{2\pi kT}}}{\sum_z e^{-z\frac{h\omega_s}{2\pi kT}}} = \left(1 - e^{-\frac{h\omega_s}{2\pi kT}}\right)e^{-z\frac{h\omega_s}{2\pi kT}} \tag{26}$$

If now the mean value given for an ellipse in equation (25) is multiplied by the probability equation (26), and the sum taken over all ellipses, *i.e.*, over z from 0 to ∞, then according to equation (22), the result will be the required value of K_s which is thus obtained in the form:

$$K_s = \left(1 - e^{-\frac{h\omega_s}{2\pi kT}}\right)\sum_z\left(1 - z\frac{h\varrho_s^2}{4\pi\mu\,\omega_s}\right)e^{-z\frac{h\omega_s}{2\pi kT}} \tag{27}$$

*The expansion is possible here as contrasted with the previously discussed instance (*Verhandl. deut. physik. Ges.*, *15*, 749 (1913)), because the amplitude of the separate movements is now extremely small (of the order of $1/N^{\frac{1}{2}}$ compared with that previously treated).

Now on the one hand:

$$\left(1 - e^{-\frac{h\omega_s}{2\pi kT}}\right) \sum_z e^{-z\frac{h\omega_s}{2\pi kT}} = 1$$

and it is easy to find that:

$$\left(1 - e^{-\frac{h\omega_s}{2\pi kT}}\right) \sum_z z\, e^{-z\frac{h\omega_s}{2\pi kT}} = \frac{1}{e^{\frac{h\omega_s}{2\pi kT}} - 1}$$

thus we can alternatively write for K_S:

$$K_s = 1 - \frac{\varrho_s^2}{2\mu\,\omega_s^2} \frac{\frac{h\,\omega_s}{2\pi}}{e^{\frac{h\omega_s}{2\pi kT}} - 1} \tag{28}$$

and with the same degree of accuracy, log K_S can be expressed as a power expansion:

$$\log K_s = -\frac{\varrho_s^2}{2\mu\,\omega_s^2} \frac{\frac{h\,\omega_s}{2\pi}}{e^{\frac{h\omega_s}{2\pi kT}} - 1} \tag{28'}$$

To find M from this according to equation (20'), the sum must be taken over all possible values of s. According to the preceding discussion, summation with respect to s is to be understood to mean: first, for each of the N points, summation over the three wave numbers associated with the point, and subsequently summation over all N points of the cube. If we now again introduce for this summation, in the interest of clarity, the symbol $\mathfrak{S}\sum_k$, and simultaneously introduce the value (21) for ρ_S, we obtain:*

$$\begin{aligned} -M = \mathfrak{S}\sum_k \frac{\varkappa^2}{2\mu\,\omega_k^2} \frac{\frac{h\,\omega_k}{2\pi}}{e^{\frac{h\omega_k}{2\pi kT}} - 1} \,[(\alpha-\alpha_0)\mathfrak{A}_k + (\beta-\beta_0)\mathfrak{B}_k \\ + (\gamma-\gamma_0)\mathfrak{C}_k]^2 \{\cos(\Omega-\delta) - \cos(\Omega'-\delta)\}^2 \end{aligned} \tag{29}$$

For the time being this expression still contains the phase δ, constant with respect to space, of the constituent oscillations. It may be assumed as certain that, in an actual crystal, each constituent oscillation does not continue indefinitely without modification, as has been assumed in the theory. Rather the oscillations will vary their phase δ irregularly in such a manner that in the mean no phase relation exists between the separate constituent oscillations. This entitles us not to consider equation (29) as the final expression for M, but to substitute for it the mean value taken with respect to δ.**

*The index k should not be confused with Boltzmann's constant k which occurs in the combination kT in the same formula.

**The mean is taken as if the movements of adjacent atoms were entirely independent of one another. This is contradictory to equation (29), according to which one single wave would have to be traced throughout the whole body. The actual behavior of the waves will lie between these two extremes and is directly related to the problem of heat conductivity. I intend to revert in the near future to the second limiting case not treated here.

If this mean value is taken, we finally obtain:

$$-M = \mathfrak{S}\sum_k \frac{\varkappa^2}{\mu\,\omega_k^2}\,\frac{\frac{h\,\omega_k}{2\pi}}{e^{\frac{h\omega_k}{2\pi kT}}-1}\left[(\alpha-\alpha_0)\mathfrak{A}_k + (\beta-\beta_0)\mathfrak{B}_k + (\gamma-\gamma_0)\mathfrak{C}_k\right]^2 \tag{30}$$

Only in one instance is this expression to be replaced by another, namely, when $\Omega = \Omega'$ *i.e.*, when in the expression originally given for the formation of the mean value the displacements u, v, w and u', v', w' relate to the same atom, or, expressed differently, if simultaneously:

$$l = l', \quad m = m', \quad n = n'$$

In this special case we obtain simply:

$$M = 0 \tag{30'}$$

We repeat the separate steps required for the computation of M:

First the "elastic constants" of the crystal A,B,C,D,E,F are written down for a wave with planes of equal phase given by:

$$l\varphi + m\psi + n\chi = \text{const.}$$

as functions of φ, ψ, χ, *then the three values of the wave numbers in* 2π *sec. compatible with these equations are calculated from equation (6); the wave numbers were designated by* ω_k *(with* $k = 1,2,3$*), and the associated coefficients by* $\mathfrak{A}_k, \mathfrak{B}_k, \mathfrak{C}_k$, *which are determined, as are the* ω_k*'s, as functions of the phase angles* φ, ψ, χ *with the assistance of the normalizing condition equation (9). If further the direction cosines of the incident primary radiation* $\alpha_0, \beta_0, \gamma_0$, *the direction cosines of the direction of observation* α, β, γ, *and the wave length of the radiation* λ *are given, the* $\sum_k$ *in equation (30) can be calculated, since:*

$$\varkappa = 2\pi/\lambda$$

Finally the summation $\mathfrak{S}$ *over all points N regularly distributed in the phase cube supplies the required value of M.*

I refer to section V for a discussion of M and the subsequent computation of the intensity J_m. Here we limit ourselves to giving the general formula; in Section IV the calculation will be carried to the corresponding point under the assumption of zero-point energy.

IV. Computation of the Required Mean Value $\mathfrak{S}$ with the Assumption of Zero-Point Energy

The consideration of the preceding section can be transferred to the present case without changes up to formula (22), since here also a separation of $\mathfrak{w}$ in $\mathfrak{w}_1$ to $\mathfrak{w}_{3N}$ is possible. A deviation from the previous course occurs only at the actual evaluation of the double integral:

$$K_s = \iint e^{i e_s Q_s}\,\mathfrak{w}_s\, dQ_s\, dP_s$$

since the $\mathfrak{w}_s$ now have to be defined differently. As is well known the difference between the new and the old assumption by Planck, as far as it concerns us here, consists in the provision that the place assigned in the previous section to the separate ellipses: energy = const., is now taken over by elliptical rings bounded by these ellipses which accordingly all have the area h. Similar to the procedure in section III, where first the mean value of $e^{i\rho_s Q_s}$ was determined on an ellipse, we now have to find this mean value for any one of the elliptical rings.

If we again put:

$$Q_s = \sqrt{\frac{2E}{\mu\omega_s^2}}\cos\varphi, \quad P_s = \sqrt{2\mu E}\sin\varphi$$

we must consider E not as constant but as variable between the limits:

$$E = z\frac{h\omega_s}{2\pi} \quad \text{to} \quad E = (z+1)\frac{h\omega_s}{2\pi}$$

In the new coordinates E and φ the elementary area in the Q,P plane is given by the expression:

$$\frac{1}{\omega_s}dE\,d\varphi$$

by introducing these coordinates we find the required mean value for an elliptical ring from the formula:

$$\frac{\iint e^{i\varrho_s\left(\frac{2E}{\mu\omega_s^2}\right)^{1/2}\cos\varphi}\,dE\,d\varphi}{\iint dE\,d\varphi}$$

where the integration is taken from $\varphi = 0$ to $\varphi = 2\pi$ and with respect to E between the limits specified above.

Just as it was permissible above, it is again permissible here to expand the exponential function in the integral in powers of the exponent and to retain only the first terms. We thus find:

$$\iint e^{i\varrho_s\left(\frac{2E}{\mu\omega_s^2}\right)^{1/2}\cos\varphi}\,dE\,d\varphi = \iint dE\,d\varphi\left[1 + i\varrho_s\left(\frac{2E}{\mu\omega_s^2}\right)^{1/2}\cos\varphi - \frac{\varrho_s^2 E}{\mu\omega_s^2}\cos^2\varphi + \ldots\right]$$

$$= 2\pi\frac{h\omega_s}{2\pi} - \frac{\pi\varrho_s^2}{2\mu\omega_s^2}\left(\frac{h\omega_s}{2\pi}\right)^2\{(z+1)^2 - z^2\}$$

further:

$$\iint dE\,d\varphi = 2\pi\frac{h\omega_s}{2\pi}$$

thus the average value computed for one of the elliptical rings becomes:

$$1 - \frac{\varrho_s^2}{2\mu\omega_s^2}\frac{h\omega_s}{2\pi}\left(z + \frac{1}{2}\right) \tag{31}$$

From the general formula (22) for K_s and in view of the deliberations leading to formulas (26) and (27) of the preceding section which apply here, we obtain here upon substitution of the mean value (31):

$$K_s = \left(1 - e^{-\frac{h\omega_s}{2\pi kT}}\right) \sum_z \left[1 - \frac{h\varrho_s^2}{4\pi\mu\omega_s}\left(z + \frac{1}{2}\right)\right] e^{-z\frac{h\omega_s}{2\pi kT}} \qquad (32)$$

The expression differs from equation (27) only in that $z + \frac{1}{2}$ instead of z appears under the summation sign; it is evaluated in exactly the same way as has been indicated in Section III following equation (27), and yields:

$$K_s = 1 - \frac{\varrho_s^2}{2\mu\omega_s^2}\left\{\frac{1}{2}\frac{h\omega_s}{2\pi} + \frac{\frac{h\omega_s}{2\pi}}{e^{\frac{h\omega_s}{2\pi kT}} - 1}\right\} \qquad (33)$$

from which follows:

$$\log K_s = -\frac{\varrho_s^2}{2\mu\omega_s^2}\left\{\frac{1}{2}\frac{h\omega_s}{2\pi} + \frac{\frac{h\omega_s}{2\pi}}{e^{\frac{h\omega_s}{2\pi kT}} - 1}\right\} \qquad (33')$$

Since the expression found here for $\log K_s$ differs from the one found before only in the structure of the expression in the brackets, and since in particular the factor ρ_s^2 occurs here in the same manner, the discussion concerning the phase angles δ in the previous section applies here. Therefore the final result M can be readily given.

One obtains:

$$-M = \mathfrak{S} \sum_k \frac{\varkappa^2}{2\mu\omega_k^2}\left\{\frac{1}{2}\frac{h\omega_k}{2\pi} + \frac{\frac{h\omega_k}{2\pi}}{e^{\frac{h\omega_k}{2\pi kT}} - 1}\right\}[(\alpha-\alpha_0)\mathfrak{A}_k + (\beta-\beta_0)\mathfrak{B}_k + (\gamma-\gamma_0)\mathfrak{C}_k]^2 \qquad (34)$$

in all instances except for the special case:

$$l = l', \; m = m', \; n = n'$$

for which again:

$$M = 0 \qquad (34')$$

For the meaning of the symbols I refer to the explanation included in the passage at the end of the preceding section.

V. General Results Regarding the Effect of the Thermal Movement

The fact that first arrests attention when examining the expression for M in both equations (30) and (34) is the non-dependence on the quantitites l, m, n, l', m', n'. An influence due to the mutual positions of the atoms, whose displacements u, v, w and u', v', w' occurred in the expression from which M was obtained by taking the mean, is thus no longer present. From this follows the result, previously obtained on the basis of a more elementary theory:

The thermal movement has no effect on the sharpness of the interference pattern.

The secondary x-ray intensity is determined by the double sum given in formula (3), section I. If attention is paid to the exceptional values for M, equation (30') or (34'), it can be computed by first eliminating from the double sum all terms for which:

$$l = l', \quad m = m', \quad n = n'$$

They contribute the total amount:

$$\frac{A^2}{r^2}\sum 1 = \frac{A^2}{r^2} N$$

All these terms, multiplied by e^M, are now subtracted from the double sum, giving:

$$\frac{A^2}{r^2} N e^M$$

to be finally added. The remainder of the double sum, except for this added expression, can be written in the form:

$$\frac{A^2}{r^2} e^M \sum_{l,m,n,l',m',n'}\sum e^{i\varkappa[(\alpha-\alpha_0)(x_0-x_0')+(\beta-\beta_0)(y_0-y_0')+(\gamma-\gamma_0)(z_0-z_0')]}$$

where M is now determined exclusively by equation (30) or (34) and because of its independence of l, m, n, l', m', n' can be put in front of the summation sign; the summations are to be taken over all values of l, m, n, l', m', n'. The last-given double sum is identical with the sum which appears in Laue's theory; it determines in known manner the location of the points in the interference pattern. Let us now introduce the abbreviation:

$$Z = N(1 - e^M) \tag{35}$$

and:

$$L = \sum\sum e^{i\varkappa[(\alpha-\alpha_0)(x_0-x_0')+(\beta-\beta_0)(y_0-y_0')+(\gamma-\gamma_0)(z_0-z_0')]} \tag{36}$$

the total intensity J_m can be expressed in terms of Z and L in the form:

$$J_m = \frac{A^2}{r^2}(Z + e^M L) \tag{37}$$

from which the preceding theorem is evident.

However, it may be asserted that:

The thermal movement has a considerable influence on the intensity of the secondary radiation.

The quantity M which contains the temperature T as a parameter is, according to equation (30) or (34), always negative. Further its absolute value increases, as will readily be seen, with increasing T to infinity. Thus for $\lim T = \infty$, $e^M = 0$, and we obtain the expression:

$$J_m = \frac{A^2}{r^2} Z = N\frac{A^2}{r^2}$$

for the intensity.

All that remains from the phenomenon in this limit is an evenly scattered intensity. On the other hand, at least in the absence of the zero-point energy, $M = 0$ for $T = 0$, and consequently:

$$J_m = \frac{A^2}{r^2} L$$

In this limiting case the intensity distribution is determined exactly by Laue's expression.

At intermediate temperatures, and also for $T = 0$, provided the zero-point energy exists, both expressions in the bracket of equation (37) have to be taken into account. The formula shows how the heat movement withdraws energy from the interference intensity, measured by L, and converts it into a scattered energy, measured by Z.

The quantity M, which measures the temperature effect, depends in a very simple manner on the wave length. Since in equation (30) as well as in equation (34) κ^2 occurs only as a factor, M is always inversely proportional to the square of the wave length of the secondary x-ray radiation. If this dependence only is considered, and if the observations are made at constant temperature and for fixed directions of the incident and the secondary beam, then the interference intensity is proportional to:

$$e^{-\frac{\text{const}}{\lambda^2}} L$$

and the scattered intensity is proportional to:

$$1 - e^{-\frac{\text{const}}{\lambda^2}}$$

Considering in conclusion the dependence on the direction of incidence and on the direction of observation, equations (30) and (34) show at once that M is a quadratic function of the differences:

$$(\alpha - \alpha_0),\ (\beta - \beta_0),\ (\gamma - \gamma_0)$$

their coefficients continuously increasing with increasing temperature. It will be seen readily that this quadratic form is not the most general form possible; in fact it contains only the squares of the above differences and not their products. If we consider, for instance, the term in $-M$ in equation (30) which is multiplied by:

$$(\alpha - \alpha_0)\ (\beta - \beta_0)$$

the associated coefficient has the value:

$$S \sum_k \frac{x^2}{\mu \omega_k^2} \frac{\frac{h\omega_k}{2\pi}}{e^{\frac{h\omega_k}{2\pi kT}} - 1} \mathfrak{A}_k \mathfrak{B}_k \tag{38}$$

Furthermore, according to section II, ω_k is to be found as a root of the third degree equation:

$$\begin{vmatrix} A-\mu\omega_k^2 & F & E \\ F & B-\mu\omega_k^2 & D \\ E & D & C-\mu\omega_k^2 \end{vmatrix} = 0$$

The coefficients $A \ldots F$ occurring therein may, for instance, be considered as determined by equation (7); it will then be readily appreciated that a reversal of the sign of any of the quantities φ, ψ, or χ does not change the determinant.* Once a definite point has been chosen in an octant of the φ, ψ, χ space and the root ω_k for it computed, then the latter is also a root of the determinant for the seven points in the seven other octants of the phase cube obtained by repeated mirror reflections on the coordinate planes. The values of the root ω_k thus repeat in each of the eight octants; the same holds also for the expression:

$$\frac{\varkappa^2}{\mu\,\omega_k^2} \; \frac{\frac{h\,\omega_k}{2\pi}}{e^{\frac{h\,\omega_k}{2\pi k T}} - 1}$$

in equation (38). Further the ratios $\mathfrak{A}_k:\mathfrak{B}_k:\mathfrak{C}_k$ follow after determination of ω_k from equation (6) in the form:

$$\left\{\begin{aligned} \mathfrak{A}_k:\mathfrak{B}_k:\mathfrak{C}_k &= \begin{vmatrix} B-\mu\omega_k^2 & D \\ D & C-\mu\omega_k^2 \end{vmatrix} : \begin{vmatrix} D & F \\ C-\mu\omega_k^2 & E \end{vmatrix} : \begin{vmatrix} F & B-\mu\omega_k^2 \\ E & D \end{vmatrix} \\ &= \Delta_k : \Delta_k' : \Delta_k'' \end{aligned}\right. \tag{39}$$

In view of the normalization equations (14) one obtains:

$$\left\{\begin{aligned} \mathfrak{A}_k &= \frac{1}{\sqrt{N}}\frac{\Delta_k}{\sqrt{\Delta_k^2+\Delta_k'^2+\Delta_k''^2}}, \quad \mathfrak{B}_k = \frac{1}{\sqrt{N}}\frac{\Delta_k'}{\sqrt{\Delta_k^2+\Delta_k'^2+\Delta_k''^2}} \\ \mathfrak{C}_k &= \frac{1}{\sqrt{N}}\frac{\Delta_k''}{\sqrt{\Delta_k^2+\Delta_k'^2+\Delta_k''^2}} \end{aligned}\right. \tag{40}$$

The determinant Δ_k does not change with a change in sign of the phase angles φ, ψ, χ. However, Δ_k' changes its sign with a change in sign of φ or ψ, and Δ_k'' behaves similarly with respect to a change in sign of φ or χ, whereas such changes in χ in the first case and in ψ in the second case have no effect. It follows that each of the combinations:

$$\Delta_k\Delta_k', \; \Delta_k'\Delta_k'', \; \Delta_k''\Delta_k$$

occurs always in four out of the eight octants with the same sign and in the other four with the opposite sign. In the summation indicated by $\mathfrak{S}$, all factors of:

$$(\alpha-\alpha_0)(\beta-\beta_0)$$

as well as the factors of:

$$(\beta-\beta_0)(\gamma-\gamma_0)$$

and of:

$$(\gamma-\gamma_0)(\alpha-\alpha_0)$$

*The symmetrical properties which we are using here retain their validity also in the case of the general assumption for the coefficients $A \ldots F$.

must vanish. It is evident that the same holds for equation (34).

A visualization of the spatial distribution of the temperature effect can be obtained as follows: A point with the coordinates α_0, β_0, γ_0 is plotted in a rectangular coordinate system, the axes of which are parallel to the edges of the elementary parallelepiped. Since these quantitites stand for the direction cosines of the incident beam:

$$\alpha_0^2 + \beta_0^2 + \gamma_0^2 = 1$$

this point is situated on a sphere with radius 1 at the point of intersection with the incident beam which traverses the origin. The temperature effect is constant (with respect to space), provided M, *i.e.*, the sum in equation (30) or (34), respectively, is constant. In view of the preceding considerations, this condition can be written in the form:

$$K_{11}(\alpha - \alpha_0)^2 + K_{22}(\beta - \beta_0)^2 + K_{33}(\gamma - \gamma_0)^2 = \text{const.} \tag{41}$$

If the family of ellipsoids represented by this equation for a series of values of the right-hand term is plotted with the point α_0, β_0, γ_0 as center, their axes extend in the direction of the coordinate axes and have lengths proportional to $K_{11}^{-\frac{1}{2}}$, $K_{22}^{-\frac{1}{2}}$, $K_{33}^{-\frac{1}{2}}$, which are functions only of temperature and wave length. The intersections of this family of surfaces with the unit sphere are curves on which the temperature effect is constant. In Figure 1 are shown schematically one such ellipsoid and the associated curve of intersection.

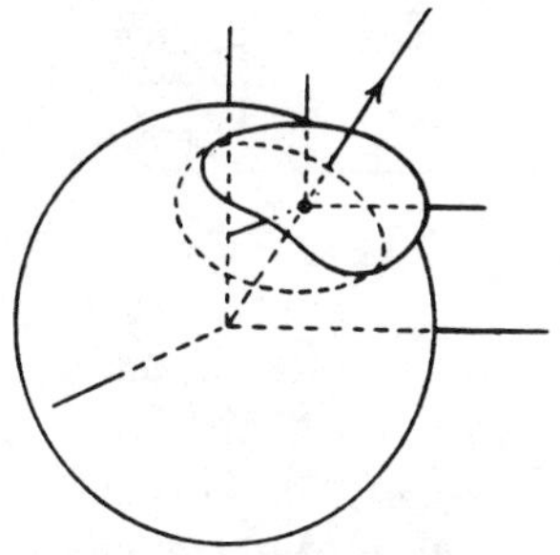

Figure 1

The behavior is much simpler for regular crystals; the ellipsoid just mentioned then becomes a sphere so that the curves of constant temperature effect are circles. The latter are located in planes extending at right angles to the direction of the incident beam and have their centers on this line. Only this special case was studied in the former approximate theory.

That the latter statement is well founded can be seen from the following. In the case of the regular crystal, the equations (7') can be regarded as definition for the quantities $A \ldots F$. If these values are introduced in equations (6), the latter equations have the properties that (1) the determinant of the system of equations remains unchanged if φ, ψ, χ are interchanged cyclically, and (2) the equations remain unchanged if φ, ψ, χ are interchanged cyclically and simultaneously $\mathfrak{A}$, $\mathfrak{B}$, $\mathfrak{C}$ are interchanged cyclically.

According to the preceding treatment it is only necessary to study more closely one octant of the phase cube; we chose the one for which φ, ψ, χ are positive. We now select an arbitrary point:

$$\varphi = \varphi, \quad \psi = \psi, \quad \chi = \chi$$

therein; then the two further points for which the determinant has not changed:

$$\varphi' = \psi, \quad \psi' = \chi, \quad \chi' = \varphi$$

and:

$$\varphi' = \chi, \quad \psi' = \varphi, \quad \chi' = \psi$$

can be plotted at once. A root ω_k belonging to the first point thus belongs to all three points. These three points have been indicated in Figure 2; further, the one octant of the phase cube has been subdivided into three pyramids by three planes including the origin and each one edge of the cube, in such a manner that for the above-mentioned cyclic interchange the system of all the points of one pyramid is transformed into the system of all the points of the succeeding pyramids.

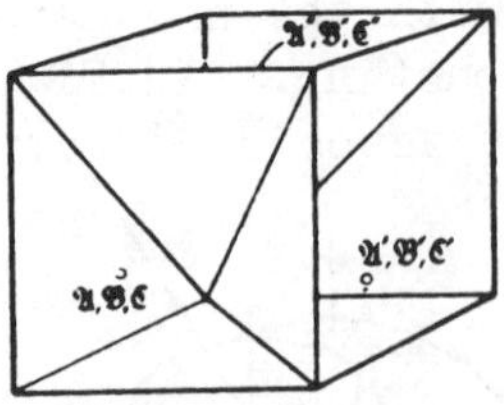

Figure 2.

With each of the three above-mentioned points — assuming a definite angular frequency ω_k— are associated three sets of values $\mathfrak{A}$, $\mathfrak{B}$, $\mathfrak{C}$, which we have designated by $\mathfrak{A}$, $\mathfrak{B}$, $\mathfrak{C}$; $\mathfrak{A}'$, $\mathfrak{B}'$, $\mathfrak{C}'$; $\mathfrak{A}''$, $\mathfrak{B}''$, $\mathfrak{C}''$, respectively. In the expressions (30) or (34) we have to form, as long as only these three points are being considered, the combinations:

$$\mathfrak{A}^2 + \mathfrak{A}'^2 + \mathfrak{A}''^2, \quad \mathfrak{B}^2 + \mathfrak{B}'^2 + \mathfrak{B}''^2, \quad \mathfrak{C}^2 + \mathfrak{C}'^2 + \mathfrak{C}''^2$$

In view of the above-mentioned second property of equations (6), we have:

$$\begin{aligned} \mathfrak{A} &= \mathfrak{A}, & \mathfrak{B} &= \mathfrak{B}, & \mathfrak{C} &= \mathfrak{C} \\ \mathfrak{A}' &= \mathfrak{B}, & \mathfrak{B}' &= \mathfrak{C}, & \mathfrak{C}' &= \mathfrak{A} \\ \mathfrak{A}'' &= \mathfrak{C}, & \mathfrak{B}'' &= \mathfrak{A}, & \mathfrak{C}'' &= \mathfrak{B} \end{aligned}$$

so that the three sums have the same common value:

$$\mathfrak{A}^2 + \mathfrak{B}^2 + \mathfrak{C}^2$$

If three points of the phase space connected by cyclic interchange are taken together in the summation indicated by $\mathfrak{S}$ in equations (30) and (34), the equality of the coefficients of:

$$(\alpha - \alpha_0)^2, \ (\beta - \beta_0)^2$$

and:

$$(\gamma - \gamma_0)^2$$

follows readily, and the above proposition has been proven.

With the above discussion all has been said which can be concluded without more detailed work on equations (30) and (34) about the influence of temperature. In the next section we will discuss the form of the temperature dependence; we start with low temperatures.

VI. The Effect of Heat at Low Temperatures

Let us confine ourselves for the time being to the case of equation (30) valid for vanishing zero-point energy; then we can replace the summation $\mathfrak{S}$ by an integration over the cube in the phase space. A volume element equal to:

$$\frac{(2\pi)^3}{N}$$

is associated with each of the N points of the cube, over which points the summation is to be taken; in the limit for $N = \infty$, the volume element can be written in the form $d\varphi d\psi d\chi$. Equation (30) can thus be transformed into:

$$\left\{ \begin{aligned} -M = \frac{N}{(2\pi)^3} \int_{-\pi}^{+\pi}\int_{-\pi}^{+\pi}\int_{-\pi}^{+\pi} d\varphi\, d\psi\, d\chi \sum_k \frac{\varkappa^2}{2\mu\omega_k^2} \frac{\frac{h\omega_k}{2\pi}}{e^{\frac{h\omega_k}{2\pi\kappa T}} - 1} [(\alpha - \alpha_0)^2 \mathfrak{A}_k^2 \\ + (\beta - \beta_0)^2 \mathfrak{B}_k^2 + (\gamma - \gamma_0)^2 \mathfrak{C}_k^2] \end{aligned} \right. \tag{42}$$

when we use the result of the preceding section, according to which the products:

$$(\alpha - \alpha_0)(\beta - \beta_0)$$

etc. do not appear in the final result.

For low temperatures, those sections of the cube in the phase space that are associated with high angular frequencies ω_k contribute only a small amount to the value of M; in this special case we may confine the derivations to long waves for which the elastic coefficients are represented by equations (7) with sufficient accuracy. An essential simplification thereby secured lies in the independence of the velocity of propagation of the elastic waves on the frequency. If the velocity corresponding to ω_k is denoted by g_k and the wave length by λ, then:

$$\omega_k = \frac{2\pi g_k}{\lambda} \tag{43}$$

If further the direction cosines of the normals to the planes of constant phase are p, q, r so that:

$$p^2 + q^2 + r^2 = 1$$

we can substitute:

$$\varphi = \frac{2\pi a}{\lambda} p, \quad \psi = \frac{2\pi b}{\lambda} q, \quad \chi = \frac{2\pi c}{\lambda} r \tag{44}$$

for φ, ψ, χ, where a, b, c are the edges of the elementary crystal parallelepiped introduced earlier in section II.

The computation of ω_k as a function of φ, ψ, χ is now replaced by the computation of g_k as a function of p, q, r. If the expressions from equations (43) and (44) are introduced in equation (6), the equation:

$$\begin{vmatrix} A_0 - \varrho g^2 & F_0 & E_0 \\ F_0 & B_0 - \varrho g^2 & D_0 \\ E_0 & D_0 & C_0 - \varrho g^2 \end{vmatrix} = 0 \qquad (45)$$

follows for g, where $A_0 \ldots F_0$, according to equation (7), are abbreviations for the following direction functions:

$$\left\{ \begin{array}{ll} A_0 = c_{11} p^2 + c_{66} q^2 + c_{55} r^2, & D_0 = (c_{23} + c_{44}) q r \\ B_0 = c_{22} q^2 + c_{44} r^2 + c_{66} p^2, & E_0 = (c_{31} + c_{55}) r p \\ C_0 = c_{33} r^2 + c_{55} p^2 + c_{44} q^2, & F_0 = (c_{12} + c_{66}) p q \end{array} \right. \qquad (46)$$

Equation (45) has three roots one of which has been designated by g_k before; the letter ρ denotes the density which originally appeared in the form μ/abc.

After selection of a definite value for g_k, the quantitites $\mathfrak{A}_k$, $\mathfrak{B}_k$, $\mathfrak{C}_k$ follow from the three linear equations from (45):

$$\left\{ \begin{array}{l} A_0 \mathfrak{A}_k + F_0 \mathfrak{B}_k + E_0 \mathfrak{C}_k = \varrho g_k^2 \mathfrak{A}_k \\ F_0 \mathfrak{A}_k + B_0 \mathfrak{B}_k + D_0 \mathfrak{C}_k = \varrho g_k^2 \mathfrak{B}_k \\ E_0 \mathfrak{A}_k + D_0 \mathfrak{B}_k + C_0 \mathfrak{C}_k = \varrho g_k^2 \mathfrak{C}_k \end{array} \right. \qquad (47)$$

with the additional condition:

$$\mathfrak{A}_k^2 + \mathfrak{B}_k^2 + \mathfrak{C}_k^2 = \frac{1}{N} \qquad (47')$$

The exponent of the exponential function in equation (42) is given by:

$$\frac{h \omega_k}{2\pi k T} = \frac{h g_k}{\lambda k T} = \xi \qquad (48)$$

Let this quantity and two coordinates of the unit sphere:

$$p^2 + q^2 + r^2 = 1$$

be introduced as new variables instead of φ, ψ, χ. The magnitude of the volume element in the phase space expressed in these new coordinates is readily found to be:

$$\left(\frac{2\pi k T}{h g_k}\right)^3 a b c \xi^2 d\xi \, d\Omega$$

where $d\Omega$ designates an elementary area of the unit sphere.

If this expression is introduced in equation (42), the latter can be reduced to:

$$\left\{ \begin{array}{l} -M = \dfrac{N}{8\pi^2} \dfrac{\varkappa^2 k^2 T^2}{h \varrho} \displaystyle\int_0^\infty \frac{\xi \, d\xi}{e^\xi - 1} \cdot \sum_k \int \frac{d\Omega}{g_k^3} [(\alpha - \alpha_0)^2 \mathfrak{A}_k^2 \\ \qquad + (\beta - \beta_0)^2 \mathfrak{B}_k^2 + (\gamma - \gamma_0)^2 \mathfrak{C}_k^2] \end{array} \right. \qquad (45)$$

Since we confine these considerations to low temperatures, the integration with respect to ξ can extend from 0 to ∞, and the **quantities** $\mathfrak{A}_k$, $\mathfrak{B}_k$, $\mathfrak{C}_k$ depend only on the coordinates of the unit sphere and not on ξ; therefore it was justified to put the integral with respect to ξ as a factor in front of the summation sign.

It is well known that:

$$\int_0^\infty \frac{\xi\, d\xi}{e^\xi - 1} = \frac{\pi^2}{6}$$

and we can write instead of equation (45):

$$\left\{ \begin{aligned} -M = \frac{N}{48}\frac{\varkappa^2 k^2 T^2}{h\varrho} \sum_k \int \frac{d\Omega}{g_k^3} [(\alpha-\alpha_0)^2 \mathfrak{A}_k^2 + (\beta-\beta_0)^2 \mathfrak{B}_k^2 \\ + (\gamma-\gamma_0)^2 \mathfrak{C}_k^2] \end{aligned} \right. \tag{45'}$$

From this we obtain the result:

At low temperatures and in the absence of a zero-point energy, the exponent of the exponential function which measures the temperature effect is proportional to T^2.

The factor of T^2 is in general a quadratic function of:

$$(\alpha - \alpha_0),\ (\beta - \beta_0),\ (\gamma - \gamma_0)$$

the coefficients of which can be calculated from equation (45) as sums over k = 1, 2, 3 of the mean values:

$$\int \frac{\mathfrak{A}_k^2}{g_k^3} d\Omega, \quad \int \frac{\mathfrak{B}_k^2}{g_k^3} d\Omega, \quad \int \frac{\mathfrak{C}_k^2}{g_k^3} d\Omega$$

which are in general different from one another.

For regular crystals, however, it was shown before that these mean values must equal one another. Each of them can thus be written in the form:

$$\frac{1}{3}\int \frac{\mathfrak{A}_k^2 + \mathfrak{B}_k^2 + \mathfrak{C}_k^2}{g_k^3} d\Omega$$

which, in view of equation (14'), simply becomes:

$$\frac{1}{3N}\sum_k \int \frac{d\Omega}{g_k^3}$$

Instead of equation (45'), we now obtain for M the expression:

$$\left\{ \begin{aligned} -M = \frac{1}{144}\frac{\varkappa^2 k^2 T^2}{h\varrho} [(\alpha-\alpha_0)^2 + (\beta-\beta_0)^2 \\ + (\gamma-\gamma_0)^2] \sum_k \int \frac{d\Omega}{g_k^3} \end{aligned} \right. \tag{46}$$

which, on introducing the angle ϑ between direction of observation and direction of incidence, reduces to the simple form:

$$-M = \frac{1}{72}\frac{\varkappa^2 k^2 T^2}{h\varrho}(1 - \cos\vartheta) \sum_k \int \frac{d\Omega}{g_k^3} \tag{46'}$$

an expression whose structure is of course in agreement with the structure given at the end of the preceding section. The sum $\sum_k$ is the one which appears in the theory of specific heats at low temperatures. According to the computations of the factor in the T^3 law for regular crystals by Born and v. Kármán[5], we have:

$$C = \frac{4\pi^4}{15} \frac{k^4 T^3}{\varrho h^3} \sum_k \int \frac{d\Omega}{g_k^3} \tag{47}$$

if C indicates the specific heat (amount of heat per gram). By combining the two formulas (46') and (47) we finally obtain:

$$-M = \frac{5}{96\pi^4} \frac{x^2 h^2}{k^2} \frac{C}{T} (1 - \cos\vartheta) \tag{48}$$

from this expression the temperature effect at low temperature can be computed independent of elastic measurements from the observed value for the specific heat.

As illustrated once more clearly by equation (48) in view of the T^3 law for the specific heat, M and thus the temperature effect vanish completely at low temperatures, provided no zero-point energy exists. At $T = 0$, the scattered radiation, denoted by Z in the preceding section, vanishes for all directions of observation, and the decrease of the interference pattern intensity with increasing angle ϑ between direction of observation and direction of incidence no longer follows an exponential law. If zero-point energy is present, we obtain another picture. In this case not equation (30) but equation (34) holds for M; thus at all temperatures M is larger compared with the value given before by the temperature-independent amount:

$$\left\{ \begin{aligned} \Delta M = \mathfrak{S} \sum_k \frac{x^2}{4\mu\,\omega_k^2} \frac{h\,\omega_k}{2\pi} [(\alpha - \alpha_0)^2 \mathfrak{A}_k^2 + (\beta - \beta_0)^2 \mathfrak{B}_k^2 \\ + (\gamma - \gamma_0)^2 \mathfrak{C}_k^2] \end{aligned} \right. \tag{49}$$

an expression which, just as has been done before, can be transformed into:

$$\left\{ \begin{aligned} \Delta M = \frac{1}{4} \frac{N}{(2\pi)^3} \iiint d\varphi\, d\psi\, d\chi \sum_k \frac{x^2}{\mu\,\omega_k^2} \frac{h\,\omega_k}{2\pi} [(\alpha - \alpha_0)^2 \mathfrak{A}_k^2 \\ + (\beta - \beta_0)^2 \mathfrak{B}_k^2 + (\gamma - \gamma_0)^2 \mathfrak{C}_k^2] \end{aligned} \right. \tag{49'}$$

The computation to be carried out to find the value of ΔM from equation (49') is of course completely determined. The three roots ω_k and the associated constants $\mathfrak{A}_k$, $\mathfrak{B}_k$, $\mathfrak{C}_k$, must be calculated from equations (6) for all points φ, ψ, χ of the phase space, and then the integration over the phase space performed. To do this rigorously is not only very tedious but impossible numerically because for this procedure all quantities $A \ldots F$ in equation (7) would have to be known with greater accuracy as functions of φ, ψ, χ than is attainable by elastic measurements. It is of primary importance that in equation (49') the effect of high frequencies is no longer neutralized by Planck's function, as was the case for the preceding computations of this section. However, we absolutely require a numerical value for ΔM, which is essential for a discussion of the question of the zero-point energy; it may therefore be permissible to apply the approximative method which proved successful for the computation of the specific heat.

If we limit ourselves to the regular crystal system, we have already found that in general the factors of:

$$(\alpha-\alpha_0)^2, (\beta-\beta_0)^2, (\gamma-\gamma_0)^2$$

appearing in equation (49') are equal to one another. We can therefore write for each of these factors:

$$\frac{1}{12}\frac{N}{(2\pi)^3}\iiint d\varphi\, d\psi\, d\chi \sum_k \frac{\varkappa^2}{\mu\,\omega_k^2}\frac{h\,\omega_k}{2\pi}(\mathfrak{A}_k^2+\mathfrak{B}_k^2+\mathfrak{C}_k^2)$$

In view of equation (14) and introducing the angle ϑ used before, we have:

$$\Delta M = \frac{1}{96\pi^4}(1-\cos\vartheta)\iiint d\varphi\, d\psi\, d\chi \sum_k \frac{\varkappa^2 h}{\mu\,\omega_k} \tag{50}$$

or:

$$\Delta M = \frac{1}{96\pi^4}(1-\cos\vartheta)\frac{\varkappa^2 h}{\mu}\sum_k \iiint \frac{d\varphi\, d\psi\, d\chi}{\omega_k} \tag{50'}$$

Up to here the computation is rigorously correct; we now introduce the new variables ω_k and two angles of the unit sphere to replace φ, ψ, χ, and can substitute:

$$\frac{a^3}{g_k^3}\omega^2\, d\omega\, d\Omega$$

for $d\varphi d\psi d\chi$, where $d\Omega$ designates an element of the solid angle, and where, as an approximation, the velocity of propagation g_k is regarded as independent of frequency and direction of propagation. We now assume, as was done in the theory of specific heat, a maximum value ω_{max} for ω and pretend that we have to integrate over the interior of a sphere from $\omega = 0$ to ω_{max};* thus:

$$\Delta M = \frac{1}{48\pi^3}(1-\cos\vartheta)\frac{\varkappa^2 h}{\varrho}\omega_{max}^2 \sum \frac{1}{g_k^3} \tag{51}$$

For the isotropic body for which at least the independence of g_k on the direction of propagation is assured, the sum can be calculated in the form:

$$\sum \frac{1}{g_k^3} = \frac{1}{g_l^3} + \frac{2}{g_t^3}$$

by means of the observed or calculated velocities of propagation g_l and g_t of the longitudinal and transverse waves, respectively. Alternatively, equation (51) can also be correlated with the previously[6] introduced characteristic temperature:

$$\Theta = \frac{h\,\omega_{max}}{2\pi k} \tag{52}$$

It had been shown[6] that for a volume V with N atoms the relation:

$$3N = \frac{V}{6\pi^2}\omega_{max}^3\left(\frac{1}{g_l^3}+\frac{2}{g_t^3}\right) = \frac{V}{6\pi^2}\omega_{max}^3 \sum \frac{1}{g_k^3} \tag{52'}$$

*In this manner M. Born and Th. v. Kármán (*Physik. Z.*, *14*, *15* (1913)) accomplished the transition from their theory to the formula for the specific heat derived by me.

holds so that, introducing the volume associated with one atom, V/N, and the characteristic temperature, we obtain from equation (51):

$$\varDelta M = \frac{3}{16\pi^2}(1-\cos\vartheta)\frac{N}{V}\frac{x^2 h^2}{\varrho k}\frac{1}{\Theta} \tag{53}$$

Introducing Avogadro's number N (number of atoms per atomic weight) and the atomic weight A, we can write for monatomic bodies:

$$\varDelta M = \frac{3}{16\pi^2}\frac{x^2 h^2}{k}\frac{N}{A\Theta}(1-\cos\vartheta) \tag{53'}$$

While, in the absence of zero-point energy, the temperature effect vanishes for $T = 0$, in the other case, the temperature function converges towards $e^{-\Delta M}$, a quantity which under certain circumstances, must be quite noticeable in experiments, as shown by the numerical discussion in section VIII.

VII. Approximate Formula for the Temperature Function Valid at all Temperatures

It is known from the theory of specific heats and has also been shown for our case in the preceding section that rigorous computations cannot be carried to the end. On the other hand investigations regarding my approximate theory of specific heat have shown that for monatomic bodies and bodies, such as for instance KCl, that are to be treated similarly, only minute differences exist between theoretical and experimental course of the curves of specific heat as function of teperature. It therefore suggests itself to develop here also a simple approximation formula for the same cases which describes as closely as possible the experimental phenomena. In the following we will attempt to attain the following end: it should be possible to use the formula as a guide for the experiments which as yet have to be carried out on this subject. Let the crystal to which our computations relate be regular; further, we disregard the zero-point energy for the time being. For this case, in view of the general result obtained at the end of section V and recalling formula (14), we can write for M in accordance with equation (42):

$$-M = \frac{1}{3}\frac{1}{(2\pi)^3}(1-\cos\vartheta)\sum_k\iiint d\varphi\, d\psi\, d\chi \frac{x^2}{\mu\,\omega_k^2}\frac{\frac{h\,\omega_k}{2\pi}}{e^{\frac{h\omega_k}{2\pi k T}}-1} \tag{54}$$

if we again introduce, as was done in paragraph 6, the angle ϑ between the direction of observation and direction of incidence. In this approximation let the velocity of propagation g_k be considered constant, and the new variables:

$$\xi = \frac{h\,\omega_k}{2\pi k T} \tag{55}$$

and two coordinates of the unit sphere be introduced instead of φ, ψ, χ, using equation (43) and (44). We then have instead of equation (54), in the terminology used in section VI:

$$-M = \frac{1}{12\pi^2}(1-\cos\vartheta)\frac{x^2 k^2 T^2}{h\,\varrho}\sum_k\iiint\frac{\xi\, d\xi}{e^{\xi}-1}\frac{d\Omega}{g_k^3}$$

Let the integration with respect to ξ be extended from 0 to a maximum value of x which corresponds to the maximum limiting frequency of the elastic spectrum and for which we write Θ/T, introducing the above mentioned characteristic temperature Θ. Further the integral over the unit sphere is calculated as it would be calculated for an isotropic body for which:

$$\iint \frac{d\Omega}{g_k^3} = 4\pi\left(\frac{1}{g_l^3} + \frac{2}{g_t^3}\right)$$

We then find for M the expression:

$$-M = \frac{1}{3\pi}(1 - \cos\vartheta)\frac{\varkappa^2 k^2 T^2}{h\varrho}\left(\frac{1}{g_l^3} + \frac{2}{g_t^3}\right)\int_0^{x=\frac{\Theta}{T}} \frac{\xi\, d\xi}{e^\xi - 1} \tag{56}$$

The sum of the reciprocal third powers of the velocities of propagation may, in accordance with equations (52) and (52'), also be written in terms of Θ; if this is done, equation (56) can be replaced by:

$$-M = \frac{3}{4\pi^2}\frac{\varkappa^2 h^2}{\mu k \Theta}(1 - \cos\vartheta)\frac{1}{x^2}\int_0^x \frac{\xi\, d\xi}{e^\xi - 1} \tag{56'}$$

if μ designates the atomic mass of the respective substance. Introducing the number of atoms per gram-molecule N and the atomic weight A, equation (56') can be written in the form:

$$-M = \frac{3}{4\pi^2}\frac{\varkappa^2 h^2}{k}\frac{N}{A\Theta}(1 - \cos\vartheta)\frac{1}{x^2}\int_0^x \frac{\xi\, d\xi}{e^\xi - 1} \tag{56''}$$

Thus the required expression for M valid for any temperature has been found.

In the next section we will present a detailed numerical discussion of this formula; the two limiting cases $T \ll \Theta$ and $T \gg \Theta$ will be treated here.

(a) $T \ll \Theta$

In this case $x = \Theta/T \gg 1$, the upper limit of the integral may then be put equal to ∞ and we find:

$$\int_0^x \frac{\xi\, d\xi}{e^\xi - 1} = \int_0^\infty \frac{\xi\, d\xi}{e^\xi - 1} = \int_0^\infty \xi(e^{-\xi} + e^{-2\xi} + e^{-3\xi} + \ldots)\, d\xi$$

$$= 1 + \frac{1}{4} + \frac{1}{9} + \frac{1}{16} + \ldots = \frac{\pi^2}{6}$$

If, further, $1/x^2$ is replaced by T^2/Θ^2, then equation (56') for instance becomes:

$$-M = \frac{1}{8}\frac{\varkappa^2 h^2}{\mu k \Theta^3}(1 - \cos\vartheta)T^2 \tag{57}$$

We again find the result, stated previously in section VI for low temperatures, that -M is proportional to the square of the absolute temperature.*

(b) $T >> \theta$

In this case $x = \theta/T << 1$ and the integral in equation (56') can be approximately calculated to:

$$\int_0^x \frac{\xi\, d\xi}{e^{\xi}-1} = \int_0^x \frac{\xi\, d\xi}{\xi} = x$$

Hence equation (56') becomes:

$$-M = \frac{3}{4\pi^2}\,\frac{\varkappa^2 h^2}{\mu k \Theta^2}(1-\cos\vartheta)\,T \qquad (58)$$

The previous approximate theory resulted in the same temperature dependence. The only difference is that now we do not have to introduce an only approximately known "quasi-elastic force" f for the atom. This may be expressed as follows: Equation (58) determines the previously introduced quasi-elastic force f. We found in the approximate theory:

$$-M = 2\frac{\varkappa^2 k}{f}(1-\cos\vartheta)\,T$$

A comparison with equation (58) furnishes the equation:

$$f = \frac{4\pi^2}{3}\,\frac{\mu k^2 \Theta^2}{h^2}$$

which determines f. In view of the definitation of θ:

$$\Theta = \frac{h\nu_{max}}{k}$$

involving the limiting frequency of the elastic spectrum, we can also write:

$$\frac{f}{\mu} = \frac{2}{3}\,4\pi^2\nu^2_{max}$$

and formulate the final result as follows:

The present improved theory furnishes, in the limit for high temperatures, the same result as the previous theory, developed without regard to the mutual bonds between the atoms; in the latter theory, a value equal to $(2/3)^{\frac{1}{2}}$ times the maximum frequency in the elastic spectrum has to be introduced for the "frequency of the atoms."

The case of existing zero-point energy can be dealt with without further computations; the change ΔM which occurs in M if this

*The identity with equation (48) immediately follows if the expression which is valid for low temperatures:

$$C = \frac{12\pi^4}{5}\,\frac{k}{\mu}\,\frac{T^3}{\Theta^3}$$

is substituted for C.

condition is taken into account was calculated already in the preceding section. In view of equations (53) and (53') we find now:

$$-M = \frac{3}{4\pi^2}\frac{\varkappa^2 h^2}{\mu k \Theta}(1-\cos\vartheta)\left[\frac{1}{4}+\frac{1}{x^2}\int_0^x \frac{\xi\, d\xi}{e^{\xi}-1}\right] \tag{59}$$

instead of equation (56'). Thus a representation valid for any temperature has been obtained also for this case. No change has to be made in the above theorem; however, the temperatures to be called "high" are much higher than for vanishing zero-point energy.

VIII. Numerical Discussion and Graphical Presentation

The temperature dependence of M is determined by the integral in equations (56') and (59). We write it in the form:

$$\frac{1}{x^2}\int_0^x \frac{\xi\, d\xi}{e^{\xi}-1} = \frac{\Phi(x)}{x}$$

the function $\Phi(x)$ can be expanded as follows:*

$$\begin{aligned}\Phi(x) &= 1-\frac{x}{4}+\frac{B_1}{3!}x^2-\frac{B_2}{5!}x^4+\frac{B_3}{7!}x^6-+\dots\\ &= 1-\frac{x}{4}+\frac{x^2}{36}-\frac{x^4}{3600}+\frac{x^6}{211680}-\frac{x^8}{10886400}+\dots\end{aligned} \tag{60}$$

and:

$$\begin{aligned}\Phi(x) = \frac{\pi^2}{6}\frac{1}{x} - e^{-x}\left(1+\frac{1}{x}\right) - e^{-2x}\left(\frac{1}{2}+\frac{1}{4x}\right)\\ - e^{-3x}\left(\frac{1}{3}+\frac{1}{9x}\right) - \dots\end{aligned} \tag{60'}$$

of which equation (60) is suitable for small values of x and equation (60') for large values of x. If equation (60) is extended only to the third term, the error is less than 1 per cent for $x = 2$; the same is true for equation (60') starting from x slightly larger than 2. The series follow in a manner analogous to that shown for the functions appearing in the theory of specific heat. The following table for $\Phi(x)$ was calculated from equations (60) and (60').

Table I

x	$\Phi(x)$	x	$\Phi(x)$	x	$\Phi(x)$	x	$\Phi(x)$
0	1	1.2	0.740	3	0.483	9	0.183
0.2	0.951	1.4	0.704	4	0.388	10	0.164
0.4	0.904	1.6	0.669	5	0.321	12	0.137
0.6	0.860	1.8	0.637	6	0.271	14	0.114
0.8	0.818	2.0	0.607	7	0.234	16	0.103
1.0	0.778	2.5	0.540	8	0.205	20	0.0822

*B_1, B_2, B_3... denote Bernouilli's numbers.

The functions Φ/x and $\frac{1}{4} + \Phi/x$ determine the temperature dependence (compare equations (62) and (62'), respectively). To obtain a clear representation we have plotted the two functions as functions of $1/x = T/\theta$ in Figure 3. They lie between the two approximations:

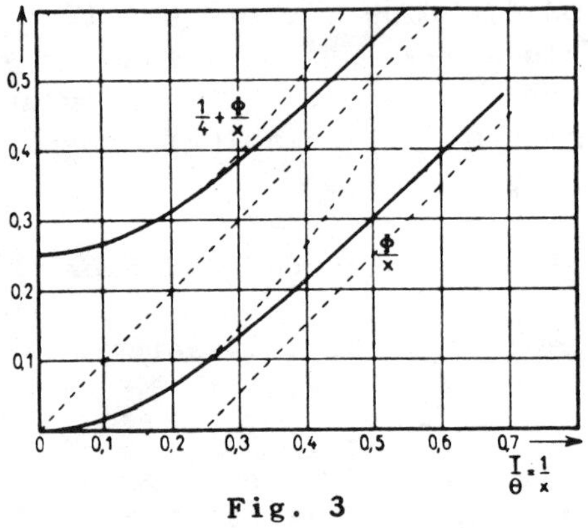

Fig. 3

$$\frac{\Phi}{x} = \frac{\pi^2}{6}\frac{T^2}{\Theta^2} \quad \text{and} \quad \frac{\Phi}{x} = \frac{T}{\Theta} - \frac{1}{4}$$

and:

$$\frac{1}{4} + \frac{\Phi}{x} = \frac{1}{4} + \frac{\pi^2}{6}\frac{T^2}{\Theta^2} \quad \text{and} \quad \frac{1}{4} + \frac{\Phi}{x} = \frac{T}{\Theta}$$

respectively, which are indicated by dashed lines.

It will be readily seen that the above approximate equation holds for low temperatures to $T/\theta = 1/8$, and the approximate equation for high temperatures holds for $T/\theta = 1.6$, with an error not exceeding 1 per cent. If an error of 10 per cent is admitted, the ranges of validity extend from $T/\theta = 0$ to $T/\theta = 1/5.5$, and from $T/\theta = 1/1.2$ to $T/\theta = \infty$.

Let the pure number responsible for the absolute magnitude of the temperature effect in x-ray interference patterns be designated by P; then from equations (56') or (56") and (59) we have:

$$P = \frac{1}{4\pi^2}\frac{\varkappa^2 h^2}{\mu k \Theta} = \frac{3}{4\pi^2}\frac{N h^2}{k}\frac{\varkappa^2}{A\Theta} \tag{61}$$

where $\varkappa$ denotes 2π times the reciprocal of the x-ray wave length, h and k the known universal constants, θ the temperature characteristic for the course of the specific heat, μ the mass of an atom, A the conventional atomic weight, and N the universal number giving the number of atoms per gram-molecule. If we introduce the wave length λ and put according to Planck: $N = 6.20 \times 10^{23}$, $k = 1.34 \times 10^{-16}$ erg, $h = 6.41 \times 10^{-27}$ erg-sec., P can be calculated from the relation:

$$P = \frac{0.571 \cdot 10^{-12}}{A\Theta\lambda^2} \tag{61'}$$

where λ is given in centimeters.

The exponent of the exponential function which controls the course of the intensity and which we previously designated by M can be written in the form:

(a) no zero-point energy:

$$M = -P(1 - \cos\vartheta)\frac{\Phi(x)}{x} \qquad (62)$$

(b) in the presence of zero-point energy:

$$M = -P(1 - \cos\vartheta)\left[\frac{1}{4} + \frac{\Phi(x)}{x}\right] \qquad (62')$$

where P is the number just defined, Φ that function of $x = \Theta/T$ which was discussed at the beginning of this section, and ϑ the angle between direction of observation and direction of incidence.

In connection with section V we recall that the scattered intensity is proportional to:

$$N\left(1 - e^{M}\right) \qquad (63)$$

while the interference intensity is measured by:

$$e^{M} L \qquad (63')$$

where L designates Laue's expression defined there, and N the number of irradiated atoms.

As the first example we chose the diamond whose strange experimental behavior initiated this theory. Here $A = 12$ and $\Theta = 1830$; we thus have for $\lambda = 10^{-8}$ cm. (compare equation (61')):

$$P = 0.26$$

for $\lambda = 5 \times 10^{-9}$ cm. P becomes four times as large, *i.e.:*

$$P = 1.04$$

Using the table for $\Phi(x)$, the exponent of the temperature function M can now readily be computed from equation, (62) or (62') for any desired temperature T and any desired angle ϑ. After computation of e^{M}, the expressions (63) and (63') result, which characterize the course of the scattered intensity and interference intensity, respectively.

The calculation was carried out for two wave lengths, $\lambda = 7.1 \times 10^{-9}$ cm. and $\lambda = 3.5 \times 10^{-9}$ cm.; the result is illustrated in Figures 4 and 5.

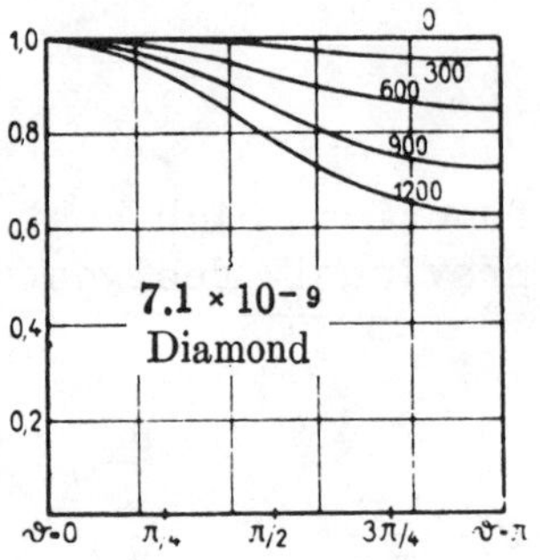

Fig. 4a.

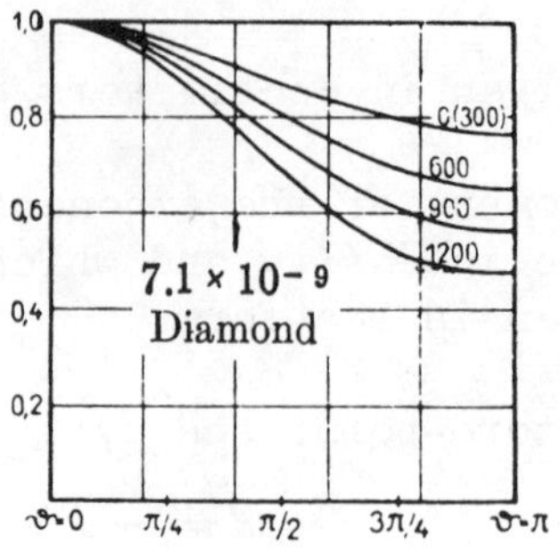

Fig. 4b.

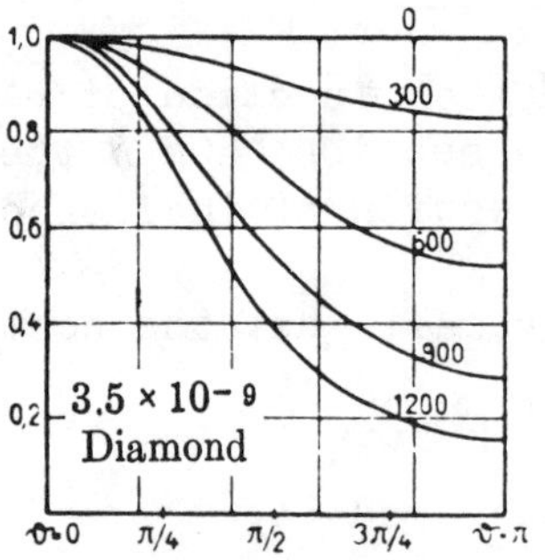

Fig. 5a.

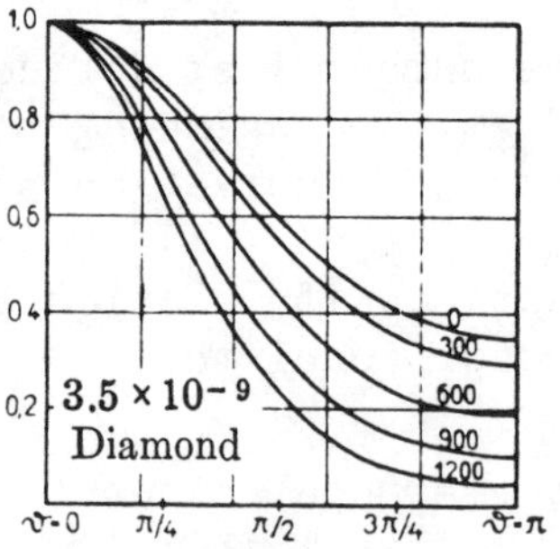

Fig. 5b.

The quantity e^M is plotted in each figure at right angles to the horizontal ϑ axis, which extends from 0 to π, for the temperatures T = 0, 300, 600, 900, 1200. The figures with index *a* relate to the case of vanishing zero-point energy; the figures with index *b* relate to the case of finite zero-point energy equation (62')). For Figures 4a and 4b, $\lambda = 7.1 \times 10^{-9}$ cm.; for Figures 5a and 5b, $\lambda = 3.5 \times 10^{-9}$ cm. The numerals next to the curves indicate the temperatures for which the respective curve was calculated.

The ordinates of the curves measure the decrease in intensity of the interference pattern; the vertical distance between curves and the upper horizontal line at the height 1 is proportional to the simultaneously present scattered intensity for the same wave length, in as much as the change is caused by the thermal movement.*

The figures illustrate the large effect of the zero-point energy, provided the observations are made at sufficiently short waves. Whereas in Figures 4a and 5a, the curve for T = 0 coincides with the horizontal line in the height 1, this is not the case in the presence of zero-point energy. As illustrated by Figures 4b and 5b, the limiting curve for T = 0 deviates considerably from the above-mentioned horizontal line, particularly for the wave length 3.5×10^{-9} . For shorter wave length, the difference would be greater still.

*For an adjustment to obtain the actually observable intensity which occurs in the actual experiment compare the second supplement at the end.

To illuminate further the special position of diamond, the same calculation has been carried out for sylvite (Figs. 6 and 7).*

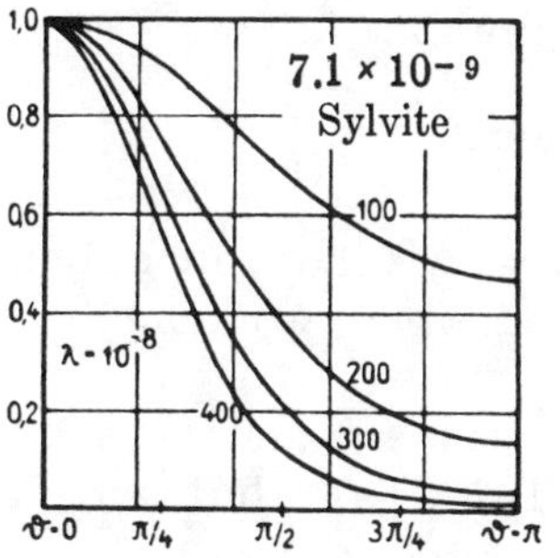

Fig. 6a.

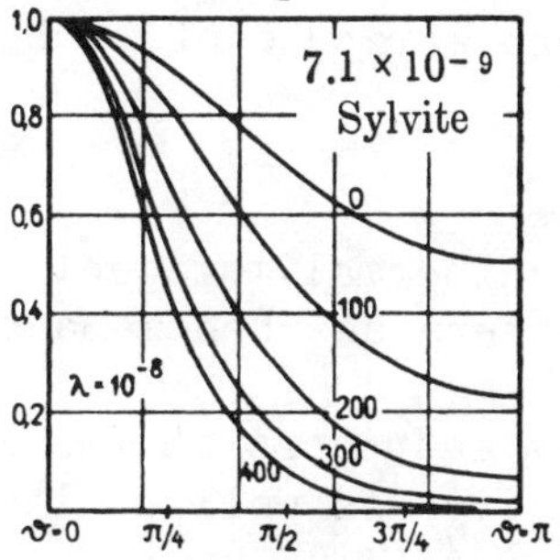

Fig. 6b.

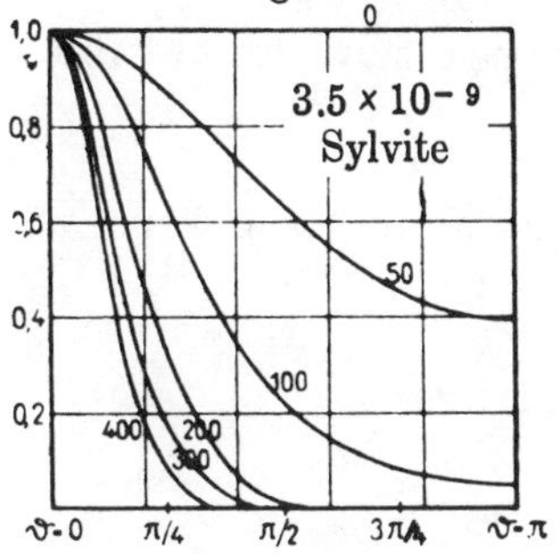

Fig. 7a.

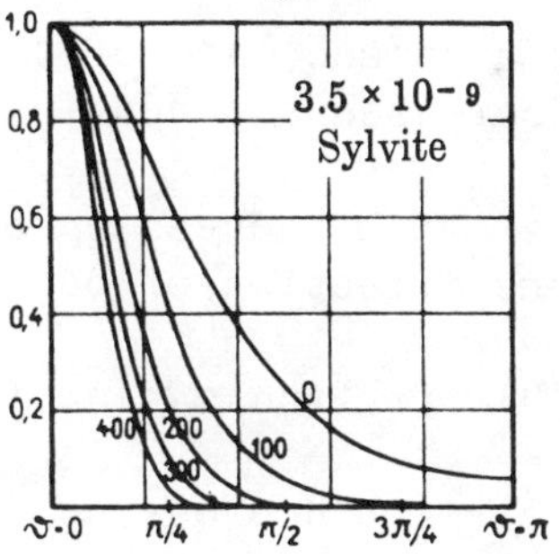

Fig. 7b.

In this instance a fairly good representation of the curve for the specific heat is obtained from my formula with $\theta = 219$. This value for θ has been assumed; further, A has been put equal to 37 (mean of K = 39 and Cl = 35). The curves have again been computed for $\lambda = 7.1 \times 10^{-9}$ cm. and $\lambda = 3.5 \times 10^{-9}$ cm. for vanishing zero-point energy (Figs. 6a and 7a) and for finite zero-point energy (Figs. 6b and 7b).

The difference between the two concepts in regard to the limiting curve for $T = 0$ is even more evident here than for diamond. Also the greater amplitude of the thermal movement is clearly demonstrated. It is in accordance herewith that we notice great changes in the temperature range below average room temperature, while there is no great difference any more between $T = 0$ and $T = 300$ for the diamond. We therefore plotted the curves for $T = 0$, 100, 200, 300, 400. It may further be pointed out that for sylvite and vanishing zero-point energy a rapid change of the intensity curve starts not before $T = 100$ approximately. We therefore entered in Figure 7a also the curve for the intermediate temperature $T = 50$.

The curves appear to present definite evidence that — assuming at least approximate correctness of the theory, — it should be possible to arrive at a definite decision for or against the hypothesis of zero-point energy on the basis of measurements of the intensity distribution as a function of temperature.

*Sylvite was chosen because the probability that this substance may be treated as monatomic is very great.

Since unfortunately no experimental material for this investiga- is at present available, I must refrain from making any presumption as to the probable result of such experiments.

Summary*

1. The thermal movement of the atoms has an essential influence on the interference phenomena observed with x-rays.

2. The sharpness of the interference maxima is not affected; their intensity, as well as its spatial distribution, however, is affected.

3. As a consequence of the thermal movement, the interference intensity decreases exponentially

(a) with increasing angle between the direction of incidence and the direction of observation;

(b) with increasing temperature;

(c) with decreasing wave length.

4. The exponent of the exponential function just mentioned vanishes at $T = 0$ in the absence of zero-point energy; it remains finite and assumes a substantial value if zero-point energy exists.

5. The exponent is always inversely proportional to the square of the wave length.

6. The interference intensity is always accompanied by a scattered intensity which has its highest intensity value where the interference intensity has its lowest value and vice versa.

7. The course of the phenomena can be approximately predicted by computations, provided data on the behavior of the specific heat as a function of temperature are available.

8. In this approximation a similarity law holds, as for the specific heat of monatomic bodies, according to which also in this instance the temperature dependence is a function only of the ratio of the characteristic temperature θ to the temperature of observation.

Supplements Appended when Reading Proofs (November 26, 1913)

1. In the meantime A. Sommerfeld informed me of a version of

*Please note the limitation in the second footnote on page 14.

the above results which permits of readily surveying the spatial distribution of the thermal effect on the spots of the interference pattern.

We found that, for instance, for vanishing zero-point energy the thermal effect is measured by an exponential function with an exponent M, which from equations (62) and (61), and introducing the directional cosines:

$$\alpha_0, \beta_0, \gamma_0, \alpha, \beta, \gamma$$

can be written in the form:

$$M = -\frac{3h^2}{2\mu k \Theta \lambda^2}\frac{\Phi(x)}{x}[(\alpha-\alpha_0)^2+(\beta-\beta_0)^2+(\gamma-\gamma_0)^2] \tag{63}$$

According to the fundamental formulas of Laue's theory:

$$\alpha-\alpha_0 = h_1\frac{\lambda}{a}, \quad \beta-\beta_0 = h_2\frac{\lambda}{a}, \quad \gamma-\gamma_0 = h_3\frac{\lambda}{a} \tag{64}$$

(valid for the regular system at any angle of incidence), we can write instead for a spot of the interference pattern:

$$M = -\frac{3h^2}{2\mu k \Theta a^2}\frac{\Phi(x)}{x}(h_1^2+h_2^2+h_3^2) \tag{63'}$$

an expression that no longer contains the wave length as such but only the integers h_1, h_2, h_3 characteristic for the spot in the interference pattern.

As has been stressed repeatedly, our initial expression (63) for M relates to a definite wave length and compares the temperature effect on the different directions ϑ in which the respective wave length can occur. In Laue's photographs the wave length is different for different spots; the expression (63'), however, permits comparison of spots for different wave lengths in a simple manner.

According to Bragg's concepts, the occurrence of the interference spot h_1, h_2, h_3 can be described as a reflection from a plane which is possible crystallographically and whose indices are in the ratio $h_1 : h_2 : h_3$. Such a plane is the less densely populated with atoms the larger the sum of the square of its indices. Thus equation (63') states that the thermal decrease in intensity of the interference spots is the more important the less densely the reflecting plane is populated with atoms.

2. During printing of the preceding note, H. A. Lorentz sent me by letter several remarks related to the computation of the actually observable intensity of the interference spots. These remarks must be given attention if an experimental decision between the two hypotheses by Planck is to be made in the manner explained before.

In reality we experiment neither with absolutely monochromatic radiation nor with a single direction of incidence. Consequently energies of different wave lengths and originating from different directions of incidence will gather in a certain direction of observation.

Let us first neglect the thermal effect; we then calculate from equation (3) the secondary intensity for waves incident within the solid angle $d\Omega_0$ and extending over the spectral range $d\kappa$, which intensity can be represented by the expression:

$$\frac{A}{r^2} d\Omega_0 \, d\varkappa \sum \sum e^{i\varkappa a[(\alpha-\alpha_0)(n_1-n_1')+(\beta-\beta_0)(n_2-n_2')+(\gamma-\gamma_0)(n_3-n_3')]} \tag{64}$$

if, for the sake of simplicity, we confine these considerations to the regular crystal system. If α_0, β_0, γ_0, α, β, γ, and κ have been so chosen that Laue's conditions are met for a point with the characteristic numbers h_1, h_2, h_3, and if subsequently α_0, β_0, γ_0 are slightly varied, these variations can be measured by three numbers ϵ_1, ϵ_2, ϵ_3, which can be introduced by means of the equations:

$$\varkappa a(\alpha - \alpha_0) = h_1 2\pi + \varepsilon_1, \qquad \varkappa a(\beta - \beta_0) = h_2 2\pi + \varepsilon_2$$
$$\varkappa a(\gamma - \gamma_0) = h_3 2\pi + \varepsilon_3$$

The actually observed intensity is obtained by a summation over $d\Omega_0 \, d\kappa$ over a finite region; recalling that in the new coordinates ϵ_1, ϵ_2, ϵ_3 the corresponding element of integration assumes the value:

$$\frac{1}{\varkappa^2 a^3} \frac{2}{[(\alpha-\alpha_0)^2 + (\beta-\beta_0)^2 + (\gamma-\gamma_0)^2]} d\varepsilon_1 \, d\varepsilon_2 \, d\varepsilon_3$$

the integrated expression can be written in the form:

$$\frac{A^2}{r^2} \frac{1}{\varkappa^2 a^3} \frac{2}{[(\alpha-\alpha_0)^2 + (\beta-\beta_0)^2 + (\gamma-\gamma_0)^2]}$$
$$\sum \sum \iiint d\varepsilon_1 \, d\varepsilon_2 \, d\varepsilon_3 \, e^{i[\varepsilon_1(n_1-n_1')+\varepsilon_2(n_2-n_2')+\varepsilon_3(n_3-n_3')]} \tag{64'}$$

The integration is to be executed between the limit:

$$-e_1 < \varepsilon_1 < +e_1, \quad -e_2 < \varepsilon_2 < +e_2, \quad -e_3 < \varepsilon_3 < +e_3$$

where e_1, e_2, e_3 are small but finite numbers. Thus we obtain:

$$\frac{A^2}{r^2} \frac{8}{\varkappa^2 a^3} \frac{2}{[(\alpha-\alpha_0)^2 + (\beta-\beta_0)^2 + (\gamma-\gamma_0)^2]}$$
$$\sum \sum \frac{\sin(n_1-n_1')e_1}{n_1-n_1'} \frac{\sin(n_2-n_2')e_2}{n_2-n_2'} \frac{\sin(n_3-n_3')e_3}{n_3-n_3'} \tag{64''}$$

If n_1, n_2, n_3 are fixed and the summation with respect to n_1', n_2', n_3' is executed, the expression:

$$\int_{-\infty}^{+\infty}\int_{-\infty}^{+\infty}\int_{-\infty}^{+\infty} \frac{\sin e_1 u_1}{u_1} \frac{\sin e_2 u_2}{u_2} \frac{\sin e_3 u_3}{u_3} du_1 \, du_2 \, du_3 = \pi^3$$

can be substituted for the partial sum in view of the small values of e_1, e_2, e_3.

If N designates the number of irradiated atoms, the same computation has to be repeated N times as n_1, n_2, n_3 are covered. The following expression is thus obtained for the observable intensity at the interference spot:

$$\frac{A^2}{r^2} \frac{8\pi^3}{\varkappa^2 a^3} \frac{2N}{[(\alpha-\alpha_0)^2 + (\beta-\beta_0)^2 + (\gamma-\gamma_0)^2]} \tag{65}$$

or in view of Laue's condition:

$$\frac{A^2}{r^2}\frac{4\pi}{a}\frac{N}{h_1^2+h_2^2+h_3^2} \qquad (65')$$

The effect of the thermal movement can be measured by the additional factor e^M. However it will be seen that as a consequence of the averaging process inherent in the experiment a further decrease occurs which is measured by the sum of the squares of the characteristic integers in the denominator of equation (65'). Furthermore the spatial radiation of a single atom may be different for different directions. In the most simple case (for instance, radiation of a bound electron) this can be taken into account by putting A^2 proportional to $1 + \cos^2\vartheta$. (ϑ denotes the angle between direction of incidence and direction of observation).

Bibliography

1. P. Debye, *Verhandl. deut. physik. Ges.*, *15*, 678, 738, 857 (1913).
2. M. v. Laue, *Sitzber. kgl. bayer. Akad. Wiss.*, 303 (1912). Compare further the survey in *Rapport du deuxieme Conseil de physique Solvay*.
3. P. Debye, *Ann. Physik*, *39*, 789 (1912).
4. M. Born and Th. v. Kármán, *Physik. Z.*, *14*, 65 (1913).
5. M. Born and Th. v. Kármán, *Physik. Z.*, *14*, 15 (1913).
6. P. Debye, *loc. cit.*

SCATTERING OF X-RAYS
(Zerstreuung von Röntgenstrahlen*)

P. Debye

Translated from
Annalen der Physik, Vol. 46, 1915, pages 809-823.

The recent developments of our views regarding the structure of the atom has forced us to recognize the possibility of electrons in motion which, in spite of large accelerations, do not radiate energy. We have to assume, for instance, the presence of two electrons in a hydrogen molecule, situated opposite one another on a circle 1.05×10^{-8} cm. in diameter and revolving with an angular velocity $\omega = 4.21 \times 10^{16}$ sec.$^{-1}$.[1] If the field generated by this motion is computed on the basis of the Maxwell-Lorentz equations of electrodynamics, the "tremendously large" value of 4.9×10^{-2} erg/sec. is obtained for the energy radiated per second. This should be compared with the kinetic energy of both electrons which is only 4.1×10^{-11} erg so that, on the basis of recognized principles, one arrives at the conclusion that a hydrogen molecule should destroy itself in approximately 10^{-8} sec. by its own radiation.+ We are therefore forced to consider the motion in the path determined by the h-hypothesis as non-radiating which is in sharp contrast to accepted principles.

On the other hand, the ordinary dispersion can be explained completely on the basis of the above model without the necessity of abandoning the known fundamentals of mechanics and electrodynamics. This proves that disturbances in the path required by the h-hypothesis behave normally in every respect. This fact becomes even clearer if the energy scattered from an incident wave is measured. It is well-known that J. J. Thomson showed that the x-ray radiation scattered by an atom can be computed according to the laws of electrodynamics and that he thereby obtained a method for determining experimentally the number of electrons per atom.

*Received February 27, 1915.

+This interval of time may appear less unlikely if we consider that 7×10^7 revolutions take place within 10^{-8} sec.

In Thomson's computations the energy scattered by one electron was evaluated and the overall effect obtained by multiplication by the number of electrons present. As long as no further details or more reliable information regarding the arrangement of the electrons in the atom as available, this procedure had to be considered satisfactory. In the meantime it became progressively clearer that the arrangement of the electrons in the atom shows regularities, as for instance in hydrogen where two electrons are located at a constant distance of 1.05×10^{-8} cm. On the other hand the wave length of the x-rays involved in these scattering experiments is also of the order of 10^{-8} cm. Consequently if atomic models with electron rings, to which hydrogen belongs as a special case, have any correspondence to reality, we must expect the atoms themselves to show occasional interference patterns when irradiated with x-rays; these interference patterns cannot be obliterated entirely even though the atoms have random orientation in space.

Exactly this phenomenon, however, is already known. The repeatedly mentioned and apparently not understood asymmetrical distribution of the scattered radiation[2], as well as the interference patterns, by Friedrich[3] should fundamentally both be interpreted in this sense.

To obtain a complete survey of the phenomena in this field, I have in the following computed the scattered energy and its distribution in space for an aggregation of atoms oriented at random and which contain each one ring with any desired number of electrons. The computation was carried through entirely on the basis of generally accepted principles. It is shown that for long waves the total scattered radiation is proportional to the square of the number of electrons and that the spatial distribution corresponds to that of a dipole. With decreasing wave length the total radiation gradually approaches a value proportional to the first power of the number of electrons, while its spatial distribution is still that corresponding to a dipole, with the exception of the region closely surrounding the direction of incidence of the primary radiation. In this direction itself the radiation is proportional to the square of the number of electrons in all cases, and in the region surrounding this direction the interference patterns photograph as rings.

If the likelihood of the following reasoning is admitted, the experimental investigation of scattered radiation, particularly of light atoms, appears to me to assume great interest, since it must be possible in this way to establish by experiment the particular arrangement of the electrons in the atoms. Such an investigation then assumes the significance of an ultra-microscopy of the interior of an atom.

1. The Energy Scattered by a Multitude of Free Electrons

The electron multitude consists of p individuals which, due to the static and dynamic forces active in the interior of the model, can remain at constant mutual distances.

Let the velocities in the interior of the system be so small that the change in position of any particle during the time of one oscillation of the incident wave will be small compared with the associated wave length. Further, let the frequency of the incident wave be large compared with the eigenfrequency of the disturbed system. These two assumptions are essential for setting up and solving the computation as easily as is done in the following. Further, they are met in practice for conventional x-rays and for light atoms.

In view of these assumptions we may now with negligible error:

(a) neglect the forces effective between electrons and between electrons and nuclei, i.e., we may consider the electrons as absolutely free. We may further

(b) neglect the movements of the electron aggregate as a whole and consider it as fixed in space, as long as the incident wave has not yet caused any disturbance.

The incident x-ray radiation is assumed to be a plane polarized wave of frequency ω propagating in the direction of the positive x-axis of a rectangular coordinate system to which the rest positions of the p electrons will be refered (by specifying each of their coordinates). Assuming the electric field of this wave to have a component of amplitude E, in the z-direction only, then the components of the electric field will be given by the known expressions:

$$\begin{cases} \mathfrak{E}_x = 0 \\ \mathfrak{E}_y = 0 \\ \mathfrak{E}_z = E e^{i\omega t} e^{-ikx} \end{cases} \qquad k = \frac{\omega}{c} \tag{1}$$

(c = velocity of light).

Before the disturbance, the electron indicated by the subscript n may have the coordinates x_n, y_n, z_n; these coordinates will change by ξ_n, η_n, ζ_n as a result of the influence of the incident wave. It follows from the above that ξ_n as well as η_n is zero, while ζ_n follows from the classical equation of mechanics:

$$\mu \frac{d^2 \zeta_n}{dt^2} = -e E e^{-ikx_n} e^{i\omega t}$$

and is equal to:

$$\zeta_n = \frac{\varepsilon}{\mu \omega^2} E e^{-ikx_n} e^{i\omega t} \tag{2}$$

where $-\varepsilon$ and μ designate the charge and mass, respectively, of an electron.

The question to be answered in this paragraph can then be expressed as follows. An electron multitude of p electrons is given. Each electron oscillates about its rest position x_n, y_n, z_n, the oscillations being described by deviations:

$$\begin{cases} \xi_n = 0, \\ \eta_n = 0, \\ \zeta_n = \frac{\varepsilon}{\mu \omega^2} E e^{-ikx_n} e^{i\omega t} \end{cases}$$

What is the intensity of the wave observed at a large distance R from the system in an arbitrarily chosen direction specified by the direction cosines α, β, γ?

For each electron this question is identical with the problem, repeatedly treated in the literature (for the first time presumably by H. Hertz), of the radiation from an oscillating dipole. The solution obtained on the basis of Maxwell's equations is given by the following statement:

Let the vector $\mathfrak{A}$ be defined by the three equations:

$$\mathfrak{A}_x = 0$$

$$\mathfrak{A}_y = 0$$

$$\mathfrak{A}_z = -\varepsilon\zeta_n\frac{e^{-ikr}}{r}$$

where r is the distance between the rest position of the n-th electron and the point of observation. The electric and magnetic field intensities of the scattered wave can then be computed from the two equations:

$$\mathfrak{E} = -k^2\mathfrak{A} - \text{grad div }\mathfrak{A}$$

$$\mathfrak{H} = -ik \text{ rot }\mathfrak{A}$$

If r is so large that $kr \gg 1$, the computation gives, for instance, for the three components of the electric field strength of the scattered wave the well-known result:

$$\mathfrak{E}_x = -\alpha\gamma k^2\varepsilon\zeta_n\frac{e^{-ikr}}{r}$$

$$\mathfrak{E}_y = -\beta\gamma k^2\varepsilon\zeta_n\frac{e^{-ikr}}{r}$$

$$\mathfrak{E}_z = (1-\gamma^2)k^2\varepsilon\zeta_n\frac{e^{-ikr}}{r}$$

Finally, considering that the observation is made at a large distance, we can substitute in place of kr the expression:

$$kr = kR - k(\alpha x_n + \beta y_n + \gamma z_n)$$

in which R designates the distance between the point of observation and the origin of the coordinate system. Hence we obtain for the n-th electron from Eq. (2):

$$\left\{\begin{aligned} \mathfrak{E}_x &= -\alpha\gamma\frac{\varepsilon^2}{\mu c^2}E\frac{e^{i(\omega t-kR)}}{R}e^{ik[(\alpha-1)x_n+\beta y_n+\gamma z_n]} \\ \mathfrak{E}_y &= -\beta\gamma\frac{\varepsilon^2}{\mu c^2}E\frac{e^{i(\omega t-kR)}}{R}e^{ik[(\alpha-1)x_n+\beta y_n+\gamma z_n]} \\ \mathfrak{E}_z &= (1-\gamma^2)\frac{\varepsilon^2}{\mu c^2}E\frac{e^{i(\omega t-kR)}}{R}e^{ik[(\alpha-1)x_n+\beta y_n+\gamma z_n]} \end{aligned}\right. \tag{3}$$

while the wave scattered from the complete system and at a large distance is described by the field strengths:

$$\left\{\begin{aligned} \mathfrak{E}_x &= -\alpha\gamma\frac{\varepsilon^2}{\mu c^2}E\frac{e^{i(\omega t-kR)}}{R}\sum_1^p e^{ik[(\alpha-1)x_n+\beta y_n+\gamma z_n]} \\ \mathfrak{E}_y &= -\beta\gamma\frac{\varepsilon^2}{\mu c^2}E\frac{e^{i(\omega t-kR)}}{R}\sum_1^p e^{ik[(\alpha-1)x_n+\beta y_n+\gamma z_n]} \\ \mathfrak{E}_z &= (1-\gamma^2)\frac{\varepsilon^2}{\mu c^2}E\frac{e^{i(\omega t-kR)}}{R}\sum_1^p e^{ik[(\alpha-1)x_n+\beta y_n+\gamma z_n]} \end{aligned}\right. \tag{4}$$

where the sums are to be taken from 1 to p, *i.e.*, over all electrons present in the system.

For the energy transport, *i.e.*, for the intensity, the square of the electric amplitude is the determining factor; it can be computed as the sum of the squares of the absolute values of $\mathfrak{E}_x$, $\mathfrak{E}_y$, and $\mathfrak{E}_z$; stated as a formula: $|\mathfrak{E}|^2 = |\mathfrak{E}_x|^2 + |\mathfrak{E}_y|^2 + |\mathfrak{E}_z|^2$

If the absolute values corresponding to Eq. (4) are substituted:

$$|\mathfrak{E}|^2 = \frac{\varepsilon^4}{\mu^2 c^4}\frac{E^2}{R^2}\{\alpha^2\gamma^2 + \beta^2\gamma^2 + (1-\gamma^2)^2\} \left|\sum_1^p e^{ik[(\alpha-1)x_n + \beta y_n + \gamma z_n]}\right|^2$$

is obtained.

In view of the relation: $\alpha^2 + \beta^2 + \gamma^2 = 1$

and by transformation of the square of the sum into a double sum, we obtain:

$$|\mathfrak{E}|^2 = \frac{\varepsilon^4}{\mu^2 c^4}\frac{E^2}{R^2}(1-\gamma^2)\sum_1^p\sum_1^p e^{ik[(\alpha-1)(x_n - x_m) + \beta(y_n - y_m) + \gamma(z_n - z_m)]} \tag{5}$$

As stressed at the beginning of this section, the expression (5) relates to the case of an incident wave polarized at right angles to the z-axis. If we had assumed a polarization at right angles to the y-axis, the factor $(1 - \gamma^2)$ would be replaced by the factor $(1 - \beta^2)$. For the case of an unpolarized wave, met in practice, the mean of the two values is obtained for $|\mathfrak{E}|^2$. Since:

$$\frac{(1-\gamma^2) + (1-\beta^2)}{2} = \frac{1+\alpha^2}{2}$$

the result for an unpolarized wave is given by:

$$|\mathfrak{E}|^2 = \frac{\varepsilon^4}{\mu^2 c^4}\frac{E^2}{R^2}\frac{1+\alpha^2}{2}\sum\sum e^{ik[(\alpha-1)(x_n - x_m) + \beta(y_n - y_m) + \gamma(z_n - z_m)]} \tag{6}$$

Remembering that E^2 is the square of the amplitude of the incident wave, the final result may be stated in the following form:

If an unpolarized wave is incident on the system of electrons under consideration, the ratio v of the intensity observed at a large distance R, with directional cosines α, β, γ, to the incident intensity can be computed from the formula:

$$v = \frac{\varepsilon^4}{\mu^2 c^4}\frac{1+\alpha^2}{2}\frac{1}{R^2}\sum\sum e^{ik[(\alpha-1)(x_n - x_m) + \beta(y_n - y_m) + \gamma(z_n - z_m)]} \tag{7}$$

2. Scattering by Very Many Molecules with Electron Rings

In an amorphous body, all orientations of the atoms (molecules) are equally probable. If a part of this body, containing N atoms, is irradiated, the intensity to be observed at an arbitrary direction α, β, γ is secured by first finding the mean intensity scattered by one atóm with respect to all possible orientations, and subsequently multiplying by N.

Without further and more specific information, the system which constitutes the basis for eq. (7) shall be assumed to form an atom. Then we shall have to find the mean of eq. (7) with respect to all possible orientations of the electron aggregate, since with the ratio v we may calculate just as well as with the intensity.

The easiest way to find the mean is to first consider a single term of the double sum and, then, considering the structure of the exponent of the exponential function, temporarily to introduce two vectors $\mathfrak{A}$ and $\mathfrak{B}$ such that in rectangular coordinates the components of $\mathfrak{A}$ are $(\alpha - 1)$, β, γ and the components of $\mathfrak{B}$ are $(x_n - x_m)$, $(y_n - y_m)$, $(z_n - z_m)$. The exponent is then equal to the scalar product $(\mathfrak{A}\mathfrak{B})$ of these two vectors. If now the atom assumes all possible orientations, any arrow rigidly connected with the atom will assume all directions in space with equal frequency. Vector $\mathfrak{B}$, for instance, may be considered to be such an arrow, while vector $\mathfrak{A}$, which is determined by the direction of incidence and the direction of observation, is constant as to direction and magnitude. If we write:

$$(\mathfrak{A}\mathfrak{B}) = AB\cos\Theta$$

introducing the lengths A and B of vectors $\mathfrak{A}$ and $\mathfrak{B}$, respectively, and the angle θ between them, we have to multiply the expression:

$$e^{ikAB\cos\Theta}$$

by the element $d\Omega$ of the spatial angle, take the sum with respect to all possible positions of $d\Omega$, and finally to divide by 4π. This computation leads to the result:

$$\frac{1}{4\pi}\int d\Omega\, e^{ikAB\cos\Theta} = \frac{1}{2}\int_0^\pi e^{ikAB\cos\Theta}\sin\Theta\, d\Theta = \frac{\sin(kAB)}{kAB} \tag{8}$$

Now

$$A^2 = (\alpha - 1)^2 + \beta^2 + \gamma^2 = 2(1 - \alpha) = 4\sin^2\frac{\vartheta}{2}$$

provided the angle between the direction of observation and the direction of incidence is designated by ϑ so that α = cos . For A itself we have:

$$A = 2\sin\frac{\vartheta}{2} \tag{9}$$

On the other hand B represents, according to its definition, the length of the line connecting the n-th with the m-th electron. Let us designate this length by s_{nm}, then after completion of the averaging process our expression (7) will be transformed into:

$$\bar{v} = \frac{\varepsilon^4}{\mu^2 c^4}\frac{1+\cos^2\vartheta}{2}\frac{1}{R^2}\sum_n\sum_m\frac{\sin\left[2ks_{nm}\sin\frac{\vartheta}{2}\right]}{\left[2ks_{nm}\sin\frac{\vartheta}{2}\right]} \tag{10}$$

In a direction deviating by an angle ϑ from the primary beam, the fraction:

$$V = \frac{N\varepsilon^4}{\mu^2 c^4}\frac{1+\cos^2\vartheta}{2}\frac{1}{R^2}\sum_n\sum_m\frac{\sin\left[2ks_{nm}\sin\frac{\vartheta}{2}\right]}{\left[2ks_{nm}\sin\frac{\vartheta}{2}\right]} \tag{11}$$

of the incident intensity will be observed at a distance R from the body, provided a total number of N atoms (molecules) is irradiated.

Equation (11) indicates that, in spite of the irregular orientation of the atoms, interference is to be expected as a consequence of the regular electron arrangement inside the atoms.

To proceed further from equation (11), assumptions regarding the arrangement of the electrons inside the atom (molecule) must be made. The computation here is carried through for the case of p electrons positioned at equal distances from one another on a ring of radius a. With regard to the hydrogen molecule, we can state perhaps with certainty that this representation is correct with

p = 2; for heavier atoms and molecules, a corresponding arrangement with larger values for p can be considered very probable. This is the reason for the special assumption made above.

If we imagine radii drawn from the center of the ring to each electron, the angle between the n-th and the m-th radius will be equal to:

$$(n - m)\frac{2\pi}{p}$$

hence the distance between the n-th and the m-th electron will be:

$$s_{nm} = 2a \sin|n - m|\frac{\pi}{p}$$

where $|n - m|$ indicates the absolute value of the difference $n - m$.

If this value for s_{nm} is substituted in the double sum of equation (11), it becomes:

$$\sum_n \sum_m \frac{\sin\left[4ka\sin\frac{\vartheta}{2}\sin|n-m|\frac{\pi}{p}\right]}{\left[4ka\sin\frac{\vartheta}{2}\sin|n-m|\frac{\pi}{p}\right]}$$

for which, as will readily be seen, the simple sum:

$$p\sum_{n=0}^{n=p-1}\frac{\sin\left[4ka\sin\frac{n\pi}{p}\sin\frac{\vartheta}{2}\right]}{\left[4ka\sin\frac{n\pi}{p}\sin\frac{\vartheta}{2}\right]}$$

may be substituted so that finally the expression:

$$V = \frac{Np}{R^2}\frac{\varepsilon^4}{\mu^2 c^4}\frac{1+\cos^2\vartheta}{2}\sum_{n=0}^{n=p-1}\frac{\sin\left[4ka\sin\frac{n\pi}{p}\sin\frac{\vartheta}{2}\right]}{\sin\left[4ka\sin\frac{n\pi}{p}\sin\frac{\vartheta}{2}\right]} \qquad (12)$$

results for the desired intensity ratio V which must be the starting point for the following discussion.

III. Discussion of the Spatial Intensity Distribution of the Scattered Radiation. Relation to Friedrich's Interferences

According to equation (12) the intensity distribution is a function of the angle between direction of observation and direction of incidence; it varies considerably with the wave length of the incident radiation. Let us consider two limiting cases.

Let the wave length of the incident and of the scattered radiation be λ. Then, according to the definition:

$$ka = 2\pi\frac{a}{\lambda}$$

(a) If λ is large compared to the radius of the electron ring, which incidentally is equal to 0.72×10^{-8} cm. for H_2, then $ka \ll 1$ and the terms of the sum in equation (12) all approach 1 in the same way as $(\sin x)/x$ approaches 1 for $x = 0$. The total sum will then equal p. and thus:

$$V = \frac{N}{R^2}\frac{\varepsilon^4}{\mu^2 c^4}\frac{1+\cos^2\vartheta}{2}p^2 \qquad (13)$$

In its dependence on the direction of observation, the energy scattered in any direction behaves exactly as if it had been radiated by a single dipole, *i.e.*, for instance, the intensity scattered at a right angle to the direction of incidence is only half the energy scattered in or opposite to the direction of incidence, at least as long as the primary radiation remains unpolarized. This is nothing but the well-known fact that an electron does not radiate in the direction of acceleration. Further the scattering is proportional to p^2, *i.e.*, proportional to the square of the number of electrons present in the atom (molecule).

(b) If on the other hand $ka >> 1$, or $\lambda << a$, then the terms in the sum from $n = 1$ to $n = p - 1$ assume values to be considered only within a small angular region in the neighborhood of $\vartheta = 0$. The discussion of this particularly interesting region will follow presently. For a sufficiently large value of ϑ, all terms of the series from $n = 1$ to $n = p - 1$ disappear as a consequence of the continously increasing denominator. Only the first term of the series for $n = 0$ is always equal to 1. Hence the sum is equal to 1 over a considerable range of ϑ-values so that:

$$V = \frac{N}{R^2} \frac{\varepsilon^4}{\mu^2 c^4} \frac{1 + \cos^2 \vartheta}{p} p \qquad (13')$$

an expression different from the one for the previously treated limiting case only in that p takes the place of p^2. The scattering is now proportional to the first power of the number of electrons. Considering the total radiation from N irradiated atoms, it is given in the first case ($\lambda >> a$) by:

$$\int V R^2 \, d\Omega = \frac{8\pi}{3} \frac{N \varepsilon^4}{\mu^2 c^4} p^2$$

and in the second case ($\lambda << a$) by:

$$\int V R^2 d\Omega = \frac{8\pi}{3} \frac{N \varepsilon^4}{\mu^2 c^4} p$$

The latter formula was used by J. J. Thomson for the determination of the number p of electrons from observations on scattered radiation. It is known that with its assistance he arrived at the conclusion that the number of electrons in the atom is approximately equal to half the atomic weight. It is evident that the choice of wave length of the primary radiation is essential in this determination; p or p^2 can be measured, according to the circumstances, as can be readily understood without much computation.

It is of course possible, by integration of the general formula equation (12), to derive the formula for the total radiation which contains the above two limiting cases. I shall not do this because the total radiation will hardly be measured in practice.

However, the dependence of the intensity on ϑ and λ shall now be illustrated by a diagram. For this purpose we first construct a preliminary diagram. To abbreviate the expression we write:

$$\varrho = 4 k a \sin \frac{\vartheta}{2} \qquad (14)$$

and plot in the preliminary diagram the sum divided by p as a function Φ of ρ. We write:

$$\Phi_p(\varrho) = \frac{1}{p}\sum_{n=0}^{n=p-1} \frac{\sin\left[\varrho \sin \frac{n\pi}{p}\right]}{\left[\varrho \sin \frac{n\pi}{p}\right]} \tag{15}$$

For hydrogen, for example, the determining quantity would be:

$$\Phi_2(\varrho) = \frac{1}{2}\left\{1 + \frac{\sin\varrho}{\varrho}\right\}$$

for an atom with 3 electrons in a ring (Li?) it would be:

$$\Phi_3(\varrho) = \frac{1}{3}\left\{1 + 2\frac{\sin\frac{\varrho}{2}\sqrt{3}}{\frac{\varrho}{2}\sqrt{3}}\right\}$$

etc. for more electrons. These various functions $\Phi_2(\rho)$, $\Phi_3(\rho)$.... will be found in Figure 1.

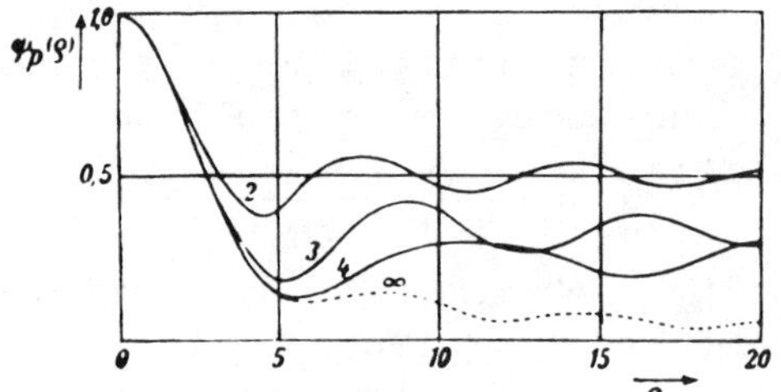

Figure 1

They all start with the value 1 at $\rho = 0$ and then approach with rapidly declining oscillations the value $1/p$ at $\rho = \infty$.

The last, dashed line in Figure 1 is the limiting function $\Phi_\infty(\rho)$; it can be represented by means of Bessel functions and can be easily computed with the assistance of the tables in Jahnke-Emde.*

*Provided $p \gg 1$, π/p is a small number. If we write, according to equation (15):

$$\Phi_p(\varrho) = \frac{1}{\pi}\sum_{0}^{p-1} \frac{\sin\left[\varrho \sin \frac{np}{p}\right]}{\left[\varrho \sin \frac{n\pi}{p}\right]}\frac{\pi}{p}$$

then, in the limit for $p = \infty$, the magnitude $n\pi/p$ may be considered as a continuous variable u, and the differential du may be substituted for π/p. Thus we obtain:

$$\Phi_\infty(\varrho) = \frac{1}{\pi}\int_0^\pi \frac{\sin[\varrho \sin u]}{[\varrho \sin u]}\,du$$

On the other hand, however:

$$\frac{\sin[\varrho \sin u]}{\sin u} = \int_0^\varrho \cos[\varrho \sin u]\,d\varrho$$

so that we can write:

$$\Phi_\infty(\varrho) = \frac{1}{\pi\varrho}\int_0^\varrho d\varrho \int_0^\pi \cos[\varrho \sin u]\,du$$

(continued on next page)

Once Figure 1 is at our disposal, it is easy to form a picture of the way in which the intensity scattered in various directions, specified by the angle ϑ, changes with decreasing wave length of the primary radiation.

With the abbreviation given in equation (15), equation (12) becomes:

$$V = \frac{N}{R^2}\frac{\varepsilon^4}{\mu^2 c^4}\frac{1+\cos^2\vartheta}{2} p^2 \Phi_p\left(4ka\sin\frac{\vartheta}{2}\right) \tag{16}$$

Provided the wave length is given and the radius of the electron path is known, then $4ka = 8\pi a/\lambda$ is known, and the quantity:

$$\varrho = 4ka\sin\frac{\vartheta}{2}$$

can be determined for any value of ϑ so that the associated value of $\Phi_p(\rho)$ can be found from Figure 1. The value of V can be calculated then from equation (16).

In Figure 2 the intensity is plotted for hydrogen as a function of ϑ, for the values:

I: $4ka=0$; II: $4ka=2{,}5$; III: $4ka=10$; IV: $4ka=40$

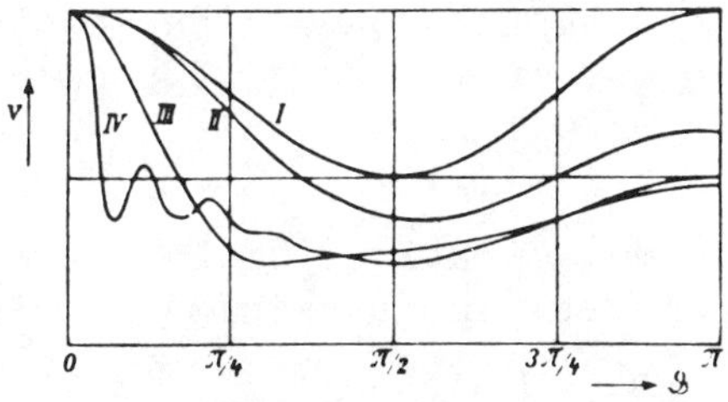

Figure 2

For $4ka = 0$, the scattered intensity is proportional to $1 + \cos^2\vartheta/2$. At $\vartheta = \pi$ it decreases continuously with decreasing wave length until it reaches half its original value. Already at comparatively large values of λ considerable deviation from the radiation of a dipole is observed. This dissymmetry in the scattering, the occurrence of which has been pointed out in the introduction, constitutes the first indication of an interference in the atom. Finally, for still shorter λ, the radiation again approaches

(con'd from preceding page)

*or, from the known definition of Bessel's function of zero order, $J_0(\rho)$:

$$\Phi_\infty(\varrho) = \frac{1}{\varrho}\int_0^\varrho J_0(\varrho)\,d\varrho$$

Finally the last mentioned integral may be expanded into a series of Bessel functions which are immediately suitable for numerical computations. The final formula reads:

$$\Phi_\infty(\varrho) = \frac{2}{\varrho}\sum_{n=0}^{n=\infty} J_{2n+1}(\varrho)$$

it can be easily evaluated from the tables found on pages 149 ff. of Jahnke-Emde.

its regular course with, however, only half the amplitude. For sufficiently small values of λ definite interferences appear in the neighborhood of the direction of incidence which should be visible on a photograph as rings of rapidly decreasing intensity. It is now suggested that these rings be related to the known observations by W. Friedrich, and to consider the respective photographs as obvious demonstration of the regular (annular) arrangement of the electrons inside the atoms. In fact not only the sensitivity of this phenomenon with regard to the hardness of the tube, stressed by Friedrich, is illustrated by Figure 2, but it is also numerically understandable how it could be that wax and paraffin, for instance, produced rings, while only a uniform darkening resulted with carbon. The materials which generated rings contain much chemically bound hydrogen, now known to have an electron path of comparatively large radius. If the bound hydrogen may be treated similarly to the free hydrogen molecule, then $2a$ would be equal to 1.05×10^{-8} and (based on Figure 1) the first ring would have to be expected at an angular distance $\vartheta = 26°$, assuming an incident wave length of 0.40×10^{-8} cm. If, on the other hand, it is assumed that there is a ring with 6 electrons inside the C atom, each revolving with the momentum $h/2\pi$ around a nucleus carrying a charge of 6_ε, $2a$ is computed to amount to 2.5×10^{-9} cm. The first ring would then, for the same wave length as above (0.40×10^{-8} cm), occur at an angular distance $\vartheta = 134°$, *i.e.*, it would not reach the plate.

The last-mentioned considerations are, of course, only a very rough approximation. However, in my opinion, it is highly probable that the electron rings furnish the correct explanation of Friedrich's interference patterns. It is immaterial for use of the measurements whether, experimentally, rings are actually photographed or a continuous deviation from the scattering law for dipoles is established, at longer incident wave length, for instance, by electrometric measurements. It appears to be essential to me that provided our view is accepted, we are in a position to measure from observations of the scattered radiation, the electron arrangement inside the atoms in centimeters. The most suitable objects for these measurements, at least at present, appear to be the first elements in the periodic system.

Bibliography

1. The Constitution of the Hydrogen Molecule. *Sitzber. d. Kgl. B. Akademie d. Wiss.*, Convention, Jan. 9, 1915.
2. See R. Pohl, *Physik der Röntgenstrahlen*, p. 58 ff., Viehweg, 1912.
3. *Physik. Z.*, *14*, 317 (1913).

X-RAY INTERFERENCE PATTERNS OF PARTICLES ORIENTED AT RANDOM. I.

(Interferenzen an regellos orientierten Teilchen im Röntgenlicht. I*)

P. Debye and P. Scherrer**

Translated from
Physikalische Zeitschrift, Vol. 17, 1916, pages 277-283.

Some time ago one of us directed attention to an experimental method which permits one to obtain information regarding the number as well as the relative positions of electrons in the atom[1]. As was then stressed, the possibility of such a measurement is based on the fact that, provided the electrons show some regularity of arrangement in the atom, this systematic arrangement can still be recognized if many atoms of random orientation occur together.

It could be shown in detail that if such a substance with the assumed regular internal electron arrangement is exposed to monochromatic x-rays the resulting secondary radiation will essentially not be radiated evenly from the substance into space, but will exhibit maxima and minima. These maxima and minima will be situated on cones, whose axes coincide with the direction of the primary radiation and whose vertices are inside the secondary radiator which is assumed to be small. It is further necessary for the appearance of the maxima and minima that the wave length of the primary radiation is of the same order of magnitude as the distances between electrons. That this second requirement is likely to be met in an experiment was then concluded from a comparison of the wave length of fluorescent x-ray radiation with the distances between electrons to be expected from Bohr's assumptions.

Experiments since then carried out by us show the expected result. However, in several instances, interference patterns of a

*Received May 31, 1916.

**A note of essentially the same content was presented to the Kgl. Ges. d. Wiss. at Göttingen on December 3, 1915. A note No. II, dated December 17, 1915, is concerned with interferences from liquids (benzol, etc.). We intend to revert to the latter in more detail.

different nature, and superposed on the expected effect, were established, which indicated definitely by the sharpness of their maxima that the regular arrangement of the presumably small number of electrons in the atom cannot be held responsible for their occurrence. The present preliminary publication will be restricted to the description and explanation of this phenomenon. In a later publication we intend to treat the electron interferences and related phenomena in more detail.

The interference maxima are sharp and consequently must be due to a phenomenon involving the cooperation of a large number of radiation centers. If this is so, then it is natural, in the instances where the interference patterns were observed, to attribute them to the crystalline structure of the substance which was penetrated by the x-rays, even though the substance, as was always the case, was used as a seemingly amorphous powder, or was even labeled as "amorphous". This point of view, adopted by us, may appear strange in view of the experimental result obtained by Friedrich, Knipping, and v. Laue in their first publication,[2] according to which a finely powdered crystal no longer showed interferences. Actually, however, this assertion, as shown by the accompanying photographs, can no longer be maintained, and further, the phenomenon follows necessarily from Laue's theory of crystal interferences, as will be shown later. With regard to our final aim, we remark that the observations reported in this note may be considered as experimental proof for the correctness of our initial statement.

If one agrees with the following reasoning, observation of the interferences in question supplies a simple method of deciding with absolute certainty about the (micro-)crystalline or amorphous state of a substance. The discussion of three photographs in section III is intended to show how it is further possible to use the photograph for an investigation of the internal structure of a single crystal. Actually the relative positions and the distances of atoms in a crystal can be determined by means of a single photograph, similar to the well-known manner in which Bragg achieved this end by electrometric investigation of reflections from the different lattice planes of a large crystal. It may perhaps be asserted with good reason that the complete problem can be solved much more easily by means of a powder of amorphous appearance than is is possible by observations on a large, perfect crystal.

If the atomic arrangement has been determined for some substance, then, conversely, the powdered substance may be used as a grating for the analysis of the incident radiation. Since, of course, there is no question of the orientation of the "amorphous" substance, and since a single line of the incident spectrum will be present as a clear narrow line at many (*e.g.*, 10) separate places on the diagram, the apparatus described later should be applicable with advantage to the determination of the wave lengths present in the incident radiation. When viewed in this way, our apparatus may be considered the simplest spectroscope of all.

I. The Experiments

The powdered substance to be investigated (among other substances, we investigated graphite, amorphous boron (from Kahlbaum), amorphous silicon, boron nitride, lithium fluoride, etc.) was pressed into the shape of a small rod approximately 2 mm. in diameter and approximately 10 mm. in length.* The rod was put upright in the center of a cylindrical camera 57 mm. in diameter, which could be closed light-tight by a cover. The x-rays were admitted to this camera in a horizontal direction through an elongated lead tube, cast into a brass tube, and provided in the middle with a drilled hole 2.5 mm. in diameter. The sharply defined beam left the camera without touching the wall, continued inside a long tube of black paper attached to the camera, and finally penetrated the bottom of this tube, also made of thin black paper. In this way no noticeable secondary radiation was produced in the camera. Control experiments, without a scattering substance, repeated several times, over the same interval of time as was used for the scattering experiment (mostly two, occasionally four hours), showed no trace of scattered radiation.

The beam impinged on the center of the rod; the secondary radiation from the rod was photographed on two pieces of film, each bent to form a half circle and in contact with the camera wall. The radiation could be registered on these films within an angular range extending from 9° to 171°, corresponding to a length of film of approximately 80 mm. The ranges missing on both sides and extending over 2 × 9° were shielded by the mechanical arrangements necessary to let the primary beam enter and leave the camera undisturbed. Occasionally, in order to obtain a complete survey of the phenomenon, a special photograph was taken with a smaller piece of film covering the exit opening.

The primary radiation was generated by means of x-ray tubes, essentially copied from a tube designed by Rausch v. Traubenberg. The tubes were connected to the pump during operation. The radiation left the tube through an aluminum window of 0.05 mm. thickness. The distance between the target and the scattering substance was approximately 12 cm.

Figure 1 of the accompanying plate shows a photograph with LiF as scattering substance.** The left-hand edge of the reproduction is the one that was positioned closest to the exit. In the following, we always count the angle starting from the center of this opening; the left-hand edge then corresponds to 9°. The film shows circular, sharply defined interference lines, which straighten out toward 90°, and gradually again assume circular shapes with increasing radius of curvature as 180° is approached. The photograph thus shows that particularly large intensities are radiated into

*Occasionally it was necessary to strengthen the small rod by coating it with a thin layer of collodion.

**The LiF forms a fine powder having the appearance of magnesia usta.

space along special cones having the LiF rod as center. The intersection of one of these cones with the cylindrically shaped film corresponds to one of the lines on the photograph. The angular space containing noticeable radiation from one space unit of the rod must be very narrow. Actually, for sufficiently hard radiation, the width of the photographic lines is identical with the thickness of the scattering rod, and can be considerably reduced for spectrographic purposes.

In this experiment, the primary radiation originated from a copper target.

In Figure 2 is reproduced a photograph which was taken with the same LiF rod but now irradiated with x-rays obtained from a platinum target.

Figure 3 shows the reproduction of the scattering from "amorphous silicon" obtained with copper radiation.

Very similar results were obtained with the other materials listed above.

Friedrich,[3] when observing passage of x-rays through wax and paraffin, first photographed rings encircling the intersection of the primary ray with the photographic plate. At that time, it was not explained what caused the rings. Although decisive experiments by Friedrich are missing, and although the structure of the substance used is not known, we believe, judging from the clearly defined rings in the reproduction, that, in this instance also, randomly arranged small crystals are responsible for the phenomenon.*

II. Theory

The theory of crystal interferences by Laue and Bragg leads, as is known, to the following two principal rules:

(1) If an x-ray beam falls on a crystal, "reflected" beams may originate which can be constructed as if they were caused by ordinary optical reflection from the net planes of the crystal.

(2) The beams so constructed are, however, actually present only when the path·difference of beams reflected from two coordinated successive net planes is equal to an integral multiple of the incident wave length.

The formulae which describe these laws will be given here for the regular system only;* they read as follows: Denoting the angle

*It should be noted that according to this statement our view regarding the observations by Friedrich is not completely identical with the view taken in the note of February 27, 1915.

**The general case can also be treated without difficulties. This requires only a suitable combination of the formulas compiled by M. v. Laue in *Enz. math. Wiss.*, Vol. 24, pp. 457 ff.

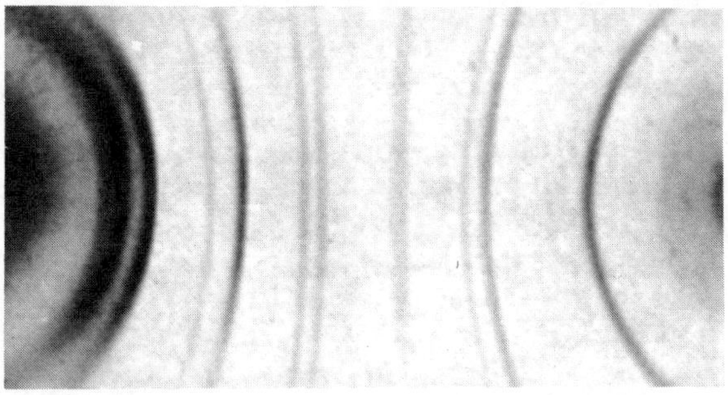

Fig. 1.

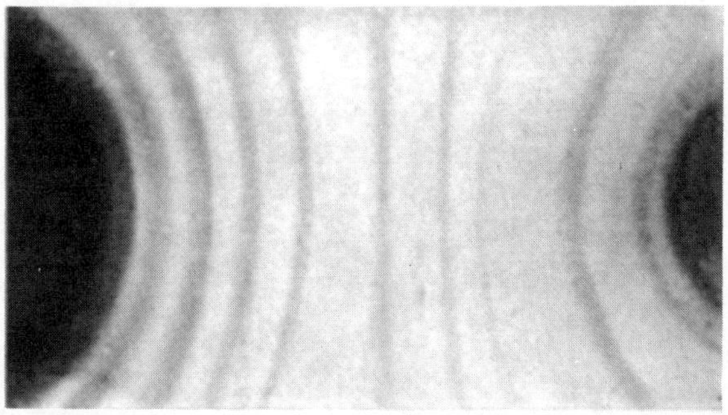

Fig. 2.

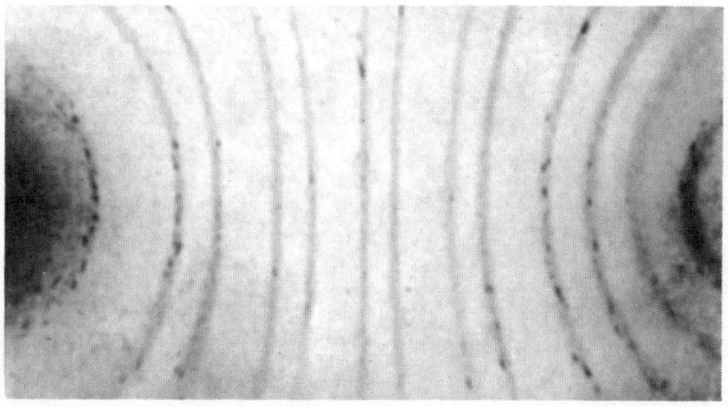

Fig. 3.

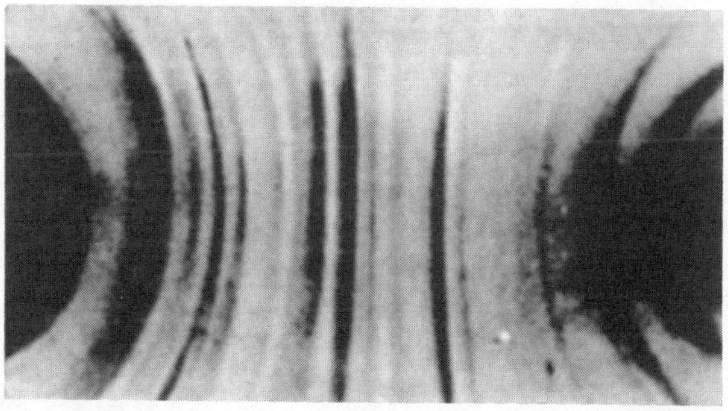

Fig. 4.

between the incident beam, which is assumed to be monochromatic, and the net plane just considered by φ, the (integral) indices of this plane by h_1, h_2, h_3, the wave length by λ, and the length of the edge of the unit cube of the cubic space lattice by a, then, as a general rule, only angles meeting the following requirement are possible:

$$\sin\varphi = \frac{\lambda}{2a}\sqrt{h_1^2 + h_2^2 + h_3^2} \qquad (1)$$

In this relation the indices 3, 2, 1, are considered to differ, for example, from 6, 4, 2 or 9, 6, 3, so that formula (1) includes higher-order interferences.

If the radiation centers are located at the corners of the space lattice only, and if they are all alike, equation (1) takes everything into consideration. If however, several radiation centers are contained in one unit cube, certain directions of reflection, possible according to equation (1), may be missing, viz., those which are destroyed by interference. This is covered by the addition to equation (1) of the statement that the intensity of the reflected beam is proportional to the square of the absolute value of the so-called structural factor S. The latter is defined by the equation:

$$S = \sum_n A_n e^{i2\pi(p_n h_1 + q_n h_2 + r_n h_3)} \qquad (2)$$

where h_1, h_2, h_3 designate the above mentioned indices of the reflecting net plane, while:

$$p_n a, q_n a, r_n a$$

are the coordinates of the nth atom within the unit cube and with respect to its corner. The factors A_n indicate the amplitudes emitted by the nth radiation center. The summation is to be taken over all these centers and thus contains as many terms as there are interlaced lattices.

We complete our remarks on the intensity by stating that it is further proportional to an expression of the form:

$$\frac{e^{-\alpha T(h_1^2 + h_2^2 + h_3^2)}}{h_1^2 + h_2^2 + h_3^2}$$

where α is a characteristic constant of the crystal measuring the mobility of the atoms, and T is the absolute temperature. This expression illustrates the decrease in the reflected intensity with an increase of the sum of the squares of the indices.[4]

If a random mixture of small crystals is irradiated with monochromatic x-rays, and only one definite net plane h_1, h_2, h_3 is considered, then, according to equation (1), reflection will occur only if it is positioned so that the angle between this plane and the incident beam has the value φ computed from equation (1). However, the orientation of the small crystals in the powder is random. The rays reflected from the plane h_1, h_2, h_3 therefore form a cone, half the vertex angle of which (called ϑ) is equal to 2φ, since the axis of the cone corresponds to the incident ray whereas each straight line on its surface is a reflected ray.

However, this conical surface carries not only radiation reflected from the single h_1-h_2-h_3-net plane just considered. All radiation reflected from net planes characterized by indices obtained from h_1, h_2, h_3 through permutation or through replacement of the positive by the negative sign will collect on the same cone. In each of these instances, the expression:

$$h_1^2 + h_2^2 + h_3^2$$

and consequently, according to equation (1), also the angle $\vartheta = 2\varphi$, remain the same. It is easy to count that there are 48 such net planes in the general case. This is the most general case; the figure may be considerably smaller. The following rules apply:

(a) If two indices are identical, the number of planes is 48/2 = 24.

(b) If three indices are identical, the number of planes is 48/(2×3) = 8.

(c) If one of the indies is zero, the result is to be multiplied by the factor $\frac{1}{2}$.

(d) If two of the indices are zero, the result is to be further multiplied by $\frac{1}{4}$.

The number of planes pertaining to the cone may exceed 48, provided two different index triplets exist, with equal value for the sum of their squares (for instance: 3,3,3 and 5,1,1). If the index triplets are arranged according to increasing values of the sum of their squares, then according to equation (1), the succession of these sums corresponds to a succession of cones with increasing angular opening. It is our contention that the lines on our photographs are the intersections of these cones with the circularly bent strip of film.

If this is so, it must be possible to establish, from measurements of the angular openings 2ϑ corresponding to single lines, that the successive values of $\sin^2(\vartheta/2)$ are proportional to integers in accordance with equation (1). Our photographs actually meet this requirement, as will be shown in the following section.

It was observed experimentally that occasionally reflections corresponding to index triplets, arranged according to increasing sums of their squares, were completely absent as lines on the photograph. This can only be understood on the basis of a special structure of the structure factor, which must disappear for these directions. Thus, conclusions can be drawn from the missing lines as to the structure of S and consequently as to the relative positions of the atoms, assumed to be the centers of radiation. This is essentially identical with Bragg's reasoning. The indicated procedure leads, for instance, to the statement (see section III) that LiF forms a lattice similar to NaCl, KCl, etc., alternate points of which are occupied by Li-atoms and F-atoms, respectively, while the so-called amorphous silicon has a lattice structure identical with that of diamond. This last statement is of particular interest, because in the periodic system Si is directly below C.

III. Discussion of Three Examples

(1) Figure 1 shows reproduction in original size of a film obtained by irradiation of a small rod pressed from very finely powdered LiF. The primary radiation consisted essentially of the *K*-series of the Cu target. All in all, 16 lines could be distinguished on the original. Measurements on the photograph gave the values entered in degrees in column 2 of Table I for ϑ, half the angular opening of the associated cone. The first column contains the designations

Table I. LiF, Copper Radiation

Density	ϑ in degrees	$\sin\frac{\vartheta}{2}$	h_1, h_2, h_3	$\frac{\sin \vartheta/2}{\sqrt{h_1^2+h_2^2+h_3^2}}$	Number of planes	Intensity
v. w.	30.0	0.259	1, 1, 1	0.150	8	---
w.	33.8	0.290	1, 1, 1	0.168	8	---
st.	37.8	0.323	1, 1, 1	0.187	8	3.85
st.	44.2	0.377	2, 0, 0	0.189	6	10.2
w.	56.2	0.472	2, 2, 0	0.167	12	---
st.	63.8	0.528	2, 2, 0	0.187	12	10.2
v. w.	67.4	0.554	3, 1, 1	0.167	24	---
v. w.	71.4	0.583	2, 2, 2	0.168	8	---
m.	76.6	0.620	3, 1, 1	0.187	24	3.15
m.	80.8	0.647	2, 2, 2	0.187	8	4.51
m.	97.8	0.753	4, 0, 0	0.188	6	1.86
w.	111.0	0.824	3, 3, 1	0.189	24	1.82
st.	116.0	0.848	4, 2, 0	0.190	24	8.10
st.	137.6	0.932	4, 2, 2	0.190	24	6.75
v. w.	153.2	0.973	4, 4. 0	0.172	12	---
st.	166.6	0.993	{3, 3, 3 / 5, 1, 1}	0.191	{8 / 24}	1.71

v. w. = very weak, w. = weak, m. = medium, st. = strong, which indicate an approximate evaluation of the line densities. According to Moseley, the largest part of the Cu radiation intensity is concentrated in the α-line (wave length 1.549×10^{-8} cm.); the line next in intensity, and belonging to the same series, is the β-line with a wave length of 1.402×10^{-8} cm. The first step in straightening out the values of $\sin(\vartheta/2)$, given in Table I, consisted in searching for ratios equal to 1.402:1.549, since, according to equation (1), if the α-line and the β-line are reflected from the same lattice plane, the associated ratios of $\sin(\vartheta/2)$ must be proportional to the ratio of the wave lengths.

Allowance being made for the intensity ratios leads to the elimination of 6 lines. The remaining figures, the underlined numerals in Table I, were considered as relating only to the α-line. It was now important to find the associated indices of the reflecting planes. It was soon established that the only structure which fitted in consisted of alternating Li atoms and F atoms arranged at equal distances. The Li atoms as well as the F atoms of such a lattice are obtained by the parallel displacement of a face-centered cube. Four Li atoms are assumed to form the basis for the elementary cube; their positions are given by:

$$\begin{array}{lll} p_1 = 0, & q_1 = 0, & r_1 = 0 \\ p_2 = 0, & q_2 = {}^1/_2, & r_2 = {}^1/_2 \\ p_3 = {}^1/_2, & q_3 = 0, & r_3 = {}^1/_2 \\ p_4 = {}^1/_2, & q_4 = {}^1/_2, & r_4 = 0 \end{array}$$

Similarly for the F atoms:

$$\begin{array}{lll} p_1 = {}^1/_2, & q_1 = {}^1/_2, & r_1 = {}^1/_2 \\ p_2 = {}^1/_2, & q_2 = 1, & r_2 = 1 \\ p_3 = 1, & q_3 = {}^1/_2, & r_3 = 1 \\ p_4 = 1, & q_4 = 1, & r_4 = {}^1/_2 \end{array}$$

If the structural factor is formed by means of this information, it is represented by:

$$S = (A_{Li} + e^{i\pi(h_1+h_2+h_3)} A_F) \qquad (3)$$
$$\left\{1 + e^{i\pi(h_1+h_3)} + e^{i\pi(h_2+h_3)} + e^{i\pi(h_1+h_2)}\right\}$$

where A_{Li} is the amplitude scattered by the Li atom and A_F that scattered by the F atom.

From this form of S it follows that:

(*a*) Lattice planes with mixed indices do not reflect.

(*b*) Lattice planes with odd indices reflect an intensity proportional to $16(A_{Li}-A_F)^2$.

(*c*) Lattice planes with even indices reflect an intensity proportional to $16(A_{Li}+A_F)^2$.

A_{Li} is certainly considerably smaller than A_F because the atomic weights of Li and F are 7 and 19, respectively. Both sets of planes mentioned in (*b*) and (*c*) will therefore reflect noticeable amounts.

If the index triplets are arranged according to increasing sums of squares, and if in view of (*a*) the mixed indices are omitted, the groupings in underlined characters in column 4 of Table I are obtained. If the model is correct, all observed lines must give the same value for:

$$\frac{\sin \vartheta/2}{\sqrt{h_1^2 + h_2^2 + h_3^2}}$$

That this is really so is shown by the figures in column 5 printed with underscoring. The small variance of the figures between 0.187 and 0.191, which amounts to only approximately 2%, is easily explained by the fact that the small rod was not positioned in the exact center of the camera.

A check is presented by carrying through the same computation with regard to the numbers not underlined in the table and relating to the β-line. These also give a good constancy of the above ratio, and thus confirm the model. The only exception is the top line in the table, which is very weak. It is not certain whether or not this line is real. It is included despite this, because we wish the table to correspond exactly to what we did. This consisted in first measuring the lines, without having worked out a theory, then developing the theory and finally checking it with the experimental facts.

Table I is completed by a 6th column in which the number of cooperating lattice planes is given for each line. In column 6 are listed the values of this number while column 7 contains values for the relative intensities calculated according to the formula mentioned in the preceding section. Since these values are to be compared with the only roughly estimated intensities of the first column, α was assumed to be zero and thus the influence of the thermal agitation ignored. A_{Li} was taken as equal to 7 and A_F as equal to 19. The figures reflect only roughly the course of the entries in column 1.

From the average value of the figures in column 5 related to the α-line, and assuming $\lambda_\alpha = 1.549 \times 10^{-8}$ cm., the value for the edge of the elementary Li cube follows from equation (1) as:

$$a = 4.11 \times 10^{-8} \text{ cm.}$$

From the β-lines of the photograph with $\lambda_\beta = 1.402 \times 10^{-8}$ cm.:

$$a = 4.17 \times 10^{-8} \text{ cm.}$$

is calculated.

Both values agree within the expected errors of observation.

(2) The second figure on the illustration shows a photograph of the scattering by the same LiF rod, this time using a platinum target in the x-ray tube. It can be observed that the lines are now closer and present in a larger number than on the previous photograph. Table II shows numerically that this is related to the smaller value for $\lambda_\alpha = 1.316 \times 10^{-8}$ cm. of the L-radiation from platinum which is effective in this case. This Table is arranged exactly like Table I and contains the lines referring to the α-radiation in underlined characters. The lines ascribed to the β-line on the basis of a wave length $\lambda_\beta = 1.121 \times 10^{-8}$ cm. are non-underlined characters.. Column 5 demonstrates the excellent constancy of the value:

$$\sin\frac{\vartheta}{2}/\sqrt{h_1^2+h_2^2+h_3^2}$$

for the α-line. All planes that are liable to radiate are represented in the table. From the average value, found from the table, of 0.1634 for the above ratio and with $\lambda_\alpha = 1.316 \times 10^{-8}$ cm., the edge of the elementary cube of LiF comes out equal to:

$$a = 4.03 \times 10^{-8} \text{ cm.}$$

Table II. LiF, Platinum Radiation

Density	ϑ in degrees	$\sin\frac{\vartheta}{2}$	h_1, h_2, h_3	$\frac{\sin\vartheta/2}{\sqrt{h_1^2+h_2^2+h_3^2}}$	Number of planes	Intensity
m.	28.6	0.247	1, 1, 1	0.143	8	---
st.	32.8	0.283	1, 1, 1	0.164	8	3.85
st.	38.2	0.327	2, 0, 0	0.163	6	10.2
m.-s.	43.6	0.372	2, 2, 0	0.132	12	---
m.	54.2	0.456	2, 2, 0	0.165	12	10.2
w.	65.8	0.543	3, 1, 1	0.164	24	3.15
w.	69.2	0.568	2, 2, 2	0.164	8	4.51
v.w.	75.8	0.614	4, 2, 0	0.137	24	---
v.w.	81.8	0.655	4, 0, 0	0.164	6	1.86
v.w.	90.6	0.710	3, 3, 1	0.163	24	1.82
m.-s.	93.6	0.729	4, 2, 0	0.163	24	8.10
m.-s.	105.8	0.797	4, 2, 2	0.163	24	6.75
w.-v.w.	109.6	0.817	5, 3, 1	0.138	48	---
w.	115.4	0.845	{3, 3, 3 / 5, 1, 1}	0.163	{8 / 24}	1.71
m.	134.4	0.922	4, 4, 0	0.163	12	2.54
m.	149.2	0.964	5, 3, 1	0.163	48	1.71
st.	155.4	0.975	{4, 4, 2 / 6, 0, 0}	0.162	{24 / 12}	5.63

The average value obtained from the previously discussed photograph of the same material with Cu radiation was:

$$a = 4.14 \times 10^{-8} \text{ cm.}$$

(3) Figure 3 shows a photograph taken with "amorphous silicon" as radiating substance. Even viewed only superficially, this photograph shows the relationship between the interference pattern and the crystal structure by the presence of small, dark points distributed over the lines. These points apparently originated at slightly larger crystals which, positioned at random, happened to be so orientated as to reflect the Cu-radiation. Thus purely qualitative observations of this photograph prove our opinion. It is shown by means of the figures in Table III that the check is also quantitative. The table is arranged in the same way as Table I. The only difference consists in that now the model used for LiF is no longer suitable. This time not only the mixed indices are missing here from the triplets arranged according to increasing sums of their squares in column 5, but also those even indices whose sum $(h_1+h_2+h_3)$ is not a multiple of 4. The inefficiency for reflection of planes of this special type is characteristic for the diamond model by Bragg. This model contains in the basic cube from which the lattice is constructed 8 elementary points with the relative coordinates:

$$p_1 = 0, \quad q_1 = 0, \quad r_1 = 0$$
$$p_2 = 0, \quad q_2 = 1/2, \quad r_2 = 1/2$$
$$p_3 = 1/2, \quad q_3 = 0, \quad r_3 = 1/2$$

Table III. Si, Copper Radiation

Density	ϑ in degrees	$\sin\frac{\vartheta}{2}$	h_1, h_2, h_3	$\frac{\sin\vartheta/2}{\sqrt{h_1^2+h_2^2+h_3^2}}$	Number of planes	Intensity
v.w.	26.0	0.225	1,1,1	0.130	8	---
st.	28.8	0.248	1,1,1	0.143	8	1.33
v.w.	43.2	0.369	2,2,0	0.130	12	---
st.	47.8	0.405	2,2,0	0.143	12	1.50
w.	51.8	0.437	3,1,1	0.132	24	---
m.-st.	56.2	0.471	3,1,1	0.142	24	1.09
v.w.	63.0	0.522	4,0,0	0.131	6	---
m.	68.6	0.563	4,0,0	1.141	6	0.275
m.	76.6	0.620	3,3,1	0.142	24	0.630
s.s.s.	81.2	0.651	4,2,2	0.133	24	---
m.-st.	87.4	0.691	4,2,2	0.141	24	1.00
m.	94.8	0.736	{3,3,3 / 5,1,1}	0.142	{8 / 24}	0.595
s.s.s.	99.0	0.760	4,4,0	0.134	12	---
s.-m.	107.2	0.805	4,4,0	0.142	12	0.375
m.-st.	114.0	0.839	5,3,1	0.142	48	0.690
m.-st.	127.4	0.896	6,2,0	0.142	24	0.600
s.s.s.	132.4	0.915	4,4,4	0.132	8	---
m.	136.0	0.927	5,3,3	0.141	24	0.278
s.	146.2	0.957	7,1,4	0.134	48	---
m.	158.8	0.983	4,4,4	0.142	8	0.167

$$\begin{aligned}
p_4 &= 1/2, & q_4 &= 1/2, & r_4 &= 0\\
p_5 &= 1/4, & q_5 &= 1/4, & r_5 &= 1/4\\
p_6 &= 1/4, & q_6 &= 3/4, & r_6 &= 3/4\\
p_7 &= 3/4, & q_7 &= 1/4, & r_7 &= 3/4\\
p_8 &= 3/4, & q_8 &= 3/4, & r_8 &= 1/4
\end{aligned}$$

Accordingly the structural factor is of the form:

$$S = \left\{1 + e^{i\frac{\pi}{2}(h_1+h_2+h_3)}\right\}\left\{1 + e^{i\pi(h_2+h_3)} + e^{i\pi(h_3+h_1)} + e^{i\pi(h_1+h_2)}\right\}$$

and shows the above mentioned characteristics required for the arrangement of the figures in the table.

From the average value of the figures printed in underlined characters in column 5 and associated with Cu_α, the edge of the basic Si cube is found to be:

$$a = 5.46 \times 10^{-8} \text{ cm.}$$

The figures not underlined referring to the β-line give the value:

$$a = 5.31 \times 10^{-8} \text{ cm.}$$

in good agreement with the above value.

(*4*) Figure 4 shows a photograph taken with graphite* (and Cu-radiation), from which it can be inferred that graphite belongs to the trigonal system and has 12 atoms in the rhombohedral elementary cell, the edge of which is evaluated at 4.69×10^{-8} cm.**

Bibliography

1. *Nachr. Kgl. Ges. Wiss. Gottingen*, Feb. 27, 1915; *Ann. Physik*, *46*, 809 (1915).
2. *Sitzber. Kgl. Bayer. Akad. Wiss.*, 1912, p. 315.
3. *Physik Z.*, *14*, 317 (1913).
4. P. Debye, *Ann. Physik*, *43*, 49 (1914).

*Flinz graphite (Bavaria), kindly placed at our disposal by Prof. Tammann, as well as Ceylon graphite from the Nürnberger Bleistiftwerken (J. Faber), and graphite from L.E.C. Hardtmuth were investigated. Here also, we thank these gentlemen for supplying the sample materials.

**More details on this subject as well as instructions for the evaluation of photographs obtained with any crystal system will be reported in the near future.

ATOMIC STRUCTURE
(Atombau)

P. Debye and P. Scherrer

Translated from
Physikalische Zeitschrift, Vol. 19, 1918, pages 474-483.

From our present day conviction that atoms are to be considered as planetary systems of moving electrically charged masses, springs our task to explore these planetary systems in such detail as do the astronomers in their field, so that on the basis of the picture obtained we can understand all manifestations of the atom. There is no doubt that the enormous amount of material, unsurpassed in accuracy, which has been collected by the spectrum analysts in the course of time, will—interpreted in the sense of the quantum theory established by Bohr - play a chief part in this connection. It would, however, be entirely unjustified to neglect, in view of this - in the pursuit of the main objective - the many other phenomena related to the structure of the atom, even if they can be evaluated quantitatively to a considerably smaller extent than is common in wave length determinations. A problem of very great importance, namely the explanation of chemical valency, is apparently hardly related to the experiences in the spectral field, which refer mainly to free atoms. It is suggested that the valencies be related to the elementary quantum; it is even possible, in one instance, to speak of this requirement as being met, if it is considered that in the model of the hydrogen molecule the hyphen connecting the atoms and indicating valency in the chemical formula is successfully replaced by an electron ring of one quantum. The question will immediately be asked whether in other cases similar electron rings can take over the part of the valency hyphen and, particularly, whether their possible existence can be established experimentally. A problem commanding our special interest arises in this connection when we consider in detail the equilibrium of forces in a crystal and therewith the reasons for the possibility of crystal structures.

*Received July 11, 1918.

From the condition that the electrostatic potential has to satisfy Laplace's equation, it follows immediately that this potential can not have a maximum or a minimum anywhere in space. This means that a crystal, for which we assume only electric forces between the atoms, never represents a system in equilibrium. It is tempting to circumvent this difficulty by resorting to electron rings with associated quantum numbers in the sense previously indicated. In conversations we have had repeated opportunity to convince ourselves that attempts to use this idea for quantitative evaluations are being made from many sides.

With crystals we are in the fortunate position to directly test the validity of the basic concept of electron rings; this fact is closely related to the single point in v. Laue's interference theory[1] which is treated there only formally. This theory starts from the assumption that an atom, when hit by radiation, emits a secondary radiation into space which has a fixed phase relationship to the incident wave, and whose amplitude and spatial distribution are also determined. No assumption is, however, made regarding the nature of the last mentioned quantitites. All this is only expressed phenomenologically by the introduction of v. Laue's undetermined radiation coefficient ψ, whose special properties are not required for the immediate purpose of v. Laue's theory. However, the importance of a more detailed determination of this quantity ψ soon became apparent. As is well known, the progress made by Bragg was rendered possible to an essential extent by the assumption that these coefficients are (at least approximately) proportional to the atomic weight of the scattering atom. It does not seem to have occurred to Bragg, and has been scarcely noted otherwise, that this assumption is contradictory to experience previously gathered on the scattering of x-rays and which was commonly considered valid. Barkla[2] had previously summarized his results regarding the scattering coefficient s, measured in the conventional manner, in a substance of density ρ (at least for substances with small atomic weight) in the formula:

$$\frac{s}{\varrho} = 0.2 \qquad (1)$$

The symbol s designates the total radiation scattered per second by 1 cm.[3] of the substance if a primary radiation of (everywhere equal) intensity 1 excites the scattering. Designating by σ the scattering coefficient of the atom (so that an atom, irradiated with the intensity J per second, scatters a total energy σJ), we have:

$$s = \frac{\sigma\varrho}{A m_H}$$

where A denotes the atomic weight and m_H the mass of a hydrogen atom. Barkla's law thus reads:

$$\sigma = 0.2\, m_H A \qquad (2)$$

i.e., the energy scattered by an atom is proportional to the atomic weight. However, Bragg's assumption contends the same proportionality for the scattered amplitude, and still both assumptions prove true. This can, of course, only be understood to indicate that both assumptions are approximations for the correct law under different external conditions.

At the time of its foundation, the law by Barkla had the advantage that a theoretical basis was available. If the scattered radiation of a single completely free electron situated in an x-ray pencil of intensity ***J***, emitted as a consequence of the movement forced upon it by the primary radiation be calculated on the basis of classical electrodynamics, then the amount:

$$\frac{8\pi}{3}\frac{\varepsilon^2}{\mu^2 c^4} J$$ [3]

is found, where ε designates the charge and μ the mass of the electron, and c the velocity of light. According to this relation an essential characteristic of the radiation scattered by an electron is that it is completely independent of the wave length of the primary radiation. If Z electrons per atom are present, we assume for the scattering coefficient σ:

$$\sigma = \frac{8\pi}{3}\frac{\varepsilon^2}{\mu^2 c^4} Z \tag{3}$$

We have only to substitute the value $A/2$ for Z, in conformity with the explanation of the periodic system of elements as revealed by x-ray investigations, to arrive at Barkla's law in the form:

$$\sigma = \frac{4\pi}{3}\frac{\varepsilon^4}{\mu^2 c^4} A$$

The numerical factor has the correct value, since with:

$$\varepsilon = 4.77 \cdot 10^{-10}, \frac{\varepsilon}{\mu} = 5.30 \cdot 10^{-17}$$

and:

$$c = 3.00 \cdot 10^{10}$$

we have:

$$\frac{4\pi}{3}\frac{\varepsilon^4}{\mu^2 c^4} = 0.27 \cdot 10^{-24}$$

whereas the factor $0.2\, m_H$ appearing in equation (2) has the value:

$$0.2\, m_H = 0.33 \cdot 10^{-24}$$

The bridge from here to Bragg's law consists, in our opinion, in a remark, published in 1915 by one of us[4], and which constituted the starting point for the investigations which we have carried out in the meantime. It has been shown there that it is permissible to retain the basic idea of the generation of the scattered radiation by electrons, since for long-wave x-rays (long compared with the distances between electrons) the electrons in the atom will all oscillate in phase, and, consequently not their scattered intensities but their scattered amplitudes will add. This means that σ is not to be computed from equation (3) but from the formula:

$$\sigma = \frac{8\pi}{3}\frac{\varepsilon^4}{\mu^2 c^4} Z^2 \tag{4}$$

With $Z = A/2$, this is Bragg's law.

Quantitative completion of this assignment (which is essentially a calculation of interferences) shows that, for any sufficiently short wave length, a solid angle around the primary beam can be specified within which Bragg's law, $\sigma \sim A^2$, holds. Its angular opening decreases with decreasing wave length of the primary radiation. Outside of this solid angle, Barkla's law $\sigma \sim A$ soon starts to be valid.

Thus it can be understood why in Bragg's experiments, where mainly small glancing angles were used, proportionality of the amplitude to the atomic weight could give a good approximation.

If the validity of the preceding reasoning is admitted, the essential result, which is of importance for the following, may be stated: Scattering of x-rays is caused only by the electrons. Intensity measurements constitute measurements of an interference phenomenon caused by the electrons, and thus it must be possible to obtain from them information regarding number and position of **the electrons.**

That fact that Bragg's as well as Barkla's law results from experiments depending on the circumstances, affords evidence that the wave length range, where the interferences under consideration play an essential part, lies within limits accessible to experiments. The special application and quantitative evaluation of this argumentation, as attempted in the following, relates to three points which will be outlined in the following three sections.

I. The Question of the Electron Rings as Bonding Elements in Crystal Structures

Since the electrons are the cause of the scattering, what is generally called coordinates of the atom in crystal interferences is not directly related to the position of the nucleus and thus, to the principal mass of the atom. Accordingly, in general, an atom, *i.e.*, in reality its "electron cloud," can not be replaced by a radiating point. Expressed differently: The atoms must demonstrate their structure in observations of interferences. A simple calculation, which we will not reproduce in this survey, shows that the effect of the atomic structure in interference experiments will manifest itself only as intensity effect, which makes it, of course, less striking. These considerations suffice to establish that, for instance, it is not admissible to speak of a contradiction if, on the one hand, superficial examination of x-ray diagrams of KCl and NaCl does not lead to a differentiation between the two structures, while on the other hand etch figure experiments on these two bodies assign them to different groups.

In special cases, however, the effect can be striking; an example would be the diamond, provided the valency hyphens are in fact to be replaced by electron rings:

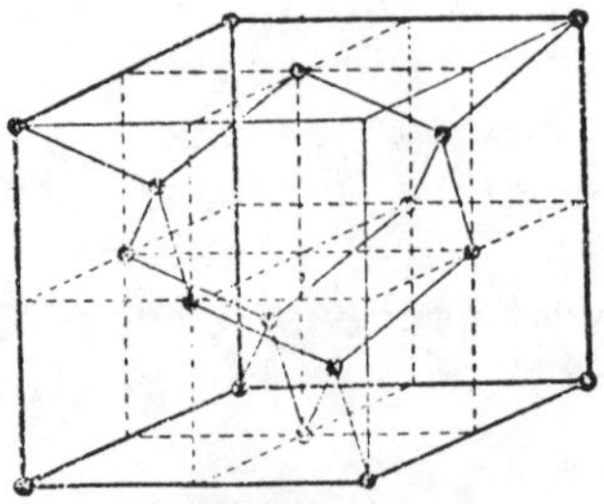

Figure 1

Assuming, for instance, the electron ring to consist of two electrons rotating in a plane at right angles to the hyphen, then each C atom would have relinquished four electrons corresponding to its four valencies. Each atom would have retained two electrons for scattering. But now the center of each valency hyphen would also have to be considered as being occupied by two electrons and thus would scatter with equal strength; it will be readily appreciated that, for sufficiently long waves, the imaginary ring has the same effect as if the two rotating electrons were situated at its center.

We thus have a new model with a new structural factor S. While, in accordance with the conventional conception, S is of the form:*

$$S = 6\left(1 + e^{i\frac{\pi}{2}(h_1+h_2+h_3)}\right)\left\{1 + e^{i\pi(h_1+h_2)} + e^{i\pi(h_2+h_3)} + e^{i\pi(h_3+h_1)}\right\} \qquad (5)$$

it will now be given by:

$$S' = 2\left[\left(1 + e^{i\frac{\pi}{2}(h_1+h_2+h_3)}\right) + e^{i\frac{\pi}{4}(h_1+h_2+h_3)}\left(1 + e^{i\frac{\pi}{2}(h_1+h_2)} + e^{i\frac{\pi}{2}(h_2+h_3)} + e^{i\frac{\pi}{2}(h_3+h_1)}\right)\right] \qquad (5')$$

$$\left\{1 + e^{i\pi(h_1+h_2)} + e^{i\pi(h_2+h_3)} + e^{i\pi(h_3+h_1)}\right\}$$

In the first column of the Table I are given the indices of the reflecting planes in the order of increasing glancing angle. In the second column are entered the squares of the absolute value of S divided by 64, which may be taken as a rough measure for the reflected intensity to be expected. In the third column are given the values of $|S'|^2/64$.

We would like to remark that all possible reflections for the wave length:

$$\lambda = 1.58 \cdot 10^{-8}\,\text{cm}$$

corresponding approximately to the Cu_α radiation are contained in Table I. There are more possible reflections for shorter waves and less for longer waves.

Table I

Indices		
(111)	18	11.6
(002)	0	0
(022)	36	4
(113)	18	0.34
(222)	0	16
(004)	36	4
(113)	18	2
(024)	0	0

The table shows that the reflections (002) and (024) are

*In accordance with our text, the amplitude is made equal to the number of electrons and not to the atomic weight.

extinguished by internal interference for the assumption of electron rings just as for the conventional model,* whereas the reflection (222), *i.e.*, the second order reflection from the octahedral plane, should appear strongly.

Bragg could not observe this reflection. A control photograph with diamond powder according to our method showed, in spite of the greater sensitivity of the photographic method, no line at the spot on the film corresponding to the reflection (222), while the adjacent line (004) was clearly present. Further, the intensity distribution in the diagram does not at all correspond to the series of figures given in the third column, as can be established by superficial inspection.

Thus we arrive at the conclusion that electron rings, analogous to those of the H_2 model, do not cause the mutual bonds between the C atoms in the diamond. It seems reasonable to presume this state of affairs for other cases also, and, in fact, we could not find one instance of electron rings acting as bonds in a crystal.

II. Counting the Number of Electrons Associated with Atoms

According to the discussion in the introduction, the scattered amplitude originating in an atom, considered as an electronic system containing z electrons, is, for sufficiently long waves, proportional to z. Bragg's assumption was the rough formulation of this law. We may expect this same proportionality for arbitrary wave length provided the scattering angle is sufficiently small, or stated precisely: The proportionality of the amplitude to z is the limiting law valid for vanishing scattering angle at arbitrary wave length. The factor in this law is theoretically known. An absolute determination of the electron number thus requires an absolute measurement of the scattered intensity. In the case of a crystal with several atoms, a relative determination of the electron numbers in the participant atoms which can be carried out with considerably less work on the basis of relative intensity measurements may replace the absolute determination.

If we take the position that the above discussed proportionality to z holds exactly for all angles of scattering, we can easily find a few cases for which this law can be relied on to decide the question of association of the electrons with the atoms without much difficulty.

Let us agree that in the following the conventional chemical symbols for the atoms designate the amplitude scattered by the atom. The structural factor for, for instance, sylvite (KCl) is then built as follows:

*The reflection (002) designates a second order reflection from the hexahedral plane (001); similarly for (024), etc.

$$S = \{1 + e^{i\pi(h_1+h_2)} + e^{i\pi(h_2+h_3)} + e^{i\pi(h_3+h_1)}\}\{K + Cl\, e^{i\pi(h_1+h_2+h_3)}\}$$

It follows from the first factor that planes with mixed indices (composed of even and odd numbers) do not reflect. The second factor states that the reflected amplitude is proportional to K + Cl for planes the sum of whose indices is even, and proportional to K - Cl for planes the sum of whose indices is odd.

Bragg investigated sylvite and established that within the limits of accuracy for his measurements planes with an odd sum of indices do not reflect. If the result is absolutely correct, it would mean:

$$K = Cl \qquad (6)$$

In the free state, K has 19 electrons, whereas Cl has 17 electrons. Thus relation (6) can not be met by ordinary atoms if the amplitude is proportional to the number of electrons. If, however, the potassium atom had given up an electron and transferred it to the chlorine atom, then K = 18 and Cl = 18, and equation (6) would be met. If we were convinced that the relation (6) is strictly confirmed by Bragg's measurements, it would follow from this statement that not the potassium and chlorine atoms but the ions are present in the crystal, and these ions carry one elementary charge, positive for the K ion and negative for the Cl ion.

The sensitivity obtained by Bragg is not high enough to make this statement, however, since it involves the question of whether reflections of intensity $(K - Cl)^2 = 2^2 = 4$ can be noticed beside reflections of the intensity $(K + Cl)^2 = 36^2 = 1296$. This question can, however, be decided by our photographic method. Among other substances we used sodium fluoride as experimental material. The photograph shows that the structure of sodium fluoride is similar to that of KCl and NaCl. Table II contains the results of an evaluation of the NaF photograph. The photograph was taken with Cu radiation. The underlined numbers refer to Cu_α, those not underlined type to Cu_β. The agreement between the figures in the last column reveals that we are dealing with cubic crystals.

The principal point which we want to consider is the following: usually Na = 11 and F = 9. If ions carrying one elementary charge are present in the crystal, then Na = 10 and F = 10, and we would expect that planes having odd sums of indices do not reflect. In the table we have indicated by asterisks three lines, for which the sum of indices is odd, according to the third column, and which exist in spite of this. Though weak, they are sufficiently distinct to be measurable, as shown by the fourth column.

Maintaining our preliminary position that the scattered amplitude is proportional to the number of electrons, it follows from the photograph that:

$$Na \neq F$$

which could be simply interpreted to mean that the atoms carry no charge.

Table II

Intensity	$\sin \frac{\vartheta}{2}$	Indices	$\frac{\sin\vartheta/2}{\sqrt{h_1{}^2+h_2{}^2+h_3{}^2}}$
s-m	0.208	002	0.149
st	0.328	002	0.164
s-m	0.417	022	0.148
st	0.465	022	0.164
s	0.515	222	0.149
*s	0.542	113	0.164
st	0.570	222	0.165
s	0.592	004	0.148
m	0.658	004	0.164
st	0.734	024	0.164
st	0.804	224	0.164
ss	0.837	044	0.148
*ss	0.853	115 333	0.164
s	0.890	006 244	0.148
m	0.930	044	0.164
ss	0.941	026	0.149
*s	0.977	135	0.165

ss = very weak, s = weak, m = medium, st = strong.

On the other hand this assumption would give rise to great difficulties. To our knowledge NaF has not been investigated as to its reflective power in the infrared; however the structural analogy to NaCl, etc., makes the presence of an absorption band in the infrared region highly probable. How such a band could possibly be explained on the basis of neutral atoms is not at all clear to us.

The discussion in the introduction readily admits another solution. It was shown there that, according to the theory, the improved law by Bragg (with the number of electrons instead of the atomic weights) is only a limiting law for small angles. Furthermore we have not taken into account another effect, *i.e.*, the temperature movement of the atoms. The latter is known to introduce a different decrease in the scattering coefficient of the atoms, provided they are subject to temperature movements of different amplitudes, which possibility can not be disregarded since the masses of the two atoms are different. From the formulas[5] published previously regarding this effect, we know it has the common property, with the internal interference effect here under consideration, to vanish for vanishing scattering angle.

This indicates that dependable results can only be obtained if the relative scattering coefficients can be successfully established by experiment for the limiting case of a vanishing scattering angle. A particularity of our photograph already points in this direction. Besides the three lines marked by asterisks, another line, the sum of whose indices is odd, namely the line (111), should be expected. The latter would have to appear at:

$$\sin\frac{\vartheta}{2} = 0.284$$

at the very beginning of the film at a very small scattering angle. We could not detect it.

Serving also as a transition to the subject treated in section 3, we will now report on intensity measurements which unequivocally determine the relative number of electrons for LiF. LiF crystallizes similarly to NaCl, KCl, NaF, etc. There is, however, a difference compared with NaF and KCl in as much as in the free state Li = 3 and F = 9, so that vanishing of the reflection from planes with odd sums of indices is not to be expected even if the "lattice points" are ions. Again only planes with mixed indices are present. For planes with an even sum of indices the structural factor is proportional to (Li+F), for planes with an odd sum of indices it is proportional to (Li - F).

A new photograph was taken by our method with LiF powder as radiating substance; care was taken that the photographic densities did not exceed the value 1.* It has been shown by Friedrich and Koch[6] that, if this is the case, proportionality exists between density and intensity, provided the density is due to x-rays. It is further essential that the lines which it is intended to compare be caused by x-rays of the same wave length; this requirement is automatically met for our photograph. All lines compared are due to the K_α radiation of copper. We found the various indices associated with the lines photometrically with a Hartmann-microphotometer in which, similarly to Koch's automatically registering instrument, the intensity to be observed is converted by means of a potassium photocell to a deflection in an electrometer. We first observe the electrometer deflection as a measure for the density of the particular spot in the line. From Hartmann we had obtained a density scale that is a photographic plate with fields of known densities which have been determined by direct measurements. With the aid of this scale the observed electrometer deflections could be converted into absolute densities, and further according to Friedrick-Koch's law into x-ray intensities. An intensity curve for each line was plotted in this manner; its area, after deduction of the background density, is directly proportional to the line intensity.

A further adjustment (based on theory) to be explained in section 3, is required to derive the structural factor from the intensity. If this adjustment has been taken care of, figures are arrived at in the present case which, in our terminology, are proportional to the quantities $(Li+F)^2$ and $(Li - F)^2$, respectively,

*As is well known, we understand by the density of a portion of the film the common logarithm of the ratio of the incident to the transmitted intensity.

for the scattering angle at which the line appears. The scattering angle is simply measured by the sum of the squares of the indices:

$$h_1^2 + h_2^2 + h_3^2 = H^2$$

of the line, since:

$$\sin^2 \frac{\vartheta}{2} = \frac{\lambda^2}{4a^2}(h_1^2 + h_2^2 + h_3^2)$$

holds if λ designates the wave length used and a designates the lattice constant. Table III contains the results of the measurements.

Table III

h_1, h_2, h_3	$H^2 = h_1^2+h_2^2+h_3^2$	$(Li + F)^2$ and $(Li - F)^2$ respectively
111	3	*107
002	4	239
022	8	112
113	11	*17.6
222	12	63.8
004	16	46.6
133	19	*7.65
024	20	28.8
224	24	17.2

Intensities associated with planes having an odd sum of indices are marked with asterisks; they are proportional to $(Li - F)^2$. All other figures belong to planes with even sum of indices, and are proportional to $(Li+F)^2$. If the figures in the table which are proportional to $(Li+F)^2$ are plotted as a function of $H^2 = h_1^2 + h_2^2 + h_3^2$ a smooth curve can be drawn through them. This was done. The three figures proportional to $(Li - F)^2$ lie, as is already shown by the table, far below this curve. The latter figures may be compared with the ordinates of the curve for the same value of the abscissa; thus, in our terminology, the ratio

$$\left(\frac{Li + F}{Li - F}\right)^2$$

can be calculated for a definite angle, *i.e.*, for a definite value of H^2. Table IV contains the result; Figure 2 illustrates the contents of the table.

Table IV

$H^2 = h_1^2+h_2^2+h_3^2$	$\frac{F + Li}{F - Li}$
3	1.72
11	2.04
19	2.06

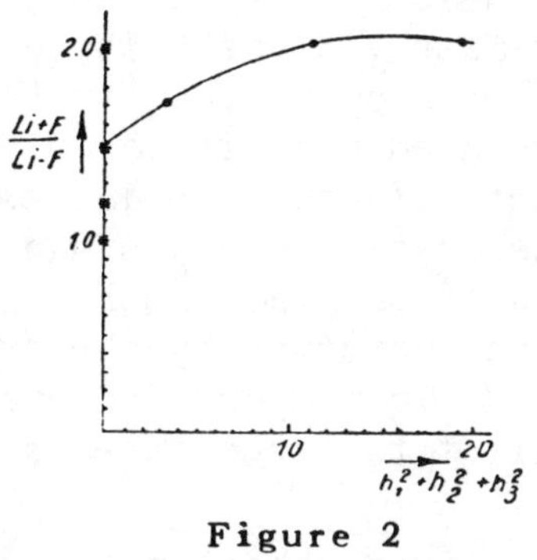

Figure 2

It will be seen that the ratio is not constant; we are interested in the limiting value to be expected for $H^2 = 0$ and, consequently, for $\vartheta = 0$. To secure this limiting value with as little arbitrary determination as possible, we draw a quadratic curve through the three points; ite equation can be easily determined to be

$$\frac{F+Li}{F-Li} = 1.523 + 0.0728\ H^2 - 0.00234\ H^4$$

Thus we have found that the experimental value

$$\frac{F+Li}{F-Li} = 1.52$$

is to be expected as the limiting value for vanishing scattering angle.

From the theory we would expect the following:

(a) neutral atoms

$$\frac{F+Li}{F-Li} = \frac{9+3}{9-3} = 2$$

(b) Li carries one positive elementary charge

$$\frac{F+Li}{F-Li} = \frac{10+2}{10-2} = 1.5$$ or

(c) Li carries two positive elementary charges

$$\frac{F+Li}{F-Li} = \frac{11+1}{11-1} = 1.2$$ or

etc.

These figures are marked by asterisks on the ordinate axis. It will be seen that the experiment agrees with case (b) within the limits of accuracy. Thus it has been shown experimentally that in the LiF crystal the Li carries a positive unit charge and the F a negative unit charge, as is familiar to us from electrolysis.

The case of LiF has been discussed as an example for the experimental procedure. We will treat other cases, especially the question in similar crystals whether univalent atoms carry a unit charge and multivalent atoms carry several unit charges according to their valency, in a more detailed report.[7]

III. Size of Atomic Electron Systems

Superficial inspection of Table III reveals a characteristic which is common to all interference observations. While H^2 increases from 4 to 24, the intensity which is proportional to $(Li+F)^2$ decreases from 239 to 17.2, *i.e.*, it is reduced by a factor of approximately 14. We know that the **temperature** effect causes a decrease, proportional to H^2, of the amplitude scattered by the atoms and responsible for the interferences. Examination of the numerical relations, however, shows that as strong a decrease as indicated by the experiment is not to be expected as a consequence of the temperature movement, even though allowance is made for the fact that the data for LiF are only estimated.

A decision on the question whether there is, besides the temperature movement, another cause for a decrease in the scattered amplitude with an increase in angle, will best be secured by investigating a substance for which the temperature movement is as small as possible. Diamond is a good example, and since the course of the specific heat is exactly known, we can determine the quantitative amount of this effect.

The diamond powder with which a photograph was made by following our method was put at our disposal by Professor Keesom of Utrecht. We wish to express our thanks for his assistance. The powder was not quite as fine as is necessary to obtain a photograph with lines that are not dotted and which can be readily evaluated by photometric methods. We could, however, overcome this difficulty by simply rotating from time to time, the small rod consisting of the powder through a small angle during the exposure and by occasionally slightly lifting and lowering it.

The finished photograph was evaluated photometrically, as described in section 2. The results obtained thus represent the actual line intensities; they are entered in column 3 of Table V; columns 1 and 2 contain the line indices and the sum of their squares H^2, respectively.

Table V

Indices	H^2	Intensity	θ	$\mathfrak{F}\theta$	$\mathfrak{F}R$	$C^2 = \dfrac{\mathfrak{F}}{R\theta}$
111	3	406	0.780	521	4.38	119
022	8	158	0.860	184	2.04	46
113	11	94	0.982	96	3.20	30
004	16	55	1.20	46	2.00	23
133	19	171	1.20	142	11.26	13

Our next assignment is to determine the scattering coefficient of the C atom from these intensities J.

First we have to make an adjustment, introduced by the geometrical arrangement of our experiment. Because of the absorption of the radiation in the rod, not every volume element of the latter is irradiated by the same primary intensity; further the scattered radiation proceeding from a volume element is partially absorbed inside the rod. In an ideal case the absorption would be negligible; the actually observed intensities are derived from the ideal ones by multiplication by a factor Θ which is a function of the direction of observation. It is represented by a double integral which cannot be evaluated without difficulty. We have therefore determined Θ as a function of ϑ by graphical evaluation of the prescribed integration, a method which is rather tedious but gives good results. The figures which represent this coefficient Θ, except for a common numerical factor (immaterial for the following), for all directions in which lines appeared, are contained in column 4 of Table V. Column 5 gives the ideal intensity corrected for absorption in the small rod.

We now have to study the relationship between this ideal intensity and the amplitude C scattered by the atom.

1. As is well known, it follows from Laue's theory that the intensity is proportional to the square of the absolute value of the structural factor S. The latter has the value:

$$S = C\left\{1 + e^{i\pi(h_1 - h_2)} + e^{i\pi(h_2 + h_3)} + e^{i\pi(h_3 + h_1)}\right\}\left\{1 + e^{i\frac{\pi}{2}(h_1 + h_2 + h_3)}\right\}$$

for diamond on the basis of the atomic arrangement proposed by Bragg.

2. According to known experiments, the scattered radiation is polarized for unpolarized primary radiation. This introduces a second factor, the polarizing factor, in the expression for the intensity, which is obtained by conventional methods, and which has the value

$$\frac{1 + \cos^2\vartheta}{2}$$

3. In our method not all reflecting planes are equally effective. Rather, for random position of the powder particles, each plane is adapted for reflection proportionally to the number of different positions in which it can terminate a crystal shape. The hexahedral plane (001), for instance, 6 times, the octahedral plane (111) eight times, etc. The intensity to be observed is also proportional to the frequency factor Z defined by the numbers 6, 8, etc.

4. Finally there is one more factor which is analogous to the one computed by H. A. Lorentz for v. Laue's arrangement.

A pencil of parallel rays of correct wave length and angle of incidence causes, in the direction for which the condition of reflection* is exactly met, an intensity proportional to the square

* $\sin\frac{\vartheta}{2} = \frac{\lambda}{2a}\sqrt{h_1^2 + h_2^2 + h_3^2}$

of the number of the irradiated atoms. There is also some intensity in the adjoining directions which decreases more rapidly the larger the number of atoms. Furthermore not only do the exactly oriented crystals reflect, but those slightly rotated with respect to this orientation also give noticeable reflection intensities. What we observe is the sum of all these effects. The summation gives a result proportional to the number of atoms with a "summation factor" determined by Lorentz to $1/H^2$ for Laue's arrangement and which in our case has the value:*

$$\frac{1}{H^2}\frac{1}{\cos\frac{\vartheta}{2}}$$

Assuming the temperature factor caused by the thermal movement to be included in the symbol C in the structural factor, we arrive at the result that the intensity is proportional to the quantity:

$$|S|^2\frac{1+\cos^2\vartheta}{2}\quad Z\frac{1}{H^2\cos\frac{\vartheta}{2}}$$

We designate it by RC^2; figures for the reducing factor R are entered in column 6 of Table V. By dividing the ideal intensity J/θ by the reducing factor R, we obtain the value of:

$$C^2=\frac{J}{R\Theta}$$

given in column 7, which are proportional to the intensity scattered by a single atom.

It will be noticed that these figures decrease comparatively fast with increasing angle.

If we disregard the discussion in the introduction, we can only hold the thermal movement of the atoms responsible for this decrease. A brief calculation immediately shows that this explanation is unsatisfactory. Let us consider—as was done in the original theory of the specific heat by Einstein—the atom as quasi-elastically bound to its rest position, then the probability that it has been displaced to a volume element specified by the coordinates:

$$\xi,\ \eta,\ \zeta,\ \xi+d\xi,\ \eta+d\eta,\ \zeta+d\zeta$$

is given by the formula:

$$\frac{1}{(2\pi)^{3/2}}\frac{1}{r^3}e^{-\frac{\xi^2+\eta^2+\zeta^2}{2r^2}}\,d\xi\,\delta\eta\,d\zeta$$

where r is a measure of the average displacement. With this assumption the value:

$$C=C_0\,e^{-\frac{4\pi^2r^2}{a^2}H^2}\tag{7}$$

is obtained for the quantity C appearing in the structural factor; a designates the lattice constant and C_0 a quantity which is an exclusive characteristic of the atom. If further m is the mass of the atom and v its vibration frequency as defined by Einstein,

*The detailed computation will be presented in an extensive publication.

and assuming that no energy exists at zero temperature, we obtain for $T = 290^\circ$ absolute

$$4\pi^2 r^2 = \frac{h}{mv}\,\frac{1}{e^{\frac{h\nu}{kT}}-1} = 7.55 \cdot 10^{-20}$$

In the presence of zero-point energy, we obtain:

$$4\pi^2 r^2 = \frac{h}{mv}\left(\frac{1}{e^{\frac{h\nu}{kT}}-1}+\frac{1}{2}\right) = 562 \cdot 10^{-20}$$

Thus the first assumption gives:

$$\frac{4\pi^2 r^2}{a^2} = 0.60 \cdot 10^{-4}$$

and the second:

$$\frac{4\pi^2 r^2}{a^2} = 4.5 \cdot 10^{-3}$$

since for diamond $a = 3.54 \times 10^{-8}$ cm.

Even the second assumption does not at all explain the observed rapid decrease of C^2, since from equation (7) we obtained:

$$C^2 = C_0^2 e^{-2\frac{4\pi^2 r^2}{a^2}} - C_0^2 e^{-9.0 \cdot 10^{-3} H^2}$$

A value of $C^2 = 102$ would thus have to be expected in the last line of Table V as a consequence of thermal movement, if the value in the first line is $C^2 = 119$; instead of which $C^2 = 13$ has been observed.

The thermal movement only plays a secondary role. Another intra-atomic reason for the observed decrease must exist.

We surmise the reason to be the finite size of the electron system associated with the atom.

It is evident at once that, where the scattering is caused by the electrons only, the finite dimension of the electron system will have an effect similar to the thermal movement of the atoms.

The discussion in the introduction led to the result that for wave lengths comparable in order of magnitude with the electron distances, a decrease in scattering intensity with increasing angle must occur.

Here, where we are dealing with the distinct interference lines of the Laue effect, the computation takes a course different from that presented in the publication cited in the introduction. Further the result depends of course— but for a first orientation to a negligible extent — on the special assumptions to be made regarding the mutual positions and movements of the electrons in the atom. As an assumption, which is very simple and essentially correct for our purpose we postulate that the electrons belonging to the atom are contained inside a sphere of radius ρ in such a manner that each volume element of the sphere is traversed equally frequently by each electron.

If this assumption is made, the computation gives:

$$C \sim 3 \frac{\sin \frac{2\pi\varrho}{a} H - \frac{2\pi\varrho}{a} H \cos \frac{2\pi\varrho}{a} H}{\left(\frac{2\pi\varrho}{a} H\right)^3} \qquad (8) \qquad (8)$$

a function which, as will be readily seen, may be approximated in its initial course by an exponential curve like the one which appears when the thermal movement is taken into consideration.

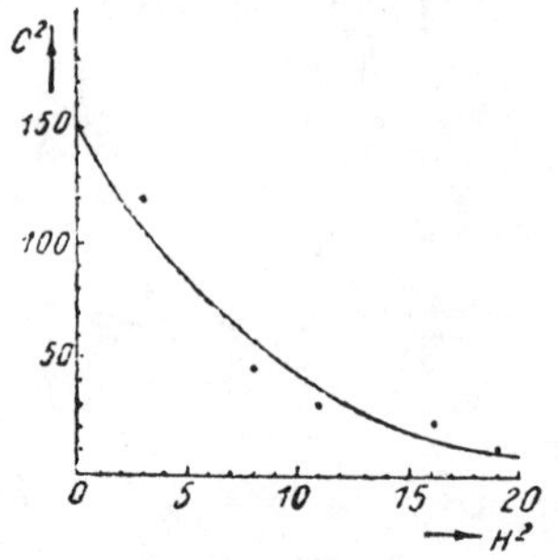

Fig. 3

As shown in Figure 3, a curve of the type determined by equation (8) can be drawn through the values observed for C^2. The curve drawn in the figure obeys the formula:

$$C^2 = 151.3 \left[3 \frac{\sin 0.761 H - 0.761 H \cos 0.761 H}{(0.761 H)^3} \right]^2$$

It represents the observations within the expected limits of accuracy as illustrated by the plotted point corresponding to observed values.

It follows from this numerical representation that:

$$\frac{2\pi\varrho}{a} = 0.761$$

i. e.:

$$\frac{\varrho}{a} = 0.12 \text{ and } \varrho = 0.43 \cdot 10^{-8} \text{ cm}$$

These values are entirely within the range of possibilities. The space occupied by the electrons, which we here derived, has a radius which is only 12% of the lattice constant or approximately one fourth of the distance between atomic centers. A ring, according to Bohr's quantum theory, of four electrons, revolving around a nucleus carrying four elementary charges, would have a radius of 0.17×10^{-8} cm.

We thus have a method to determine the actual size of the planetary system consisting of electrons belonging to the atom. We would like to emphasize that our discussion has been carried out on the basis of classical theory.

No experience is known to us which would invite doubts as to the correctness of the conclusions drawn on the basis of this theory in the range of wave lengths contemplated. Observations of absorption for considerably shorter wave lengths, in the region of γ rays, however, appear to indicate new aspects.

Bibliography

1. See, for instance, the article "Wellenoptik" by M. v. Laue, *Enzyklopadie*, Vol. V, No. 24, p. 459.
2. C. G. Barkla, *Phil. Mag.*, *7*, 534 (1904); *21*, 648 (1911).
3. I. I. Thomson, Conduction of Electricity Through Gases, Cambridge, 1903, p. 271.
4. P. Debye, *Ber. d. Köngl. Ges. d. Wiss.*, Göttingen, Feb. 27, 1915.
5. P. Debye, *Ann. d. Physik, 43*, 49 (1914).
6. W. Friedrich and P. P. Koch, *Ann. d. Physik, 45*, 399 (1914).
7. Born, with whom we discussed these investigations several months ago, has since evaluated experiments with infrared radiation in order to estimate the charges in the lattice, *Berl. Ak.*, 1918, p. 604, similar to an attempt by W. Dehlinger, *Physik. Z.*, *15*, 276 (1914).

X-RAY SCATTERING AND QUANTUM THEORY
(Zerstreuung von Röntgenstrahlen und Quantentheorie*)

P. Debye

Translated from
Physikalische Zeitschrift, Vol. 24, 1923, pages 161–166

(1) It is known that scattered x-ray radiation is polarized and varies in intensity with the direction of scattering. If the radiation emitted by a free electron is computed in accordance with the laws of electrodynamics, the theoretical result, as shown by J. J. Thomson, is in qualitative agreement with experiments. Thus the polarization observed by Barkla is disclosed by the computation; also more energy is scattered in the direction of the primary beam and in the opposite direction thereto than is scattered at right angles to the beam. If θ is the angle between the primary and the secondary beam, the formula:

$$i_s \sim \frac{1 + \cos^2 \Theta}{2}$$

follows for the dependence of the scattered intensity i_s on the angle. At comparatively long wave lengths of the primary radiation, however, a substantial discrepancy exists between the observed and the calculated intensity distribution. In practice much more is radiated in the forward direction ($\theta = 0$) than backward ($\theta = \pi$). It can be shown that this effect is caused by the fact that the distances between the electrons in the atom are of the same order of magnitude as the wave length of the x-rays used for these particular experiments. Consequently the secondary rays originating at the different electrons of one single atom interfere with one another, and the scattered intensity at average wave length is computed to be proportional to z^2 for $\theta = 0$, and proportional to z for $\theta = \pi$, where z indicates the number of electrons in the atom.[1]

If we now proceed to very short waves, the interference effect should, according to the calculations, contract to a progressively smaller angular region surrounding $\theta = 0$, and the intensity outside

*Received March 14, 1923.

this region should be represented by Thomson's formula. All experimental evidence with short waves is contradictory to this assumption. In the following I wish to present some thoughts referring exclusively to the short-wave region.

Four points seem to me to deserve particular attention:

1. The intensity of the scattered radiation is considerably higher in the direction of the primary radiation ($\theta = 0$) than in the opposite direction ($\theta = \pi$) in contradiction to the mathematically derived proportionality to $(1 + \cos^2\theta)$.

2. It now appears to be certain that the radiation scattered in the direction of the primary beam is harder than that scattered in the opposite direction. Thus the wave length is changed, again in contradiction to the results of the above mentioned computation.

3. The total energy of the scattered radiation sinks below the limiting value corresponding to Thomson's calculation. However this limiting value is in agreement with the experimental value of 0.2, found by Barkla, for the ratio of scattering coefficient s to density ρ, under the assumption, which cannot be doubted today, that for light elements the number of electrons is equal to the atomic number.

4. Each scattering is accompanied by electron emission. The shorter the wave length, the more the electrons appear to be ejected in the direction of the primary beam.

Not all experimental results are so unequivocal that the assertions 1 to 4 can be considered as absolutely confirmed by experiments. However I recently gained the impression from a survey by A. H. Compton[2] that it is highly probable that they are correct. I will therefore hesitate no longer to present for discussion an explanation of these effects based on quantum theory which occurred to me as a possibility quite some time ago.

(2) Let us assume that classical electrodynamics fails also in the computation of the energy scattered by a free electron excited by a primary radiation, and must be replaced by a quantum concept. In particular the following picture holds. The primary x-ray radiation of frequency ν_0 transfers to a free electron* in an elementary process the energy $h\nu_0$. This energy is transformed quantitatively, serving, first, to generate a secondary ray of frequency ν and energy $h\nu$, and, second, to impart a velocity v to the electron.

The secondary radiation may be considered as "needle radiation" in Einstein's interpretation. On the basis of these premises, a highly detailed picture of the process can be secured, provided it is assumed that (a) the law of conservation of energy and (b) the principle of conservation of momentum hold also in this instance. The two theorems are sufficient without any additional hypothesis.

The velocities of the electrons will be of the order of the velocity of light c; we therefore use the relativistic formulas,

*Compare also section (4) on the assumption of free electrons.

and have accordingly for the kinetic energy of an electron of rest mass m moving with the velocity v the expression:

$$E = mc^2\left(\frac{1}{\sqrt{1-\beta^2}} - 1\right) \tag{1}$$

if we put:

$$\beta = \frac{v}{c} \tag{1'}$$

in the conventional manner.

For the momentum I (in the direction of the velocity) we have:

$$I = \frac{mv}{\sqrt{1-\beta^2}} \tag{2}$$

Application of the law of conservation of energy gives:

$$h\nu_0 = mc^2\left(\frac{1}{\sqrt{1-\beta^2}} - 1\right) + h\nu \tag{I}$$

Let us assume that the secondary ray be emitted at an angle Θ with respect to the primary ray, and that the electron is ejected at an angle ϑ. Both directions are in one plane with the primary ray. Application of the principle of conservation of momentum affords two equations, one for the components of the momentum in the direction of the primary ray and another for the components at right angle thereto. They read:

$$\frac{h\nu_0}{c} = \frac{mv}{\sqrt{1-\beta^2}}\cos\vartheta + \frac{h\nu}{c}\cos\Theta \tag{II}$$

$$0 = \frac{mv}{\sqrt{1-\beta^2}}\sin\vartheta + \frac{h\nu}{c}\sin\Theta \tag{III}$$

since, as is well known, an impulse u/c in the direction of propagation is associated with the radiating energy $u = h\nu$.

Insight is best obtained by introducing the following dimentionless quantities:

$$\mu = \frac{\nu}{\nu_0} \tag{3}$$

i.e., the ratio of the secondary frequency to the primary frequency, and:

$$x = \frac{mc^2}{h\nu_0} \tag{4}$$

The quantity mc^2/h has the dimension of a frequency; it will be denoted by N. Its numerical value is given by:

$$N = 1.23 \cdot 10^{20} \frac{1}{\text{sec}}$$

corresponding to a wave length:

$$\Lambda = 0.0243 \cdot 10^{-8}\,\text{cm}$$

and is universal as indicated by its definition.

Introduction of μ and x yields the basic equations in the form:

$$1 - \mu = x\left(\frac{1}{\sqrt{1-\beta^2}} - 1\right) \tag{I'}$$

$$1 - \mu\cos\Theta = x\frac{\beta}{\sqrt{1-\beta^2}}\cos\vartheta \tag{II'}$$

$$-\mu\sin\Theta = x\frac{\beta}{\sqrt{1-\beta^2}}\sin\vartheta \tag{III'}$$

According to equation (4), $x = \infty$ corresponds to the primary frequency $\nu_0 = 0$, and $x = 0$ to the primary frequency $\nu_0 = \infty$.

If now a secondary ray exists in the direction θ, the three equations (I'), (II'), (III') permit calculation of the three quantities μ, β and ϑ. For a given x, i.e., for a given wave length of the primary radiation, we find, first, according to equation (3), from μ the frequency ν of the secondary ray; second, according to equation (I'), from β the velocity of the secondary electron; and third, the direction ϑ in which the latter is ejected.

(3) A survey of these relationships may be obtained as follows. Eliminating ϑ from equations (II') and (III') results in:

$$1 + \mu^2 - 2\mu\cos\Theta = x^2 \frac{\beta^2}{1-\beta^2}$$

and substituting the value of $1 - \beta^2$ which can be derived therefrom into equation (I') yields:

$$1 + \frac{1-\mu}{x} = \sqrt{1 + \frac{1+\mu^2 - 2\mu\cos\Theta}{x^2}}$$

Squaring this equation and putting:

$$1 + \mu^2 - 2\mu\cos\Theta = (1-\mu)^2 + 2\mu(1-\cos\Theta)$$

gives:

$$1 - \mu = \frac{\mu}{x}(1-\cos\Theta)$$

or:

$$\mu = \frac{1}{1 + \frac{1}{x}(1-\cos\Theta)} = \frac{1}{1 + \frac{2}{x}\sin^2\frac{\Theta}{2}} \qquad (5)$$

Since:

$$\mu = \frac{\nu}{\nu_0}$$

Equation (5) states that the frequency of the secondary radiation is always lower than the frequency of the primary radiation, except in the case where the secondary ray is the continuation of the primary ray. Furthermore the decrease in frequency is larger the smaller x, i.e., according to equation (4), the shorter the wave length of the primary radiation. The effect is considerable only if the primary wave length λ is comparable to or smaller than the universal wave length Λ.

The right-hand side of equation (I') is the ratio of the energy of the ejected electrons:

$$E = mc^2\left(\frac{1}{\sqrt{1-\beta^2}} - 1\right)$$

to the energy of the primary quantum $h\nu_0$. Thus it follows, using equation (5), that:

$$\left.\begin{aligned}\frac{E}{h\nu_0} = x\left(\frac{1}{\sqrt{1-\beta^2}} - 1\right) = 1 - \mu = \\ = \frac{\frac{2}{x}\sin^2\frac{\Theta}{2}}{1 + \frac{2}{x}\sin^2\frac{\Theta}{2}}\end{aligned}\right\} \qquad (6)$$

a formula which readily supplies the energy of the secondary electrons as a function of the direction of the associated secondary x-ray. If this ray constitutes the continuation of the primary ray ($\theta = 0$), then $E = 0$; if it is propagated in the opposite direction ($\theta = \pi$), then E is a maximum given by:

$$E = \frac{1}{1 + \frac{x}{2}}$$

The smaller x, i.e., the shorter the wave length of the primary ray, the larger is the fraction of the primary radiation quantum found in the electron beam, so that as the wave length decreases the electron energy is increasingly preferred at the expense of the secondary radiation.

Finally the direction of the velocity of the released electron must be determined. Dividing equation (III') by equation (II') yields:

$$\operatorname{tg} \vartheta = -\frac{\mu \sin \Theta}{1 - \mu \cos \Theta}$$

for which we can write in view of equation (5):

$$\operatorname{tg} \vartheta = -\frac{x}{1 + x} \frac{1}{\operatorname{tg} \frac{\Theta}{2}} \tag{7}$$

If θ varies from 0 to π, the right-hand side of equation (7) assumes all values between $-\infty$ to 0. Since ϑ is limited to the range between 0 and π, this angle simultaneously varies between $\pi/2$ and π. Equation (6) determines only β^2 and not β, so that the choice of the sign is still open. However if equations (II') and (III') each are to be satisfied by values of ϑ between $\pi/2$ and π, the choice of the negative sign for β is imperative. This means that the electrons are always hurled forward. If the direction of the secondary radiation deviates in an upward direction from the direction of the primary ray — the latter being assumed horizontal — the electron takes a downward course. For long primary waves (x large), we have in the limit:

$$\operatorname{tg} \vartheta = -\frac{1}{\operatorname{tg} \frac{\Theta}{2}}$$

and thus simply:

$$\vartheta = \frac{\pi}{2} + \frac{\Theta}{2}$$

However the smaller the primary wave length, the smaller the factor $x/(1 + x)$; except for small values of θ the right-hand side of equation (7) is then always small, i.e., ϑ is close to π. For very hard primary radiation, the electrons are ejected approximately in the direction of the primary beam. Figure 1 illustrates the theoretical conditions for the case $x = 1$, i.e., for:

$$\lambda_0 = \Lambda = 0.0243 \cdot 10^{-8}\,\text{cm}$$

The upper part of the figure shows a half-circle indicated by a dashed line with radius $h\nu_0$; further a series of arrows bounded by a full-line curve are drawn. Their lengths indicate the magnitude

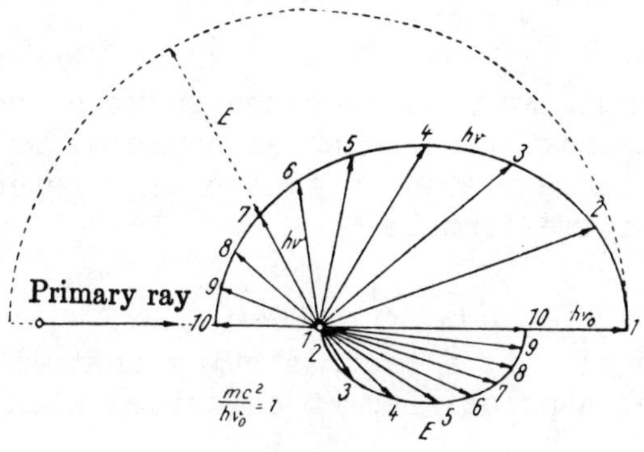

Figure 1

of the secondary radiation quanta $h\nu$ in accordance with equation (5).* The section of the radius extending between this curve and the circle measures the respective absolute value of the energy E of the electron, since it is required by the law of the conservation of energy that $h\nu$ and E add up to $h\nu_0$, (compare equation (6)). The lower part of the figure shows a series of arrows also bounded by a full-line curve. The length of these arrows represents the respective amount of the electron energy E, and they point in the direction of electron velocity. The arrows for the radiation quanta $h\nu$ as well as the arrows for E are numbered in such a manner that those belonging together have the same number. Their directions have been calculated from equation (7). It will be seen that the lower arrows are confined to an angular region of 90°, whereas the directions of the secondary radiation quanta cover a range of 180°.

(4) The consequences listed under (3) agree qualitatively with the first, second and last of the points made at the beginning. I had thought that the frequency shift called for was too large, equation (5) requiring, at a primary wave length λ = 0.708 A and

*As equation (5) shows, the curve is an ellipse, the radiation center being situated in its left-hand focus. The horizontal and vertical semi-axes are:

$$a = h\nu_0 \frac{x+1}{x+2}$$

and:

$$b = h\nu_0 \sqrt{\frac{x}{x+2}}$$

respectively; the center of the ellipse is displaced by:

$$\frac{h\nu_0}{1+x}$$

from the radiation center.

for an angle of observation of 90°, a frequency ratio of:

$$\mu = \frac{\nu}{\nu_0} = \frac{1}{1 + \frac{1}{x}} \cong 1 - \frac{1}{x} + \ldots = 1 - 0.0330$$

However Compton[3] seems to have established by direct observations, at λ_0 = 0.708 (molybdenum, $K\alpha$), an increase in wave length of 3.4% on radiation scattered from graphite at an angle of 90° to the primary radiation. This is in agreement with the calculated value of μ. Furthermore he gives the formula:

$$\frac{\nu}{\nu_0} = 1 - \frac{h}{mc\lambda_0}$$

for the computation of this effect, stating that it can be derived on the basis of Doppler's principle; it coincides with the expansion given above for:

$$\mu = \frac{\nu}{\nu_0}$$

However the question is still open whether and how this frequency change becomes manifest in reflections from crystals, particularly whether the angles of successive reflections of increasing order show anything of this effect.

The totality of the measurements on crystal reflections conveys the impression that nothing can be observed of the frequency shift under discussion. It thus is of primary interest to confront Compton's observations with these measurements, and to establish whether two entirely different effects are indeed present.

The position that the preceding reasoning takes in regard to Einstein's assumption for the photoelectric effect:

$$E = h\nu_0$$

is also of interest. This assumption does not apply to the interaction of radiation with a single free electron. Assuming that, in the sense of this equation, an electron hurled in the direction of the primary beam be generated at the expense of the radiation energy $h\nu_0$ to the exclusion of any other effect, we would have (for constant mass m):

$$\frac{mv^2}{2} = h\nu_0$$

i.e., a momentum of magnitude:

$$mv = \sqrt{2mh\nu_0}$$

would be generated. The momentum:

$$\frac{h\nu_0}{c}$$

would have disappeared with the radiation. In general these two amounts do not at all compensate one another. For light waves and x-rays that are not too hard, the generated momentum would be larger than the one that disappeared. We would have a contradiction to the principle of conservation of momentum. However, Einstein's equation holds in practice in these regions, and it must be concluded that the bond between electron and atom plays a part. As a consequence of this bond, the remainder of the atom may absorb some of the momentum, and then compensation is possible; the small momentum $h\nu_0/c$, which disappears with the radiation, appears as the difference between the momentum of the electron and of the remainder of the atom, directed in opposite directions, without necessitating a decrease of the electron energy noticeably below $h\nu_0$. For entirely free electrons, the compensation was assured above by the simultaneous emission of a secondary ray. In this case, as shown by equation (6), a complete transfer of the radiation quantum onto the electron occurs only in the limit for infinitely short wave lengths. There are indications that the conventional photoelectric effect as interpreted by Einstein, for which, as has just been shown, the internal forces of the atom are of determining influence, is progressively replaced, as the wave length of the primary radiation decreases, by an effect of the nature described above, where the electron is almost free and the internal forces no longer play a part. This transition should occur at longer wave lengths for lower atomic weights of the irradiated body, since the forces inside the atom diminish with diminishing atomic weight.

(5) The third of the experimental results mentioned in the beginning has not as yet been discussed. In the terms of the present status of quantum theory, this problem involving intensity can be solved only on the basis of a statement regarding the probability of an elementary process. In our case, the probability to be given is that — for a known intensity i of the primary radiation — a secondary ray is generated which points into a solid angle $d\Omega$, whose central line deviates by an angle θ from the primary direction. Relying on the correspondence principle, i.e., requiring that in the limit for long waves the results of electrodynamics are valid, the expression:

$$\text{const.}\, i \frac{1 + \cos^2 \Theta}{2} d\Omega$$

must hold for the probability, where const. denotes the constant in Thomson's formula.

In optics Bohr used the expedient of postulating: the expression for the probability valid at infinitely long waves shall hold with sufficient accuracy also in the optical region. If this is done here, the energy scattered by an electron is proportional to:

$$\int_0^\pi \frac{1 + \cos^2 \Theta}{2} \frac{\sin \Theta d\Theta}{1 + \frac{1}{x}(1 - \cos \Theta)}$$

while it is proportional to the integral:

$$\int_0^\pi \frac{1 + \cos^2 \Theta}{2} \sin \Theta d\Theta$$

in the limit for long waves ($x = \infty$). The ratio of these two integrals is the factor f by which the limiting value for the scattering has to be multiplied to find the actual value at a desired wave length λ_0. The factor f is a function of:

$$x = \frac{mc^2}{h\nu_0} = \frac{\lambda_0}{\Lambda}$$

Evaluation yields:

$$f(x) = \frac{3}{4} x^3 \left\{ \left(1 + \frac{2}{x} + \frac{2}{x^2}\right) \ln\left(1 + \frac{1}{x}\right) - \frac{1}{x}\left(1 + \frac{3}{2}\frac{1}{x}\right) \right\} \qquad (8)$$

For large values of x this becomes:

$$f(x) = 1 - \frac{7}{16}\frac{1}{x} + \cdots$$

for small values of x we obtain:

$$f(x) = \frac{3}{2} x \left[\ln \frac{1}{x} - \frac{3}{4} + \cdots \right.$$

At $\lambda_0 = 0$ this factor $f = 0$ increases with increasing wave length and reaches unity at $\lambda_0 = \infty$. The important variation of f is of course concentrated in the region surrounding:

$$\lambda_0 = \Lambda = 0.0234 \text{ Å}$$

Thus the theory in fact demands that the scattering decrease below the electrodynamic limiting value to zero. It should be noted, however, that the reduction in scattering in this representation is associated with a simultaneously occurring increase in electron energy by such an amount that, as regards energy, the two effects together correspond to the electrodynamic limiting value at any wave length. If it were established beyond doubt that for hard radiations the total amount of scattered radiation and electron energy sinks below this limiting value, it would be of particular interest in view of an improvement in the extrapolation (from $\lambda_0 = \infty$) used in the correspondence principle, since appreciable confusion still prevails in this region. It may be briefly mentioned that the correspondence principle can be used in a similar manner to describe the polarization phenomena in terms of quantum theory.

In conclusion I wish to stress that I would like the preceding discussion to be considered as nothing but an attempt to derive as detailed conclusions as possible from the two assumptions: "energy quanta" and "radiation quanta," taking recourse only to general laws: "law of conservation of energy" and "principle of conservation of momentum." By means of experiments which reveal characteristic deviations from this scheme, we may hope to secure deeper insight into the laws of quantum theory, particular as regards their relation to physical optics.

III. Bibliography

1. P. Debye, *Ann. d. Physik, 46*, 809 (1915); compare also the summarizing presentation in the *Naturwissenschaften*, 1922, No. 16.
2. *Bulletin of the National Research Council, 4*, Part 2, No. 20, October, 1922 (National Academy of Sciences, Washington, D.C.). I wish to refer to this publication for all detailed literature references.
3. See reference 2, p. 19.

NOTE ON THE SCATTERING OF X-RAYS

By P. Debye[1]

I. The phenomena observed by v. Laue and Bragg give us evidence of two fundamental facts:

1. They show that X-rays, scattered by an atom, must have, at least partly, phase relations with the primary rays,[2] producing the well-known interference effects.

2. They show that the intensity of scattering depends on the inner properties of the atoms inducing scattering.

There can be no question that not only in crystals, but also in the molecules of substances in the liquid or gaseous state, the atoms occupy definite places. It is true that in general the distances of the atoms will not be quite constant, but certain types of inner vibrations will exist. The values for the specific heat at constant volume for more complicated molecules give evidence of this heat motion. It seems, however, reasonable to admit that these inner vibrations will only have a disturbing effect of a second order of magnitude, as has been found in the case of the temperature effect in crystal reflection.

Moreover, there seems to be no reason why the atoms in a molecule should act in another manner than the atoms in a crystal. This is obvious if the crystal is considered as one large molecule, the picture of the lattice constituting its chemical structure formula, and if it is borne in mind that the interatomic forces in the crystal are of the same kind as the interatomic forces which help to build up the molecule.

Therefore it should be possible to detect the interference effects corresponding to the geometrical arrangement of the atoms in the molecule.

In the theory of v. Laue it is assumed that each atom acts as

[1] Professor of Physics at the Eidgenössische Technische Hochschule, Zürich, Lecturer at the Department of Physics, Massachusetts Institute of Technology, Winter Term, 1924–1925.

[2] Experiments of G. E. M. Jauncey and C. H. Eckart, Nature, 112, Sept., 1923; H. Kulenkampff, Zeits. f. Physik, 19, 1923; and Y. H. Woo, Proc. Nat. Acad., 10, 145, 1924, seem to show that the scattered radiation with shifted wave-length, detected by Compton, is not able to give interference effects of the kind necessary for the crystal X-ray reflection.

*Reprinted from *Journal of Mathematics and Physics*, *4*, 133–147 (1925).

a resonator, scattering the primary radiation in all directions but with an amplitude and phase which may depend, in general, on the constitution of the atom and the wave-length of the primary rays. Laue himself does not make any attempt to find a connection between the atomic properties and their scattering coefficient, this coefficient being introduced as a quantity only phenomenologically defined. In the later work of C. G. Darwin, P. Debye and A. H. Compton, an attempt is made to give a theory of the properties of this coefficient, assuming with J. J. Thomson that each electron in the atom is the seat of the fundamental scattering effect, and the whole atomic effect is really the result of a summation extending over all the electrons included in the atom. The discussion of some early experimental work of Debye and Scherrer on the scattering by diamond, and the later, more finished work of W. L. Bragg on rock salt, is based on this theory.

It is not an easy matter to derive the atomic scattering coefficient from observations of the intensity of the reflected beam, the only thing really measured. It can only be done with the aid of a complicated and very elaborate theory consisting of a summation extending over the elementary wavelets issuing from the single atoms. This theory itself is not beyond question; very recently P. Ewald[3] has been led to a different final result, taking account of the interaction of the wavelets more decidedly than in the first theory of v. Laue.

At first glance it seems that we may avoid this difficulty if the experiments are not performed with substances in the crystalline state. If, for instance, the scattering of a liquid has been measured, one might possibly think that the properties of the scattering function, *i.e.*, the scattered intensity plotted as a function of the angle between the secondary and primary X-ray, only depend on the dimensions and the form of the atomic frame constituting the molecule. From this point of view P. Debye and P. Scherrer in 1916 discussed the scattering of benzene.[4] In the meantime it became evident, however, that a very large number of different liquids yield diffraction-patterns which show only slight differences. Therefore, at the meeting of the German Physical

[3] P. Ewald, Phys. Zeitschr., 25, 29, 1925.
[4] Gött. Nachr., 16, 1916.

Society in 1920 at Jena, Debye stated that the principal maximum of the scattering function (*i.e.*, the first interference ring surrounding the primary ray) must be due to interferences between the different molecules of the liquid. Experiments carried out by W. H. Keesom[8] which led him to the same statement — especially the fact that he was able to photograph the ring even with a monatomic gas in the liquid state — are conclusive.

The theory of this effect seems to be as difficult as the theory of the correction for the dimensions of the molecules in the equation of state. Nevertheless, I believe that a first approximation, which can readily be obtained for the limiting case that the total volume of the molecules is small compared with the volume occupied by the gas, may be of some interest. It will be shown that, even if the molecules are comparable with hard spheres and do not interact in any other way than to prevent each other from entering into the domain defined by this sphere, this fact alone is sufficient to cause a scattering function, exhibiting a maximum. This maximum occurs at an angle defined by the quotient of the wave-length and the diameter of the sphere substituting the molecule.

In this way it seems proved, both experimentally and theoretically, that it will only be possible, even in the case of iiquids, to arrive at the scattering function characteristic of the molecule and its atomic frame, if we succeed in freeing the primary experimental result of the undesired intermolecular interferences. Further, it seem improbable that it will be possible to perform theoretical calculations which will give a reliable formula for this process in the case of dense liquids. Therefore the only possible way to find the interferences due to the interaction of the atoms constituting the molecule, seems to be the performance of scattering experiments with gases of different densities. If such experiments are done, then, with the help of theoretical considerations of the kind given below, it should be possible to eliminate the effect of molecular interaction, and to arrive at the direct measurement of the atomic distances in the molecule.

[8] W. H. Keesom and J. de Smedt, Journ. de Phys. Ser. VI, 4, 144, 1923.

II. Suppose a beam of primary X-rays, traveling in such a direction that the cosines of the angles with the three axes of a rectangular system of coördinates are α_0, β_0, γ_0 and define this

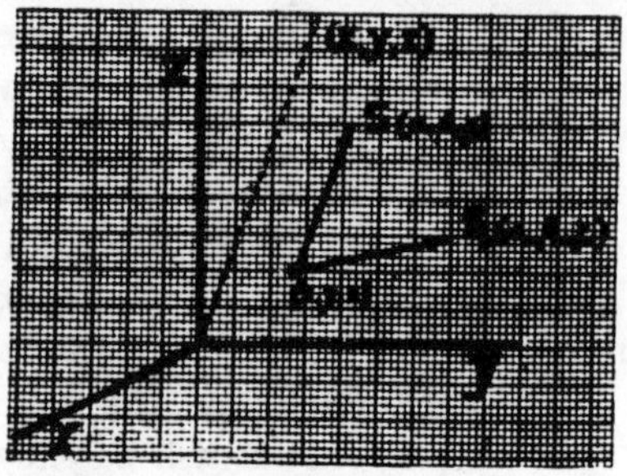

Fig. 1

direction by a vector S_0 of unit length and components α_0, β_0, γ_0. Let this ray strike a particle at a point x, y, z and let us assume that this particle will give rise to a scattered radiation in all directions with amplitude and phase defined in the usual way by a scattering factor Ψ. If the amplitude of the primary ray equals 1, the electric force at any point x, y, z, and any time t is equal to

$$e^{i\omega t}e^{-ik(\alpha_0 x+\beta_0 y+\gamma_0 z)}$$

if ω is the frequency and

$$k=\frac{\omega}{c}=\frac{2\pi}{\lambda}$$

is defined by the frequency ω and the velocity of propagation c or by the wave length λ. The particle at the point x, y, z will therefore emit a secondary wave with amplitude and phase given by the factor

$$\Psi e^{-ik(\alpha_0 x+\beta_0 y+\gamma_0 z)}.$$

The electric force of this scattered wave in a point X, Y, Z at a relatively large distance l, from the particle will be equal to

$$\Psi e^{-ik(\alpha_0 x+\beta_0 y+\gamma_0 z)}\frac{e^{-ikl}}{l}.$$

Suppose now the point X, Y, Z at so large a distance from the scattering medium that, introducing the distance R of this point from the origin, and the cosines α, β, γ of the radius vector R with the axes, we may use the approximation

$$l = R - (\alpha x + \beta y + \gamma z).$$

The amplitude of the scattered wave will then be equal to

$$\Psi \frac{e^{-ikR}}{R} e^{ik[(\alpha-\alpha_0)x + (\beta-\beta_0)y + (\gamma-\gamma_0)z]}.$$

Introducing a vector S of unit length and the components α, β, γ, and the vector r with the components x, y, z, defining the place of the particle, this expression can be written in the form

$$\Psi \frac{e^{-ikR}}{R} e^{ik(S-S_0).r}$$

if $(S-S_0).r$ is the scalar product of the two vectors $S-S_0$ and r.

Suppose now, that the scattering medium consists of a large number of particles l . . . n . . . N, at different places, characterized at a certain moment by the vectors r_l . . . r_n . . . r_N. At this moment the amplitude of the scattered radiation at the point X, Y, Z will then be equal to

$$\Psi \frac{e^{-ikR}}{R} \sum_n e^{ik(S-S_0).r_n}. \tag{1}$$

But the intensity corresponding to this amplitude cannot really be observed if the particles of the medium have a heat motion, as for instance, the molecules of a gas or a fluid. The only thing which can be measured is the mean value of this intensity during an appreciable time, very large in comparison with the time necessary for the positions of the particles to become thoroughly interchanged.

According to (1) we will find the intensity I at the moment under consideration by multiplying this expression by the conjugate value, finding

$$I = \frac{\Psi^2}{R^2} \sum_m \sum_n e^{ik(S-S_0).(r_n-r_m)}, \tag{2}$$

if we treat Ψ as a real value. The intensity to be measured I_m, will be given by the mean value of this expression, an operation which will be indicated by the letter M, so that

$$I_m = \frac{\Psi^2}{R^2} \sum_m \sum_n M[e^{ik(S-S_0).(r_n-r_m)}]. \tag{3}$$

III. If the number of particles of the scattering gas is equal to N, the sum in (3) consists of N^2 terms. The N terms of this sum, in which $n=m$, each have the value 1; the $N(N-1)$ remaining terms depend on the relative positions of two particles n and m. We now have to calculate the mean value

$$M[e^{ik(S-S_0)\cdot(r_n-r_m)}]=M_{mn}$$

giving the particles all possible places in the gas. We introduce the assumption that the gas considered is not very dense, so that, dealing with two particles m and n, we need not take into account the other particles. Moreover, it will not be necessary to make any difference between the real volume of the gas and the free volume in which the molecules can move. Calling the whole volume occupied V and an element of this volume dV, the probability that the center of the particle n will be in dV_n and the center of the particle m in dV_m is equal to

$$\frac{dV_n}{V}\,\frac{dV_m}{V}\,.$$

Multiplying the expression $e^{ik(S-S_0)\cdot(r_n-r_m)}$ by $\frac{dV_n}{V}\,\frac{dV_m}{V}$ and integrating over all the possible situations will yield the desired mean value M_{mn}.

If we fix the molecule n and allow the molecule m to reach every point not forbidden by our fundamental assumption that each molecule will act as a hard sphere with a certain radius a, the integration according to dV_m must be performed over the whole volume, with the exception of a sphere of the radius $2a$ surrounding the molecule n. This integration achieved, we will have to move dV_n to all the places included in V. Instead of performing the integration in this way, we may also integrate first by taking into account all the elements of volume dV_n and dV_m, of V without making any exception, corresponding to the molecular volume. We will then have to subtract the result issuing from the integration with respect to dV_m, now moving dV_m only in the interior of the sphere, with radius $2a$, followed by an integration with respect to dV_n over all the elements contained in V. This is the method we will follow.

The result therefore assumes the form:

$$M_{mn} = K'_{mn} - K''_{mn} \tag{4}$$

with

$$K'_{mn} = \int\limits_{\text{Vol.}} \int\limits_{\text{Vol.}} \frac{dV_n}{V} \frac{dV_m}{V} e^{ik(S-S_0)\cdot(r_n-r_m)} \tag{5}$$

and

$$K''_{mn} = \int\limits_{\text{Vol.}} \int\limits_{\text{Sphere}} \frac{dV_n}{V} \frac{dV_m}{V} e^{ik(S-S_0)\cdot(r_n-r_m)} \tag{5'}$$

the suffixes vol. and sphere at the integrals referring to the domains over which the integration has to be performed.

IV. Putting $S-S_0=s$, $r_n-r_m=r$ and introducing the angle θ between the vectors $S-S_0$ and r_n-r_m, the inner integral in (5′) assumes the form

$$\frac{2\pi}{V} \int\limits_{r=0}^{r=2a} r^2 dr \int\limits_{\theta=0}^{\theta=2\pi} e^{iksr\cos\theta} \sin\theta \, d\theta.$$

The result is

$$\frac{4\pi}{V} (2a)^3 \frac{\sin 2ksa - 2ksa \cos 2ksa}{(2ksa)^3},$$

which we will write in the form

$$\frac{1}{V} \frac{4\pi}{3} (2a)^3 \phi\,(2ksa),$$

introducing the function

$$\phi(u) = \frac{3}{u^3} (\sin u - u \cos u) \tag{6}$$

being equal to 1 for $u=0$.
The calculation of the outer integral in (5′) means merely the multiplication with the factor 1. Therefore we arrive at

$$K''_{mn} = \frac{1}{V} \frac{4\pi}{3} (2a)^3 \phi\,(2ksa). \tag{7}$$

From a purely mathematical point of view the calculation of

K'_{mn} is very difficult, because the expression will depend upon the geometrical form of the volume V. From a physical point of view it is, however, evident that this geometrical form can practically be of no importance at all. To find the order of magnitude of the value designated by K'_{mn} we will calculate the integral for a volume of definite form, *i.e.*, a cube of side d.

Let the coördinates of r_m and r_n be called x_m, y_m, z_m and x_n, y_n, z_n, the coördinates of S and S_0 being α, β, γ and α_0, β_0, γ_0. If the corner of the cube coincides with the origin of the system of coördinates and its sides are parallel to the axes, then we find, according to (5)

$$K'_{mn} = \frac{1}{d^6}\int_0^d \dots$$

$$\int_0^d e^{ik[(\alpha-\alpha_0)(x_n-x_m)+(\beta-\beta_0)(y_n-y_m)+(\gamma-\gamma_0)(z_m-z_n)]}\, dx_n dy_n dz_n dx_m dy_m dz_m.$$

Now we have at once

$$\int_0^d\int_0^d e^{ip(\xi_1-\xi_2)} d\xi_1 d\xi_2 = 2\frac{1-\cos pd}{p^2}$$

so that

$$K'_{mn} = 8\frac{1-\cos kd(\alpha-\alpha_0)}{k^2d^2}\;\frac{1-\cos kd(\beta-\beta_0)}{k^2d^2}\;\frac{1-\cos kd(\gamma-\gamma_0)}{k^2d^2} \quad (8)$$

It is obvious that K'_{mn} will be equal to 0 for very small values of the angle between the scattered and the primary ray because $kd = 2\pi\frac{d}{\lambda}$ is always an exceedingly large number. On the other hand, K''_{mn} depends on the value of ksa. The diameter of the molecule a being of the same order of magnitude as the wave length, this expression will have an appreciable value even for large angles.[6] We can therefore practically put

$$K'_{mn} = 0 \quad (8')$$

[6]As is readily seen the quotient K'_{mn}/K''_{mn} is of the order of magnitude volume of one molecule / total volume of the gas.

V. With the help of (7) and (8′), we can calculate

$$M_{mn} = -\frac{1}{V}\frac{4\pi}{3}(2a)^3\,\phi\,(2ksa).$$

M_{mn} not depending on the suffixes, and entering $N(N-1)$ times in the sum (3), expressing I_m, we find

$$I_m = \frac{\Psi^2}{R^2}\left[N - \frac{N(N-1)}{V}\frac{4\pi}{3}(2a)^3\phi(2ksa)\right].$$

The product

$$N\frac{4\pi}{3}(2a)^3 = \Omega$$

is the total volume of all the "spheres of action" of all the molecules, each sphere having a radius equal to the diameter of one molecule. Moreover, N being always very large, we may put $N-1=N$. The intensity of the scattering, therefore, is

$$I_m = N\frac{\Psi^2}{R^2}\left[1 - \frac{\Omega}{V}\phi\,(2ksa)\right]. \qquad (9)$$

We put $|S-S_0|=s$; if now the angle between the secondary and the primary ray is called θ, then

$$s^2 = (S-S_0)^2 = S^2 + S_0^2 - S.S_0 = 2(1-\cos\theta) = 4\sin^2\frac{\theta}{2}.$$

The argument of our function ϕ, defined by (6) can, as a result, be written in the form

$$2ksa = 8\frac{\pi a}{\lambda}\sin\frac{\theta}{2}$$

remembering that $k=\frac{2\pi}{\lambda}$. According to (9), the intensity of scattering, therefore, depends on the angle θ, even for the case underlying our calculation, that the molecules only act as hard spheres and show no association at all to double molecules. This remark seems necessary because it has sometimes been claimed that the existence of the interference ring has to be connected with the existence of double molecules, no other cause seeming to be available for the understanding of interference effects.

In Fig. 2 the function

$$\phi(u) = \frac{3}{u^3} (\sin u - u \cos u)$$

is plotted as an ordinate against u. For $u=0$ it is found $\phi(u)=1$, with increasing argument the function ϕ becomes 0, performing oscillations of decreasing amplitude. If the scattering of each

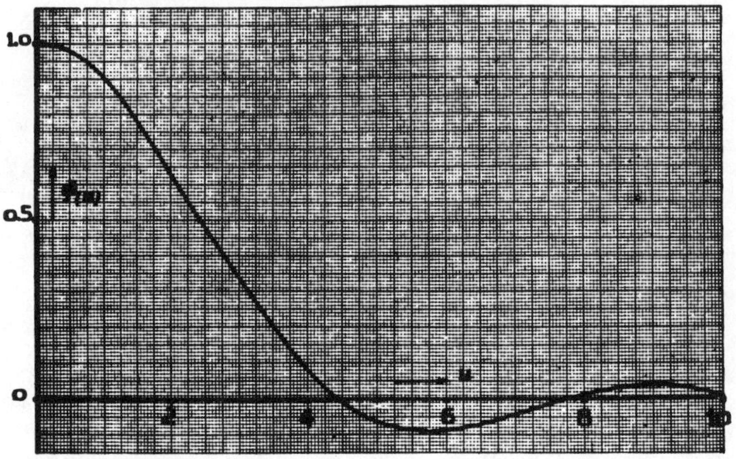

Fig. 2

particle were the same as the scattering of a single resonator, Ψ^2 would have to be proportional to $\frac{1+\cos^2\theta}{2}$ because of the fact that the primary radiation is unpolarized and the effect of scattering yields a polarization of the scattered radiation which is complete for an angle of $\theta = 90°$. To visualize the effect of the interference here considered, a curve is drawn in Fig. 3, giving the function

$$\frac{1+\cos^2\theta}{2}\left[1-\tfrac{1}{2}\phi\left(12\,\pi\,\sin\frac{\theta}{2}\right)\right]$$

This corresponds to

$$\frac{2a}{\lambda} = 3$$

meaning a diameter of $2a = 2.1 \times 10^{-8}$ cm., if a wave length of $\lambda = 0.7 \times 10^{-8}$ is used. Moreover, the factor $\frac{\Omega}{V}$ has been put equal to $\frac{\Omega}{V} = \frac{1}{2}$. Under these assumptions, a ring would be observed with a maximum of intensity at an angle of about $\theta = 16°$.

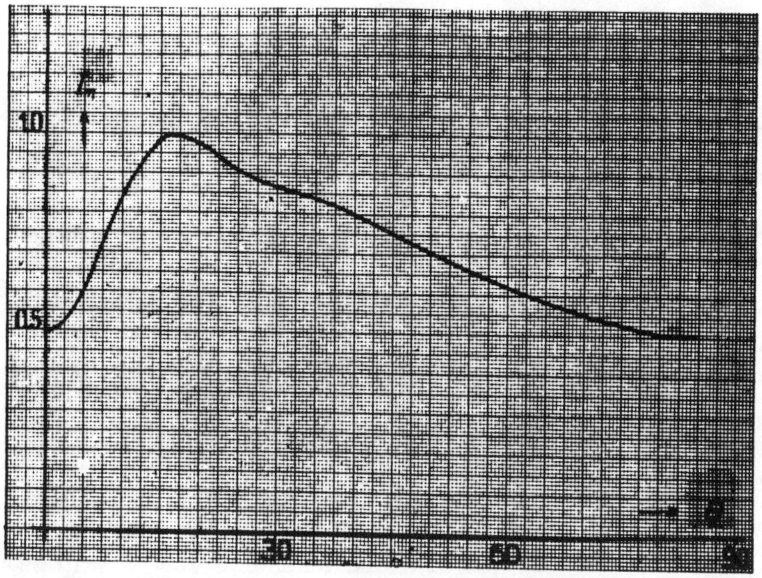

Fig. 3

In general, the molecule will not act as a single resonator, but it is to be expected that there will exist an interference effect corresponding to the shape of the atomic frame. The whole scattering curve must now show a superposition of the two kinds of interference effects which may be briefly called the "inner" and the "outer" effect. Now the outer effect will, as in (9'), always be proportional to $\frac{\Omega}{V}$. In making measurements on a gas under different pressures, it should therefore be possible to eliminate the outer effect which tends to vanish with decreasing density, and to arrive at a measurement of the inner effect. Should such measurements prove to be practically possible, then we may, in the case of nitrogen for instance, expect as a result the direct measurement of the distance of the two N-atoms.

VI. It will, perhaps, be of some interest to find the scattering function also for this more complicated case. Consider spherical molecules of radius a, each containing two scattering centers acting as simple resonators at a mutual distance l. The line connecting these two centers is supposed to pass through the center of the sphere so that this center is at equal distance $\frac{l}{2}$ from the two resonators. The molecules themselves will be numerated in the same manner as before with the numbers $1 \ldots n \ldots N$. The resonators may be designated by dashes, in such a way that, for instance, r_n' means the radius-vector from the origin to one of the resonators of the particle n, and r_n'' means the radius-vector from the origin to the second resonator of the same molecule. The undashed letter r_n will be reserved for the vector drawn from the origin to the center of the sphere. Moreover, we shall put

$$r_n' = r_n + v_n' \quad \text{and} \quad r_n'' = r_n + v_n''$$

v_n' and v_n'' meaning the radius-vector drawn from the center of the sphere to the first or second resonator. According to our previous reasoning, (cf. (1)), the amplitude of the scattered radiation is now equal to

$$\frac{\Psi}{R} \sum_n \left(e^{ik(S-S_0)\cdot r_n'} + e^{ik(S-S_0)\cdot r_n''}\right)$$

and the mean intensity I_m can be calculated by the equation

$$I_m = \frac{\Psi^2}{R^2} \sum_m \sum_n M_{mn} \tag{10}$$

with

$$M_{mn} = M\left[\left\{e^{ik(S-S_0).r_n'} + e^{ik(S-S_0).r_n''}\right\}\left\{e^{ik(S-S_0).r_m'} + e^{ik(S-S_0).r_m''}\right\}\right]. \tag{10'}$$

Now the mean values of each of the four terms of this product must be calculated. Each of them will give another result according as to whether $n = m$ or $n \neq m$. Assuming that all the possible directions of the line connecting the two resonators and therefore of the vectors v_n' and v_n'' are equally probable, this calculation however, offers no difficulty,[7] and leads to the following formula:

(a) $m=n$

$$M[e^{ik(S-S_0).(r_n'-r_m')}]=1$$

$$M[e^{ik(S-S_0).(r_n'-r_m'')}]=\frac{\sin ksl}{ksl}$$

$$M[e^{ik(S-S_0).(r_n''-r_m'')}]=\frac{\sin ksl}{ksl}$$

$$M[e^{ik(S-S_0).(r_n''-r_m'')}]=1$$

(b) $m \neq n$

$$M[e^{ik(S-S_0).(r_n'-r_m')}]=-\frac{1}{N}\frac{\Omega}{V}\phi(2ksa)\left(\frac{\sin\frac{ksl}{2}}{\frac{ksl}{2}}\right)^2$$

and the same value for the three remaining terms. Remembering that in N terms of the sum (10) the condition $n=m$ is fulfilled and the number of the remaining terms in $N(N-1)$, we find first

$$n=m:\ M_{nm}=2\left[1+\frac{\sin ksl}{ksl}\right]$$

$$n\neq m:\ M_{nm}=-\frac{4}{N}\frac{\Omega}{V}\phi(2ksa)\left(\frac{\sin\frac{ksl}{2}}{\frac{ksl}{2}}\right)^2$$

[7]Calculating $M\ e^{ik(S-S_0).(r_n'-r_m')}$ we put

$$r_n'=r_n+v_n' \quad , \quad r_m'=r_m+v_m'$$

The mean value M to be found is then the result of the following integration

$$M=\int\int\frac{dV_n}{V}\frac{dV_m}{V}e^{ik(S-S_0).(r_n-r_m)}\int\int\frac{d\Omega_n}{4}\frac{d\Omega_m}{4}e^{ik(S-S_0).(v_n'-v_m')}$$

if $d\Omega_n$ and $d\Omega_m$ denote elements of the solid angles defining the directions of the vectors v_n' and v_m'. Remembering that the length of either of these vectors is $l/2$, and introducing the angle Θ between S–S_0 and v_n', one of the last integrals assumes the form

$$\frac{1}{2}\int_0^\pi e^{\frac{ikls}{2}\cos\Theta}\sin\Theta\, d\Theta=\frac{\sin\frac{ksl}{2}}{\frac{ksl}{2}}$$

The first integral has been calculated in the preceding part. The combination of these results yields the formula under (b) of the text.

and on introducing these values in (10):

$$I_m = 4N\frac{\Psi^2}{R^2}\left\{\tfrac{1}{2}\left[1+\frac{\sin ksl}{ksl}\right]-\frac{\Omega}{V}\left(\frac{\sin\frac{ksl}{2}}{\frac{ksl}{2}}\right)^2 \phi(2ksa)\right\}. \quad (11)$$

It may be remembered that,

$$ksl = 4\pi\frac{l}{\lambda}\sin\frac{\theta}{2}$$

$$2ksa = 8\pi\frac{a}{\lambda}\sin\frac{\theta}{2}$$

and ϕ is the function defined by (6) and given in the diagram Fig. 2.

With decreasing density the influence of the second term in (11) vanishes and the scattering function finally assumes the limiting form:

$$I_m = 4N\frac{\Psi^2}{R^2}\ \frac{1}{2}\left[1+\frac{\sin ksl}{ksl}\right], \quad (11')$$

agreeing with a general formula previously calculated.[8]

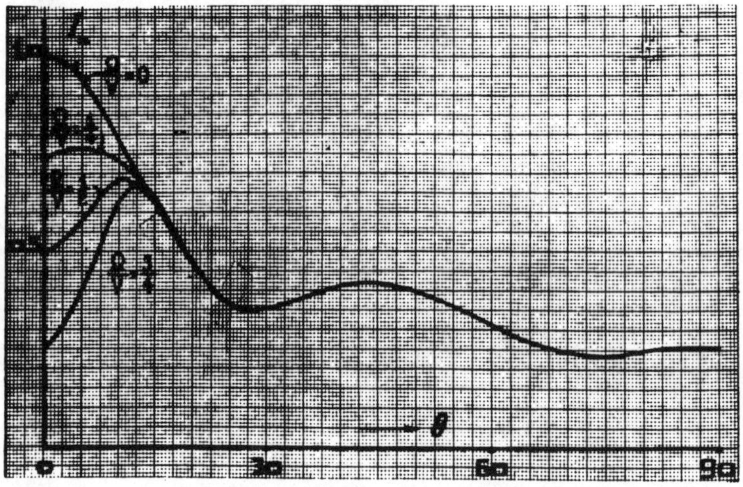

Fig. 4

[8]P. Debye, Ann. d. Phys. 46, 809, 1915.

To illustrate formula (11′) the intensity I_m has been calculated according to the assumptions $\frac{2a}{\lambda}=3$ and $l=a$, corresponding, for instance, at a given wave length of, say $\lambda=0.7\times10^{-8}$ cm. to a diameter of the sphere, substituting the molecule of 2.1×10^{-8} cm. and a distance of the resonators substituting the atoms of 1.05×10^{-8} cm. The scattering function moreover depends on the density, measured by the quotient Ω/V. Therefore four curves are given in Fig. 3, corresponding to the values $\frac{\Omega}{V}=0$, $\frac{\Omega}{V}=\frac{1}{4}$, $\frac{\Omega}{V}=\frac{1}{2}$ and $\frac{\Omega}{V}=\frac{3}{4}$. The diagram shows how the first maximum, occurring at about $\theta=12°$ and corresponding to intermolecular interference, disappears with decreasing density, leaving a second maximum at $\theta=45°$ undisturbed, because this maximum corresponds to the interatomic interference of the two atoms constituting the molecule.

March, 1925.

SCATTERING OF X-RAYS BY SINGLE MOLECULES
(Zerstreuung von Röntgenstrahlen an einzelnen Molekeln)

P. Debye, L. Bewilogua, and F. Ehrhardt*

Translated from
Physikalische Zeitschrift, Vol. 30, 1929, pages 84-87

(1) A number of years ago one of us, together with Scherrer, performed experiments on the scattering of x-rays by liquids in the hope of producing interference due to the interaction of elementary waves stemming from the individual atoms of free molecules. It soon became evident that the interference rings observable due to such experiments could be explained for the most part by the interference of the components of the scattered radiation arising from the different molecules as entities. A clear separation of the two effects, presumably by theoretical calculations, is impossible, because a sufficiently exact theory of the interference of arbitrarily and densely spaced molecules does not exist; and if it is at all possible it is very difficult. Consequently, a result of the kind looked for could be reached only by experimental elimination of the molecular interference and this could be accomplished only in one way: by sufficiently enlarging the separation between molecules, i.e., observing the diffraction on gases or vapors. In this case it can be shown that the effect to be eliminated can only have a relative intensity of the order of magnitude: ratio of volume of molecule to total volume of gas.[1] With gases at atmospheric pressure the number determined by this ratio is so small compared to 1 (about 10^{-3}), that experiments under such circumstances must result in an interference pattern which for all practical purposes can only be accounted for by possible intramolecular interference. During the past year the necessary apparatus was installed at the Leipzig Institute and experiments were conducted with it. The following is a report about experiments which we believe constitutes a proof of the expected interference phenomenon.

*Received December 22, 1928.

If one considers each atom as a scattering center, as is successfully done in the explanation of interference in crystals, then it is clear that a fixed molecule, since it represents an atom lattice with fixed spatial configuration of the constituent atoms, will produce a characteristic distribution of intensities of the scattered radiation. This distribution, if considered as a function of the angle θ between the primary and secondary ray, will generally show maxima and minima. The position of the maxima depends on the spatial configuration of the atom in a manner easily specified. The entire interference pattern represents therefore, if properly explained, a model of the configuration of the atoms. In a gas, however, it is impossible to orient all molecules, for instance, parallel. The question arises therefore whether the entire interference effect is completely obliterated, due to the fact that all possible orientations occur. Calculations show that this does not occur, and that maxima and minima should appear even with completely random orientation.[2] In the special case, e.g., in which one considers only identical radiating atoms, the formula for the intensity I of the scattered radiation contains the factor (which alone is important to us at this point):

$$\sum_i \sum_j \frac{\sin x_{ij}}{x_{ij}}$$

where x_{ij} is an abbreviation for the expression:

$$x_{ij} = 4\pi \frac{l_{ij}}{\lambda} \sin\frac{\Theta}{2}$$

and λ denotes the wavelength, θ the angle between primary and secondary ray and l_{ij} the distance between two atoms in the molecule designated by indices i and j. We shall simply set:

$$I = \sum_i \sum_j \frac{\sin x_{ij}}{x_{ij}}$$

omitting all other less important factors. The omitted factors vary, in accordance with all experiences, only relatively slowly and, what is most important, decrease continuously with increasing angle θ.[3] If, for instance, the scattering molecule consists of four identical atoms at the corners of a regular tetrahedron of side a, I becomes:

$$I = \sum_{i=1}^{i=4} \sum_{j=1}^{j=4} \frac{\sin x_{ij}}{x_{ij}} = 4\left[1 + 3\frac{\sin x}{x}\right]$$

where:

$$x = 4\pi \frac{a}{\lambda} \sin\frac{\Theta}{2}$$

A first maximum of I is situated near $x = 5(\pi/2)$, i.e., at an angle θ_m which can be calculated from the formula:

$$\sin\frac{\Theta_m}{2} = \frac{5}{8}\frac{\lambda}{a}$$

(2) The experiments which are reported here were conducted in the following manner. An x-ray bundle is collimated by a suitable screening system. It has a diameter of about 1 mm. measured at the next to the last diaphragm. The x-ray enters into a cell which is filled with the gas to be investigated, and goes through it for a distance of 2 cm. It is then absorbed by a little brass rod, which

is in the form of a slide. The front wall of the cell of circular cross section is covered half way by a semi-circular brass plate, the upper part of which consists of this slide. The other half is closed by an aluminum sheet of 0.01 mm. thickness. The back wall is provided with a heating coil. The individual dimensions can be taken from Figure 1, based on the statement that the inner diameter of the cell

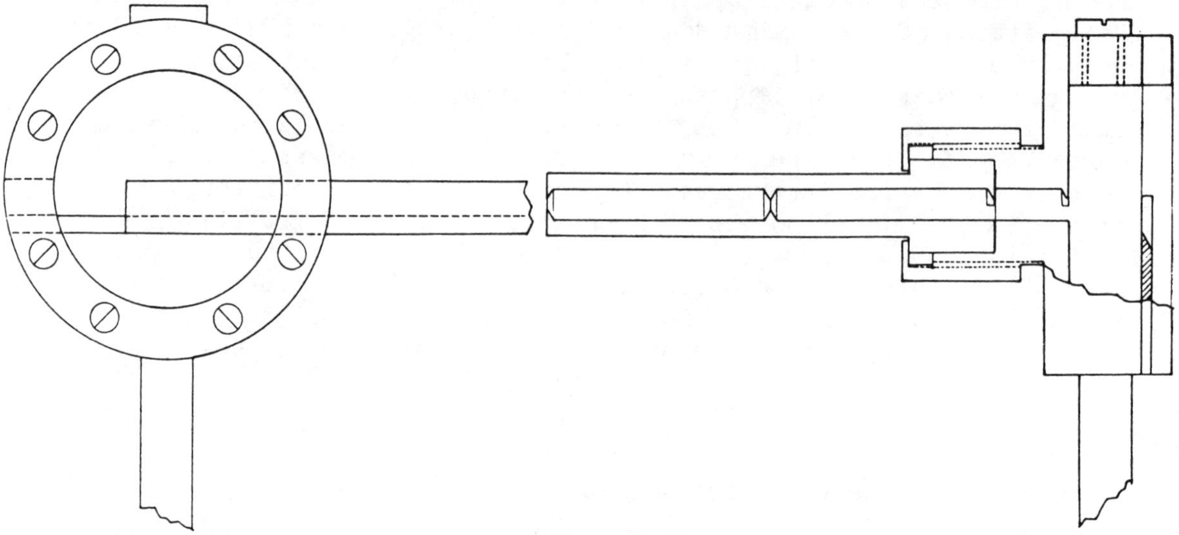

Fig. 1.

is 3 cm. The whole apparatus is arranged so that no secondary radiation from metal parts touched by the primary ray can leave the cell through the aluminum sheet. The scattered radiation, due to the gas, is caught after going through the sheet at a few centimeters distance on a photographic film which is contained in a cassette. The front of the cassette consists of black paper, while the back wall on which the film is placed forms a shallow vessel, which is cooled by running water. How far it has been possible to eliminate the scattering by the metal parts is shown in Figure 2, the reproduction of a 15-hour exposure, during which the air was removed from the cell which was itself within a vacuum. As a source of radiation, an ionic x-ray tube made of metal, according to Hadding, with a copper anti-cathode was used. It was operated at a voltage of 40 kV and consumed 15 mA corresponding to a power of 9.6 kW. No darkening of the plate was to be seen. Since the cell was found suitable by this control exposure, pictures were taken with air. Figure 3 is a reproduction of such a picture, which was obtained by an exposure of 20 to 30 hours. The scattering is present, but the decrease of blackening toward larger angles is uniform.[4] If one thinks of a nitrogen molecule as **consisting** of 2 scattering centers, the intensity would have to be calculated, according to the above-mentioned formula, by:

$$I = 2\left(1 + \frac{\sin x}{x}\right)$$

Fig. 2.

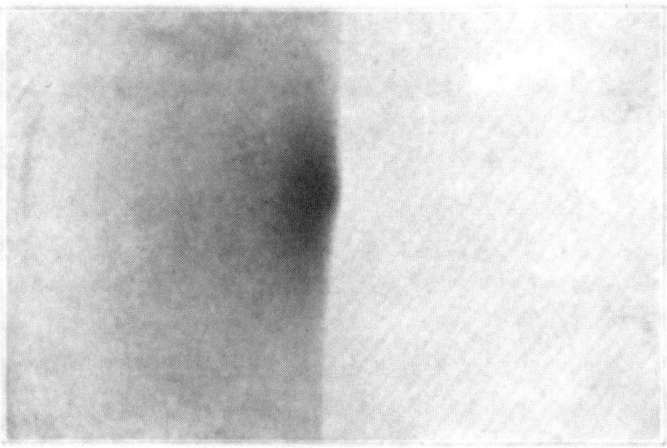

Fig. 3.

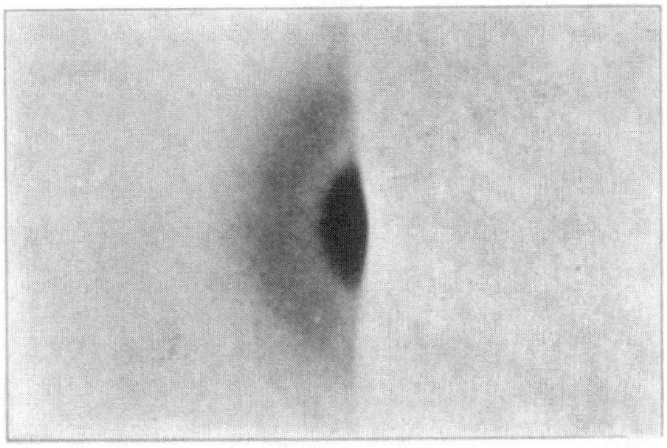

Fig. 4.

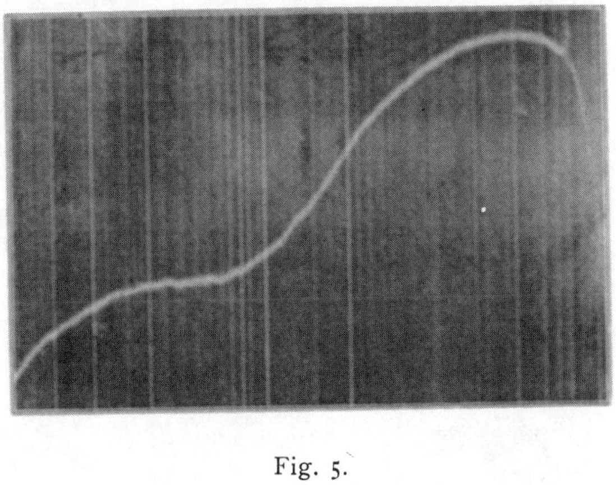

Fig. 5.

An indistinct ring would thus be possible. However, the atoms are not to be regarded as points, and a certain amount of thermal motion of the atoms is present. To this is to be added the fact that the gas volume, in which the scattering takes place, has a noticeable extension. All these circumstances tend to decrease the expected interference phenomenon. The formula is, therefore, under no circumstance to be interpreted as a true representation of the variation of intensity with angle; it can only give an indication of what one has to look for.

An experiment with CCl_4 seemed much more promising. According to organic chemistry it has four heavy chlorine atoms situated at the corners of a regular tetrahedron at constant distances a from one another. In addition, the small amount of scattering, due to the light central atom, can be neglected compared to the scattering by the four outer atoms. The theoretical intensity formula is the one given above; the term which leads to maxima and minima of intensity is 3 times as big in the case of di-atomic molecules.

The experiment was conducted in such a way that the cell was constantly connected with a boiler, provided with a cooler, in which the carbon tetrachloride boiled continuously. The boiling point of the liquid is at 76.7° C.; the cell was kept at a higher temperature, 85° C., by the use of a heating coil. Before beginning the experiment, the cell was opened by a screw situated at the top; and one waited long enough to permit the air to escape together with the escaping vapor. The above-mentioned cooling of the film, situated a few centimeters from the front side of the cell, was carried out in order to prevent harm to the sensitive layer by heat. A picture of the scattering by CCl_4 vapor is reproduced in Figure 4. The time of exposure was 20 hours, the distance from the front side of the cell to the film was 1.5 cm. Now a clearly recognizable interference ring occurs. Figure 5 shows a photographic density recording across the picture obtained from a Koch recording photometer. In spite of the fact that deception due to the liquid condensed on the walls of the cell seems unavoidable, not only because of the temperature of the cell which is kept above the boiling point, but also because the only place where such condensed material could lie in the primary ray is screened from the film by brass, various control experiments were made. At one time the cell was kept at 120° C. instead of at 85° C. This changed nothing. During a different test a lattice of copper wires was placed in front of the aluminum sheet (outside of the cell) with the purpose of securing a sufficiently high temperature of the thin aluminum sheet itself. In this picture the ring is again clearly visible between the shadow of the wires.

(3) We believe we have the right to conclude from the above results that the observed interference ring is indeed due to the inner interferences within the individual CCl_4 molecule. In order to determine the angle at which the ring appears somewhat more exactly, a second picture was taken with a larger distance from cell to film. From these exposures it could also be seen that the domain from which the

scattered radiation originates lies in the cell within the volume of vapor. For the angle at which the ring appears, the value θ_m = 34° with an error of about ± 10% was found. If one substitute this value in the formula given at the beginning of this article:

$$\sin\frac{\Theta_m}{2} = \frac{5}{8}\frac{\lambda}{a}$$

and set λ = 1.54 A (K_α - line of copper radiation), one obtains: a = 3.3 A. This number represents, according to our ideas, the distance between two chlorine atoms in a CCl_4 molecule, determined here using a direct interferometric method. For comparison it may be mentioned that Wasastjerna obtains the value 1.72 A for the radius of a Cl ion, based on his ideas about the volume occupied by atoms in a crystal. Two such ions would therefore have a distance of 3.44 A between them when touching. It is also of interest to make a comparison with the result which one obtains upon radiating liquid carbon tetrachloride. According to the statements of Wyckoff,[5] a ring appears at an angle θ = 18° in the latter case when using copper radiation. The maximum found by us is situated at θ = 34° , and therefore has no relalation to the liquid ring. In addition to the main maximum, however, a weak secondary maximum has been observed with the liquid at θ = 31°. It would not be impossible that this secondary maximum is identical to the vapor ring. If this supposition should be true, our pictures on vapor would give the proof for the old assumption that the pictures obtained by scattering from liquids represent a superposition of two effects of which the one of main interest here is obscured to a large extent by intermolecular interferences. According to Katz[6] a second ring always appears in the case of liquids if in the molecule a central atom is linked to 2 or 3 atomic groups. This might be an effect of intramolecular interference.

The experiments are being continued with other gases and vapors*; at the same time other designs of the apparatus are being considered with the aim of securing better pictures extending over a larger range of angles. We not only hope first to measure atomic distances in the molecule, assuming the chemical structural formula to be correct, but also to prove or disprove the validity of such a geometrical picture by comparison with the observed intensity distribution of the scattered radiation.

*In the meantime we have observed one interference maximum with chloroform vapor. In this case only 3 atoms cooperate, accordingly the maximum is less pronounced than with carbon tetrachloride.

Bibliography

1. P. Debye, *Physik. Z. 28*, 135 (1927).
2. P. Debye, *Ann. Physik, 46*, 809 (1915).
3. This can be deduced theoretically on the basis of the wave-mechanical distribution of charge in the atom; it follows, however, also definitely from the experimental work of W. L. Bragg and his students. This fact was directly proved for the Hg atom by the experiments of P. Scherrer and A. Stäger, *Helv. Phys. Acta, 1*, 518 (1928).
4. C. S. Barrett, *Phys. Rev. 32*, 22 (1928) reports on measurements of scattering by H_2, He, N_2, O_2, A, and CO_2; for which, also, no interference maxima were found.
5. R. W. G. Wyckoff, *The Structure of Crystals*, Reinhold, New York, 1924, page 383; *Am. J. Sci. 5*, 455 (1923).
6. J. R. Katz and J. Selman, *Z. Physik, 46*, 392 (1928). Other interesting experiments with liquids have been performed by C. W. Stewart (*Phys. Rev.*, 1927 and 1928) and in Raman's laboratory by C. M. Sogani, *Indian J. Phys., 1*, 357 (1927). Compare also the survey of C. Drucker, *Physik. Z., 29*, 273 (1928).

INTERFERENCE OF X-RAYS BY ISOMERIC MOLECULES
(Röntgeninterferenzen an isomeren Molekülen*)

P. Debye
(Experiments performed with L. Bewilogue and F. Ehrhardt)

Translated from
Physikalische Zeitschrift, Vol. 31, 1930, page 142

Preliminary Report

Some time ago it was reported that one can recognize interferences in the scattered radiation of vapors which are caused by the interaction of some few atoms in a single molecule (*Physik. Z.*, *30*, 84-87 (1929); *30*, 524-525 (1929) (*Züricher Vorträge*); *Ber. Sächs. Akad. Leipzig, Math.phys. Kl.*, *81*, 29-37 (1929)). In the meantime the first three interference maxima were observed and photometrically measured in the case of CCl_4, using copper radiation. The maximum of fourth order was also seen when using molybdenum radiation. Only two atoms in a molecule still give recognizable interference if essentially monochromatic x-rays are used and if the atoms themselves are heavy enough compared to the rest of the molecule. These facts result in the possibility of measuring spacings in the molecule interferometrically. They were here used in order to investigate two interesting isomeric cases.

(*1*) The two dichloroethylenes:

cis trans (1)

can be obtained as liquids which boil at 60° and 48°, respectively. The structure of the molecules is visualized in accordance with chemical experience and according to I. Wislicenus (because of the suppression of the so-called free rotation by the double bond)

*Received January 22, 1930.

by the above formulae. That the structure in its geometrical aspect actually corresponds to these formulae was already proved by the polarity measurements which were carried out by J. Errera. (Only the cis form has an electric moment.) In addition, the cis form should evidently have a smaller distance a from Cl to Cl atom than the trans form. Photographs of the scattering by the two substances in gaseous form gave the expected interferences. The photographs appear essentially different. The distances:

$$a_{cis} = 3.6 \text{ A.}; \quad a_{trans} = 4.1 \text{ A.}$$

follow from the **angles** at which the first interference maxima appear. The distance is therefore indeed larger in the trans form.

(2) The two dichloroethenes:

$$\begin{array}{cc}
\begin{array}{c}
\text{H} \\
\text{H} - \overset{|}{\text{C}} - \text{H} \\
\text{H} - \overset{|}{\underset{|}{\text{C}}} - \text{Cl} \\
\text{Cl}
\end{array}
&
\begin{array}{c}
\text{H} \\
\text{H} - \overset{|}{\text{C}} - \text{Cl} \\
\text{H} - \overset{|}{\underset{|}{\text{C}}} - \text{Cl} \\
\text{H}
\end{array}
\end{array} \qquad (2)$$

1, 1-compound   1, 2-compound

should also have different distances a of the Cl atoms in accordance with their chemical formulae. Here, also, interference maxima were found, and the results, in agreement with expectation, are:

$$a_{1,1} = 3.4 \text{ A.}; \quad a_{1,2} = 4.4 \text{ A.}$$

Moreover, the interference pattern of the 1, 2-compound is most probably not compatible with an unrestricted rotation around the C--C bond; an equilibrium position seems to exist, which is similar to the trans form with the double bond discussed above.

The discussion of these results formed a part of a report at the meeting of the Gauverein of the Phys. Ges. at Breslau on the 11th and 12th of January, 1930.

X-RAY INTERFERENCE AND ATOMIC SIZE*
(Röntgeninterferenzen und Atomgrösse)

P. Debye**

Translated from
Physikalische Zeitschrift, Vol. 31, 1930, pages 419–428.

(1) The interferences observed in radiation scattering by gases show in their dependency of the intensity on the angle characteristic deviations from calculations according to the point theory, in which the scattering centers are considered to be of negligible dimensions. If, for instance, one has a diatomic molecule with identical atoms of distance l between them, then except for a constant factor the point theory gives for the intensity I the expression,

$$I = \frac{1 + \cos^2 \vartheta}{2} 2 \left[1 + \frac{\sin x}{x} \right] \tag{1}$$

with:

*The investigations reported on here began in 1915 in a paper entitled "Scattering of X-Rays" (P. Debye, *Ann. Physik*, *46*, 809 (1915)). This note contained the basic formulae used here, to demonstrate the essential point that random orientation cannot destroy the appearance of interference maxima. It is dated February 15, 1915 and is marked by the editors as "received February 27, 1915". By a strange coincidence, a paper on the same subject by P. Ehrenfest was presented by H. A. Lorentz and H. Kamerlingh Onnes (*Amst. Akad.*, *23*, 1132 (1915)) was given at the meeting of the Amsterdam Academy on Saturday, February 27, 1915. The title of the paper is: "About Interference Phenomena to be Expected when X-rays Pass Through a Diatomic Gas." The two atoms are considered as two scattering points with a fixed distance between them, and it is shown that for this special case, also, interference effects should be observable in spite of random orientation.

**Received March 6, 1930.

$$x = 4\pi \frac{l}{\lambda} \sin \frac{\vartheta}{2}$$

where λ is the wave length of the primary radiation and ϑ the angle between the direction of scattered radiation and the direction of the primary ray. The factor:

$$\frac{1+\cos^2\vartheta}{2}$$

measuresthe influence of the polarization which occurs during the scattering, and which decreases the intensity at $\vartheta = 90°$ to one half compared to that in the direction of $\vartheta = 0°$ and $\vartheta = 180°$. According to this formula, one would have to expect only a relatively small decrease of radiation intensity with increasing angle. Thus, in our special case, when l is a few times as big as λ, the ratio I_{180}/I_0 would, according to equation (1), be just about $\frac{1}{2}$.

In all practical cases, the intensity decreases much more rapidly with increasing angle, as any photogtaph will show at once. A similarly large decrease in intensity with increasing glancing angle occurs also if x-rays are deflected from crystals. In this case the effect is successfully explained by the fact that the wave length which is used is comparable in size with the dimensions of the atoms. The result is that, especially at large angles of scattering, the waves originating from various parts of the atom destroy each other by interference. Under these circumstances, it seems obvious that at least the main part of the observed effect in the case of scattering by gases can be explained in the same way. However, certain special effects occur in the gas case, about which we shall report in the following:

(2) We want to start from the assumption that the electron density in each atom depends only on the corresponding nuclear distance, and hence disregard for the present the asymmetry which the chemical binding must cause, especially in the distribution of the outer electrons. At a distance r from the nucleus let the density be $\nu(r)$, so that in a spherical shell between r and $r+dr$ we find:

$$4\pi\, r^2\, \nu\,(r)\, dr$$

electrons. Calculations which are similar to those in the point theory then give for a molecule which is built up by the atoms $1 \ldots i \ldots n$ a scattering intensity

$$I = \frac{1}{R^2}\frac{e^4}{m^2 c^4}\frac{1+\cos^2\vartheta}{2}\sum_i\sum_j \psi_i \psi_j \frac{\sin x_{ij}}{x_{ij}} \tag{2}$$

where:

$$x_{ij} = 4\pi \frac{l_{ij}}{\lambda} \sin \frac{\vartheta}{2}$$

Here l_{ij} signifies the distance from atom i to atom j, while ψ_i (or ψ_j) is an amplitude factor which is based on the distribution of the electrons in the atom. In contrast to the original formula (1), the constant factor is here also represented, so that expression (2) gives the absolute intensity of the scattered radiation for the primary intensity=1. The letters e, m, and c signify electronic charge, electron mass, and speed of light, respectively, while R is the distance from the molecule to the point of observation.

The important thing is the influence of the amplitude factor ψ. For this one obtains the expression:

$$\psi_i = \int_0^\infty \nu_i \frac{\sin k s r}{k s r} 4 \pi r^2 d r \tag{3}$$

in which $k=\frac{2\pi}{\lambda}$ and $s= 2 \sin \frac{\vartheta}{2}$. Always the quantity $2 \sin \frac{\vartheta}{2}$ and not the angle ϑ itself is the variable which is of importance and which recurrs in all interference problems. It corresponds to twice the sine of the glancing angle in the case of crystal reflection. At the beginning of our experiments we obtained for our private use a general orientation concerning the effects to be expected by assuming a special electron distribution such that calculations were easy and still the main features were not distorted too much. We assumed:

$$\nu_i = \frac{1}{\pi^{3/2}} \frac{z_i}{a_i^3} e^{-\frac{r^2}{a_i^2}} \tag{4}$$

thus distributing the electrons of the ith atom in accordance with a Gaussian error function, which with the aid of the characteristic length a_i determines the extension of the atom. The quantity z_i signifies the total number of electrons contained within the atom, while the numerical factor is so chosen that by integrating ν_i over the volume the total number z_i is actually obtained. With this distribution function one obtains, by carrying through the integration suggested in equation (3):

$$\psi_i = z_i e^{-\frac{k^2 s^2 a_i^2}{4}} \tag{5}$$

The amplitude factor, therefore, decreases with increasing angle ϑ ($s = 2 \sin \frac{\vartheta}{2}$) the faster the larger a_i is, that is, the bigger the electron domain of the atom is. In Figure 1 a series of scattering curves are drawn for the case of a (fictitious) continuous series of diatomic molecules. The distance between nuclei is kept constant but the size of the electron domain increases from one curve to the next. In order to visualize this, each curve is accompanied by a drawing of a molecular model in which the atoms are indicated as circles which were drawn with a radius $r = a\sqrt{2}$. According to equation (4) the electron density on this circle would have fallen to a fraction $\frac{1}{e^2} = \frac{1}{7.39}$ of its value in the center. The ordinates y of the functions represented are (in our case of the diatomic molecule), according to equations (2), (3), and (5):

$$y = 2 e^{-\frac{k^2 s^2 a^2}{2}} \left[1 + \frac{\sin x}{x}\right] = 2 e^{-\frac{a^2}{l^2} \frac{x^2}{2}} \left[1 + \frac{\sin x}{x}\right]$$

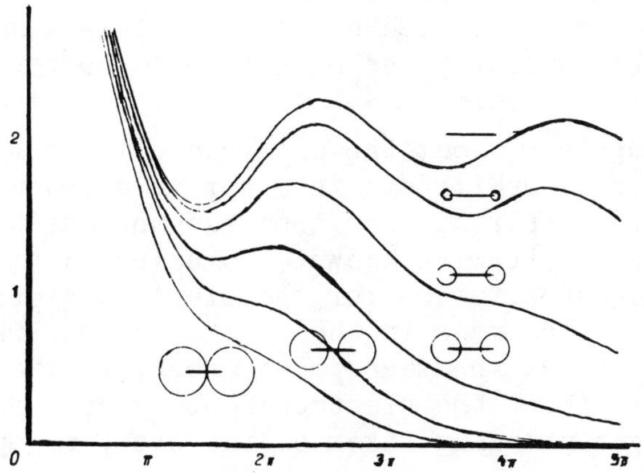

Figure 1

We have omitted the factor:

$$\frac{1 + \cos^2\vartheta}{2}$$

as insignificant for our purpose. On the horizontal axis the variable:

$$x = ksl = 4\pi\frac{l}{\lambda}\sin\frac{\vartheta}{2}$$

is represented.

It appears that the ratio a/l, *i.e.*, the ratio of size of atom to distance between atoms, is of a special importance. The bigger this ratio is, the more rapidly the scattering curve will decrease with increasing angle. At the same time it is seen that this ratio has an influence upon the interpretation of the molecular interference. As long as a/l is small, two atoms making up a molecule will still give recognizable interference maxima and minima. If, however, this ratio becomes too big, the maxima and minima disappear, and only a practically smooth curve remains. In fact, it has been shown that molecules such as O_2 and N_2, if investigated with copper radiation, will not give a recognizable first maximum, while, on the other hand, two chlorine atoms connected to a carbon atom will show the first interference maximum quite clearly.

Similarly, the scattering curve for CO_2 is essentially smooth, while CS_2 gives clear maxima and minima. Further on in this report, we shall once more return to these cases, when our ideas have been improved as to their quantitative representation. Nothing will change qualitatively. Because of this a second remark can be made here, which is of importance for the exact determination of atomic distances from interference pictures. It is seen in Fig. 1 how the position of the first maximum changes with increasing atomic size and constant atomic distance; it moves toward smaller values of x. Thus, with relatively big atoms one may be led erroneously to assume too large an atomic distance. We shall further on return to corrections which are necessitated by this displacement.

(3) The assumptions about the electron distribution which we just used were entirely arbitrary; in order to advance from here, it will be necessary to substitute for them assumptions which rest on better foundations. As is well known, Thomas and Fermi[1] have independently found an approximation for the electron distribution in which they treat the electrons in the atom as a gas, where the energy of the atom appears as the zero point energy in the customary sense. According to this, the electron density at a place with the potential V, due account being taken of the electron spin, must be equal to:

$$\nu = \frac{8\pi}{3}\left(\frac{2\,m\,e\,V}{h^2}\right)^{3/2} \tag{6}$$

On the other hand, Poisson's equation:

$$\Delta V = 4\pi\nu e \tag{7}$$

must hold for the potential V, since the charge of an electron is equal to $-e$. By combining equations (6) and (7), one obtains for the potential the equation:

$$\Delta V = \frac{32\pi^2}{3}\,e\left(\frac{2\,m\,e}{h^2}\right)^{3/2} V^{3/2} \tag{8}$$

If V is determined according to equation (8), assuming the proper boundary conditions, the desired electron density ν follows according to equation (6). It may be remarked that the method corresponds exactly to that which is used in order to determine the ion cloud in electrolytic solutions. The only difference is that for the solutions the usual Boltzmann statistics may be used, whereas with the electrons in the atom and in the temperature region which prevails here, complete degeneracy has occurred.

It is practical to go over to a dimensionless representation, measuring the distance r to the nucleus as a multiple x of a basic length a, with:

$$a = \frac{1}{z^{1/3}}\left(\frac{3}{32\pi^2}\right)^{2/3}\frac{h^2}{2\,m\,e^2} \tag{9}$$

If one introduces numerical values, the length a has a value

$$a = \frac{0{,}47}{z^{1/3}}\,\text{Å} \tag{9'}$$

so that a decreases proportionally to the third root of the electron number. The potential V is best measured as a multiple of a fundamental potential:

$$V_0 = \frac{z\,e}{a} \tag{10}$$

Now introducing the function φ by the definition (with $\frac{r}{a} = x$):

$$\frac{V}{V_0} = \frac{\varphi(x)}{x} \tag{11}$$

φ must satisfy the equation:

$$\frac{d^2\varphi}{dx^2} = \frac{\varphi^{3/2}}{x^{1/2}} \tag{12}$$

with the added boundary conditions:

$$\varphi = 1 \text{ for } x = 0$$

and

$$\int_0^\infty \varphi^{3/2} x^{1/2}\, dx = 1$$

The electron density ν is to be calculated from φ with the aid of the relation:

$$\nu = \frac{z}{4\pi a^3}\left(\frac{\varphi}{x}\right)^{3/2} \tag{13}$$

Fermi gives a table of φ as a function of x. One can now say that by this one curve, which represents the values of his table, the electron distribution of all atoms is given at the same time. It is represented by equation (13) when the basic length has been chosen according to equation (9) or (9'). Fig. 2 represents φ as a function of x in accordance with Fermi's table.

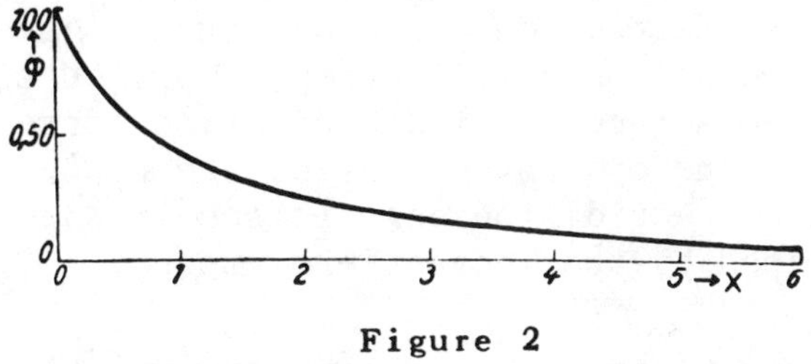

Figure 2

In order to find out what the amplitude factor for x-ray scattering of an atom is we must substitute this representation of ν in equation (3) and carry through the integration. Always using the same dimensionless variable x, we obtain:

$$\psi = \int_0^\infty \nu \frac{\sin ksr}{ksr} 4\pi r^2\, dr = \frac{z}{ksa}\int_0^\infty \frac{\varphi^{3/2}}{x^{1/2}} \sin ksax\, dx$$

and hence the amplitude factor as a function of $s = 2\sin\vartheta/2$. If one uses the abbreviation:

$$u = ksa = 4\pi\frac{a}{\lambda}\sin\frac{\vartheta}{2} \tag{14}$$

and defines a function $\Phi(u)$ by the equation:

$$\Phi(u) = \frac{1}{u} \int_0^\infty \frac{q^{3/2}}{x^{1/2}} \sin u x \, d x \tag{15}$$

the amplitude factor ψ becomes:

$$\psi = z \Phi(u) \tag{16}$$

The integration prescribed in equation (15) was carried out graphically for various values of u. Figure 3 represents the function $\Phi(u)$ as ordinate to the abscissa u, based on these calculations. At the same time another curve (the lower curve) for Φ^2, since the scattered intensity is proportional to Φ^2. Remembering equation (14) and using the definition (9) or (9') for the length a, we have represented by one single curve the scattering function of all atoms as a function of the angle and the wave length.*

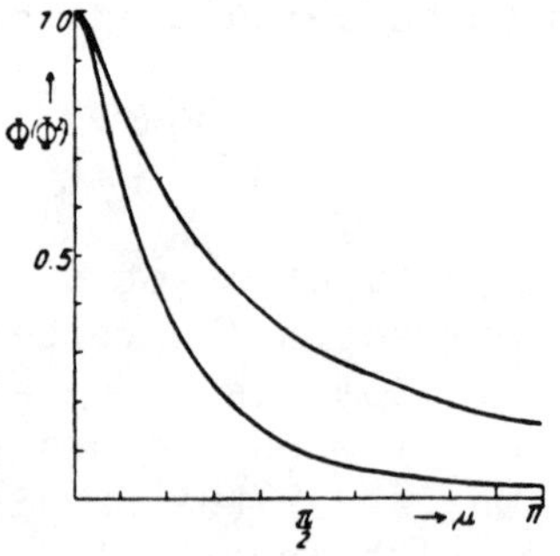

Figure 3

The Thomas-Fermi method does not, of course, give an exact representation of the electron distribution. Hartree's numerical calculations[2] have given a more exact distribution curve for some atoms, in which, above all, the concentrations of the *K*, *L*, etc. electrons appear at definite nuclear distances. It seems, however, as if the approximation obtained by the Thomas-Fermi method suffices for our purposes.

Another point is to be considered when using the calculated scattering curve. It is calculated for a single atom; if it is used for the discussion of the **results** of scattering by complex molecules, the change of position in those electrons which take care of the chemical bonds in the molecule has obviously been neglected. The binding electrons, however, belong to the outer electrons, and therefore give only small contributions to the total scattering.**

*After finishing this paper we found a similar remark by W. L. Bragg and S. West, *Z. Krist.*, *69*, 118 (1929).

**This remark is well illustrated by some calculations of R. W. James, G. W. Brindley, and R. G. Wood (*Proc. Roy. Soc.*, *A125*, 401 (1929)) about the scattering by aluminum ions of various ionization.

Thus it can be expected that by putting together atoms with central symmetry of their electron distribution a **sufficiently good** approximation **will** result. At the same time, I must, however, emphasize that the experimental determination of deviations from the calculated curve would be interesting, especially because in this way we could probably find out something about the charge distribution of the binding electrons. It is possible to solve Fermi's potential quotation for more complicated cases, perhaps for two centers.* Once this has been done, we shall also have a theoretical basis for the discussion of spatial charge distribution of the binding electrons.

(4) As a first application of the above ideas, we want to discuss the scattering by CCl_4 molecules as completely as is possible on the basis of the experimental results which we have at present. At first, we assumed the Cl **atoms** to be points without extension and disregarded the scattering by the C atoms. This gave for the scattering intensity to be expected:

$$I = \frac{1 + \cos^2\vartheta}{2} 4 \left| 1 + 3\frac{\sin x}{x} \right| \qquad (17)$$

where:

$$x = k s l = 4\pi \frac{l}{\lambda} \sin\frac{\vartheta}{2} \qquad (17')$$

in which the scattering amplitude of a Cl atom is set equal to 1 and where l signifies the distance from Cl atom to Cl atom. Of course, one should really consider the C atom also. If one does this, still from the point of view of the point theory, then one obtains for the intensity:

$$I = \frac{1 + \cos^2\vartheta}{2} \left\{ 4 \left[1 + 3\frac{\sin x}{x} \right] + 8 \cdot \frac{6}{17} \frac{\sin x'}{x'} + \left(\frac{6}{17}\right)^2 \right\} \qquad (18)$$

because one must now consider, in addition to the interferences between the Cl atoms, also the interferences between the C atoms and the Cl atoms. The factor **6/17 measures the ratio of the scattering amplitudes** of the two atoms C and Cl, which is taken to be equal to the number of electrons, 6 and 17, respectively. The quantity x' is an abbreviation for:

$$x' = k s l' = 4\pi \frac{l'}{\lambda} \sin\frac{\vartheta}{2} \qquad (18')$$

where l' is the distance C-Cl, which is equal to:

$$l' = \frac{l}{2}\sqrt{\frac{3}{2}}$$

based on a tetrahedral picture, so that the relationship:

$$x' = \frac{x}{2}\sqrt{\frac{3}{2}} \qquad (18'')$$

*Compare the final remark concerning this point by E. Fermi in Leipzig lectures, loc. cit.

exists. For purposes of visualization, the factor in braces from equation (18) (after dividing by 4) is represented as a function of x/π in Fig. 4. One thus obtains the upper curve. For very large values of x/π, the ordinates of this curve would, therefore, converge to the value $1 + \frac{1}{4}(6/17)^2 = 1.03$. The difference between this and equation (17) is due only to the added interference of the C scattering with the Cl scattering, which is measured by the second member in the brace. Disregarding a small displacement of the maxima and minima, the effect limits itself to an increase of the second maximum.

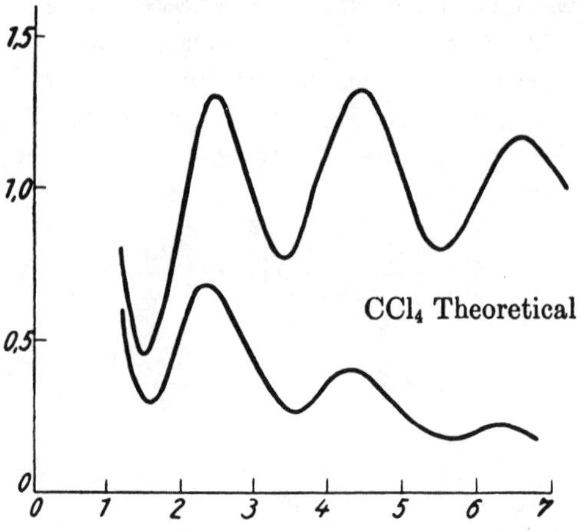

Figure 4

The main point, however, is the influence of the atomic dimensions, and this must now be considered. This is done in accordance with the above ideas not by just taking the scattering amplitude of an atom of electron number z to be proportional to z, but in substituting for it, by taking into consideration the electron distribution, the expression:

$$\psi = z\Phi(ksa)$$

according to equations (16) and (14). The characteristic length a of the atom is, according to equation (9) or (9'):

$$a = \frac{0.47}{z^{1/3}}\,\text{Å}$$

and $\Phi(u)$ is the universal curve which is given in Fig. 3 as a function of u. If we call the amplitude factor of the Cl atom ψ_1, and of the C atom ψ_2, one obtains for the scattering intensity the expression:

$$I = \frac{1+\cos^2\vartheta}{2}\left\{4\,\psi_1^2\left[1 + 3\frac{\sin x}{x}\right] + 8\,\psi_1\psi_2\frac{\sin x'}{x'} + \psi_2^2\right\} \tag{19}$$

in which x and x' have the same meaning as before. To be explicit, we have for the Cl atom with 17 electrons:

$$\psi_1 = 17\,\Phi\left(ks\frac{0.47}{17^{1/3}}\right)$$

and for the C atom:

$$\psi_2 = 6\Phi\left(k s \frac{0.47}{6^{1/3}}\right)$$

in which the wave length is measured in A.

The curve which is represented by equation (19) is also drawn in Fig. 4 (the lower curve). In order to make it comparable with the curve which was calculated according to the point theory (18), we have not drawn the value of the expression in the brace, but that which is obtained if this value is multiplied by the factor $(1/4)(1/17^2)$. **The abscissa are, as before:**

$$x = ksl$$

In order to be able to carry out the drawings, it was necessary to assume a ratio of atom radius to atom distance. This was done by determining the distance l from Cl to Cl approximately as $l = 3.1$ A as derived from a comparison of the point theory with the measurements. If we call a_1 the characteristic length of the Cl atom, the argument of the Φ **function** belonging to Cl is:

$$k s a_1 = k s l \frac{a_1}{l} = x \frac{a_1}{l}$$

In this:

$$a_1 = \frac{0.47}{17^{1/3}} \text{Å}$$

was substituted for a_1 and l was assumed to be 3.1 A. After this substitution the scattering functions ψ could be drawn as a function of x. The same procedure was used for ψ_2. It is seen from Fig. 4 that the total intensity is decreased significantly by the internal interferences, and the more so the bigger x is, *i.e.* the bigger the angle between primary and secondary ray is. At the same time, the maxima of various orders are flattened out, the more so the higher the order. Since we have now obtained a theoretical scattering curve for the $C\,Cl_4$ molecule in which the atomic dimensions are duly considered, we can approach the problem of determining the distance Cl-Cl. The abscissa of the curve is, as already mentioned, the quantity $x = ksl$; by choosing l appropriately, we can now try to stretch the curve in a horizontal direction, so that the maxima and minima occur for those values for which they were observed.

In the meantime, observations have been made on CCl_4 vapor with three different instruments, all of the type formerly described (*Ber. Sächs. Akad., Math.-naturwiss. Kl.*, *81*, 29 (1929)). The first apparatus was especially designed for small scattering angles down to $\vartheta = 8°$; the second one for medium angles and the third one for large scattering angles. The diagrams were measured with a Zeiss recording photometer.*

*The last curves were made at the Leipzig Institute with a Zeiss instrument. I wish to thank the Notgemeinschaft at this time for its use.

Such a curve,* which contains the first two maxima, is shown in Fig. 5.

At the same time, reference marks of intensity were recorded which were photographed using a rotating sector. It was, therefore, possible to evaluate the recorded deflections of the electrometer in terms of intensities. The result is shown in Fig. 6.**

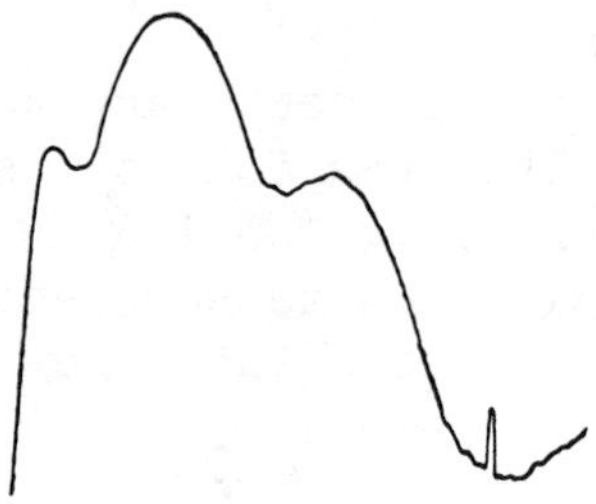

Fig. 5.

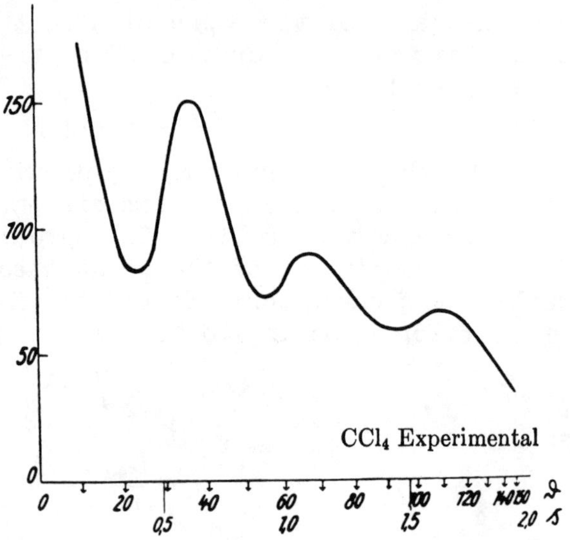

Fig. 6.

The intensity curves of the various instruments had common domains and fit together very well. The quantity $s = 2 \sin(\vartheta/2)$ was chosen as abscissa. We can cover the region $0 < s < 2$; one notes that the observations indeed cover the entire region, except for two small stretches at the beginning and end. The interference maxima of the first, second, and third order are clearly recognizable. If one compares the experimental curve in detail with the theoretical curve drawn to about the same scale (see Fig. 7), the main difference exists in the fact that the maxima of the theoretical curve are somewhat sharper than the ones of the experimental curve. To a small degree, this may be due to the fact that in the experiment we do not obtain pure monochromatic radiation by our filter, and that the secondary bundle of rays have a finite opening. The most important point, however, is probably the fact that we did not consider the thermal motion of the atoms in the molecule in the theoretical curve.

*The photographing of the scattered radiation itself took an exposure time of about 5 hours.

**The curve is already corrected both for absorption in the camera and with respect to polarization.

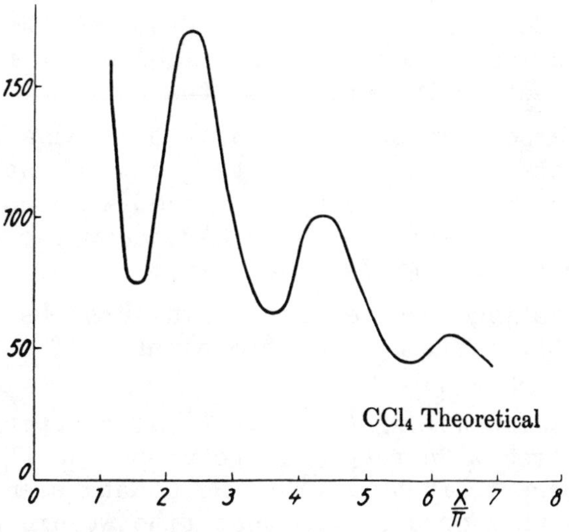

Figure 7

In the experimental curve, we observe 6 points of special importance, three minima and three maxima; the experimental value of $s = 2\sin(\vartheta/2)$ belonging to these were determined. Thus we obtained six independent determinations of the distance l. In Table I we have in the fourth column the values of x/π, taken from the theoretical curve; the fifth column contains the observed values of s; the last column contains the distance l calculated from s. As an example for the calculation, we may take the following:

Table I

	Theoretical x/π			s	l
	Cl_4 (·)	CCl_4 (·)	CCl_4 (o)	obs.	(A)
1st Min.	1.43	1.50	1.60	0.411	3.00
1st Max.	2.47	2.48	2.38	0.618	2.97
2nd Min.	3.47	3.42	3.60	0.93	2.98
2nd Max.	4.48	4.44	4.30	1.10	3.01
3rd Min.	5.48	5.50	5.70	1.46	3.00
3rd Max.	6.48	6.56	6.30	1.64	2.96

For the first maximum we have, according to the fourth column, $x/\pi = 1.60$; on the other hand, according to definition, $x = ksl = 2\pi s(l/\lambda)$. According to observation, $s = 0.411$; therefore, with $\lambda = 1.539$ A., l follows, as given in the last column:

$$l = \frac{1.60}{2 \cdot 0.411} 1.539 = 3.00 \text{ Å}$$

In order to be able to judge the influence of the atomic dimensions on the position of the maxima and minima, I have given those values of x/π in the third column which are obtained from the theoretical curve drawn in accordance with the point theory. This is indicated in the table by printing in the heading a dot next to CCl_4, while the values calculated under consideration of finite atom dimensions are indicated by a circle in the heading. Finally, the table contains in the second column the values of x/π, which follow from the point theory disregarding the scattering effect of the central C atom altogether.

If the six experimental values of l are compared with one another, it is seen that the agreement is quite good and that it can be claimed that the distance Cl-Cl has been determined with an accuracy of about ± 1% as l = 2.99 A. Formerly, we obtained the value l = 3.1 A., using the point theory (*Sächs. Akad., loc. cit.*).

It is interesting to note that, using Bragg's radii (0.7 A. for C and 1.05 A. for Cl, one can calculate a value of l = 2.98 A. based upon the tetrahedron model.

Earlier (*Sächs. Akad., loc. cit.*), the scattering by $CHCl_3$ and CH_2Cl_2 were compared with the CCl_4 pictures, and it was concluded that with the decrease in the number of Cl atoms an increase in the distance between them occurs. At that time we did not consider the corrections based on the finite size of the atom. Now it has been found that even when making these corrections, the former assertion still holds true. The percentage differences are, however, somewhat smaller than the value given there (marked as preliminary). A complete report on these questions will be written by Mr. Bewilogua.*

(5) We shall now discuss some cases which at present are still in the laboratory stage, so that a detailed discussion as above cannot be given here. I do that in order to show that the predictions which can be made on the basis of the scattering curve calculated according to Thomas and Fermi correspond in every case to the qualitative aspects of the pictures obtained andare hence valuable in anticipating experimental results. Mr. Gajewski at the Leipziger Institute has taken a series of pictures of diatomic gases, or such

*In the meantime Messrs. Mark and Wierl have succeeded in performing a beautiful experiment (*Naturwiss. 18*, Heft 9, 205 (1930)) in which the interferences here discussed are obtained with cathode rays, the DeBroglie wave length of which is about 30 times as small as the wave length of the x-rays we used. They give as a result for their measurements on CCl_4 the distance l = 3.14 A. This value, however, should be corrected, as we did with our earlier value of l = 3.1 A., with regard to the angular decrease in intensity which occurs with β-rays just as with x-rays. Based on the Schrödinger equation, this correction can be calculated, at least for sufficiently fast electrons, in a similar manner using the Fermi distribution.

Mr. Mark remarks that Mr. Bothe suggested the possibility of using β-rays instead of x-rays at the occasion of a lecture I gave at Zürich. I may add to this that Mr. Pringsheim has made a similar remark after an earlier lecture which I gave at the Berliner Physikalische Gesellschaft.

gases as, with respect to scattering, can be considered as diatomic. According to the point theory, one would expect, for instance in the case of N_2, and using Cu radition, a first interference maximum at $\vartheta = 120°$, if for the distance N-N the value l = 1.10 A. is substituted, which Rasetti has calculated from the moment of inertia[3] determined by the Raman effect. Our experiment shows no such maximum. If, however, the extension of the atoms is considered, a scattering curve can be calculated, which, at least qualitatively, corresponds to the experimental results. The large angular decrease in intensity which is caused by the finite extension of the atoms flattens the maximum so much that it cannot be recognized as such. This is represented in Fig. 8; the upper curve has been calculated for atoms of point dimensions; the lower curve takes accont of the finite size of the electron shell. Here, as in all the following curves, the factor $(1 + \cos^2\vartheta)/2$ is omitted; the effect of polarization of the scattered radiation was therefore omitted. The abscissa is always $s = 2 \sin(\vartheta/2)$. With increasing ing atomic weight, the density of the electronic charge concentrates itself around the nucleus, in accordance with Fermi's theory (the characteristic length $a = 0.47/z^{1/3}$ becomes smaller; on the other hand, the size of the atoms increases (size in the sense of occupied volume). If, therefore, different diatomic molecules with atoms of different atomic weights are compared with each other, the interference maxima should become more pronounced with increasing atomic weight.

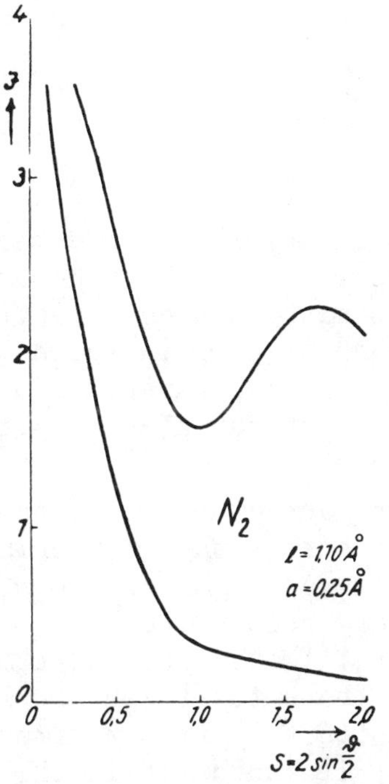

Figure 8

Figs. 9 and 10 are an attempt to visualize this. The first figure corresponds to oxygen, the second to chlorine. The upper curves in these figures again represent the result of the point theory, while the lower curves consider the extension of the atoms. The interference maximum which is barely indicated in the case of O_2 is much more pronounced in the case of Cl_2, which has a much larger interatomic distance and a much higher concentration of electrons around the nuclei.

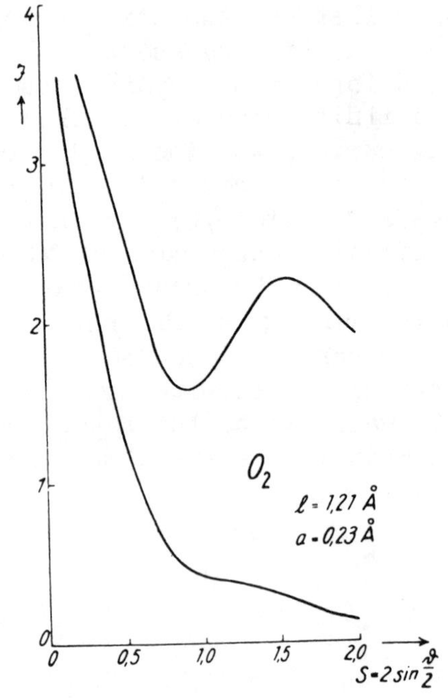

Fig. 9.

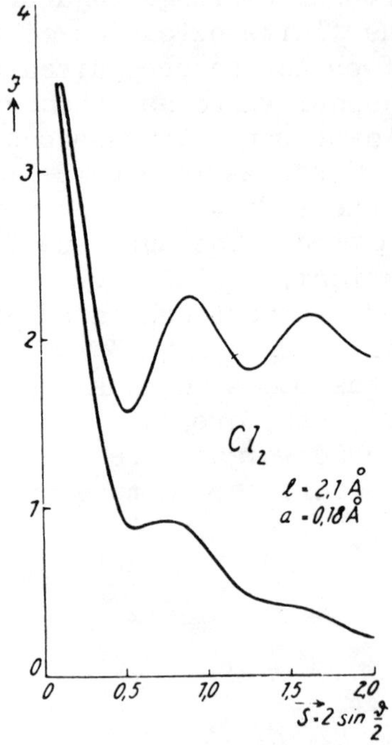

Fig. 10.

The experiments which Gajewski conducted on N_2 and O_2 gave results which coincide, according to all appearances, with these expectations. Cl_2 itself has not yet been investigated; our experience, however, with the compound CH_2Cl_2, in which two Cl atoms are connected by one C atom, corresponds to expectations and permits us to expect that the scattering picture calculated for Cl_2 will agree well with experiment.

In a similar manner, our discussion can be illustrated by experimental results obtained for the scattering by the molecules CO_2 and CS_2. Carbonic acid gives a general decrease of intensity with angle without a definitely recognizable first maximum, whereas for carbon disulfide the first two interference maxima are visible. Theoretical results for these two cases are represented in Figs. 11 and 12. The point at which the first maximum should occur, according to the point theory, is indicated in the CO_2 curve (Fig. 11) by an arrow. Two curves are drawn in Fig. 12, which corresponds to CS_2. The curve with the more pronounced maxima was obtained by neglecting

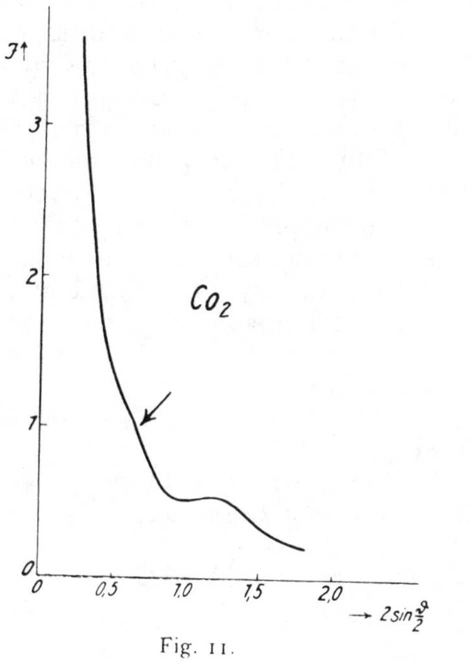

Fig. 11.

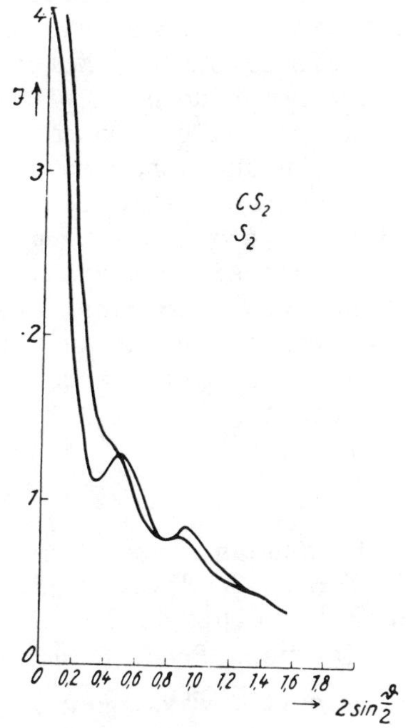

Fig. 12.

the scattering from the C atom; in the smoother curve all three atoms are considered. The experiments with CS_2 had a result which perhaps is more similar to a curve in between. A final decision will not be possible until more exact intensity curves have been obtained.

Finally, in Fig. 13, curves are drawn representing what is to be theoretically expected for the two cis-trans-isomeric ethylene derivatives $C_2H_2Cl_2$ about which we reported recently (P. Debye, *Physik. Z.*, *31*, 142 (1930)). As mentioned in this report, first maxima for the

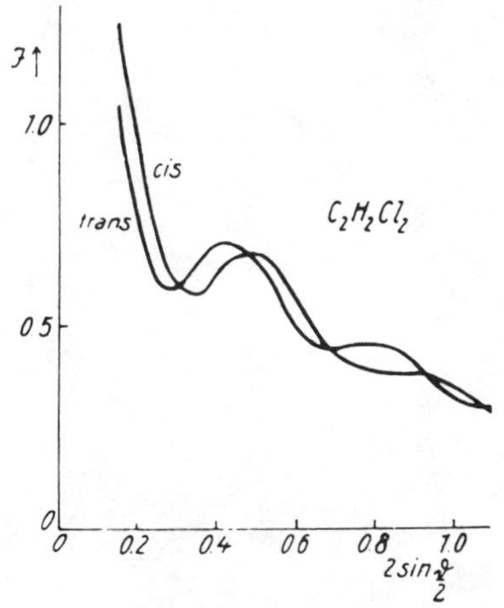

Figure 13

two isomers occur at different angles, the maximum for the trans compound (with the bigger distance Cl-Cl) at a smaller angle than the one for the cis compound. Moreover, it is found that the second maximum is much more pronounced with the trans compound than with the cis compound, which would also correspond to the theory as represented by Fig. 13. On the other hand, it is situated closer to the first maximum than on the theoretical curve, in which, however, the scattering due to the C atoms has been neglected. Mr. Ehrhardt is expected to discuss these experimental results in more detail. All in all, it is clear that with due consideration of atomic dimensions (based on the Thomas-Fermi distribution) the interference theory of molecular scattering gives results which coincide in large measure with our experimental results.

Bibliography

1. L. H. Thomas, *Proc. Cambridge Phil. Soc.*, *23*, 524 (1927); E. Fermi, *Z. Physik*, *48*, 73 (1928); *Leipzig lectures, 1928*, Hirzel, Leipzig.
2. J. Hartree, *Proc. Cambridge Phil. Soc.*, *24*, 89 (1928).
3. F. Rasetti, *Phys. Rev.*, *34*, 367 (1929).

REFRACTION OF X-RAYS BY LIQUIDS AND GASES
(Röntgenzerstreuung an Flüssigkeiten und Gasen*)

P. Debye

Translated from
Physikalische Zeitschrift, Vol. 31, 1930, pages 348-350

The lecture given at the Breslau meeting had reference to the differentiation of isomers by x-ray interference experiments as expressed by **its** title. In the introduction to this address I explained the principles of the measurement and showed by an example how differently liquids and gases behave. The results with regards to the isomers have already been published in a preliminary report (*Physik. Z.*, *31*, 142 (1930)). I intend, therefore, to use this opportunity for a summary of the Breslau lectures in order to present those introductory remarks **alone,** as is also indicated by a change in the title.

If one irradiates a liquid two effects will be important for the distribution of its scattered radiation: the intermolecular interference from molecule to molecule, and the intramolecular interference from atom to atom within a single molecule. As long as the latter effect had not been observed, one could not differentiate between the two effects with certainty; one only knew that intermolecular interference played an important role in the case of liquids. Since now intramolecular interference has been shown to exist for substances in gaseous form (*Physik. Z.*, *30*, 84 (1929); *Sächs. Akademie-Berichte*, *81*, 29 (1929); *Physik. Z.*, *30*, 524 (1929)), I would like to show, using CCl_4, how the gas diagrams and the liquids diagrams behave when compared to each other.

It has now been possible to observe the first four maxima in the curve of the scattered radiation from CCl_4 vapor; by using molybdenum $K\alpha$-radiation, filtered by a thin **zirconium** sheet.** When

*Received February 11, 1930.

**I wish to thank at this place the laboratory of Philips at Eindhoven for their kindness in lending me a small sheet of zirconium metal.

using copper *K*-radiation (filtered by a sheet of nickel) there is only room for three maxima in the angular region from 0° to 180°. The pictures were taken by using scattering apparatus, the construction of which was already explained by a sketch (*Sächs. Akademie-Berichte, loc. cit.*). This apparatus was used up to the middle of the angular region; a new scattering camera was constructed for larger angles with the primary ray. The photographs were measured photometrically* and the obtained photographic densities reverted to intensities. The result is represented by Figure 1. The upper curve represents the scattered intensity by CCl_4 gas as a function of the angle between primary and secondary ray. The lower curve gives the intensity as a function of the quantity $x = 2\sin(\vartheta/2)$, which is the essential variable always occurring in interference problems. The usable range of values of x is from $x = 0$ to $x = 2$. One notices how practically the entire range is covered, which contains the three maxima to be expected as the result of the interference of the secondary radiation of the four Cl-atoms. In the presentation of the intensity as a function of x, the abscissa of the minima should, in a first approximation (*i.e.*, if the atoms are considered to be points with no extension), be in the ratio of 3:7:11; and the abscissa of the maxima in the ratio of 5:9:13, which can be verified in Figure 1. In

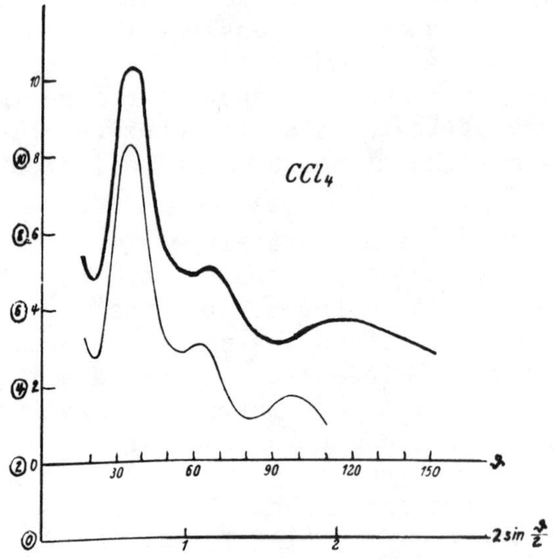

Figure 1

addition to the gas scattering curves, various exposures of the scattering by liquid CCl_4 were made. For this purpose an apparatus was constructed in which the primary radiation fell upon an

*We thank Dr. Hansen, who made the Zeiss Photometer in Jena available to us. Some curves were obtained from an instrument in the Physics Laboratory of the University in Jena, for which we thank its director Dr. M. Wien.

open surface of the fluid, and the scattered radiation was recorded on a film bent in a circle, the center of which coincided with the spot of the surface of the fluid on which the primary radiation fell. The method is similar to the one used by Prins (*Z. Physik*, 56, 617 (1929)). Prins considers it necessary to oscillate the x-ray tube back and forth together with the film-cassette which is firmly connected to it. We believe that this oscillation is entirely unnecessary and kept the tube and film at rest. The pictures were taken for two angles of incidence of 5° and 15°.* The films obtained were measured photometrically, and the photographic density converted to intensity. An error in the fundamental intensity curve is caused by absorption of the radiation in the liquid. If α gives the glancing angle of the primary radiation with the surface of the liquid, one finds that the absorption of the intensity in the direction ϑ introduces a factor of the value:

$$\frac{1}{\mu}\frac{\sin(\vartheta-\alpha)}{\sin\vartheta+\sin\alpha}$$

where μ is the absorption coefficient of the radiation in the liquid. The intensity curves were corrected for this absorption. In this way the full-line curve in Figure 2 was obtained. Two pronounced maxima

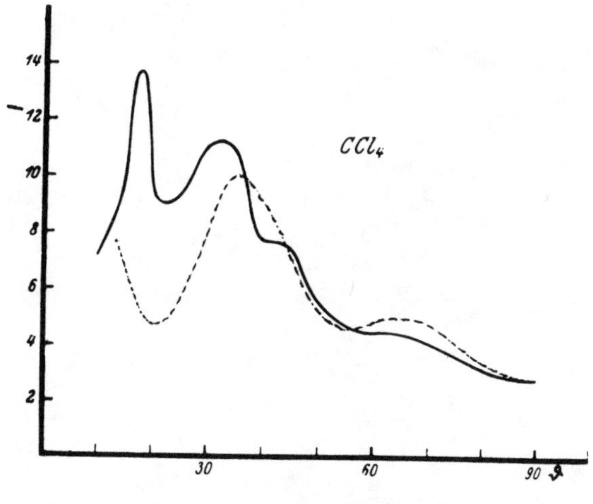

Figure 2

occur at ϑ = 17.5° and at ϑ = 32°, and two less pronounced elevations are visible at about ϑ = 44° and ϑ = 66°. The curve for CCl_4 gas is also presented; it is the dotted curve in Figure 2. The first three elevations of the liquid curve are probably essentially due to the intermolecular interferences characteristic of the liquid; the fourth elevation, which coincides with the second maximum of the gas curve, might on the other hand be due to intramolecular interference. It occurs at such a large angle that the intermolecular interference effects should have died down. One can correct such intensity curves for liquids for the effects of intramolecular interference and obtain a reduced curve which should represent the purely intermolecular

*They were carried through at the Leipzig Institute by R. Kaiser and H. Mencke.

interferences. From such a reduced curve it should then be possible to obtain, using a simple procedure, a curve which represents the distribution probability of the molecules. In this way the experiment would lead to a picture of the molecular arrangement in the liquid. We intend to report about this procedure in the near future.

DETERMINATION OF THE INNER STRUCTURE OF LIQUIDS BY X-RAYS
(Bestimmung der inneren Struktur von Flüssigkeiten mit Röntgenstrahlen)

P. Debye and H. Menke

Translated from
Physikalische Zeitschrift, Vol. 31, 1930, pages 797-798

When radiating liquids by x-rays, one obtains a scattering picture representing the superposition of two different interference phenomena. If the intermolecular part can be separated from the intramolecular part, conclusions about the structure of the liquid can be drawn. Up to now the method used was first to find out by calculation something about the relative positions of the molecules, and to compare this theoretical result, even if only qualitatively, with the experimental results.[1] Even with the simplest assumptions possible, namely, that the molecules behave like hard spheres, it has not been possible to carry out such calculations exactly. It therefore seemed indicated to us to carry out the procedure in the opposite direction, *i.e.*, to put the experimental results of the scattering first and to deduce from them, without any prejudice, the structure of the liquid. We reported about this procedure and gave some tentative results at the meeting (Gautagung) in Halle.

Mercury, which, as a monatomic liquid, should give especially simple results, was first investigated.

If a volume which contains N mercury atoms is irradiated and if the scattered radiation of the different single atoms did not interfere with one another, one would observe an intensity:

$$J = \frac{1 + \cos^2 \vartheta}{2} N \psi^2 \qquad (1)$$

ψ^2 is a function of the angle ϑ between primary and secondary ray, which characterizes the scattering of the single atom (atom-form factor), while the factor $(1 + \cos^2\vartheta)/2$ was introduced in order account for the polarization occurring during scattering. The angular distribution of ψ^2 can either be calculated according to Fermi[2] or obtained from experimental data about the scattering of mercury vapor.[3] It is, of course, assumed in the above equation that the correction due to the absorption in the irradiated volume has already been carried out. This correction becomes especially simple if one permits the primary ray to fall on a free surface of the liquid and investigates the radiation emitted by this surface.[4] This is the reason why we have chosen this method of observation. The result of this experiment is not in agreement with equation (1). As is well known, certain maxima and minima of intensity occur which should not occur according to equation (1). The angle between primary and secondary ray was originally used as the variable; it is, however, more convenient to introduce not ϑ itself, but the quantity:

$$s = 2 \sin \frac{\vartheta}{2} \tag{2}$$

as the variable. One may then represent the result of the experiment in a purely experimental fashion by assuming:

$$J = \frac{1 + \cos^2 \vartheta}{2} N \psi^2 E(s) \tag{3}$$

in place of equation (1). Here $E(s)$ is a function of s which can be determined purely by experiment. This function is represented in Fig. 1 in accordance with our experiments; for higher values of s, E approaches unity. (That $E(s)$ could be represented for values which surpass $s = 2$ is based on the fact that the experiments, which were originally carried out with Cu-K_α-radiation, were later completed by experiments with Mo-K_α-radiation of shorter wave length.)

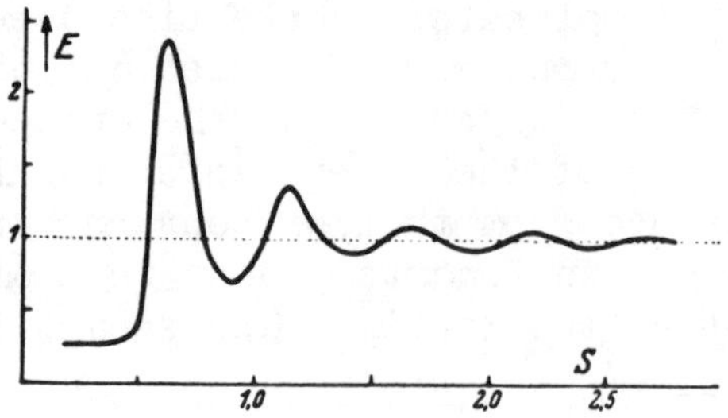

Figure 1

The problem is to determine the structure of the liquid from the experimental curve $E(s)$. The structure can be characterized in the following manner. If two atoms within the liquid are marked and one thinks of oneself as moving along with one atom, which at the end of a distance r carries with it a volume element dS, the probability for the second atom being found with its center within dS can be written as:

$$W \frac{dS}{V} \tag{4}$$

If V is the entire volume which may be used, $W = 1$ would mean that all distances r are of equal probability. In general, however, W will depend on r. If one succeeds in determining this probability function, the structure of the fluid will be determined. Let us first assume W as known. It can then be shown that the function $E(s)$ can be calculated according to the formula:

$$s[1-E(s)] = 2\frac{\lambda^3}{d^3}\int_0^\infty \varrho[1-W(\varrho)]\sin 2\pi s\varrho\, d\varrho \quad (5)$$

We have chosen for the variable not the distance r itself, but the ratio of distance to wave length:

$$\varrho = \frac{r}{\lambda} \quad (6)$$

The quantity d is defined by the equation:

$$Nd^3 = V \quad (6')$$

d^3 therefore is the volume occupied on the average by one atom. However, matters are just reversed. W is not known as a function of ρ, but E is known as a function of s. Using Fourier's theorem, one can invert equation (5) and one obtains:

$$\varrho[1-W(\varrho)] = 2\frac{d^3}{\lambda^3}\int_0^\infty s[1-E(s)]\sin 2\pi\varrho s\, ds \quad (7)$$

In order to find the value of the probability function W from the observed curve $E(s)$ for any distance ρ measured in wave lengths, one only needs to carry through the integration suggested in equation (7). Zernike and Prins[5] have already suggested this possibility; they, however, had no intensity curves which they could have used. The result, obtained for W by actually carrying out the reversion in the case of mercury is shown in Figure 2, in which the distance from the central atom is given in A. First, it is clear that certain distances, about 3 A., 5.6 A., 8.1 A., etc., are preferred, and that in-between distances are avoided if possible by the atom. The curve thus shows that even in the liquid stage there exists a quasi-crystaline structure and defines this structure quantitatively, using the probability curve.

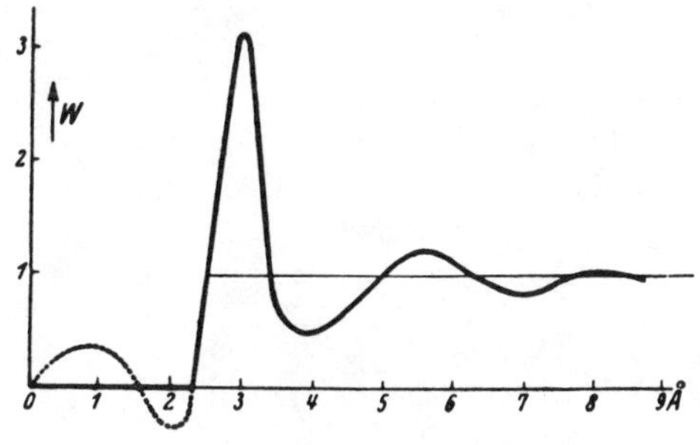

Figure 2

Thus, for instance, one can read from the curve that the distance 3 A. is about 6 times as probable as the distance 4 A. For small distances one should expect $W = 0$. The integration, according to equation (7), gave, instead, the dotted part of the curve. The deviation is explained by the fact that for the calculation of the W curve for small distances the shape of the E curve for high values of s is of importance. The intensity distribution in the maxima and minima of higher orders, however, cannot be determined very accurately. In the meantime we are trying to improve the accuracy of the intensity curve as well as to extend the method to other liquids, especially those with molecules of a greater number of atoms.

Bibliography

1. J. A. Prins, *Z. Physik.*, *56*, 617 (1929). Further literature is cited there.
2. P. Debye, *Physik. Z.*, *31*, 419 (1930).
3. P. Scherrer and A. Stäger, *Helv. Phys. Acta*, *1*, 518 (1928).
4. P. Debye, *Physik. Z.*, *31*, 348 (1930).
5. F. Zernike and J. A. Prins, *Z. Physik*, *41*, 184 (1927). See especially the remarks near the end of this paper.

DIPOLE MOMENTS

VAN DER WAALS' COHESIVE FORCES
(Die van der Waalsschen Kohäsionskräfte)

P. Debye

Translated from
Physikalische Zeitschrift, Vol. 21, pages 178-187, 1920.

As is well known the great success of van der Waals' equation of state is based essentially on the assumption of attractive forces between the molecules. These forces cause, in addition to the external pressure, an internal pressure which is proportional to the square of the density. According to van der Waals, these forces of attraction exist between molecules of any kind, and constitute a general property of matter. It appears, therefore, of particular interest to consider the origin of this universal attraction.

Today we know with absolute certainty that the molecule is a system of electric charges, and we are led to an attempt to find an electric origin for van der Waals' forces. It will certainly be unnecessary to consider details of molecular structure. A property of matter as general as the van der Waals' attraction can require for its explanation nothing but structural features common to all molecules. We shall show in the following that it is in fact sufficient only to know that the molecules are electrical systems in which the charges are not rigidly bound to their rest positions. A relation between van der Waals' constant of attraction on the one hand and the index of refraction and the widening of spectral lines on the other hand will be derived on the basis of this hypothesis.

1. Van der Waals' Equation

We start by presenting a few known relations which will be used subsequently.

Let us consider a gas of actual volume V at pressure p and at the absolute temperature T. If its mass is G and its molecular weight M, and if R designates the universal gas constant

(R = 8.31 × 10^7 erg), then according to van der Waals' equation of state, we have:

$$\left(p + \frac{A}{V^2}\right)(V - B) = G\frac{R}{M}T \qquad (1)$$

Here B designates four times the actual volume of the molecules contained in the volume V, while A is a measure for the mutual attraction of the molecules.

The relations:

$$p_k = \frac{1}{27}\frac{A}{B^2}, \; V_k = 3B, \; \frac{GR}{M}T_k = \frac{8}{27}\frac{A}{B} \qquad (2)$$

for the critical point are obtained from this equation in the usual way.

The total energy of the gas is the sum of its kinetic and potential energies. Its dependence on volume is easily secured from the thermodynamic relation:

$$\frac{\partial U}{\partial V} = T^2\frac{\partial}{\partial T}\left(\frac{p}{T}\right) \qquad (3)$$

Insertion of the equation of state into the right-hand term of this relation gives:

$$\frac{\partial U}{\partial V} = \frac{A}{V^2} \qquad (4)$$

or, integrated,

$$U = \Psi(T) - \frac{A}{V} \qquad (4')$$

In order to compute van der Waals' constant A, it will only be necessary to determine the potential energy of the gas, given in the formula by $-A/V$.

In addition to the form of the equation just given, which refers to an arbitrary amount G of the gas, we shall use another form referring to one mole. Calling:

$$\mathfrak{v} = \frac{MV}{G}$$

the molar volume, and introducing instead of A and B the constants a and b, defined by:

$$a = \frac{M^2}{G^2}A, \; b = \frac{M}{G}B \qquad (5)$$

van der Waals' equation reads:

$$\left(p + \frac{a}{v^2}\right)(v - b) = RT \tag{6}$$

and the critical variables are given by:

$$p_k = \frac{1}{27}\frac{a}{b^2}, \; v_k = 3b, \; RT_k = \frac{8}{27}\frac{a}{b} \tag{7}$$

For the molar energy $u = MU/G$, we obtain the expression:

$$u = \psi(T) - \frac{a}{v} \tag{8}$$

II. Cause of the Cohesion

If we imagine the molecules to be rigid electrical systems, then there will, of course, always be a force acting between two such systems which will change sign and magnitude with their mutual orientation. Since all possible orientations occur in a gas, the average over these orientations must be taken in order to compute the attraction term appearing in the equation of state.

In general, in carrying through this averaging process, the probability of an arbitrary orientation would have to be determined on the basis of the Boltzmann-Maxwell principle. The higher the temperature, however, the less important is the dependence on the mutual energy. In the limit for high temperatures, all orientations will have to be considered equally probable. Obviously, van der Waals' assumption requires that the characteristic cohesion introduced into the equation persist in this limiting case.

It can be easily shown that two rigid electrical systems do not, in the mean, exert a force on one another. The potential which is generated at a distant point by a molecule can be considered as originating from a series of concentric spheres covered with a layer of electric charges of constant surface density. If the molecule assumes all possible orientations in space, each charge occupies, on the average, all points of a sphere with equal frequency. Since it is known that a sphere with a charge of constant surface density affects points outside of the sphere as if its total charge were concentrated at its center, and since the molecule carries no total charge, the average of the potential at the point considered will be zero. Thus, no force is effective, on the average, between two rigid molecules.

The situation is immediately and essentially changed if we

consider molecules that are not completely rigid. The fact that each gas has a refractive index different from unity is proof of the mobility of the separate charges in the molecule. Taking this into consideration, it will be clear that a given molecule assumes an electric moment in the field $\mathfrak{E}$ of another molecule, which moment is proportional to the field $\mathfrak{E}$. Thereby a mutual energy arises between the two molecules which is proportional to the product of field strength times moment, i.e., proportional to the square of $\mathfrak{E}$. Thus the average of the corresponding force cannot vanish. Further it will be readily seen that the force is always one of attraction. Hence we may conclude that we have found in* this force the origin of van der Waals' universal attraction.

*The situation can be illustrated by the following example. Two dipoles are situated opposite to each other.

(a) In a position I. Here the main effect is repulsive. As a consequence of the action, the field $\mathfrak{E}$ on the elastically bound charges, the latter are displaced in such a way that the electric moments are reduced. Thereby the repulsive force decreases; in other words, an attractive force appears as a secondary effect.

(b) In position II. Here the main effect is attractive. The field strength now displaces the charges so that the moments increase. The main effect is now increased, or expressed differently: again an attractive force has been added as secondary effect.

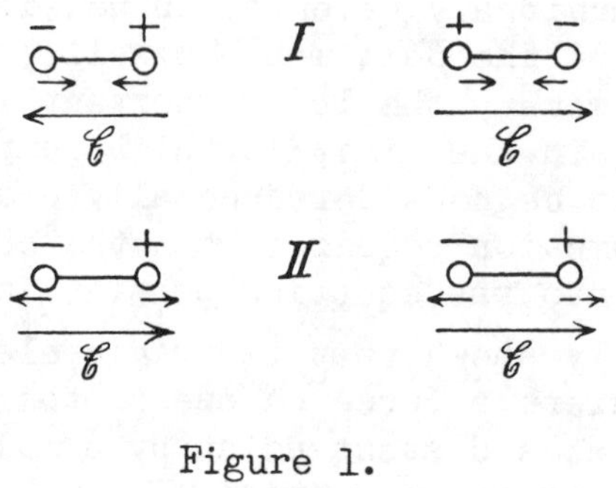

Figure 1.

The main effect vanishes when the average over all possible orientations is taken; the secondary effect is always positive and thus can never vanish.

We add the comment, that in terms of our concept van der Waals' force has the same origin as the forces effective in the basic experiment of electrostatics. There a charged body attracts other bodies irrespective of whether they are conductors or insulators i.e., forces appear which drive the mobile particles from places of lower to places of higher density of lines of force. Similarly the molecules are subject to forces which tend to bring the particles nearer together and thus bring them into as strong a field as possible.

Once this has been recognized, it is only necessary to compute the potential energy of the gas. For this it suffices to know how much work must be expended in order to move a molecule from inside the gas across the gas surface to a sufficiently distant point. The computation of this work is best carried out in two steps.

The first question to be answered is: How large is the amount of work which must be done in order to move a polarizable molecule whose electric moment can be expressed by:

$$\mathfrak{m} = \alpha \mathfrak{E} \tag{9}$$

from a point of zero field strength to a point of field strength $\mathfrak{E}$?

It can easily be shown that the amount of work involved is:

$$-\frac{\alpha}{2}\mathfrak{E}^2 \tag{10}$$

which is independent of the path over which the molecule has been moved.*

The second question to be answered is: How large is the mean square of the electric field strength at a point inside the gas? By combining the two answers, the desired amount of work can readily be found.

III. Mean Square Value of the Electric Field Strength

Consider a molecule which consists of electric charges e_k situated at points with the coordinates ξ_k, η_k, ζ_k. The potential at a point located at a distance r from the origin with the coordinates x, y, z, expanded in negative powers of r, will be:

$$\left.\begin{aligned}\varphi = \frac{\Sigma e_k}{r} + \frac{1}{r^2}\Big[\frac{x}{r}\Sigma e_k \xi_k + \frac{y}{r}\Sigma e_k \xi_k + \\ + \frac{z}{r}\Sigma e_k \zeta_k\Big] + \frac{1}{r^3}\Big[\frac{1}{2}\Big(3\frac{x^2}{r^2} - 1\Big)\Sigma e_k \xi_k^2 + \\ \frac{1}{2}\Big(3\frac{y^2}{r^2} - 1\Big)\Sigma e_k \eta_k^2 + \frac{1}{2}\Big(3\frac{z^2}{r^2} - 1\Big)\Sigma e_k \zeta_k^2 + \\ + \frac{3xy}{r^2}\Sigma e_k \xi_k \eta_k + \frac{3yz}{r^2}\Sigma e_k \eta_k \zeta_k + \\ + \frac{3zx}{r^2}\Sigma e_k \zeta_k \xi_k + \Big] + \ldots\end{aligned}\right\} \tag{11}$$

*The corresponding result for a magnetic field is involved, for instance, in connection with the determination of the susceptibility of weakly magnetizable bodies by means of the force which is exerted on them in an inhomogeneous field (Curie).

Since the molecule carries no charge, the first term vanishes.

It can happen that the center of gravity of the positive charges doe not coincide with the center of gravity of the negative charges. Then the second term has a finite value; it is said that the molecule carries an electric dipole. Whether or not a molecule belongs to this category is decided, for instance, by means of the dielectric constant δ and the index of refraction n. A dipole is associated with the molecule if Maxwell's law $\delta = n^2$ does not hold. Another indication of the presence of dipole moments is anomalous temperature dependence of δ, such that δ decreases with increasing temperature at constant density.[1] Alternatively, a dipole is associated with a molecule if Drude's anomalous dispersion exists.[2]

Provided the electric moment is sufficiently large, gases of this type exhibit definite association phenomena,* and show, simultaneously, comparatively large deviations from the simple van der Waals' equation. We will, therefore, exclude such gases from the present considerations; their behavior shall be explained in more detail in the near future.

For our purposes, the potential is, in this case, represented by the third term of equation (11). The molecules then behave like electrical quadrupoles. As is known we can orient the coordinate system with respect to the molecule so that:

$$\Sigma e_k \xi_k \eta_k = \Sigma e_k \eta_k \zeta_k = \Sigma e_k \zeta_k \xi_k = 0$$

and the complete molecule can be characterized by the three basic electrical moments of inertia:

$$\Theta_1 = \Sigma e_k \xi_k^2, \ \Theta_2 = \Sigma e_k \eta_k^2, \ \Theta_3 = \Sigma e_k \zeta_k^2 \quad (12)$$

If $\theta_1 = \theta_2 = \theta_3$, the third term also vanishes, and the next term of the series expansion will have to be used. It is probable that this condition is not met for most gases, and we will, therefore, carry out our computations with the expression for the quadrupole.

*For problems related to the equation of state, substances are classified as normal and anomalous (J. P. Kuenen, *Zustandsgleichung und Kontinuitätstheorie*, Braunschweig, 1907, p. 139). I am of the opinion that these categories express the division of molecules into those which do not and those which do carry dipoles. The differences between the two groups are, however, not always sharply defined. For molecules with relatively small moments (CO, CO_2), it must depend on the temperature range concerned whether the substance is to be classified as belonging to the first or to the second group.

From the expression for the potential, let us find the electric field strength at a point located at a distance r by means of the relation:

$$\mathfrak{E} = -\operatorname{grad}\varphi$$

compute its square, and take the average over all possible orientations of the molecule. As a result of this elementary computation we obtain for the mean square value a quantity dependent only on a combination of the three basic moments of inertia; this combination is designated by τ^2; its value is given by:

$$\tau^2 = \Theta_1^2 + \Theta_2^2 + \Theta_3^2 - (\Theta_1\Theta_2 + \Theta_2\Theta_3 + \Theta_3\Theta_1) \tag{13}$$

With this abbreviation, the average value is equal to:

$$\frac{3\tau^2}{r^8} \tag{14}$$

If n molecules are present per centimeter, the mean square value of the electric field strength at a point in the gas is obtained by evaluating:

$$\overline{\mathfrak{E}^2} = \int_{r=d}^{\infty} n\, 4\pi r^2\, dr \frac{3\tau^2}{r^8} = \frac{12\pi}{5} \frac{n\tau^2}{d^5} \tag{15}$$

Here the integration has been taken from a lower limit d to infinity. The quantity d indicates the diameter of a molecule; it is equal to the shortest distance to which two centers of molecules can approach one another.*

IV. Computation of van der Waals' Constant of Cohesion

The combination of equations (10) and (15) would give:

$$-\frac{\alpha}{2}\overline{\mathfrak{E}^2} = -\frac{6\pi}{5}\frac{n\alpha\tau^2}{d^5} \tag{16}$$

*For the determination of $\overline{\mathfrak{E}^2}$ the probability formula developed by J. Holtsmark, *Physik. Z.*, 20, 162 (1919), cannot be used. The diameter of the molecules was there assumed to be zero, for which case from Holtsmark's formulas lead to an infinite value of $\overline{\mathfrak{E}^2}$. We mention in this connection that it would be desirable in the problem treated by Holtsmark to take account of the finite diameter, the more so as this should affect the dependence on pressure of the widening effect.

for the potential energy of a molecule inside the gas, provided the electric field is produced by only the molecules initially present, and provided the introduced test molecule is considered as polarizable only and not as field producing. In fact, however, it is surrounded by its own electric field, and, consequently, polarizes all previously present particles. Accordingly, we readily see that the work done in introducing a molecule into the gas will be double the amount given by equation (16).

If Z molecules are contained in a volume V so that $n = Z/V$, and if dZ molecules are then added, the additional amount of work done is given by:

$$-\frac{12\pi}{5}\frac{\alpha\tau^2}{d^5}\frac{ZdZ}{V}$$

The total potential energy of a gas having N molecules in a volume V thus amounts to:

$$-\frac{6\pi}{5}\frac{\alpha\tau^2}{d^5}\frac{N^2}{V}$$

On the other hand, in paragraph 1 we found for this energy the value $-A/V$ so that we finally obtain:

$$A = \frac{6\pi}{5}\frac{\alpha\tau^2}{d^5}N^2 \qquad (17)$$

The total energy is, as expected, inversely proportional to the volume. If, instead of the constant A referring to an arbitrary volume, we desire the constant a referring to a molar volume, we obtain:

$$a = \frac{6\pi}{5}\mathsf{N}^2\frac{\alpha\tau^2}{d^5} \qquad (18)$$

where N ($N = 6.06 \times 10^{23}$) designates Loschmidt's number. The constant α can be readily related to experimental quantities. It is common practice to define the molecular refraction P of a substance of molecular weight M and density ρ in terms of its index of refraction n by means of the equation:

$$\frac{\mathrm{n}^2 - 1}{\mathrm{n}^2 + 2}\frac{M}{\varrho} = P \qquad (19)$$

The limiting value of this quantity for zero frequency is, according to the theory of dispersion, the quantity α we have been seeking multiplied by $(4\pi/3)N$, so that we obtain:*

$$P_0 = \frac{4\pi}{3} N \cdot \alpha \qquad (20)$$

Consequently the formula for a may alternatively be written as:

$$a = \frac{9}{10} P_0 \frac{N \cdot \tau^2}{d^5} \qquad (21)$$

The combination τ of the principal electrical moments of inertia, which we shall call "average electrical moment of inertia", could be found from the widening of the spectral lines produced by the addition of other gases,[3] if sufficient measurements were available in this field. Unfortunately this is not the case, so that we can only say that the quantity τ must be of the order of magnitude 5×10^{-26} (charge of an electron times square of the diameter of a molecule).

The quantity d follows from van der Waals' constant b, since, as is well known,

$$b = 4N \frac{4\pi}{3} \left(\frac{d}{2}\right)^3 = \frac{2\pi}{3} N d^3$$

On the other hand, b also follows from the critical molecular volume, $v_k = 3b$, or if this has not been observed, as is the case in most instances, from the critical pressure and the critical temperature according to the formula:

$$\frac{RT_k}{p_k} = 8b \qquad (22)$$

Table I is presented to demonstrate to what extent the formula derived for van der Waals' constant presents a picture of actual conditions.

In the first column are listed the chemical formulas of the gases, and in the second and third columns are given the observed values for critical temperature and critical pressure (the latter in atmospheres). The fourth column contains the first van der Waals' constant a, and the fifth the second van der Waals' constant b calculated from these values. From the latter follow the diameters d entered in the sixth column. The seventh column contains the molecular refractions derived from observed indices of refraction.

The values of a, P_0, and d are now inserted into equation (21), and the mean electrical moment of inertia τ is computed.

*We are obviously justified in using the limiting value of zero for frequency. The proper frequencies of the electrons essential for dispersion are of the order of magnitude 10^{15} sec.$^{-1}$. We have to compare this frequency with the number of times a molecule travels the distance of one molecular diameter at average temperatures. The latter figure is of the order of 10^{12} sec.$^{-1}$.

Table I

	T_k	p_k (atm.)	a dyne cm.⁴	b cm.³	d cm.	P_0 cm.⁻³	τ gr.½cm.7/2sec.⁻¹
He	5	2.75	0.026×10^{12}	18.6	2.44×10^{-8}	0.52	2.84×10^{-26}
A	151	48.0	1.37	32.2	2.93	4.17	11.2
Kr	210	54.3	2.34	39.6	3.14	6.25	14.5
Xe	288	57.2	4.17	51.6	3.42	10.2	18.9
H_2*	32	19.4	0.15	16.6	2.36	2.03	3.20
N_2	128	33.6	1.40	39.0	3.12	4.35	13.3
O_2	154	50.8	1.34	31.0	2.90	3.98	11.2
Cl_2	419	93.5	5.39	45.8	3.30	11.6	18.3
CH_4	191	54.9	1.91	35.6	3.04	6.36	11.9
C_2H_6	305	48.9	5.47	63.9	3.68	11.3	25.0
C_5H_{12}	470	33.0	9.2	145	4.85	25.7	60.7
C_2H_4	283	51.1	4.50	56.7	3.54	9.42	22.1

*From the known value for hydrogen, we obtain $\tau = 2.07\times10^{-26}$.

For our theory to deserve confidence, values of τ of the order of magnitude 5×10^{-26} must result. The last column of the table shows that this is really the case. Further we should expect an increase in the values of τ with increasing complexity of the molecule. This behavior is indicated by the last substances mentioned in the table.

V. Surface Tension

If we may now assume to be in possession of the actual law of forces between the molecules* we can define exactly what is to be understood by the radius of the sphere of action of a molecule. This quantity is of essential importance for the computation of surface tension, and we may, therefore, hope to be in a position to calculate the observed values for the surface tension directly from van der Waals' cohesion constant. A more exact theory of surface tension would have to consider thermal movement. We will here confine ourselves to a theory which does not take thermal movement into account, and which can only supply values for the surface tension extrapolated to $T = 0$.[4]

*In this note only the mean square of the electric field strength in the gas is required, a quantity which can be considered as derived from the law for the force. It follows readily from our assumptions that for the normal gases here considered the attraction is inversely proportional to the ninth power of the distance between two molecules. For anomalous dipole gases, it would be inversely proportional to the seventh power. It will be seen that the theory conforms with the familiar conclusion that van der Waals' attraction decreases rapidly with increasing distance.

We measure the z coordinate at right angles to the surface of the liquid; its value on the surface is zero, its values outside the liquid positive, and its values inside the liquid negative. Considering a reference point in the vicinity of the surface, there will be no more molecules in the positive region; the mean square value of the field strength will decrease continuously on a path from the inside of the liquid traversing the boundary layer toward the space containing no matter. The dependence of $\overline{\mathfrak{E}^2}$ on the z coordinate can be determined by an elementary computation. The result is given by the following formulas and represented in Figure 2.

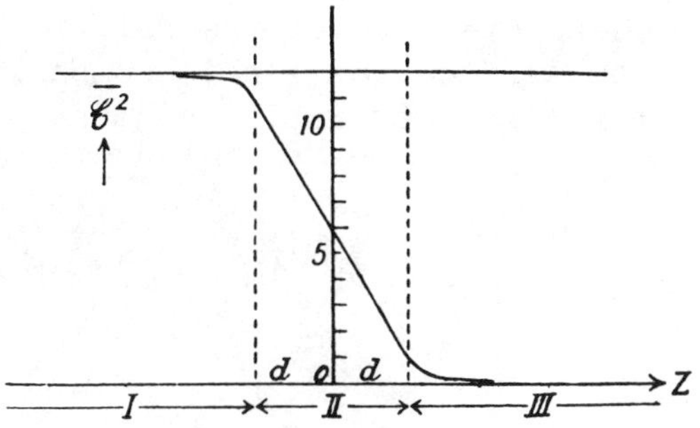

Figure 2.

$$\text{I)} \quad \overline{\mathfrak{E}^2} = \frac{12\pi}{5}\frac{n\tau^2}{d^5} + \frac{\pi}{5}\frac{n\tau^2}{z^5}$$

$$\text{II)} \quad \overline{\mathfrak{E}^2} = \frac{\pi}{5}\frac{n\tau^2}{d^5}\left(6 - \frac{5z}{d}\right)$$

$$\text{III)} \quad \overline{\mathfrak{E}^2} = \frac{\pi}{5}\frac{n\tau^2}{z^5}$$

Inside the liquid at a sufficient distance below the surface, the original formula for $\overline{\mathfrak{E}^2}$ holds according to formula (I). As the surface is approached, an amount inversely proportional to the fifth power of the distance from the surface must be deducted. This rule holds up to the distance d below the surface. From there on, as is shown by formula (II),the mean square of the field strength decreases proportional to z up to a point situated at a distance d above the surface. Here formula (III) starts, according to which $\overline{\mathfrak{E}^2}$ decreases inversely proportional to the fifth power of the distance from the surface with increasing coordinate.

We consider a liquid with volume V and area F. Considering the changes of $\overline{\mathfrak{E}^2}$ in the vicinity of the surface, it would be a mistake to put the total potential energy simply proportional to the volume. Rather an amount proportional to the surface area must be added. If n again designates the number of molecules per cubic centimeter, the value:

$$-\frac{6\pi}{5}\alpha\frac{n^2\tau^2}{d^5}V+\frac{3\pi}{8}\alpha\frac{n^2\tau^2}{d^4}F \qquad (23)$$

for the total potential energy will be found easily on the basis of formulas I, II, and III. The surface energy may be defined as the energy per unit area of the surface; hence:

$$\sigma=\frac{3\pi}{8}\alpha\frac{n^2\tau^2}{d^4} \qquad (24)$$

A more readily comprehensible form of the relation is obtained if this expression for the surface tension is compared with the previous expression (18) for van der Waals' constant a; in this way the relation:

$$\frac{\sigma}{a}=\frac{5}{16}d\left(\frac{\rho}{M}\right)^2 \qquad (25)$$

is obtained.

To what extent a correct value for σ can be calculated by this formula is demonstrated by Table II.

Table II

	T_k	p_k atm.	a dyne cm.^{4}	b cm.^{3}	d cm.	ρ/M cm.$^{-3}$	σ calc. dyne cm.$^{-1}$	σ obs'd. dyne cm.$^{-1}$
$C_4H_{10}O$	467	35.6	17.5×10^{12}	134	4.72×10^{-8}	0.96×10^{-2}	24	46
C_6H_6	561	47.9	18.9	120	4.56	1.12	34	59
C_6H_{14}	508	29.6	24.9	175	5.15	0.77	24	42
CS_2	548	76	11.3	73.5	3.86	1.66	38	72

In the first three columns of this table are given the chemical formulas, the critical temperature, and the critical pressure of the substance, respectively. In the fourth, fifth, and sixth columns follow the values of a, b, and d derived from the critical data by means of van der Waals' formulas. The seventh column contains the reciprocal of the molar volume (ρ/M). The column preceding the last contains the value of the surface tension computed from these data by means of formula (25). To enable comparison with experiments, it is necessary to extrapolate the observed values for the surface tension to $T = 0$. This is done by means of Eötvös' law. The values entered in the last column were obtained in this manner. It will be seen that only the order of magnitude is correctly given by the theoretical formula. In this connection it should be remembered that as far as numerical agreement is concerned van der Waals' equation never gives complete satisfaction.*

The relation of the surface tension to van der Waals' constant a can also be illustrated in the following way: According to Eötvos' law we have:

$$\left(\frac{M}{\varrho}\right)^{2/3} \sigma = 2.1\,(T_k - T) \qquad (26)$$

or in the limit for $T = 0$:

$$\left(\frac{M}{\varrho}\right)^{2/3} \sigma = 2.1\,T_k \qquad (26')$$

Our considerations lead to equation (25) by means of which we find:

$$\left(\frac{M}{\varrho}\right)^{2/3} \sigma = \frac{5}{16}\left(\frac{\varrho}{M}\right)^{1/3} a d$$

Further, according to van der Waals' equation the volume occupied by a substance in the limit for $T = 0$ is equal to b, and $b = (2\pi/3)\mathrm{N}d^3$. Hence we can substitute:

$$d = \left(\frac{3}{2\pi}\frac{b}{\mathrm{N}}\right)^{1/3}$$

for the quantity d. If this is done, the formula for the molar surface energy assumes the form:

$$\left(\frac{M}{\varrho}\right)^{2/3} \sigma = \left(\frac{\varrho}{M}\right)^{1/3} \frac{5}{16}\, d\, a = \frac{5}{16}\left(\frac{3}{2\pi}\right)^{1/3} \frac{1}{\mathrm{N}^{2/3}}\frac{a}{b}$$

*We recall, for instance, that according to van der Waals' equation:

$$\frac{RT_k}{p_k v_k} = \frac{8}{3} = 2.67$$

while for normal substances we have on the average the experimental value:

$$\frac{RT_k}{p_k v_k} = 3.7$$

We also know that:

$$\frac{a}{b}=\frac{27}{8}RT_k=\frac{27}{8}\mathrm{N}\cdot kT_k$$

introducingBoltzmann's constant, $k = 1.37 \times 10^{-16}$ erg instead of the universal gas constant R. By making use of this relation involving the critical temperature, we obtain:

$$\left(\frac{M}{\varrho}\right)^{2/3}\sigma=\frac{135}{128}\left(\frac{3}{2\pi}\right)^{1/3}\mathrm{N}^{2/3}kT_k \qquad (27)$$

Thus the theory gives in fact a proportionality between the molar surface energy and the critical temperature. For Eötvös' constant, we find a multiple of $\mathrm{N}^{2/3}k$, which, obviously, is to be expected, and which was also found by Madelung, as well as by Born and Courant, in their theory of the temperature coefficient in Eötvös' law. Inserting numerical values, the constant of this law becomes equal to 0.82, with $k = 1.37 \times 10^{-16}$ and $\mathrm{N} = 6.06 \times 10^{23}$. Proportionality with the critical temperature is found, while the numerical factor, as might be expected, deviates considerably from the experimental factor.

In order to calculate the surface tension and to evaluate Eötvös' constant another line may be followed, which does not require as extensive extrapolations as were made above. A more reliable value of a can be calculated for the liquid state, if, instead of critical quantities, only observations relating to the liquid state itself are utilized. To vaporize a mass M of a liquid, an amount of work has to be done which is measured by the molar heat of vaporization Λ. The potential energy involved in separating all the molecules of the mass M to a distance where molecular forces are negligible, is, according to the preceding calculations, equal to:

$$\frac{6\pi}{5}\frac{N^2\alpha\tau^2}{d^5}\frac{\varrho}{M}$$

Disregarding minor corrections (for the expenditure of external work and the energy difference of the thermal motion in gases and liquids, we have:

$$\Lambda=\frac{6\pi}{5}\frac{N^2\alpha\tau^2}{d^5}\frac{\varrho}{M} \qquad (28)$$

On the other hand:

$$a=\frac{6\pi}{5}\frac{\mathrm{N}^2\alpha\tau^2}{d^5}$$

so that:

$$a = \Lambda \frac{M}{\varrho} \qquad (29)$$

In Table III we have, for the same substances that were mentioned before, in the first column the chemical formula, in the second column the observed latent heat of vaporization, in the third

Table III

	Λ obs'd. cal.	a dyne cm.^{4}	σ calc'd. dyne cm.$^{-1}$	σ obs'd. dyne cm.$^{-1}$
$C_4H_{10}O$	6900	30×10^{12}	41	46
C_6H_6	7800	29	52	59
C_6H_{14}	7700	42	40	42
CS_2	6800	17	57	72

column van der Waals' constant of attraction calculated therefrom by equation (29), and in the fourth column the value of the surface tension calculated by equation (25). The last column contains the observed values for comparison. Good agreement is now obtained.

Continuing this line of thought, an attempt can also be made to calculate a more satisfactory value for Eötvös' constant. It is known that the actual value for the limiting density considerably exceeds the value calculated from van der Waals' law; the average value of the ratio of these two quantities is found* to be equal to 1.82. If we use the actual density, we therefore obtain an additional factor equal to $(1.82)^{4/3}$ in Eötvös' constant. The latter then assumes the value 1.8, instead of the previously calculated value of 0.82. This comes considerably closer to the experimental figure of 2.1.

VI. Concluding Remarks

Considering the situation as it now presents itself, one is, I believe, compelled to admit the probability of our concept of van der Waals' cohesive forces and, thus, to accept their electric origin. Under these circumstances, a relation between

*This figure was obtained from a combination of results regarding the relationship between the experimental values for critical density and limiting density and the corresponding theoretical values in J. P. Kuenen, *loc. cit.*, pp.60 and 83.

optical constants and van der Waals' constants is indicated, and we should be in a position to use the extensive experimental material on the equation of state for the determination of the principal electrical moments of inertia. I do not believe that the figures given for this quantity in Table I are more than an indication of the order of magnitude. The reason for this is, I believe, not so much the basis of the theory but its execution which, quantitatively, is as deficient as van der Waals' equation with a and b constant. There is, however, no reason to stop at this point of the survey which as given here can only suffice as a first orientation. The right course, it appears to me, would be to write the equation of state in the form eventually preferred by Kamerlingh-Onnes:

$$p\mathfrak{v} = \mathfrak{A} + \frac{\mathfrak{B}}{\mathfrak{v}} + \frac{\mathfrak{C}}{\mathfrak{v}^2} + \cdots$$

with virial coefficients $\mathfrak{A}, \mathfrak{B}, \mathfrak{C}$, etc. In fact a rigorous method based on Boltzmann-Maxwell's law exists for the computation of these coefficients as functions of temperature against which no objections can be raised. It has already been used in several papers by Keesom. It only remains to add to the forces between the molecules, introduced by Keesom, those considered here. Work of this kind applying to anomalous dipole gases is being undertaken. Among other things, I hope, it will lead to another method for the determination of molecular electric moments and thus establish a quantitative relation between observations concerning the equation of state and observations concerning the temperature dependence of the dielectric constant.*

It will also be necessary to follow more closely the surface phenomena and particularly their dependence on temperature. It should be possible to explain in terms of molecular theory the peculiar effects of small quantities of added substances on the surfacetension. Thus we arrive at the subject of mixtures, where without stressing surface phenomena - the formula for van der Waals' adhesive constant for mixtures, readily obtained from the theory, should be useful.

The main objects to be attained through an exact theory, it appears to me, are quantitatively correct figures for the electric moment and the electric moments of inertia, which can, in turn, be made the basis for the testing and setting up of detailed molecular models.

*Computations along these lines have been made in the meantime by Falkenhagen; they confirmed my expectations.

Supplement Added in Proof

My endeavors to find a second, independent means for the determination of the average electric moment of inertia τ have since led me to consider the temperature dependence of viscosity. The only formula which represents this dependence satisfactorily for higher temperatures is the one proposed by Sutherland:

$$\eta = \frac{C\sqrt{T}}{1 + \frac{\Theta}{T}}$$

where η is the coefficient of viscosity and C and Θ designate constants characteristic for the gas. C is inversely proportional to the square of the molecular diameter, and essentially measures the length of the free path which would correspond to the case of no forces between the molecules; Θ is a temperature characteristic for the gas and equal to the potential energy of two molecules in contact.

It would not be possible to use Sutherland's formula correctly for our purposes, unless the numerical coefficients were known with mathematical certainty. The computation required for this determination has been carried out in the meantime in a thesis by Enskog under the direction of Oseen. The following result is pertinent to our problem:

If the attractive force K between two molecules is proportional to the νth power of the reciprocal of the distance, so that:

$$K = \frac{c}{r^\nu}$$

then we obtain Sutherland's formula for the internal friction with:

$$\Theta = \frac{217}{205}\frac{2\Delta}{3}\frac{Nc}{(\nu - 1)d^{\nu-1}R}$$

and:

$$C = 1{,}015\frac{5}{16\sqrt{\pi}}\frac{\sqrt{MR}}{Nd^2}$$

Here 2Δ is a numerical factor which depends on the power in the law for the force and for which Enskog presents a table.*

*Enskog designates the numerical factor by 2δ; we write 2Δ instead, because the letter δ has already been used for the dielectric constant.

According to our assumption, the potential energy of two moleculcs at a distance r from one another equals:

$$-\alpha\mathfrak{E}^2 = -\frac{3\alpha\tau^2}{r^8}$$

where the mean value at a distance r is inserted for $\mathfrak{E}^2$ according to equation (14). Hence the force K is given by:

$$K = \frac{24\alpha\tau^2}{r^9}$$

Furthermore Enskog gives for $r = 9$, $2\Delta = 7/15$. Introducing the molar refraction $P_0 = (4\pi/3)N\alpha$ instead of α, we obtain therefore:

$$\Theta = \frac{217}{205}\frac{7}{20\pi}\frac{P_0\tau^2}{Rd^8}$$

$$C = 1{,}015\frac{5}{16\sqrt{\pi}}\frac{\sqrt{MR}}{Nd^2}$$

Thus the molecular diameter d follows from the experimentally determined values for C; Sutherland's temperature θ should then furnish the mean electric moment of inertia τ. In Table IV we

Table IV

	θ	C g. cm.$^{-1}$sec.$^{-1}$	d cm.	τ g.$\frac{1}{2}$cm.7/2sec.$^{-1}$	d cm.	τ g.$\frac{1}{2}$cm.7/2sec.$^{-1}$
He	75	1.46×10^{-5}	1.92×10^{-8}	4.33×10^{-26}	2.44×10^{-8}	2.84×10^{-26}
A	162	2.04	2.88	11.4	2.93	11.2
Kr	188	2.38	3.20	15.5	3.14	14.5
Xe	252	2.46	3.53	20.6	3.42	18.9
H_2	82	0.666	2.40	5.60	2.36	3.20
N_2	113	1.43	3.15	13.4	3.12	13.3
O_2	137	1.76	2.94	11.7	2.90	11.2
C_2H_4	226	1.04	3.70	24.4	3.54	22.1

have listed in the first column the chemical formula for the gas, in the second the values of θ, and in the third those of C as determined from the observations. The fourth column contains the diameter d calculated from C and the fifth the moment of inertia τ calculated from θ and C. For comparison with the preceding section, a sixth and a seventh column have been added which contain the values of d and τ as found before from the critical data.

Agreement between the values of d and τ found in these two ways is (with the exception of the figures for He and H_2) much better than I dared to expect. Apparently our law for the forces and its interpretation correspond entirely to reality.*

References

1. P. Debye, *Physik. Z.*, *13*, 97 (1912); M. Jona, *ibid.*, *20*, 14 (1919).
2. P. Debye, *Verhandl. deut. physik. Ges.*, *15*, 777 (1913).
3. P. Debye, *Physik. Z.*, *20*, 160 (1919); J. Holtsmark, *ibid.*, *20*, 162 (1919); J. Holtsmark, *Ann. Physik*, *58*, 577 (1919).
4. In the works of E. Madelung, *Physik. Z.*, *14*, 729 (1913), and of M. Born and R. Courant, *ibid.*, *14*, 731 (1913), a theory of the temperature dependence is presented although no explanation of the absolute value is attempted. The latter is the main objective here.

*Our formula for θ leads also to the conclusion that Sutherland's temperature θ is proportional to the critical temperature T_k. The proportionality factor obtained is $\theta/T_k = \frac{217}{205}\cdot\frac{7}{8} = 0.925$, while from the experimental data for the above mentioned gases (with the exception of H_2 and He) we find the average θ/T_k = 0.904.

THE THEORY OF ANOMALOUS DISPERSION IN THE REGION OF LONG-WAVE ELECTROMAGNETIC RADIATION

(Zur Theorie der anomalen Dispersion im Gebiete der langwelligen elektrischen Strahlung)

P. Debye

Translated from
Berichte der deutschen physikalischen Gesellschaft, Vol. 15, No. 16, pages 777-793, 1913.

Various substances, such as water and alcohol, show a striking variation of both refractive index and absorption coefficient even in the region of long electric waves. Let us consider the refractive index first; the characteristics of the curve which represents it as a function of frequency can be described as follows:

The curve shows (a) an overall continuous decrease of refractive index with increasing frequency from a rather high value, corresponding to the static dielectric constant, to a value which does not differ much from unity and which appears to be of the same order of magnitude as the optical index of refraction. Superimposed are (b) secondary maxima which, it seems, are not responsible for the main course of the absorption curve.

It struck my attention that in particular the characteristic anomalous dispersion mentioned under (a) appears to occur in substances in which the temperature dependence of the dielectric constant suggests the existence of rigid, constant electric moments in the interior of the molecule. I therefore attempted, partly in co-operation with R. Ortvay, to understand the quantitative aspects of the behavior outlined above on the basis of permanent moments.

I would like to show in the following how the behavior mentioned under (a) can be explained on the basis of this hypothesis. I plan to report later with Ortvay[2] on the maxima mentioned under (b), which we are inclined to attribute to structures formed by the association of molecules and which are capable of oscillation.

The qualitative aspects of the explanation suggested in the succeeding pages may be summarized briefly as follows: If an electric field is suddenly applied to a dielectric, the molecules will align the axis of their electrical moments within a short but finite interval of time in such a way that each cubic centimeter has assumed an electric moment. This moment depends on temperature and field strength, and will subsequently be subject to only small fluctuations due to molecular motion. This final moment is essentially the origin of the high dielectric constants which one observes in a static field. Moreover, electrons displaced from their rest positions contribute to the dielectric constant; in the static case, however, their contribution is considerably smaller than that due to the effect of permanent moments. If the field disappears after a very short time, or, more practically, if we use an alternating electric field, in which the electric force oscillates many times per second between a positive and a negative maximum value, then one quarter of a period of the electrical force will be insufficient time for the molecules to orient themselves to the extent that the full static moment per cubic centimeter is reached. If we proceed to very rapid oscillations, the molecules will be unable to align themselves at all, and only the electrons can contribute to the dispersion phenomena. Consequently the essential step in the quantitative computation is the calculation of the "relaxation time" just mentioned; from it we may readily deduce the frequency at which the dispersion curve assumes an anomalous behavior. Rather than restricting the calculation to the time of relaxation above, we shall consider the more general problem, by setting up a differential equation, the solution of which will give the mean rate of change of the orientation of the axes of the moments as a function of the variable driving force. The average moment of one cubic centimeter can then be derived by a simple calculation, whence we obtain the refractive index and the absorption coefficient as a function of frequency. The differential equation is a generalization of Einstein equation for the determination of the mean square of the displacement of a particle with Brownian movement.

I. Differential Equation for the Distribution Function of the Moment Axes

In order to simplify the folowing reasoning as much as possible and to establish a direct connection with Einstein's considerations, we will-for the purposes of this note- replace the actual medium by an imaginary one. In the real medium, the axes of the moments can point in all possible direction of space; in our model, however, we shall assume that the axes of all moments can rotate only in one plane, the one which contains the direction of the electric force. The direction of a given axis can then, at any instant, be defined by means of a single angle α, which will be measured from the positive direction of the electric force. The dielectric state of the substance is then determined by the number of molecules for which the direction of the axis of the moment lies within the angle $d\alpha$. We designate this number by

$$dz = F d\alpha$$

F is the unknown distribution function; it depends on α and on the time t, since we are considering non-static conditions. In the course of a short interval, Δt, F will have hanged by ΔF, and the number of dipoles in the interval $d\alpha$ is changed by:

$$\Delta dz = \Delta F d\alpha$$

Both the variation indicated by d as well as the one indicated by Δ are considered as essentially finite, and only as "physically infinitely small."

The problem then is to calculate F. For this purpose we shall use Einstein's method[3] for the computation of Brownian movement.

Let us imagine an α axis, and mark the beginning of the interval $d\alpha$; then we can compute the number of molecules whose moments "traverse the point α from left to right" during an interval Δt. The quantity in question is made up of two contributions.

First a torque of the magnitude:

$$-m\mathfrak{K}\sin\alpha$$

is exerted on each molecule by the external field, where m is the electrical moment of the dipole and the electric force on the dipole. The latter can be calculated according to Lorentz, from the electric field and the polarization by means of the formula:

$$\mathfrak{K} = \mathfrak{E} + \frac{4\pi}{3}\mathfrak{P}$$

If the coefficient of friction of the liquid under investigation is η, a moment due to fictional force:

$$-8\pi\eta a^3 \frac{d\alpha}{dt} = -\varrho\frac{d\alpha}{dt}$$

acts on the molecule, which is imagined to be a sphere of radius a; during the time interval Δt, as a consequence of the external torque, an average change in angle equal to:

$$\Delta\alpha = -\frac{m\mathfrak{K}}{\varrho}\sin\alpha\,\Delta t$$

is to be expected. To this change corresponds a number of molecules equal to:

$$F\Delta\alpha = -\frac{m\mathfrak{K}\sin\alpha}{\varrho}F\Delta t \qquad 1)$$

whose moments traverse the point α from left to right.

Second, the angle α will fluctuate by an amount ε as a result of collisions. If the mean square of these fluctuations is designated by ε_m^2, then, as a consequence:

$$-\frac{\varepsilon_m^2}{2}\frac{\partial F}{\partial \alpha} \qquad 2)$$

molecules will traverse the point α during each interval Δt.

We now obtain from equations (1) and (2) the number of moments which traverse the point α from left to right during the time interval Δt:

$$-\left[\frac{m\mathfrak{R}\sin\alpha}{\varrho}F\Delta t+\frac{\varepsilon_m^2}{2}\frac{\partial F}{\partial \alpha}\right] \qquad 3)$$

This number would be equal to the increase in molecules in the interval $d\alpha$, if other molecules would not simultaneously leave the interval at the point $\alpha + d\alpha$. The latter number can be computed by expanding expression (3); this gives:

$$-\left[\frac{m\mathfrak{R}\sin\alpha}{\varrho}F\Delta t+\frac{\varepsilon_m^2}{2}\frac{\partial F}{\partial \alpha}\right]-\frac{\partial}{\partial\alpha}\left[\frac{m\mathfrak{R}\sin\alpha}{\varrho}F\Delta t+\frac{\varepsilon_m^2}{2}\frac{\partial F}{\partial \alpha}\right]d\alpha \qquad 3')$$

By subtracting equation (3') from (3), we obtain as the required change, during the time interval Δt, in the number of moment axes situated in $d\alpha$.

$$\Delta F d\alpha=\frac{\partial}{\partial\alpha}\left[\frac{m\mathfrak{R}\sin\alpha}{\varrho}F\Delta t+\frac{\varepsilon_m^2}{2}\frac{\partial F}{\partial \alpha}\right]d\alpha \qquad 4)$$

If this were a stationary state, i.e., if $\mathfrak{R}$ were independent of t, we could compute from equation (4), by means of Einstein's relation, the mean squre of the variation ε_m^2, since F would in this case be determined by Maxwell-Boltzmann's law. According to the latter, F must have the form:

$$F=\mathrm{Const}\,e^{\frac{m\mathfrak{R}\cos\alpha}{kT}}$$

where $k = 1.346 \times 10^{-16}$ erg is Boltzmann's constant and T the absolute temperature. On the other hand, the expression in the bracket of equation (4) must vanish for the stationary case, and hence we find in agreement with Einstein:

$$\frac{\varepsilon_m^2}{2}=\frac{kT}{\varrho}\Delta t$$

This value of the mean square of the fluctuation, characteristic for Brownian motion, will be at least approximately correct for states which are not stationary. We therefore substitute this value for $\varepsilon_m^2/2$ in equation (4) and obtain:

$$\Delta F d\alpha = \frac{\partial}{\partial\alpha}\left[\frac{m\mathfrak{K}\sin\alpha}{\varrho}F + \frac{kT}{\varrho}\frac{\partial F}{\partial\alpha}\right]\Delta t\, d\alpha$$

By dividing by Δt and substituting $\partial F/\partial t$ for $\Delta F/\Delta t$, we obtain, upon rearranging, the required differential equation for F in the form:*

$$\varrho\frac{\partial F}{\partial t} = \frac{\partial}{\partial\alpha}\left[m\mathfrak{K}\sin\alpha\, F + kT\frac{\partial F}{\partial\alpha}\right] \qquad 5)$$

from which F is determined, when $\mathfrak{K}$ is given, as a function of t and α.

We recall further that ρ is given by:

$$\varrho = 8\pi\eta a^3 \qquad 6)$$

provided we accept in our molecular case the validity of the usual coefficient of friction η and consider the molecule as a sphere of radius a. It is evident that under these assumptions we can arrive at only an approximate value for ρ; from the manner of its derivation, equation (5) has more general validity, provided the value of ρ is not prescribed.

II. Computation of the Distribution Function for a Periodically Alternating Field

It is known that at average temperatures the torques exerted on molecular dipoles by the external field are much too weak to be able to cause any significant dissymmetry in the arrangement of the dipole axes. As a consequence, the moment is proportional to the generating electric field strength to a close approximation. The assumption of proportionality is even better justified for the case of alternating fields, where the final state is not reached during a quarter period. We may then, in order to obtain a practical solution of equation (5), develop F in powers of the amplitude of $\mathfrak{K}$ which we designate by K, and drop powers of K beyond the first. We assume $\mathfrak{K}$ to be a periodic function of the form:

$$\mathfrak{K} = Ke^{i\omega t}$$

where ω denotes the number of oscillations in 2π seconds, and set F equal to:

$$F = F_0 + KF_1$$

*The meaning of equation (5) may be more evident if it is considered that the first term in the bracket measures the variation of the density of the distribution of the moment axes due to the external force, and the second term the analogous variation due to an "osmotic torque" proportional to temperature and density.

Inserting this value for F in equation (5) and comparing the leading terms on both sides of the equation, we obtain the two equations:

$$\left.\begin{aligned} \varrho\frac{\partial F_0}{\partial t} - kT\frac{\partial^2 F_0}{\partial \alpha^2} &= 0. \\ \varrho\frac{\partial F_1}{\partial t} - kT\frac{\partial^2 F_1}{\partial \alpha^2} &= \frac{m\mathfrak{K}}{K}\frac{\partial}{\partial \alpha}(F_0 \sin\alpha) \end{aligned}\right| \qquad 7)$$

for the determination of F_0 and F_1.

Let us first consider F_0. We may state *a priori* that F_0 must be a periodic function of α, i.e., we can expand F_0 in a sine or cosine series of multiples of α. If we write, for instance, a single term of this series in the form:

$$\Theta_n \cos n\alpha$$

where Θ_n is a function of t yet to be determined, this function must satisfy the equation:

$$\varrho\frac{d\Theta_n}{dt} + n^2 kT\Theta_n = 0$$

From this it would follow for Θ_n:

$$\Theta_n = \text{const}\, e^{-\frac{n^2 kT}{\varrho}t}$$

i.e., all terms of the Fourier series which correspond to an index n different from zero, are attenuated by an exponential function of time. Therefore, they do not play any part in the final state; therefore, F_0 must have a constant value A:

$$F_0 = A \qquad 8)$$

Inserting this value of F_0 in the second equation of (7), and replacing $\mathfrak{K}/K$ by $e^{i\omega t}$, we obtain the differential equation:

$$\varrho\frac{\partial F_1}{\partial t} - kT\frac{\partial^2 F_1}{\partial \alpha^2} = mA\cos\alpha\, e^{i\omega t}$$

for the determination of F_1.

The particular solution of this equation corresponding to the right hand term has the form:

$$\frac{mA}{kT + i\omega\varrho}\cos\alpha\, e^{i\omega t}$$

From the other solutions which are possible if the right hand term vanishes, only $F_1 = B =$ const. remains, for the same reason as stated above, so that we obtain:

$$F_1 = B + \frac{mA}{kT + i\omega\varrho} \cos\alpha\, e^{i\omega t} \tag{9}$$

According to equations (8) and (9), F can now be represented by:

$$F = A\left[1 + \frac{mKe^{i\omega t}}{kT + i\omega\varrho}\cos\alpha\right] + KB$$

For very small values of the angular frequency ω, we have on the basis of Maxwell-Boltmann's law:

$$F = A e^{\frac{m\Re}{kT}\cos\alpha} = A\left[1 + \frac{mKe^{i\omega t}}{kT}\cos\alpha\right]$$

Accordingly we must set $B = 0$, and finally arrive at:

$$F = A\left[1 + \frac{mKe^{i\omega t}}{kT + i\omega\varrho}\cos\alpha\right] \tag{10}$$

for arbitrary values of ω.

With the determination of the distribution function by equation (10), we have reached the object of this paragraph; from it we can compute the average polarization per cubic centimeter as a function of time, and calculate the index of refraction and the absorption coefficient. Before proceeding to these computations, we shall make a calculation showing the quantity which plays the part of relaxation time τ. For this purpose we pose the problem of how a systematic arrangement of dipole axes set up by an external force reverts to a random arrangement of the force suddenly vanishes.

While the force K is effective, we can put:

$$F = A\left[1 + \frac{mK\cos\alpha}{kT}\right]$$

in accordance with statistical mechanics where the symbols designate the same quantitities as before. For $K = 0$, i.e., upon the sudden removal of the external field, F must satisfy the equation:

$$\varrho\frac{\partial F}{\partial t} = kT\frac{\partial^2 F}{\partial\alpha^2}$$

which results from the basic equation (5) for $\Re = 0$. It can readily be seen that the solution of this partial differential equation, corresponding to the above initial condition for $t = 0$, has the form:

$$F = A\left[1 + \frac{mK\cos\alpha}{kT} e^{-\frac{kT}{\varrho}t}\right]$$

Thus the systematic arrangement, represented by the second term, vanishes exponentially with time so that after a relaxation time:

$$\tau = \frac{\varrho}{kT}$$

its effect will have been reduced by the factor $1/e$.

If we compare the result obtained here for the relaxation time τ with formula (10), it will be seen that the freqency of the exciting field will cause significant variations in the distribution of the dipole axes when $\omega\rho/kT$, i.e., $\omega\tau$, assmes values of the order of magnitude of unity. This means, as had to be expected, that the influence of the frequency becomes noticeable when the duration of an electrical oscillation and the relaxation time assume comparable values.

III. Computation of Indices of Refraction and Absorption as Functions of Frequency

The number of molecules whose dipoles lie within the angular interval $d\alpha$, is, according to equation (10):

$$F\,d\alpha = A\left[1 + \frac{mKe^{i\omega t}}{kT + i\omega\varrho}\cos\alpha\right]d\alpha$$

The component of this electrical moment in the direction of the exciting force, which alone is of interest to us, has the value $m\cos\alpha$.

The electrical moment per cubic centimeter thus becomes:

$$\int_0^{2\pi} Fm\cos\alpha\,d\alpha = A\int_0^{2\pi}\left[1 + \frac{mKe^{i\omega t}}{kT + i\omega\varrho}\cos\alpha\right]m\cos\alpha\,d\alpha$$

or, since the mean value of $\cos\alpha$ is equal to zero,

$$A\frac{m^2Ke^{i\omega t}}{kT + i\omega\varrho}\int_0^{2\pi}\cos^2\alpha\,d\alpha$$

The number of molecules per cubic centimeter is given by:

$$\int_0^{2\pi} F\,d\alpha = A\int_0^{2\pi} d\alpha$$

hence we have for the mean observable moment m of a molecule the expression:

$$\bar{m} = \frac{\int_0^{2\pi} F m \cos\alpha \, d\alpha}{\int_0^{2\pi} F d\alpha} = \frac{m^2 K e^{i\omega t}}{kT + i\omega\varrho} \frac{\int_0^{2\pi} \cos^2\alpha \, d\alpha}{\int_0^{2\pi} d\alpha}$$

or:*

$$\bar{m} = \frac{m^2 K}{2kT} \frac{e^{i\omega t}}{1 + \frac{i\omega\varrho}{kT}} \qquad 11)$$

since the mean value of $\cos^2 \alpha$ is equal to $\frac{1}{2}$.

The further course of the computation follows a known method. According to Lorentz' assumption, $\mathfrak{K} = Ke^{i\omega t}$ is identified with $\mathfrak{E}(4\pi/3)\mathfrak{P}$. Here, the polarization $\mathfrak{P}$ is calculated as the mean electrical moment of a cubic centimeter, and the dielectric displacement $\mathfrak{D}$ is derived therefrom in accordance with the formula:

$$\mathfrak{D} = \mathfrak{E} + 4\pi\mathfrak{P}$$

The right hand term can now be expressed as a multiple of $\mathfrak{E}$; the factor of $\mathfrak{E}$ is the desired square of the complex index of refraction $\mathfrak{n}$. The calculation will be carried to this point in the present paragraph. In the next paragraph, we shall separate into its two components, ordinary index of refraction n and absorption coefficient κ, and discuss these two experimental quantities as functions of frequency. If the number of molecules per cubic centimeter is N, then, according to equation (11), we have:

$$\mathfrak{P} = \frac{Nm^2}{2kT} \frac{Ke^{i\omega t}}{1 + \frac{i\omega\varrho}{kT}} = \frac{Nm^2}{2kT} \frac{\mathfrak{K}}{1 + \frac{i\omega\varrho}{kT}}$$

or assuming the Lorentz field:

*It follows from equation (11) that in the limit for $\omega = 0$

$$\bar{m} = \frac{m^2 K}{2kT}$$

If the simplifying assumption that all dipole axes are located in a plane had not been introduced, the factor 1/3 would appear instead of the factor 1/2.

$$\mathfrak{P} = \frac{Nm^2}{2kT} \frac{1}{1 + \frac{i\omega\varrho}{kT}} \left(\mathfrak{E} + \frac{4\pi}{3}\mathfrak{P}\right)$$

This result would be correct, provided the total dielectric polarization were derived exclusively from the dipoles. However, the influence of the displaced electrons must also be taken into consideration. Assuming, as an example, one such electron with a charge e and a quasi-elastic constant f to be available per molecule, then these electrons add to $\mathfrak{P}$ the contribution:

$$\mathfrak{P} = \frac{Ne^2}{f}\left(\mathfrak{E} + \frac{4\pi}{3}\mathfrak{P}\right)$$

If more electrons are present per molecule and if the strength of their bonds is different, we substitute $\Sigma e^2/f$ for e^2/f. The above expression is correct only if the frequency of the exciting wave is not near the resonant frequency of one of the electrons. The present discussion is limited to the long-wave region of the spectrum, where we can measure the influence of the electrons by the expression just given.

We thus obtain for the total polarization:

$$\mathfrak{P} = \left(\frac{Nm^2}{2kT} \frac{1}{1 + \frac{i\omega\varrho}{kT}} + \frac{Ne^2}{f}\right)\left(\mathfrak{E} + \frac{4\pi}{3}\mathfrak{P}\right)$$

If we substitute as a temporary abbreviation:

$$\frac{Nm^2}{2kT} \frac{1}{1 + \frac{i\omega\varrho}{kT}} + \frac{Ne^2}{f} = \frac{3}{4\pi}\beta \qquad 12)$$

we obtain for $4\pi\,\mathfrak{P}$ the value:

$$4\pi\mathfrak{P} = \frac{3\beta}{1-\beta}\mathfrak{E}$$

Furthermore the dielectric displacement $\mathfrak{D}$ can be calculated from:

$$\mathfrak{D} = \mathfrak{E} + 4\pi\mathfrak{P}$$

so that we obtain the relationship:

$$\mathfrak{D} = \mathfrak{E} + \frac{3\beta}{1-\beta}\mathfrak{E} = \frac{1+2\beta}{1-\beta}\mathfrak{E} \qquad 13)$$

The factor $(1+2\beta)/(1-\beta)$ is the square of the complex index of refraction n, for which we obtain from equations (12) and (13) the explicit formula:

$$n^2 = \frac{1 + 2\left(\frac{4\pi}{3}\frac{Ne^2}{f} + \frac{4\pi}{3}\frac{Nm^2}{2kT}\frac{1}{1+\frac{i\omega\varrho}{kT}}\right)}{1 - \left(\frac{4\pi}{3}\frac{Ne^2}{f} + \frac{4\pi}{3}\frac{Nm^2}{2kT}\frac{1}{1+\frac{i\omega\varrho}{kT}}\right)} \qquad 14)$$

Instead of the combinations Ne^2/f and $Nm^2/2kT$ we can introduce the square of the index of refraction for very rapid oscillations or for very slow oscillations, respectively. We designate the first by ε_∞, the second by ε_0. The symbol ∞ in ε_∞ is to be understood to mean that this quantity refers to oscillations for which the effect of the dipoles has vanished and not to oscillations which are in fact infinitely fast. As shown by numerical calculation to be presented later, this will occur for wave lengths of approximately 1 mm. to 0.1 mm. (measured in air).

Equation (14) gives for very high frequencies:

$$n^2 = \varepsilon_\infty = \frac{1 + 2\frac{4\pi}{3}\frac{Ne^2}{f}}{1 - \frac{4\pi}{3}\frac{Ne^2}{f}}$$

and for very low frequencies:

$$n^2 = \varepsilon_0 = \frac{1 + 2\frac{4\pi}{3}\left(\frac{Ne^2}{f} + \frac{Nm^2}{2kT}\right)}{1 - \frac{4\pi}{3}\left(\frac{Ne^2}{f} + \frac{Nm^2}{2kT}\right)}$$

From these two formulas we obtain:

$$\frac{4\pi}{3}\frac{Ne^2}{2f} = \frac{\varepsilon_\infty - 1}{\varepsilon_\infty + 2}$$

and:

$$\frac{4\pi}{3}\frac{Nm^2}{2kT} = \frac{\varepsilon_0 - 1}{\varepsilon_0 + 2} - \frac{\varepsilon_\infty - 1}{\varepsilon_\infty + 2}$$

so that we can write instead of equation (14):

$$n^2 = \frac{\left(1 + 2\frac{\varepsilon_\infty - 1}{\varepsilon_\infty + 2}\right) + \frac{2}{1+\frac{i\omega\varrho}{kT}}\left(\frac{\varepsilon_0 - 1}{\varepsilon_0 + 2} - \frac{\varepsilon_\infty - 1}{\varepsilon_\infty + 2}\right)}{\left(1 - \frac{\varepsilon_\infty - 1}{\varepsilon_\infty + 2}\right) - \frac{1}{1+\frac{i\omega\varrho}{kT}}\left(\frac{\varepsilon_0 - 1}{\varepsilon_0 + 2} - \frac{\varepsilon_\infty - 1}{\varepsilon_\infty + 2}\right)}$$

The final formula follows from this by a simple rearrangement:

$$n^2 = \frac{\frac{\varepsilon_0}{\varepsilon_0 + 2} + \frac{i\omega\varrho}{kT}\frac{\varepsilon_\infty}{\varepsilon_\infty + 2}}{\frac{1}{\varepsilon_0 + 2} + \frac{i\omega\varrho}{kT}\frac{1}{\varepsilon_\infty + 2}} \qquad 14')$$

IV. Numerical Computation and Graphical Representation

Before proceeding to the separation of n into refractive index and absorption coefficient, we shall estimate the numerical value to be expected for the quantity $\rho\omega/kT$, which determines the course of the dispersion curve.

According to equation (6), we have:

$$\varrho = 8\pi\eta a^3$$

For $a = 10^{-8}$ cm. and $\eta = 0.0106$, (corresponding to water at 18°) and introducing the number of oscillations v per second instead of ω, we find that the ratio $\rho\omega/kT$ equals unity at 18° for a frequency v, which from:

$$\nu = \frac{kT}{2\pi\varrho} = \frac{kT}{16\pi^2\eta a^3}$$

turns out to be:

$$\nu = 2.35 \cdot 10^{10}$$

Thus, according to equation (14'), anomalous dispersion would be appreciable at a wave length:

$$\lambda = 1.28 \text{ cm}$$

measured in air. If the molecular radius is estimated to be 2×10^{-8} cm. instead of 1×10^{-8} cm., the wave length becomes 8 times as long, i.e., it becomes:

$$\lambda = 10.2 \text{ cm}$$

It is remarkable that the position of anomalous dispersion in the spectrum can be calculated in advance, given the coefficient of friction and the molecular dimensions based on the theory of gases. In future experiments relating to this theory, it is particularly recommended that attention be given to the temperature dependence of dispersion and absorption, because the coefficient of friction η^4 is so sensitive to temperature.

Qualitatively, the formula for the relaxation time can be confirmed independently of dispersion measurements by the known decrease of the Kerr constant for electrical double refraction in the electric spectrum with increasing frequency. The theory of this phenomenon can be developed in an analogous way to the theory given here for electrical dispersion, using Langevin's suggestion that the orientation of molecules with dissymmetrically bound electrons is responsible for the Kerr effect. It may suffice to point out that experimental results by H. Abraham and J. Lemoine give relaxation times of the same order of magnitude as appear in our case.

Always keeping in mind that the dispersion discussed here is limited to high frequency electrical waves, the formulas for the index of refraction n and for the absorption coefficient κ, following from equation (14'), will now be discussed in more detail.

By definition, we have:

$$\mathfrak{n} = n(1 - i\varkappa) \qquad 15)$$

so that it is only necessary, for the determination of n and κ, to separate the real and imaginary parts of the square root of the right-hand side of equation (14'). We can write for the numerator:

$$\frac{\varepsilon_0}{\varepsilon_0+2} + \frac{i\omega\varrho}{kT}\cdot\frac{\varepsilon_\infty}{\varepsilon_\infty+2} = \left[\left(\frac{\varepsilon_0}{\varepsilon_0+2}\right)^2 + \frac{\omega^2\varrho^2}{k^2T^2}\left(\frac{\varepsilon_\infty}{\varepsilon_\infty+2}\right)^2\right]^{1/2} e^{i\varphi_1}$$

with the abbreviation:

$$\operatorname{tg}\varphi_1 = \frac{\omega\varrho}{kT}\,\frac{\varepsilon_0+2}{\varepsilon_\infty+2}\,\frac{\varepsilon_\infty}{\varepsilon_0} \qquad 16)$$

Similarly the denominator can be transformed into:

$$\left[\left(\frac{1}{\varepsilon_0+2}\right)^2 + \frac{\omega^2\varrho^2}{k^2T^2}\left(\frac{1}{\varepsilon_\infty+2}\right)^2\right]^{1/2} e^{i\varphi_2}$$

with:

$$\operatorname{tg}\varphi_2 = \frac{\omega\varrho}{kT}\,\frac{\varepsilon_0+2}{\varepsilon_\infty+2} \qquad 16')$$

Hence:

$$\mathfrak{n}^2 = n^2(1-i\varkappa)^2 = \left[\frac{\left(\frac{\varepsilon_0}{\varepsilon_0+2}\right)^2 + \frac{\omega^2\varrho^2}{k^2T^2}\left(\frac{\varepsilon_\infty}{\varepsilon_\infty+2}\right)^2}{\left(\frac{1}{\varepsilon_0+2}\right)^2 + \frac{\omega^2\varrho^2}{k^2T^2}\left(\frac{1}{\varepsilon_\infty+2}\right)^2}\right]^{1/2} e^{-i(\varphi_2-\varphi_1)}$$

where, according to equations (16) and (16'), the difference $\varphi_2 - \varphi_1$ is defined by the equation:

$$\operatorname{tg}(\varphi_2 - \varphi_1) = \frac{\operatorname{tg}\varphi_2 - \operatorname{tg}\varphi_1}{1 + \operatorname{tg}\varphi_2 \operatorname{tg}\varphi_1} = \frac{\omega\varrho}{kT}\,\frac{\varepsilon_0 + 2}{\varepsilon_\infty + 2}\,\frac{1 - \frac{\varepsilon_\infty}{\varepsilon_0}}{1 + \frac{\omega^2\varrho^2}{k^2T^2}\left(\frac{\varepsilon_0 + 2}{\varepsilon_\infty + 2}\right)^2 \frac{\varepsilon_\infty}{\varepsilon_0}} \tag{17}$$

Finally, separation of real and imaginary parts gives:

$$\left.\begin{aligned} n &= \left[\frac{\left(\frac{\varepsilon_0}{\varepsilon_0 + 2}\right)^2 + \frac{\omega^2\varrho^2}{k^2T^2}\left(\frac{\varepsilon_\infty}{\varepsilon_\infty + 2}\right)^2}{\left(\frac{1}{\varepsilon_0 + 2}\right)^2 + \frac{\omega^2\varrho^2}{k^2T^2}\left(\frac{1}{\varepsilon_\infty + 2}\right)^2}\right]^{1/4} \cos\frac{\varphi_2 - \varphi_1}{2} \\ \varkappa &= \operatorname{tg}\frac{\varphi_2 - \varphi_1}{2} \end{aligned}\right\} \tag{18}$$

In numerical evaluation, the angle $\varphi_2 - \varphi_1$ is first found from equation (17), and then the required quantities are calculated by means of equation (18).

A discussion in qualitative terms immediately reveals the expected course of n and κ. It is obvious from equation (17) that the angle $(\varphi_2-\varphi_1)$ vanishes for small as well as for large values of ω. Hence, according to equation (18), κ must have a maximum and must vanish for very long and very short waves. Since, furthermore, in these limiting cases $\cos(\varphi_2-\varphi_1)/2$ has the value 1, the first equation in (18) shows that for increasing values of ω the index of refraction n will decrease from $n = \sqrt{\varepsilon_0}$ to $n = \sqrt{\varepsilon_\infty}$, thus indicating anomalous dispersion.

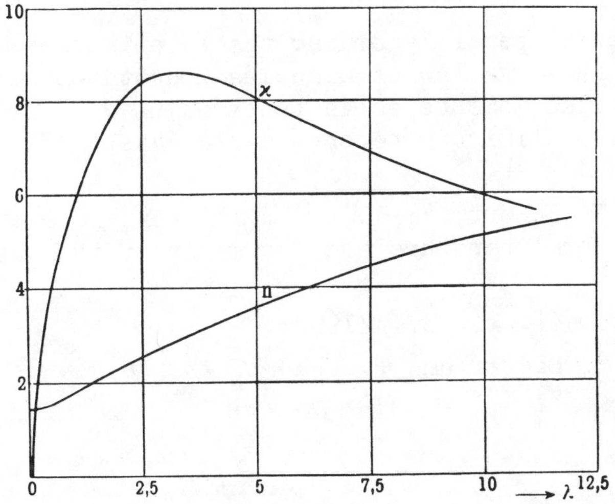

The figure serves to illustrate this behavior. It was plotted under the assumption that $\omega\rho/kT$ has the value unity at a wave length λ equal to 1 cm. measured in air; then $1/\lambda$ can be substituted for $\omega\rho/kT$. Further we put $\omega_0 = 80$ and $\omega_\infty = 2$. With these numerical values, $\varphi_2 - \varphi_1$ can be calculated from the equation:

$$\operatorname{tg}(\varphi_2 - \varphi_1) = \frac{1}{\lambda} \frac{20.0}{1 + 10.5\left(\frac{1}{\lambda}\right)^2}$$

and, subsequently, n and κ follow, according to equation (18) from:

$$\left.\begin{aligned} n &= 8.94 \left[\frac{1 + 0.262\left(\frac{1}{\lambda}\right)^2}{1 + 420\left(\frac{1}{\lambda}\right)^2}\right]^{1/4} \cos\frac{\varphi_2 - \varphi_1}{2} \\ \kappa &= \operatorname{tg}\frac{\varphi_2 - \varphi_1}{2}. \end{aligned}\right\}$$

The result of the numerical computations gives the figure in which the curve n for the refractive index and the curve κ for the absorption coefficient are plotted as functions of the wave length, measured in centimeters. The legend of the ordinate axis refers to the refractive index, while κ is represented on a tenfold enlarged ordinate scale. It appears superfluous to describe the figure in more detail.

Even though the results of the above discussion require quantitative improvements, we believe that it represents a sound theoretical treatment of anomalous dispersion and absorption as a function of temperature and of the frequency of the alternating field. The hypothesis that molecules contain permanent dipole moments is the essential basis of the theory.

Bibliography

1. A comprehensive paper regarding the relation between the deviations from the law of Clausius-Mosotti and the hypothesis of fixed moments is in preparation. It follows the lines of my preliminary report in the *Physik. Z.*, *13*, 97 (1912).
2. I refer to F. Eckert, *Verhandl. deut. physik. Ges.*, *15*, 307 (1913) for literature and a summary of the experimental results.
3. *Ann. Physik*, [4], *19*, 371 (1906).
4. Compare O. v. Baeyer and F. Eckert, *loc. cit.*

SOME RESULTS OF A KINETIC THEORY OF INSULATORS
(Einige Resultate einer kinetischen Theorie der Isolatoren)

P. Debye

Translated from
Physikalische Zeitschrift, Vol. 13, No. 3, pages 97-100, February 1, 1912.

I. Introduction

The currently accepted structure of a dielectric assumes that the electrons in its molecules are bound to their rest positions by a force proportional to their displacement. Then the effect of an electric field will be to displace the electron from their rest positions, producing a dielectric polarization of the material. This hypothesis has successfully accounted for dispersion and for the Zeeman effect. Nevertheless, we shall show that it cannot account for all the known facts. The shortcomings of this hypothesis appear, for example, if we consider the problem of the temperature dependence of the dielectric constant. The few data available[1-5] which bear on this question show that the dielectric constant ε increases rapidly with decreasing temperature. Thus, for instance for ethyl alcohol, $\varepsilon = 25.8$ at an absolute temperature $T = 293$, while at a temperature lower by 140° ($T = 153$) $\varepsilon = 54.6$, which is already more than double the original value. In contrast, an attempt to base a kinetic theory of insulators on the above hypothesis leads to the result that the dielectric constant would be entirely independent of temperature.

Two explanations of the contradiction are possible. On the one hand, we might doubt the validity of a computation carried out in conformity with statistical mechanics, and from the discrepancy conclude as would no longer be surprising at this time - that statistical mechanics has again failed. On the other hand, however, we might try to modify or improve the basic hypothesis concerning the structure of insulators. We have taken this second course; its success proves it to be correct. The assumption that *the interior of dielectrics contains not only elastically bound* *electrons but also permanent dipoles of constant electric moment* enables us to explain in a completely satisfactory way the temperature dependence of the dielectric constant, using classical statistical mechanics as the method of computation.* This is at least true for the observations available to date. We are, however, justified in doubting the results for lower temperatures, in analogy to the deviations from Curie-Langevin's law which Perrier found for magnetic susceptibility.

II. Results of the Computation

We proceed from the hypothesis stated in the introduction according to which every molecule (a) contains electrons bound to their rest positions by elastic forces, and (b) further possesses a constant electric moment of magnitude m.

If the polarization, P, produced by an electric field of strength, E, is calculated, an expression with two additive terms is found. The first term of this expression measures the effect of the "displaced electrons" and is independent of temperature. We represent it by $(\varepsilon_0-1)E$. The second term has the Curie-Langevin form $E(a/T)$, where a is a constant and T the absolute temperature. For gases we then obtain:

$$P = E(\varepsilon_0 - 1) + E\frac{a}{T} \qquad (1)$$

so that the dielectric displacement D will be given by:

$$D = P + E = E\left(\varepsilon_0 + \frac{a}{T}\right) \qquad (2)$$

and we obtain for the dielectric constant:

$$\varepsilon = \varepsilon_0 + \frac{a}{T} \qquad (3)$$

Equation (1) holds, provided the temperature is changed at constant volume. If the observations refer to constant pressure, the variation in volume must be considered. For ideal gases, for instance, we would have for the difference $\varepsilon - 1$ an expression of the form:**

*The theory was carried through assuming that the electric moment shows no dependence on temperature. To what extent this can be justified will be discussed in the more detailed publication.

**For the more detailed discussion we refer here as well as in the following to a subsequent more detailed publication.

$$\varepsilon - 1 = \frac{\alpha}{T} + \frac{\beta}{T^2} \qquad (3')$$

If we attempt to apply equation (3) to the observations of Abegg-Seitz on liquids, no satisfactory result is obtained. We must consider not only the "external" electric field strength as a measure of the effective force inside the substance, but the effect of the induced molecular field must also be taken into account. If the force[6] produced by the molecular field is measured by $P/3$, the following formula, as a first approximation for small field strength, is found for P:

$$P = \left(E + \frac{P}{3}\right)\Theta \qquad (4)$$

where θ is a function of the temperature of the form:

$$\Theta = 3b + \frac{3a}{T} \qquad (5)$$

and a and b are constants which will be dealt with later on. From equations (4) and (5) together with the formula $D = E + P = \varepsilon E$ the dielectric constant ε can be calculated thus:

$$D = E + P = \varepsilon E$$

$$\varepsilon = \frac{1 + 2\frac{\Theta}{3}}{1 - \frac{\Theta}{3}} \qquad (6)$$

In view of equation (5) we can write instead:

$$\frac{\varepsilon - 1}{\varepsilon + 2} = \frac{\Theta}{3} = \frac{a}{T} + b \qquad (7)$$

or finally:

$$\frac{\varepsilon - 1}{\varepsilon + 2} T = a + bT \qquad (7')$$

Stated in words, the product of the characteristic expression $(\varepsilon - 1)/(\varepsilon + 2)$ and the absolute temperature T is a linear function* of T.

*We use the theorem in the form given because we shall only consider fluids at relatively low temperatures, where the temperature expansion can be neglected. Strictly, the theorem reads: The product of the Clausius-Mosotti expression $(\varepsilon-1)/(\varepsilon+2)\rho$ (ρ designates the density) times the absolute temperature T is a linear function of the latter.

III. Comparison with Experiment

Discussion of formula (3) for gases will be reserved for a later detailed publication; only the results of a comparison of formula (7') with the data of Seitz-Abegg will be given here. Observations were made on five alcohols (methyl, ethyl, propyl, isobutyl, and amyl alcohol) and on ethyl ether. The temperatures at which the observations were made cover a range of from $T = 153$ to $T = 293$. If the quantity:

$$\frac{\varepsilon - 1}{\varepsilon + 2} T$$

is computed[7] and the figures thus obtained are plotted as ordinates against the temperature as abscissa, the diagram in the figure is obtained. We see at a glance that the observed values

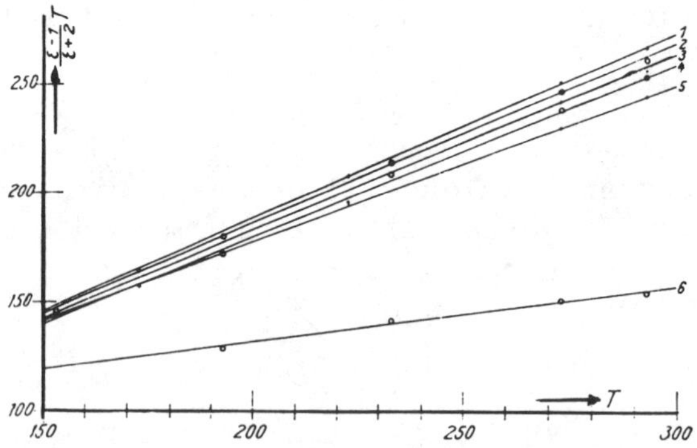

do not deviate essentially from straight lines,* *i.e.*, formula (7') is confirmed by the observations. The constant a is equal to the ordinate at $T = 0$, and b is equal to the slope. Table I, which should be self-explanatory, summarizes these values:

Table I

		a	b
1	Methyl Alcohol	18	0.85
2	Ethyl Alcohol	20	0.82
3	Propyl Alcohol	21	0.80
4	Isobutyl Alcohol	29	0.76
5	Amyl Alcohol	32	0.72
6	Ethyl Ether	80	0.25

*The numbers 1, 2, etc. in the figure refer to the different substances in the order given above. The observed values are indicated, alternately, by circles and crosses.

If formula (7) is extrapolated to such low values of T that $\theta/3 = 1$, *i.e.*, so far that T becomes equal to a critical temperature defined by:

$$\frac{a}{T_k} + b = 1$$

or:

$$T_k = \frac{a}{1 - b} \qquad (8)$$

it would follow that $\varepsilon = \infty$. Actually this extrapolation is unjustified. Nevertheless, the critical temperature T_k has a real meaning. If the temperature is decreased below T_k, the detailed theory shows that polarization in the substance may occur without an external field, *i.e.*, a residual dielectric polarization could be observed. As far as I know, this has not been achieved in practice. Before $T = T_k$ is reached, liquids solidify, and the validity of relations (7) and (7'), respectively, ceases, since the molecules are no longer free to rotate. For what can be said about solids, I refer again to the more detailed publication, remarking only here that the critical temperature T_k plays a part analogous to the so-called Curie point for ferromagnetic substances.

IV. Physical Meaning of Constants a and b. Calculation of Elementary Electric Moments

The deviation from the horizontal of the straight lines in the figure is caused by the presence of "displacement electrons." If these only were present, we would have $a = 0$ and instead of equation (7') the formula:

$$\frac{\varepsilon - 1}{\varepsilon + 2} = b \qquad (9)$$

i.e., as has already been pointed out in the introduction, ε would be independent of the temperature, in contradiction to the observation.

If the charge of the "displacement electrons" is e, and there are N_1 of them which are drawn back towards their rest positions by a force f_1 per unit displacement, N_2 by a force f_2, etc., then b can be expressed as a function of these quantities by:

$$3b = \sum_p \frac{N_p e^2}{f_p} \qquad (10)$$

The quantity $1 + 3b$ would then be equal to the dielectric constant, provided (a) displacement electrons only were present, and (b) the mutual effect of the molecules on one another could be neglected.

Here we are primarily concerned with constant a. Assuming that all molecules have the same (fixed) electric moment m, and designating by N the number of molecules per cubic centimeter, we obtain for a the relation:

$$a = \frac{N m^2}{9 k} \qquad (11)$$

where k is the universal constant with which the natural logarithm of the probability is to be multiplied to obtain the entropy. Its value is, according to Planck[8], 1.346×10^{-16} erg. The number of molecules N per cubic centimeter can be calculated from the density of the substance, the molecular weight, and the known mass of the hydrogen atom[8] (1.63×10^{-24} g.) Using the observed values for the density at room temperature, we calculated the values for m from the formula:

$$m = \frac{3}{\sqrt{4\pi}} \sqrt{\frac{k a}{N}}$$

which follows from equation (11). The factor $1\sqrt{4\pi}$ on the right was introduced in order to convert to conventional electrostatic units because the previous formulas referred to rational units in which system the unit of electricity is $1\sqrt{4\pi}$ times that of the conventional electrostatic unit. The moments m have the dimensions $g.^{1/2}cm.^{5/2}sec.^{-1}$. They are given in Table II for the fluids mentioned above as well as for a few more.

Table II

Name	Chemical formula	m
Methyl Alcohol	$CH_3.OH$	$3.4.10^{-19}$
Ethyl Alcohol	$C_2H_5.OH$	$4.3.10^{-19}$
Isobutyl Alcohol	$C_3H_7.OH$	$5.0.10^{-19}$
	$(CH_3)_2.CH.CH_2OH$	$6.5.10^{-19}$
Amyl Alcohol	$C_5H_{11}.OH$	$7.9.10^{-19}$
Water	$H.O.H$	$5.7.10^{-19}$
Ethyl Ether	$C_2H_5.O.C_2H_5$	$11.8.10^{-19}$
Acetone	$CH_3.CO.CH_3$	$3.4.10^{-19}$
Toluol	$C_6H_5.CH_3$	$5.1.10^{-19}$
Nitrobenzene	$C_6H_5.NO_2$	$5.7.10^{-19}$

While these figures do not claim to be very accurate, they show clearly that the kind of molecule has no essential effect on the magnitude of m. As shown by the chemical formulas, the oxygen appears, for instance, in the hydroxy group, serves in another case (ethyl ether) for bonding two alkyl radicals, and in acetone has a double bond to a C atom. Furthermore, two cyclic compounds are included; we note that toluene contains no oxygen at all, while the other, nitrobenzene, contains oxygen in the nitro group, which is essentially different from those mentioned above. Also the values of the dielectric constants at room temperature vary within wide limits, *i.e.*, between 2.33 for toluene to 37.8 for nitrobenzene and 80 for water. I consider it too early yet to draw more general conclusions in regard to the numerical values of m. We shall revert to this point in the more detailed publication using more extensive data.

One other item should be emphasized here. If we imagine the moment m formed by two elementary electric charges e (of opposite sign) separated by a distance l, then with $e = 4.69 \times 10^{-10}$ electrostatic units[8], and assuming for the moment an average value $m = 5.4 \times 10^{-19}$, we obtain $l = 1.1 \times 10^{-9}$ cm.

Since, on the other hand, we estimate the diameter of a molecule to be a multiple of 10^{-8} cm., we see that considerable space is available inside a molecule for such a dipole. By presenting this consideration we do not want to imply that the electric moment must be thought of as being constructed in this manner.

References

1. R. Abegg and W. Seitz, *Z. phys. Chem.*, *29*, 242, 491 (1899).
2. F. Hasenoehrl, *Wiener Ber.*, *105*, 460 (1896).
3. K. Baedeker, *Z. phys. Chem.* *36*, 305 (1901).
4. K. Tangl, *Ann. Phys.*, *10*, 748 (1903).
5. F. Ratz, *Z. phys. Chem.*, *19*, 94 (1896).
6. Regarding the computation of this force see, for example, H. A. Lorentz, "*The Theory of Electrons*", Teubner, 1909, p. 303, note 54.
7. The figures given in Landolt-Boernstein, p. 764 et seq., were used.
8. M. Planck, *Theorie der Waermestrahlung*, Leipzig, 1906, p. 162.

MOLECULAR FORCES AND THEIR ELECTRICAL INTERPRETATION
(Molekularkräfte und ihre elektrische Deutung)

P. Debye

Translated from Physikalische Zeitschrift, Vol. 22, pages 302-308, 1921.

(1) About a year ago[1] I showed that a mutual attraction must exist between the molecules of all substances as a consequence of the polarization of one molecule in the electric field of the other molecule. Whereas the "polarizability" of a molecule may be considered as known experimentally, when the index of refraction (and sometimes the dispersion) has been measured, the generation of the field must be based on the (average) arrangement of the charges in the molecule.

Depending on whether the center of gravity of the positive charges fails to or does coincide with the center of gravity of the negative charges, the field at a fairly great distance is derived from a dipole moment associated with the molecule or from three main electrical moments of inertia defining a quadrupole. (In the first instance the electrical properties of the molecule are geometrically characterized by a vector and in the second by a surface of the second order). The division, based thereon, of molecules into two large main groups - those belonging to the first group being usually capable of generating considerably stronger electric fields in their surrounding than those belonging to the second group - seems to me to correspond very well to the conspicuous differences observed in their behavior (association, dissociating ability for electrolytes, dielectric behavior,[2] anomalous dispersion at long wave length, etc.[3]). While the differentiation between the two main groups and the quantitative determination of dipole moments is comparatively easy (for instance, by means of measurements on the temperature dependence of the dielectric constant, a method similar to the determination of the corresponding magnetic moments according to Curie-Langevin) proof of existence and an absolute determination of the electrical moments of inertia present greater difficulties.

A homogeneous electric field exerts a torque on a dipole molecule; a quadrupole molecule, under the same circumstances, is not affected. Quadrupole moments therefore are not observable with conventional measurements carried out by means of homogeneous fields. It would obviously be of interest to establish the existence of quadrupoles by a purely electrical method. This seems to be possible in the following manner.

(2) The indifference of the quadrupole to the applied electrical field no longer exists if the latter is inhomogeneous. In this case, quadrupoles must have a tendency for orientation similar to that of dipole moments in a homogeneous field. In order to establish such an orientation experimentally, an optical method suggests itself. Although the molecular orientation will be strongly disturbed by thermal motion, double refraction of the substance should be induced by an inhomogeneous field. However, another type of double refraction is already known to occur in homogeneous fields (Kerr effect). According to the theory of Langevin,[4] which seems to be essentially correct, this effect is based on the fact that molecules have different polarizabilities in different directions, and, consequently, tend to orient themselves in a homogeneous field. To prevent the Kerr effect from obscuring the presumed new effect, observations could be made at a point in the field where the field strength itself is zero while its spatial variation is as large as possible. An arrangement is sketched in the accompanying Figure 1, in the center of which

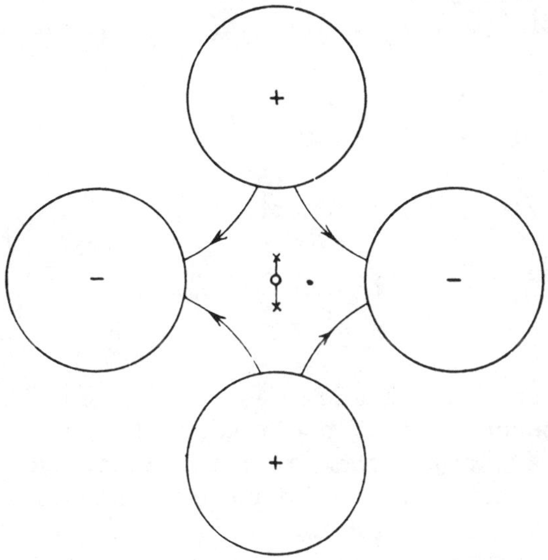

Figure 1

these requirements are met; the orientation of a simple quadrupole molecule, consisting of two negative charges marked by crosses and one positive charge marked by a circle, is also shown. A more detailed computation indicates that the effect might be just observable; experiments will have to be postponed to a later date.

(3) In principle, however, any phenomenon due to the electric field present inside the material which is generated by the molecules or atoms, furnishes a method. In the publication previously cited, quadrupole moments have been derived from van der Waals' adhesive forces as determined from observations on the equation of state.

The derivation of an exact equation of state from clearly prescribed premises concerning the forces effective between molecules is known to be an extremely difficult task, which has not as yet been solved. We must content ourselves with approximations such as van der Waals' equation. I already had the opportunity to refer to this point and to stress that the values for the mean electrical moment of inertia τ, calculated from van der Waals' constant of attraction a, can only be correct in order of magnitude. I also considered it advisable to point to a better method introduced by Kamerlingh-Onnes and developed in several publications by Keesom, which obtains the equation of state in the form of a systematically arranged series expansion. If the equation of state is written in the form:

$$p\mathfrak{v} = \mathfrak{A} + \frac{\mathfrak{B}}{\mathfrak{v}} + \frac{\mathfrak{C}}{\mathfrak{v}^2} + \dots \qquad (1)$$

we must compute the "virial coefficients" $\mathfrak{A}$, $\mathfrak{B}$, $\mathfrak{C}$... as functions of temperature (usually in the form of series arranged in decreasing powers of T). Written in this form, van der Waals' equation:

$$\left(p + \frac{a}{\mathfrak{v}^2}\right)(\mathfrak{v} - b) = RT \qquad (2)$$

would supply the virial coefficients:

$$\left.\begin{aligned} \mathfrak{A} &= RT \\ \mathfrak{B} &= RT\left(b - \frac{a}{RT}\right) \\ \mathfrak{C} &= RTb^2 \\ &- - - - - - \end{aligned}\right\} \qquad (2')$$

In his publications, Keesom has carried out a rigorous computation of the second virial coefficient $\mathfrak{B}$, on the assumptions that the molecules carry either a rigid electrical dipole or else rigid electrical quadrupole, and further behave like hard spheres as far as repulsion is concerned. Thus he arrives at a series expansion of the form:

$$\mathfrak{B} = RT\left(c_0 + \frac{c_2}{T^2} + \frac{c_3}{T^3} + \dots\right)$$

He has recently summarized the results of his computations.[5] It is characteristic for his expansions that the terms of the series in the first power of $1/T$ are consistently missing. This is a consequence of the rigidity of the electrical arrangement in the molecule, assumed by Keesom, and would mean, in opposition to van der Waals' equation and the form for $\mathfrak{B}$ following therefrom (compare Equation 2'), that the mutual attraction between molecules vanishes in the limit of high temperatures. I have tried to show in the previously cited paper that a universal attraction effective at all temperatures can only be understood if the assumption of rigidity of the arrangement is abandoned, and "polarizability" of the molecule is introduced in accordance with optical experience.

In the publication cited, Keesom recognizes this point of view as justified, and attempts to improve the series expansion for $\mathfrak{B}$, along the lines of my suggestion, by introducing the "polarizing forces." The term in the first power of $1/T$ now appears, as it should, (assuming correctness of the computation) with exactly the coefficient which follows from my value of a (see Equation 2').

By comparison of his improved expanion with the experimental values for hydrogen, Keesom concludes that, in the temperature range considered, the effect of the polarizing forces, though present, is of little practical importance compared with the forces which he previously discussed. Accordingly, he prefers different and smaller values than those given by me for the mean electrical moment of inertia (for instance, the value 3.55×10^{-26} for O_2 instead of 11.2×10^{-26}). Though I do not and did not attribute a high degree of accuracy to my figures any more than to van der Waals' equation, I feel compelled to doubt the recent figures by Keesom and to consider them too low, at least those for O_2 and N_2. The difference of opinion between Keesom and myself thus relates exclusively to the numerical values and to the gases which he uses for the comparison; we are of the same opinion as to fundamentals. For monatomic gases the orientation theory loses its meaning anyway (compare under section 5).

(4) The reason why I believe greater values of the mean electrical moment of inertia τ to be more probable is the following. Based on a purely experimental point of view and using only fundamental theory, Kamerlingh-Onnes established a mean empirical reduced equation of state based on experimental data on hydrogen, oxygen, nitrogen, ethyl ether, and isopentane. This equation is in practice highly superior to van der Waals' equation. Kamerlingh-Onnes writes this equation of state using the reduced variables p/p_k, $\mathfrak{v}/\mathfrak{v}_k$, and T/T_k in the form corresponding to equation (1):

$$\frac{p}{p_k}\frac{\mathfrak{v}}{\mathfrak{v}_k}=A+\frac{B}{\mathfrak{v}/\mathfrak{v}_k}+\frac{\Gamma}{(\mathfrak{v}/\mathfrak{v}_k)^2}+\cdots \qquad (3)$$

The "reduced virial coefficients" are again represented in the form:

$$\left.\begin{aligned} A &= K\frac{T}{T_k} \\ B &= K^2\frac{T}{T_k}\left\{\beta_0 + \beta_1\frac{T_k}{T} + \beta_2\left(\frac{T_k}{T}\right)^2 + \ldots\right\} \\ \Gamma &= K^3\frac{T}{T_k}\left\{\gamma_0 + \gamma_1\frac{T_k}{T} + \gamma_2\left(\frac{T_k}{T}\right)^2 + \ldots\right\} \\ & - - - - - - - - - - - - - \end{aligned}\right\} \quad (3')$$

with:

$$K = \frac{RT_k}{p_k v_k} \quad (3'')$$

The coefficients of expansion β_0, $\beta_1 \ldots, \gamma_0, \gamma_1, \ldots$, now have the dimension of pure numbers. Of these 5 × 5 = 25 are calculated from the experimental material and given, for instance, in the article by H. Kamerlingh-Onnes and W. H. Keesom (*Enz. d. mathemat. Wiss.*, *5*, 10, 730). We are here interested only in β_0 and β_1 for which Kamerlingh-Onnes finds the values:

$$\beta_0 = 0.117796, \; \beta_1 = -0.228038 \quad (4)$$

The original van der Waals' equation can, of course, also be written in the above reduced form. If the virial coefficients thus obtained are designated by A_w, B_w, ... we find:

$$\left.\begin{aligned} A_w &= K\frac{T}{T_k}, \\ B_w &= K^2\frac{T}{T_k}\left[\frac{p_k}{RT_k}b - \frac{p_k}{R^2T_k^2}a\frac{T_k}{T}\right] = \\ &= K^2\frac{T}{T_k}\left[\frac{1}{8} - \frac{27}{64}\frac{T_k}{T}\right] \\ \Gamma_w &= K^3\frac{T}{T_k}\left(\frac{p_k}{RT_k}\right)^2 b^2 = K^3\frac{T}{T_k}\frac{1}{64} \\ & - - - - - - - - - - - - - \end{aligned}\right\} \quad (5)$$

It is especially important for the theory of polarizing forces that in the empirical reduced equation of state the second virial coefficient B actually contains a negative term proportional to the first power of $1/T$ (β_1 negative). This proves strictly empirically the existence of polarizing forces.

A comparison of equation (5) with equations (3') and (4) shows immediately that the mean reduced equation of state required slightly different values for b and a than does van der Waals' equation; the new value for b is obtained by multiplying van der Waals' value by:

$$8\beta_0 = 8 \cdot 0.117796 = 0.942$$

and the new value for a by multiplying van der Waals' value by:

$$-\frac{64}{27}\beta_1 = \frac{64}{27} 0.228038 = 0.540$$

Previously, the average electrical moment of inertia was computed from the formula (equation 21, *loc. cit.*, page 182):

$$\tau^2 = \frac{10}{9} \frac{1}{NP_0} a d^5$$

which is still valid. If now not van der Waals' classical equation, but instead the practically excellent mean reduced equation of state is used, the new values for τ will follow from the previously given values by multiplication by the factor:

$$0.540^{1/2}\, 0.942^{5/6} = 0.699$$

Adaptation of the theory to the empirical equation of Kamerlingh-Onnes therefore reduces the electrical moments of inertia by 30%. The following table contains the values of τ for the three gases (H_2, O_2, N_2); values based on van der Waals' constants are given in the second column, those calculated by Keesom in the third, and those based on the attraction constant in the empirical, reduced equation of state in the fourth.

Table A

Values for $\tau \times 10^{26}$

H_2	3.20	2.03	2.14
O_2	11.2	3.55	7.84
N_2	13.3	3.86	9.30

Only for hydrogen, investigated with special care by Keesom, is the agreement satisfactory; we find the electrical moment of inertia for O_2 and N_2 to be considerably greater.

To summarize, the combination of my theory with the empirical reduced equation of state provides the simple relation:

$$\tau^2 = 0.00209 \frac{RT_k}{NP_0} \left(\frac{RT_k}{p_k}\right)^{5/3} \qquad (6)$$

for the calculation of the mean electrical moment of inertia τ. (R = 8.31 × 10^7 = gas constant, N = 6.06 × 10^{23} = Loschmidt's number, Γ_k = critical temperature, p_k = critical pressure, P_0 = molar refraction.) The figures in the last column of the preceding table may be considered as having been computed by means of this formula.

(5) It may seem surprising that I prefer the relation expressed by equation (6) involving the empirical reduced equation of state. It is not just because this relation is so simple; my primary reason is that I doubt the strict validity of Keesom's series expansion which cannot be mended even by the introduction of polarizing forces. There are three principal points to which I want to draw attention.

1. Only for large distances is it possible to describe the average electric field of a molecule by that of a quadrupole. For distances which are not large compared to the distances between the separate charges in the molecule it is necessary to include higher terms of the expansion of the potential. Such small distances occur at every collision.

2. Keesom's series expansion is obtained by application of the principles of statistical mechanics. Thus it contains implicitly the statement, for instance, that the specific heat (at constant volume) of hydrogen has the value $5R/2$ at all temperatures, whereas Eucken's measurements[6] are known to show that the quantum deviations from the law of equipartition become noticeable at temperatures slightly below 0°C. This discrepancy is even more pronounced in the case of monatomic gases. Their atoms do not have any rotational energy, which is temperature dependent, which in quantum theory is a consequence of their small moment of inertia. Accordingly, Keesom's series expansion is void of any meaning for monatomic gases, and the only forces which can explain the existing molecular attraction are the polarization forces.

3. The series expansion used by Keesom treats colliding molecules as hard spheres. It is most improbable that this assumption corresponds to reality. No objection can be made to using this assumption as a first approximation. If, however, the accuracy of the calculation is such that the fourteenth term is retained in the expansion, as Keesom does, there is, in my opinion, no guaranty for its reliability.

Considering the importance of these three remarks with regard to their theoretical evaluation, it will be seen immediately that remark 3 is the most essential. It will be clear that the criticism contained in the first two remarks can possibly be eliminated. Neither the more exact description of the electric field surrounding a molecule nor the transition from classical statistics to quantum theory presents fundamental difficulties today. If it is admitted that the attractive forces between the molecules are exclusively electrical in origin (and I see no reason to doubt this any more), the necessary completion of the theory can be expected from future work.

The circumstances concerning the third remark are entirely different. It deals with the problem of whether it is possible to understand not only the attractive but also the repulsive forces between the molecules and to interpret them as a consequence of the assumed electrical structure. This question appears to me to be of primary importance, the more so as it plays an essential part not only in the equation of state but also with regard to solids. Thus for instance, the fundamental instability of the atomic arrangement appearing in the beautiful investigations by Born on the forces in the crystal lattice could be overcome if only the theoretical necessity of repulsive forces of a special kind not provided for in the static theory of the potential could be revealed. In conclusion, I would like to show how this can be done without introducing more calculations than are required for the understanding of the essential points.

(6) All experiments indicate that a universal repulsion exists between molecules. A molecule is a complicated structure; the first question to be answered in the search for an explanation of the universal repulsion is whether the complicated structure as such is responsible for the repulsion. Previous experiments by Lenard[7] on the penetration of cathode rays through matter and the experiments by J. Franck and co-workers[8] have proved conclusively that a single charge, an electron, is also in general repelled by a molecule if sufficiently close. It is known that even the movements of slow electrons in helium could be interpreted by the assumption that the helium atoms behave as hard spheres when colliding with an electron. Consequently, we are in a position to simplify our problem and to search for a repulsive force between a free electron and a molecule.

If we consider a molecule to be a structure consisting of charges but neutral in its entirety, the repulsion demanded by experiment appears to be missing. If a sphere of radius r is drawn around such a structure, and the surface integral of the potential $\int \varphi d\sigma$ on this sphere is determined, the value:

$$\int \varphi d\sigma = 4\pi r \sum e$$

is found, where Σe designates the total sum of all enclosed charges. Thus if the molecule is neutral, the surface integral will vanish. Next to regions where the potential assumes negative values, to which an electron cannot be moved without doing work, there will always be regions with positive potential. If the electron is moved here, work is gained so that within the realm of these considerations, a molecule or atom would have to behave differently than is indicated by the experiment.

Admittedly, for all preceding deliberations the assumption that a molecule consists of charges at rest was sufficient. In reality, however, the charges are in motion and generate a field surrounding the molecule varying continuously in direction and magnitude. The assumption of a system of charges at rest amounts to calculations based on the average value of the field strength with respect to time, as seems justified by the rapid rate of change of the field. The question now arises as to whether, as a consequence of this variation with time and in spite of the short duration of a period, additional forces may not appear which could account for the observed repulsion. From this point of view it is of interest to compare the instantaneous values of the field strength in the neighborhood of a molecule with their time average. It is immediately seen that in general the instantaneous values greatly exceed their average.

A hydrogen atom is illustrated in Figure 2, in which an

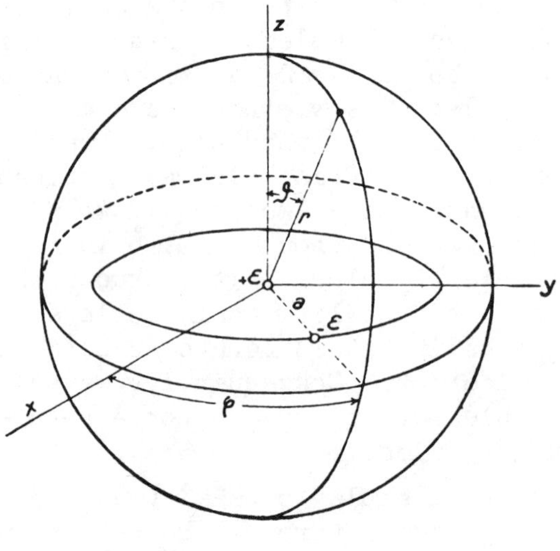

Figure 2

electron carrying a charge $-\varepsilon$ revolves with angular velocity ω in a circle with radius a around a nucleus carrying a charge $+\varepsilon$ in the first quantum orbit. A sphere of radius r is drawn with the nucleus as a center, and the position of a point thereon is specified by the angles ϑ and φ. In first approximation, such an atom will behave as an alternating dipole and generate a field which can be derived from the potential:

$$\Phi_1 = -\frac{\varepsilon a}{r^2} \sin\vartheta \cos(\varphi - \omega t)$$

Only in second approximation do we obtain a potential which is constant in time. This part has the value:

$$\Phi_2 = \frac{\varepsilon a^2}{r^3}\left(\frac{1}{2} - \frac{3}{4}\sin^2\vartheta\right)$$

and corresponds to the field of a nucleus surrounded by a ring which carries a uniform layer of negative electricity.

The mean square of the field strength $\overline{\mathfrak{E}_1^2}$ calculated from Φ_1 is equal to:

$$\overline{\mathfrak{E}_1^2} = \frac{\varepsilon^2}{a^4}\left(\frac{a}{r}\right)^6\left(\frac{3}{2}\sin^2\vartheta + 1\right)$$

and from Φ_2:

$$\overline{\mathfrak{E}_2^2} = \frac{\varepsilon^2}{a^4}\left(\frac{a}{r}\right)^8 \frac{9}{4}\left(1 - 2\sin^2\vartheta + \frac{5}{4}\sin^4\vartheta\right)$$

At a distance $r = 10a = 5.3 \times 10^{-8}$ cm. we thus obtain for the maximum value of $\sqrt{\overline{\mathfrak{E}_2^2}}$ 8×10^6 volt per centimeter and for the maximum value of $\sqrt{\overline{\mathfrak{E}_2^2}}$ 7.6×10^5 volt per centimeter. In this example the amplitude of the field strength variations is approximately ten times the time average (incidentally, the large absolute values are of interest). Having established that the varying part of the field strength preponderates (and the more so the greater the distance from the molecule), we now consider the average force exerted on a free electron by a field varying in time and space.

We shall divide the field into a constant part and a part variable in time in such a manner that the average value of the field strength $\mathfrak{E}$ associated with the latter vanishes. An electron of mass μ and charge $-\varepsilon$ will then be displaced in this variable field by an amount σ given by the equation:

$$\mu \frac{d^2\sigma}{dt^2} = -\varepsilon\mathfrak{E} \qquad (7)$$

When σ can be assumed sufficiently small, the instantaneous force $\mathfrak{K}$ on the electron is found from Taylor's expansion of the field in the neighborhood of the rest position. We thus obtain:*

$$\mathfrak{K} = -\varepsilon\mathfrak{E} - \varepsilon(\sigma\nabla)\mathfrak{E}$$

In general, however, we have for any two vectors $\mathfrak{A}$ and $\mathfrak{B}$:

$$\nabla(\mathfrak{A}\mathfrak{B}) = (\mathfrak{A}\nabla)\mathfrak{B} + (\mathfrak{B}\nabla)\mathfrak{A} + [\mathfrak{A}\,\mathrm{rot}\,\mathfrak{B}] + [\mathfrak{B}\,\mathrm{rot}\,\mathfrak{A}]$$

*∇ is the known symbolic vector with the components: $\partial/\partial x$, $\partial/\partial y$, $\partial/\partial z$.

Putting $\mathfrak{A} = \sigma$ and $\mathfrak{B} = \mathfrak{E}$, and recalling that rot $\mathfrak{E} = 0$ so that from equation (7) rot $\sigma = 0$, we obtain:

$$\nabla(\sigma\mathfrak{E}) = (\sigma\nabla)\mathfrak{E} + (\mathfrak{E}\nabla)\sigma$$

If we now take the mean value with respect to time, we obtain in view of equation (7):

$$\overline{(\sigma\nabla)\mathfrak{E}} = -\frac{\mu}{\varepsilon}\overline{(\sigma\nabla)\frac{d^2\sigma}{dt^2}} = -\frac{\mu}{\varepsilon}\overline{\left(\frac{d^2\sigma}{dt^2}\nabla\right)\sigma} = \\ = \overline{(\mathfrak{E}\nabla)c}$$

so that finally:

$$\overline{(\sigma\nabla)\mathfrak{E}} = \tfrac{1}{2}\nabla\overline{(\sigma\mathfrak{E})}$$

Since $\overline{\mathfrak{E}} = 0$, we find for the time average of the force exerted on the electron:

$$\overline{\mathfrak{K}} = -\nabla\frac{\varepsilon}{2}\overline{(\sigma\mathfrak{E})} \tag{8}$$

Thus the mean force is represented as the gradient of a potential, *i.e.*, the electron behaves as if it had a potential energy U amounting to:

$$U = \frac{\varepsilon}{2}\overline{(\sigma\mathfrak{E})} \tag{9}$$

relative to the molecule generating the field. This potential energy has a simple meaning. By scalar multiplication of both sides of the equation of motion with σ, formation of the time average (integration with respect to t and division by the time interval θ), and partial integration, we arrive at:

$$\varepsilon\overline{(\sigma\mathfrak{E})} = -\mu\frac{1}{\Theta}\int_0^\Theta dt\left(\sigma\frac{d^2\sigma}{dt^2}\right) = \\ = -\mu\frac{1}{\Theta}\left[\left(\sigma\frac{d\sigma}{dt}\right)\right]_0^\Theta + \mu\frac{1}{\Theta}\int_0^\Theta\left(\frac{d\sigma}{dt}, \frac{d\sigma}{dt}\right)dt$$

Since the first term disappears for a sufficiently long period of time θ, this reduces to:

$$\varepsilon\overline{(\sigma\mathfrak{E})} = 2\overline{T}$$

where $\overline{T}$ designates the mean kinetic energy of the electron. Thus we conclude that: the pulsating electron behaves as if it possessed a potential energy with respect to the molecule, equal to the kinetic energy of its pulsations:

$$U = \overline{T} \qquad (9')$$

Since the latter energy is necessarily positive, the apparent potential energy is also positive, and since the alternating field increases as the molecule is approached, this is only another way of stating the fact that a universal repulsion exists between an electron and a molecule which is greater the closer the electron approaches the molecule.*

(7) Here I am only concerned with the more fundamental features of this repulsive force. It may therefore suffice to draw attention briefly to the fact that for the case of the hydrogen atom the order of magnitude of this force is adequate to explain the molecular diameter observed in the kinetic theory of gases. If the electron approaches the H atom in the direction of the Z axis (see Figure 2), the value:

$$\overline{T} = \frac{1}{2}\frac{\varepsilon^2}{a}\left(\frac{a}{r}\right)^6$$

is found for the mean kinetic energy of the pulsations caused by the alternating field, where r denotes the distance between electron and nucleus. Hence the electron possesses a positive potential energy with respect to the atom, given by:

$$U = \frac{1}{2}\frac{\varepsilon^2}{a}\left(\frac{a}{r}\right)^6 \qquad (10)$$

as a consequence of its pulsations.

On the other hand it is attracted statically more strongly by the nucleus than it is repelled by the rotating electron further away, and in a first approximation, these forces correspond to a negative potential energy U' amounting to:

$$U' = -\frac{1}{2}\frac{\varepsilon^2}{a}\left(\frac{a}{r}\right)^3 \qquad (10')$$

*An illustrative interpretation for the repulsive forces results, for instance, from a consideration of the following simple, special case. A center generates in its vicinity a radial field periodic with time. An electron will execute radial oscillations in this field in such a manner that (in view of the fundamental laws of mechanics), seen from its central position, it is closer to the center of force during the interval of the repulsive forces, and further away from the center of forces during the interval of attraction. If the amplitude of the alternating field increases with decreasing distance from the center, the time average repulsion will prevail in view of the spatial variations.

If it were permissable to calculate with these values of the energies for short distances, then:

$$U + U' = 0$$

would hold for $r = a = 0.50 \times 10^{-8}$ cm., which value would represent the "radius" of the H atom, since the kinetic energy of the thermal movement is negligibly small compared with the energy ε^2/a.

It will naturally depend on the structure of the molecule whether the positive potential energy exceeds the negative potential energy in all directions of approach, or whether there exists a direction of approach for which the negative potential energy continuously controls. Accordingly, we will speak, in experimental terms, either of atoms (molecules) without electron affinity which behave as hard spheres at impacts, or else we will assign an electron affinity to the atom (molecule) which will be the larger the larger the (solid) angle within which the negative potential energy predominates. In conclusion, we mention that these arguments may be extended to forces between entire atoms and molecules, thus suggesting a possibility of completing the potential forces of Born's lattice by those discussed here. In this way, an explanation of the stability of crystal structure on the basis of purely electrical forces seems possible.

Bibliography

1. P. Debye, *Physik. Z.*, *21*, 178 (1920).
2. P. Debye, *ibid.*, *13*, 97 (1912); M. Iona, *ibid.*, *20*, 14 (1919).
3. P. Debye, *Verhandl. deut. physik. Ges.*, *15*, 777 (1913); J. Timmers, *Dissertation*, Utrecht, 1914.
4. P. Langevin, *Le Radium*, *7*, 233, 249 (1910).
5. W. H. Keesom, *Physik. Z.*, *22*, 129 (1921).
6. A. Eucken, *Berl. Ber.*, *1912*, 141.
7. Lenard, *Quantitatives über Kathodenstrahlen aller Geschwindigkeiten*, Heidelberg, 1918.
8. J. Franck and G. Hertz, *Physik. Z.*, *17*, 409, 430 (1916); *20*, 132 (1919) (summarizing presentation).

THE BASIC LAWS OF ELECTRIC AND MAGNETIC EXCITATION FROM THE POINT OF VIEW OF THE QUANTUM THEORY
(Die Grundgesetze der elektrischen und magnetischen Erregung vom Standpunkte der Quantentheorie)

P. Debye

Translated from
Physikalische Zeitschrift, Vol. 27, pages 67–74, 1926.

(1) The law established by P. Curie[1] and interpreted by P. Langevin[2] according to which the paramagnetic (mass) susceptibility χ_m is inversely proportional to the absolute temperature:

$$\chi_m = \frac{C}{T} \qquad (1)$$

may be called the basic law of paramagnetic excitation. In my opinion, the law I suggested[3] according to which the dielectric (mass) susceptibility χ_e can be represented in the form:*

$$\chi_e = A + \frac{B}{T} \qquad (2)$$

may be called its analog for electrical excitation. Both laws hold only as long as molecular interaction may be neglected; in this region both laws have been confirmed experimentally. The deviations which occur when this condition is not met, and also a possible interpretation of Weiss' molecular field, will be dealt with in the last section of this note.

*The dielectric (mass) susceptibility is defined as the electric moment which the unit of mass assumes in a field of strength unity. For a medium of dielectric constant D and density ρ, we obtain:

$$\chi_e = \frac{D-1}{4\pi}\frac{1}{\rho}$$

The slight difference in form of the two laws is entirely superficial. As a matter of fact, an additive constant should also be added, in general, to the right side of Curie's equation to allow for diamagnetic magnetization which is independent of temperature. In the magnetic case, however, the term C/T predominates in most practical cases and the constant therefore is frequently omitted. The two laws are directly related, the first with the Zeeman effect, the second with the Stark effect: this relationship appears in the statement of the two effects which involves not only splitting which is linear in the field, but also quadratic. Apparently it is not commonly recognized that the temperature law (2) constitutes a very general consequence of the energetic molecular behavior which is measured by these effects. It therefore seems worthwhile to make a few remarks on this subject despite their simplicity. Their value lies primarily in the fact that they seem to offer a direct way to the comprehension of deviations from the basic laws.

In order to be able to start from first principles, complicated as little as possible with unnecessary additional hypotheses, it seems best to proceed from a general statistical formulation of the problem. To avoid any misunderstanding and to make clear the kind of statistics which are being used, I start with its explicit formulation.

(2) Consider a system which may occur in s states $1 \ldots i \ldots s$, each associated with definite values for the energy $u_1 \ldots u_i \ldots u_s$ and definite statistical weights $\rho_1 \ldots \rho_i \ldots \rho_s$. If the system is observed at Z instants of time, or if a number Z of such systems are observed simultaneously, and if we find that $z_1 \ldots z_i \ldots z_s$ of these systems are in the states $1 \ldots i \ldots s$, respectively, then, according to Boltzmann, the probability of this distribution is given by:*

$$W = Z! \prod_1^s \frac{\varrho_i^{z_i}}{z_i!} \qquad (1)$$

If we next seek that distribution for which W is a maximum, for a given number of systems with a given energy, ZU, we find the familiar result that z_i must assume the definite values $z_i = \overline{z_i}$, represented by the formula:

*If the weights are so defined that:

$$\sum_1^s \rho_i = 1$$

W corresponds to the conventional definition of probability and is always less than unity. The sum of all probabilities of all possible distributions is equal to unity, since W is simply a term in the known expansion of:

$$(\varrho_1 + \ldots + \varrho_i \ldots + \varrho_s)^{z_1 + \ldots + z_i + \ldots + z_s} = 1^z$$

$$\overline{z}_i = \alpha \varrho_i e^{-\beta u_i} \qquad (2)$$

The two constants α and β are determined from Z and ZU by solving equations:

$$\left.\begin{aligned} Z &= \alpha \sum_1^s \varrho_i e^{-\beta u} \\ ZU &= \alpha \sum_1^s \varrho_i u_i e^{-\beta u_i} \end{aligned}\right\} \qquad (3)$$

These steps complete the solution of the formal statistical problem. But physical significance is only obtained through the use of Boltzmann's principle, according to which the associated entropy ZS is defined by the expression:

$$ZS = k \log \overline{W} \qquad (4)$$

where $\overline{W}$ is that value of W according to equation (1) which is obtained* when $z_i = \overline{z_i}$. From equation (1) this definition is equivalent to the formula:

$$S = k(\log Z - \log \alpha + \beta U)$$

or if the sum:

$$\sigma = \sum_1^s \varrho_i e^{-\beta u_i} \qquad (5)$$

is designated by σ, also equivalent to the representation:

$$S = k\left(\log \sigma - \frac{\beta}{\sigma} \frac{\partial \sigma}{\partial \beta}\right) \qquad (6)$$

Since from equation (6):

$$\frac{\partial S}{\partial \beta} = k\beta \frac{\partial}{\partial \beta}\left(-\frac{1}{\sigma}\frac{\partial \sigma}{\partial \beta}\right) \qquad (6')$$

and from equation (3):

$$U = -\frac{1}{\sigma}\frac{\partial \sigma}{\sigma \beta} \qquad (6'')$$

*Classical thermodynamics may consistently be defined as the relations obtaining in the limit for $Z = \infty$. It is known that this also leads to equation (4) if the distribution experiment is assumed to be repeated many times and the entropy is defined as the average value of all log W which appear.

it follows from the thermodynamic definition of entropy that in general:

$$\beta = \frac{1}{kT} \qquad (7)$$

(T = absolute temperature) must hold.*

Ordinarily the result is summarized by inferring from equations (6), (3), and (7), as follows:

To find the free energy $F = U - TS$, the "sum of state":

$$\sigma = \sum_1^s \varrho_i e^{-\frac{u_i}{kT}} \qquad (8)$$

must first be calculated, and then the free energy is simply:

$$F = U - TS = -kT \log \sigma \qquad (9)$$

This course of reasoning is obviously quite independent of any special assumptions regarding the possible states, in agreement with the Maxwell-Boltzmann distribution law, expressed by the familiar relation:

$$\bar{z}_i = a \varrho_i e^{-\frac{u_i}{kT}}$$

Thus the difference between the points of view of the ergodic hypothesis and quantum theory first becomes apparent when the statistical weights ρ_i must be determined.

In our cases of electric and magnetic excitation, the statements presented in equations (8) and (9) must be amplified to

*If the terms in equation (5) are arranged in the order of increasing u_i so that u_1 represents the lowest energy value present, we obtain in the limit for T = 0:

$$\sigma = \varrho_1 e^{-\beta u_1}$$

and from equation (6):

$$S = k \log \varrho_1$$

This representation satisfies Nernst's theorem of heat in its limited form, since S remains finite for T = 0. It is, of course, only necessary that the possible states be distributed discretely.

The postulate that the same statistical weight ρ_1 is associated with the different systems at the state of the lowest energy would correspond to Nernst's theorem of heat in its general form. Whether the ρ_1 are then normalized so that ρ = 1, and thus S = 0, appears to be immaterial.

some extent. If a body is located in a magnetic field H and has acquired a magnetic moment M, it will have a potential energy $-MH$ with respect to the field. Designating by U the *internal* energy of the body, not merely U alone but the *total* energy $(U - MH)$ must be inserted into the second equation (3). The expressions for u_i will contain the field strength H as parameter. If, as in equation (6'), the quantity $\partial S/\partial\beta$ is calculated, $(\partial S/\partial\beta)d\beta$ is the differential of the entropy for constant field strength. Furthermore the left-hand side of the equation corresponding to equation (6') does not contain U but $(U - MH)$, the differential of which, under the same assumption, is $(dU - HdM)$. A combination of these two equations gives:

$$dS = k\beta(dU - HdM)$$

and since the differential of the magnetic work done on the body is equal to HdM, we have in accordance with thermodynamics,

$$dS = \frac{1}{T}(dU - HdM)$$

Thus the relation (7):

$$\beta = \frac{1}{kT}$$

remains valid.

Since, however, the left-hand side of equation (6') contains $(U - MH)$ instead of U, we obtain from equation (6) not the ordinary free energy but rather the thermodynamic potential:

$$\Phi = U - TS - HM \qquad (10)$$

as:

$$\Phi = -kT \log \sigma \qquad (11)$$

All this amounts to the statement that the logarithm of the sum of state equation (8) now defines the thermodynamic potential in accordance with equations (10) and (11) instead of the free energy.

If Φ is found,

$$S = -\frac{\partial \Phi}{\partial T}, \quad M = -\frac{\partial \Phi}{\partial H} \qquad (12)$$

follow at once from equation (10). The latter quantity M is the one which is of interest to us. The essential quantity is therefore the sum of state defined by equation (8), which determines the thermodynamic potential in accordance with equation (11) and from which the magnetic moment is derived by differentiation in accordance with equation (12).

Similar considerations hold for electric excitation, where electrical field strength E must be substituted for H in order to find the electrical moment M of the body.

(3) Let us now consider a system consisting of N molecules, assumed entirely independent of one another, which are subjected to an external electric or magnetic influence. In this case, the sum of state equation (8) will be equal to the Nth power of a sum relating to a *single* molecule only. Consequently, according to equation (11), the thermodynamic potential will assume a value equal to N times an elementary value which is computed from the sum for a single molecule. Thus we may speak of the sum of state and the thermodynamic potential of a single molecule, and obtain the corresponding quantities for the total system by raising them to the Nth power by multiplication by N, respectively. In order to obtain the sum of state σ in the magnetic case for a single molecule, we must know the energy levels u_i which the molecule can assume in the magnetic field. Let W designate the energy level in zero field and assume that s levels are available for a field H where the ith level is assigned the energy w_i dependent on H. Under these conditions, the sum of state referred to the single molecule becomes, according to equation (8),

$$\sigma = e^{-\frac{W}{kT}} \sum_1^s \varrho_i e^{-\frac{w_i}{kT}} \quad (13)$$

It follows from equation (11) that:

$$w_i = -\mu_i H - \frac{\alpha_i}{2} H^2 + \ldots \quad (14)$$

and from equation (12) that:

$$m = -\frac{\partial \Phi}{\partial H} = kT \frac{\partial}{\partial H} \log \sum_1^s \varrho_i e^{-\frac{w_i}{kT}} \quad (13'')$$

if m denotes the mean magnetic moment which a single molecule assumes in the field H. We now introduce the assumption that the energy values w_i of magnetic origin can be expanded in powers of the field strength as follows:

$$\Phi = W - kT \log \sum_1^s \varrho_i e^{-\frac{w_i}{kT}} \quad (13')$$

It would be meaningless to include higher powers in the field strength, as will be seen by inspection of equation (13"), as long as it is found experimentally that m is proportional to the field strength. If the sum in equation (13) is expanded in powers of H, it follows that:

$$\Sigma \varrho_i e^{-\frac{w_i}{kT}} = \Sigma \varrho_i + \frac{H}{kl} \Sigma \varrho_i \mu_i + \\ + \frac{H^2}{2kT} \Sigma \varrho_i \left(\alpha_i + \frac{\mu_i^2}{kT}\right)$$

Hence, from equation (13'),

$$\Phi = W - kT\left\{\log \Sigma \varrho_i + \frac{H}{kT}\frac{\Sigma \varrho_i \mu_i}{\Sigma \varrho_i} + \right.$$
$$+ \frac{H^2}{2kT}\frac{\Sigma \varrho_i \alpha_i}{\Sigma \varrho_i} + \frac{H^2}{2k^2T^2}\left[\frac{\Sigma \varrho_i \mu_i^2}{\Sigma \varrho_i} - \right.$$
$$\left.\left. - \left(\frac{\Sigma \varrho_i \mu_i}{\Sigma \varrho_i}\right)^2\right]\right\}$$

and finally from equation (13"):

$$m = \frac{\Sigma \varrho_i \mu_i}{\Sigma \varrho_i} + H\left\{\frac{\Sigma \varrho_i \alpha_i}{\Sigma \varrho_i} + \frac{1}{kT}\left[\frac{\Sigma \varrho_i \mu_i^2}{\Sigma \varrho_i} - \right.\right.$$
$$\left.\left. - \left(\frac{\Sigma \varrho_i \mu_i}{\Sigma \varrho_i}\right)^2\right]\right\} \qquad (15)$$

According to the result, equation (15), a magnetic moment exists in general even for $H = 0$. It vanishes only if the statistical weights are so distributed that $\Sigma\rho_i \mu_i$ also vanishes. This is apparently the usual case. Assuming:

$$\Sigma \varrho_i \mu_i = 0$$

we obtain instead of equation (14):

$$m = H\left\{\frac{\Sigma \varrho_i \alpha_i}{\Sigma \varrho_i} + \frac{1}{kT}\frac{\Sigma \varrho_i \mu_i^2}{\Sigma \varrho_i}\right\} \qquad (15')$$

Thus it has been proved for the general case that susceptibility is a linear function of $1/T$, provided the approximation in equation (14') for the energy as a quadratic term is sufficient, and provided the molecules may be considered independent of one another. An equivalent proposition obviously holds for electric excitation.

Our next problem is the determination of the expansion coefficients in equation (14'). Here we may use Bohr's theory of spectra by selecting the term which gives the energy for the normal state in an electric or magnetic field. The first term in the expression (14) for the energy, which leads to the temperature dependent term in equation (15'), is to be interpreted as the energy of a magnetic or electric moment whose orientation is quantized. The second term in equation (14'), which, as the first term in equation (15') suggests, is connected with the polarizability of the molecule, is to be interpreted as a distortional energy of the electron orbits. The method of derivation, however, implies that the relations are completely independent of the models which may be associated therewith. In the term representation the linear term in equation (14') corresponds to the linear

Zeeman or Stark effect and the quadratic term to the quadratic effect. Furthermore the treatments by Pauli,[4] Epstein,[5] Gerlach,[6] and Sommerfeld[7] of Curie's constant are based on a relation which is identical with equation (15'), provided $\Sigma\rho_i\alpha_i = 0$. This is evidently permissible for the magnetic case; in the electric case, however, when discussing the dielectric constant, neglect of the polarizability, *i.e.*, the quadratic Stark effect, is not justifiable.[8]

Cases can be imagined for practically attainable field strengths where the dependence on field strength expressed by equation (14') no longer suffices. Let us consider, using an example which has been treated mathematically in detail, the hydrogen atom in a high quantum state. In accordance with Sommerfeld's fine structure it occurs in various modifications whose energy differences are so small that they may all be present simultaneously in comparable quantities.* If such atoms are exposed to an electric field, a shift of energy levels proportional to the square of the field strength will, according to computations by Kramers,[9] occur first. This will gradually change with increasing field strength to a shift which is a linear function of the field and corresponds to the ordinary Stark effect. These phenomena all appear within the experimentally accessible range of field strengths. For such atoms, the complicated law for w_i, which expresses this shift in energy levels, would have to be inserted into equation (13), and the mean moment would have to be found from equation (13"). This would correspond to the strange case that for increasing field strength the electric susceptibility increases from an amount expressed by the first term in equation (15'), through a transitional region, to an amount expressed by the second term of equation (15'). It is hardly to be expected that experiments with hydrogen atoms in the higher quantum states will be possible within the near future. But it does not seem to be excluded *a priori* that other atoms might exist for which similar conditions are within the realm of experiments.

The magnetic analog to these conditions is correspondingly related to the Paschen-Back effect. It is interesting, in connection with this point of view, that Woltjer and Kamerlingh-Onnes[10] found in some instances very strange curves for the relationship between moment and field strength at low temperatures; for instance, $COCl_2$ and $NiCl_2$ exhibit a transition from a smaller to a larger susceptibility at 2.3° absolute, when the field is increased from zero to 10,000 gauss. However, further experiments are necessary before an analogy of these experiments to the Paschen-Back effect can be established.

*The order of magnitude of the difference in wave length is 0.1Å.; this corresponds to a difference in energy of approximately 10^{-16} erg. Setting this energy equal to kT leads to a comparison temperature of 1 degree.

It is not certain whether an atom in the normal state can be found which possesses an electric moment in the sense used in equation (14). However, such moments have been established for molecules consisting of more than one atom. It should be noted that the above discussion does not apply rigorously to the latter case. So far it has been assumed permissible to proceed from a single normal state with energy W. This is correct to sufficient approximation only if the other possible states differ from W by energies large compared with kT. This is not the case for molecules because the different possible ground states differ only by the rotational energy of the whole structure; since the moments of inertia are rather large, the orresponing energy levels differ only by amounts which are usually considerably smaller than kT. A possibility of retaining the formula corresponding to equation (15') for the electrical case exists when the term splitting is the same for all rotational energy levels. If the electric field produces a series of additional energies w_i $(i = 1 \ldots s)$ for each rotational energy level W_j, the sum of state is represented in the form:

$$\sigma = \sum_{(j)} \left[e^{-\frac{W_j}{kT}} \sum_1^s \varrho_i e^{-\frac{w_i}{kT}} \right]$$

which can only be written as a product:

$$\sigma = \sum_{(j)} e^{-\frac{W_j}{kT}} \cdot \sum_1^s \varrho_i e^{-\frac{w_i}{kT}}$$

if the w_i are independent of the numeral j specifying the rotational energy level. An entirely different possibility, which strangely enough leads to the same temperature law, is discussed by Pauli[11] for special molec les of the HCl type. The energy of each level is a quadratic function of the field strength, and therefore the moment corresponding to a level is independent of temperature. This moment, however, decreases with increasing energy of the rotational energy level: At the same time the distribution of the molecules with respect to the levels is, of course, dependent on temperature. Together these two relations create the old temperature law for the average excited moment, which is amply confirmed experimentally.[12] The energy levels of the rotation manifest themselves in band spectra; the temperature dependence of the dielectric constant is thus directly related to the Stark effect of the band lines.[13]

(4) The last remark is related to a possible interpretation of the deviations from Curie's law observed in liquids and solids. P. Weiss, on the basis of his theory of the molecular field, derived the law:

$$\chi_m = \frac{C}{T - \Theta} \qquad (16)$$

to replace equation (1); it has, in fact, been shown that in many instances $1/\chi_m$ is a linear function of temperature in which an additive constant θ appears. It should, however, be noted that in many instances θ is negative which, in the interpretation given by Weiss, corresponds to a negative molecular field, *i.e.*, to a field opposed to that of the polarization. Furthermore, the field is so strong that it is impossible to account for it as a consequence of the mutual magnetic effects. Moreover, law (16) no longer holds and the temperature dependence becomes more complicated, a phenomenon called the kryomagnetic anomaly by Kamerlingh-Onnes. For these and other reasons, it is doubtful whether one should attempt to derive formula (16) on the basis of the hypothesis of Weiss.[14] It may be that the law in the form of equation (16) constitutes only an approximation of a more general temperature law, similar to the volume correction used by van der Waals in his equation of state.[15]

In fact, formulas of this type for the temperature dependence are obtained provided a somewhat more general assumption than that in (3) is made regarding the energy states in the magnetic field. There it was assumed that the energy of the separate particle can be represented by:

$$W + w_i \qquad (17)$$

where w_i was to be calculated from equation (14). The different values for w_i are imagined to be caused by the different orientations of the electron system with respect to the external field. If, however, the particle is not completely free, but other atoms are nearby, as in liquids and crystals, it appears almost obvious that different energies should be associated with the different orientations even for zero field. We will try to express this by assigning the energy:

$$W + \varepsilon_i + w_i \qquad (17)$$

to a quantum state i, where ε_i is the orientation energy just mentioned, while w_i is given by equation (14) as before.

If equation (17') is used, the sum of state becomes:

$$\sigma = e^{-\frac{W}{kT}} \sum^{s} \varrho_i e^{-\frac{\varepsilon_i}{kT}} e^{-\frac{w_i}{kT}}$$

or expanded in powers of H:

$$\sigma = e^{-\frac{W}{kT}} \left\{ \sum \varrho_i e^{-\frac{\varepsilon_i}{kT}} + \frac{H}{kT} \sum \varrho_i \mu_i e^{-\frac{\varepsilon_i}{kT}} + \frac{H^2}{2kT} \sum \varrho_i e^{-\frac{\varepsilon_i}{kT}} \left(\alpha_i - \frac{\mu_i^2}{kT} \right) \right\} \qquad (18)$$

From equation (11), we now obtain:

$$\Phi = W - kT\left\{\log \sum \varrho_i e^{-\frac{\varepsilon_i}{kT}} + \frac{H}{kT}\frac{\sum \varrho_i \mu_i e^{-\frac{\varepsilon_i}{kT}}}{\sum \varrho_i e^{-\frac{\varepsilon_i}{kT}}} + \frac{H^2}{2kT}\frac{\sum \varrho_i \alpha_i e^{-\frac{\varepsilon_i}{kT}}}{\sum \varrho_i e^{-\frac{\varepsilon_i}{kT}}} + \right.$$
$$\left. + \frac{H^2}{2k^2T^2}\left[\frac{\sum \varrho_i \mu_i^2 e^{-\frac{\varepsilon_i}{kT}}}{\sum \varrho_i e^{-\frac{\varepsilon_i}{kT}}} - \left(\frac{\sum \varrho_i \mu_i e^{-\frac{\varepsilon_i}{kT}}}{\sum \varrho_i e^{-\frac{\varepsilon_i}{kT}}}\right)^2\right]\right\} \qquad (18')$$

and finally from equation (12) the average moment:

$$m = \frac{\sum \varrho_i \mu_i e^{-\frac{\varepsilon_i}{kT}}}{\sum \varrho_i e^{-\frac{\varepsilon_i}{kT}}} + H\left\{\frac{\sum \varrho_i \alpha_i e^{-\frac{\varepsilon_i}{kT}}}{\sum \varrho_i e^{-\frac{\varepsilon_i}{kT}}} + \frac{1}{kT}\left[\frac{\sum \varrho_i \mu_i^2 e^{-\frac{\varepsilon_i}{kT}}}{\sum \varrho_i e^{-\frac{\varepsilon_i}{kT}}} - \left(\frac{\sum \varrho_i \mu_i e^{-\frac{\varepsilon_i}{kT}}}{\sum \varrho_i e^{-\frac{\varepsilon_i}{kT}}}\right)^2\right]\right\} \qquad (18'')$$

If $m = 0$ for $H = 0$,

$$\sum \varrho_i \mu_i e^{-\frac{\varepsilon_i}{kT}} = 0 \qquad (19)$$

must hold for all values of the temperature T. This condition is not as stringent as might appear on first glance. To satisfy it, it is only necessary that the term splitting be symmetrical and that to an energy level with positive μ_1 be associated the same value of ε_i as for the corresponding level with negative μ_1. If the μ_1 are imagined as the projections of the magnetic moment onto the direction of the field, we thus require that the positive and negative projections involve the same energy ε_i with respect to the surroundings. Since the positive projection is transformed into the negative projection if the direction of rotation of the electron generating the moment is reversed, and since the energy ε_i is to be attributed in all probability to the electrical forces effective between the electron orbit and the surroundings, it is obvious that relation (19) will be satisfied in most instances. If this is the case, equation (18") assumes the form:

$$m = H\left\{\frac{\sum \varrho_i \alpha_i e^{-\frac{\varepsilon_i}{kT}}}{\sum \varrho_i e^{-\frac{\varepsilon_i}{kT}}} + \frac{1}{kT}\frac{\sum \varrho_i \mu_i^2 e^{-\frac{\varepsilon_i}{kT}}}{\sum \varrho_i e^{-\frac{\varepsilon_i}{kT}}}\right\} \qquad (19)$$

The correction due to diamagnetic excitation can in most instances be neglected. The first term in equation (19') then drops out, and we obtain:

$$m = \frac{H}{kT}\frac{\sum \varrho_i \mu_i^2 e^{-\frac{\varepsilon_i}{kT}}}{\sum \varrho_i e^{-\frac{\varepsilon_i}{kT}}} \qquad (19')$$

Curie's law is no longer obeyed; the temperature dependence is more complicated. Only in the limit of high temperatures do we again have the previous result:

$$m = \frac{H}{kT} \frac{\sum \varrho_i \mu_i^2}{\sum \varrho_i}$$

If we put in general:

$$m = \frac{H}{kT} \frac{\sum \varrho_i \mu_i^2}{\sum \varrho_i} \tau(T) \qquad (20)$$

the correction fnction τ is given by:

$$\tau = \frac{\sum \varrho_i \mu_i^2 e^{-\frac{\varepsilon_i}{kT}} \Big/ \sum \varrho_i \mu_i^2}{\sum \varrho_i e^{-\frac{\varepsilon_i}{kT}} \Big/ \sum \varrho_i} \qquad (21)$$

Expansion of τ (T) in negative powers of the temperature leads to the expression:

$$\tau = 1 + \frac{1}{kT}\left(\frac{\sum \varrho_i \varepsilon_i}{\sum \varrho_i} - \frac{\sum \varrho_i \varepsilon_i \mu_i^2}{\sum \varrho_i \mu_i^2}\right) + \cdots \qquad (21')$$

as a first approximation. If, on the other hand, the law of Weiss is considered along the lines explained before, it leads to the expansion:

$$\tau = 1 + \frac{\Theta}{T} + \cdots \qquad (22)$$

If the law of Weiss is indeed an approximation, the characteristic temperature θ should have the significance which follows from the equation:

$$\left.\begin{aligned} k\Theta &= \frac{\sum \varrho_i \varepsilon_i}{\sum \varrho_i} - \frac{\sum \varrho_i \varepsilon_i \mu_i^2}{\sum \varrho_i \mu_i^2} = \\ &= \sum \left(\frac{\varrho_i}{\sum \varrho_i} - \frac{\varrho_i \mu_i^2}{\sum \varrho_i \mu_i^2}\right) \varepsilon_i \end{aligned}\right\} \qquad (23)$$

No difficulty regarding the sign of θ is encountered, since from equation (23) θ may obviously be positive or negative.

(5) It seems desirable to illustrate the general treatment in section (4) by several specific examples.

Let us first assume the simplest possible case of a moment which can only point either in the direction of the field or in the opposite direction. Both positions have the same statistical weight ρ and are assigned the same energy ε. According to equation

(19') we have for this case the rigorous result:

$$m = \frac{H}{kT}\mu^2$$

i.e., Curie's law is valid without correction.

Let us assume for the second case an elementary moment which may have the four projections $+2\mu$, -2μ, $+\mu$, $-\mu$ in the direction of the field. Let the first two positions be characterized by the energy ε_1, and the last two by the energy ε_2 and assume the statistical weight to be the same for all four positions. In this case we have from equation (19"):

$$m = \frac{H}{kT}\frac{5}{2}\mu^2 \frac{\frac{4}{5}e^{-\frac{\varepsilon_2}{kT}} + \frac{1}{5}e^{-\frac{\varepsilon_1}{kT}}}{\frac{1}{2}e^{-\frac{\varepsilon_2}{kT}} + \frac{1}{2}e^{-\frac{\varepsilon_1}{kT}}}$$

Thus in this instance Curie's law is not obeyed because the correction function:

$$\tau = \frac{\frac{4}{5}e^{-\frac{\varepsilon_2}{kT}} + \frac{1}{5}e^{-\frac{\varepsilon_1}{kT}}}{\frac{1}{2}e^{-\frac{\varepsilon_2}{kT}} + \frac{1}{2}e^{-\frac{\varepsilon_1}{kT}}} = 1 + \frac{3}{5}\mathfrak{T}g\frac{\varepsilon_1 - \varepsilon_2}{2kT}$$

appears. It follows from this representation or equally well from equation (23) that this statement is in first approximation idential with an expression derived by Weiss, according to which the characteristic temperature θ is given by the relation:

$$k\Theta = \frac{3}{10}(\varepsilon_1 - \varepsilon_2)$$

Whether a positive or a negative value for θ results depends solely on whether ε_1 is greater or smaller than ε_2, *i.e.*, on whether energy is gained or lost when the moment is shifted from the oblique position to the parallel. The order of magnitude of the energy difference in question presents no difficulties, since it is equal to the equipartition energy of a linear oscillator at the temperature. This is, of course, extremely small compared to the energy levels usually present inside the atom, and may easily be due to external causes.

Bibliography

1. P. Curie, *Ann. Chim. Phys.*, (7), *5*, 289 (1895); *Oeuvres Paris*, *1908*, 2.
2. P. Langevin, *J. Physik* (4), *4*, 678 (1905); *Ann. Chim. Phys.* (8), *5*, 70 (1895).
3. P. Debye, *Physik. Z.*, *13*, 97 (1912). Compare also the review article in *Handb. Radiologie*, 6, 597 ff. (1924).
4. W. Pauli, *Physik. Z.*, *21*, 615 (1920).
5. P. Epstein, *Science*, *57*, 532 (1923).
6. W. Gerlach, *Physik. Z.*, *24*, 275 (1923).
7. A. Sommerfeld, *Z. Physik*, *19*, 221 (1923).
8. According to experiments by R. Ladenburg (*Z. Physik*, *28*, 51 (1924); *Physik, Z.*, *26*, 685 (1925); *Berl. Akad.*, *5*, 420 (1925) there is now in prospect a test of the relation between the quadratic Stark effect and the dielectric constant.
9. H. A. Kramers, *Z. Physik*, *3*, 199 (1920).
10. H. R. Woltjer, *Versl. Amst. Akad.*, *34*, 494 (1925); H. R. Woltjer and H. Kamerlingh-Onnes, *Versl. Amst. Akad.*, *34*, 502 (1925).
11. W. Pauli, *Z. Physik*, 6, 319 (1921).
12. To the papers mentioned in my article in *Handb. Radiologie*, 6, 597 ff. (1924), the beautiful experiments by C. T. Zahn, *Phys. Rev.*, *24*, 400 (1924), and by C. P. Smyth and C. T. Zahn, *J. Am. Chem. Soc.*, *47*, 2501 (1925) have recently been added. The law is confirmed for HCl, HBr, HI, and $CH_2{:}CH.CH_2{\cdot}CH_3$.
13. A theory of this effect is given by G. Hettner, *Z. Physik*, *2*, 349 (1920).
14. For a detailed discussion I refer to my article in the *Handb. Radiologie*, 6, 597 ff. (1924).
15. B. Cabrera has recently (*J. Physique*, 6, 241, 273 (1925)) presented a series of reasons of an experimental nature explaining why an interpretation on the basis of the hypothesis of Weiss is unlikely.

THE SYMMETRICAL TOP IN WAVE MECHANICS

P. Debye and C. Manneback

Reprinted from
Nature, Vol. 119, page 83, January 15, 1927

In a recent issue of Nature (Dec. 4, 1926, p. 805) there appeared a letter of Messrs. R. de L. Kronig and I. I. Rabi, in which they gave, on the basis of the new wave mechanics of Schrödinger, an expression for the energy of a symmetrical rotator, *i.e.* a rigid polyatomic molecule having two equal moments of inertia.

A like result has also been obtained by F. Reiche (*Zeit. f. Phys.*, 39, 444, 1926) using the wave mechanics. Furthermore, under the assumption that the molecule possesses a permanent electric moment along the direction of its figure axis, Reiche derived to first order approximation the addition to the energy expression caused by placing the molecule in an external electric field. Independently of Reiche, also using the wave mechanics, we have carried the calculation to the second order of approximation and have thus been able to compute the dielectric constant. We find for the total energy $W^*_{j,m,n}$ of the molecule in the presence of an electric field of strength F:

$$W^*_{j,m,n} = W_{j,n} - \mu F \frac{mn}{j(j+1)} + \frac{(\mu F)^2}{h^2/8\pi^2 A} 4(\Phi_{j,m,n} - \Phi_{j+1,m,n})$$

where $W_{j,n}$ is the energy of the molecule without electric field, as already given by D. M. Dennison (*Phys. Rev.*, 28, 318, 1926) using the matrix mechanics. μ represents the permanent dipole moment, A the moment of inertia about an axis perpendicular to the figure axis, and j, m, and n three quantum numbers. The first of these may take all positive integral values not including zero, while the others may take both positive and negative integral values including zero, subject to the restriction that the absolute value of each shall not exceed the value of j. The function

$\Phi_{j,m,n}$ is a numerial factor depending only upon the quantum numbers:

$$\Phi_{j,m,n} = \frac{(j^2 - m^2)(j - n^2)}{(2j - 1)(2j)^2(2j + 1)}$$

$\Phi_{j+1,m,n}$ is the same expression, where only $j + 1$ is substituted in place of j.

From the energy expression given above, the dielectric constant of a perfect gas is found to have the following value at high temperatures:

$$1 + \frac{4\pi}{3}\frac{N\mu^2}{KT}$$

where T is the absolute temperature, N the number of molecules per unit volume, and k the Boltzmann constant. This result is in complete agreement with the value of the dielectric constant of such molecules already found by Kronig (*Proc. U. S. Nat. Acad. Sci.*, *12*, 608, 1926) using the matrix mechanics, and it means that at high temperatures the dielectric constant obeys the law of Langevin-Debye. We have found that at usual temperatures the departure from this law cannot exceed a few per cent.

It may be remarked that the second term in the energy expression given above predicts the existence of a Stark effect of the first order in the band spectra of symmetrical molecules, whereas for diatomic molecules an effect only of the second order is to be expected. The separation of the lines in the Stark effect of the first order for symmetrical molecules depends upon the magnitude of the dipole-moment and the field strength, but *not* upon the moments of inertia of the molecule. The intensity of the lines, on the other hand, is dependent upon the moments of inertia. The separation of the lines in the Stark effect of the band spectra of polyatomic molecules, which a simple calculation shows to be of a sufficient magnitude to be measured experimentally, thus provides a means of finding the dipole strength of such molecules. One finds $\Delta\lambda/\lambda = 22 \times 10^{-6}$ with $F = 50{,}000$ volts/cm. and $\mu = 1 \times 10^{-18}$ C.G.S. units.

A detailed paper covering the work outlined here will appear shortly by one of us in the *Physikalische Zeitschrift*.

Reprinted from pages 589–594 of
THE JOURNAL OF CHEMICAL PHYSICS VOLUME 19, NUMBER 5 MAY, 1951

Electric Moments of Polar Polymers in Relation to Their Structure*

P. DEBYE AND F. BUECHE
Department of Chemistry, Cornell University, Ithaca, New York

(Received December 21, 1950)

From the observed angular dissymmetry of the light scattered by polymer solutions it has been possible to draw conclusions on the average size of the polymer coil. In general, the actual coil diameter found in this way is much larger than that calculated from a model with free rotation. (For polystyrene in benzene the factor is about 3.5.) This increase in size can satisfactorily be attributed to hindering of rotation around the chemical bonds. In order to obtain information about structural details of this hindered rotation, however, the size determination will have to be combined with evidence derived from other sources than light scattering. As such, the dipole moment, as derived from measurements of the dielectric constant of polymer solutions, obviously is a good choice. If a number of units are connected to each other in a chain, the contribution of the whole chain may be larger or smaller than would be the case if the units were free, depending on whether positions of parallelism or antiparallelism of the elementary electric moments are preponderant, taking the average over all the possible forms the coiling molecule can assume. A theory of this effect has been worked out, and experiments have been carried out. One of the examples is polyparachlorostyrene, for which the average contribution to the polarization of one chain element is substantially smaller in the polymer than that of the corresponding monomer.

VISCOSITY and light scattering measurements have shown that coiling chain polymer molecules occupy a volume much larger than is predicted by simple theories. The explanation for this which has often been given states that, owing to steric hindrances, the polymer chain cannot take on all configurations allowed by free rotation about the chain bonds. It is generally agreed that a very large part of this extra extension of the chain is due to hindering of rotation along the polymer chain.

The nature of this hindered rotation is now a dominant problem in polymer behavior. Unfortunately, viscosity and light scattering data can do little more than tell us that hindering exists. Infrared investigations have been made on very small molecules to solve just such problems as the one at hand, but the extension of this method to polymer molecules except in a few cases would seem to be excessively tedious if not impossible. We are therefore led to search for still another method for determining polymer behavior in solution-*a* method which will give a more detailed picture of the configurations assumed by the polymer molecule. Such a more detailed picture is provided by an analysis of the dipole moments of polar polymer molecules in solution.

To illustrate the sensitivity of the dipole moment of a polymer molecule to hindering of rotation, let us first consider a particularly simple type of polymer chain. It is one in which the valence bond angle is 180° and therefore the chain is a straight line of atoms. Consider further that dipole groups of moment μ come off the chain in some periodic fashion in a direction normal to the chain.

Now it is well known that a molecule's dipole contribution to the dielectric constant of a medium is proportional to μ^2. If our simple chain of N units is assumed to be completely stiff, then its contribution to the dielectric constant will either be zero or of the order of $(N\mu)^2$, depending upon whether the fixed dipoles on the stiff chain are so oriented that the dipoles cancel or add. This is to be compared with the value of $N\mu^2$ which would be true if the units of the chain were perfectly free to rotate. Therefore, since N is very large in practice, we see that the dipole moment contribution to the dielectric constant is extremely sensitive to the polymer configurations and to the hindering of internal polymer rotation.

Let us now proceed to an actual and general polymer molecule. It may be shown (see Appendix I) that for any polymer molecule, the average dipole contribution to the molecular polarization is equal to

$$\left\langle P\sum_{m=1}^{N_0}\sum_{n=1}^{N_0}(\mu_n f)(\mu_m f)\right\rangle_{Av}\frac{E}{3kT},$$

where μ_n is the vector magnitude of the nth dipole of the chain and f is a unit vector in the direction of the applied electric field E. The sums extend over all of the N_0 dipoles on the chain, and the average is to be taken over all of the possible chain configurations and orientations. P is the probability of occurrence of any particular chain configuration.

Therefore, we come to the result that the polymer molecule will act exactly like a single molecule of permanent moment $\bar{\mu}$ if we define $\bar{\mu}$ by the relation

$$\frac{\bar{\mu}^2}{3}=\left\langle P\sum_{m=1}^{N_0}\sum_{n=1}^{N_0}(\mu_n f)(\mu_m f)\right\rangle_{Av}.$$

We shall call $\bar{\mu}$ the average moment of the polymer molecule.

It is shown in Appendix II that in the case of a carbon-carbon chain with a dipole of moment μ coming off every other chain atom at an angle γ with the preceding $c-c$ chain bond and angle β with the following $c-c$ chain bond, the average turns out to be

$$\tfrac{1}{3}\mu^2N_0[1+2p\cos\beta\cos\gamma/(1-p^2)],$$

* The work reported in this paper was done under a contract with the Reconstruction Finance Corporation, Office of Rubber Reserve.

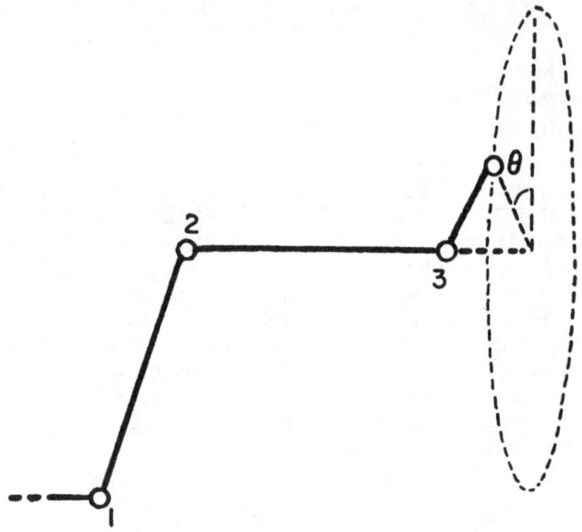

FIG. 1(a). A schematic diagram showing a series of four atoms along a polymer chain.

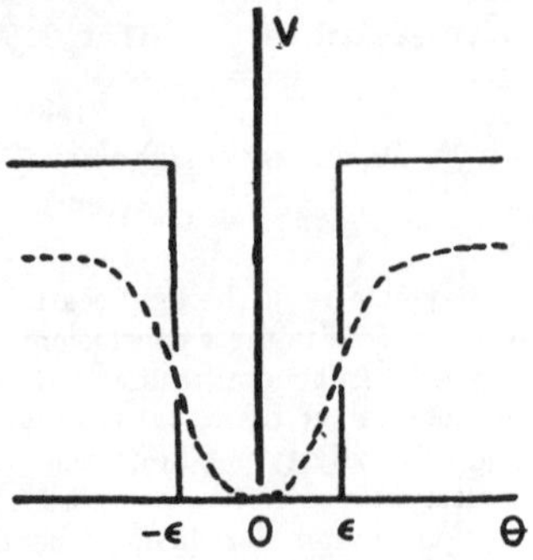

FIG. 1(b). An assumed potential for chain atom 4.

providing free rotation is assumed. By free rotation, we mean that each chain carbon rotates freely on the valence cone subject only to the restrictions that the $c-c$ bond distance and angle must be preserved. In the last expression, p is the cosine of the valence angle, i.e., $\frac{1}{3}$ for a $c-c$ chain.

For a molecule such as polyparachlorostyrene we have also that $\cos\beta = -\cos\gamma = p = \frac{1}{3}$. In that case, again assuming free rotation, we find the average dipole moment to be given by

$$\bar{\mu}^2 = 11N_0\mu^2/12,$$

where N_0 is the number of dipoles on the chain and μ is the moment of each of these dipoles.[1]

Therefore, we have obtained the result that if our assumption of free rotation were true, polyparachlorostyrene and similar polymers should have an average moment equal to $(0.92N_0)^{\frac{1}{2}}$ times the dipole moment of each chain unit. It is shown in Appendix IV that the quantity $\bar{\mu}^2/N_0$ is given by a single dipole moment measurement.

Since μ, the dipole moment of an individual chain group, is generally known, a check is easily obtained for our derived equation. Work in our laboratory (see Appendix IV) has shown that the theoretical factor 0.92 is too large. The correct factor as found experimentally is about 0.56 for p.p. chlorostyrene. The similar compound, polyvinylchloride is also found to give a low value for this factor, namely, about 0.75.[2]

We are therefore confronted with additional evidence that real polymer molecules do not possess complete freedom of internal rotation in dilute solution. It will be the purpose of the remaining sections of this paper to show how these observed values of the average dipole moment may be combined with light scattering data in order to determine the nature and extent of the hindering of rotation in polymer molecules.[3]

Hindering of rotation must now be considered along with its consequences upon the calculation of the average dipole moment. To do this it becomes necessary to postulate a probable form of hindering to free rotation. This may be done as follows:

Let us consider any four adjacent chain atoms such as are shown schematically in Fig. 1(a). At any instant atoms 1, 2, and 3 define a plane. Consider now the possible movement of atom 4. If free rotation exists, this atom will be free to rotate on the dotted circle shown. We shall call this the valence circle.

In general, however, we have seen that atom 4 will be constrained in such a way that all positions of the valence circle are not equally probable. We can express this fact by stipulating that this atom experiences a certain potential at each point on this valence circle and, in general, the potential varies from point to point on the circle.

If θ is the angle of rotation of atom 4 out of the plane of atoms 1, 2, and 3 we can then say that this atom experiences a potential as a function of θ. The hindering to free rotation should then be expressible in terms of this potential, $V(\theta)$. For our calculations, it will be sufficient to approximate the actual potential by a deep rectangular potential well. An example of such an approximation is shown in Fig. 1(b).

Effectively then, we stipulate that there is a certain range of values of θ_4 in which atom 4 is free to rotate and that it is excluded from all other angles.

Such an approximation may seem at first thought to be quite a serious departure from reality. However, at a fixed temperature, it can be shown that any real potential may be represented to a fair accuracy by one or more such potential wells. This representation does, however, sacrifice an explicit temperature dependence of the dipole moment.

It may be easily seen that temperature effects will become apparent in the variation of the width of the

[1] R. M. Fuoss and J. G. Kirkwood, J. Am. Chem. Soc. **63**, 385 (1941), have derived an expression similar to this. However, owing to a mistake in their averaging process, the answers do not agree.

[2] R. M. Fuoss, J. Am. Chem. Soc. **63**, 2410 (1941), and also results obtained in our laboratory.

[3] This subject has been treated in another manner by W. Kuhn, Helv. Chim. Acta **31**, 1092 (1948). Unfortunately, his method of treatment does not lend itself well to the determination of the hindering process.

potential well. Since we shall express the width in terms of the so-called angle of free rotation, 2ϵ, we shall expect ϵ to vary with temperature. For example, if we were dealing with a simple potential such as is shown in Fig. 1(b), ϵ should increase with rising temperature. In what is to follow we shall assume ϵ and therefore the temperature to be constant.

Various hindering potentials have been used in the past to explain the size of molecules as found from light scattering and viscosity measurements. When any explicit form has been used, it has usually been similar to that shown in Fig. 1(b). Many others can be reduced to such a form. Furthermore, it is usually assumed that the center of this well is situated at such a position on the valence circle that its center coincides with $\theta=0$. That is to say, if the potential well were of near zero width, the chain would be nearly completely stiff and all its chain links would lie in a saw tooth arrangement in a single plane.

Owing to the wide use of such a model, we shall first calculate the dipole moment to be expected for polyparachlorostyrene if its hindering were of this form. Such calculations have been made (see Appendix III) and the results are plotted in Fig. 2. On this graph we have plotted the ratio $\bar{\mu}^2/N_0\mu^2$ *vs* $\sin\epsilon/\epsilon$. For completely free rotation $\sin\epsilon/\epsilon$ is zero, and for a stiff chain it is unity.

For this calculation it has been assumed that the dipoles occur on alternate chain carbon atoms. These dipoles come off at the tetrahedral angle. But now it is possible to obtain two isomers because of the fact that the dipole group can occur on either one of two positions on the chain carbon.

Curve A in Fig. 2 is for the case where the dipoles appear on the same position on each chain unit. Curve B is for the case where the dipole position alternates. The latter structure probably applies in the case of polyvinyl chloride.[4] Although the structure of p.p.

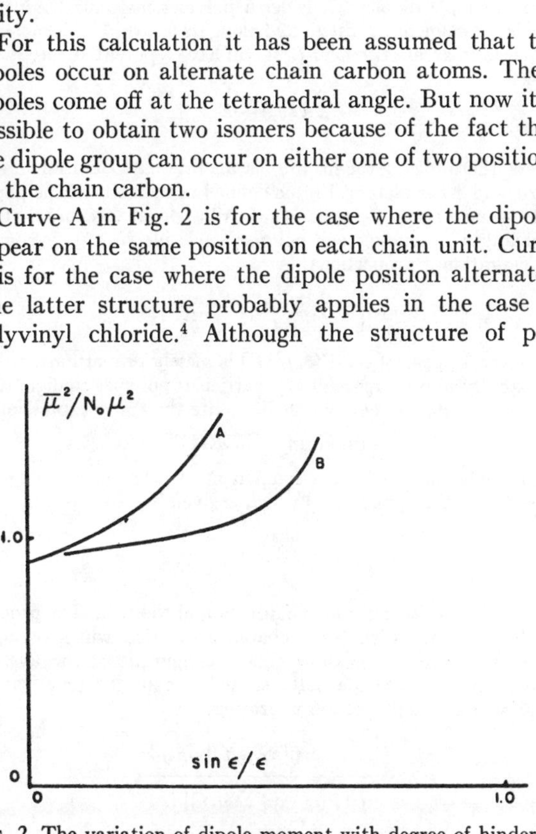

FIG. 2. The variation of dipole moment with degree of hindering for two normally planar molecular forms.

[4] C. W. Bunn, *Advances in Colloid Science* (Interscience Publishers, Inc., New York, 1946), Vol. 11.

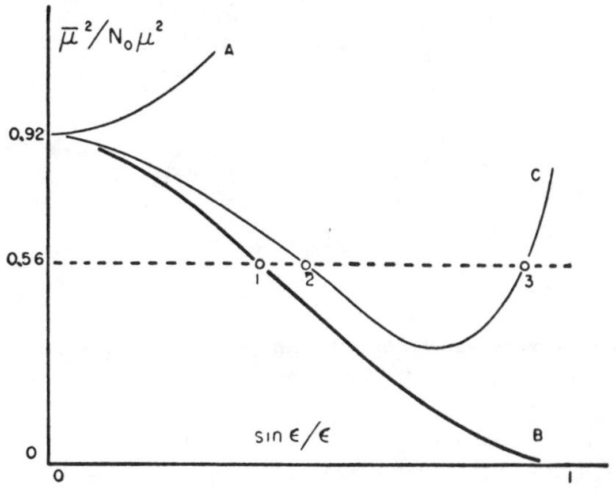

FIG. 3. Three typical curves showing the variation of dipole moment with degree of hindering.

chlorostyrene is not accurately known, it is probable that it is very much like polyvinyl chloride, and we shall assume this to be true.

In any event, both isomers are seen to give rise to curves which start out at $\bar{\mu}^2/N_0\mu^2$ values of 0.92 (i.e., the free rotation value) and rise rather rapidly as the chain becomes stiffer. Obviously, since the experimental value is 0.56, we see that it is impossible for this particular model for the hindering of free rotation in p.p. chlorostyrene to be correct. We must therefore conclude that symmetrical rotations of this type about a planar polymer configuration do not represent the true picture of this molecule.

We are, therefore, confronted with the problem of finding a potential for bond rotations which will give rise to an average dipole moment for the molecule equal to that observed. It need not be true that each chain bond experience the same potential as a function of θ; and, therefore, a large number of possible models exist. However, owing to obvious steric hindrances and with the known fact that the dipoles tend to cancel each other, this large number of possibilities may be reduced without actually calculating each one.

This process leaves a certain number of models as definite candidates for the true structures. It now becomes necessary to take definite models for the hindering and calculate how the dipole moment of our polymer molecule will depend upon the degree of stiffening of the chain for that particular model. In general, one obtains three different types of dependence. They are illustrated in Fig. 3.

Curve A is like the ones obtained previously for the traditional model. It starts at the free rotation value, 0.92, and rises more or less rapidly as the hindering increases. In such a model the dipoles do not oppose each other enough to reduce the value to the 0.56 found experimentally.

It is then logical to assume a model such that, if it were completely stiff, the dipoles exactly cancel each other. Such a model must obviously give a value of zero

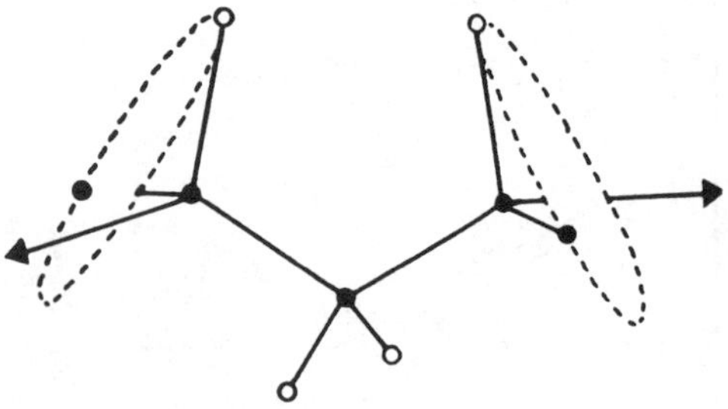

FIG. 4. A model which illustrates the probable zero configuration for polyparachlorostyrene.

when the chain is stiff. Curve B is typical of such a model. Since this curve passes through the experimental value, this model is a possible choice for the true polymer molecule.

A molecular configuration which has its dipoles nearly opposed when the chain is stiff gives rise to a curve such as curve C. The rapid rise near $\sin\epsilon/\epsilon = 1$ is due to the fact that the stiff molecule must have a moment of $(N_0 f\mu)^2$, where f is the fraction of each dipole which remains uncompensated. It should be mentioned that this curve is extremely sensitive to small alterations in the assumed model. The position of the minimum rises very rapidly as the compensation of dipoles decreases.

We are therefore confronted with two possible types of models which are capable of giving the observed dipole moment. To proceed further we need the results of some other experiment. Fortunately, the size of the polymer molecule found from light scattering is sufficient to resolve the ambiguity.

For each of these models it is possible to calculate an expected size for the molecule. This may be done in a straightforward manner by simple modifications of known methods. These methods have been described elsewhere.[5] For each model we then obtain two values for $\sin\epsilon/\epsilon$, one which will explain the observed dipole moment and one which will explain the observed size.[6]

Points 1 and 2 of Fig. 3 are found to be at far lower values of $\sin\epsilon/\epsilon$ than is needed to explain the observed polymer size. Therefore, we are left with only one type of model which is capable of explaining both experimental facts. However, since this curve is so very sensitive to the particular model chosen for calculation, it is somewhat simpler to obtain the true value of $\sin\epsilon/\epsilon$ from the size of the molecule once the model has been approximately determined. This is true, since the size at a given $\sin\epsilon/\epsilon$ is relatively insensitive to the exact model chosen.

From such considerations we propose the following approximate picture of a polyparachlorostyrene molecule in solution.

If each chain link were at the center of the potential well in which it rotates, the chain would be rather extended and periodic in structure, repeating every four chain links. A typical segment of the chain may be illustrated schematically by use of Fig. 4. The actual model is thought to be a slight distortion of this form. Owing to the fact that the potential well has a definite width, the atoms rotate about this configuration through a certain free angle. This angle is about 65 degrees.

The model which we have postulated above is not the only one capable of satisfying the experimental facts. However, this model is probably the simplest reasonable one which is in agreement with experiment. The present method is very similar to x-ray crystal structure methods in that it is not capable of determining one and only one molecular model. Others may be possible. However, given any other postulated model, this method of attack can very easily determine if such a model is consistent with experimental facts.

APPENDIX I

The general equation for the dipole moment of a polymer molecule may be derived in a manner similar to that used in the case of simple dipoles. Consider a polymer molecule having N_0 dipoles of moment μ_n distributed along its length. If this molecule is placed in an electric field E, it will have a potential energy

$$-\sum_1^{N_0} V(\theta_n) - \sum_1^{N_0} \mu_n E,$$

where the added term in $V(\theta)$ is an internal potential due to steric and other factors. $V(\theta)$ is defined in the body of the paper.

If we now apply Boltzmann's distribution law, we find that the probability of finding our polymer molecule in any particular configuration is proportional to

$$\exp\left[-\sum_1^{N_0}\frac{V(\theta_n)}{kT} - \sum_1^{N_0}\frac{\mu_n E}{kT}\right].$$

However, since $\exp[-\Sigma V(\theta_n)/kT]$ is merely proportional to the probability of occurrence of any particular polymer configuration without an applied field we shall rewrite the above expression as

$$P(c)\exp[-\Sigma\mu_n E/kT].$$

The polymer will have a component of its moment along the direction of the applied field which is given by

$$\sum_1^{N_0} \mu_n f,$$

where f is a unit vector in the direction of the field. The product of this quantity with the probability function will give, upon integration over all possible chain configurations, the average contribution of a single polymer molecule to the polarization. This is, after the proper normalization,

$$\frac{\int\{P(c)\exp[x\Sigma\mu_n f]\}\Sigma\mu_n f d\tau}{\int\{P(c)\exp[x\Sigma\mu_n f]\}d\tau},$$

where x is E/kT and the integral is to be extended over all possible configurations.

This expression may be considerably simplified by expanding the exponential and dropping the higher terms so as to make the

[5] See, for example, a paper by W. J. Taylor, J. Chem. Phys. **16**, 257 (1948), which contains one method of calculation and gives references to others.

[6] We have found in our laboratory that polyparachlorostyrene molecules are about the same size as polystyrene molecules. Therefore, $R^2 = 25Na^2$. (A. M. Bueche, J. Am. Chem. Soc. **71**, 1452 (1949).)

average moment proportional to the electric field. Such an approximation is justified for all practical cases in which we can define a dielectric constant and no saturation effects are observable. The result for the average dipole polarization is $x\langle P\{\sum\mu_n f\}^2\rangle_{Av}$.

But this quantity is equal to $\bar{\mu}^2E/3kT$, where $\bar{\mu}$ is the average dipole moment of the whole polymer molecule. We therefore have finally for the average dipole moment of the polymer molecule

$$\bar{\mu}^2=3\left\langle P\left\{\sum_1^{N_0}\sum_1^{N_0}(\mu_n f)(\mu_m f)\right\}\right\rangle_{Av}.$$

APPENDIX II

To calculate the average dipole moment for a polymer molecule having free rotation we must evaluate $\langle\sum\sum(\mu_n f)(\mu_m f)\rangle_{Av}$. Now from a well-known geometrical relation we can obtain at once that

$$\cos(\mu_n,f)\cos(\mu_m,f)=\cos(\mu_n,\mu_m)\cos^2(\mu_n,f)+\sin(\mu_n,f)\cos(\mu_n,f)\sin(\mu_n,\mu_m)\cos\phi,$$

where ϕ is as shown in Fig. 5.

Since all directions of the field are equally probable, we obtain upon averaging that

$$\langle\cos(\mu_n,f)\cos(\mu_m,f)\rangle_{Av}=\tfrac{1}{3}\langle\cos(\mu_n,\mu_m)\rangle_{Av}.$$

The average for $\cos(\mu_n,\mu_m)$ may be easily found by a method similar to one first suggested by Eyring.[7]

If we consider the diagrams in Fig. 6, it is clear that we may write

$$\langle\cos(\mu_m,\mu_n)\rangle_{Av}=\langle\cos(a_k,\mu_m)\cos\beta+\sin(a_k\mu_m)\sin\beta\cos(\phi_k)\rangle_{Av}.$$

Now if free rotation exists, ϕ_k takes on all values from zero to 2π, and so the average of the second term is zero. We are then left with the relation

$$\langle\cos(\mu_n,\mu_m)\rangle_{Av}=\cos\beta\langle\cos(a_n,\mu_m)\rangle_{Av}.$$

The same process may be repeated again giving for the average

$$\cos\beta\cos a\langle\cos(a_{n-1},\mu_m)\rangle_{Av}.$$

Finally, we obtain for the average

$$\cos\beta(\cos a)^{(n-m-1)}\langle\cos(a_{m+1},\mu_m)\rangle_{Av},$$

and then

$$\langle\cos(\mu_n,\mu_m)\rangle_{Av}=\cos\beta\cos\gamma(\cos a)^{n-m-1}.$$

In general, then if $K=(n-m)$ we obtain for $n\neq m$ the result

$$\langle\cos(\mu_n,\mu_m)\rangle_{Av}=\cos\beta\cos\gamma\cos^{k-1}a.$$

If we use this fact, the series in the expression for the average dipole is readily evaluated. The result is

$$\bar{\mu}^2=\mu^2N_0[1+2p\cos\beta\cos\gamma/(1-p^2)],$$

where $p=\cos a$. The approximation has been made that $p^N\ll 1$. This is true in all practical cases.

APPENDIX III

To calculate the average dipole moment when hindering is present we proceed as in Appendix II down to the point where

$$\langle P(c)\cos(\mu_n,f)\cos(\mu_m,f)\rangle_{Av}=(1/3\mu^2)\langle P(c)\mu_n\mu_m\rangle_{Av}.$$

We now define unit vectors i_k, j_k, and h_k as shown in Fig. 6. We must therefore calculate $\langle P(c)i_k i_0\rangle_{Av}$. Now for the particular model under consideration it is clear that $P(c)$ is zero except for those values of θ_k which are inside the potential well. In these regions $P(c)$ will be unity. Therefore, because of the particular form assumed for the polymer molecule, it is true that as i_k rotates in its free angle, it will rotate in a symmetrical fashion so as to give an average value which is in the plane of h_k and h_{k-1}. It is therefore possible to express the average value of h_k and h_{k-1} and to write

$$\langle i_k i_0\rangle_{Av}=\langle Ah_k i_0+Bh_{k-1}i_0\rangle_{Av},$$

[7] H. Eyring, Phys. Rev. **39**, 746 (1932).

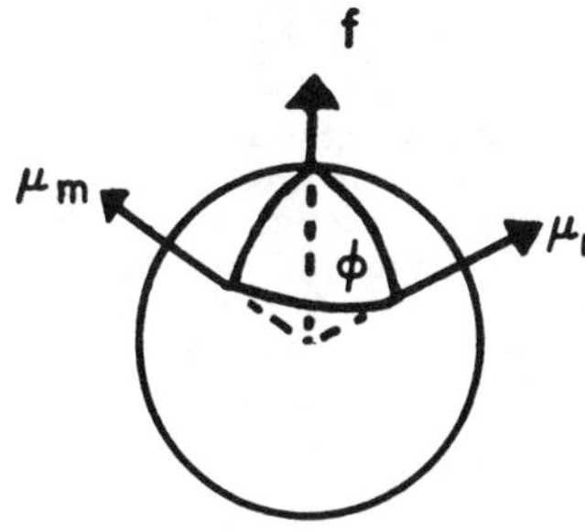

FIG. 5. A diagram illustrating the symbols used in the text.

where A and B are known constants which depend upon the function $\sin\epsilon/\epsilon$.

But the average of $h_k i_0$ can be related to the average of $h_{k-1}i_0$ and $h_{k-2}i_0$ in a similar manner. Therefore, we are able to arrive at a simple recurrence formula for the $h_k i_0$ and $h_{k-1}i_0$ averages. In this particular case this recurrence relation is sufficiently simple so that a compact closed form may be found quite readily for $\langle h_k i_0\rangle_{Av}$. After this is obtained, the result is substituted in the equation above, and all $\langle i_k i_0\rangle_{Av}$ are then determined. Using this relation, the series for $\bar{\mu}^2$ can then be summed. The above procedure was used to obtain the curves shown in Fig. 2.

More complicated potentials may be treated in the same manner. However, for twisted chain configurations it is often found that the recurrence relations become rather complicated. In such cases, the average is best found by reducing the recurrence relations down to a form such that each contains only the h's, i's, or j's along with i_0. The averages may then be calculated by numerical substitution in these relations. This latter procedure was followed in calculating the curves shown in Fig. 3.

APPENDIX IV

The experimental determination of the average dipole moment of poly p. chlorostyrene was done in the usual way by measuring

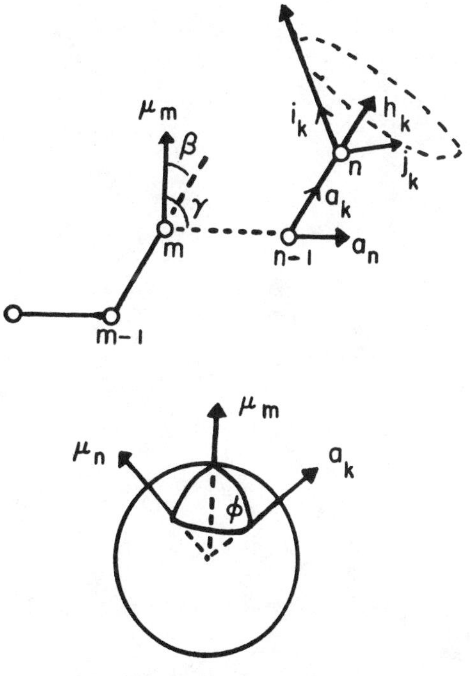

FIG. 6. Diagram showing the coordinates chosen for the calculations.

the dielectric constant of several concentrations of the polymer in benzene solutions. The cell was composed of three concentric metal cylinders enclosed in a glass tube. This cell has been used for a number of years in this laboratory for the determination of the dipole moments of low molecular weight compounds. Its capacitance was measured by the heterodyne beat method at a frequency of 1.5 Mc, which is well outside the dispersion range.

If M is the molecular weight of each chain unit, that is to say, the molecular weight of the polymer molecule is N_0M, then, providing N_0 is large compared with unity, it may be shown that

$$\overline{\mu^2}/N_0=(9kT/4\pi A)[P_\infty - M(n^2-1)/d_1(n^2+2)],$$

where A is Avogadro's number, n is the index of refraction of the solute, and P_∞ is the value at infinite dilution of

$$P=M\frac{100-x}{x}\left[\frac{\epsilon_{12}-1}{\epsilon_{12}+2}\frac{100}{(100-x)d_{12}}-\frac{\epsilon_0-1}{\epsilon_0+2}\frac{1}{d_0}\right].$$

In this expression, x is the weight percent of polymer, ϵ is the dielectric constant and d is the density. The subscripts 0, 1, and 12 refer to solvent, polymer, and solution, respectively.

Complete dielectric measurements were taken on two different batches of polymer. The average results obtained were 1.45D at 30°C and 1.46D at 50°C with an estimated error of about 1 percent.[8]

To find the quantity desired, $\overline{\mu^2}/N_0\mu^2$, we need yet the moment of the free chain unit. This was assumed to be the same as that for p-chlorotoluene, and a value of 1.93D was used. Although the values given in the literature do not agree too well with each other, this value was taken as an average of what appears to be the most reliable values. Assuming this value, we then obtain an experimental value of 0.56 for $\overline{\mu^2}/N_0\mu^2$.

[8] The monomer, parachlorostyrene, was kindly furnished by the Dow Chemical Company. Their analysis showed it to be 99.6 percent of the para isomer. The dipole values stated have been corrected for a small amount of chlorine lost during polymerization.

ELECTROLYTES

ON THE THEORY OF ELECTROLYTES. I. FREEZING POINT DEPRESSION AND RELATED PHENOMENA.*

(Zur Theorie der Elektrolyte. I. Gefrierpunktserniedrigung und verwandte Erscheinungen)

P. Debye and E. Hückel**

Translated from

Physikalische Zeitschrift, Vol. 24, No. 9, 1923, pages 185-206

I. Introduction

It is known that the dissociation hypothesis by Arrhenius explains the abnormally large values of osmotic pressure, freezing point depression, etc., observed for solutions of electrolytes, by the existence of free ions and the associated increase in the number of separate particles. The quantitative theory relies on the extension, introduced by van't Hoff, of the laws for ideal gases to diluted solutions for the computation of their osmotic pressure. Since it is possible to justify this extension on the basis of thermodynamics, there can be no doubt regarding the general validity of these fundamentals.

*Submitted February 27, 1923.

**The present considerations were stimulated by a lecture by E. Bauer on Ghosh's works, held at the Physikalische Gesellschaft. The general viewpoints taken as the basis for the computation of the freezing point depression as well as of the conductivity lead me, among other things, to the limiting law involving the square root of the concentration. I could have reported on this during the winter of 1921 at the "Kolloquium." With the active assistance of my assistant, Dr. E. Hückel, a comprehensive discussion of the results and their collection took place during the winter of 1922.

However, for finite concentrations we obtain smaller values for freezing point lowering, conductivity, etc. than one would expect on first consideration, in the presence of a complete dissociation of the electrolyte into ions. Let P_k, for instance, be the osmotic pressure resulting from the classical law by van't Hoff for complete dissociation, then the actually observed osmotic pressure will be smaller, so that:

$$P = f_0 P_k$$

where, according to Bjerrum,[1] the "osmotic coefficient" f_0 thus introduced is intended to measure this deviation independent of any theory-and can be observed as a function of concentration, pressure, and temperature. In fact, these observations do not relate directly to the osmotic pressure but to freezing point lowering, and boiling point rise, respectively, which can both be derived on the basis of thermodynamics, and by means of the same osmotic coefficient f_0, from their limiting value following from van't Hoff's law for complete dissociation.

The most evident assumption to explain the presence of the osmotic coefficient is the classical assumption, according to which not all molecules are dissociated into ions, but which assumes an equilibrium between dissociated and undissociated molecules which depend on the over-all concentration, as well as on pressure and temperature. The number of free, separate particles is thus variable, and would have to be made directly proportional to f_0. The quantitative theory of this dependence, as far as it relates to the concentration, relies on the mass action law of Guldberg-Waage; the dependence on temperature and pressure of the constant of equilibrium appearing in this law can be determined thermodynamically, according to van't Hoff. The complete aggregate of dependencies, including the Guldberg-Waage law, can be proved by thermodynamics, as is shown by Planck.

Since, the electric conductivity is determined exclusively by the ions, and since, according to the classical theory the number of ions follows immediately from f_0, the theory requires the well known relation between the dependence on the concentration of the conductivity on the one hand and of the osmotic pressure on the other hand.

A large group of electrolytes, the strong acids, bases, and their salts, collectively designated as "strong" electrolytes, exhibits definite deviations from the dependencies demanded by the classical theory. It is especially noteworthy that these deviations are the more pronounced the more the solutions are diluted.* Thus, as was recognized in the course of developments and following

*A summary presentation of this subject was given by L. Ebert, "Forschungen ueber die Anomalien starker Elektrolyte," *Jahrb. Rad. u. Elektr.*, *18*, 134 (1921).

the classical theory, it is possible only with a certain degree of approximation to draw a conclusion from f_0 as to the dependence of the conductivity on the concentration. Moreover the dependence of the osmotic coefficient f_0 on the concentration is also represented entirely incorrectly. For strongly diluted solutions, f_0 approaches the value 1; if $1-f_0$ is plotted as a function of the concentration c, classial theory requires for binary electrolytes, such as KCl, that this curve meets the zero point with a finite tangent (determined by the constant of equilibrium, K). In the general case, provided the molecule of the electrolyte splits into ν ions, we obtain from the law of mass action for low concentrations:

$$1-f_0=\frac{\nu-1}{\nu}\frac{c^{\nu-1}}{K}$$

so that in cases where splitting into more than two ions occurs, the curve in question should present even a higher order of contact with the abscissa. The complex of these dependencies constitutes Ostwald's dilution law.

Actually observations on strong electrolytes show an entirely different behavior. The experimental curve starts from the zero point at a right angle (cf. Figure 2) to the abscissa, independent of the number of ions, ν. All proposed, practical interpolation formulas attempt to represent this behavior by assuming $1-f_0$ to be proportional to a fractional power (smaller than 1, such as 1/2 or 1/3) of the concentration. The same remark holds with regard to the extrapolation of the conductivity to infinite dilutions which, according to Kohlrausch, requires the use of the power $\frac{1}{2}$.

It is clear that under these circumstances the classical theory can not be retained. All experimental material indicates that its fundamental starting point should be abandoned, and that, in particular, an equilibrium calculated on the basis of the mass action law does not correspond to the actual phenomena.

W. Sutherland,[2] in 1907, intended to build the theory of the electrolytes on the assumption of a complete dissociation. His work contains a number of good ideas. N. Bjerrum[3] is, however, the first to have arrived at a distinct formulation of the hypothesis. He clearly stated and proved that, for strong electrolytes, no equilibrium at all is noticeable between dissociated and undissociated molecules, and that, rather, convincing evidence exists which shows that such electrolytes are completely dissociated into ions up to high concentrations. Only in considering weak electrolytes, undissociated molecules reappear. Thus the classical explanation as an exclusive basis for the variation of, for instance, the osmotic coefficient, has to be abandoned and the task ensues to search for an effect of the ions, heretofore overlooked, which explains, in the absence of association, a decrease in f_0 with an increase in concentration.

Recently, under the **influence** of Bjerrum, the impression gained strength that consideration of the electrostatic forces, exerted by the ions on one another and of considerable importance because of the comparatively enormous size of the elementary electric charge, must supply the desired explanation. Classical theory does not discuss these forces, rather, it treats the ions as entirely independent elements. A new interaction theory has to be analogous in some respects to van der Waals' generalization of the law of ideal gases to the case of real gases. However, it will have to resort to entirely different expedients, since the electrostatic forces between ions decrease only as the square of the distance and thus are essentially different from the intermolecular forces which decline much more rapidly with an increase in distance.

Milner[4] computed the osmotic coefficient along such lines. His computation can not be objected to as regards its outline, but it leads to mathematical difficulties which are not entirely overcome, and the final result can only be expressed in the form of a graphically determined curve for the relation between $1-f_0$ and the concentration. From the following it will further emerge that the comparison with experience, carried through by Milner, supposes the admission of his approximations for concentrations which are much too high and for which, in fact, the individual properties of the ions, not taken into account by Milner, already play an important part. In spite of this, it would be unjust to discard Milner's computation in favor of the more recent computations by Ghosh[5] on the same subject. We shall have to revert, in the following, to the reason why we can not agree to Ghosh's calculations, neither in their application to the conductivity nor in their more straightforward application to the osmotic pressure. We will even have to reject entirely his calculation of the electrostatic energy of an ionized electrolyte, which is the basis for all his further conclusions.

The circumstances to be considered in the computation of the conductivity are very similar to those for the osmotic coefficient. Here also the new interaction theory has to make an attempt at understanding the mutual electrostatic effect of the ions with regard to its influence on their mobility. An earlier attempt was made in this direction by Hertz.[6] He transcribes the methods of the kinetic theory of gases, and, in fact, finds a mutual interference of the ions. However, the transcription of this classical method, and particularly the use of concepts like that of the free path length of a molecule in a gas for the case of free ions surrounded by the molecules of the solvent, does not seem to be very reliable. The final result obtained by Hertz cannot, in fact, be reconciled with the experimental results.

In this first note, we shall confine ourselves to the "osmotic coefficient f_0" and to a similar "activity coefficient f_a," used by

Bjerrum[7] and stressed in its significance. Even for such (weak) electrolytes, where a noticeable number of undissociated molecules is present, the equilibrium cannot simply be determined by the Guldberg-Waage formula in its classical form:

$$c_1^{\mu_1} c_2^{\mu_2} \ldots c_n^{\mu_n} = K$$

(c_1, c_2,...c_n, are the concentrations, K the constant of equilibrium). It will be necessary, in view of the mutual electrostatic forces between the ions, to write:

$$f_a K$$

instead of K, introducing the activity coefficient* f_a. This coefficient, just as f_0, will depend on the concentration of the ions. Though, according to Bjerrum, a relation to be proved by thermodynamics exists between f_a and f_0, their dependence on the concentration is different for the two coefficients.

The detailed treatment of conductivity shall be reserved for a later note. This division seems justified, since the determination of f_0 and f_a requires solely a consideration of reversible processes, whereas the computation of mobilities has to do with essentially irreversible processes for which no direct relation to the fundamental laws of thermodynamics exists.

II. Fundamentals

As is well known, it is shown in thermodynamics that the properties of a system are completely known, provided one of the many possible thermodynamic potentials is given as a function of the correctly chosen variables. In view of the form in which the terms based on the mutual electric effects will appear we chose the quantity:**

*The activity coefficient f_a introduced here is not identical with that introduced by Bjerrum. Bjerrum splits our coefficient f_a in order to give a produce of coefficients each of which is associated with a separate ion type. (Compare section 8).

**The potential G differs from Helmholtz' free energy $F = U - TS$ only by the factor $-1/T$. This difference is not essential at all; we define it as in the text: to have immediate connection with Planck's thermodynamics.

$$G = S - \frac{U}{T} \qquad (1)$$

(S = entropy, U = energy, T = absolute temperature) as basic functions. As variables in this case (besides the concentration) we have, naturally, volume and temperature, since:

$$dG = \frac{p}{T} dV + \frac{U}{T^2} dT \qquad (1')$$

The computations which follow differ from the classical computations in that the electrical effects of the ions are taken into account. Accordingly, we divide U into two parts, a classical part U_k and an additional electrical energy U_e:

$$U = U_k + U_e$$

If we consider that, according to equation (1):

$$T^2 \frac{\partial G}{\partial T} = U \qquad (2)$$

and divide the potential G also into two parts:

$$G = G_k + G_e$$

we find that, according to equation (2):

$$G_e = \int \frac{U_e}{T^2} dT \qquad (3)$$

It is our main assignment to determine the electric energy U_e of an ionic solution. For practical evaluations, however, the potential G is not as suitable as the function:

$$\Phi = S - \frac{U + pV}{T} \qquad (4)$$

also preferred by Planck. As shown by the differential form of this definition:

$$d\Phi = -\frac{V}{T} dp + \frac{U + pV}{T^2} dT \qquad (4')$$

the variables pertaining to the potential Φ are pressure and temperature, and since a large majority of the experiments are carried through at constant pressure (and not at constant volume), Φ is preferable. A comparison between (4) and (1) results in:

$$\Phi = G - \frac{pV}{T} \tag{5}$$

if, according to the above, G is known, it remains to find the additional term $-pV/T$ as a function of p and T, and to add it. In view of (1') we can conclude that:

$$\frac{p}{T} = \frac{\partial G}{\partial V} = \frac{\partial G_k}{\partial V} + \frac{\partial G_e}{\partial V} \tag{6}$$

and so have obtained the equation of state which relates pressure, volume, and temperature for the ionic solution. It can be interpreted by assuming that, as a consequence of the electric effect of the ions, an additional, electric pressure p_e, is added to the external pressure p; the electric pressure is to be computed from the relation:

$$p_e = -\frac{\partial G_e}{\partial V} \tag{6'}$$

Later we shall, incidentally,* have occasion to determine this electric pressure p_e; it amounts to approximately 20 atmospheres for an aqueous solution of, for instance, KCl at a concentration of 1 mole per liter. Strictly speaking, it is incorrect to use the classical expression for V (as function of p and T) without regard to the electric effect of the ions, since the pressure p_e causes also a change in volume. In view of the low compressibility of water which results in a relative change of volume of 0.001 for 20 atmospheres, the electric addition to V (as function of p and T) can be neglected for most applications. In the light of this remark, we shall also divide Φ in a classical part and an additional electric part:

$$\Phi = \Phi_k + \Phi_e \tag{7}$$

and can put, according to equation (3):

$$\Phi_e = G_e = \int \frac{U_e}{T^2}\, dT \tag{7'}$$

The classical part, Φ_k, has, according to Planck, the form:

$$\Phi_k = \sum_0^s N_i (\varphi_i - k \log c_i) \tag{7''}$$

*Compare with footnote, page 237.

where:

$$N_0,\ N_1, \ldots N_i, \ldots N_s$$

designate the number of separate particles in the solution, and N_0 refers to the solvent.* Further the thermodynamic potential referred to a single particle is equal to:

$$\varphi_i = s_i - \frac{u_i + p v_i}{T}$$

a quantity independent of the concentration; k is Boltzmann's constant, $k = 1.346 \times 10^{-16}$ erg, and c_i denotes the concentration of the i type particle so that:

$$c_i = \frac{N_i}{N_0 + N_1 + \ldots + N_i + \ldots + N_s}$$

of which the relation:

$$\sum_0 c_i = 1$$

is a consequence.

Upon completion of this thermodynamic introduction, we proceed to the discussion of the principal assignment: computation of the electric energy U_e.

At a first glance it appears as if this energy could be obtained directly in the following manner. In a solvent with dielectric constant D are located two electric charges of magnitudes ε and $-\varepsilon$, respectively, at a distance r, so that their energy is given by:

$$-\frac{1}{D}\frac{\varepsilon^2}{r}$$

For simplicity, in these general considerations, a binary electrolyte, such as KCl, is kept in mind which is completely split into

*Our relation deviates from that given by Planck in as much as we do not use the number of moles but rather the number of actual particles, which appears to be more suitable for our purpose. This corresponds to the appearance of Boltzmann's constant k instead of the gas constant R. An essential difference compared with Planck is, of course, not caused by this formulation.

ions so that in a volume V of the solution $N_1 = N$ K ions with the charge $+\varepsilon$ and an equal number $N_2 = N$ Cl ions with the charge $-\varepsilon$ are present. It can then be imagined that the average distance r, which enters into the computation of energy, equals the average distance between the ions, and since the volume associated with one ion is equal to $V/2N$, we write:

$$r = \left(\frac{V}{2N}\right)^{1/3}$$

In using this value for r, the electric energy of the solution would be estimated to:

$$U_e = -N\frac{\varepsilon^2}{D}\left(\frac{2N}{V}\right)^{1/3}$$

In fact, Ghosh[5] proceeds in this manner. This consideration, however, is wrong, and the complete theory based on it (practically characterized by the introduction of the cube root of the concentration), is to be rejected.

The (negative) electric energy of an ionic solution originates from the fact that, considering any one ion, we shall find on the average more dissimilar than similar ions in its surroundings, an immediate consequence of the electrostatic forces effective between the ions. There is some similarity to the case of crystals, such as NaCl, KCl, etc., in which, according to Bragg's investigations, each atom (here also present as ion) has dissimilar ones as nearest neighbors. Though it is correct (in agreement with the exact calculations by Born) to estimate the electric energy of the crystal by inserting the distance of two neighboring dissimilar atoms, it is a mistake to overestimate the analogy, and to use the same average distance $(V/2N)^{1/3}$ in the case of solutions. In fact, here an entirely different length is of fundamental importance, since now the ions can move freely, and, consequently, the desired length can only follow on the basis of an evaluation of the difference in probability for the period of time spent by similar and dissimilar ions in the same volume element in the vicinity of a specified ion. From this alone it already follows that the temperature movement will be of importance in the calculation of U_e.

From dimensions alone, we can only conclude the following. Assuming that the size of the ions does not have to be taken into account* for high dilutions, then one energy is the expression given above:

*In the following it will be shown that this assumption actually holds.

$$\frac{\varepsilon^2}{D}\left(\frac{2N}{V}\right)^{1/3}$$

Another energy, that of the temperature movement, measured by kT, is of equal importance. Thus it is to be expected that U_e will take the form:

$$U_e = -N\frac{\varepsilon^2}{D}\left(\frac{2N}{V}\right)^{1/3} f\left(\frac{\varepsilon^2}{D}\left(\frac{2N}{V}\right)^{1/3}/kT\right) \quad (8)$$

in which f is a function of the ratio of the two energies about which we can not make any *a priori* statements.*

A consideration of the limiting case of high temperatures leads to the same conclusion. If the energy of the temperature movement is large, and if we consider a volume element in the neighborhood of an ion singled out for this consideration, the probability of finding in it a similar ion is equal to the probability to find a dissimilar one. In the limit for high temperatures, U_e must disappear, i.e., the expression for U_e has T as essential parameter also for medium temperatures.

III. Computation of the Electric Energy of an Ionic Solution of a Uni-Univalent Salt

In a volume V, N molecules of a uni-univalent salt (for instance, KCl) are present disintegrated into ions; if the absolute value of the charge of an ion is ε (4.77×10^{-10} electrostatic units), the dielectric constant of the solvent is D. We follow one of these ions with the charge $+\varepsilon$, and we intend to ascertain its potential energy u with respect to the surrounding ions. Direct calculation, as attempted by Milner, who considers each possible arrangement of ions and lets it enter into the computation with the probability corresponding to Boltzmann's principle, proved too difficult mathematically. We therefore replace it by another consideration, where the computation is, from the beginning directed toward the average of the electric potential generated by the ions.

At a point P in the surroundings of the specified ion, the average value of the electric potential with respect to time be ψ; to transport a positive ion there, the work expended is $+\varepsilon\psi$; for a negative ion, however, the work expended is $-\varepsilon\psi$. Therefore, in

*The considerations by O. Klein are in agreement with this discussion on dimensions. *Meddel. från. K. Vetenskapsakad. Nobelinstitut*, 5, No. 6 (1919) (Article to celebrate the 60th birthday of S. Arrhenius).

a volume element dV at this location, we shall find, as an average value with respect to time, according to Boltzmann's principle,

$$ne^{-\frac{\varepsilon\psi}{kT}}dV$$

positive and:

$$ne^{+\frac{\varepsilon\psi}{kT}}dV$$

negative ions, putting $n = N/V$. In fact, in the limit for $T = \infty$, the ion distribution must become uniform, so that the multiplying factor of the exponential function must be put equal to N/V, i.e., equal to the number of ions of one kind per cubic centimeter of solution. With these statements however, nothing can as yet be obtained, since the potential ψ of the point P is still unknown. However, according to Poisson's equation, this potential must satisfy the condition:

$$\Delta\psi = -\frac{4\pi}{D}\varrho$$

if the electricity is distributed with a density ρ in a medium of dielectric constant D. On the other hand, from the above:

$$\varrho = n\varepsilon\left(e^{-\frac{\varepsilon\psi}{kT}} - e^{+\frac{\varepsilon\psi}{kT}}\right) = -2n\varepsilon\,\mathfrak{Sin}\frac{\varepsilon\psi}{kT} \qquad (9)$$

so that ψ can be determined as a solution of the equation:

$$\Delta\psi = \frac{8\pi n\varepsilon}{D}\,\mathfrak{Sin}\frac{\varepsilon\psi}{kT} \qquad (10)$$

The further we go from the specified ion, the smaller will be the potential ψ. For large distances we can then replace, with sufficientapproximation, $\sinh(\varepsilon\psi/kT)$ by $\varepsilon\psi/kT$. If this is done, equation (10) assumes the much simpler form:*

$$\Delta\psi = \frac{8\pi n\varepsilon^2}{DkT}\psi \qquad (10')$$

*We have also investigated the effect of the successive terms in the expansion of $\sinh(\varepsilon\psi/kT)$, and could establish that their effect on the final result is rather small. This computation is not presented.

In this equation, the factor of ψ on the right hand side has the dimension of the reciprocal of the square of a length. We put:

$$\kappa^2 = \frac{8\pi n\varepsilon^2}{DkT} \qquad (11)$$

so that κ is the reciprocal of a length, and equation (10') becomes:

$$\Delta\psi = \kappa^2\psi \qquad (12)$$

The length, introduced in this way:

$$\frac{1}{\kappa} = \sqrt{\frac{DkT}{8\pi n\varepsilon^2}}$$

is the essential quantity in our theory and replaces the average distance between ions in Ghosh's consideration. If numerical values are inserted (see later) and the concentration is measured, as usual, in moles per liter solution, then, if this concentration is denoted by γ,

$$\frac{1}{\kappa} = \frac{3{,}06}{\sqrt{\gamma}} 10^{-8}\,\text{cm}$$

for water at 0°C. The characteristic length reaches molecular dimensions for a concentration of $\gamma = 1$ (1 mole per liter).

We shall now interrupt the course of the consideration in order to investigate the physical interpretation of our characteristic length.

If an electrode is immersed in an electrolytic solution of potential *o*, the surface of the electrode compared with the solution have a potential difference ψ. The transition from ψ to *o* will then take place within a layer of finite thickness which is given by the above considerations. Using equation (12) and designating by z a coordinate at right angles to the surface of the electrode, we obtain:

$$\psi = \Psi e^{-\kappa z}$$

a function which satisfies equation (12). Since the right term of equation (12), from Poisson's equation, stands for $-4\pi\rho/D$, the charge density related to the given potential is:

$$\rho = -\frac{D\kappa^2}{4\pi}\psi e^{-\kappa z}$$

According to this formula, $1/\kappa$ measures the length within which the charge density of the ion atmosphere reduces to one eth part. Our characteristic length is a measure for the thickness of the ion atmosphere (i.e., of the well-known double layer by Helmholtz); according to equation (11), it depends on concentration, temperature, and dielectric constant of the solvent.*

Having clarified the significance of $1/\kappa$, we shall now use equation (12) to determine the potential distribution and the density distribution in the neighborhood of the singled-out ion with charge $+\varepsilon$. We designate the distance from this ion by r, and introduce spatial polar coordinates in equation (12). Equation (12) then becomes:

$$\frac{1}{r^2}\frac{d}{dr}\left(r^2\frac{d\psi}{dr}\right)=\varkappa^2\psi \qquad (12')$$

and this equation has the general solution:

$$\psi=A\frac{e^{-\varkappa r}}{r}+A'\frac{e^{\varkappa r}}{r} \qquad (13)$$

Since ψ disappears at infinity, A' must equal zero; the constant A, however, will have to be found from the conditions prevailing in the neighborhood of the ion. This determination shall be carried through in two steps, (a) and (b), assuming for (a) that the dimensions of the ion have no effect, and under (b) take the size of the ion into consideration. The trend of thought under (a) will then give the limiting law for high dilutions, while under (b) will fall the changes to be made in order to adapt this limiting law to larger concentrations.

(a) Ion Diameter Negligible

The potential of a single point charge ε in a medium of dielectric constant D would be:

$$\psi=\frac{\varepsilon}{D}\frac{1}{r}$$

assuming no other ions in the medium. Our potential, equation (13), must agree with this expression for infinitely small distances, consequently we must put;

*Agreement of the above results on the double layer with computations by M. Gouy, *J. Physik.* (4), 9, 457 (1910) on the theory of the capillary electrometer was subsequently established. We may, perhaps, point out that in this instance the original equation (10) permits a simple solution.

$$A = \frac{\varepsilon}{D}$$

and the desired potential becomes:

$$\psi = \frac{\varepsilon}{D}\frac{e^{-\varkappa r}}{r} = \frac{\varepsilon}{D}\frac{1}{r} - \frac{\varepsilon}{D}\frac{1 - e^{-\varkappa r}}{r} \qquad (14)$$

We split the potential into two parts, the first part representing a potential which is undisturbed by the surrounding ions, and the second part representing the potential derived from the ion atmosphere. For small values of r, the value of this latter potential is equal to:

$$-\frac{\varepsilon}{D}\varkappa$$

the potential energy u of the singled-out ion $+\varepsilon$ with respect to its surroundings amounts to:*

$$u = -\frac{\varepsilon^2}{D}\varkappa \qquad (15)$$

If we have a series of charges e_i, and if the potential at the respective location of each charge is ψ_i, then, according to the laws of electrostatics, the total potential energy:

$$U_e = \tfrac{1}{2}\sum e_i \psi_i$$

In our case, where N positive ions are present, each of which at a potential difference $-\varepsilon\kappa/D$ against its surrounding, and further N negative ions with a potential difference of $+\varepsilon\kappa/D$, the desired potential energy** will be:

*Besides the graphical result mentioned in the preface, the article by Milner contains a footnote (*Phil. Mag.*, *24*, 575, 1912), according to which in the case of the above text and in our terminology:

$$u = -\frac{\varepsilon^2}{D}\kappa(\pi/2)^{\frac{1}{2}}$$

A derivation of this formula is not included. It differs from our result by the factor $(\pi/2)^{\frac{1}{2}}$.

**Since we are only concerned with the mutual potential energy, we have to take for ψ_i not the value of the total potential but only the part caused by the surrounding charges, always calculated for the point at which the charge e_i is located.

$$U_e = \frac{N\varepsilon}{2}\left(-\frac{\varepsilon\varkappa}{D}\right) - \frac{N\varepsilon}{2}\left(+\frac{\varepsilon\varkappa}{D}\right) = -\frac{N\varepsilon^2\varkappa}{D} \tag{16}$$

Since $\varkappa$ is given by equation (11) as a function of the concentration, the potential energy of the ion solution is proportional to the square root of the concentration and not, as according to Ghosh, to the cube root of this quantity.

(b) Ion Diameter is Finite

We observed, previously, that the characteristic length $1/\varkappa$ reaches the magnitude of molecular dimensions for concentrations of 1 mole per liter. At suh concentrations, it is therefore inadmissible to replace an ion of finite, molecular size by a point charge, as was done under (a). It would not be within the nature of our calculations, based on Poisson's equation, to introduce a detailed concept of the distances to which the ions approach one another. Rather we shall in the following visualize ions as spheres of radius a, the interior of which is to be treated as a medium of dielectric constant D, and in the center of which is located a point charge of value $+\varepsilon$ or $-\varepsilon$. Then the magnitude a, obviously, does not measure the radius of the ion but a length which constitutes the mean value for the distance to which the surrounding ions, positive as well as negative, can approach the singled-out ion. Correspondingly, for positive and negative ions of equal size, for instance, a would be expected to be of the order of magnitude of the ion diameter. In general, the ion diameter is not to be considered the diameter of the actual ion, since, most likely, the ions have to be imagined as surrounded by a firmly attached layer of water molecules. It is obvious that the introduction of such a length a cannot be expected to be more than a rough approximation. The discussion of practical cases (compare later) will show that, in practice, this approximation is rather good.

As before, we express the potential surrounding a singled-out ion by:

$$\psi = A \cdot \frac{e^{-\varkappa r}}{r} \tag{17}$$

only the constant A must now be determined differently. According to our assumptions, we have to write:

$$\psi = \frac{\varepsilon}{D}\frac{1}{r} + B \tag{17'}$$

for the interior of the ion sphere (for a positive ion). Constants A and B must be determined from the boundary conditions on the surface of the sphere. There, that is, for $r = a$, the potentials ψ as well as the field strengths $-d\psi/dr$ must be continuous. From this it follows that:

$$\left.\begin{aligned} A\frac{e^{-\varkappa a}}{a} &= \frac{\varepsilon}{D}\frac{1}{a} + B \\ A \cdot e^{-\varkappa a}\frac{1+\varkappa a}{a^2} &= \frac{\varepsilon}{D}\frac{1}{a^2} \end{aligned}\right\} \quad (18)$$

hence:

$$A = \frac{\varepsilon}{D}\frac{e^{\varkappa a}}{1+\varkappa a}, \quad B = -\frac{\varepsilon\varkappa}{D}\frac{1}{1+\varkappa a} \quad (18')$$

The value of B represents the potential generated by the ion atmosphere in the center of the ion sphere; from this we obtain for the potential energy of a positive ion with respect to its surroundings the expression:

$$u = -\frac{\varepsilon^2\varkappa}{D}\frac{1}{1+\varkappa a} \quad (19)$$

As shown by comparison with equation (15), the effect of the size of the ion is expressed by the factor $1/(1 + \kappa a)$ only. For low concentrations (n small) κ is also small, according to equation (11), and the energy approaches the value given previously for infinitely small ions. For large concentrations (κ large), however, u gradually approaches the value:

$$-\frac{\varepsilon^2}{Da}$$

so that our characteristic length, $1/\kappa$, loses its effect in favor of the new length a which measures the size of the ions.

By means of equation (19) we obtain, as under (a), for the total electric energy of the ion solution the expression:

$$U_e = -\frac{N}{2}\frac{\varepsilon^2\varkappa}{D}\left[\frac{1}{1+\varkappa a_1} + \frac{1}{1+\varkappa a_2}\right] \quad (20)$$

provided - as appears to be indicated - the positive ions are characterized by a radius a_1 and the negative ions by another radius a_2. We could use (16) or (20) directly for the determination of the thermodynamic function in accordance with the discussion in section II. However, we shall first derive the expression for the energy, corresponding to (20), for any ionic solution, by eliminating the restriction to uni-univalent salts introduced in the interest of brevity.

IV. The Potential Fnergy of an Arbitrary Ion Solution

A solution contains:

$$N_1 \ldots N_i \ldots N_s$$

different ions with charges:

$$z_1 \ldots z_i \ldots z_s$$

such that the integers $z_1 \ldots . z_i \ldots . z_s$ measure the valencies and may assume positive as well as negative values. Since the total charge is equal to zero,

$$\sum N_i z_i = 0$$

must hold. In addition to the total numbers N_i, the number of ions per cubic centimer:

$$n_1 \ldots n_i \ldots n_s$$

be introduced.

Again any one of the ions is singled out, and the potential in its surroundings is determined from Poisson's equation:

$$\Delta\psi = -\frac{4\pi}{D}\varrho$$

From Boltzmann's principle, the density of the ith ion type is given by:

$$n_i e^{-z_i \frac{\varepsilon\psi}{kT}}$$

so that:

$$\varrho = \varepsilon \sum n_i z_i e^{-z_i \frac{\varepsilon\psi}{kT}}$$

and the fundamental equation is given by:

$$\Delta\psi = -\frac{4\pi\varepsilon}{D}\sum n_i z_i e^{-z_i \frac{\varepsilon\psi}{kT}} \qquad (21)$$

If we use again the expansion of the exponential function, as in the previous paragraph, the equation:

$$\Delta\psi = \frac{4\pi\varepsilon^2}{DkT}\sum n_i z_i^2 \psi \qquad (21')$$

instead of (21) will be the basic equation, since, because of the condition:

$$\sum n_i z_i = 0$$

the first term of the expansion disappears. Thus, in the general case, the square of the characteristic length $1/\kappa^2$ is to be defined by the equation:*

$$\kappa^2 = \frac{4\pi\varepsilon^2}{DkT}\sum n_i z_i^2 \qquad (22)$$

while the equation for the potential retains its previous form:

$$\Delta\psi = \kappa^2\psi$$

Again an ion shall be singled out, and the potential ψ in its vicinity be determined. In concordance with the discussion in the preceding paragraph,

$$\psi = A\frac{e^{-\kappa r}}{r}$$

is obtained for the field outside the ion.

Provided the ion carries a charge $z_i\varepsilon$ and a distance of approach to it equal to a_i is to be considered, then we have for the interior of the ion sphere:

$$\psi = \frac{z_i\varepsilon}{D}\frac{1}{r} + B$$

while the constants A and B are evaluated to:

$$A = \frac{z_i\varepsilon}{D}\frac{e^{\kappa a_i}}{1+\kappa a_i}, \quad B = -\frac{z_i\varepsilon\kappa}{D}\frac{1}{1+\kappa a_i}$$

To the given value of B corresponds the potential energy:

$$u = -\frac{z_i^2\varepsilon^2\kappa}{D}\frac{1}{1+\kappa a_i}$$

of the singled-out ion with respect to its ion atmosphere, while the total electric energy of the ion solution, as will readily be seen, assumes the value:

$$U_e = -\sum\frac{N_i z_i^2}{2}\frac{\varepsilon^2\kappa}{D}\frac{1}{1+\kappa a_i} \qquad (23)$$

*Since for uni-univalent salts $n_1 = n_2 = n$ and $z_1 = -z_2 = 1$, the general expression (22) for κ^2 agrees with the one (compare equation 11) given for this special case.

The inverse length κ is, in the general case, defined by equation (22).*

V. The Additional Electric Term to the Thermodynamic Potential

In section II we arrived at the result that the additional term to the thermodynamic potential:

$$G = S - \frac{U}{T}$$

resulting from the mutual effect of the ions, should be found from the equation:

$$G_e = \int \frac{U_e}{T^2} dT$$

To take care of the general case, let us take the expression (23) for U_e, then, when integrating, we have to consider that, according to (22), the reciprocal length in this expression depends on the temperature. The computation becomes clearer, if we first conclude from (22) that:

$$2\varkappa d\varkappa = -\frac{4\pi\varepsilon^2}{Dk}\sum n_i z_i^2 \frac{dT}{T^2}$$

where D is assumed independent of temperature,** and then use κ and not T as variable of integration. Thus results:

$$G_e = \frac{k}{4\pi\sum n_i z_i^2}\sum N_i z_i^2 \int \frac{\varkappa^2 d\varkappa}{1+\varkappa a_i} \quad (24)$$

If we introduce the abbreviation:

$$\varkappa a_i = x_i \quad (25)$$

find:

*From the expression for U_e, we can immediately derive the heat of dilution. We established that the theoretical value agrees with the experiments.

**In fact a direct, kinetic theory of the osmotic pressure, reported in Recueil des travaux chimiques des Pays-Bas et de la Belgique , proves the validity of the final expression for G_e independent of this assumption. For a discussion of the thermodynamic computation we may refer to B. A. M. Cavanagh, *Phil. Mag.*, *43*, 606 (1922).

$$\int\limits^{x=x} \frac{x^2 dx}{1+x a_i} = \frac{1}{a_i^3} \int\limits^{u=x_i} \frac{u^2 du}{1+u} = \frac{1}{a_i^3} \left\{ \text{const} + \log(1+x_i) - 2(1+x_i) + \tfrac{1}{2}(1+x_i)^2 \right\}$$

The constant of integration must be so determined that in the limit for infinite dilution the electrical addition G_e to the total potential disappears. Since according to (22), $\varkappa$ is proportional to:

$$\sqrt{\sum n_i z_i^2}$$

$\varkappa = 0$ corresponds to infinite dilution. Consequently the constant in the bracket must be so determined that for $x_i = 0$ the expression in the bracket is also equal to zero. Since at this limit:

$$\log(1+x_i) - 2(1+x_i) + \tfrac{1}{2}(1+x_i)^2$$

assumes the value -3/2, our constant is 3/2. Then:

$$\int \frac{x^2 dx}{1+x a} = \frac{1}{a_i^3} \left\{ \tfrac{3}{2} + \log(1+x_i) - 2(1+x_i) + \tfrac{1}{2}(1+x_i)^2 \right\}$$

and:

$$G_e = \frac{k}{4\pi \sum n_i z_i^2} \sum \frac{N_i z_i^2}{a_i^3} \left\{ \tfrac{3}{2} + \log(1+x_i) - 2(1+x_i) + \tfrac{1}{2}(1+x_i)^2 \right\} \qquad (26)$$

If expanded with respect to powers of x_i the function in the bracket takes the form:

$$\tfrac{3}{2} + \log(1+x_i) - 2(1+x_i) + \tfrac{1}{2}(1+x_i)^2 = \\ = \frac{x_i^3}{3} - \frac{x_i^4}{4} + \frac{x_i^5}{5} - \frac{x_i^6}{6} + \cdots$$

If we, therefore, introduce the abbreviation:

$$\chi_i = \chi(x_i) = \\ = \frac{3}{x_i^3} \left\{ \tfrac{3}{2} + \log(1+x_i) - 2(1+x_i) + \tfrac{1}{2}(1+x_i)^2 \right\} \qquad (27)$$

χ will, for small concentration, approach unity, and can be expanded into:

$$\chi_i = 1 - \tfrac{3}{4} x_i + \tfrac{3}{5} x_i^2 - \ldots \qquad (27')$$

By introduction of this function and consideration of the equation (22) defining κ^2, the addition to the thermodynamic potential can now be presented in the form:*

$$G_e = \sum N_i \frac{z_i^2 \varepsilon^2}{DT} \frac{\varkappa}{3} \chi_i \qquad (28)$$

where, for clarity, the expression (22) for κ according to which:

$$\varkappa^2 = \frac{4\pi\varepsilon^2}{DkT} \sum n_i z_i^2$$

be repeated explicitly.

For small concentrations, therefore, an amount of G_e proportional to κ, i.e., proportional to the square root of the concentration, is associated with each ion. If the finite dimensions of the ions are neglected, then, according to (27') and (25), χ_i would be equal to 1 throughout, and this dependency would appear to be valid for all concentrations. The dependence on the size of the ions, which accounts for the individual properties of the ions, is, then, measured by the function χ, given by (27) or (27'). In the limit for high dilutions, however, this influence disappears, and the ions can only be distinguished if their valencies are different.

VI. Osmotic Pressure, Vapor Pressure Depression, Freezing Point Depression, Boiling Point Increase

In accordance with the discussion of section II and in view of equations (7), (7'), and (7''), the thermodynamic function Φ of the solution is given by the expression:

$$\Phi = \sum_0^s N_i (\varphi_i - k \log c_i) + \sum_1^s N_i \frac{z_i^2 \varepsilon^2}{3D} \frac{\varkappa}{T} \chi_i \quad (29)$$

Here equation (28) has been used for the additional electric term to Φ, where:

$$\chi_i = \chi(x_i) = \chi(\kappa a_i)$$

is given by (27), and, as explained in the preceding paragraph, approaches unity in the limit for infinitely small concentrations. κ is our characteristic reciprocal length, defined by equation (22),

*The additional electric pressure p_e, mentioned in section II, equation (6'), results from this formula. The numerical value given there was computed in this way.

$$\varkappa^2 = \frac{4\pi\varepsilon^2}{DkT}\sum n_i z_i^2$$

By the method followed in Planck's *Lehrbuch der Thermodynamik*, the rules for the description of the phenomena mentioned in the heading can all be derived by differentiation of equation (29). The condition for equilibrium of a transition of a quantity δN_0 molecules of the solvent from the solution to the appropriate other phase is, as is well known,

$$\delta\Phi + \delta\Phi' = 0$$

where Φ' designates the thermodynamic potential of the second phase. We put:

$$\Phi' = N_0'\varphi_0' \qquad (30)$$

We wish to carry out the computations for the case of equilibrium between the solution and the frozen solvent, in view of the fact that the most extensive and the most reliable measurements are available for the freezing point depression as a function of concentration. We let N_0 vary by the amount δN_0 and N_0' by the amount $\delta N_0'$, and obtain immediately:

$$\delta(\Phi + \Phi') = \varphi_0'\delta N_0' + (\varphi_0 - k\log c_0)\delta N_0 + \\ + \sum_1^s N_i \frac{z_i^2\varepsilon^2}{3DT}\frac{d(\varkappa\chi_i)}{d\varkappa}\frac{\partial\varkappa}{\partial N_0}\delta N_0 \qquad (31)$$

since, as will readily be seen:

$$\sum_0^s N_i\delta\log c_i = \sum_0^s N_i\frac{\partial\log c_i}{\partial N_0}\delta N_0$$

assumes the value zero.

Since:

$$\delta N_0' = -\delta N_0$$

the condition for equilibrium reads:

$$\varphi_0' - \varphi_0 = -k\log c_0 + \sum_1^s N_i\frac{z_i^2\varepsilon^2}{3DT}\frac{d(\varkappa\chi_i)}{d\varkappa}\frac{\partial\varkappa}{\partial N_0} \qquad (32)$$

As presented here, it may be applied to all the phenomena mentioned in the heading, and it constitutes a relation between pressure, temperature, and concentrations.

In the definition of κ, n_i designates the number of ions of the ith type per unit volume so that:

$$n_i = \frac{N_i}{V}$$

and, as in Planck's treatment, the method is based on the linear relation for the volume:

$$V = \sum_0^s n_i v_i = n_0 v_0 + \sum_1^s n_i v_i$$

According to equation (22),

$$2\varkappa \frac{\partial \varkappa}{\partial N_0} = -\frac{4\pi\varepsilon^2}{DkT} \sum z_i^2 \frac{N_i v_0}{V^2} =$$
$$= -\frac{4\pi\varepsilon^2}{DkT} \frac{v_0}{V} \sum n_i z_i^2$$

Again using the equation defining κ, we obtain:

$$\frac{\partial \varkappa}{\partial N_0} = -\frac{\varkappa}{2} \frac{v_0}{V}$$

and our condition for equilibrium assumes the form:

$$\varphi_0 - \varphi_0' = k \log c_0 + v_0 \sum_1^s n_i \frac{z_i^2 \varepsilon^2}{6DT} \varkappa \frac{d(\varkappa \chi_i)}{d\varkappa} \quad (32')$$

The function of the concentration:

$$\frac{d(\varkappa \chi_i)}{d\varkappa}$$

characterizing the phenomena considered, can be computed easily from equation (27). Let us designate it by σ_i, then, retaining the abbreviation:

$$x_i = \varkappa a_i$$

we obtain:

$$\sigma_i = \frac{d(\varkappa \chi_i)}{d\varkappa} =$$
$$= \frac{3}{x_i^3} \left[(1 + x_i) - \frac{1}{1 + x_i} - 2 \log(1 + x_i) \right] \quad (33)$$

For small values of x_i, the expansion:

$$\sigma_i = 1 - \tfrac{3}{2} x_i + \tfrac{9}{5} x_i^2 - 2 x_i^3 + \ldots = \\ = \sum_{\nu=0}^{\nu=s} 3 \frac{\nu+1}{\nu+3} x_i^{\nu}$$

is valid so that σ_i approaches unity for small concentrations; for large concentrations σ_i approaches zero as $3/x_i^2$. Table I contains numerical values for σ as a function of $x = \kappa a$.

Table I

x	$\sigma(x)$	x	$\sigma(x)$	x	$\sigma(x)$	x	$\sigma(x)$
0	1.000	0.4	0.598	0.9	0.370	3.0	0.1109
0.05	0.929	0.5	0.536	1.0	0.341	3.5	0.0898
0.1	0.855	0.6	0.486	1.5	0.238	4.0	0.0742
0.2	0.759	0.7	0.441	2.0	0.176	4.5	0.0628
0.3	0.670	0.8	0.403	2.5	0.136	5.5	0.0540

In Figure 1 a plot of this function is presented.

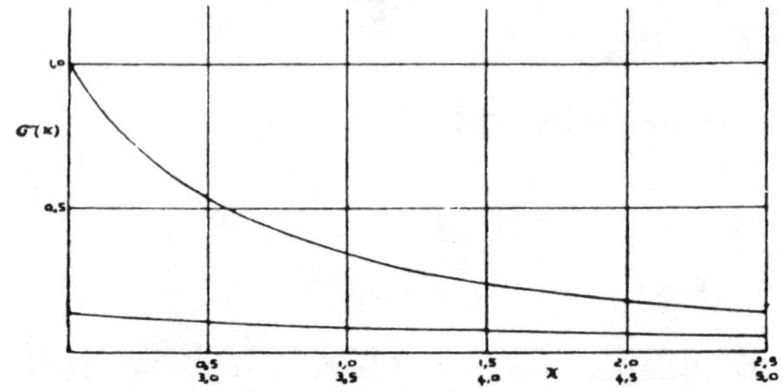

Figure 1

Since we shall have occasion to treat the freezing point depression of more concentrated solutions, it is advisable to calculate the amount of this depression from equation (32'), without introducing all simplifications permissible for highly diluted solutions. Let the freezing point of the pure solvent be T_0, the freezing point of the solution $T_0 - \Delta$, the heat of fusion of the frozen solvent q, the specific heat at constant pressure of the liquid solvent c_p, and the same quantity for the frozen solvent c_p', where the three last mentioned quantities be referred to an actual molecule so that they constitute the conventional quantities for one mole divided by Loschmidt's number. According to the equation defining φ, we then have:

$$\varphi_0 - \varphi_0' = -\frac{\Delta}{T_0}\frac{q}{T_0} + \frac{\Delta^2}{T_0^2}\left[(c_p - c_p') - \frac{2q}{kT_0}\right]$$

For c_0 we can put:

$$c_0 = 1 - \sum_1^s c_i$$

Since further we abbreviated:

$$\frac{d(\varkappa\chi_i)}{d\varkappa} = \sigma_i$$

it finally follows that:

$$\frac{\Delta}{T_0}\frac{q}{kT_0} - \frac{\Delta^2}{T_0^2}\left(\frac{c_p - c_p'}{2k} - \frac{q}{kT_0}\right) = \\ = -\log(1 - \sum_1^s c_i) - \frac{\varepsilon^2\varkappa}{6DkT}\sum_1^s v_0 n_i z_i^2 \sigma_i$$

If Loschmidt's number is designated by N,

$$Nq = Q$$

the heat of fusion of one mole,

$$Nk = R$$

the gas constant, and:

$$Nc_p = C_p \text{ e.g. } Nc_p' = C_p'$$

the specific heat per mole of the liquid and solid solvent, respectively, so that we can also write:

$$\left.\begin{aligned}&\frac{\Delta}{T_0}\frac{Q}{RT_0} - \frac{\Delta^2}{T_0^2}\left(\frac{C_p - C_p'}{2R} - \frac{Q}{RT_0}\right) = \\ &= -\log(1 - \sum_1^s c_i) - \frac{\varepsilon^2\varkappa}{6DkT}\sum_1^s v_0 n_i z_i^2 \sigma_i\end{aligned}\right\} \quad (34)$$

For low concentrations we can (1) neglect Δ^2/T_0^2 compared with Δ/T_0, (2) put:

$$-\log(1 - \sum_1^s c_i) = \sum_1^s c_i$$

and (3) the total volume can be identified with the volume of the water, considering the number of the dissolved ions as infinitely small compared with the number of the water molecules. This corresponds to:

$$v_0 n_i = \frac{v_0}{V} N_i = \frac{N_i}{N_0} = \frac{N_i}{N_0 + \sum_1^s v_i N_i} = c_i$$

With these approximations, we obtain:*

$$\frac{\Delta}{T_0} \frac{Q}{RT_0} = \sum_1^s c_i \left(1 - \frac{\varepsilon^2 \varkappa}{6DkT} z_i^2 \sigma_i\right) \qquad (35)$$

whereas the classical formula for the same assumptions reads:

$$\frac{\Delta}{T_0} \frac{Q}{RT_0} = \sum_1^s c_i \qquad (35')$$

VII. Freezing Point Depression of Diluted Solutions

The characteristics of the electric effect of the ions are particularly clear in the limit for highly diluted solutions as expressed in equation (35). We shall therefore treat the laws for this limiting case separately. The formula (35) applies to the general case of a mixture of several electrolytes which, moreover, may be only partially dissociated into ions. We consider the special case of one type of molecule in solution. The molecule be completely dissociated into ions, and consist of s types of ions, numbered $1, \ldots\ldots i, \ldots\ldots\ldots s$, so that:

$$\nu_1, \quad \ldots \nu_i, \quad \ldots \nu_s$$

ions of the type $1, \ldots\ldots i, \ldots\ldots s$ constitute the molecule. If the charges associated with each of these ions are:

$$z_1 \varepsilon, \quad \ldots z_i \varepsilon, \quad \ldots z_s \varepsilon$$

(For H_2SO_4, dissociated into the ions H and SO_4, for instance,

$$\nu_1 = 2, \quad \nu_2 = 1, \quad z_1 = +1, \quad z_2 = -2$$

*It is not necessary to distinguish between ions and neutral molecules; if both are present, we imply have to put $z_i = 0$ for the latter. If all separate particles are neutral, then, naturally, equations (35) and (35') become identical.

provided subscript 1 refers to the H ions and subscript 2 to the SO_4 ions.)

Since the molecule as a unit carries no charge, we have:

$$\sum_1^s \nu_i z_i = 0$$

The solution thus consists of N_0 molecules of the solvent and N molecules of the added electrolyte, where N may be considered small compared with N_0. Then:

$$c_i = \frac{N_i}{N_0 + \sum_1^s N_i} = \frac{N_i}{N_0}$$

If we take into account that:

$$N_i = \nu_i N$$

and designate by c the concentration referred to the dissolved type of molecule so that, in the approximation here used,

$$c = \frac{N}{N_0}$$

then:

$$c_i = \nu_i c$$

Equation (35) for the freezing point depression then becomes:

$$\frac{\Delta}{T_0}\frac{Q}{RT_0} = f_0 \Sigma c_i = f_0 c \Sigma \nu_i \qquad (36)$$

where:

$$f_0 = 1 - \frac{\varepsilon^2 \varkappa}{6DkT}\frac{\Sigma \nu_i z_i^2 \sigma_i}{\Sigma \nu_i} \qquad (37)$$

The quantity f_0 is the osmotic coefficient mentioned in the introduction, since $f_0 = 1$ represents the transition to classical theory as indicated by equation (35'). If Δ_k designates the freezing point depression calculated in accordance with classical theory, then:

$$\frac{\Delta}{\Delta_k} = f_0$$

or:

$$1 - f_0 = \frac{\Delta_k - \Delta}{\Delta_k}$$

Relation (37) indicates qualitatively that the actual freezing point depression should be smaller than that expected from classical theory, a result which is consistently confirmed for diluted electrolytic solutions. The quantities κ and s which appear in equation (37) are determined by formulas (22) and (33) (the latter one with associated table). As explained in the preceding paragraph, σ_i is a measure for the effect of the finite size of the ions which disappears for very small concentrations, since σ then approaches unity. Therefore, if we first consider the limiting laws valid for highly diluted solutions, we have for this limiting case:

$$f_0 = 1 - \frac{\varepsilon^2 \varkappa}{6DkT} \frac{\Sigma \nu_i z_i^2}{\Sigma \nu_i} \tag{38}$$

Further, from equation (22):

$$\varkappa^2 = \frac{4\pi\varepsilon^2}{DkT} \Sigma n_i z_i^2$$

and since:

$$n_i = \nu_i \frac{N}{V} = \nu_i n$$

introducing the volume concentration, n, of the dissolved molecules, we have:

$$\varkappa^2 = \frac{4\pi n \varepsilon^2}{DkT} \Sigma \nu_i z_i^2$$

It follows that for very low concentrations:

$$f_0 = 1 - \frac{\varepsilon^2}{6DkT} \sqrt{\frac{4\pi\varepsilon^2}{DkT} n \Sigma \nu_i} \left(\frac{\Sigma \nu_i z_i^2}{\Sigma \nu_i} \right)^{3/2} \tag{38'}$$

where $n\,\Sigma\nu_i$ designates the number of total ions per cubic centimeter in the solution, and:

$$w = \left(\frac{\Sigma \nu_i z_i^2}{\Sigma \nu_i} \right)^{3/2} \tag{39}$$

shall be called the "valency factor," since it measures the effect of the **valencies, z_i, on the phenomena.** It is best not to consider f_0, but its deviation from unity, and thus to write for very low concentrations:

$$1 - f_0 = w \frac{\varepsilon^2}{6DkT} \sqrt{\frac{4\pi\varepsilon^2}{DkT} n \Sigma \nu_i} \tag{40}$$

First, this formula expresses the dependence of $1 - f_0$ on concentration, stating in this regard:

Law No. 1

For all electrolytes and in the limit for small concentrations, the percentage deviation of the freezing point depression from its classical value is proportional to the square root of the concentration.

It is possible to state this law as a general law, because, in highly diluted solutions, all electrolytes can be considered as completely dissociated into ions. However, the region of complete dissociation is, in practice, only reached by strong electrolytes.

Second, equation (39) makes a statement on the effect of the valencies of the ions which may be formulated as follows:

Law No. 2

If the dissolved molecule dissociates into $\nu_1, . \nu_i . . . \nu_s$ different ions $1, . . . i, . . . s$ with the valencies $z_1 . . . z_i . . . z_s$, then, for low concentrations, the percentage deviation of the freezing point depression from its classical value is proportional to a valency factor, w, which can be computed from:

$$w = \left(\frac{\Sigma \nu_i z_i^2}{\Sigma \nu_i}\right)^{3/2}$$

As an example for the calculation of this valency factor Table II is presented, where the type of the salt is determined by the example given in the left column, and the value of w is given in the right column:

Table II

Type	Valency factor, w
KCl	$1 * 1$
$CaCl_2$	$2\sqrt{2} = 2.83$
$CuSO_4$	$4\sqrt{4} = 8$
$AlCl_3$	$3\sqrt{3} = 5.20$
$Al_2(SO_4)_3$	$6\sqrt{6} = 16.6$

Thus the influence of the ions increases considerably with increasing valency which also is in accordance with the qualitative prediction.

Third, the solvent also has an effect in keeping with the well known suggestion by Nernst intended to explain the ionizing force of solvent with high dielectric constant. According to equation (40), we have:

Law No. 3

For low concentrations the percentage deviation of the freezing point depression from the classical value is inversely proportional to the 3/2th power of the dielectric constant of the solvent.

The other constants appearing in equation (40) are the elementary charge $\varepsilon = 4.77 \times 10^{-10}$ e.s.u., Boltzmann's constant $k = 1.346 \times 10^{-16}$ erg, and the temperature, T, which latter is present explicitly and implicitly, since the dielectric constant, D, varies with T.

If we deal with diluted solutions in the conventional sense, σ can no longer be replaced by unity, and equation (37) applies, which reads explicitly:

$$1 - f_0 = w\frac{\varepsilon^2}{6DkT}\sqrt{\frac{4\pi\varepsilon^2}{DkT}n\sum\nu_i}\frac{\Sigma\nu_i z_i^2 \sigma_i}{\Sigma\nu_i z_i^2} \quad (41)$$

As shown by Table I, as well as by the formula (33) from which the table is derived, σ_i continuously decreases with increasing concentration and finally decreases as:

$$\frac{3}{x_i^2} = \frac{3}{\varkappa^2 a_i^2}$$

i.e., inversely proportional to the concentration, since κ is proportional to the square root of this quantity. According to (41) the deviation $1 - f_0$ increases proportional to the square root of the concentration for very small concentrations, then, for higher concentrations, in view of the effect of σ, i.e., in view of the finite diameter of the ions, the deviation will reach a maximum, and finally decrease inversely proportionally to the square root of the concentration. Even though this statement contains an extrapolation to higher concentrations of equation (41) which is not entirely justified, it seems to describe the behavior of concentrated solutions (compare section IX). In fact, measurements show a maximum of $1- f_0$ as a characteristic of the curves for the freezing point depression. However, we believe that the phenomenon of hydration (compare the last section) also contributes considerably to the formation of the maximum. A numerical comparison of theory and experiments will be given in section IX.

VIII. Dissociation Equilibrium

Not limiting our considerations to strong electrolytes only, a dissociation equilibrium will exist between undissociated molecules and ions. However, the equilibrium is not to be computed by means of the classical formula, because also in this instance the mutual electric forces have a disturbing influence. How this can be taken into consideration in accordance with our theory shall now be computed. We start again with the expression (29) for the thermodynamic potential Φ of the solution:

$$\Phi = \sum_0^s N_i(\varphi_i - k \log c_i) + \sum_1^s N_i \frac{z_i^2 \varepsilon^2}{3D} \frac{\varkappa}{T} \chi_i$$

some of the particles present in the solution will carry an electric charge, while **others will be electrically neutral. For** the latter, we will have simply $z_i = 0$. The solvent shall be indicated by the subscript o. We now undertake, in a well-known procedure, the variation of the numbers N_i, and calculate the associated changes in the potential. This leads to:

$$\delta\Phi = \sum_{i=0}^{i=s} \delta N_i(\varphi_i - k \log c_i) + \sum_{i=1}^{i=s} \delta N_i \frac{z_i^2 \varepsilon^2}{3DT} \varkappa \chi_i$$
$$+ \sum_{i=1}^{i=s} N_i \frac{z_i^2 \varepsilon^2}{3DT} \frac{d(\varkappa\chi_i)}{d\varkappa} \sum_{j=1}^{j=s} \frac{\partial\varkappa}{\partial N_j} \delta N_j$$

If it is taken into account that, according to the defining equation (22):

$$\varkappa^2 = \frac{4\pi\varepsilon^2}{DkT} \sum_{l=1}^{l=s} n_l z_l^2 = \frac{4\pi\varepsilon^2}{DkT} \sum_{l=1}^{l=s} \frac{N_l z_l^2}{V}$$

the quantity κ may depend on all numbers $N_1 \ldots N_s$. If in the third sum the indices of summation, i and j, are interchanged, $\delta\Phi$ may be written in the form:

$$\delta\Phi = \delta N_0(\varphi_0 - k \log c_0) + \sum_{i=1}^{i=s} \delta N_i \left[\varphi_i - k \log c_i + \frac{\varepsilon^2}{3DT}\left(z_i^2 \varkappa \chi_i + \sum_{j=1}^{j=s} N_j z_j^2 \frac{d(\varkappa\chi_j)}{d\varkappa} \frac{\partial\varkappa}{\partial N_i}\right)\right]$$

However, $\partial\kappa/\partial N_i$ can be calculated from the definition of κ. We obtain, provided the linear relation regarding the volume is retained,

$$\frac{\partial\varkappa}{\partial N_i} = \frac{\varkappa}{2\sum_1^s n_l z_l^2} \cdot \frac{z_i^2 - v_i \sum_1^s n_l z_l^2}{V}$$

If the conventional assumption is made that a chemical reaction may take place in the solution, where the proportions:

$$\delta N_1 : \delta N_2 : \ldots \delta N_i : \ldots : \delta N_s = \\ = \mu_1 : \mu_2 : \ldots \mu_i : \ldots \mu_s$$

hold, the condition of equilibrium follows from the expression

$$\sum_{i=1}^{i=s} \mu_i \log c_i = \sum_{i=1}^{i=s} \frac{\mu_i \varphi_i}{k} + \frac{\varepsilon^2 \varkappa}{6DkT} \sum_{i=1}^{i=s} \left\{ 2\mu_i z_i^2 \chi_i + \mu_i (z_i^2 - v_i \sum_1^s n_j z_j^2) \frac{\sum_{j=1}^{j=s} n_j z_j^2 \frac{d(\varkappa \chi_j)}{d\varkappa}}{\sum_{j=1}^{j=s} n_j z_j^2} \right\} \quad (42)$$

for the variation of the potential. This condition is distinguished from the classical condition by the additional term on the righthand side. If the activity coefficient f_a is introduced, as was done in the introduction, by putting:

$$\sum_1^s \mu_i \log c_i = \log (f_a K)$$

where K represents the classical constant of equilibrium, the activity coefficient is defined by the relation:

$$\log f_a = \frac{\varepsilon^2 \varkappa}{6DkT} \sum_{i=1}^{i=s} \left\{ 2\mu_i z_i^2 \chi_i + \mu_i (z_i^2 - v_i \sum_1^s n_j z_j^2) \frac{\sum_{j=1}^{j=s} n_j z_j^2 \frac{d(\varkappa \chi_j)}{d\varkappa}}{\sum_{j=1}^{j=s} n_j z_j^2} \right\} \quad (43)$$

According to this formula it is, of course, possible to provide a special activity coefficient for each atom or molecule taking part in the reaction by putting:

$$\log f_a = \mu_1 \log f_a^1 + \ldots \\ \mu_i \log f_a^i + \ldots \mu_s \log f_a^s \quad (44)$$

with:

$$\log f_a^i = \frac{\varepsilon^2 \varkappa}{6DkT} \left\{ 2 z_i^2 \chi_i + (z_i^2 - v_i \sum_1^s n_j z_j^2) \frac{\sum_1^s n_j z_j^2 \frac{d(\varkappa \chi_i)}{d\varkappa}}{\sum_1^s n_j z_j^2} \right\} \quad (44')$$

Then, however, as equation (44') indicates by the appearance of κ, this coefficient referred to a definite type of molecule does not solely depend on quantities which are related to this type of atom.

Here again simplifications are possible by limitation to lower concentrations. In this case:

$$v_i \sum_1^s n_j z_j^2$$

is negligible compared with z_i^2; if this is done, the volume of the dissolved substance is considered negligible compared with the total volume. Thus:

$$\log f_a{}^i = \frac{\varepsilon^2 \varkappa}{6DkT} z_i^2 \left\{ 2\chi_i + \frac{\sum_1^s n_j z_j^2 \frac{(d\varkappa\chi_i)}{d\varkappa}}{\sum_1^s n_j z_j^2} \right\} \quad (45)$$

Finally we can find the limiting value of the activity coefficient for **increasingly diluted solutions.** In this limit, where the effect due to the dimension of the ion vanishes, we can put $\chi = 1$, and obtain:

$$\log f_a{}^i = \frac{\varepsilon^2 \varkappa}{2DkT} z_i^2 \quad (45')$$

Since κ depends on the properties of all ions (is affected by their valency), the special coefficient $f_a{}^i$ is, not even in this limiting case, exclusively a function of the properties of the ith ion. We shall not discuss this limiting law in detail, and only observe that here again in the limit proportionality exists between log f_a and the square root of the concentration.

IX. Comparison of Freezing Point Depression with Experimental Results

Figure 2 is a representation of the characteristic behavior of

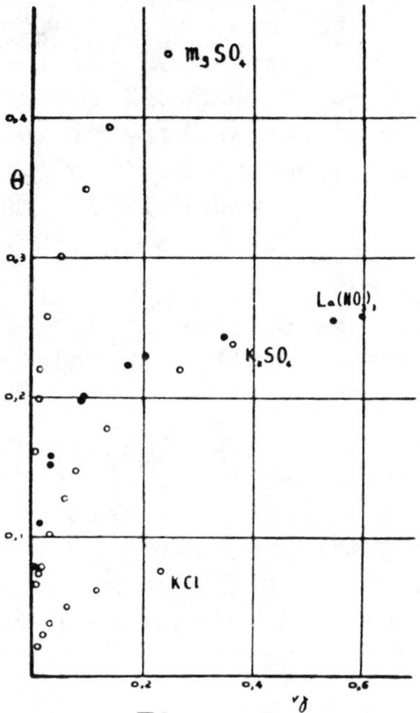

Figure 2

strong electrolytes. A magnitude $\nu\gamma$ which measures the ion concentration is plotted on the horizontal axis, where γ, as indicated before, designates the concentration of the electrolyte in moles per liter,* while $\nu = \Sigma\nu_i$ represents the number of ions into which the molecule of the salt dissociates. Four representatives KCl, K_2SO_4, $La(NO_3)_3$ and $MgSO_4$ were chosen from four types differing in the valencies of their ions. KCl dissociates into two univalent ions, K_2SO_4 in two univalent and one bivalent ion, $La(NO_3)_3$ in three univalent and one trivalent ion, and $MgSO_4$ in two bivalent ions. If we designate the freezing point depression expected from classical theory for complete dissociation with Δ_k, and the observed freezing point depression with Δ, the expression:

$$\Theta = \frac{\Delta_k - \Delta}{\Delta_k} \qquad (46)$$

i.e., the percentage deviation from the classical value was found and plotted as ordinate. According to section VII we can also put:

$$\Theta = 1 - f_0 \qquad (46')$$

thus represented, Θ measures the deviation of the osmotic coefficient from its limiting value 1. Since in a solution with water as solvent:

$$\Delta_k = \nu\gamma \cdot 1.860^0 \qquad (47)$$

a point on the abscissa corresponds, for all electrolytes, to a concentration which should produce the same freezing point depression provided the mutual forces are disregarded. We plotted the observed values and omitted to connect corresponding points by a curve in order to avoid any preconceived interpretation. This method, however, is possible only because recent and excellent measurements by American research workers of the freezing point depression at low concentrations are available. The measurements in Figure 2 are taken from Adams and Hall and Harkins.[8]

It is evident that the deviation Θ does not increase for low concentrations with the first or even higher powers of the concentration as required by the law of mass action. Further the curve demonstrates the strong effect of the ion valency.

*For the salts K_2SO_4, $La(NO_3)_3$, $MgSO_4$ the concentration γ' in moles per 1000 g. water is substituted for γ, as given by the authors cited below, since in the absence of density measurements for these salt solutions at 273^0, a conversion to moles per liter could not be carried out; this means only an insignificant deviation for the low concentrations considered here.

Our theory requires that, for very low concentrations, the percentage deviation Θ be proportional to the square root of the concentration and that the factor of proportionality depend strongly on the valency of the ions. According to equations (39) and (40) we have (assuming the molecule dissociates into $\nu_1 \ldots \nu_i \ldots \nu_s$ ions with the valencies $z_1 \ldots z_i \ldots z_s$):

$$\Theta = 1 - f_0 = w \frac{\varepsilon^2}{6DkT} \sqrt{\frac{4\pi\varepsilon^2}{DkT} n \Sigma\nu_i} \quad (48)$$

with the valency factor:

$$w = \left(\frac{\Sigma\nu_i z_i^2}{\Sigma\nu_i}\right)^{3/2} \quad (49)$$

First we want to express the number of ions n per cc. as the concentration γ measured in moles per liter. We take the value 6.06×10^{23} for Loschmidt's number; then:

$$n = 6.06 \cdot 10^{20} \gamma$$

It is further assumed that $\varepsilon = 4.77 \times 10^{-10}$ e.s.u., $k = 1.346 \times 10^{-16}$ erg, and since the following deals with the freezing point depression of solutions having water as a solvent, $T = 273$. The dielectric constant of water is calculated from the interpolation formula[9] given by Drude. We find for 0°C.:

$$D = 88.23$$

Using these figures, we get (with $\Sigma\nu_i = \nu$):

$$\sqrt{\frac{4\pi\varepsilon^2}{DkT} n\nu} = 0.231 \cdot 10^8 \sqrt{\nu\gamma} \frac{1}{\text{cm}}$$

and hence:

$$\Theta = 0.270 w \sqrt{\nu\gamma} \quad (50)$$

Our quantity κ becomes with the above numerical values:

$$\kappa = 0.231 \cdot 10^8 \sqrt{\nu\gamma} \sqrt{\frac{\Sigma\nu_i z_i^2}{\nu}} \frac{1}{\text{cm}} \quad (51)$$

In Figure 3 observed values[10] of Θ have been plotted against a new abscissa $\sqrt{\nu\gamma}$, the experimental points have been interconnected by straight lines. Further, four straight lines starting at the origin are presented, which illustrate the limiting law expressed

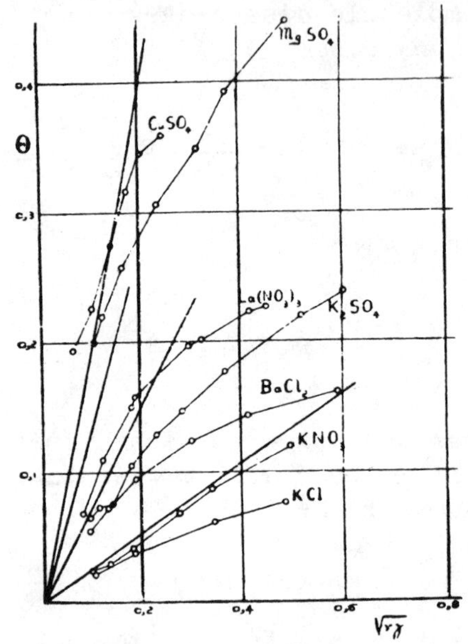

Figure 3

by equation (50). The four types of salts in the figure have the valency factors:

$$w = 1, \quad w = 2\sqrt{2}, \quad w = 3\sqrt{3}, \quad w = 8$$

the straight lines correspond to these values. It will be seen that for low concentrations the straight lines are actually approximated, so that, apparently, the limiting law involving the square root of the concentration corresponds to the facts. Further the absolute values of the slope - computed by means of the dielectric constant equal to 88.23, and distinguished theoretically only by the valency factor (as expressed in equation (50) by the factor 0.270 w) - are confirmed by experiment. However, Figure 3 indicates that early deviations from the limiting law take place. This is in agreement with the considerations in section III and equation (51), according to which, even for uni-univalent electrolytes, the characteristic length $1/\kappa$ is of the order of magnitude of the ion diameter already for $\gamma = 1$, so that it is no longer permissible to neglect it. We have further based our theory on the simplified version, equation (21'), of the potential equation. This also may have some effect. However, we pointed out (see footnote on page 227) that this latter effect is theoretically comparatively insignificant. The experimental results also indicate that the deviations from the limiting law are caused by the individual properties of the ions. To show this, we present Figure 4. Here observations

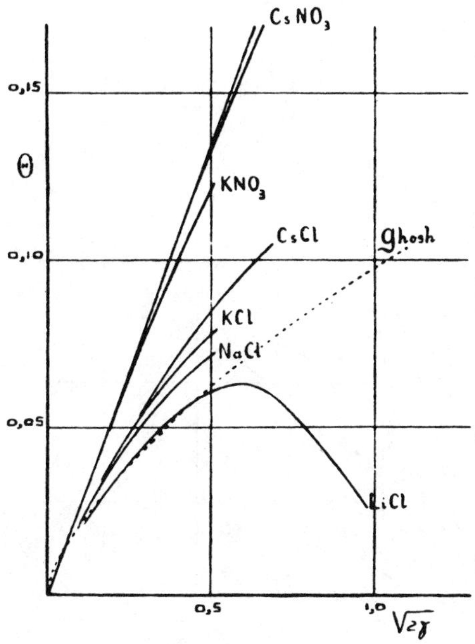

Figure 4

of uni-univalent ions, exclusively, are entered* as a function of $\sqrt{2\gamma}$ (since, here, $\nu = 2$). The straight line represents the limiting law discussed above; all curves approach this line for low concentrations. The deviations vary greatly in magnitude, and, it should be noted, are in the order Cs, K, Na, Li for the salts of chlorine. This is the same order as is obtained if the alkali ions are arranged according to decreasing mobility, an order which is in contradiction with the assumed dimension of the ions, and which was correlated only recently by Born[11] with the relaxation time of water for electric polarization following from dipole theory. To afford orientation with regard to the work by Ghosh, the curve for Θ, as evaluated by this theory, is given in the figure by a dashed line. It should be valid for all salts, and, moreover, has a vertical tangent at the origin.

*Besides the measurements already cited, we used here measurements by H. Jahn, *Z. phys. Chem.*, *50*, 129 (1905); *59*, 31 (1907) (LiCl, CsCl); E. W. Washburn and MacInnes, *J. Am. Chem. Soc.*, *33*, 1686 (1911) (LiCl, $CsNO_3$); W. H. Harkins and W. A. Roberts, *ibid.*, *38*, 2658 (1916)(NaCl) (concentration partly in moles per liter, partly moles per 1000 g. water).

The question now is: to what extent can our theory, improved with regard to the ion dimensions, account for the individual deviations. The conditions are illustrated in Figure 5. We have

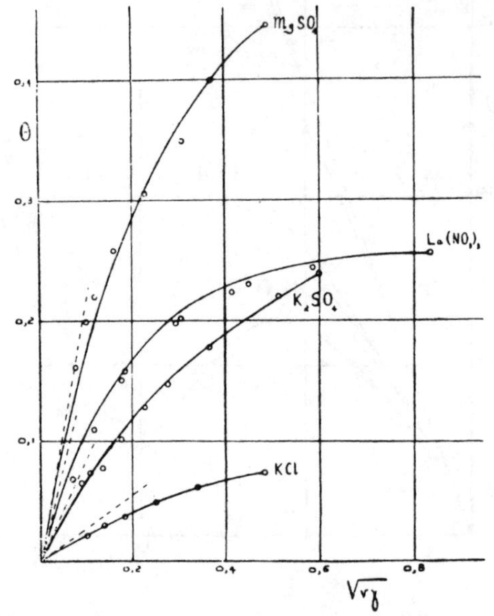

Figure 5

again chosen the four electrolytes of the four types previously mentioned, and plotted the observed values of Θ as a function of $\sqrt{\nu\gamma}$. According to equation (41) we obtain, taking into account the dimensions of the ions (upon introduction of the numerical values):

$$\Theta = 1 - f_0 = 0.270\, w \sqrt{\nu\gamma}\, \frac{\Sigma \nu_i z_i^2 \sigma_i}{\Sigma \nu_i z_i^2} \qquad (52)$$

where σ_i designates the function of the argument $x_i = \kappa a_i$ tabulated in Table I and given by formula (33), and where a_i denotes the length which measures the size of the ith ion with respect to its surroundings. At the present state of affairs, it did not appear advisable to us to study the separate ion dimensions, but to calculate with an average diameter a equal for all ions of one electrolyte. Then all σ_i become equal, and the expression:

$$\Theta = 0.270\, w \sqrt{\nu\gamma}\, \sigma(\kappa a) \qquad (53)$$

is obtained for Θ. For the determination of the magnitude of a, we chose, for the time, only one observed value, the one corresponding to the highest concentration, and then plotted the curve-resulting from the theoretical formula (53) with the a thus obtained-in the figure. Four dashed straight lines radiating from the origin (tangents to the curves) represent the limiting law (50) for strongly diluted solutions. Agreement with the observations is very good, particularly in view of the determination of constants from a single observed value.* The figure is supplemented by the following tables:

Table III.

KCl (a = 3.76 × 10^{-8}cm)

2γ	$\sqrt{2\gamma}$	Θ observed	Θ calculated
0.0100	0.100	0.0214	0.0237
0.0193	0.139	0.0295	0.0313
0.0331	0.182	0.0375	0.0392
0.0633	0.252	0.0485	0.0499
0.116	0.341	0.0613	0.0618
0.234	0.484	0.0758	---

In each first column is entered the ion concentration $\nu\gamma$, in the second column the value of the abscissa $\sqrt{\nu\gamma}$ in Figure 5, in

*The method for the determination of a is explained in detail for $La(NO_3)_3$ as an example. For γ' = 0.17486, Θ' = 0.2547 was observed; since ν = 4, the abscissa becomes $(\nu\gamma')^{\frac{1}{2}}$ = 0.836. According to the limiting law (50) for extreme dilution, we would obtain with $w = 3\sqrt{3}$ (corresponding to equation 49 for $\nu_1 = 1$, $\nu_2 = 3$, $z_1 = 3$, $z_2 = -1$) a value of Θ = 1.173, the actually observed value is obtained from this limiting value by multiplication with 0.216. According to equation (53) this factor is equal to σ. From Figure 1 we find that an abscissa $x = \kappa a$ = 1.67 belongs to the ordinate σ = 0.216; further from equation (51) by substituting $\sqrt{\nu\gamma'}$ = 0.836, the value of $\kappa = 0.336 \times 10^{-8}\text{cm}^{-1}$. Consequently, a diameter

$$a = x/\kappa = 4.97 \times 10^{-8} \text{ cm.}$$

corresponds to the observed data.

(For the salts K_2SO_4, $La(NO_3)_3$, $MgSO_4$, the concentration γ' is given in moles per 1000 g. water and used for the determination of Θ which is, therefore, designated by Θ . For the low concentrations considered here, the resulting deviations are very small; a conversion of γ' to γ would not give a noticeable change in the values for Θ' observed, Θ' calculated, and a.)

Table IV.

$K_2SO_4 (a = 2.69 \times 10^{-8}$ cm)

$3\gamma'$	$\sqrt{3\gamma'}$	Θ observed	Θ calculated
0.00722	0.0906	0.0647	0.0612
0.0121	0.110	0.0729	0.0724
0.0185	0.136	0.0776	0.0871
0.0312	0.176	0.101	0.108
0.0527	0.229	0.128	0.132
0.0782	0.280	0.147	0.152
0.136	0.369	0.178	0.183
0.267	0.516	0.220	0.217
0.361	0.600	0.238	---

Table V.

$La(NO_3)_3 (a = 4.97 \times 10^{-8}$ cm)

$4\gamma'$	$\sqrt{4\gamma'}$	Θ observed	Θ calculated
0.00528	0.0728	0.0684	0.0828
0.0142	0.119	0.110	0.121
0.0322	0.179	0.151	0.157
0.0343	0.185	0.158	0.161
0.0889	0.298	0.197	0.204
0.0944	0.308	0.201	0.207
0.173	0.418	0.223	0.230
0.205	0.453	0.229	0.235
0.346	0.588	0.243	0.248
0.599	0.836	0.255	---

Table VI.

$MgSO_4 (a = 3.35 \times 10^{-8}$ cm)

$2\gamma'$	$\sqrt{2\gamma'}$	Θ observed	Θ calculated
0.00640	0.0800	0.160	0.147
0.0107	0.103	0.199	0.179
0.0149	0.122	0.220	0.203
0.0262	0.162	0.258	0.248
0.0534	0.231	0.306	0.311
0.0976	0.312	0.349	0.368
0.138	0.372	0.392	0.400
0.242	0.493	0.445	---

the third column the observed value of Θ,* and in the fourth column the value of the same quantity computed from equations (53) and (51). The figure corresponding to the highest concentration is not entered here, since from it, in each instance, the average diameter a, given in the title of the tables, was computed.

Finally, in Figure 6, is given a representation of the theory

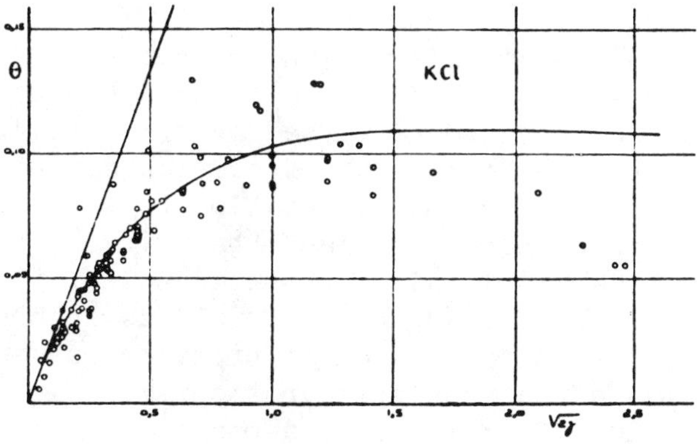

Figure 6

and observation of KCl solutions with water as solvent. In discussing this figure, it is our principal intention to present a few indications regarding the behavior of concentrated solutions; incidentally, we intend to show how large the discrepancies are between the separate results given in the literature in spite of high accuracy claimed by the individual observers. For this purpose the figure contains all observations on KCl solutions since 1900 that we found.[12] As abscissa is chosen, as before, $\sqrt{2\gamma}$, where γ denotes, according to our definition, the concentration in moles per liter solution. All information referring to concentrations measured differently, is here calculated for these concentrations by means of the measured densities[13] of KCl solutions. The ordinate is again designated by Θ, it does not, however, exactly represent the previous expression:

$$\frac{\Delta_k - \Delta}{\Delta_k}$$

In fact, not even the classical theory prescribes proportionality between freezing point depression and concentration for concentrated solutions. First, this is so because log $(1-c)$ and not the concentration itself appears in the classical equation. Second, the difference between the thermodynamic potentials of ice and water is no longer given with sufficient accuracy by the first

*See previous footnote.

term of the Taylor expansion, proportional to Δ, the second term, involving Δ^2, must be retained. Accordingly, in this instance, we have to use the complete equation (34). For KCl we have $n_1 = n_2 = n$ and $z_1 = z_2 = 1$, further we shall again replace the two ion diameters a_1 and a_2 by an average value a. Then equation 34 may be rearranged as follows:

$$\frac{1}{2nv_0}\left[\frac{\Delta}{T_0}\frac{Q}{RT_0} - \frac{\Delta}{T_0^2}\left(\frac{C_p - C_p'}{2R} - \frac{Q}{RT_0}\right) + \log(1 - 2c)\right] = -\frac{\varepsilon^2 \varkappa}{6DkT}\sigma \qquad (54)$$

The term on the left-hand side was now computed for different concentrations. For this purpose $C_p - C_p'$ was put equal to 3.6, corresponding to an approximate value for $C_p' = 14.4$, extrapolated from Nernst's measurements[14] of the specific heat of ice at 273°. It is further required for the computation to know the relation between the molar concentration c and the volume concentration γ. By means of the observed density of the solution, this relation can be readily given, however, here as well as in the derivation of the equation, the molecular weight of water has a certain effect. Though this effect vanishes in the first approximation, it can not be eliminated from the second-order terms. Inasmuch as the effect is of second order only, its influence is considerably reduced; we have, therefore, used the simple molecular weight 18, throughout. Finally the quantity $2nv_0$ in the denominator can be put equal to:

$$2nv_0 = 2\Omega_0 \frac{\gamma}{1000}$$

where Ω_0 designates the molar volume of water. Provided no mutual electric effect of the ions was present, the left-hand term should give zero for substitution of the observed freezing point depression. Actually it gives a finite value, and we designate this finite value by $-\Theta$. Then, according to the theory, this difference Θ must be represented by the right-hand term so that:

$$\Theta = \frac{\varepsilon^2 \varkappa}{6DkT}\sigma(\varkappa a) = 0.270\sqrt{2\gamma}\,\sigma(\varkappa a) \qquad (55)$$

should hold. It can be ascertained that the definition for Θ obeyed here is, in the limit, identical with the one given above for low concentrations.

The points entered in the figure have the ordinates calculated from the observations in the manner indicated. The curve in the figure represents the right-hand term of equation (55), under the assumption that $a = 3.76 \times 10^{-8}$ cm. This value of a was determined from one observation by Adams, according to which the experimental value $\Theta = 0.0758$ is associated with $\gamma = 0.117$. The straight line which is also given in the figure again represents the limiting law for extreme dilution corresponding to $\sigma = 1$. It may be stated that, up to concentrations of approximately one mole per liter, the observations are well represented by the curve. For higher concentrations, the observations show a maximum for Θ. The theoretical curve also has a maximum; this is so flat, however, that it is hardly indicated, as shown by the figure. We are inclined to consider this discrepancy at high concentrations as factual, and wish to present a few pertinent remarks in the next section.

X. General Remarks

From the preceding discussion it may be concluded that it is inadmissible from a theoretical as well as from an experimental point of view to consider the electric energy of an ionic solution to be essentially determined by the average mutual distance of the ions. Rather, a quantity which measures the thickness of the ion atmosphere or, to connect with something known better, the thickness of a Helmholtz double-layer proves to be a characteristic length. In view of the fact that this thickness depends on the concentration of the electrolyte, the electric energy of the solution also becomes a function of this quantity. The fact that this thickness is inversely proportional to the square root of the concentration is responsible for the characteristic appearance of the limiting laws for highly diluted solutions. Though we must decline to talk in terms of a lattice structure of the electrolyte in the conventional sense, and though, as shown by the development of the subject, taking this image too literally leads to inadmissible mistakes, it still contains a grain of truth. To make this clear, the following two imaginary experiments are carried out. First, we take an element of space, and consider it placed, repeatedly, at arbitrary positions in the electrolyte. It is clear that, in a binary electrolyte, we shall find therein positive and negative ions with equal frequency. Second, we take the same spatial element, and again place it repeatedly in the electrolyte, now not arbitrarily, but always such that it is, for instance, located at a definite distance (of several angstrom units) from an arbitrarily selected positive ion. Now we shall not find positive and negative charges with equal frequency, the negative charges will prevail in number. In that the oppositely charged ions, on the average, prevail in number in the immediate surroundings of each ion, we can see, correctly, an analogy to the crystal structure of the NaCl type, where each Na ion is immediately surrounded by 6 Cl ions and each Cl ion by 6 Na ions. However, it is to be considered an essential characteristic of the electrolytic solution that the measure for this order is determined by the thermal equilibrium between attracting forces and temperature movement, while it is definitely predetermined for the crystal.

The computations and comparison with experience were carried out by taking the conventional dielectric constant for the surrounding solvent. The success justifies this assumption. Though this procedure is justifiable for low concentrations, it should cause mistakes for higher concentrations. In fact, it follows from dipole theory that for high field intensities, dielectrics must show saturation phenomena similar to the known magnetic saturation. The recent experiments by Herweg[15] may be taken as an experimental confirmation of this theoretical requirement. Since at a distance of 10^{-7} cm. from a singly charged ion, a field intensity of approximately 200,000 volt/cm. is to be expected, we should be prepared to observe something of these saturation phenomena. It would, of course, be very interesting if an attempt to separate this effect in its consequences from the observations were successful, the more so that nature puts at our disposal field intensities of a magnitude hardly attainable otherwise with conventional experimental means.

In another respect concentrated solutions should show a special behavior. If many ions are present in the surroundings of each single ion, this can be regarded as a change of the surrounding medium with respect to its electrical properties, an effect which has not been taken into account in the preceding theory. The manner in which this may become effective may be indicated by the following considerations. Let us consider one fixed ion and another mobile ion, oppositely charged, and investigate the amount of work required to remove the mobile ion. This work may be regarded as composed of two parts: (1) the ion will require a certain amount of work for its removal, and (2) we shall gain work by filling the space, previously taken up by the ion, with solvent. Experiments concerning the heat of dilution actually provide an indication of the existence of such conditions. Let us take, for example, a HNO_3 solution of initially low concentration and dilute it with a large quantity of water (i.e., so much that further dilution would not cause any heat effect), cooling will take place, i.e., work must be done in the sense of the previous considerations to separate the ions from one another. If the initial solution has a higher concentration, then, in the same experiment, heat is generated, i.e., work is obtained, if the surrounding of each ion is freed of a sufficient number of other ions which are replaced by water molecules. In conventional language, it is said that a predominant hydration of the ions occurs, and that this is to be regarded as an exothermic process. Obviously the above considerations intend an explanation of this so-called hydration on a purely electric basis. In fact an approximate computation can be carried through which gives theoretically Berthelot's rule, valid in this connection for the dependence of the heat of dilution from the initial concentration, and which makes plausible the order of magnitude of the experimentally determined numerical coefficient of this rule. These considerations have some bearing on the freezing point observations inasmuch as they suggest the possibility of computing why and to what extent the curves found for the percentage deviation Θ (compare the case of KCl) bend downward for higher concentrations and may even cross the abscissa provided the concentration is high enough. In this instance, the freezing point depression exceeds the one expected from classical theory (also, as may be stated explicitly, if the classical theory is used in its unabbreviated form). Until now, one has been resigned, in such cases, to talk about hydration.

However, before conditions for concentrated solutions can be investigated, it must be shown that the irreversible process of electric conduction in strong electrolytes can also be understood quantitatively from our point of view. We reserve the detailed presentation of this subject for a future article. Here only the basic ideas, which will be discussed more thoroughly in that paper, may be indicated. If an ion moving in a liquid is subjected to the influence of an external field, the surrounding ions will have to move constantly in order to form the ion atmosphere. If we now assume for a moment that a charge is suddenly generated in the electrolyte, an ion atmosphere will have to appear which requires a certain time of relaxation for its formation. Similarly, for a moving ion, the surrounding atmosphere will not attain its equilibrium distribution and thus cannot be computed on the basis of the Boltzmann-Maxwell principle. However, the determination of its charge distribution can be carried through on the basis of an obvious interpretation of the equations for the Brownian movement. It can be estimated qualitatively in which direction this effect, caused by the presence of a finite relaxation time, will be operative. At a point in front of the moving ion (i.e., a point toward which it moves) the electric density of the ion atmosphere must increase with time; it must decrease for a point behind the ion. As a consequence of the relaxation time, the density in front of the ion will be slightly smaller than its value at equilibrium; behind it, however, it will not yet have decreased to its equilibrium value. Consequently, during the movement there always exists a slightly larger electrical density of the ion atmosphere behind the ion than in front of it. Since charge density in the atmosphere and charge of the central ion always carry opposite signs, a force braking the ion movement will occur, independent of its sign, and obviously this force will increase with increasing concentration.

This is one effect which operates in the same sense as a decrease in dissociation calculated on the basis of Ostwald's dilution law. However, still another effect is present which must be taken into consideration. In the vicinity of an ion are predominently ions of the opposite sign, which under the influence of the **external** field will, of course, move in the opposite direction. These ions will, to a certain degree, drag along the surrounding solvent, thus causing the considered single ion not to move relative to a stationary solvent but relative to a solvent moving in the opposite direction. Since, apparently, this effect increases with increasing concentration, we have a second effect operating in the same sense as a decrease in dissociation. The effect can be calculated quantitatively according to the principles used by Helmholtz for the treatment of electrophoresis.

The common factor of the two effects just mentioned consists, as is shown by the computations, in the fact that both are closely related to the thickness of the ion atmosphere, and that, therefore, the generated forces are proportional to the square root of the concentration of the electrolyte, at least in the limit for very low concentrations. Thus we obtain a law, found by Kohlrausch[16] according to which for low concentrations the percentage deviation of the molecular conductivity from its limiting value at infinite dilution is proportional to the square root of the concentration. Also the proportionality factor thus finds a molecular interpretation.

Anticipating the detailed representation of electrolytic conductivity in prospect for a following article, we can state as an over-all result that the view, according to which strong electrolytes are completely dissociated, is entirely supported.

Bibliography

1. N. Bjerrum, *Z. Elektrochem.*, *24*, 231 (1918).
2. W. Sutherland, *Phil. Mag.*, *14*, 1 (1907).
3. Proceedings of the Seventh International Congress of Applied Chemistry, London, May 27, to June 2, 1909, Section X: "A New Form for the Electrolytic Dissociation Theory."
4. Milner, *Phil. Mag.*, *23*, 551 (1912); *25*, 743 (1913).
5. J. C. Ghosh, *J. Chem. Soc.*, *113*, 449, 627, 707, 790 (1918); *Z. phys. Chem.*, *98*, 211 (1921).
6. P. Hertz, *Ann. Phys.* (4), *37*, 1 (1912).
7. N. Bjerrum, *Z. Elektrochem.*, *24*, 231 (1918); *Z. anorg. Chem.*, *109*, 275 (1920).
8. L. H. Adams, *J. Am. Chem. Soc.*, *37*, 481, 1915 (KCl); L. E. Hall and W. D. Harkins, *ibid.*, *38*, 2658, 1916 (K_2SO_4, $La(NO_3)_3$, $MgSO_4$).
9. *Ann. Phys.*, *59*, 61 (1896).
10. L. H. Adams, *J. Am. Chem. Soc.*, *37*, 481, 1915 (KNO_3, KCl): R. E. Hall and W. D. Harkins, *ibid.* *38*, 2658 1916 (K_2SO_4, $La(NO_3)_3$, $MgSO_4$, $BaCl_2$); T. G. Bedford, *Proc. Roy. Soc.*, *A83*, 454, 1909 ($CuSO_4$) (concentration in moles per liter for KCl, $CuSO_4$; in moles per 1000 g. water for KNO_3, $BaCl_2$, K_2SO_4, $La(NO_3)_3$).

Bibliography (con'd)

11. M. Born, *Z. Physik* , *1*, 221 (1920).

12. J. Barnes, *Trans. Nova Scotia Inst. Science*, *10*, 139 (1900); C. Hebb, *ibid.*, *10*, 422 (1900); H. J. Jones, J. Barnes, and E. P. Hyde, *J. Am. Chem. Soc.*, *27*, 22 (1902); H. B. Jones and C. G. Caroll, *ibid.*, *28*, 284 (1902); W. Biltz, *Z. phys. Chem.*, *40*, 185 (1902); T. W. Richards, *ibid.*, *44*, 563 (1903); S. W. Young and W. H. Sloan, *J. Am. Chem. Soc.*, *26*, 919 (1904); H. Jahn, *Z. phys. Chem.*, *50*, 129 (1905); T. G. Bedford, *Proc. Roy. Soc.*, *A83*, 454 (1909); F. Fluegel, *Z. phys. Chem.*, *79*, 577 (1912); L. H. Adams, *J. Am. Chem. Soc.*, *37*, 481 (1915); W. H. Rodebusch, *J. Am. Chem. Soc.*, *40*, 1204 (1918).

13. Baxter and Wallace, *J. Am. Chem. Soc.*, *38*, 18 (1916).

14. W. Nernst, *Berl. Ber.*, *1*, 262 (1910).

15. *Z. Physik*, *3*, 36 (1920), and *8*, 1 (1922).

16. F. Kohlrausch and L. Holborn, *Das Leitungsvermoegen der Elektrolyte*. Second Ed. Leipzig, 1916, pages 108 and 112.

ON THE THEORY OF ELECTROLYTES.
II. LIMITING LAW FOR ELECTRIC CONDUCTIVITY
(Zur Theorie der Electrolyte. II. Das Grenzgesetz für die elektrische Leitfahigkeit)

By P. Debye and E. Hückel

Translated from
Physikalische Zeitschrift, Vol. 24, 305-325, 1923.

It has been mentioned in the introduction to our first article on electrolytes[1] that also the curve for the conductivity as a function of concentration cannot be explained by the law of mass action. Let us introduce a "conductivity coefficient" f_λ which we will use in the following discussion - defined similarly to the previous osmotic coefficient f_0, so that:

$$\Lambda = f_\lambda \Lambda_0$$

where Λ designates the molar conductivity for an arbitrary con-concentration and Λ_0 the same quantity in the limit for vanishing concentration. Then, provided we have a binary electrolyte of concentration c, and assuming the known decrease in molar conductivity with increasing concentration to be a consequence of the relative decrease in the number of ions required by the law of mass action, we can compute that for small concentrations:

$$1 - f_\lambda = \frac{c}{K}$$

should hold. K denotes the constant of equilibrium determining the supposed equilibrium between ions and molecules. The experiments with strong electrolytes are in disagreement with this consequence. The curves, representing Λ as a function of the concentration - several of which are plotted in the book by Kohlrausch and Holborn - show a strong curvature at low concentrations, and do not at all exhibit a linear approach to the limit for zero concentration, as required by the law of mass action. Rather they reach it with a vertical tangent. Kohlrausch attempted to find a law which approximately describes the course of the curve over a fairly wide range. He found the formula[2] to be suitable. Usually

$$1 - f_\lambda \sim c^{1/3}$$

we find this formula mentioned. However, Kohlrausch remarked that it has the value of an interpolation formula only and fails for large as well as for small concentrations-this is of special importance. In this last case he finds another expression[3] much more suitable:

$$1 - f_\lambda \sim c^{1/2}$$

In view of the fact that frequently very little attention is paid to these circumstances (to such an extent that the law involving the $\frac{1}{2}$ power is occasionally forgotten, and theoretical deductions are based on the interpolation formula involving the 1/3 power) we want to quote the words with which Kohlrausch accompanied the derivation of the limiting law involving the $\frac{1}{2}$th power.[3]

"If in accordance herewith the course of the conductivity with increasing dilution assumes so distinctly such a simple behavior, I believe it to be very probable that this behavior constitutes the law. Of course, this law does not comply with the wish that it be deductible from the theory of dissociation."

In the following an attempt is made to compute, assuming complete dissociation, in which manner the mutual electric forces of the ions affect the conductivity. At the end of our first article, we presented already the qualitative discussion which shall be carried through here quantitatively. It may be permitted to formulate here the two reasons, mentioned there, for the decrease in molecular conductivity with an increase in concentration. In the pure solvent, an ion moves under the exclusive influence of the external field strength and reaches a velocity at which the frictional force to which it is submitted is equal to the external force. If the solution reaches a certain concentration, then, first, a supplementary force in addition to the external force and increasing with concentration will be generated, which supplementary force is caused by the mutual Coulomb's forces of the ions. Second, the frictional force, to which the ion is subjected, will be modified by the presence of the other ions. Both causes are effective in the direction of decreasing conductivity. The magnitudes of these effects, the computation of which is attempted in sections A and B, are related to the average thickness of the ion atmosphere. This thickness is, for small concentrations, inversely proportional to the square root of the concentration. Thus it will be understood that the square root will appear, and the theory actually explains Kohlrausch's law:

$$1 - f_\lambda \sim c^{1/2}$$

Since even the limiting case of low concentrations needs a lengthy discussion, we have limited ourselves, in the following, consistently to this case, and have not attempted to generalize the formulas in order to make them applicable to higher concentrations.

In section C results of the former sections A and B are combined to yield a formula for the conductivity coefficient. In section D, finally, theory and experiment are compared.

As already mentioned in our first article, two calculations with regard to the same subject are available: One by P. Hertz,[4] the second by J. C. Ghosh.[5] Though we are of the opinion that the computation by P. Hertz probably is too highly idealized and therefore is not able to furnish an explanation of Kohlrausch's law, it is of special value as a consistent attempt to comprehend the mutual hindrance of the ions. We are under the impression that the later papers by J. C. Ghosh on the subject are of less importance. They, also, do not explain Kohlrausch's law. Further, as opposed to Hertz, the mechanism of hindrance is not discussed. The premise is made that with an increase in concentration a certain portion of the ions of the completely dissociated electrolyte are no longer to be considered as free. This portion is computed by the alleged use of Boltzmann's principle. We do not believe that, here, the principle has been applied as it should be.

A. Ionic Forces

I. Basic Equation for the Determination of the Distribution Function

Assume that an ion is moved in the x-direction with constant velocity v. The ion atmosphere, surrounding the ion, can no longer be computed from Boltzmann-Maxwell's principle, because this is not a static problem, and we have to reach further for an equation which can determine the distribution function in such a case. This is afforded by the completed and suitably interpreted equation for the Brownian movement, which, in a similar way, P. Debye made the basis for an explanation of the anomalous dipole dispersion.

In a space with volume elements dS a distribution of particles with the instantaneous distribution function f may exist. The number of particles $f\,dS$, present in the volume element dS, will then change for two reasons.

During a short time of observation τ the particles will dissipate, in accordance with a certain probability function w, because of their Brownian movement. Consequently, the element dS loses all its particles $f\ dS$ except for an amount of minor magnitude; in exchange it obtains from each element dS' of its surroundings a number of particles

$$f'dS'wdS$$

The total increase in the number of particles due to the Brownian movement, therefore, during the time interval τ for which the probability function w is formulated, amounts to:

$$-fdS + dS \int f'wdS'$$

If τ is chosen sufficiently small so that only the immediate surroundings of dS are to be taken into account, then, according to Taylor's theorem

$$f' = f + \left(x\frac{\partial f}{\partial x} + \cdots + \cdots\right) + \\ + \frac{1}{2}\left(x^2\frac{\partial^2 f}{\partial x^2} + \cdots + \cdots + 2xy\frac{\partial^2 f}{\partial x\,\partial y} + \cdots\right)$$

and only the mean values

$$\overline{x} = \int xwdS', \;\ldots\ldots,\; \overline{x^2} = \int x^2wdS', \;\ldots\ldots \\ \overline{xy} = \int xywdS' \ldots\ldots$$

have to be determined. Now we have

$$\overline{x} = \overline{y} = \overline{z} = 0 \\ \overline{xy} = \overline{yz} = \overline{zx} = 0 \\ \overline{x^2} = \overline{y^2} = \overline{z^2} = s^2$$

where, for the time being, s^2 remains indeterminate. Thus the desired increase in the number of particles due to the Brownian movement is

$$\frac{s^2}{2}\left(\frac{\partial^2 f}{\partial x^2} + \frac{\partial^2 f}{\partial y^2} + \frac{\partial^2 f}{\partial z^2}\right)dS = \frac{s^2}{2}\,\mathrm{div\,grad}\, f \cdot dS$$

A second cause for the change in the number of particles will have to be looked for in the effect of forces which are present. Let us assume that a force $\mathfrak{K}$ acting on the particle causes a velocity $\mathfrak{v}$, according to the equation

$$\mathfrak{K} = \rho\mathfrak{v}$$

such that ρ is the coefficient of friction associated with the particle. Then the increase in the number of particles due to $\mathfrak{K}$ within dS can be represented by

$$-\tau \operatorname{div} f \frac{\mathfrak{K}}{\rho} \cdot dS$$

A third cause for the change in the number of particles will consist in the motion of the liquid in which the particles are suspended. However, this effect will be negligible in the approximation which we do not intend to exceed here. Accordingly the total increase in the number of particles will be

$$\tau \frac{\partial f}{\partial t} dS = \frac{s^2}{2} \operatorname{div} \operatorname{grad} f \cdot dS - \tau \operatorname{div} \frac{f \mathfrak{K}}{\rho} \cdot dS$$

i.e., the distribution function as a function of time and space is to be found from the equation

$$\tau \frac{\partial f}{\partial t} = \operatorname{div} \left(\frac{s^2}{2} \operatorname{grad} f - f \frac{\mathfrak{K}}{\rho} \tau \right)$$

From the fact that, for the static case, the solution of this equation must correspond to Boltzmann-Maxwell's principle, we can conclude - as shown by Einstein - that

$$\frac{s^2}{2\tau} = \frac{kT}{\rho}$$

(k = Boltzmann's constant). Thus the basic equation assumes the form

$$\rho \frac{\partial f}{\partial t} = \operatorname{div} (kT \operatorname{grad} f - f \mathfrak{K}) \qquad (1)$$

II. The General Conditions for the Determination of the Ion Atmosphere

Assume that ions of the kinds $1, \ldots i, \ldots s$, are present in the solution with charges $e_1 \ldots e_i, \ldots e_s$, respectively. A certain ion is selected and forced to move in the x-direction with the velocity v. Then in a spatial element dS, the numbers of ions $n_1 dS, \ldots n_i dS, \ldots n_s dS, \ldots$ will be present on the average.

The solution is traversed by a current, generated by a constant electric field strength $\mathfrak{E}$, also directed in the x-direction.

Thus a force

$$\mathfrak{K}' = c_i \mathfrak{E}$$

will be exerted on an ion of the ith kind.

Similarly to the static case, an ion distribution surrounding the selected ion will arise so that, at least in its immediate neighborhood, oppositely charged ions will predominate. The ion atmosphere which is thereby created will, on the average, give rise to an electric potential ψ, and if - for reasons similar to those in the static case - we perform our calculations starting with this mean potential ψ, a second force amounting to

$$-e_i \operatorname{grad} \psi$$

will be exerted on the ion of the ith kind. Using the basic equation (1), we are now in a position to present the equation

$$\varrho_i \frac{\partial n_i}{\partial t} = \operatorname{Div}(kT \operatorname{Grad} n_i - n_i c_i \mathfrak{E} + n_i e_i \operatorname{Grad} \psi) \quad (2)$$

for the determination of the number of particles n_i. Here ρ_i is the coefficient of friction for particles of the ith kind; further we introduced Div and Grad instead of div and grad, inasmuch as we intend to distinguish between a coordinate system fixed in space, to be designated by capital letters, and another coordinate system, to be designated by lower case letters, which moves with the selected ion. There are as many equations (2) as there are kinds of ions; they would suffice for the determination of the n_i's, *i.e.*, of the ion distribution, but for the appearance of the unknown potential. However, ψ must satisfy Poisson's equation; we can, therefore, complete the system by adding the equation

$$\operatorname{Div} \operatorname{Grad} \psi = -\frac{4\pi}{D} \sum n_i e_i \quad (2')$$

where D designates the dielectric constant of the solvent.

It is evident that the numbers of particles n_i as well as the potential are stationary in regard to a coordinate system moving with the selected particle. Thus if we put

$$X - vt = x, \quad Y = y, \quad Z = z$$

the unknown quantities depend only on x, y, and z. Accordingly

$$\frac{\partial}{\partial t} = -v \frac{\partial}{\partial x}, \quad \operatorname{Div} = \operatorname{div}, \quad \operatorname{Grad} = \operatorname{grad}$$

and the system

$$\left.\begin{array}{l} \cdots\cdots\cdots \\ -\varrho_i v \dfrac{\partial n_i}{\partial x} = \mathrm{div}(kT\,\mathrm{grad}\,n_i - n_i e_i \mathfrak{E} + n_i e_i \,\mathrm{grad}\,\psi) \\ \cdots\cdots\cdots \\ \mathrm{div}\,\mathrm{grad}\,\psi = -\dfrac{4\pi}{D}\sum n_i e_i \end{array}\right\} \quad (3)$$

obtains.

III. Approximations and Their Physical Interpretation

If, in equation (3), we put $v = 0$ and let $\mathfrak{E}$ vanish, it would be possible to satisfy the first s equations by

$$n_i = \mathrm{const}\, e^{-\frac{e_i \psi}{kT}}$$

since they would then assume the form

$$\mathrm{div}\,(kT\,\mathrm{grad}\,n_i + n_i e_i \,\mathrm{grad}\,\psi) = 0 \qquad (4)$$

This would represent, in accordance with our previous discussion, the Boltzmann-Maxwell assumption. Previously we did not use the complete expression but expanded the exponential function and retained only the first term of the expansion. This procedure may be justified, independently, by means of equation (4), as follows. At fairly large distances from the selected ion, n_i will approach a constant value $\bar{n}_i$ which may be identified with the mean density of the ith kind of ions in the liquid. Provided grad ψ is sufficiently small, and this will hold better with increasing dilution, we can write

$$n_i e_i \,\mathrm{grad}\,\psi$$

instead of

$$\bar{n}_i e_i \,\mathrm{grad}\,\psi$$

Hence equation (4) becomes

$$\mathrm{div}\,(kT\,\mathrm{grad}\,n_i + \bar{n}_i e_i \,\mathrm{grad}\,\psi) = 0$$

and will be satisfied by the expression

$$n_i = \bar{n}_i - \frac{\bar{n}_i e_i}{kT}\psi \qquad (5)$$

This expression, however, is nothing but the beginning of the expansion for Maxwell-Boltzmann's relation which was the basis for all developments in the static case. Thus we stay within the same boundaries if we replace

$$-n_i e_i \mathfrak{E} + n_i e_i \operatorname{grad} \psi$$

by

$$-\overline{n_i} e_i \mathfrak{E} + \overline{n_i} e_i \operatorname{grad} \psi$$

in the first s equations of (3). Here, as can be justified, the "external" field strength is assumed not to exceed the order of magnitude of grad ψ. The modified equations (3) for the moving ion are still not satisfied by expression (5), since the movement must cause a deviation from the Maxwell-Boltzmann distribution. We therefore make the assumption that

$$n_i = \overline{n_i} - \frac{\overline{n_i} e_i}{kT} \psi + \mu_i \qquad (6)$$

where μ_i measures the deviation from the distribution at equilibrium. It will not be necessary to take into account terms of higher than the first order in v. Since μ_i vanishes with v, terms of the form, for example, $v\mu_i$ may be neglected. Taking all these conditions into account, we can with sufficient accuracy proceed from (3) to the following system:

$$\left.\begin{array}{l} \cdots\cdots\cdots \\ \varrho_i v \dfrac{\overline{n_i} e_i}{kT} \dfrac{\partial \psi}{\partial x} = \operatorname{div}(kT \operatorname{grad} \mu_i) = kT \operatorname{div} \operatorname{grad} \mu_i \\ \cdots\cdots \\ \operatorname{div} \operatorname{grad} \psi = \dfrac{4\pi}{DkT} \Sigma \overline{n_i} e_i^2 \psi - \dfrac{4\pi}{D} \Sigma \mu_i e_i \end{array}\right\} (7)$$

considering that

$$\operatorname{div} \mathfrak{E} = 0$$

and

$$\Sigma \overline{n_i} e_i = 0$$

The quantity

$$\frac{4\pi}{DkT} \Sigma \overline{n_i} e_i^2$$

already appeared in the static case. We designated it by κ^2 which we will do here also. It is the square of the reciprocal length measuring the thickness of the ion atmosphere. If we apply the operation

$$\Delta = \text{div grad}$$

to the last equation of (7), the first s equations may be immediately used for the elimination of the μ_i, and we obtain

$$\Delta(\Delta\psi - \varkappa^2\psi) = -\frac{4\pi}{DkT}\frac{v}{kT}\sum \overline{n}_i e_i^2 \varrho_i \frac{\partial\psi}{\partial x}$$

We now introduce an average coefficient of friction ρ by putting

$$\varrho = \frac{\sum \overline{n}_i e_i^2 \varrho_i}{\sum \overline{n}_i e_i^2} \qquad (8)$$

and, further, designate the quantity $\rho v/kT$, which is proportional to the velocity, by

$$\frac{\varrho v}{kT} = \omega \qquad (9)$$

With these abbreviations, the equation for the potential becomes

$$\Delta(\Delta\psi - \varkappa^2\psi) = -\varkappa^2\omega\frac{\partial\psi}{\partial x} \qquad (10)$$

here each of the quantities κ and ω has the dimension of a reciprocal length. We now have to find a solution of equation (10) which satisfies the conditions imposed by the problem.

IV. Potential Distribution Surrounding a Moving Ion

The potential to be determined from equation (10) need not be found accurately. It suffices to establish a solution up to the first order of v, *i.e.*, of ω. We shall make use of a well known method by temporarily putting the right hand term equal to zero, and determining ψ of zero order. Then we shall insert this value of ψ in the right hand side, and compute ψ from the equation in a first order approximation.

At zero approximation the potential:

$$A \cdot \frac{e^{-\varkappa r}}{r}$$

of the stationary case, evidently, deserves consideration; for the time being we do not enter any value for the constant A; r is the distance from the selected ion. Hence the potential for the moving ion is determined with sufficient accuracy by the equation

$$\Delta(\Delta\psi - \varkappa^2\psi) = -\varkappa^2\omega\frac{\partial}{\partial x}\left(A\frac{e^{-\varkappa r}}{r}\right) \quad (10')$$

Now note that

$$\Delta\left(A\frac{e^{-\varkappa r}}{r}\right) = \varkappa^2\left(A\frac{e^{-\varkappa r}}{r}\right)$$

So we can obtain a special solution of equation (10') from the equation

$$\Delta\psi - \varkappa^2\psi = -\omega\frac{\partial}{\partial x}\left(A\frac{e^{-\varkappa r}}{r}\right) =$$
$$= -\omega\frac{d}{dr}\left(A\cdot\frac{e^{-\varkappa r}}{r}\right)\cos\vartheta$$

provided the angle between x and r is denoted by ϑ. Since further, if we put

$$\psi = R\cos\vartheta$$

the relation

$$\Delta\psi - \varkappa^2\psi = \left(\frac{1}{r^2}\frac{d}{dr}r^2\frac{dR}{dr} - \frac{2R}{r^2} - \varkappa^2 R\right)\cos\vartheta$$

holds, it will be readily understood that

$$R = -\frac{\omega A}{2}e^{-\varkappa r}$$

Thus the special solution corresponding to the right hand term is

$$\psi = -\frac{\omega A}{2}e^{-\varkappa r}\cos\vartheta \quad (11)$$

To this solution may be added any expression which satisfies the equation

$$\Delta(\Delta\psi - \varkappa^2\psi) = 0$$

We take this expression also proportional to cos ϑ and obtain

$$\psi = \left(A'\frac{d}{dr}\frac{e^{-\varkappa r}}{r} + A''\frac{d}{dr}\frac{e^{\varkappa r}}{r} + \frac{B'}{r^2} + B''r\right)\cos\vartheta \quad (11')$$

Evidently terms in this expression which do not vanish at infinity must be rejected. Hence

$$A'' = 0 \text{ and } B'' = 0$$

Thus we finally obtain from equation (11) and equation (11')

$$\psi = A \cdot \frac{e^{-\varkappa r}}{r} + \left(- \frac{\omega A}{2} e^{-\varkappa r} + A' \frac{d}{dr} \frac{e^{-\varkappa r}}{r} + \right.$$
$$\left. + \frac{B'}{r^2} \right) \cos \vartheta \qquad (12)$$

Now follows the determination of the constants. Investigating the supplementary potential originating through the movement in the vicinity of the origin, and expanding for this purpose in powers of r, we obtain

$$\left[(-A' + B') \frac{1}{r^2} + \frac{-\omega A + \varkappa^2 A'}{2} + \right.$$
$$\left. \left(\frac{\omega A}{2} - \frac{\varkappa^2 A'}{3} \right) \varkappa r + \cdots \right] \cos \vartheta$$

We will consider the dimensions of the selected ion to be negligible compared with $1/\kappa$. As pointed out in the preface, this corresponds to restricting ourselves to the limiting laws for highly diluted solutions. Then the first term in the square bracket vanishes so that

$$B' = A' \qquad (13)$$

otherwise the selected ion would appear to carry a fixed dipole at the origin whose potential, as is well known, is proportional to $\cos \vartheta / r^2$. The second term also must vanish which makes

$$\varkappa^2 A' = \omega A \qquad (14)$$

For if we compute

$$\text{div grad} = \Delta$$

of the supplementary potential this term would contribute an amount proportional to $\cos \vartheta / r^2$; this would mean, in accordance with Poisson's equation, that the density of the charge distribution associated with the supplementary potential would become infinite in the neighborhood of the selected ion. Since this is inadmissible, equation (14) is justified. Denoting the charge of the selected ion by e_j, the constant A has the value

$$A = \frac{e_j}{D} \qquad (15)$$

and, summarizing, we obtain

$$\psi = \frac{e_j}{D}\frac{e^{-\varkappa r}}{r} - \omega\frac{e_j}{D}\left(\frac{e^{-\varkappa r}}{2} - \right.$$
$$\left. - \frac{d}{d(\varkappa r)}\frac{e^{-\varkappa r}}{\varkappa r} - \frac{1}{\varkappa^2 r^2}\right)\cos\vartheta \qquad (16)$$

for the potential of the moving ion.

V. Field and Charge Distribution Surrounding a Moving Ion

By means of Poisson's equation we obtain from the formula (16) for the potential the mean electric density of the ion atmosphere. We find for it the expression

$$-\frac{e_j\varkappa^3}{4\pi}\frac{e^{-\varkappa r}}{\varkappa r} + \frac{\omega}{2}\frac{e_j\varkappa^2}{4\pi}e^{-\varkappa r}\cos\vartheta$$

The first term corresponds to the static case ($\omega = 0$), and describes a centrally symmetric density distribution with a sign opposite throughout to that of the ion charge e_j. The second term gives the change which occurs as a consequence of the movement. In front of the ion ($\vartheta = 0$), the sign of the charge density agrees with that of the selected ion; behind the ion ($\vartheta = \pi$), it has the opposite sign. It will be seen that - in agreement with the results of the qualitative considerations in our first article - forces will arise which have a retarding effect. To convey a picture, the lines of force are plotted in Figure 1, representing

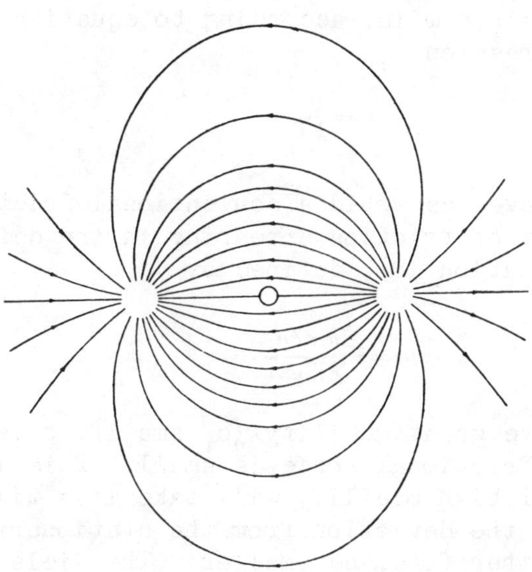

Figure 1

the lines of force such as they will be caused by the supplementary potential generated by the movement. It will be seen that, in the vicinity of the origin, where the ion with the charge e_j should be assumed, the supplementary field becomes homogeneous. This is confirmed and at the same time the value of the supplementary field strength at this point is found by expanding the supplementary potential in regard to powers of r. It follows for the total potential ψ:

$$\psi = \frac{e_j}{D}\frac{e^{-\varkappa r}}{r} + \frac{\omega\varkappa}{6}\frac{e_j}{D} r\cos\vartheta + \cdots \quad (17)$$

Thus in the vicinity of the origin the supplementary potential is equal to

$$\frac{\omega\varkappa}{6}\frac{e_j}{D} r\cos\vartheta = \frac{\omega\varkappa}{6}\frac{e_j^2}{D} x$$

corresponding to a field strength of the value

$$\mathfrak{E} = -\frac{\omega\varkappa}{6}\frac{e_j}{D}$$

The force arising as a consequence of the finite relaxation time of the ion atmosphere is thus

$$e_j\mathfrak{E} = -\frac{\omega\varkappa}{6}\frac{e_j^2}{D} \quad (18)$$

It is always retarding (because of the minus sign), independently of the sign of e_j. Secondly it is proportional to the velocity v of the moving ion, since ω is, according to equation (9), an abbreviation for the expression

$$\omega = \frac{\varrho v}{kT}$$

Consequently it behaves as would a conventional frictional force. The mean coefficient of friction appearing in the definition of ω is, according to equation (8), defined by

$$\varrho = \frac{\sum \overline{n}_i e_i^2 \varrho_i}{\sum \overline{n}_i e_i^2}$$

Thus if the ions have great mobility (ρ_i small), ρ is also small, and the additional frictional force is small. This is to be expected since ions of high mobility will take less time to build up the ion atmosphere; the deviation from the stationary potential distribution will, therefore, be smaller. The dielectric constant D in the denominator of equation (18) measures in a familiar way the redction in the forces between charges which occurs inside a dielectric medium. Figure 2 serves to illustrate the course of

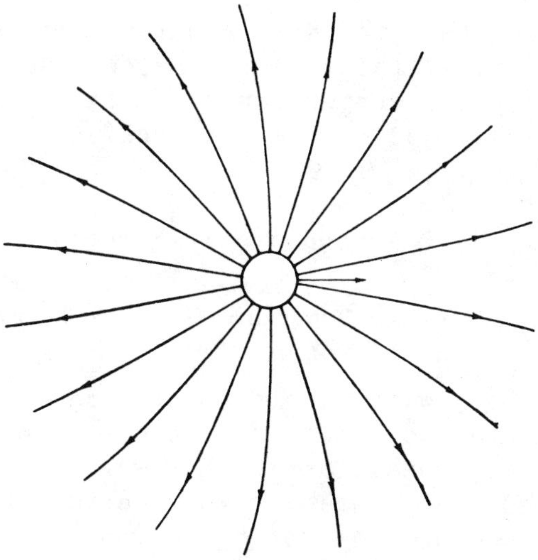

Figure 2

the field lines in the complete field. In the static case the field lines would radiate radially and in straight lines from the ion. The movement causes them to bend as if they could follow the movement of the ion only with a certain retardation.

To obtain some orientation regarding the magnitude of the supplementary frictional force caused by the movement, let all ρ_i be assumed equal. Then the conventional frictional force to which the ion would be subject in pure water would be ρ_v. Thus the ratio of the supplementary force equation (18) to this initial frictional force will be

$$\frac{e_j^2 \varkappa}{6\,DkT}$$

Assuming a monovalent ion so that $e_j = \varepsilon = 4.77 \times 10^{-10}$ and putting (at 0°C.) $D = 88.23$, $T = 273$ and $k = 1.346 \times 10^{-16}$, we have for the ratio

$$\frac{e_j^2 \varkappa}{6\,DkT} = 1.17 \cdot 10^{-8} \varkappa$$

For a solution of a uni-univalent salt with a concentration γ in moles per liter, we found previously the reciprocal length

$$\varkappa = 0.326 \cdot 10^8 \sqrt{\gamma}$$

In this case we would have

$$\frac{e_j^2 \varkappa}{6DkT} = 0.382 \cdot \sqrt{\gamma}$$

Thus it will be seen that the supplementary forces, arising as a consequence of the final relaxation time, become already appreciable compared with the conventional frictional forces of the ions at comparatively small concentrations.

B. Electrophoretic Forces

VI. Equations of Motion for the Solvent

It is known that, with a certain approximation, it is permissible to calculate the frictional forces to which the ions are subject during their transit as if they were small spheres moving in a viscous liquid, and thereby give rise to currents in accordance with the macroscopic laws of hydrodynamics for viscous fluids. From the movement of the liquid calculated in this manner, Stokes' law results, according to which the arising fictional force is

$$6\pi\eta b v$$

in which η designates the coefficient of viscosity and b the radius of the moving sphere. We want to start with the image of the moving sphere to establish whether and how the flow of the liquid is changed by the presence of the ions. Our interest therefore is centered not so much on Stokes' law itself, but on the changes which will be found as a consequence of the presence of the ions. It may appear doubtful, in view of the small dimensions of the ions, whether the calculation based on the fundamental equations of hydrodynamics still applies. We feel that these doubts should not be given too much weight, though it is, *a priori*, evident that the radii, which will have to be introduced, will correspond only in their order of magnitude to the actual dimensions of the ions. Application of the following method of computation to the mobility of colloidal particles and the effect of added electrolytes would not be affected by this objection.

The ion with the charge e_j, considered as a sphere with a radius b, is assumed to move with the velocity v in the x-direction of a coordinate system. It will carry along its ion atmosphere where the electricity is distributed on the average with a certain density which can be derived from a potential ψ. For this potential we make the assumption

$$\psi = \frac{e_j}{D}\frac{e^{-\varkappa r}}{r} \qquad (19)$$

ψ is thus given by the expression for the static case and for the limit for highly diluted solutions; it satisfies the equation

$$\Delta\psi - \varkappa^2\psi = 0 \qquad (20)$$

The motion of the ion with the velocity v takes place by virtue of the electric field strength $\mathfrak{E}$, which generates the current, which shall have a component in the x-direction only, namely $\mathfrak{E}_x = X$. The difference with Stokes' case consists in that, now, the electric charges of the ion atmosphere are present in each volume element of the liquid with the result that the external field strength $\mathfrak{E}$ gives rise to a force acting on the volume which will modify the flow. Since the charge in the atmosphere is opposite to that of the ion, it follows immediately that the flow of the liquid will be modified in such a way as to increase the friction. According to Poisson's equation, the charge density is

$$-\frac{D}{4\pi}\Delta\psi$$

so that the force acting per unit volume has the value

$$\mathfrak{F} = -\frac{D\mathfrak{E}}{4\pi}\Delta\psi = -\frac{D\mathfrak{E}}{4\pi}\varkappa^2\psi \qquad (21)$$

It has been explained in Section I that the potential surrounding the moving ion can no longer be presented accurately by equation (19). If we are using equation (19) in spite of this, we do it because we consider only strongly diluted solutions, as pointed out in the preface. For the same reason it is possible to separate the problem as is being done in sections A and B. Strictly the fluid flow would have had to be taken into account when determining the distribution function (as indicated in A), and the unknown distribution function on the potential, when determining the flow of the fluid. Accordingly, strictly speaking, the following equations form one system with the preceding equations which is to be taken as a unit. The separation into two systems is only a consequence of our limitation to small concentrations.

If the velocity of the fluid stream is denoted by $\mathfrak{v}$, then, according to Stokes, the hydrodynamic equations read

$$\left.\begin{aligned} \eta \operatorname{rot}\operatorname{rot}\mathfrak{v} &= -\operatorname{grad} p + \mathfrak{F} \\ \operatorname{div}\mathfrak{v} &= 0 \end{aligned}\right\} \qquad (22)$$

where $\mathfrak{F}$ is to be substituted from equations (21) and (19), and p designates the pressure.

Our immediate assignment is the determination of the fluid velocity $\mathfrak{v}$ as a function of space-coordinates. We intend to solve the problem assuming the ion sphere to rest and the liquid at great distances to have a velocity $-v$ in the x-direction. The actual case can then be obtained by superposing a constant velocity $+v$ in the x-direction, which does not affect the flow relative to the ion sphere. Once the velocity distribution is known, the stresses on the surface of the sphere are calculated, and conclusions drawn from these stresses as to the force exerted on the ion.

It will be observed that the problem is set up in a way similar to that followed by Helmholtz when he computed the phenomena of electrophoresis. A difference exists only in that we determine in detail by the potential ψ the structure of the ion atmosphere which corresponds to a part of Helmholtz' double layer.

VII. General Formulation of Velocity and Pressure Distribution

Forming the div of the first equation of (22), it follows, in view of equation (21) and the assumption made concerning the direction of $\mathfrak{F}$, that

$$\Delta p = \operatorname{div} \mathfrak{F} = -\frac{DX}{4\pi} \Delta \frac{\partial \psi}{\partial x} \qquad (23)$$

A particular solution is given by

$$p = -\frac{DX}{4\pi} \frac{\partial \psi}{\partial x}$$

to which an arbitrary solution of

$$\Delta p = 0$$

has to be added. We now introduce polar coordinates (see figure 3) and put

$$p = \left(A_0 + \frac{B_0}{r}\right) + \\ + \left(A_1 r + \frac{B_1}{r^2} - \frac{DX}{4\pi} \frac{d\psi}{dr}\right) \cos \vartheta \qquad (24)$$

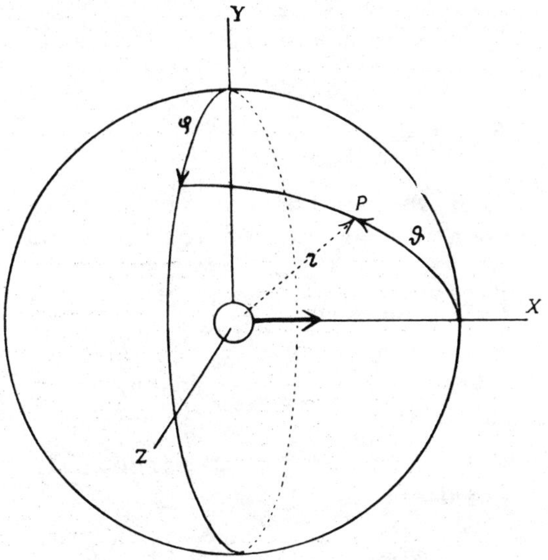

Figure 3

in which A_0, B_0, A_1, B_1 designate arbitrary constants and the relation

$$\frac{\partial \psi}{\partial x} = \frac{d\psi}{dr}\cos\vartheta$$

has been used.

We now form the rot of the first equation of (22), and focus our attention on the circulation

$$\mathfrak{w} = \operatorname{rot}\mathfrak{v} \qquad (25)$$

We obtain for the determination of $\mathfrak{w}$:

$$\eta \operatorname{rot}\operatorname{rot}\mathfrak{w} = \operatorname{rot}\mathfrak{F} \qquad (26)$$

In accordance with the assumptions of the problem, the flow is identical in each meridian plane containing the x-axis. Thus $\mathfrak{v}_r$ and $\mathfrak{v}_\vartheta$ exist, while $\mathfrak{v}_\varphi = 0$. Since in general the three components of $\mathfrak{w}$ in the r, ϑ, and φ directions can be determined explicitly from (25) by means of the equations

$$\mathfrak{w}_r = \frac{1}{r\sin\vartheta}\frac{\partial}{\partial\vartheta}\sin\vartheta\,\mathfrak{v}_\varphi - \frac{1}{r\sin\vartheta}\frac{\partial}{\partial\varphi}\mathfrak{v}_\vartheta$$

$$\mathfrak{w}_\vartheta = \frac{1}{r\sin\vartheta}\frac{\partial}{\partial\varphi}\mathfrak{v}_r - \frac{1}{r}\frac{\partial}{\partial r}r\mathfrak{v}_\varphi$$

$$\mathfrak{w}_\varphi = \frac{1}{r}\frac{\partial}{\partial r}r\mathfrak{v}_\vartheta - \frac{1}{r}\frac{\partial}{\partial\vartheta}\mathfrak{v}_r$$

there exists only one component of $\mathfrak{w}$ in the φ direction (the lines of circulation form circles around the x-axis), *i.e.*,

$$\mathfrak{w}_r = 0, \qquad \mathfrak{w}_\vartheta = 0, \qquad \mathfrak{w}_\varphi = \mathfrak{w}_\varphi$$

We now have to find the vector rot rot $\mathfrak{w}$ from vector $\mathfrak{w}$ which also has only a φ component, as can be verified easily, given by

$$(\text{rot rot } \mathfrak{w})_\varphi = -\frac{1}{r^2}\frac{\partial^2}{\partial r^2} r \mathfrak{w}_\varphi - \\ -\frac{1}{r^2}\frac{\partial}{\partial \vartheta}\frac{1}{\sin\vartheta}\frac{\partial}{\partial\vartheta}\sin\vartheta\, \mathfrak{w}_\varphi$$

Similarly, since $\mathfrak{F}$ points in the X-direction, rot $\mathfrak{F}$ also has a φ component only which is equal to

$$\text{rot}_\varphi \mathfrak{F} = -\varkappa^2 \frac{DX}{4\pi}\sin\vartheta \frac{d\psi}{dr}$$

The sole existing φ component of the circulation must satisfy the equation

$$\frac{1}{r}\frac{\partial^2}{\partial r^2} r\mathfrak{w}_\varphi + \frac{1}{r^2}\frac{\partial}{\partial\vartheta}\frac{1}{\sin\vartheta}\frac{\partial}{\partial\vartheta}\sin\vartheta\,\mathfrak{w}_\varphi = \\ = -\varkappa^2\frac{DX}{4\pi\eta}\sin\vartheta\frac{d\psi}{dr} \qquad (27)$$

Putting $\mathfrak{w}_\varphi$ proportional to sin ϑ and considering the differential equation which ψ must satisfy, we obtain the solution

$$\mathfrak{w}_\varphi = \left(C_1 r + \frac{D_1}{r^2} - \frac{DX}{4\pi\eta}\frac{d\psi}{dr}\right)\sin\vartheta \qquad (28)$$

where C_1 and D_1 are constants to be determined.

The two results, equation (24) for the pressure and equation (28) for the circulation, were obtained from the basic equations (22) upon suitable differentiation of the same (formation of div and rot). Consequently $\mathfrak{w}_\varphi$ and p must now be inserted in the basic equations to ascertain whether the expressions are compatible, specifically, which relations must hold between the constants A_0, B_0, A_1, B_1, C_1, D_1 to assure compatibility.

The equation

$$\eta \text{ rot rot } \mathfrak{v} = -\text{grad } p + \mathfrak{F} = \\ = -\text{grad } p - \frac{D\mathfrak{E}}{4\pi}\varkappa^2\psi$$

can be divided into three equations for the components in the r, ϑ and φ **directions,** respectively. It can be established that the third equation, relating to the φ direction, assumes the form $0 = 0$, corresponding to the condition that $\mathfrak{E}$ has $\mathfrak{E}_r$ and $\mathfrak{E}_\vartheta$ components only. The first two equations read

$$\left.\begin{aligned}
&\left(2\eta C_1 + 2\eta\frac{D_1}{r^3} - \frac{DX}{4\pi}\frac{2}{r}\frac{d\psi}{dr}\right)\cos\vartheta = \\
&= \frac{B_0}{r^2} + \left(-A_1 + \frac{2B_1}{r^3} + \right.\\
&\left. + \frac{DX}{4\pi}\frac{d^2\psi}{dr^2}\right)\cos\vartheta - \frac{DX}{4\pi}\varkappa^2\psi\cos\vartheta \\
&\left(-2\eta C_1 + \eta\frac{D_1}{r^3} + \frac{DX}{4\pi}\frac{1}{r}\frac{d}{dr}r\frac{d\psi}{dr}\right)\sin\vartheta = \\
&= \left(A_1 + \frac{B_1}{r^3} - \frac{DX}{4\pi}\frac{1}{r}\frac{d\psi}{dr}\right)\sin\vartheta + \\
&+ \frac{DX}{4\pi}\varkappa^2\psi\sin\vartheta
\end{aligned}\right\} \tag{29}$$

From this we conclude

$$B_0 = 0, \quad A_1 + 2\eta C_1 = 0, \quad B_1 - \eta D_1 = 0$$

or

$$B_0 = 0, \quad C_1 = -\frac{A_1}{2\eta}, \quad D_1 = \frac{B_1}{\eta} \tag{30}$$

whereas the terms containing X cancel out in view of the differential equation which holds for ψ. Summarizing we have from equation (24) and (28) and using equation (30)

$$\left.\begin{aligned}
p &= A_0 + \left(A_1 r + \frac{B_1}{r^2} - \frac{DX}{4\pi}\frac{d\psi}{dr}\right)\cos\vartheta \\
\mathfrak{w}_r &= 0 \\
\mathfrak{w}_\vartheta &= 0 \\
\mathfrak{w}_\varphi &= \left(-\frac{A_1}{2\eta}r + \frac{B_1}{\eta}\frac{1}{r^2} - \frac{DX}{4\pi\eta}\frac{d\psi}{dr}\right)\sin\vartheta
\end{aligned}\right\} \tag{31}$$

It remains to find the velocity $\mathfrak{v}$ from the expression obtained for the circulation

$$\mathfrak{w} = \operatorname{rot}\mathfrak{v}$$

considering, of course, the condition for incompressibility div $\mathfrak{v} = 0$. The two equations in question read:

$$\left.\begin{aligned} \mathfrak{w}_\varphi &= \operatorname{rot}_\varphi \mathfrak{v} = \frac{1}{r}\frac{\partial}{\partial r} r\mathfrak{v}_\vartheta - \frac{1}{r}\frac{\partial}{\partial \vartheta}\mathfrak{v}_r = \\ &= \left(-\frac{A_1}{2\eta} r + \frac{B_1}{\eta}\frac{1}{r^2} - \frac{DX}{4\pi\eta}\frac{d\psi}{dr}\right)\sin\vartheta \\ \operatorname{div}\mathfrak{v} &= \frac{1}{r^2}\frac{\partial}{\partial r} r^2\mathfrak{v}_r + \frac{1}{r\sin\vartheta}\frac{\partial}{\partial\vartheta}\sin\vartheta\,\mathfrak{v}_\vartheta = 0 \end{aligned}\right| \qquad (32)$$

If we put

$$\mathfrak{v}_r = R_1 \cos\vartheta, \qquad \mathfrak{v}_\vartheta = R_2 \sin\vartheta \qquad (33)$$

where R_1 and R_2 are two functions of r only which are to be determined later, equation (32) can be replaced by

$$\left.\begin{aligned} &\frac{1}{r}\frac{d}{dr} r R_2 + \frac{R_1}{r} = -\frac{A_1}{2\eta} r + \frac{B_1}{\eta}\frac{1}{r^2} - \frac{DX}{4\pi\eta}\frac{d\psi}{dr} \\ &\frac{1}{r^2}\frac{d}{dr} r^2 R_1 + 2\frac{R_2}{r} = 0 \end{aligned}\right|$$

By considering r^2R_1 and rR_2 as the functions to be determined, this system can be replaced by the following:

$$\left.\begin{aligned} &\frac{d^2}{dr^2}(r^2 R_1) - \frac{2}{r^2}(r^2 R_1) = \\ &= \frac{A_1}{\eta} r^2 - \frac{2B_1}{\eta}\frac{1}{r} + \frac{DX}{2\pi\eta} r \frac{d\psi}{dr} \\ &(r R_2) + \frac{1}{2}\frac{d}{dr}(r^2 R_1) = 0 \end{aligned}\right| \qquad (34)$$

The solution corresponding to the right hand term of the first equation in (33) assumes the form

$$r^2 R_1 = \frac{A_1}{\eta}\frac{r^4}{10} + \frac{B_1}{\eta} r + \frac{DX}{2\pi\eta\varkappa^2} r \frac{d\psi}{dr}$$

the correctness of the last term, proportional to X, follows readily in view of the **differential** equation satisfied by ψ. To this result may be added a solution of the homogeneous equation so that we obtain:

$$\begin{aligned} r^2 R_1 = \frac{A_1}{\eta}\frac{r^4}{10} + \frac{B_1}{\eta} r + A_3 r^2 + B_3 \frac{1}{r} + \\ + \frac{DX}{2\pi\eta\varkappa^2} r \frac{d\psi}{dr} \end{aligned}$$

introducing the two new constants A_3 and B_3. The second equation of (34) immediately gives

$$rR_2 = -\frac{A_1}{2\eta}\frac{2}{5}r^3 - \frac{B_1}{2\eta} - A_3 r + \\ + \frac{B_3}{2}\frac{1}{r^2} - \frac{DX}{4\pi\eta\varkappa^2}\frac{d}{dr}r\frac{d\psi}{dr}$$

Considering the formulation of the expressions (33) and taking p from equation (31), we finally obtain the following velocity and pressure distributions satisfying the hydrodynamic equations:

$$\left.\begin{aligned} \mathfrak{v}_r &= \left[\frac{A_1}{\eta}\frac{r^2}{10} + \frac{B_1}{\eta}\frac{1}{r} + A_3 + \right. \\ &\quad \left. + B_3\frac{1}{r^3} + \frac{DX}{2\pi\eta\varkappa^2}\frac{1}{r}\frac{d\varphi}{dr}\right]\cos\vartheta \\ \mathfrak{v}_\vartheta &= \left[-\frac{A_1}{\eta}\frac{r^2}{5} - \frac{B_1}{2\eta}\frac{1}{r} - A_3 + \right. \\ &\quad \left. + \frac{B_3}{2}\frac{1}{r^3} - \frac{DX}{4\pi\eta\varkappa^2}\frac{1}{r}\frac{d}{dr}r\frac{d\varphi}{dr}\right]\sin\vartheta \\ \mathfrak{v}_\varphi &= 0 \\ p &= A_0 + \left[A_1 r + \frac{B_1}{r^2} - \frac{DX}{4\pi}\frac{d\psi}{dr}\right]\cos\vartheta \end{aligned}\right\} \quad (35)$$

VIII. Satisfying the Boundary Conditions. Total Force

According to our assumption, at great distances the velocity is parallel to the x-axis and its value is $-v$. This means that $\mathfrak{v}_r$ and $\mathfrak{v}_\vartheta$ asymptotically approach

$$\mathfrak{v}_r = -v\cos\vartheta, \qquad \mathfrak{v}_\vartheta = v\sin\vartheta$$

Comparison with (35) shows that consequently

$$A_1 = 0 \quad \text{and} \quad A_3 = -v$$

Except for A_0—a constant essential only for the absolute determination of the pressure, and which may therefore be made zero without restricting the generality—B_1 and B_3 can still be determined arbitrarily. They are determined by the condition that $\mathfrak{v}_r$ and $\mathfrak{v}_\vartheta$ must vanish on the surface of the sphere (radius = b), and can be computed from the two equations:

$$\left.\begin{aligned}
&\frac{B_1}{\eta}\frac{2}{b}+B_3\left(\frac{2}{b}\right)^3=\\
&\qquad=v-\frac{DX}{2\pi\eta\varkappa^2}\left(\frac{1}{r}\frac{d\psi}{dr}\right)_{r=b}\\
&\frac{B_1}{\eta}\frac{2}{b}-B_3\left(\frac{2}{b}\right)^3=\\
&\qquad=2v-\frac{DX}{2\pi\eta\varkappa^2}\left(\frac{1}{r}\frac{d}{dr}r\frac{d\psi}{dr}\right)_{r=b}
\end{aligned}\right\} \quad (36)$$

It is found, taking into account the differential equation for ψ

$$\left.\begin{aligned}
&\frac{B_1}{\eta}\frac{2}{b}=+\frac{3}{2}v-\frac{DX}{4\pi\eta}(\psi)_{r=b}\\
&B_3\left(\frac{2}{b}\right)^3=-\frac{1}{2}v+\frac{DX}{4\pi\eta}\left(\frac{d^2\psi}{d(\varkappa r)^2}\right)_{r=b}
\end{aligned}\right\} \quad (36')$$

Thus all available constants have been determined, and the stresses on a surface element can now be computed. The force exerted on the sphere will be directed in the x-direction, and will be exerted on the surface elements of the sphere having a normal r. We therefore want the components of stress which, in the conventional terminology, be designated by p_{rx}.

We have

$$p_{rx}=p_{xx}\frac{x}{r}+p_{yx}\frac{y}{r}+p_{zx}\frac{z}{r}$$

and further

$$\begin{aligned}
p_{xx}&=-p+2\eta\frac{\partial \mathfrak{v}_x}{\partial x}\\
p_{yx}&=\eta\left(\frac{\partial \mathfrak{v}_y}{\partial x}+\frac{\partial \mathfrak{v}_x}{\partial y}\right)\\
p_{zx}&=\eta\left(\frac{\partial \mathfrak{v}_z}{\partial x}+\frac{\partial \mathfrak{v}_x}{\partial z}\right)
\end{aligned}$$

finally

$$\begin{aligned}
\mathfrak{v}_x&=\mathfrak{v}_r\cos\vartheta-\mathfrak{v}_\vartheta\sin\vartheta\\
\mathfrak{v}_y&=(\mathfrak{v}_r\sin\vartheta+\mathfrak{v}_\vartheta\cos\vartheta)\cos\varphi\\
\mathfrak{v}_z&=(\mathfrak{v}_r\sin\vartheta+\mathfrak{v}_\vartheta\cos\vartheta)\sin\varphi
\end{aligned}$$

Thus by means of a rather lengthy but elementary computation p_{rx} can be determined on the basis of the expressions for $\mathfrak{v}$ and p given in equation (35). Thus, with $A_0 = 0$ and using the differential equation for ψ, we have

$$\left.\begin{array}{l} p_{rx} = \Bigg[\left(B_3 \frac{3}{r^4} + \frac{DX}{4\pi\eta\varkappa^2}\frac{d^3\psi}{dr^3}\right) - \\ -\left(\frac{B_1}{\eta}\frac{3}{r^2} + B_3\frac{9}{r^4} - \frac{6\,DX}{4\pi\eta\varkappa^2}\left(\frac{1}{r}\frac{d^2\psi}{dr^2} - \right.\right. \\ \left.\left. -\frac{1}{r^2}\frac{d\psi}{dr}\right)\right)\cos^2\vartheta\Bigg] \end{array}\right\} \quad (37)$$

The force in the x-direction on a sphere of radius r is obtained by integration of p_{rx} over the surface of the sphere with the surface element $d\sigma$ and furnishes

$$\int p_{rx}\,d\sigma = 4\pi\eta\left[-\frac{B_1}{\eta} + \frac{DX}{4\pi\eta}r^2\frac{d\psi}{dr}\right] \quad (38)$$

The value of the constant B_1 from equation (36') can now be substituted, and r made equal to b. We thus find:

$$\int p_{rx}\,d\sigma = -6\pi\eta v b + DXb\left[\frac{d}{dr}(r\psi)\right]_{r=b} \quad (38')$$

Finally the expression previously found for the potential ψ surrounding an ion with the charge e_j:

$$\psi = \frac{e_j}{D}\frac{e^{-\varkappa r}}{r}$$

which holds for low concentrations, can be used for the computation. Thus we obtain for the force K_x exerted on the ion of radius b in the x-direction:

$$K_x = -6\pi\eta v b - e_j X \varkappa b \qquad (39)$$

It is seen that, in addition to Stokes' force

$$6\pi\eta v b$$

a force caused by the eletrophoresis appears which increases with increasing $\varkappa$, *i.e.*, with increasing concentration. The ratio of this supplementary force to the force e_jX acting on the ion in a pure solvent is given by

$$\varkappa b$$

equal to the ratio of the ion radius to the average thickness of the ion atmosphere. We have found before for a uni-univalent salt dissolved in water that

$$\varkappa = 0.326 \cdot 10^8 \sqrt{\gamma}$$

were γ designates the concentration in moles per liter. Hence the ratio can be computed for this case as

$$\varkappa b = 0.326 \cdot 10^8 b \sqrt{\gamma}$$

and since b is of the order of 10^{-8} cm., it will be appreciated that the electrophoretic force, as well as the ionic force before, is already of importance at small concentrations. To the fact that both forces are proportional to the square root of the concentration (if we restrict ourselves to the limiting case of low concentrations) is to be attributed the role which the square root plays for the conductivity case. This proportionality is a consequence of the fact that the thickness of the ion atmosphere is inversely proportional to the square root of the concentration, and this in turn followed essentially from Coulomb's law, involving the square of the distance.

C. Conductivity

IX. Coefficient of Conductivity

The external field strength X is supposed to act in the x-direction on an ion with the charge e_j This be in a solution where each cm.³ contains $n_1, .. n_i, .. n_s$ ions* with charges $e_1, .. e_i, .. e_s$. As a consequence of the finite relaxation time, required by the ion atmosphere being rebuilt continuously around the moving ion, the amount

$$-\frac{\omega_j \varkappa}{6} \frac{e_j^2}{D}$$

must be added to the force $e_j X$, in accordance with equation (18). The total force

$$e_j X - \frac{\omega_j \varkappa}{6} \frac{e_j^2}{D} \tag{40}$$

tends to set the ion in motion and causes a velocity such as to establish equilibrium with the total frictional force. As a consequence of the electrophoretic effect, however, the total frictional force exceeds that calculated from Stokes' formula, and, according to equation (39), is equal to

$$-6\pi\eta b_j v_j - e_j X \varkappa b_j \tag{41}$$

*In section A we designated the same quantitites by $\bar{n}_i$; in the following, however, we may omit the bar without ambiguity.

if the velocity of the ion is designated by v_j and its radius with b_j. The velocity can be determined from the condition that the sum of the forces be zero, *i.e.*, from the equation

$$e_j X - \frac{\omega_j \varkappa e_j^2}{6\ D} - 6\pi\eta b_j v_j - e_j X \varkappa b_j = 0 \quad (42)$$

Here, according to section A, the quantity ω_j is an abbreviation for

$$\omega_j = \frac{\varrho v_j}{kT}$$

Further the average coefficient of friction ρ is derived from the individual coefficients ρ_i by the formula

$$\varrho = \frac{\Sigma n_i e_i^2 \varrho_i}{\Sigma n_i e_i^2} \quad (43)$$

while, finally, κ has its previous meaning, according to which

$$\varkappa^2 = \frac{4\pi}{DkT} \Sigma n_i e_i^2 \quad (44)$$

For infinitely small concentrations, $\kappa = 0$, thus

$$e_j X - 6\pi\eta b_j v_j = 0$$

so that the velocity is to be computed from the conventional equation where, because of the meaning given to ρ_j

$$6\pi\eta b_j = \varrho_j$$

Taking this into consideration it follows from equation (42) as a second approximation, *i.e.*, for finite concentrations

$$v_j = \frac{e_j X}{\varrho_j}\left[1 - \frac{\varrho}{\varrho_j}\frac{e_j^2 \varkappa}{6\,DkT} - b_j \varkappa\right] \quad (45)$$

The specific conductivity λ (expressed in electrostatic units) is given by the formula

$$\lambda = \frac{1}{X} \Sigma n_j e_j v_j$$

Thus, in view of equation (45), it follows that

$$\lambda = \Sigma \frac{n_j e_j^2}{\varrho_j} - \varkappa\left[\Sigma \frac{n_j e_j^4}{6\,DkT}\frac{\varrho}{\varrho_j^2} + \Sigma \frac{n_j e_j^3}{\varrho_j} b_j\right] \quad (46)$$

The specific conductivity at infinite dilution be designated by λ_0, then

$$\lambda_0 = \Sigma \frac{n_j e_j^2}{\varrho_j}$$

Further the coefficient of conductivity f_λ be defined by

$$\frac{\lambda_0 - \lambda}{\lambda_0} = f_\lambda$$

so that it measures, as usual, the relative change in conductivity with an increase in concentration. In view of the equations (43) and (44), defining the average values for ρ and κ and replacing the summation index j by i throughout, we obtain

$$1 - f_\lambda = \sqrt{\frac{4\pi}{DkT}\Sigma n_i e_i^2} \left[\frac{\Sigma n_i e_i^2 \varrho_i}{\Sigma n_i e_i^2} \frac{\Sigma \frac{n_i e_i^4}{6\,DkT} \frac{1}{\varrho_i^2}}{\Sigma \frac{n_i e_i^2}{\varrho_i}} + \frac{\Sigma \frac{n_i e_i^2}{\varrho_i} b_i}{\Sigma \frac{n_i e_i^2}{\varrho_i}} \right] \quad (47)$$

No special assumption is made in equation (47) regarding the nature of the dissolved substances; the formula thus also applies, for instance, to mixtures. It occurs frequently that n molecules of a particular salt are dissolved in one cm.[3]. Provided each of these molecules dissociates into

$$\nu_1, \dots \quad \nu_i, \dots \quad \nu_s$$

ions with valencies

$$z_1, \dots \quad z_i, \dots \quad z_s$$

and the electronic charge is designated by ε, equation (47), for this particular case, assumes the form

$$1 - f_\lambda = \sqrt{\frac{4\pi\varepsilon^2}{DkT}\nu n} \left[\frac{\varepsilon^2}{6\,DkT} w_1 + b w_2 \right] \quad (48)$$

Here

$$\nu = \Sigma \nu_i \quad (48')$$

denotes the total number of ions into which a molecule dissociates, while w_1 and w_2 are numbers which we call valency factors, defined by the equations:

$$\left.\begin{aligned} w_1 &= \frac{\Sigma \nu_i z_i^2 \varrho_i}{\sqrt{\nu \Sigma \nu_i z_i^2}} \cdot \frac{\Sigma \frac{\nu_i z_i^4}{\varrho_i^2}}{\Sigma \frac{\nu_i z_i^2}{\varrho_i}} \\ w_2 &= \sqrt{\frac{\Sigma \nu_i z_i^2}{\nu}} \end{aligned}\right\} \quad (48'')$$

Further, an average ion diameter b has been introduced instead of the ion diameter b_i by the defining formula

$$b = \frac{\Sigma \frac{\nu_i z_i^2}{\varrho_i} b_i}{\Sigma \frac{\nu_i z_i^2}{\varrho_i}} \qquad (48''')$$

X. Discussion of the Conductivity Coefficient

According to equation (48) $1 - f_\lambda$ is proportional to $\sqrt{n}$, *i.e.*, proportional to the square root of the concentration. The theory thus confirms the Kohlrausch's law, mentioned in the introduction.

Equation (48) further shows that $1 - f_\lambda$ is associated with the dielectric constant of the solvent. The dependence is not simple, since the expression for $1 - f_\lambda$ is composed of two additive terms, one of them being proportional to $D^{-3/2}$ and the other to $D^{-1/2}$. We can only establish that for a given concentration, the theory requires the larger deviations from the limiting conductivity, the smaller the dielectric constant of the solvent. This is in qualitative agreement with the experimental results of Walden, whose rule shall be treated in more detail in section D.

For water as a solvent and at a temperature of 18°C., we have, according to Drude's formula,[6] in general (t = temperature in °C.),

$$D = 88.23 - 0.4044\,t + 0.001035\,t^2$$

from which it follows that at 18°C.

$$D = 81.29$$

If we further put $\varepsilon = 4.77 \times 10^{-10}$, $k = 1.346 \times 10^{-16}$, $T = 291$, and designated by γ the concentration of the salt in moles per liter so that

$$n = 6.06 \cdot 10^{20}\,\gamma$$

we can then compute the factors in equation (48), and we obtain for solutions with water as solvent

$$1 - f_\lambda = [0.278\, w_1 + 0.233 \cdot 10^8\, b w_2]\sqrt{\nu\gamma} \quad (49)$$

It is of interest to compare the coefficient of condutivity f_λ with the osmotic coefficient f_0 found previously. Referred to water at 18°C. we obtain from previous statements

$$1 - f_0 = 0.278\, w \sqrt{\nu\gamma} \quad (50)$$

where the valency factor essential for the osmotic coefficient was given by

$$w = \left(\frac{\Sigma \nu_i z_i^2}{\nu}\right)^{3/2} \quad (50')$$

It will be seen that indeed a relation exists between f_λ and f_0. However, it is not simple, and we cannot easily draw a conclusion from f_λ as to the value of f_0 except for the order of magnitude. The two reasons which prevent this are that (*a*) the valency factor in f_0 can be determined from the valencies exclusively, whereas the valency factor w_1 in f_λ, according to equation (48"), also contains the ratio of the mobilities; (*b*) as shown for example by equation (49), the average ion diameter b is necessary for the calculation of f_λ which - assuming the relation

$$\varrho_i = 6\pi\eta b_i$$

derived from Stokes' formula to be correct - amounts to the fact that for the computation of the second term of f_λ the mobilities again are of importance. We believe that, from a physical point of view, the assertion to the contrary by Ghosh, according to which f_0 and f_λ are mutually dependent as to their magnitude, is improbable from the very beginning.

The characteristic difference between the behavior of f_0 and f_λ may be illustrated as follows. Since the expression for f_0 contains only the number of dissociated particles ν_i and the valencies z_i, all salts, which dissociate similarly, must, in the limit for small concentrations, show the same deviations from the osmotic pressure calculated by classical theory. However, since f_λ further depends on the mobilities, the deviation of the conductivity from its limit will show individual differences, even in the limit for small concentrations and for similarly dissociated salts, In section D we shall be in a position to prove by experimental data that this is actually so and that the assertion to the contrary by Ghosh's theory does not hold in practice.

A few data for the valency factors w_1 and w_2 shall now be presented for the three types of salts KCl, K_2SO_4, and $MgSO_4$. The following table contains the values for w_2.

Table I

Type of Salt	w_2
KCl.....	$\sqrt{1} = 1$
K_2SO_4...	$\sqrt{2} = 1.414$
$MgSO_4$...	$\sqrt{4} = 2$

The valency factor w_1 is a function of the mobilities also; it contains, as will be readily seen, only their ratios.

(*a*) For a monovalent salt (of the type KCl) we have

$$w_1 = \frac{\varrho_1 + \varrho_2}{2} \frac{\frac{1}{\varrho_1^2} + \frac{1}{\varrho_2^2}}{\frac{1}{\varrho_1} + \frac{1}{\varrho_2}} \tag{51}$$

If we put

$$\sigma_1 = \frac{\varrho_1}{\varrho_1 + \varrho_2}, \quad \sigma_2 = \frac{\varrho_2}{\varrho_1 + \varrho_2} \tag{52}$$

so that

$$\sigma_1 + \sigma_2 = 1$$

and the σ's have the dimension of a number, then

$$w_1 = \frac{1}{2} \frac{\sigma_1^2 + \sigma_2^2}{\sigma_1 \sigma_2} = \frac{1}{2}\left(\frac{\sigma_1}{\sigma_2} + \frac{\sigma_2}{\sigma_1}\right) \tag{53}$$

For equal coefficients of friction, $\sigma_1 = \sigma_2$ and w_1 is equal to 1. If the coefficients of friction are different, then $w_1 > 1$. Thus the larger the difference between the mobilities, the larger will be the first term in $1 - f_\lambda$. An orientation regarding the values of w_1 for different ratios σ will be gained from table II for uni-univalent salts.

Table II

Type of Salt KCl.

σ_1	σ_2	w_1
0.1	0.9	4.55
0.2	0.8	2.12
0.3	0.7	1.38
0.4	0.6	1.08
0.5	0.5	1
0.6	0.4	1.08
0.7	0.3	1.38
0.8	0.2	2.12
0.9	0.1	4.55

The minimum is obtained for $\rho_1 = \rho_2$.

(*b*) For bi-univalent salts (of the type K_2SO_4) we obtain in a similar manner

$$\left.\begin{aligned} w_1 &= \frac{2\rho_1 + 4\rho_2}{3\sqrt{2}} \frac{\dfrac{2}{\rho_1^2} + \dfrac{16}{\rho_2^2}}{\dfrac{2}{\rho_1} + \dfrac{4}{\rho_2}} = \\ &= \frac{\sqrt{2}}{3} \frac{\sigma_1 + 2\sigma_2}{2\sigma_1 + \sigma_2}\left(8\frac{\sigma_1}{\sigma_2} + \frac{\sigma_2}{\sigma_1}\right) \end{aligned}\right\} \quad (53')$$

Here subscript 1 refers to the monovalent and subscript 2 to the bivalent ion. Table III contains numerical values.

Table III

Type of Salt K_2SO_4.

σ_1	σ_2	w_1
0.1	0.9	8.05
0.2	0.8	4.24
0.3	0.7	3.55
0.4	0.6	3.68
0.5	0.5	4.24
0.6	0.4	5.22
0.7	0.3	6.87
0.8	0.2	10.1
0.9	0.1	18.8

The minimum, $w_1 = 3.53$ is reached at $\rho_1 = 0.48\rho_2$.

(*c*) Finally, for the bi-bivalent salts (type $MgSO_4$) the following formula holds

$$w_1 = 4(\rho_1+\rho_2)\frac{\frac{1}{\rho_1^2}+\frac{1}{\rho_2^2}}{\frac{1}{\rho_1}+\frac{1}{\rho_2}} = 4\left(\frac{\sigma_1}{\sigma_2}+\frac{\sigma_2}{\sigma_1}\right) \quad (53'')$$

so that in this instance the magnitudes of the valency factors are simply 8 times the values given in Table II. The minima for the valency factors w_1 for the three types of salts, respectively, are 1, 3.53, and 8.

The steep increase of $1 - f_\lambda$ with increasing valency, required by the theory, is in qualitative agreement with the finding that, for equal equivalent concentration, the deviations from the limiting conductivity are the greater, the greater the valency of the dissolved ions.

D. Comparison with Experimental Results

X1. Uni-univalent Salts Dissolved in Water

A comparatively large and carefully investigated - which we owe to years of work by Kohlrausch - body of material is available for testing the limiting law for uni-univalent salts. It is known that it was this material from which Kohlrausch derived his law[7] according to which $1 - f_\lambda \sim \sqrt{\gamma}$.

Kohlrausch's law is impressively demonstrated by the diagram, presented in the book by Kohlrausch and Holborn, in which the equivalent conductivity is plotted as a function of the square root of the concentration. It is therefore superfluous to make any further remarks beyond the statement that by virtue of the preceding developments the law may now be considered to be supported by theory. Moreover, the numerical value of the factor of proportionality should be of interest.

In the following we will use throughot the so-called molar conductivity and not the equivalent conductivity. Let the first be defined as the quotient of the specific conductivity and the number of moles dissolved per cm.^{3}, $\gamma/1000$. For uni-univalent salts, molar **conductivity** and equivalent conductivity are equal. To define the conductivity coefficient f_λ, we put

$$\lambda = \lambda_0 f_\lambda$$

where λ_0 was that conductivity which the solution would have if the ions moved as they do at infinite dilution. Therefore, we had generally

$$\lambda_0 = \sum \frac{n_i e_i^2}{\varrho_i}$$

If n molecules of a salt are dissolved in one cm.^{3}, each of which dissociates into

$$\nu_1, \ldots \nu_i, \ldots \nu_s$$

ions with the valencies

$$z_1, \ldots z_i, \ldots z_s$$

then

$$\lambda_0 = n\varepsilon^2 \sum \frac{\nu_i z_i^2}{\varrho_i}$$

where ε is the electronic charge. Further

$$n = N \frac{\gamma}{1000}$$

where $N = 6.06 \times 10^{23}$ is Loschmidt's number. Designating the molar conductivity by Λ, we have

$$\Lambda = \frac{1000\,\lambda}{\gamma} = f_\lambda N \varepsilon^2 \sum \frac{\nu_i z_i^2}{\varrho_i} \qquad (54)$$

the molecular conductivity for infinite dilution be denoted by Λ_0; it is equal to

$$\Lambda_0 = N\varepsilon^2 \sum \frac{\nu_i z_i^2}{\varrho_i} \qquad (54')$$

It is not usual to express Λ and Λ_0, as in these formulas, in electrostatic units; rather, the unit ohm^{-1} cm.^{2} has been chosen. If in these practical units the conductivities be designated by $\overline{\Lambda}$ and $\overline{\Lambda}_0$, then

$$\overline{\Lambda} = \frac{\Lambda}{9 \cdot 10^{11}} \text{ und } \overline{\Lambda}_0 = \frac{\Lambda_0}{9 \cdot 10^{11}}$$

the values in the following tables refer throughout to these practical units.

The square root law is only a limiting law, and the deviations therefrom increase with increasing concentration. Therefore, we selected fro the values of $\bar{\Lambda}$ given by Kohlrausch[8] the 6 lowest concentrations available, namely, γ = 0.0001, 0.0002, 0.0005, 0.001, 0.002, and 0.005, and represented $\bar{\Lambda}$ by the formula

$$\bar{\Lambda} = \bar{\Lambda}_0 - \alpha \sqrt{2\gamma} + \beta (2\gamma) \qquad (55)$$

where the coefficients $\bar{\Lambda}_0$, α, and β were each determined by the method of least squares. For the uni-univalent salts agreement is very good, as will be seen from the accompanying table which relates to KCl solutions:

Table IV

2γ	$\sqrt{2\gamma}$	$\bar{\Lambda}$calc.	$\bar{\Lambda}$obs.	Δ
0.0002	0.014142	129.09	129.07	-0.02
0.0004	0.020000	128.75	128.77	+0.02
0.0010	0.031623	128.08	128.11	+0.03
0.0020	0.044721	127.34	127.34	0.00
0.0040	0.063246	126.32	126.31	-0.01
0.0100	0.100000	124.39	124.41	+0.02
0.0200	0.14142	122.36	122.43	+0.07
0.0400	0.20000	119.75	119.96	+0.21
0.1000	0.31623	115.51	115.75	+0.24

The first and second columns contain 2γ and $\sqrt{2\gamma}$, respectively, the third column the computed and the fourth column the observed molar conductivity; the last column contains the difference between the two last-mentioned quantities. The table is divided into two parts by a horizontal line; the region below the line was not used for the calculation of the three coefficients $\bar{\Lambda}_0$, α, β. The formula, resulting from the method of least squares, which is the basis for the values of $\bar{\Lambda}_{calc.}$, reads

$$\bar{\Lambda} = 129.93 - 59.94 \sqrt{2\gamma} + 45.3 (2\gamma)$$

The following table V contains in the first column the chemical formulas of the 18 salts for which the computations were carried through. In the second, third and fourth columns are entered the values found for $\bar{\Lambda}_0$, α, and β. The fifth column contains the ratio $\alpha/\bar{\Lambda}_0$. The values are valid for a temperature of 18°C.

Table V

Salt	$\bar{\Lambda}_0$	α	β	$\alpha/\bar{\Lambda}_0$	$0.278w_1$	$10^8 b$
LiCl	98.93	57.35	71.4	0.580	0.342	1.02
$LiIO_3$	67.35	48.33	36.6	0.718	0.278	1.89
$LiNO_3$	95.24	56.27	71.5	0.591	6.332	1.11
NaF	90.05	50.42	23.1	0.557	0.278	1.20
NaCl	108.89	54.69	34.9	0.502	0.301	0.85
$NaIO_3$	77.42	51.39	34.2	0.664	0.286	1.62
$NaNO_3$	105.34	58.27	52.7	0.553	0.295	1.11
KF	111.29	55.88	44.9	0.502	0.292	0.90
KCl	129.93	59.94	45.3	0.461	0.278	0.79
KBr	132.04	62.17	55.9	0.471	0.278	0.83
KI	130.52	51.53	-16.6	0.395	0.278	0.50
KIO_3	98.41	54.18	19.6	0.551	0.338	0.92
$KClO_3$	119.47	58.16	14.4	0.487	0.281	0.88
KNO_3	126.46	65.67	59.3	0.519	0.278	1.04
KCNS	121.04	54.10	10.9	0.445	0.281	0.71
CsCl	133.08	53.75	-26.4	0.404	0.278	0.54
$AgNO_3$	115.82	62.35	43.2	0.558	0.281	1.19
$TlNO_3$	127.55	63.40	-14.1	0.497	0.279	0.94

It is further of interest to compare Kohlrausch's measurements with those of another observer. Figure 4 was designed for this

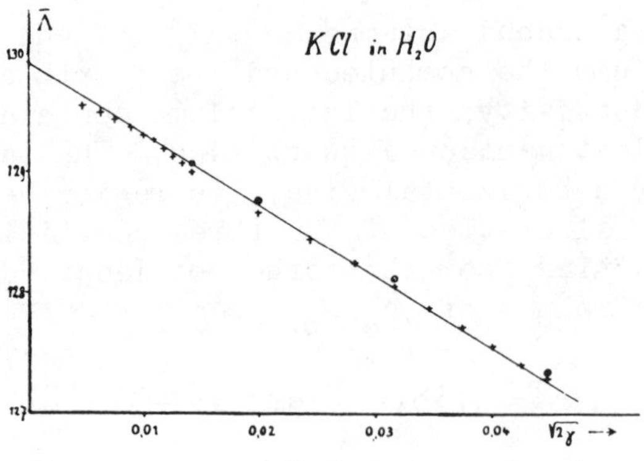

Figure 4

purpose. It shows, plotted as a function of $\sqrt{2\gamma}$, the conductivities of KCl observed by Kohlrausch indicated by circles, and the limiting straight line from table V:

Further observations by Weiland,[9] extending from $\gamma = 10^{-5}$ to $\gamma = 10^{-3}$, are entered as crosses. One gains the impression that Kohlrausch's values are excellent.

The ratio $\alpha/\bar{\Lambda}_0$ in the fifth column is of special interest to us. It follows from the definition of f_λ that

$$1 - f_\lambda = \frac{\alpha}{\bar{\Lambda}_0}\sqrt{2\gamma}$$

The table shows the variation of $\alpha/\bar{\Lambda}_0$ between 0.395 for KI and 0.718 for $LiIO_3$, there is no doubt that the limiting law is not the same for all uni-univalent salts, as would be expected from Ghosh's theory. According to the theory presented here, and from equation (49), the following relation should hold

$$1 - f_\lambda = [0.278\, w_1 + 0.233 \cdot 10^8 b]\sqrt{2\gamma}$$

where w_2 was taken to be 1 from table I, while the valency factor w_1 has to be computed from equation (51) or equation (53), and consequently depends on the ratio of the two mobilities. We define the mobility L_i as the amount in equation (54') associated with one ion so that

$$L_i = N\varepsilon^2 \frac{z_i^2}{\varrho_i} \qquad (56)$$

Thus

$$\Lambda_0 = \sum \nu_i L_i. \qquad (57)$$

In practical units, the mobilities are again designated by $\bar{L}$, so that

$$\bar{L}_i = \frac{L_i}{9 \cdot 10^{11}}$$

Hence, according to equation (53),

$$w_1 = \frac{1}{2}\left(\frac{L_2}{L_1} + \frac{L_1}{L_2}\right) \qquad (58)$$

We thought it advisable to use the same material whenever feasible, and we, therefore, calculated the mobilities on the basis of the values for the limiting conductivity $\bar{\Lambda}_0$ listed in table V. The ratio of the transport numbers[10] for K and Cl, 0.497:0.503, was assumed to be absolutely accurate; then the mobilities of the ions were so computed that the sum of the squares of the deviations of the limiting conductivities, calculated from these mobilities from the values of $\bar{\Lambda}_0$ given in table V, became a minimum. Thus we obtained

	Li	*Na*	*K*	*Cs*	*Ag*	*Tl*
$\bar{L}$ =	33.46	43.49	64.61	67.69	43.99	65.72

	F	*Cl*	*Br*	*I*	ClO_3	IO_3	NO_3	*CNS*
$\bar{L}$ =	46.62	65.39	67.43	65.91	54.86	33.87	61.83	56.43

Table VI affords information regarding these values of $\bar{\Lambda}_0$ computed on the basis of these mobilities. In the last column are listed the percentage deviations Δ. Where the computed values of $\bar{\Lambda}_0$ are in parentheses, this indicates that the salt contains an ion that occurs only once in the table; under these circumstances, evidently, a value for Δ does not exist.

Table VI

	$\bar{\Lambda}_{obs.}$	$\bar{\Lambda}_{calc.}$	Δ in %
LiCl	98.93	98.85	−0.08
$LiIO_3$	67.35	67.33	−0.03
$LiNO_3$	95.24	95.29	+0.05
NaF	90.05	90.11	+0.07
NaCl	108.89	108.88	−0.01
$NaIO_3$	77.42	77.36	−0.08
$NaNO_3$	105.34	105.32	−0.02
KF	111.29	111.23	−0.05
KCl	129.93	130.00	+0.05
KBr	132.04	(132.04)	---
KI	130.52	(130.52)	---
KIO_3	98.41	98.48	+0.07
$KClO_3$	119.47	(119.47)	---
KNO_3	126.46	126.44	−0.02
KCNS	121.04	(121.04)	---
CsCl	133.08	(133.08)	---
$AgNO_3$	115.82	(115.82)	---
$TlNO_3$	127.55	(127.55)	---

In the table, $\bar{\Lambda}_{o\,obs.}$ indicates that the value was obtained from the experimental data by means of equations (55), while $\bar{\Lambda}_{o\,calc.}$ designates the value derived by addition - according to equation (57) - of the values for $\bar{L}$ given above.

We now compute the valency factor w_1 for each salt from equation (58). The first additive term in $1 - f_\lambda$ is, as given above, equal to 0.278 w_1, the numerical values for this quantity are entered in the sixth column of table V. It will be noted that they are always smaller than the experimental gradient divided by $\bar{\Lambda}_0$ or $\alpha/\bar{\Lambda}_0$. This means that the ionic effect does not account for the total decrease in conductivity, we will attribute the discrepancy to the second term in $1 - f_\lambda$ caused by electrophoresis. If this is done, the descrepancy must be equal to $0.233 \times 10^8 b$, and the values for $10^8 b$ listed in the seventh column of table V are thus obtained. Since the electrophoresis was computed by means of the conventional hydrodynamic equations, these values for the radius of the spheres will not be considered of great importance. However, it agrees with expectations that the values for the b's are all of the order of 10^{-8} cm.

Finally, it should be remarked that from the group of salts measured by Kohlrausch, the NH_4 salts were not considered, because the measurements are less accurate. Further TlF was omitted, because the experimental values show a maximum which could not be represented by our limiting law. The maximum is always found, more or less distinct, for acids and bases. It is highly probable that the maximum, at least in this instance, does not correspond to reality, and Kohlrausch himself is of this opinion. A strong support for this interpretation is provided by more recent measurements by Kraus and Parker,[11] who investigated HIO_3 in vessels of glass, pyrex glass and quartz, and, in this order, observed the maximum less and less distinctly. However, it had not disappeared even for the quartz vessel; but from these experiments it appears doubtful whether these very difficult measurements for strongly diluted solutions can be considered conclusive.

XII. Salts with Multivalent Ions Dissolved in Water

(*a*) Uni-bivalent salts. Again measurements by Kohlrausch shall be considered; seven salts are studied. The molecular conductivity (in this instance double the equivalent conductivity) shall be discussed by the same method as before. We put

$$\bar{\Lambda} = \bar{\Lambda}_0 - \alpha\sqrt{3\gamma} + \beta(3\gamma)$$

and from observations for $\gamma = 0.00005$, $\gamma = 0.0001$, $\gamma = 0.00025$, $\gamma = 0.0005$ and $\gamma = 0.0010$, we determined the coefficients $\bar{\Lambda}_0$, α, and β for each salt by the method of least squares. In table VII, an example is given for the case of $Ba(NO_3)_2$ for which the formula

$$\bar{\Lambda} = 233.90 - 262.23\sqrt{3\gamma} - 187.0\,(3\gamma)$$

resulted.

Table VII

3γ	$\sqrt{3\gamma}$	$\bar{\Lambda}$calc.	$\bar{\Lambda}$obs.	Δ
0.00015	0.012274	230.66	230.64	−0.02
0.00030	0.017320	229.30	229.30	0.00
0.00075	0.027386	226.58	226.60	+0.02
0.00150	0.038730	223.46	223.44	−0.02
0.00300	0.054772	218.98	219.00	+0.02
0.00750	0.086603	209.79	210.58	+0.79
0.01500	0.122474	198.87	201.92	+3.05
0.03000	0.173205	186.87	191.32	+4.45
0.07500	0.273861	148.06	173.62	+5.56

A horizontal line is drawn in the table; the formula of interpolation relates to the region above the line. It will be seen that the figures are well represented in this region, while outside of it fairly large deviations soon appear. The steeper course of the curve for $\bar{\Lambda}$ is the reason why the certainty with which α can be determined is smaller. Further Kohlrausch himself considers these measurements less reliable than those for mono-monovalent salts.[12] We would like to mention in this connection that where, as i the case of the $Ba(NO_3)_2$, the coefficient α is given by a number with five digits, this is only so because, when applying the method of least squares, the experimental figures were taken to be absolutely accurate. The number with five digits, however, is by no means to be considered as given with this accuracy by the experiments.

Table VIII contains in the first four columns the chemical

Table VIII

	$\bar{\Lambda}_0$	α	β	$\alpha/\bar{\Lambda}_0$	$0.278w_1$	$10^8 b$
$Ba(NO_3)_2$	233.90	262.23	−187.0	1.121	0.986	0.41
$Sr(NO_3)_2$	226.83	276.26	313.2	1.218	0.986	0.71
$CaCl_2$	233.38	248.69	110.7	1.066	0.992	0.23
$Ca(NO_3)_2$	227.11	275.50	327.7	1.213	0.982	0.70
$MgCl_2$	221.75	235.55	53.1	1.062	1.026	1.09
$Pb(NO_3)_2$	246.11	379.51	543	1.542	0.980	1.71
K_2SO_4	264.46	229.04	−1488	0.866	0.980	---

formula of the salt and the values of $\bar{\Lambda}_0$, α, and β. In the fifth column is entered the value of $\alpha/\bar{\Lambda}_0$ (the limiting slope of the curve divided by $\bar{\Lambda}_0$). The entries refer to 18°C.

From the experiments we have in the limit for small concentrations

$$1 - f_\lambda = \frac{\alpha}{\bar{\Lambda}_0} \sqrt{3\gamma}$$

theoretically we found equation (49)

$$1 - f_\lambda = [0.278\, w_1 + 0.233 \cdot 10^8 b\, w_2] \sqrt{3\gamma}$$

since here $\nu = 3$. Further, from the general formula for w_1, equation (53'), we obtain for the special case of the uni-bivalent salts

$$w_1 = \frac{2\rho_1 + 4\rho_2}{3\sqrt{2}} \frac{\frac{2}{\rho_1^2} + \frac{16}{\rho_2^2}}{\frac{2}{\rho_1} + \frac{4}{\rho_2}}$$

First the mobility of the Ba, Sr, Ca, Mg, Pb, and SO_4 ions, respectively, was determined from the limiting conductivities $\bar{\Lambda}_0$ given in table VIII also using the figures for NO_3 and Cl given in section XI. The Ca ion appears twice; its mobility found from $Ca(NO_3)_2$ is $\bar{L}$ = 103.45, while that found from $CaCl_2$ is $\bar{L}$ = 102.60; the mean value of these two figures was assumed to be correct. It follows that

	Ba	Sr	Ca	Mg	Pb	SO_4
$\bar{L}$ =	100.24	103.17	103.03	90.97	122.45	135.24

According to equation (56)

$$L_i = N\varepsilon^2 \frac{z_i^2}{\rho_i}$$

and since

$$\bar{L}_i = \frac{L_i}{9 \cdot 10^{11}}$$

so that the quantity $z_i^2/\bar{L}_i$ may simply be substituted for ρ_i in computing w_1. This was done and from it w_1 was calculated, and the value of $0.278\omega_1$ entered into the sixth column of table VIII. The electrophoretic term was held responsible for the difference between $\alpha/\bar{\Lambda}_0$ and $0.278w_1$, and th radius $10^8 b$ computed therefrom. The figures in the seventh column were arrived at in this manner. In the line for K_2SO_4 no value for $10^8 b$ is given. In this case

$0.278w_1$ would be larger than the value of $\alpha/\bar{\Lambda}_0$ derived from the experiments; thus a negative value for b would result from the computations which does not make sense. It will be noted that for this salt the coefficient β is negative and that it assumes an unusually large value. This means that the experimental curve for the conductivity has a large curvature in the direction opposite to the one which is commonly found. It may therefore be justified to consider the value of $\alpha/\bar{\Lambda}_0 = 0.866$ as not very certain. Negative signs for β occurred several times, even for uni-univalent salts. Therefore, it would seem advisable, not to overestimate the accuracy for the figures giving the slope, and to consider the three digits in the figures not as an indication of the actual accuracy.

(b) Bi-bivalent Salts. The computations for four different bi-bivalent salts, again from observations by Kohlrausch, were carried through. Not only are the observations less good, the determination of the tangent is also less accurate since the concentrations at which the observations were made are, for this case of higher valencies, too high. We put

$$\bar{\Lambda} = \bar{\Lambda}_0 - \alpha\sqrt{2\gamma} + \beta(2\gamma)$$

and determined the three coefficients from observations with $\gamma = 0.00005$, $\gamma = 0.0001$, $\gamma = 0.00025$, $\gamma = 0.00050$ and $\gamma = 0.00100$. The results are presented in table IX.

Table IX

	$\bar{\Lambda}_0$	α	β	$\alpha/\bar{\Lambda}_0$	$0.278w_1$	$10^8 b$
$MgSO_4$	229.40	970.7	1067	4.23	2.42	3.8
$ZnSO_4$	230.85	1065.1	1145	4.61	2.40	4.7
$CdSO_4$	231.62	1200.5	1922	5.18	2.39	6.0
$CuSO_4$	230.80	1082.6	684	4.69	2.40	4.9

The large values for b are striking; however, at present, we do not feel them to be sufficiently well founded to warrant further conclusions.

XIII. Solvents Other Than Water

Nonaqueous solutions are of interest because it is to be expected that the conductivity curve will reveal the effect of the dielectric constant. It is known that we owe it to work by Walden[13] in particular that a survey of this field is available. Walden formulated two laws describing all observations. According to the first, the product

$$\bar{\Lambda}_0 \eta = K_1$$

where K_1 is a constant. This law holds the better the larger the ions are and the less tendency the solvent has to associate.* This rule is plausible provided the frictional force exerted on the ions can be computed from the hydrodynamic equations as interpreted by Stokes, and it is evident that this is the more so the larger the ions are. Since, on the other hand, the tendency for association is apparently related to the presence of permanent electric moments of the molecules of the solvent which, in the interpretation of Born, cause a disturbance in Stokes' flow, the other effect is also to be expected.

Walden's second rule reads as follows: For solutions of a salt of equal concentration in different solvents, the difference $\bar{\Lambda}_0 - \bar{\Lambda}$ is inversely proportional to the product of dielectric constant D and coefficient of friction η of the solvent, or

$$\bar{\Lambda}_0 - \bar{\Lambda} = \frac{K_2}{D\eta}$$

Combining the two rules, it follows for the conductivity coefficient f_λ that

$$1 - f_\lambda = \frac{\bar{\Lambda}_0 - \bar{\Lambda}}{\bar{\Lambda}_0} = \frac{K_2}{K_1} \frac{1}{D}$$

According to the present theory, the dependence (for a given concentration) on the dielectric constant is slightly different. From equation (48), we have**

$$1 - f_\lambda = \frac{\text{const}}{D^{3/2}} + \frac{\text{const}}{D^{1/2}}$$

where the first term represents an ionic effect and the second term is due to electrophoresis. Qualitatively the theory makes the same statement as Walden's rule: $1 - f_\lambda$ decreases with increasing dielectric constant. To find whether the theoretical law is justified by the observed numerical coefficients, we carried through some computations which are summarized in the

*Water, it is known, deviates the most from this rule.

**The factors of $D^{-3/2}$ and $D^{-1/2}$ are constants, as long as the ratio of the mobilities is the same for different solvents.

following tables.* In table X, the name of the solvent is entered in the first column, the chemical formula in the second column, and the dielectric constant in the third column.

Table X

Name	Formula	D	S_1	S_2	$S_1 + S_2$
Nitromethane	CH_3NO_2	38.8	0.81	1.33	2.14
Acetonitrile	CH_3CN	36	0.91	1.38	2.29
Nitrobenzene	$C_6H_5NO_2$	35.5	0.93	1.39	2.32
Methanol	CH_3OH	35.4	0.94	1.57	2.51
Propionitrile	C_2H_5CN	27.5	1.37	1.62	2.99
Benzonitrile	C_6H_5CN	26	1.48	1.64	3.12
Ethyl Alcohol	C_2H_5OH	25.4	1.54	1.74	3.28
Epichlorhydrin	$CH_2Cl\,\underset{\diagdown O \diagup}{CHCH_2}$	22.6	1.83	1.79	3.62
Acetone	$(CH_3)_2CO$	21.2	2.01	1.79	3.80
Acetophenone	$C_6H_5COCH_3$	18	2.58	1.95	4.53
Benzaldehyde	C_6H_5COH	17	2.82	2.01	4.83

The theoretical formula equation (48) reads, if the concentration γ in moles per liter is introduced instead of n,

$$1 - f_\lambda = \left[\frac{\varepsilon^2}{6DKT} \sqrt{\frac{4\pi\varepsilon^2}{DKT} \frac{N}{1000}} w_1 + b \sqrt{\frac{4\pi\varepsilon^2}{DKT} \frac{N}{1000} w_2} \right] \sqrt{\nu\gamma} \qquad (59)$$

where N designates Loschmidt's number. The salts used in our computations are all uni-univalent, so that $\nu = 2$. Further we are assuming the mobilities of the ion in the valency factor w_1 to be equal; then $w_1 = 1$. Finally an average value shall be inserted throughout for b; we chose $b = 4 \times 10^{-8}$ cm. Then the two additive terms in the bracket can be evaluated numerically. We designate the first one by S_1, the second one by S_2; their values are listed in columns 4 and 5 of table X, their sum in column 6.

The experimental values are now plotted as a function of $\sqrt{2\gamma}$, and the tangent at the origin is drawn graphically for each salt in each solvent. Thus $\overline{\Lambda}_0$ and α are determined in the relation

$$\overline{\Lambda} = \overline{\Lambda}_0 - \alpha \sqrt{2\gamma}$$

*Those measurements were selected where the concentrations were chosen sufficiently small.

and $\alpha/\bar{\Lambda}_0$ can be found from the slope of the tangent. Then the name of the solvent is entered in column 1 of table XI, and in column 2 is inserted the factor $S_1 + S_2$ from table X. In the following column are listed the values of $\alpha/\bar{\Lambda}_0$ obtained graphically by the method just outlined and which, according to the theory, are to be compared with the values of $S_1 + S_2$.

Table XI

Solvent	$S_1 + S_2$	$\alpha/\bar{\Lambda}_0$			
		KI	NaI	$N(C_2H_5)_4I$	$N(C_3H_7)_4I$
Nitromethane	2.14	2.1[15]	---	2.2[14]	---
Acetonitrile	2.29	2.4[14]	2.3[14]	1.9[14]	2.2[14]
Nitrobenzene	2.32	---	---	1.9[14]	---
Methanol	2.51	2.5[18]	---	2.3[14]	---
Propionitrile	2.99	---	---	3.5[14]	---
Benzonitrile	3.12	---	3.3[16]	3.7[14]	3.2[14]
Ethyl Alcohol	3.28	2.5[14]	---	3.2[14]	---
Epichlorhydrin	3.62	---	---	3.2[14]	3.2[14]
Acetone	3.80	3.5[16]	4.2[16]	---	---
Acetophenone	4.53	---	4.7[17]	---	---
Benzaldehyde	4.83	---	---	5.3[14]	---

The superscripts refer to the Bibliography giving the names of the observers. The observations do not permit a very accurate determination of the slope of the tangent, and further show fairly large discrepancies for different observers. For instance, Philipp and Courtmann find a course for tetraethyl ammonium iodide in nitromethane from which $\alpha/\bar{\Lambda}_0 = 1.4$ follows, whereas the value of 2.2 given in the table follows from the observations of Walden. The conditions are illustrated in Figure 5 where observations of

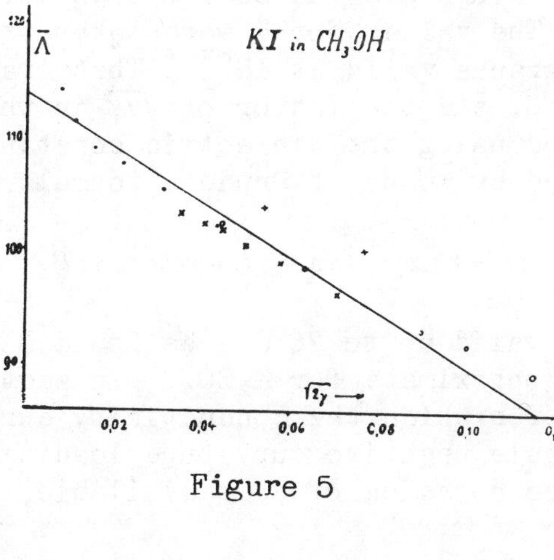

Figure 5

KI in methanol are plotted as a function of $\sqrt{2\gamma}$. The dotted values were measured by Kreider and Jones,[18] the straight crosses by Philipp and Courtmann,[15] the tilted crosses by Jones, Bringham and McMaster,[19] the circles by Fischler.[20] In view of these conditions we contented ourselves with the above approximate method of computation. In spite of this it seems possible to establish that the essential points of the theory are justified by the observations.

XIV. Temperature Dependence of the Conductivity Coefficient

According to equation (48), the conductivity coefficient contains the temperature only in the combination DT, provided the ratios of the mobilities are considered as independent of temperature. In order to investigate to what extent the theory is capable of representing the temperature dependence, we could use a few observations at 100° made by Noyes and Coolidge,[21] and Noyes, Melcher, Cooper and Eastman,[22] which are also mentioned in Kohlrausch and Holborn (p. 210). The observations are plotted and the tangent determined graphically; the designations used here are the same as those in sections XI and XII. Table XII contains in the first column the chemical formulas of the salts studied (two uni-univalent salts, two uni-bivalent salts, one bi-bivalent salt). In the second column are listed the numerical values found graphically for $\alpha/\bar{\Lambda}_0$.

Table XII

	$(\alpha/\bar{\Lambda}_0)_{obs.}$	$(\alpha/\bar{\Lambda}_0)_{calc.}$
$NaCl$	0.554	0.550
$AgNO_3$	0.573	0.608
$Ba(NO_3)_2$	1.17	1.27
K_2SO_4	1.30	1.12
$MgSO_4$	4.40	4.64

In the third column are values for $\alpha/\bar{\Lambda}_0$ derived by the following computation: The values for b were taken from the figures in the previous paragraphs valid at 18°C. These radii were considered valid also at 100°C. and the factor of $\sqrt{\nu\gamma}$ in the expression for $1 - f_\lambda$ was computed using the dielectric constant for 100°C. The latter was obtained by means of Drude's formula.[23]

$$D = 88.23 - 0.4044\,t + 0.001035\,t^2$$

though it is only valid up to 76°C. We found $D = 58.14$. The calculation is very approximate for K_2SO_4. In section XII, the case of K_2SO_4 was the one for which the conductivity curve showed a very strong and improbable negative curvature leading to a negative value for b. Since no value of b is available, we put $b = 0$, which, evidently, must give a computed value for the slope which is too small. So far as an opinion is possible, the theoretical temperature dependence appears to agree fairly well with experiments.

Bibliography

1. *Physik. Z.*, *24*, 185 (1923).
2. F. Kohlrausch, *Gesamm. Abhandl.*, *II*, 1911, p. 360, Leipzig; *Wiedem. Ann. Vol.*, *26*, 161 (1885).
3. F. Kohlrausch, *Gesamm. Abhandl.*, *II*, pp. 1127, 1132 ff.; *Z. Elektrochem.*, *13*, 333 (1907).
4. *Ann. Physik (4)*, *37*, 1 (1912).
5. *J. Chem. Soc.* *113*, 449, 627, 707, 790 (1918); *Z. phys. Chem.*, *98*, 211 (1921).
6. P. Drude, *Ann. Physik*, *9*, 61 (1896).
7. F. Kohlrausch, *Gesamm. Abhandl.*, *II*, 1127 ff., *Z. Elektrochem.*, *13*, 333 ff. (1907).
8. F. Kohlrausch, *Gesamm. Abhandl.*, *II*, 943, 1266; *Berl. Ber.*, *1900*, (I), 581; *1900* (II), 1002.
9. *J. Am. Chem. Soc.*, *40*, 138 (1918).
10. F. Kohlrausch, *Gesamm. Abhandl.*, *II*, *906*; *Wiss. Abhan. Phys. Techn. Reichsanstalt*, *3*, 156-227 (1900).
11. Ch. A. Kraus and H. C. Parker, *J. Am. Chem. Soc.*, *44*, 2429 (1922).
12. F. Kohlrausch, *Gesamm. Abhandl.*, *II*, 1134 ff.; *Z. Elektrochem.*, *13*, 333 (1907).
13. *Z. anorg. allgem. Chem.*, *115*, 49 (1920). Compare also the previous works cited here by the same author.
14. Walden, *loc. cit.* (ref. 13), and *Z. phys. Chem.*, *54*, 129 (1906); *54*, 183 (1906); *78*, 275 (1911); *Bull. Akad. Soc. St. Petersburg*, *1913*, 577.
15. Philipp and Courtmann, *J. Chem. Soc.*, *97*, 1268 (1910).
16. Dutoit and Levrier, *Chem. Chim. Ph.*, *3*, 547, 628 (1905); *Z. Elektrochem.*, *12*, 643 (1906).
17. Nicollier and Dutoit, *Z. Elektrochem.*, *12*, 643 (1906). Thèse, Lausanne, 1906.
18. Kreider and Jones, *Am. Chem. J.*, *45*, 282 (1911); *46*, 576 (1911).
19. Jones, Bringham and McMaster, *Z. phys. Chem.*, *57*, 193 (1907).
20. Fischler, *Z. Elektrochem.*, *19*, 127 (1913).
21. Noyes and Coolidge, *Z. phys. Chem.*, *46*, 323 (1903).

Bibliography (continued)

22. Noyes, Melcher, Cooper, and Eastman, *Z. phys. Chem.*, *70*, 335 (1910).

23. P. Drude, *Ann. Physik* *59*, 1896. In *Z. phys. Chem.*, *98*, 217 (1921) and *J. Chem. Soc.*, *113*, 449 (1918) by J. C. Ghosh is given a temperature formula for *D* which is said to be by Drude, and from which a value of *D* equal to 52.6 is derived for 100°C. We could not find this formula anywhere in Drude, and presume that the numerical coefficients resulted from an error in copying from the tables by Landolt-Boernstein.

ON IONS AND THEIR ACTIVITY
(Over Ionen en Hun Activiteit)

P. Debye

Translated from
Chemisch Weekblad, 20, 1923, pages 562-568

1. Whenever it was found that the osmotic pressure, the lowering of the freezing point, or any of the other colligative properties of a water solution of a salt such as KCl were not proportional to the number of ions into which the dissolved salt can dissociate, then one could readily give an explanation based on the familiar concept of incomplete dissociation. One reasoned that not all the KCl molecules are dissociated; and calling γ_1 the concentration of the K ion (in gram molecules per liter), γ_2 the concentration of the Cl ion, and γ_3 the concentration of the undissociated KCl molecule, one calculated the equilibrium by means of the Goldberg-Waage law from the formula

$$\frac{\gamma_1 \gamma_2}{\gamma_3} = K$$

where K is the equilibrium constant. If the number of free particles is determined in this way--in the special case mentioned above it will be proportion to $\gamma_1 + \gamma_2 + \gamma_3$--it is then assumed that the osmotic pressure is proportional to this number, and completely independent of the type and the specific properties of these particles.

Of late it has become more and more obvious that this very simple reasoning yields a satisfactory approximation only for compounds which dissociate slightly. The greater the degree of dissociation, the less well the theory fits. Since, in general, most common salts dissociate to a great degree, one was forced to introduce a special category: ***strong salts*** (as well as acids and bases) or, more generally, ***strong electrolytes***.

In order to obtain an overall view of the particular properties of strong electrolytes, two courses have been followed: first, a purely phenomenological, empirical interpretation of the experimental results; and second, a more theoretical approach directed towards the investigation of the causes of the phenomena.

The first course, followed especially by American researchers,[1] has led to a definition of various ionic coefficients, especially osmotic coefficient, activity coefficient, and coefficient of conductivity. The definitions were based on the following type of argument. The osmotic pressure is not necessarily proportional to the number of free particles. It could be possible that the particles, for some unspecified reason, act in a manner that is not proportional to their concentration, although they must be considered as free according to the usual concept. This lack of freedom is computed by multiplying the actual concentration γ by an osmotic coefficient f_o. In the particular case of KCl, the practical results would be expressed by assuming the osmotic pressure as proportional to:

$$f_o^{(1)} \gamma_1 + f_o^{(2)} \gamma_2 + f_o^{(3)} \gamma_3$$

Similarly, it was suggested that what may be called the "active mass" in the mass action law need not be proportional to the concentration. Here also a particular sort of freedom-deficiency is postulated which can be measured by writing $f_a \gamma$ instead of γ, where f_a is the activity coefficient. The mass action law has then (with KCl again as an example) the form:

$$\frac{f_a^{(1)} \gamma_1 \, f_a^{(2)} \gamma_2}{f_a^{(3)} \gamma_3} = K$$

Since there is no reason why the activity coefficient should equal the osmotic coefficient, two different coefficients, f_o and f_a, are introduced here. In order to simplify the following presentation, coefficients for irreversible processes such as conductivity will not be considered.

It is obvious that it will be possible, in this manner, to derive formulas which will express the experimental data. It can, of course, not be denied that the computation of the coefficients f_o and f_a for the single ions from the experimental data is not a straightforward task. This does not change the fact that the introduction of ionic coefficients constitutes in itself an important step. It makes clear that the usual laws, as they are still presented in most textbooks, really have to be modified in essential points. (In the *Theoretische Chemie* of Nernst, 8-10th ed. of 1921, 3 pages out of 896 deal with the discrepancies and their explanations.)

The second course takes as its starting point the hypothesis that the cause for the interaction, measured, e.g., as above by the coefficients f_o and f_a, is the effect of the electric charges of the ions. If this path is followed it can be considered immediately as a good sign that the different particular effects of strong electrolytes depend indeed very characteristically on the valence of the ions, which is proportional to the electric charge.[2] The task is now to express this electrical interaction in a formula. Milner[3] made such calculations for the osmotic coefficient, and P. Hertz[4] for the conductivity coefficient. Later, Ghosh[5] developed a theory intended to be completely general.

N. Bjerrum,[6] on the basis of several interesting considerations, defended most successfully the idea that the electric interaction is indeed the important point in the case of strong electrolytes. I tried recently[7] to explain why it is necessary to go further than Milner, and why the theory of Ghosh is not correct in an important point. A good article on the theory of Ghosh, by Ada Prins,[8] appeared in the Chemisch Weekblad. Therefore, I will recapitulate only what can be said against the theory of Ghosh (for reversible phenomena) and restrict myself rather to the discussion of the activity coefficients as derived from an improved theory. I refer to the publications in Physikalische Zeitschrift for a detailed discussion of the osmotic coefficient.

2. The easiest way to arrive at a theory is to calculate one of the thermodynamic potentials whereby, of course, only the changes of this potential that result are of importance because we are now taking into consideration the electric forces between the ions. One has then the advantage that all the different phenomena can be very easily calculated by differentiation. For this reason I determined first — in my publication in Physikalische Zeitschrift — the "thermodynamic potential at constant pressure and at constant temperature." One can see easily that the main task is the determination of the potential electric energy of a solution of ions. Ghosh also derived an expression for this energy. He calculates an average distance r between the ions, assuming that the ions in a solution will have about the same arrangement as found in the crystal by Bragg's experiments. For the potential energy between two oppositely charged ions of charge e, he puts:

$$-\frac{1}{D}\frac{e^2}{r}$$

where D is the dielectric constant of the solvent. By multiplying this expression by the number of ions, he assumes to have found, in essence, the electric energy of the solution. In an electrolyte solution it will certainly be true that in the immediate vicinity of an ion, as a consequence of the Coulomb forces, oppositely charged ions will be present, on the average, during a longer time than ions of the same sign. This is really an analogy to the arrangement of the ions in crystal-line KCl, NaCl, etc. But the distribution of the ions in a solution is essentially determined by the dynamic equilibrium between the action of the electric forces, which are trying to establish a regular arrangement of ions of alternating signs, and the irregular thermal motion counteracting these forces. Therefore the distance, which is of importance in the energy calculation, must be essentially dependent on the temperature, and Ghosh goes too far if he substitutes for it the average distance between neighboring ions, as calculated from the concentration alone.

It is, however, possible to reason as follows. Around each ion the other ions will occupy all possible positions in the course of time. It is then possible to speak of an average charge density in the vicinity of the ion; of course, this electric "atmosphere" has the opposite sign, because oppositely charged ions remain, on the average, longer in its vicinity than the others. What we must find is the "thickness" of this atmosphere, that is to say, the distance from the ion at which the electric density has decreased to a given fraction (for instance, to one-half). This distance, and not the average distance, will then be the one to be used in the calculation of the energy. One can see that the calculation of the ionic atmosphere is similar to the analogous derivations in the theory of the well-known ionic layer which Helmholtz assumes to exist at electrodes, that is, at places where the electric potential undergoes apparent sudden change. Let λ be the distance (which will be calculated more precisely later) characteristic of the thickness of the ionic atmosphere; then, the theory gives for the electric energy of a very dilute solution the expression:

$$(1) \qquad U = -\sum_{i=1}^{i=s} N_i \frac{e_i^2}{2D\lambda}$$

if a mixture of $N_1 \ldots N_i \ldots N_s$ ions of type 1...i...s and of charge $e_1 \ldots e_i \ldots e_s$ is present. Further, if:

$$(2) \qquad \frac{1}{\lambda} = \varkappa$$

then $\varkappa$ can be calculated according to the formula:

$$(3) \qquad \varkappa^2 = \frac{4\pi}{DkT} \sum_{i=1}^{i=s} n_i e_i^2$$

in which κ is the Boltzmann constant ($\kappa = 1.346 \times 10^{-16}$ erg), T the absolute temperature, and n_i the number of ions of the type i per cc.

Practically speaking, the difference between these results and those of Ghosh makes itself felt, for instance, in the following way: since κ^2 is proportional to the concentration, the energy will be inversely proportional to the *square root* of the concentration. On the other hand, according to Ghosh, the energy is inversely proportional to r, that is, to the *cube root* of the concentration. Immediately the following question arises. Whatever might be said against the theory of Ghosh, it cannot be denied that his calculation gives the order of magnitude of the ion effect. How is this possible? It can easily be seen that this agreement is purely accidental. With solutions of KCl, taking, for example, 0.001 and 1 normal, and assuming complete dissociation, one has according to (2) and (3) (calculated with D = 88.23 at 0°C.):

$$\lambda = 96.7 \times 10^{-8} \text{ cm. and } \lambda = 3.06 \times 10^{-8} \text{ cm.}$$

whereas the average distance between ions is:

$$r = 9.37 \times 10^{-8} \text{ cm. and } r = 0.937 \times 10^{-8} \text{ cm.}$$

Thus, the energies derived by either formula will be of the same order of magnitude, but there is nevertheless a very considerable difference.

It can be seen from the numerical values just given for λ that an expression like (1) for the energy can be exact only for very dilute solutions, since λ is not so very much greater than the dimensions of the ions. Hence we must try also to take into account the influence of these dimensions. This can be done roughly with a formula of the form

$$U = -\Sigma N_i \frac{e_i^2 \varkappa}{2D} \frac{1}{1 + \varkappa a_i} \tag{4}$$

in which a_i is the radius that has to be assigned to the ion of the kind i in the particular surroundings in which it exists.

3. With the help of the expression for the energy it is immediately possible to calculate, for instance, the heat of solution. This expression has first to be introduced into the thermodynamic potentials in order to lead to a theory of the ionic coefficients. I do not wish to enter here into this development and only give the results for the above named coefficients.

In the first place it is found that the electric action between the ions really leads to the definition of osmotic coefficients, and for dilute solutions one finds:

$$f_o^i = 1 - \frac{e_i^2 \varkappa}{6\,DkT} \sigma_i \tag{5}$$

in which σ_i is again a function of the ratio:

$$x_i = \frac{a_i}{\lambda} = \varkappa a_i \tag{5'}$$

This function is expressed as follows:

$$\sigma_i = \frac{3}{x_i^3}\left[(1 + x_i) - \frac{1}{1 + x_i} - 2 \ln (1 + x_i)\right] = \tag{5''}$$

$$= 1 - \frac{3}{2} x_i + \frac{9}{5} x_i^2 - + \ldots$$

The behavior of the osmotic coefficients for very dilute solutions is of special importance. In this case, λ becomes so large that we are allowed to put

$$x^i = \frac{a_i}{\lambda} = o$$

so that σ_i becomes equal to 1. The peculiar result is that for completely dissociated solutions the osmotic coefficient, at a given temperature, is not determined by the concentration alone, but also by the valence of the ions in solution.

We consider the special case that only a single electrolyte is dissolved, of which the molecules dissociate into $\nu_1 \ldots \nu_i \ldots \nu_s$ ions, with the valences $z_1 \ldots z_i \ldots z_s$, so that:

$$e_1 = z_1 \varepsilon, \ldots . \ e_i = z_i \varepsilon, \ldots . \ e_s = z_s \varepsilon$$

in which ε is the electronic charge. Then, with the help of (3), we can write for the osmotic coefficient:

$$(6) \quad f_o^i = 1 - \frac{z_i^2 \varepsilon^2}{6DkT} \sqrt{\frac{4\pi n \varepsilon^2}{DkT} \Sigma \nu_i z_i^2}$$

if n molecules per cc. are dissolved.

If the concentration is called γ and is measured as before in gram moles per liter, and if we take $\varepsilon = 4.77 \times 10^{-10}$, $D = 88.23$, $k = 1.346 \times 10^{-16}$, then, for practical use, we can write:

$$(6') \qquad f_o^i = 1 - 0.270 z_i^2 \sqrt{\Sigma \nu_i z_i^2} \sqrt{\gamma}$$

The osmotic coefficient is thus not an ionic coefficient in the sense that it can be calculated separately for each ion, without considering the other ions in the solution. On the contrary, in adding other ions, the osmotic coefficient will be changed. For molecules without a charge, $f_o = 1$, because $z = 0$. It is evident that as a next step in the theory one could take into consideration the electric field of the uncharged molecules which would have to be considered as dipoles or quadrupoles. This will, naturally, change the results given above. I believe, however, it is at present too early to undertake such an extension.

The application of (6') for very dilute solutions, or of (5) for moderately dilute solutions, will not be discussed further here. However, it has to be pointed out that this discussion, which can be found elsewhere (see references above), makes the conclusion inevitable that the ordinary strong electrolytes are probably completely dissociated, even at relatively great concentrations. What was interpreted in former times as an incomplete dissociation of strong electrolytes is only a consequence of the Coulomb forces between free ions.

In the second place, it is found that the electric interaction also leads to the definition of activity coefficients. To start, we consider the case of an electrolyte which is not completely dissociated, but sufficiently dissociated so that the electric forces have to be taken into consideration. For dilute solutions, the formula for the activity coefficients is not very simple, if one takes into account the different radius a_i of the various ions. However, in discussing the data for the lowering of the freezing point it is found that one has reasonably good success if one takes (as a somewhat crude approximation, it is true) a single average radius a, which is supposed to be the same for all ions. Because the radius is of little importance for very dilute solutions, I will indicate here only formulas which operate with the average radius a. Under these circumstances, one finds for the general case:

$$(7) \qquad \ln f_a^i = -\frac{e_i^2 \varkappa}{2\,DkT}\,\frac{1}{1+\varkappa a}$$

Therefore, for very dilute solutions, the theory predicts the relation:

$$(8) \qquad \ln f_a^i = 3\,(1 - f_o^i)$$

between the activity coefficient and the osmotic coefficient. In the same approximation and for practical use (at 0°C.), (7) can again be written as:

$$(9) \qquad \ln f_a^i = -\,0.810\, z_i^2 \sqrt{\Sigma \nu_i z_i^2}\,\sqrt{\gamma}$$

if only a single salt is dissolved, that dissociates completely into:

$$\nu_1 \ldots \nu_i \ldots \nu_s$$

ions of different kind. Of greater importance, however, is the case of mixed electrolytes because one obtains an expression for the activity coefficient which is useful in the discussion of the influence of added salts on the solubility. If, as above, γ_i is again the concentration of the ions of type i in gram moles (or gram atoms) per liter, then one has for very dilute solutions:

$$(10) \qquad \ln f_a^i = -\,0.810\, z_i^2 \sqrt{\Sigma z_i^2 \gamma_i}$$

whereas for moderately dilute solutions it follows from (7), after substitution of the numerical values, that:

$$(10') \ln f_a^i = -\,0.810\, z_i^2 \sqrt{\Sigma z_i^2 \gamma_i}\,\frac{1}{1 + 0.231\, a \sqrt{\Sigma z_i^2 \gamma_i}}$$

where a is measured in A ($=10^{-8}$ cm.).

4. As an example I will discuss the case of a saturated solution, where an excess of undissolved salt is always present. If then another salt in various concentrations is added to the solution, and if further it is supposed, as a case (a), that the two salts have no common ions, it is found that more of the first salt goes into solution. In order to explain this phenomenon, it is usually reasoned as follows. Let us take the particular case of Ag_2SO_4 as the first salt whose solubility is influenced by the addition of KNO_3. In the solution there are initially Ag ions, SO_4 ions, and undissociated Ag_2SO_4. When KNO_3 is added besides K ions, NO_3 ions, and undissociated KNO_3, we also expect undissociated K_2SO_4, and undissociated $AgNO_3$. This undissociated $AgNO_3$ withdraws Ag ions from the solution, and because Ag_2SO_4 is in its dissociation equilibrium, Ag_2SO_4 molecules will begin to dissociate. However, since undissolved Ag_2SO_4 is present there will be no equilibrium between the solid Ag_2SO_4 and the Ag_2SO_4 molecules in the

solution until new Ag_2SO_4 goes into solution.

However, there exists another quite different but no more complex explanation. In a solution of ions, work against the electric forces must be done in order to remove an ion from the solution, in accordance with the negative value of U given in (1). An ion is thus electrically attracted by an ion solution. When, in the example under discussion, KNO_3 is added and is completely dissociated, then the Ag and SO_4 ions, which want to leave the crystal in order to go into solution, will be more strongly attracted by the solution than before the addition. The work required to free Ag and SO_4 ions from the crystal has become smaller, so that more Ag_2SO_4 will go into solution in the form of ions.

Actually both reasonings are not contradictory and there will certainly be cases in which both mechanisms are simultaneously present. However, since it became evident in the discussion of the lowering of the freezing point of strong electrolytes that practically the consequences of the electric interaction alone are visible, and that not even a small number of undissociated molecules can be found in solutions, it can be conjectured that the second explanation is also the better one in our present case. A definite conclusion can only be drawn from a quantitative discussion of the experiments. I do not wish to enter into the well-known theoretical formulation of the first explanation with the help of the simple mass action law; rather, I wish to show the kind of formulas to which the second explanation leads.

Let us suppose that the concentration of the pure saturated Ag_2SO_4 solution is $\overline{\gamma}$, then, with complete dissociation, as is assumed, the concentration of the Ag ions is $2\overline{\gamma}$, and that of the SO_4 ions $\overline{\gamma}$. In solution each of these ions has an activity coefficient, which we will call $\overline{f_{Ag}}$ and $\overline{f_{SO_4}}$. According to a well-known application of the mass action law there exists the following relation between the dissolved and the undissolved salt:

$$(\overline{f}_{Ag}\, 2\overline{\gamma})^2\ (\overline{f}_{SO_4}\overline{\gamma}) = K$$

where the concentrations have now been multiplied (as is clearly necessary) by their activity coefficients. When KNO_3 is added, the Ag_2SO_4 concentration will have changed from $\overline{\gamma}$ to γ; the concentration of the added KNO_3 will be called γ. In the changed solution the activity factors have also changed; I write them now without the bar, f_{Ag}, f_{SO_4}, etc. Again, through application of the improved mass action law, it follows that:

$$(f_{Ag}\, 2\,\gamma)^2\ (f_{SO_4}\gamma) = K$$

When these two relations are divided by each other and the logarithm taken, it follows:

$$3 \ln \frac{\gamma}{\overline{\gamma}} = -\ln \frac{f_{Ag}^2\, f_{SO_4}}{\overline{f}_{Ag}^2\, \overline{f}_{SO_4}} \tag{11}$$

Now, it only remains to substitute for the activity coefficients their values according to (10) or (10'). Since $z_{Ag} = 1$, $z_{SO_4} = -2$, $z_K = 1$, $z_{NO_3} = -1$, one can, for instance, for very dilute solutions, immediately calculate from (10) the following values:

$$\begin{cases} \ln \bar{f}_{Ag} = -0.810 \, . \sqrt{6\bar{\gamma}} \\ \ln \bar{f}_{SO_4} = -0.810 \, . \, 4\sqrt{6\bar{\gamma}} \\ \ln f_{Ag} = -0.810 \, . \sqrt{6\gamma + 2\gamma'} \\ \ln f_{SO_4} = -0.810 \, . \, 4\sqrt{6\gamma + 2\gamma'} \end{cases}$$

As a consequence, (11) changes to:

$$(12) \qquad 3 \ln \frac{\gamma}{\bar{\gamma}} = 0.810 \left[6\sqrt{6\gamma + 2\gamma'} - 6\sqrt{6\bar{\gamma}} \right]$$

This equation enables us to calculate for a given concentration $\bar{\gamma}$ of the pure Ag_2SO_4 solution the concentration γ, after addition of KNO_3, as a function of the concentration γ of the added KNO_3. The noteworthy fact in this equation is that it can be set up with the help of the valence numbers alone. This implies that it should be possible to describe by the same equation the change in solubility of any salt dissociating into two monovalent ions and one bivalent ion (like Ag_2SO_4) caused by the addition of a second salt dissociating into two monovalent ions (like KNO_3).

For moderately dilute, instead of very dilute, solutions, (10') has to be used for the calculation of the activity coefficients. It is then found that:

$$(12') \quad 3 \ln \frac{\gamma}{\bar{\gamma}} = 0.810 \left[\frac{6\sqrt{6\gamma + 2\gamma'}}{1 + 0.231 \, a\sqrt{6\gamma + 2\gamma'}} - \frac{6\sqrt{6\bar{\gamma}}}{1 + 0.231 \, a\sqrt{6\bar{\gamma}}} \right]$$

5. Before I finally compare these theoretical conclusions with the experiment, the case (b) must still be mentioned, in which both salts have one or more ions in common. We consider again a particular case, e.g., Ag_2SO_4, which is influenced by $AgNO_3$. It is well known that Ag_2SO_4 becomes less soluble when $AgNO_3$ is added to the solution. The usual explanation is as follows. In the solution there are at first undissociated Ag_2SO_4 molecules, as well as Ag ions and SO_4 ions. When I add $AgNO_3$ the number of Ag ions will become larger, there will thus be a better opportunity for the formation of Ag_2SO_4 molecules. Since the concentration of the Ag_2SO_4 molecules is uniquely determined by their equilibrium with the solid salt, Ag_2SO_4 must precipitate.

Using the new explanation, however, one reasons in the following way: In adding $AgNO_3$, which dissociates completely, the number of Ag ions becomes larger, so that there will be a better opportunity for Ag ions and SO_4 ions to combine and to form solid salt -- making Ag_2SO_3 less soluble. At the same time, however, the ion concentration in the solution has increased by this addition, and Ag and SO_4 ions will therefore be more strongly attracted into solution, so that Ag_2SO_4 becomes more soluble through this effect. What will be the net result cannot be determined *a priori*; it depends on whether the first or the second effect will be stronger. According to the first explanation, the solubility of the poorly soluble salt must always decrease. I can probably assume as well known that there are numerous deviations from this rule,[9] which then have to be explained in an indirect way. It seems of interest to me that in some cases the application of the theory of electric interaction also gives the possibility of a direct explanation.

[The theory allows us to treat the general case in the following way: Suppose that one molecule of the primary salt dissociates into:

$$\nu_1 \;.\;.\;.\;.\; \nu_\sigma,\; \nu_{\sigma+1} \;.\;.\;.\;.\; \nu_s$$

ions with the valence:

$$z_1 \;.\;.\;.\;.\; z_\sigma,\; z_{\sigma+1} \;.\;.\;.\;.\; z_s$$

Suppose further that one molecule of the added salt dissociates into:

$$\nu_1' \;.\;.\;.\;.\; \nu_\sigma',\; \nu_{\sigma+1}' \;.\;.\;.\;.\; \nu_t'$$

ions with the valence:

$$z_1' \;.\;.\;.\;.\; z_\sigma',\; z_{\sigma+1}' \;.\;.\;.\;.\; z_t'$$

Suppose further that the ions of the types $1 \ldots \sigma$ are the same in both salts, so that:

$$z_1 = z_1' \;.\;.\;.\;.\;.\;.\;.\; z_\sigma = z_\sigma'$$

One then calculates:

$$\nu = \nu_1 + \;.\;.\;.\; + \nu_\sigma + \nu_{\sigma+1} + \;.\;.\;.\; + \nu_s$$

$$\nu^1 = \nu_1' + \;.\;.\;.\; + \nu_\sigma'$$

$$s = \nu_1 z_1^2 + \;.\;.\; \nu_\sigma z_\sigma^2 + \nu_{\sigma+1} z_{\sigma+1}^2 + \;.\;.\; + \nu_s z_s^2$$

$$s^1 = \nu_1' z_1'^2 + \;.\; + \nu_\sigma^1 z_\sigma'^2 + \nu'_{\sigma+1} z'_{\sigma+1} + \;+\;.\;.\; + \nu_t' z_t'^2$$

As above, $\bar{\gamma}$ is the concentration of the primary salt in pure water, γ the concentration after addition of the second salt, γ' the concentration of the added salt.

Introducing:

$$\frac{\gamma}{\bar{\gamma}} = 1 + y \text{ and } \frac{\gamma'}{\bar{\gamma}} = x_1$$

it follows for small values of x that:

$$y = -\frac{\nu' - \sqrt{s'}}{\nu \quad \sqrt{s}}\, x$$

Here Γ is an abbreviation for the expression:

$$\Gamma = 0.405 \frac{\sqrt{s_\gamma}}{(1 + 0.231\, a \sqrt{s_\gamma})^2}$$

The equation gives the tangent to the solubility curve. The primary salt becomes more soluble as soon as:

$$\Gamma s' > \nu'$$

A rough estimate is possible when we assume an average value for a; we can put approximately 0.231a = 0.6.]

If, in case (b) the concentration of the added $AgNO_3$ is γ', and if the same notation as in case (a) is used, we have the two equations:

$$[\bar{f}_{Ag}\, 2\bar{\gamma}]^2\, [\bar{f}_{SO_4}\, \bar{\gamma}] = K,$$
$$[f_{Ag}\, (2\gamma + \gamma')]^2\, [f_{SO_4}\, \gamma] = K$$

from which follows:

$$(13) \qquad \ln \frac{(2\gamma + \gamma')^2 \gamma}{(2\bar{\gamma})^2 \bar{\gamma}} = - \ln \frac{f^2{}_{Ag} f_{SO_4}}{\bar{f}^2{}_{Ag} \bar{f}_{SO_4}}$$

Furthermore, since the activity coefficients are determined by the same formulas as in case (a), we have for very dilute solutions:

$$(14) \ \ln \frac{(2\gamma + \gamma')^2 \gamma}{(2\bar{\gamma})^2 \bar{\gamma}} = 0.810\, [6\sqrt{6\gamma + 2\gamma'} - 6\sqrt{6\bar{\gamma}}]$$

and for moderately dilute solutions:

$$(14') \quad \ln \frac{(2\gamma + \gamma')^2 \gamma}{(2\bar{\gamma})^2 \bar{\gamma}} = 0.810 \left[\frac{6\sqrt{6\gamma + 2\gamma'}}{1 + 0.231\, a\sqrt{6\gamma + 2\gamma'}} - \frac{6\sqrt{6\bar{\gamma}}}{1 + 0.231\, a\sqrt{6\bar{\gamma}}} \right]$$

6. I will show by a single example how the experiments are confirmed by the above formulas. In Figure 1 the concentration γ of the added salt is plotted as the abscissa, and the concentration of γ of the salt in excess (in our case Ag_2SO_4) as the ordinate. The experiments represented here were performed by Harkins.[10] When salts with common ions are added, the solubility of Ag_2SO_4 is reduced much more by the monomonovalent $AgNO_3$ than by $MgSO_4$ or K_2SO_4, with the bivalent SO_4 ion. If the solubility of Ag_2SO_4 in pure water were slightly larger, $AgNO_3$ would produce, according to the detailed explanation in section 5, a reduction of the solubility; $MgSO_4$, on the other hand, would give an increase. The two salts with foreign ions, KNO_3 and $Mg(NO_3)_2$, produce an increase which is greater for $Mg(NO_3)_2$, with the bivalent Mg ion, than for KNO_3, with only monovalent ions. The value of a (called $\bar{a}$ in the figure) was calculated according to the formulas previously mentioned, from the results of a single experiment; the curves, calculated with this value of a, are reproduced in the figure, and fit the experimental results extremely well.* In the figure, a is given in centimeters and γ in millimoles, while in the text we give a in A. and γ in moles per liter.

*This graph and the following figures were obtained by O. Schärer; they are part of an unpublished article on solubility.

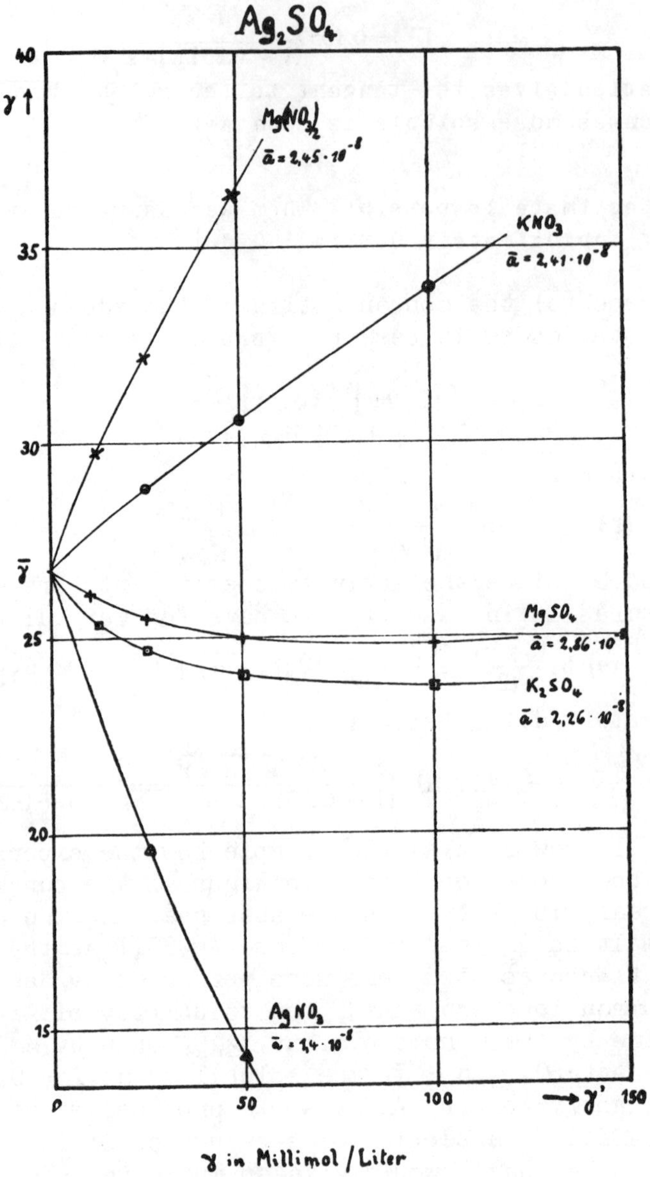

Figure 1

In Figures 2 and 3 the same experiments are represented in a slightly different manner. Let the concentrations of the ions present in the primary salt be:

$$\gamma_1 \ldots . \gamma_i \ldots . \gamma_s$$

and let one molecule dissociate into:

$$\nu_1 \ldots . \nu_i \ldots . \nu_s$$

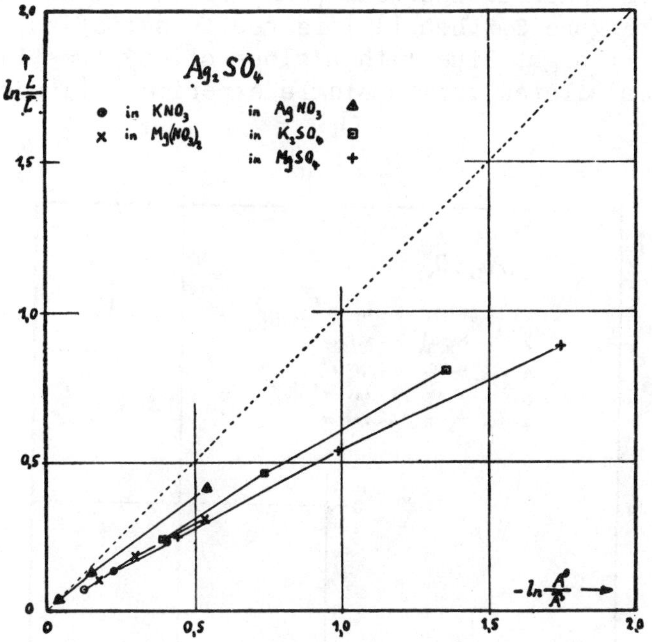

Figure 2

ions, then the condition of equilibrium is:

$$f_1^{\nu_1}\gamma_1^{\nu_1} \dots . \; f_i^{\nu_i}\gamma_i^{\nu_i} \dots . \; f_s^{\nu_s}\gamma_s^{\nu_s} = \text{const.}$$

where the f_i's are again the activity coefficients. We have called the product:

$$\gamma_1^{\nu_1} \dots \gamma_i^{\nu_i} \dots \gamma_s^{\nu_s} = L$$

the "solubility product," and the product:

$$f_1^{\nu_1} \dots f_i^{\nu_i} \dots f_s^{\nu_s} = A$$

the "activity product." If we let the values of these products for the solution in pure water be $\bar{L}$ and $\bar{A}$, and, after the addition of the second salt, L and A, then, according to the condition of equilibrium, the relation:

$$\ln \frac{L}{\bar{L}} = - \ln \frac{A}{\bar{A}} \qquad (15)$$

must exist.

In a graph with $-\ln A/\bar{A}$ as the abscissa and $\ln L/\bar{L}$ as the ordinate, a straight line with a slope of 45° must result, independent of whether or not the added salt has ions in common with the primary salt.

In Figure 2, relation (15) is plotted for the case that a is zero. Therefore, A_0 is written instead of A. It is evident that the relation (15) is not satisfied. However, if the activity coefficients and A are calculated from the improved formula in which the influence of a is taken into account, and if the values of a are used, as indicated in Figure 3, then (15) is really satisfied, for now the points lie on a straight line with a slope of 45° (see Fig. 3). The value of a was calculated from a single experiment (indicated in the figure by

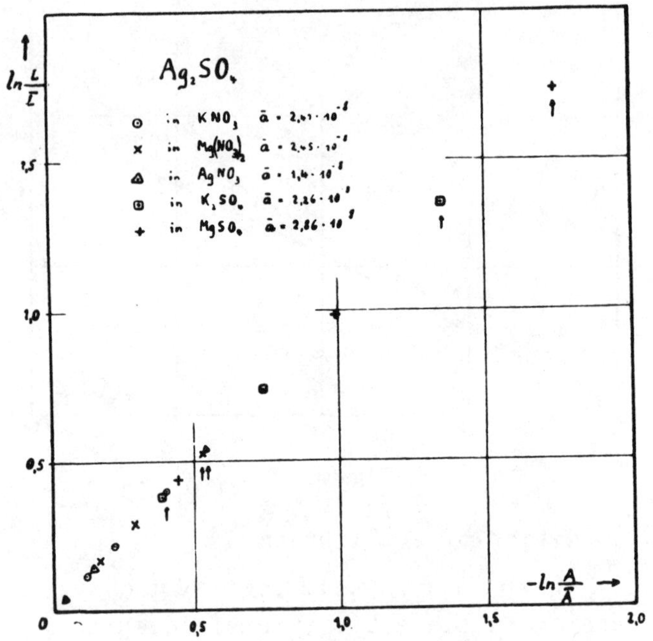

Figure 3

an arrow). Finally, it must be noted that the experiments were carried out at 25°C. and not at 0°C. Consequently, the numerical factors in the formulae for the activity coefficient (see formulae 10 and 10') are not 0.810 and 0.231, but 0.849 and 0.235.

Bibliography

1. Cf., e.g., William C. Mc. C. Lewis, *A System of Physical Chemistry*, Longmans, Green and Co., London, 1920, Vol. 2, p. 202. The first definition of "activity" appeared in G. N. Lewis, *Proc. Am. Acad.*, *43*, 259 (1907) or *Z. physik. Chem.*, *61*, 129 (1908). The "osmotic coefficients" are encountered for the first time in N. Bjerrum, *Z. Elektrochem.*, *24*, 259 (1907). A good review, both with respect to applications and additional literature, can be obtained from the following four papers: J. N. Brönsted, Studies on Solubility, *J. Am. Chem. Soc.*, *42*, 76, 1448 (1920); *43*, 2266 (1921); *44*, 877 (1921).

Bibliography (Con'd)

2. Cf., e.g., J. N. Brönsted and Agnes Petersen, *J. Am. Chem. Soc.*, *43*, 229 (1921). Also 10 years earlier a similar remark by A. A. Noyes and W. C. Bray, *J. Am. Chem. Soc.*, *33*, 1643 (1911). This paper also points out the example of the $PbCl_2$ solution which presents a very good demonstration experiment. At ordinary temperatures $PbCl_2$ is precipitated from a saturated solution by the addition of some drops of concentrated KCl solution. If, however, a few drops of concentrated $Pb(NO_3)_2$ solution are added, then there results only a barely visible $PbCl_2$ precipitate. In the first case the monovalent Cl ions are common, in the second case the bivalent Pb ions. It will be discussed later in this article what the theory can say about such experiments.

3. *Phil. Mag.*, *35*, 214, 354 (1918).

4. *Ann. Physik*, (4) *37*, 1 (1912).

5. *J. Chem. Soc.*, *113*, 449, 627, 707, 790 (1918); *Z. physik. Chem.*, *98*, 211 (1921).

6. *Proc. 7th intern. congress appl. chem.*, London, May 21 to June 2, 1909, Sect. X; *Z. Elektrochem.*, *24*, 231 (1918); *Z.f. anorg. Chem.*, *109*, 275 (1920).

7. Address before the "Natuur- en Geneeskundig Congres," Maastricht, 1923, to appear in the proceedings of the congress. Also: *Physik. Z.*, *24*, 185 (1923). There are also in both places remarks on conductivity and the theory of P. Hertz, but for more detail see *Physik. Z.*, *24*, 305 (1923).

8. *Chem. Weekblad*, *20*, 237 (1923).

9. A characteristic case is represented by $Ba(OH)_2$ or $Sr(OH)_2$, which become more soluble through the addition of $Ba(NO_3)_2$ or $Sr(NO_3)_2$. These cases were investigated by C. L. Parsons and H. P. Carson, *J. Am. Chem. Soc.*, *32*, 1383 (1910).

10. Harkins, *J. Am. Chem. Soc.*, *33*, 1817 (1911).

OSMOTIC EQUATION OF STATE AND ACTIVITY OF DILUTED STRONG ELECTROLYTES
(Osmotische Zustandsgleichung und Activität verdünnter starker Elektrolyte)

P. Debye *

Translated from
Physikalische Zeitschrift, Vol. 25, 1924, No. 5, pages 97–107

I. Aims and Restrictions

The theory of strong electrolytes published some time ago[1] was presented in a manner which permits considerable simplification, as I have since found out. The new representation, which I want to publish in the following pages, is much more obvious than the old one. I consider it, therefore, justified to examine the matter once again. Besides, I have in the meantime come across some laws on the activity of strong electrolytes which G. N. Lewis discovered in a purely experimental way. I am glad to have this opportunity to emphasize the special importance of these fine investigations, the more so since the laws of Lewis can be explained very easily by the proposed theory.**

I start with the law of osmotic pressure, the osmotic equation of state, and I assume this law to be completely known to us for the limiting case of vanishing molecular forces. This limiting case is exactly defined by the known equation for the thermodynamic potential of diluted solutions. We will call an ideal solution that which would follow this equation for any concentration. It represents a perfect analog to the ideal gas. The osmotic law governing such a solution shall be termed the osmotic equation of state. Incidentally, for many purposes it is sufficient to specialize this ideal equation of state for diluted solutions. It then becomes identical with van't Hoff's equation:

$$\overline{P} = nkT$$

where, as in the following, $\overline{P}$ represents the ideal osmotic pressure, n the number of dissolved single particles contained in one cc., k Boltzmann's constant, and T the absolute temperature.[2]

* Received January 28, 1924
** Compare section VI.

The behavior of the electrolytes becomes the more peculiar the greater the dissociation calculated by the classical methods and the higher the valency of the ions. For this reason it is wise to consider as the first molecular effect merely the Coulomb effect between the ions. This alone shall be discussed in the following. It is obvious that, by proceeding in this way, we can advance only one step beyond the classical laws. Expressed mathematically, only the first of all the supplementary terms required by the osmotic law, the one involving the lowest power of the concentration, is accurately taken into account. In the earlier paper an extension of the theory (to a certain approximation) was attempted by taking into account the size of the ions—that is, by the assumption of forces counteracting the mutual approach of the ions. It was intimated, however, that not everything had been accomplished that was desired, and the direction for future work was indicated. In this note I want to confine myself strictly to the first step. We gain thereby a clarification of the different parts of our consideration. We have to accept the drawback that the resulting laws are valid only in the limiting case of great dilution. In a later note I plan to discuss the effect of the other molecular forces more accurately than was done in my first publication.

II. Methods

We have a given solution of volume V, consisting of N_0 molecules of solvent, in which:

$$N_1, N_2 \cdots N_i \cdots N_s$$

single particles (atoms, ions, or molecules) of the kinds:

$$1, 2, \cdots i \cdots s$$

are dissolved. The osmotic equation of state for this solution shall be determined under the assumption that the different single particles carry the charges:

$$e_1, e_2 \cdots e_i \cdots e_s$$

which may in part be zero.

To determine the osmotic law we start with a greatly diluted solution which also contains the single particles $N_1 \ldots N_s$ but much more solvent. The above mentioned solution shall be obtained from this extremely diluted one in two different ways, which, however, shall both be chosen isothermally. According to the basic laws of thermodynamics, the same amount of work is performed in either case.

The first way is represented by the isothermal movement of a semi-permeable piston which permits the passage only of the solvent but not of the dissolved materials. At each instant of the process the (unknown) osmotic pressure, P, acts on the piston. The work performed in the change of volume dV equals $-PdV$. The total work, A, is consequently:

$$A = -\int_{\Omega}^{V} P\, dV$$

Ω being the initial volume.

The second process takes the following course. First, the charges e_i of the single particles of the extremely diluted solution are removed infinitely slowly.* After this has been done, the system has become an ideal solution in the sense of section I. With the help of a semi-permeable piston, we now adjust its concentration to the required extent. Finally, we recreate therein the charges e_i infinitely slowly. The total work performed in this process consists of two parts. First, the work required by the (isothermal) compression is:

$$\bar{A} = -\int_{\Omega}^{V} \bar{P}\, dV$$

in which $\bar{P}$ is the ideal osmotic pressure in the sense of section I. Second, however, the removal and the reintroduction of the charges requires electrical work, which we will call W, so that the total work amounts to:

$$\bar{A} + W$$

It is required by the laws of thermodynamics that:

$$A = \bar{A} + W \qquad (1$$

The equation of state follows from eq. (1) by a simple additional consideration. We can repeat the two processes with the variation that we now compress only to $V + dV$ and not to the volume V. For this second process we formulate the thermodynamic condition corresponding to eq. (1). Subtraction of the two equations immediately furnishes the desired osmotic pressure. Mathematically, this amounts to a differentiation of eq. (1) with respect to V. If we carry this out, we obtain, considering the meaning of A and $\bar{A}$:

$$P = \bar{P} - \frac{\partial W}{\partial V} \qquad (2$$

*Physically this process is impossible since, at least at present, we have no means of subdividing the elementary charge. Notwithstanding, there is no objection to its use just as there is no objection to a semi-permeable membrane which can not be realized in practice.

Relation (1) shows at once that the actual osmotic pressure P is smaller than the ideal one, provided work is done during the discharge and charge processes, as is the case for small concentrations. The reverse can be observed with large concentrations. Incidentally, there is a close connection between the total energy consumed, as given by eq. (1), and the free energy of the solution. In section V we shall make use of this latter relation.

III. Osmotic Equation of State

The electric work W to be computed now is composed of two parts. The first part is the work required to remove the charges of the single particles from an infinitely diluted solution; the second part constitutes the work required to recreate the charges in a solution of prescribed concentration. We assume the charging and discharging processes to be conducted in such a manner that at any stage of the processes the charges are equal to the product $\lambda\epsilon_i$. First, the function of time λ slowly decreases from 1 to 0 and, subsequently, it increases again from 0 to 1. The amount of work:

$$\frac{e^2}{2Da}$$

is required to charge a sphere of radius a to a total charge e in a medium of dielectric constant D.

The amount of work for a single particle in an infinitely diluted solution will be given by a similar expression with opposite sign, provided the dielectric constant of the pure solvent is equal to D. If the charge is recreated in the second phase of the process, the solution has in the meantime assumed its final concentration. The surrounding of the particle is thus changed as regards its dielectric properties, and an amount of work:

$$\frac{e^2}{2D'a}$$

will now have to be expended for the charging process. If only diluted solutions are considered, as is done here, D' can be put equal to D for terms of the first order in the concentration. Since we shall establish in the following that the work due to the mutual Coulomb forces is proportional to the square root of the concentration, the difference between D' and D is a term of higher order with respect to the last mentioned principal term in the expression for the work. In the limit, $-e^2/2Da$ therefore cancels against $e^2/2D'a$. Therefore, only the difference between these two amounts of work will have to be treated in the following part of this paper. There, an attempt will be made to prove that this difference is the cause of the increase in the activity coefficient beyond unity observed for highly concentrated solutions.

To calculate the amount of work due to Coulomb forces, we take the following result from the previous presentation. If a particle (with index i) is singled out, all other particles create a potential ψ_i of the amount:

$$\psi_i = -\frac{e_i \varkappa}{D} \qquad (3)$$

at the location of this particle, where κ designates the reciprocal of the "thickness of the ion atmosphere." The relation:

$$\varkappa^2 = \frac{4\pi}{DkT} \sum n_j e_j^2 \qquad (4)$$

holds for the computation of κ, in which D is the dielectric constant of the solvent, k Boltzmann's constant, and T the absolute temperature, while:

$$n_1, n_2 \cdots n_j \cdots n_s$$

separate particles of the types:

$$1, 2 \cdots j \cdots s$$

are dissolved in the unit volume; the charges of these particles are:

$$e_1, e_2 \cdots e_j \cdots e_s$$

respectively.

The result may be stated as follows: in a diluted solution, the potential of all ions in regard to one ion is identical with the potential which would be created by another single ion of equal and opposite charge situated at a distance equal to a length $1/\kappa$, which is a measure of the extension of the ion atmosphere. If, during their creation, the charges have reached the values:

$$\lambda e_1, \lambda e_2 \cdots \lambda e_i \cdots \lambda e_s$$

then, according to equations (3) and (4), the potential will have reached the value:

$$\lambda^2 \psi_i$$

The increase in the singled-out charge during a time element amounts to:

$$d(\lambda e_i) = e_i \, d\lambda$$

so that during this infinitely short time, the work:

$$e_i \psi_i \lambda^2 d\lambda$$

has to be done for a particle of the type i. This work is equal to the instantaneous potential $\lambda^2\psi_i$ times the change in charge $e_i d\lambda$. The total electric work is derived by integration with respect to λ from 0 to 1 and summation over all particles present in the liquid. We thus find:

$$W = \Sigma N_i \frac{e_i \psi_i}{3} \qquad (5)$$

or, according to equation (3):

$$W = -\Sigma N_i e_i^2 \frac{\varkappa}{3D} \qquad (5')$$

In order to obtain the osmotic pressure P, according to equation (2), the expression for the work W must be differentiated with respect to the volume V, which is implicitly contained in κ' (compare eq. 4), since the number of particles n_j refers to the unit of volume, *i.e.*, since $n_j = N_j/V$. In view of this we have from equation (4):

$$\frac{\partial \varkappa}{\partial V} = -\frac{\varkappa}{2V}$$

and hence from equation (5'):

$$\frac{\partial W}{\partial V} = \Sigma n_j e_i^2 \frac{\varkappa}{6D}$$

From equation (2) it follows that the osmotic pressure is given by:

$$P = \overline{P} - \Sigma n_i e_i^2 \frac{\varkappa}{6D} \qquad (6)$$

This expression agrees with our previous statements. Further, it is very easy to verify by the above method the results obtained by taking into account the finite dimensions of the ions.

IV. Osmotic Coefficients

It shall now be assumed that the osmotic pressure of the ideal solution is represented by van't Hoff's expression; thus:

$$P = \sum n_i k T$$

From equation (6) it then follows that for the actual solution:

$$P = \sum \left(1 - e_i^2 \frac{\varkappa}{6DkT}\right) n_i k T \qquad (7)$$

i.e., the osmotic pressure is derived by calculating not with actual but with fictitious (volume) concentrations which are obtained from the actual concentrations by multiplication with the factors:

$$g_i = 1 - e_i^2 \frac{\varkappa}{6DkT} \qquad (8)$$

These factors, which (within the range of validity of the foregoing calculation) are different from unity only for ions (with finite charge), may be termed "osmotic coefficients" following Bjerrum.*

Originally Bjerrum defined, for a salt which can dissociate into a total of ν ions and which does so in an infinitely diluted solution, an osmotic coefficient g, relating to the complete salt, by the expression:

$$P = g \nu n k T \qquad (9$$

where n designates the number of salt molecules dissolved per cubic centimeter. Provided a molecule also dissociates at higher concentrations completely in:

$$\nu_1 + \nu_2 + \cdots + \nu_i + \cdots + \nu_s = \nu$$

ions of the kind 1 ...s, then:

$$\nu g = \sum \nu_i g_i \qquad (10$$

Only g is measured experimentally. However, from the interpretation of this coefficient for strong electrolytes, it appears justifiable and desirable to separate the special ion coefficients g_i according to equations (10) and (8).

Since, according to equation (4) κ is proportional to the square root of the concentration, it is confirmed that, in diluted solutions, the work due to Coulomb forces is of primary importance. The reader is referred to the previous publication for a numerical discussion.

* I use the designations g_i and g instead of f_0^i and f_0 to avoid aggregation of indices.

V. Activity Coefficients

According to the fundamental laws of thermodynamics, the work done on a system at constant temperature is equal to the increase in free energy, where the latter, in accordance with the definition by Helmholtz, is a thermodynamic function given by:*

$$F = U = TS$$

U is the energy and S the entropy of the system. The total work:

$$A = \bar{A} + W$$

given by equation (1) has been added to the original system of volume Ω by an isothermal and reversible process; it thus constitutes the increase in free energy of the system during the change in concentration of the solution caused by the displacement of the semi-permeable piston.

A free energy, the increase of which is measured by the work $\bar{A}$, would correspond to an ideal solution in the terminology of section I. Since at infinite dilution no difference exists between the ideal solution and the actual solution, W indicates the amount to be added to the ideal free energy of the system in order to find the free energy of the actual system. The total system consists of the solution below the semi-permeable piston and the separated pure solvent above the piston. Obviously, W is also the difference existing between the actual and the ideal free energy of the solution alone. If the first is termed F and the latter $\bar{F}$, we thus have:

$$F = \bar{F} + W$$

As is well known, it is, according to Gibbs, usually much simpler to proceed from the thermodynamic potential Z instead of from the free energy F. The potential Z differs from F only by the addition of pV and is adapted to observations made at constant pressure p and constant temperature T; thus

$$Z = F + pV$$

A variation at constant temperature and constant pressure (indicated by a preceding Δ) corresponds to a variation of Z equal to:

$$\Delta Z = A + p\Delta V$$

* The free energy by Helmholtz and the thermodynamic potential by Gibbs shall be used instead of the previous potentials preferred by Planck, since these quantities permit a simpler expression for the conditions of work.

The work done on the system by the external pressure p is given by $-p\Delta V$; further we had:

$$\Delta F = A$$

Provided the work done by the external pressure is negligibly small, what has been stated with respect to the relation between the real and ideal free energy holds to a high degree of approximation for the relation between the actual thermodynamic potential Z and the ideal thermodynamic potential $\bar{Z}$; hence:

$$Z = \bar{Z} + W \qquad (11)$$

We are indeed using F or Z effectively without distinguishing between these two thermodynamic functions; this amounts to considering the systems studied as incompressible.

As is well known the ideal solution is defined by the expression:

$$\bar{Z} = \sum_0 N_i(\zeta_i + kT \log c_i)$$

where it is assumed that $N_1 \ldots N_s$ molecules (atoms, ions) are added to N_0 molecules of solvent. The ζ_i values are functions of pressure and temperature only and do not contain the concentrations; the concentrations c_i are defined by the expression:

$$c_i = \frac{N_i}{\sum_0^s N_i}$$

Thus, for diluted solutions, the actual thermodynamic function Z is, according to equations (11) and (15'):

$$Z = \sum_0^s N_i\left(\zeta_i + kT \log c_i - \frac{e_i^2 \varkappa}{3D}\right) \qquad (12)$$

It is significant that the separate terms in the addition to $\bar{Z}$ are proportional to the quantity $N_i\kappa$, *i.e.*, according to equation (4), to the 3/2 power of the number of molecules. It will now be understood why previous attempts, which regarded the expression for $\bar{Z}$ as the beginning of a series expansion in integral powers of the N_i, and which completed it accordingly, could not be successful in explaining the properties of strong electrolytes.[3]

The change in the thermodynamic function at constant pressure and constant temperature caused by a variation of the numbers N_i as computed according to eq. (12) is:

$$\delta Z = \sum_{i=0}^{i=s} \delta N_i \left[\zeta_i + kT \log c_i - \frac{e_i^2 \kappa}{3D} - \sum_{j=0}^{j=s} \frac{e_j^2}{3D} N_j \frac{\partial \kappa}{\partial N_i} \right]$$

From the definition of κ it readily follows that for diluted solutions:

$$\sum_j \frac{e_j^2}{3D} N_j \frac{\partial \kappa}{\partial N_i} = \frac{e_i^2 \kappa}{6D}$$

provided the index i refers to the dissolved particles, *i.e.*, assumes any of the values 1...*s*. If, however, the index i is equal to zero, *i.e.*, if the variable refers to the solvent, it follows that:

$$\sum_j \frac{e_j^2}{3D} N_j \frac{\partial \kappa}{\partial N_0} = - v_0 \sum_j n_j \frac{e_j^2 \kappa}{6D}$$

where v_0 designates the space occupied by one molecule of the solvent. Hence:

$$\delta Z = \delta N_0 \left[\zeta_0 + kT \log c_0 + v_0 \sum_j n_j \frac{e_j^2 \kappa}{6D} \right] + \sum_1^s \delta N_i \left[\zeta_i + kT \log c_i - \frac{e_i^2 \kappa}{2D} \right] \quad (13)$$

Gibbs called the multiplying factors of δN_i the chemical potentials μ_i. In the ideal case we have:

$$\mu_i = \bar{\mu}_i = \zeta_i + kT \log c_i \quad (14)$$

Thus equation (13) demonstrates the change in the ideal chemical potential caused by the electrostatic effects of the ions. The change in the chemical potential may be measured formally by introducing into eq. (14) a fictitious concentration $f_i c_i$ instead of the actual concentration c_i, which is obtained from c_i by multiplication by a numerical factor f_i. Then:

$$\mu_i = \zeta_i + kT \log f_i c_i \quad (14')$$

The numerical factor f_i is called by Lewis the "activity coefficient of the particular kind of molecule" and the product $f_i c_i$ the "activity."

According to eq. (13) we find for the activity coefficient of an ion of arbitrary kind i, in the limit for diluted solutions, the relation:*

$$\log f_i = -\frac{e_i^2 \varkappa}{2DkT} \qquad (15)$$

At the same time the activity of the solvent is also affected by the dissolved ions and is characterized by an an activity coefficient f_0 following from the relation:

$$\log f_0 = v_0 \sum_j n_j \frac{e^2{}_j \varkappa}{6DkT} \qquad (15')$$

Thus in the terminology just introduced, the activity of the ions is decreased, that of the solvent increased, as a consequence of the electrostatic effect between ions.

Let us consider an arbitrary reaction between the dissolved particles, characterized by changes $\delta N_1 \ldots \delta N_s$ corresponding to an assumed reaction; the thermodynamic condition for equilibrium, as is well known, then reads:

$$\delta Z = 0$$

If the expression for δZ corresponding to the ideal case is inserted, Guldberg-Waage's mass action law results. In general it follows that in the law of mass action the activities $f_i c_i$ should replace the concentrations c_i so that the activity coefficients may be defined also by means of the corrected law of mass action. These coefficients were defined in a similar though not identical manner in our first article; we then introduced as activity coefficients those numbers which must be divided by the concentrations to yield the respective activities. I consider it preferable to accept the above definition; we then have the advantage of agreeing with previous work, particularly that by Lewis.

It may further be pointed out that occasionally and for practical reasons the activity coefficient f of all of the dissolved salt is considered instead of the separate activity coefficients f_i. For a completely dissociated salt whose molecules split into:

$$\nu = \nu_1 + \cdots + \nu_i + \cdots + \nu_s$$

ions, the total activity coefficient has been defined by the expression:

$$\nu \log f = \sum \nu_i \log f_i \qquad (16)$$

* The correction of this formula for the dimension of the ion yields the remarkably simple formula:

$$\log f_i = -\frac{e_i^2 \varkappa}{2DkT} \frac{1}{1 + \varkappa a}$$

as established in the previous article.

which we are also adopting.[4] The definition is analogous to that of the osmotic coefficient in eq. (10).

VI. Experimental Results on the Activity Coefficients

Since 1907 G. N. Lewis has been contending[5] that the difficulties arising in various fields upon application of the classical laws of van't Hoff and Guldberg-Waage on the theory of Arrhenius would be overcome best by a systematic extension of the thermodynamic fundamentals. This extension was formally accomplished by the introduction of the activities instead of the concentrations, a method which will be associated with the discussions of this note in the next section. From the preceding discussion it will be seen that by this introduction primarily Gibb's chemical potentials are transformed into a new, less definite, and thus more readily adaptable form. Since, as is well known, all phenomena of varied appearance in the domain of thermodynamics can be derived from these potentials, it had become an experimental problem to determine the activity coefficients, and it was evident from the beginning that, in principle, quantitive observation of the course of any arbitrary phenomenon would be adequate. In the work thus started, speculations as to the cause and the molecular interpretation of the activity coefficients were consistently avoided. A summary of the results[6] dated 1921, which in the meantime has come to my attention, leads, on the basis of extensive experimental material, to the formulation of a theorem which Lewis calls the "hypothesis of the independent activity coefficients of the ions." It reads[7]:

"In diluted solutions the activity coefficient of an arbitrary ion depends exclusively on the total ionic strength of the solution."

It had been defined previously[8] what was to be understood by ionic strength. In our terminology Lewis designates by ionic strength a magnitude proportional to:

$$\sum n_i e_i^2$$

He multiplies, for each kind of ion, the number of moles of the ions of this kind contained in one liter by the square of the valency, takes the sum of these products for all kinds of ions and divides by 2. The actual number of ions n_i in one cubic centimeter is proportional to the number of moles per liter, the charge e_i of an ion is $z_i\varepsilon$, where z_i designates the valency and ε the electronic charge. The above mentioned sum is thus proportional to the ionic strength.

In a footnote to his hypothesis, Lewis explains what is to be understood by "diluted solutions." He prefers this term to be so interpreted that his theorem becomes a limiting law, or, expressed differently: his theorem will be followed the closer, the more diluted the solution, and it will be absolutely true for infinitely diluted solutions. Let us compare this statement with our theoretical result. According to eq. (15) we have the limiting law:

$$\log f_i = -\frac{e_i^2 \varkappa}{2DkT}$$

where κ', according to eq. (4), is defined by the expression:

$$\varkappa^2 = \frac{4\pi}{DkT} \sum n_i e_i^2$$

Hence the theory yields a formula corresponding to Lewis' theorem.

It is further to be stressed that Lewis and Linhart arrived at the experimental result[9] that, at low concentrations, the logarithm of the activity coefficient of a salt varies proportional to a power of the concentration which was approximately determined as 1/2. Lewis believes 1/2 to be correct[10], and the theory in fact leads to a proportionality to κ, *i.e.*, to the square root of the concentration.

In view of our theoretical deliberations, the cause of the relations found experimentally by Lewis must be seen in the fact that the thickness of the ion atmosphere is a determining factor for the energy of a solution and that the thickness of the ion atmosphere, for low concentrations, may be assumed proportional to the square root of the concentration and independent of the individual properties of the ions except for their valency. Based on the hypothesis of the independent activity coefficients, assumed to hold also for somewhat greater concentrations, and proceeding from the assumed equality of the activities of K and Cl ions, Lewis gives[11] a table of these coefficients for the separate ions as a function of the ionic strength. He notes that the figures were obtained from sketchy material and thus require revision. According to the theory, the hypothesis can hold only for strongly diluted solutions, and as long as it holds, the properties of an ion are determined exclusively by the valencies of the ions present in the solution. However certain differences are found between different monovalent ions; this can also be observed in the table by Lewis. The assumption that the activity of an ion depends only on the ionic strength of the solution is thus not quite justified, as far as the table goes; as a consequence this uncertainty is introduced into the calculations. To be consistent one would have to complete the limiting laws by considering the individual properties, and the introduction of the mean ion radius in our previous article may be taken as the first step in this direction. In spite of these objections, we will apply the limiting law, eq. (15), quantitatively and compare it with Lewis' figures. In view of the extrapolation mentioned above which cannot be entirely justified, this comparison does not attain the precision which could be attemped in the prior article discussing measurements on freezing point depression. If the number of moles of ions of the kind i dissolved in one liter is γ_i and the valency z_i, then:

$$J = \tfrac{1}{2} \sum z_i^2 \gamma_i$$

is the quantity called by Lewis the ionic strength of the solution. Introducing J, eq. (4) yields:

$$x^2 = \frac{8\pi\varepsilon^2}{DkT}\frac{N}{1000}J$$

where ε is the electronic charge and N Loschmidt's number. Further, from eq. (15):

$$\log f_i = -\frac{\varepsilon^2}{2DkT}\sqrt{\frac{8\pi\varepsilon^2}{DkT}\frac{N}{1000}} \cdot z_i^2\sqrt{J} = -B_e z_i^2\sqrt{J}. \qquad (17)$$

Throughout this article we are using the natural logarithm, and therefore the proportionality factor is designated by B_e. Expressed in common logarithms we write:

$$\log_{10} f_i = -B_{10} z_i^2 \sqrt{J} \qquad (17')$$

The following table lists the theoretical values for B_e and B_{10} obtained from eq. (17) for aqueous solutions under the assumption that it is permissible to calculate the dielectric constant at the temperature $t\,^\circ$C. from Drude's formula:[12]

$$D = 88.23 - 0.4044t + 0.001035t^2$$

and for $k = 1.371 \times 10^{-16}$, $\varepsilon = 4.77 \times 10^{-10}$, $N = 6.06 \times 10^{23}$.

Table I

t	B_e	B_{10}
0	1.12	0.485
20	1.15	0.499
40	1.19	0.519
60	1.23	0.535
80	1.27	0.553
100	1.31	0.568

For 20°C. the activity coefficient is, according to eq. (17'), for univalent ions in solutions of ionic strength $J = 0.01$ and $J = 0.1$, equal to $f_i = 0.89$ and $f_i = 0.70$, respectively. The values in the table by Lewis vary from $f_i = 0.92$ to $f_i = 0.85$ for $J = 0.01$ and from $f_i = 0.84$ to $f_i = 0.61$ for $J = 0.1$. At the same temperature we obtain for bivalent ions in solutions of ionic strength $J = 0.01$ and $J = 0.1$, again according to eq. (17'), $f_i = 0.63$ and $f_i = 0.23$, respectively.

In the table by Lewis, the various bivalent ions are grouped together. At the above mentioned ionic strengths $J = 0.01$ and $J = 0.1$, the values $f_i = 0.60$ and $f_i = 0.34$, respectively, are given for this group, while $f_i = 0.56$ and $f_i = 0.26$ are given for the SO_4 ion.

The formulas (17) and (17') contain no constants which can be adapted. Therefore, and in view of the previously discussed limitations, the agreement between the theory and the values obtained by Lewis from experiments over a period of years without knowledge of the theory may be considered satisfactory.

VII. General Remarks on Activity

In concluding I wish to incorporate the previous considerations into the system of general thermodynamics in order to facilitate understanding of this survey for those who have not yet had close contact with the present complex of problems.

It is evident that in general the expression for the thermodynamic potential of an ideal solution:

$$\overline{Z} = \Sigma N_i(\zeta_i + kT \log c_i) \qquad (18$$

will be adapted to actual circumstances by an additional term such as:

$$W = \Sigma N_i w_i \qquad (19$$

where the function w_i will depend on pressure and temperature as well as on concentrations. In the particular instance treated in the preceding paragraphs, W was the electric work which was determined by special considerations based on molecular theory. If the additional term W is written in the form:

$$W = \Sigma N_i kT \log h_i \qquad (19'$$

which implies that:

$$h_i = e^{\frac{w_i}{kT}}$$

the potential of the actual solution becomes:

$$\overline{Z} = Z + W = \Sigma N_i(\zeta_i + kT \log h_i c_i) \qquad (20$$

This can be summarized in the statement:

The potential of an actual solution is obtained by substituting for the real concentrations c_i fictitious concentrations $h_i c_i$, secured from the real concentrations by multiplication with certain factors h_i. Once these factors h_i are known, all properties of the solution can be derived by thermodynamic methods. It was the object of the work by Lewis, reported in the previous section, to secure information on the factors h_i by experimental methods; we have tried to attain the same objective by theoretical means. The statements and deliberations in the literature, however, are not worded in the above terminology. They use activities and activity coefficients. These can be derived from the h_i by differentiation; I would like, therefore, to take the liberty of calling the h_i activity potentials.

The variation of the thermodynamic potential corresponding to a change in the molecular numbers N_i (at constant pressure and constant temperature) is given by:

$$\delta Z = \sum_i{}' \delta N_i \left[\zeta_i + kT \log c_i + kT \log h_i + kT \sum_j{}' N_j \frac{\partial \log h_j}{\partial N_i} \right] \qquad (21)$$

since in general the activity potentials will depend on all concentrations. This expression can be transformed to correspond to eq. (20) by putting:

$$\delta Z = \sum_i \delta N_i [\zeta_i + kT \log f_i c_i] \qquad (22)$$

where the new coefficients f_i are related to the activity potentials by the equation:

$$\log f_i = \log h_i + \sum_j{}' N_j \frac{\partial \log h_j}{\partial N_i} \qquad (22')$$

As is well known, factors multiplying the δN_i in eq. (22) are called, following Gibbs, the chemical potentials of the respective constituents. In an ideal solution the chemical potential $\overline{\mu}_i$ of the ith constituent is represented by:

$$\overline{\mu}_i = \zeta_i + kT \log c_i$$

The new expression again differs therefrom by replacement of the actual concentration c_i by the fictitious concentration $f_i c_i$, thus:

$$\mu_i = \zeta_i + kT \log f_i c_i$$

The coefficients f_i introduced in this way are called activity coefficients; their deriviation from the activity potentials is given by eq. (22').

A reaction considered possible is characterized by certain relative values of the δN_i; the correlated equilibrium follows from the condition $\delta Z = 0$. If the ideal value $\overline{Z}$ has been chosen for Z, the mass action law (in logarithmic form) is obtained. In the general case, represented by eq. (22), this law is modified only inasmuch as the quantities $f_i c_i$ replace the quantities c_i. Therefore $f_i c_i$ is called the activity of the respective constituent, and it can be stated that: In ideal solutions the activity is identical with the concentration; actually, however, the activity and the concentration differ by the activity coefficient, and the law of mass action is correct only if the activities are inserted instead of the concentrations.

Thus a scheme has been outlined that completely corresponds to the one outlined by Lewis in 1907, and an experimental or theoretical attempt can now be made to obtain an understanding of the cause of the hindrance or furtherance of the effectiveness measured by the activity coefficients.

In practice we can distinguish between processes where the number of molecules of the solvent is changed and those where the number of molecules of the dissolved substance is changed.

In the first category belong the phenomena of osmotic pressure, freezing point depression, boiling point increase, vapor pressure depression. Thus during the displacement of a semi-permeable piston, permitting δN_0 molecules of the solvent to pass through, an amount of work $Pv_0 \delta N_0$ is done, where v_0 is a volume associated with one molecule of the solvent. The change in potential, caused by the loss of the solution of δN_0 molecules of the solvent and by the generation of δN_0 molecules of pure solvent, is given by:

$$\delta N_0 [\zeta_0 - (\zeta_0 + kT \log f_0 c_0)]$$

Hence, as is well known:

$$Pv_0 = -kT \log f_0 c_0 \qquad (23)$$

is the osmotic equation of state, from which van't Hoff's equation of state follows as a first approximation for low concentrations. The phenomena of the first group thus provide facilities for the determination of the activity coefficient f_0 of the solvent.

To the second group belong the phenomena related to the law of mass action, solubility and effects on solubility, electromotive force, etc. For a completely dissociated electrolyte, for instance, whose molecule splits into:

$$\nu = \nu_1 + \dots + \nu_i + \dots + \nu_s$$

particles of the kind $1 \dots s$, an increase of the solution by one molecule of the dissolved salt causes an increase in potential by:

$$\sum_{i=1}^{i=s} \nu_i(\zeta_i + kT \log f_i c_i) = \sum_{i=1}^{i=s} \nu_i \bar{\mu}_i + kT \sum_{i=1}^{i=s} \nu_i \log f_i$$

where $\bar{\mu}_i$ designates the chemical potential for the ideal case.

If as before the activity coefficient f of the salt is defined by the expression:

$$\nu \log f = \Sigma \nu_i \log f_i \tag{24}$$

the phenomena of the second group yield, in this terminology, the activity coefficient of the salt.

In view of these general thermodynamic relations it is superfluous to introduce osmotic coefficients in addition to the activity coefficients of the solvent and of the dissolved substance. If this is done, nevertheless, as Bjerrum suggests, and as has been done for diluted solutions in section IV, then a relation between the activity coefficient and the osmotic coefficient gains practical significance. This relation has been discussed by Bjerrum.[13] If N molecules of a (completely dissociated) salt are dissolved in N_0 molecules of water, the concentration of the salt is measured by:

$$C = \frac{N}{N_0}$$

The above osmotic equation of state becomes for sufficiently low concentrations:*

$$\frac{P v_0}{kT} = \nu C - \log f_0$$

* We have

$$c_0 = 1 - \Sigma c_i = 1 - \nu C$$

and log c_0 expanded gives:

$$\log c_0 = \log(1 - \nu C) = -\nu C$$

provided the number of the dissolved particles is assumed small compared with the number of the molecules of the solvent.

The additional term -log f_0 thus measures the change in the osmotic law due to the molecular forces; the osmotic coefficient g is defined by Bjerrum by the expression:

$$\frac{Pv_0}{kT} = g\nu C$$

Hence:

$$g = 1 - \frac{1}{\nu C} \log f_0 \qquad (25)$$

Both f_0 as well as f, the latter coefficient by means of its definition eq. (24), follow from the activity potential eq. (22'). Considering that these potentials can depend only on the concentrations $c_1 \ldots c_i \ldots c_s$ and writing in a permissible approximation:

$$c_1 = \frac{N_1}{N_0}, \ldots c_i = \frac{N_i}{N_0}, \ldots c_s = \frac{N_s}{N_0}$$

as is sufficiently accurate for diluted solutions, we obtain:

$$\frac{\partial}{\partial N_0} = -\frac{1}{N_0} \sum_i c_i \frac{\partial}{\partial c_i}$$

and:

$$\frac{\partial}{\partial N_i} = \frac{1}{N_0} \frac{\partial}{\partial c_i}$$

It is further assumed permissible, as it is according to the deliberations based on molecular theory presented in the preceding paragraphs, that the activity coefficient of the solvent h_0 have the value 1. Then according to eq. (22'):

$$\left. \begin{aligned} \log f_0 &= -\sum_i \sum_j c_i c_j \frac{\partial \log h_j}{\partial c_i} = \\ &= -C^2 \sum_i \sum_j \nu_i \nu_j \frac{\partial \log h_j}{\partial c_i} \end{aligned} \right| \qquad (26)$$

Further, also in accordance with eq. (22'):

$$\left.\begin{aligned}\nu \log f = \sum_i \nu_i \log f_i = \sum_i \nu_i \log h_i + \\ + C \sum_i \sum_j \nu_i \nu_j \frac{\partial \log h_i}{\partial c_i}\end{aligned}\right| \quad (26')$$

If the activity potential h of the salt is defined by the expression:

$$\nu \log h = \sum \nu_i \log h_i$$

it can be stated that the double sum in eqs. (26) and (26') follows directly from this activity potential by differentiation with respect to the concentration C. In fact:

$$\begin{aligned}\frac{d}{dC}(\nu \log h) = \sum_i \nu_i \frac{\partial}{\partial c_i}(\nu \log h) = \\ = \sum_i \sum_j \nu_i \nu_j \frac{\partial \log h_j}{\partial c_i}\end{aligned}$$

Hence:

$$\left.\begin{aligned}\log f_0 = -C^2 \frac{d}{dC}(\nu \log h) \\ \nu \log f = \nu \log h + C \frac{d}{dC}(\nu \log h)\end{aligned}\right| \quad (27)$$

These relations permit derivation of the activity coefficient of the solvent as well as the activity coefficient of the dissolved salt from the activity potential. Elimination of this potential yields either:

$$\frac{d}{dC} \log f = -\frac{1}{\nu C} \frac{d}{dC} \log f_0 \qquad (28)$$

or, if f_0 is replaced by the osmotic coefficient in accordance with eq. (25):

$$\frac{d}{dC} \log f = \frac{g - 1}{C} + \frac{d}{dC}(g - 1) \qquad (28')$$

The latter is the relation given by Bjerrum. If it follows, for instance, from observations of the osmotic pressure that in practice:

$$g = 1 - \alpha C^{1/2}$$

with an experimentally determined constant α, it follows from eq. (28') that:

$$\log f = -3\alpha C^{1/2}$$

Bjerrum took advantage of this relation in a similar manner.

I wish to stress that the discussion in this last section contains no results which cannot be found in the literature on the subject. However, it is hoped that this systematic survey will prove helpful.

Bibliography

1. P. Debye, "De moderne ontwikkeling van de elektrolyt-theorie," *Handelingen van het 19^e Natuur- en Geneeskundig-Congres*, gehouden April 1923; P. Debye and E. Hueckel, *Physik. Z.*, *24*, 185, 305 (1923).
2. This presentation agrees with the opinion of J. J. van Laar, *Sechs Vorträge uber das thermodynamische Potential*, Braunschweig, 1906.
3. H. Jahn, *Z. phys. Chem.*, *37*, 490 (1901); *38*, 125 (1901); *41*, 257 (1902); *50*, 129 (1905); W. Nernst, *Z. phys. Chem.*, *38*, 484 (1901).
4. I published a short report on the relationship between the activity and the observations on solubility: *Chem. Weekblad*, *20*, 562 (1923). A thesis by O. Schaerer to be published presently treats the subject in more detail.
5. *Proc. Am. Acad.*, *43*, 259 (1907); *Z. phys. Chem.*, *61*, 109 (1907); *70*, 212 (1909).
6. G. N. Lewis and M. Randall, *J. Am. Chem. Soc.*, *43*, 1112 (1921). During the printing of this article I had access to the publication (G. N. Lewis and M. Randall, *Thermodynamics and the Free Energy of Chemical Substances*, McGraw-Hill, New York, 1923), in which the methods used by Lewis and his co-workers are described in great detail.
7. *J. Am. Chem. Soc.*, *43*, 1145 (1921).
8. *J. Am. Chem. Soc.*, *43*, 1140 (1921).
9. *J. Am. Chem. Soc.*, *41*, 1951 (1919).
10. *J. Am. Chem. Soc.*, *43*, 1121 (1921).
11. *J. Am. Chem. Soc.*, *43*, 1147 (1921).
12. P. Drude, *Ann. Physik*, *59*, 61 (1896).
13. N. Bjerrum, *Z. Elektrochem.*, *24*, 325 (1918); *Z. phys. Chem.*, *104*, 406 (1923).

REMARKS ON A THEOREM ON THE CATAPHORETIC MIGRATION VELOCITY OF SUSPENDED PARTICLES

(Bemerkungen zu einem Satze über die kataphoretische Wanderungsgeschwindigkeit suspendierter Teilchen)

P. Debye and E. Hückel*

Translated from
Physikalische Zeitschrift, Vol. 25, 1924, No. 3, pages 49-52.

It is well known that small particles which are suspended in a liquid show a migration velocity V if an electric field (whose field strength will be denoted by X) is applied; the velocity is proportional to the field strength. It is generally assumed, in accordance with Helmholtz, that such particles have a potential difference $\bar{\psi}$ with respect to the liquid and carry a corresponding surface charge which exerts electrostatically attractive and repellent forces on the ions of the surrounding liquid. Consequently, an ion atmosphere with space charges of opposite sign is created in the vicinity of the surface of such particles, the structure of which has been treated already by Gouy[1] and Chapman[2]; we had to deal with it recently to explain the properties of strong electrolytes.[3] For many discussions, such as that of Helmholtz, it is unnecessary to know the distribution of charges in this ion atmosphere; it is sufficient to know that it extends to a noticeable amount only over very short distances. If the electric field X is applied, an electric force, for instance to the right, is exerted on the particle. Consequently, the particle will start moving to the right. If no ion atmosphere were present, the stationary velocity would be proportional to the total charge of the particle, since the resulting current of the liquid would be independent of this charge. In fact, however, this atmosphere is present, and since it carries an opposite charge relative to the particle, the field strength X will cooperate with the volume element of the liquid to drive it toward the left. Thus, as a consequence of the increased friction, the attainable stationary velocity V is smaller than in the fictitious case of the missing ion atmosphere. Based on considerations by Smoluchowski,[4] we find consistently[5] a very simple formula for the computation of V which is supposed to be valid independently of the distribution of charges in the ion atmosphere and even of the shape of the suspended particles. It reads:

* Received December 12, 1923

$$V = \frac{1}{4\pi} \frac{D\overline{\psi}}{\eta} X \tag{1}$$

where V designates the velocity, D the dielectric constant, η the coefficient of friction, $\overline{\psi}$ the potential of the particle with respect to the liquid, and X the driving electric field strength.

If the ion atmosphere is ignored, a velocity is obtained that is proportional to the charge of the particles, whereas it is now proportional to the potential. It would be entirely incorrect to assume this potential to be simply proportional to the charge of the particle so that replacing the charge by the potential would amount only to the omission of a factor of proportionality. In fact, the presence of the ion atmosphere considerably complicates the relation between the charge and the potential, so that doubling of the charge does not result in doubling of the potential. Still more surprising is the statement that the factor of proportionality always has the numerical value $1/4\pi$ independently of the shape of the particle. These considerations caused us to study the basis for the formula (1) in the case of the cataphoretic migration velocity. We thereby arrived at the following result. It can be proved that a formula:

$$V = C \frac{D\overline{\psi}}{\eta} X \tag{1'}$$

is valid, independently of the distribution of charges in the ion atmosphere, but only provided that the thickness of this atmosphere is small (in the limit infinitely small) compared to the linear dimensions of the particle. This is an assumption already made by Smoluchowski and which may be conceded in most instances. However, it is generally incorrect to put the numerical constant C equal to $1/4\pi$. C is, in fact, dependent on the form of the particles; the exact value of this factor can only be arrived at by a detailed analysis of the movement of the liquid in a special case. In the following we would like to set forth the considerations that lead to equation (1'). We shall not present special calculations of C. It may suffice to state that for spherical particles* C is given by:

$$C = \frac{1}{6\pi}$$

whereas for a cylinder:

$$C = \frac{1}{4\pi}$$

* It may be mentioned that for the special case of a spherical particle, the migration velocity is proportional to the potential, independently of the assumption of a thin double layer; the factor of proportionality is $1/6\pi$.

It is assumed that in a liquid with coefficient of friction η the charge density is ρ, so that under the influence of an external field strength $\mathfrak{E}$ a force:

$$\mathfrak{F} = \rho\mathfrak{E} \tag{2}$$

is exerted on each volume element. Further the potential associated with the charge distribution ρ is denoted by ψ. Then:

$$\Delta\psi = -\frac{4\pi\rho}{D}$$

and we can write for $\mathfrak{F}$:

$$\mathfrak{F} = -\frac{D\mathfrak{E}}{4\pi}\Delta\psi \tag{2'}$$

A particle of arbitrary shape is at rest in the liquid. The liquid passes the particle in such a manner that at great distances the lines of flow are parallel to the x axis of a rectangular coordinate system and the velocity has the value $-V$. It is further assumed that the field strength extends in the x direction, so that the components of $\mathfrak{E}$ have the values $X, 0, 0$. Provided the velocity is $\mathfrak{v}$ at any point in the liquid and the pressure at the same point is p, the equations of motion hold:

$$\left.\begin{aligned} \eta\,\mathrm{rot}\,\mathrm{rot}\,\mathfrak{v} + \mathrm{grad}\,p &= \mathfrak{F} \\ \mathrm{div}\,\mathfrak{v} &= 0 \\ \mathfrak{F} &= -\frac{D\mathfrak{E}}{4\pi}\Delta\psi \end{aligned}\right\} \tag{3}$$

From these equations $\mathfrak{v}$ is to be determined in such a way that the components $-V, 0, 0$ are reached at infinity, while $\mathfrak{v}$ vanishes at the surface of the particle.

First a current can be segregated which corresponds to the effect of $\mathfrak{F}$, but which violates the boundary conditions. These must then be satisfied subsequently by an additional current which has the advantage, from the computational point of view, of involving no force. This is obtained by the following tentative solutions:

$$\left.\begin{aligned} p &= -\frac{DX}{4\pi}\frac{\partial\psi}{\partial x} + p^* \\ \mathfrak{v}_x &= \frac{DX}{4\pi\eta}\psi - \frac{DX}{4\pi\eta}\frac{\partial\Pi}{\partial x} + \mathfrak{v}_x^* \\ \mathfrak{v}_y &= \qquad\quad - \frac{DX}{4\pi\eta}\frac{\partial\Pi}{\partial y} + \mathfrak{v}_y^* \\ \mathfrak{v}_z &= \qquad\quad - \frac{DX}{4\pi\eta}\frac{\partial\Pi}{\partial z} + \mathfrak{v}_z^* \\ \Delta\Pi &= \frac{\partial\psi}{\partial x} \end{aligned}\right\} \tag{4}$$

It can be proved that the quantities p^* and $\mathfrak{v}^*$ which are marked with asterisks, must satisfy the equations:

$$\begin{rcases} \eta \operatorname{rot} \operatorname{rot} \mathfrak{v}^* + \operatorname{grad} p^* = 0 \\ \operatorname{div} \mathfrak{v}^* = 0 \end{rcases} \tag{5}$$

involving no force, provided a new potential function Π is determined from the given potential on the basis of the last equation in (4) and the tentative assumptions defined by the other equations in (4) are made for p and $\mathfrak{v}$.

(3) At this time we will disregard the current $\mathfrak{v}^*$ and determine what forces are exerted on the particle by the remaining part of the current. The potential function Π is the solution of the equation:

$$\Delta \Pi = \frac{\partial \psi}{\partial x}$$

Provided the ion atmosphere extends only for a short distance from the surface of the particle into the liquid,

$$\frac{\partial \psi}{\partial x}$$

assumes values of interest only inside a thin layer surrounding the particle. We will now consider a surface element $d\sigma$ of the body and define a rectangular coordinate system s_1, s_2, n such that n points in the direction of its normal (away from the body), while s_1 and s_2 extend tangentially to the surface of the particle. In the limit, we then have:

$$\frac{\partial \psi}{\partial s_1} = \frac{\partial \psi}{\partial s_2} = 0$$

so that ψ is variable only in the direction of the normal n. The condition for the validity of this assumption is, obviously, that the smallest radius of curvature of the particle surface at the point considered may be assumed large compared with the thickness of the ion atmosphere. By assuming that only:

$$\frac{\partial \psi}{\partial n}$$

is to be considered, we assumed the limit of a very small thickness of the ion atmosphere. We further designate the directional cosines of the normal n in the x, y, z coordinate system by α, β, γ; then:

$$\frac{\partial \psi}{\partial x} = \alpha \frac{d\psi}{dn}$$

and Π can be found from the equation:

$$\frac{d^2\Pi}{dn^2} = \alpha \frac{d\psi}{dn}$$

It follows that:

$$\Pi = \alpha\left[\int_0^n \psi\, dn - \int_0^\infty \psi\, dn\right] \tag{6}$$

if the constant of integration is chosen so that Π vanishes at greater distances from the surface.

If we designate by p' and v' the amounts to be added to p^* and v^* in equation (4), it follows from equation (6) that:

$$\left.\begin{aligned} p' &= -\frac{DX}{4\pi}\frac{\partial\psi}{\partial x} \\ v_x' &= \frac{DX}{4\pi\eta}(1-\alpha^2)\psi \\ v_y' &= -\frac{DX}{4\pi\eta}\alpha\beta\psi \\ v_z' &= -\frac{DX}{4\pi\eta}\alpha\gamma\psi \end{aligned}\right\} \tag{7}$$

We obtain for the x component of stress, p_{nx}, exerted on the surface element with normal, n, by the liquid:

$$p_{nx} = \alpha p_{xx} + \beta p_{yx} + \gamma p_{zx} \tag{8}$$

Further:

$$\begin{aligned} p_{xx} &= -p + 2\eta \frac{\partial v_x}{\partial x} \\ p_{yx} &= \eta\left(\frac{\partial v_y}{\partial x} + \frac{\partial v_x}{\partial y}\right) \\ p_{zx} &= \eta\left(\frac{\partial v_z}{\partial x} + \frac{\partial v_x}{\partial z}\right) \end{aligned} \tag{8'}$$

Thus, it follows from equation (7), that:

$$p_{nx} = \frac{DX}{4\pi}\left[\alpha\frac{\partial\psi}{\partial x} + (1-\alpha^2)\left(\alpha\frac{\partial\psi}{\partial x} + \beta\frac{\partial\psi}{\partial y} + \gamma\frac{\partial\psi}{\partial z}\right)\right]$$

Since:

$$\frac{\partial\psi}{\partial x} = \alpha\frac{d\psi}{dn},\quad \frac{\partial\psi}{\partial y} = \beta\frac{d\psi}{dn},\quad \frac{\partial\psi}{\partial z} = \gamma\frac{d\psi}{dn}$$

we obtain the simple expression:

$$p_{nx}' = \frac{DX}{4\pi}\frac{d\psi}{dn} \tag{9}$$

The total force exerted in the x direction by the liquid as a consequence of the current $\mathfrak{v}'$ is derived from equation (9) by integration over the surface of the particle. According to the fundamental principles of electrostatics:

$$\int \frac{d\psi}{dn} d\sigma = -\frac{4\pi e}{D}$$

where e is the charge of the particle. Thus, the liquid current exerts the force $-eX$, while the force $+eX$ is exerted on the charge. The direct force is thus just compensated by the frictional force of the current $\mathfrak{v}'$.

(4) The only force left is the one due to the current $\mathfrak{v}^*$, involving no force and obeying equations (5); the boundary conditions are essential for its determination. These relate to the surface of the particle. There the velocity $\mathfrak{v}^*$ must be of such value as to give zero when added to the velocity $\mathfrak{v}'$. If we designate the values on the surface of the particle by a bar, the conditions:

$$\left.\begin{aligned} \overline{\mathfrak{v}_x^*} &= -\frac{DX}{4\pi\eta}(1-\alpha^2)\overline{\psi} \\ \overline{\mathfrak{v}_y^*} &= \frac{DX}{4\pi\eta}\alpha\beta\overline{\psi} \\ \overline{\mathfrak{v}_z^*} &= \frac{DX}{4\pi\eta}\alpha\gamma\overline{\psi} \end{aligned}\right| \tag{10}$$

must hold according to equation (7).

Further, for great distances, it is required that:

$$\left.\begin{aligned} \mathfrak{v}_x^* &= -V \\ \mathfrak{v}_y^* &= 0 \\ \mathfrak{v}_z^* &= 0 \end{aligned}\right| \tag{11}$$

i.e., that the velocity of the current be directed in the x direction and of predetermined magnitude $-V$.

The velocity $\mathfrak{v}^*$ satisfies the equations (5) which involve no force. It may also be considered as made up of two parts. One current then corresponds to Stoke's law, which meets conditions (11) at infinity and prescribes zero velocity on the surface of the particle. Since the equations are linear, the velocities associated with this case will be everywhere proportional to V. The second current may meet conditions (10) at the surface of the particle and vanish at infinity. The velocities of this second current will be proportional to:

$$\frac{DX}{4\pi\eta}\overline{\psi}$$

The current made up of these two partial currents meets the prescribed conditions (10) and (11).

The total force K transferred by the current $\mathfrak{v}^*$ is derived on the basis of the stress components computed from equation (8') by integration over the surface of the particle so that a result of the form:

$$K = -A_1\eta V + A_2 DX\overline{\psi} \tag{12}$$

is obtained.

The quantities A_1 and A_2 both have the dimension of a length and no general expression can be given for them.

According to equation (12) it is feasible that the total force (12) exerted on the particle vanishes. If this is intended, V must be adjusted so that:

$$V = \frac{A_2}{A_1}\frac{DX}{\eta}\overline{\psi} \tag{13}$$

The ratio A_2/A_1, a pure number, is designated by C. Then for $K = 0$:

$$V = C\frac{DX}{\eta}\overline{\psi} \tag{14}$$

Another possible case - where, different from the preceding case, the liquid is at rest at infinity - is obtained by adding everywhere the velocity constant V, in the positive x direction. Then the particle moves with velocity V in such a manner that the total force exerted thereon vanishes. Since this constitutes the condition for the stationary case, equation (14) gives the cataphoretic velocity with which a particle of potential $\overline{\psi}$ moves in a liquid, having a coefficient of friction η and a dielectric constant D, under the influence of a field strength X. As we stated previously, the conventional general expression for the dependence is confirmed; however, it is established that the numerical factor cannot be given readily and will depend on the shape of the particle.

Bibliography

1. M. Gouy, *J. phys.*, *(4)*, *9*, 457 (1910).
2. D. L. Chapman, *Phil. Mag.*, *25*, 475 (1919).
3. P. Debye and E. Hückel, *Physik. Z.*, *24*, 185, 305 (1923).
4. M. v. Smoluchowski, *Anz. Akad. Wiss.*, *Krakau*, 182 (1903).
5. See for instance Freundlich, *Kapillarchemie*, 2nd ed., p. 331.

THE ELECTRIC FIELD OF IONS AND NEUTRAL SALT ACTION
(Das elektrische Feld der Ionen und die Neutralsalzwirkung)

P. Debye and J. McAulay*

Translated from
Physikalische Zeitschrift, Vol. 26, 1925, pages 22-29.

Strong electrolytes are typified not only by the fact that they show characteristic deviations from classical laws when they are present alone in solution, but they also cause a group of interesting phenomena when nonelectrolytes are added to an electrolytic solution. The cause of the pecularities of strong electrolytes is the large agglomeration of free electric charge resulting from their almost complete dissociation. Therefore, in the following we will seek to understand first qualitatively and finally quantitatively the phenomena mentioned in the title, known by the name of "neutral salt effects," as consequences of the Coulomb fields emanating from the ions.

If we limit ourselves to equilibrium states, then the above neutral salt effects are of two different types. These are differentiated by the nature of the reactions involved. To the first type belong, in general, the phenomena in which the solvent experiences changes in concentration; to the second, those in which the dissolved substance experiences changes in concentration. Osmotic pressure, freezing point depression, etc., belong in the first category; the law of mass action, factors affecting solubility, electromotive force, etc., to the second[1].

A phenomena of the first type is the mutual effect of electrolyte and nonelectrolyte on the lowering of the freezing point (first observed by Tammann[2] and by Abegg[3]). Suppose that (*a*) an experiment shows that the lowering of the freezing point of a solution of concentration γ of a nonelectrolyte is Δ_N . Further, suppose that (*b*) a second experiment shows that a solution of γ' moles per liter of a (strong) electrolyte produces a freezing point lowering Δ_E . Now suppose that (*c*), in a third experiment in which γ moles of the nonelectrolyte and γ' moles of the electrolyte are dissolved together so that again a liter of solution is formed, a freezing point lowering Δ is observed. Then, for the most part, the result is that the freezing point lowering Δ is greater than the sum $\Delta_N + \Delta_E$.

To the second category belongs the well known salting-out effect: the solubility of a nonelectrolyte is generally diminished by the addition of a strong electrolyte.

Both phenomena are closely connected from a thermodynamic standpoint, and it is sufficient to understand an observation of one type in order to be able to deduce from it the other. At the same time, this is the requirement which must be placed on a correct explanation from the beginning. The neutral salt effect can be most easily visualized in

* Received December 10, 1924.

the case of osmotic pressure; we wish to discuss therefore, why, for example, the osmotic pressure of a mixture of salt and sugar is greater than the sum of the osmotic pressures of the same amounts of salt and sugar alone. In this we wish to limit ourselves from the first to dilute solutions, in order to get away from additional considerations of secondary importance which must be taken account of at higher concentrations.

Now we imagine a solution of sugar and salt at a certain concentration obtained with the help of a piston impenetrable for the two types of dissolved molecules but penetrable for water. If the piston is advanced a little farther, the volume is changed by dV, and the work done is $-P\,dV$, where P is the osmotic pressure. Disregarding the charge of the ions, we have

$$P = P_N + P_E$$

where P_N and P_E are the osmotic pressures which the nonelectrolyte and electrolyte, respectively, would exert if they were present alone. However, as a result of its charge, each ion of the electrolyte is surrounded by an electric field which has a certain energy content per unit volume in a medium of dielectric constant D. This energy content density can be calculated in a well known way and is

$$\frac{DE^2}{8\pi} \quad \text{or} \quad \frac{(DE)^2}{8\pi D}$$

if the field strength prevailing in the volume element is E. Since in the process being considered the charge of the ions remains unchanged, so that the dielectric displacement DE, which according to the basic laws of electrostatics is proportional to the charge, also remains constant, then movement of the piston requires an additional amount of work if a decrease of the dielectric constant is produced by the increase in concentration resulting from the advance of the piston. Now it is a well known fact that most nonelectrolytes in dissolving in water lower the dielectric constant in just this way. Therefore, we can try to explain the mutual effect of electrolyte and nonelectrolyte which manifests itself as an increase of osmotic pressure, as a result of the work required by the increase of the electric energy of the ionic field.

If we turn to phenomena of the second type, then, for example, we can sketch the following picture of the salting-out effect in immediate connection with the foregoing. Since the mixing of nonelectrolyte and water is connected with a change of the dielectric constant, then in an electric field variable in space the mixing ratio is not independent of the field strength to which a volume element of the mixture is subjected. Now, if by adding additional ions we increase the electric field strength considerably, then such a separation results that the dielectric constant increases and leads in most cases to a salting out of the nonelectrolyte, since the latter, as already mentioned, usually lowers the dielectric constant. Less accurately this can be expressed as follows: wherever additional electric fields are generated, particularly in the neighborhoods of the ions, the molecules of greater polarizability accumulate at the cost of those with lesser polarizability (just as in our atmosphere, gases with heavy molecules have accumulated near the earth's surface due to the effect of gravity).

As these qualitative and incomplete considerations demonstrate, the purpose of the following explanations show the relationship between the neutral salt effects and the changes of dielectric constant occurring when nonelectrolytes are dissolved. It must be remarked in this connection that cases occur, though rare in practice, in which the dielectric constant changes in the direction opposite to that assumed above. If the expected relation exists, then in these exceptional cases the phenomena should occur with the opposite sign. Thus, for example, the osmotic pressure of the mixture is smaller than the sum of the osmotic pressures of the individual solutions, and the nonelectrolyte is not salted out, but its solubility is increased in the electrolytic solution.

In order to decide whether or not the above suggestions are admissible, we must pursue their consequences quantitatively and see whether or not values of the molecular constants resulting from them really are of the orders of magnitude known already. In the form in which the theory is presented in the following, the size of the ions play an important role, and in fact it is shown that ionic radii of the correct order of magnitude, 10^{-8} cm., are necessary to describe the effects quantitatively. It must be emphasized at this point that characterizing the solution by means of the macroscopic dielectric constant is not much more than a poor substitute for the real, as yet unproduced, molecular theory of the processes involved. However, the great simplification which is brought about by this procedure may excuse the form of the presentation.

In analogy with earlier analyses, the extra work of electrical nature, which causes a change in the classical expression for the free energy of a solution, will be calculated here. Then, according to general rules, the improved expression for the free energy gives the activity coefficients of the individual components of the solution. It will be shown, for example, that in the ordinary case of a decrease of the dielectric constant upon addition of a nonelectrolyte, an increased activity of the nonelectrolyte in the mixture results, which is only another way of expressing the existence of a salting-out effect. Equilibrium with the precipitate exists at a given temperature for a definite fixed value of the activity of the dissolved substance. However, since the activity is equal to the product of the activity coefficient and concentration, an increase in the activity coefficient diminishes the solubility.

Although only the equilibrium case is dealt with in the following, and even this is discussed in detail only with reference to the effect found by Tammann, still a remark may be in order here about the reaction velocity.

As is well known, it was found by Spohr[4] and Arrhenius[5] that, for example, the rate of inversion of sugar, which at small concentrations of acid (which acts catalytically in the inversion) is proportional to the H-ion concentration, increases more rapidly at higher concentrations. Arrhenius found that a similar effect occurs when neutral salts are added at constant acid concentration. According to the current view, it is to be expected that when KNO_3, for example, is added during an inversion by means of nitric acid, then the dissociation of nitric acid is repressed, so that the number of H-ions diminishes and thus the reaction velocity decreases. Actually, just the opposite is the case. Moreover, other salts without a common ion[6] behave entirely similarly. Now our analysis will be carried out under the assumption that the electrolytes concerned are practically completely dissociated. No diminishing of the H-ion concentration is possible. However, we did conclude that the activity of sugar is increased by the addition of an electrolyte. If we accept the conclusion of Brönsted[7] and Bjerrum[8] that the inversion rate is not proportional to the concentration, but to the activity of the sugar, the surprising experimental result is demanded by our theory.

I. Osmotic Pressure and Activity in Dilute Mixtures

In order to find the addition to the free energy required by the charge of the ions, the following method has been used earlier.[9] Remove the charges from the ions of a solution infinitely diluted by the addition of water and replace them on the ions of a solution of a prescribed concentration. The work, W, necessary to carry out this imaginary process is the additional energy we want to calculate. Now if there are $N_1 \ldots N_i \ldots N_s$ ions of types $1 \ldots i \ldots s$ with charges $e_1 \ldots e_i \ldots e_s$ in the solution, then according to the results of the reference cited above, this work W is:

$$W = -\sum N_i \frac{e_i^2}{2D_0 a_i} + \sum N_i \frac{e_i^2}{2D a_i} - \sum N_i e_i^2 \frac{\varkappa}{3D} \qquad (1)$$

Here each ion of any type i is considered to be a sphere of radius a_i, whose charge is removed from it when it is in pure water of dielectric constant D_0 and later returned to it when the solution has been brought up to a prescribed concentration at which the dielectric constant is D.[10] The third term in equation (1) represents the work which is done against the forces which the ions exert on each other. It gives this amount of work only to a first approximation, as is explained in detail in the reference cited above. The quantity κ is the reciprocal thickness of the ionic atmosphere and is defined by:

$$\varkappa^2 = \frac{4\pi}{DkT} \sum n_j e_j^2 \qquad (2)$$

in which $n_1 \ldots n_j \ldots n_s$ are the number of ions of types $1 \ldots j \ldots s$ per cc.

If we content ourselves throughout with a first approximation, it is possible to use a linear interpolation formula for the dielectric constant D of the mixture:

$$D = D_0 [1 - \alpha n - \beta n'] \qquad (3)$$

In this formula, n and n', respectively, refer to the number of molecules of nonelectrolyte and electrolyte per cc., while α and β are two constants which must be determined experimentally. In the expression for D, we have written the constants α and β with negative signs because experience shows that ordinarily an addition lowers the dielectric constant. For nonelectrolytes, this has never been questioned, but for electrolytes, on the other hand, experiments performed up to now are perhaps not sufficient to decide this definitely, although recent experiments appear to support it. However, in the following it is shown that the constant β plays no role in first approximation. For a discussion of what goes on in pure electrolytes, we might refer to a paper of E. Hückel in which the problem in question is explained in detail and the theory is not limited to a first approximation as it is here.

Now if in the second term of eq.(1) $1/D$ is expanded in powers of n and n' and only the first term of this expansion is retained, and if in the third correction term D is replaced by its value D_0 for pure matter, we arrive at the result:

$$\left. \begin{aligned} W = \alpha n \sum N_i \frac{e_i^2}{2D_0 a_i} + \beta n' \sum N_i \frac{e_i^2}{2D_0 a_i} - \\ - \sum N_i \frac{e_i^2 \varkappa_0}{3D_0}, \end{aligned} \right\} (4)$$

with:

$$\varkappa_0^2 = \frac{4\pi}{D_0 kT} \sum n_i e_i^2 \qquad (4')$$

(*a*) First we will use expression (4) for W to find an improved osmotic equation of state. In the reference cited it is shown that the osmotic pressure can be obtained by adding to the classical expression $\bar{P}$ the term $-\partial W/\partial V$, where V, the volume of the solution, is imagined changed by means of a semipermeable piston by an amount dV. The first two terms of eq. (4) depend on V since:

$$n = N/V \qquad \text{and} \qquad n' = N'/V$$

if N and N', respectively, are the number of molecules of the nonelectrolyte and electrolyte present in the entire solution. The third term of eq. (4) depends on V, since $\varkappa_0$ is defined by eq. (4') and:

$$n_i = N_i/V$$

If this is borne in mind, it is easy to see that:

$$\frac{\partial \varkappa_0}{\partial V} = -\frac{\varkappa_0}{2V}$$

and $\partial W/\partial V$ is:

$$\left.\begin{aligned} -\frac{\partial W}{\partial V} &= \alpha n \sum \frac{n_i e_i^2}{2D_0 a_i} + \\ &+ \beta n' \sum \frac{n_i e_i^2}{2D_0 a_i} - \sum \frac{n_i e_i^2 \varkappa_0}{6D_0} \end{aligned}\right\} \quad (5)$$

On the other hand, the classical expression for the osmotic pressure is:

$$\overline{P} = nkT + \Sigma n_i kT \qquad (5')$$

The sum of eqs. (5) and (5') is the true osmotic pressure P to the approximation considered here.

If a molecule of the electrolyte decomposes into $\nu_1 \ldots \nu_i \ldots \nu_s$ ions of types I... $i \ldots s \ldots$, with valencies $z_1 \ldots z_i \ldots z_s$, then $n_i = \nu_i n'$ and $e_i = z_i \varepsilon$, where ε is the elementary charge. After substitution of these relations, which are also used in the foregoing publications, eq. (5) plus (5') becomes:

$$\begin{aligned} P = nkT + \nu n' kT \Big[1 - \varkappa_0 \frac{\varepsilon^2}{6D_0 kT} \frac{\Sigma \nu_i z_i^2}{\nu} + \\ + \beta n' \frac{\varepsilon^2}{2D_0 kT} \frac{1}{\nu} \sum \frac{\nu_i z_i^2}{a_i} \Big] + \\ + \alpha n \nu n' \frac{\varepsilon^2}{2D_0} \frac{1}{\nu} \sum \frac{\nu_i z_i^2}{a_i} \end{aligned} \qquad (6)$$

The quantity ν is an abbreviation for:

$$\nu = \nu_1 + \ldots + \nu_i + \ldots \nu_s$$

and thus is the total number of ions into which one molecule of the electrolyte decomposes. Now the expression (6) for the osmotic pressure can be interpreted as follows. P is made up of 3 parts: the first part is the osmotic pressure P_N which would be generated by the nonelectrolyte if it were present by itself in the solution. The second part is the osmotic pressure P_E which the $\nu n'$ ions of the electrolyte would exert if they were present alone. It must be understood that these two terms in eq. (6) are the beginning of a series expansion of P_N and P_E, respectively. To these terms must be added still a third term, which we will call p, in order that:

$$P = P_N + P_E + p \qquad (7)$$

Hence:

$$p = \alpha n \nu n' \frac{\varepsilon^2}{2D_0} \frac{1}{\nu} \sum \frac{\nu_i z_i^2}{a_i} \tag{8}$$

Again, in agreement with our treatment of the problem, this expression for p is to be thought of only as the first term of a series expansion; therefore, no more can be expected than that expression (8) represents a limiting law for the anomalous change of osmotic pressure.

We want to put the result in a somewhat clearer form by first introducing an average radius a by means of the expression:

$$\sum \frac{\nu_i z_i^2}{a_i} = \frac{\Sigma \nu_i z_i^2}{a} \tag{9}$$

and second, comparing the resulting value of p with the value of the total osmotic pressure $\bar{P}$ following from a classical calculation leading to the well known formula:

$$\bar{P} = nkT + \nu n'kT \tag{10}$$

The result for the ratio $p/\bar{P}$ is:

$$\frac{p}{\bar{P}} = \alpha \frac{n\nu n'}{n+\nu n'} \frac{\Sigma \nu_i z_i^2}{\nu} \frac{\varepsilon^2}{2D_0 a kT} \tag{11}$$

that is, the ratio $p/\bar{P}$ is proportional to the quotient of the product and the sum of the concentrations of the two constituents, proportional to a valence factor $\Sigma\nu_i z_i^2/\nu$, proportional to the ratio of the electric energy $\varepsilon^2 2D_0 a$ of the central ions to the thermal energy kT, and finally proportional to a constant α, which according to equation (3) measures how strongly the dielectric constant of water is altered by addition of the non-electrolyte. According to our theory, whether or not an anomalous increase or decrease of the osmotic pressure occurs depends on the sign of α. If α is positive, that is, if the dielectric constant decreases with addition of the nonelectrolyte, then the anomalous pressure change p is positive in agreement with the qualitative considerations in the introduction.

(*b*) Second, equation (4) for the additional work W will be used to find the effect of electrolytes on the activity of nonelectrolytes when both are present in solution. The general plan of the method is indicated in the analyses of the reference frequently cited above. First, the activity potential h_i is defined there as follows. If the additional work W is written in the form:

$$W = \Sigma N_i w_i$$

then

$$\log h_i = \frac{w_i}{kT}$$

According to equation (4) for W:

$$\left.\begin{aligned} \log h_i &= \alpha n \frac{e_i^2}{2D_0 a_i kT} + \\ &+ \beta n' \frac{e_i^2}{2D_0 a_i kT} - \frac{e_i^2 \varkappa_0}{3D_0 kT} \end{aligned}\right\} \tag{12}$$

so that the activity potential of nonelectrolytes is equal to 1, and only the activity potential of ions is different from 1.

According to the general methods of the reference cited above, the activity coefficient f_i can be found by carrying out the differentiations indicated in the following equation:

$$\log f_i = \log h_i + \Sigma N_j \frac{\partial \log h_j}{\partial N_i} \qquad (13)$$

If the quantities referring to nonelectrolytes are written without indices, then $h = 1$; that is, $\log h = 0$. Further, in the sum in eq. (13) the differentiation is with respect to the number of molecules N of the nonelectrolyte and this occurs in eq. (12) only in the first term as $n = N/V$. If this is borne in mind, then we find immediately for the natural logarithm of the activity coefficient of the nonelectrolyte the equation:

$$\log f = \alpha \sum \frac{n_j e_j^2}{2 D_0 a_j k T}. \qquad (14)$$

which upon introduction of the average radius a defined by eq. (9) and substitution of $z_j \varepsilon$ for e_j reduces to the form*:

$$\log f = \alpha \nu n' \frac{\Sigma \nu_i z_i^2}{\nu} \frac{\varepsilon^2}{2 D_0 a k T} \qquad (14')$$

The activity coefficient is the factor by which the actual concentration must be multiplied in the law of mass action in order to obtain the active concentration or activity. Thus, equation (14) shows that in first approximation if an electrolyte is added the activity of the nonelectrolyte increases proportionally to the number of added ions, provided α is positive, that is, according to eq. (3), provided adding nonelectrolyte to the solvent decreases the dielectric constant. However, since the activity at which the solution is in equilibrium with the precipitate is fixed, this increase of the activity means that even a smaller concentration suffices to sustain solution equilibrium, and this means that the nonelectrolyte is salted out. If its solubility before the electrolyte is added is L_0, and its solubility after the electrolyte is added is L, then

$$L/L_0 = 1/f$$

so that a theoretical limiting law of the neutral salt effect follows from eq. (14).

II. Comparison with Experiments on the Freezing Point Lowering of Mixtures

In order to determine whether the theoretical formulae can predict the salt effect, in the following some observations on the freezing point lowering of mixtures of sugar and different electrolytes will be examined in the light of our theory. The numerical data are from experiments of Rivett.[11] The method of calculation may be explained using for the example data of mixtures of sucrose and KCl. In the first two columns of Table I are recorded the concentrations γ' and γ in moles per liter of the electrolyte and nonelectrolyte of the mixture. As in the introduction, let the freezing point lowering of the mixture be Δ, the corresponding quantity for the electrolyte alone Δ_E, and for nonelectrolyte alone Δ_N. From the observed values of these three quantities, the difference:

$$\delta = \Delta - (\Delta_E + \Delta_N)$$

*In eq. (14') i is used again as summation index in place of j.

was calculated and recorded in the third column of the table. δ is invariably positive, just as Tammann found originally. Further, the freezing point lowering of the mixture following from the classical **van't Hoff** law and denoted by $\bar{\Delta}$ is recorded in the fourth column. The fifth column contains the ratio $\delta/\bar{\Delta}$, which can be considered equal to the ratio $p/\bar{P}$ obtained in the preceding section I. Thus, according to eq. (11), in the limit of small concentrations:

$$\frac{\delta}{\bar{\Delta}} = \frac{p}{\bar{P}} = \alpha \frac{n\nu n'}{n+\nu n'} \frac{\Sigma \nu_i z_i^2}{\nu} \frac{\varepsilon^2}{2D_0 akT} \quad (15)$$

Table I

Cane Sugar and Potassium Chloride

γ' (Salt)	γ (Sugar)	δ	$\bar{\Delta}$	$\delta/\bar{\Delta}$	$\dfrac{\delta/\bar{\Delta}}{2\gamma\gamma'/(\gamma+2\gamma')}$
0.1402	0.3717	0.038	1.213	0.0313	0.196
0.2359	0.6108	0.116	2.014	0.0576	0.216
0.3690	0.8571	0.272	2.967	0.0917	0.231
0.4667	0.7562	0.282	3.144	0.0897	0.215
0.7030	0.5114	0.269	3.566	0.0756	0.201
0.9414	0.2639	0.177	3.993	0.0443	0.191

In order to have a direct comparison with experiment, we substitute instead of the actual number n of molecules per cc. the number of moles per liter, γ. These are related as follows:

$$n = \frac{N}{1000}\gamma$$

denoting by N Loschmidt's number. Further, according to eq. (3), for a pure sucrose solution, for example, α is defined by:

$$D = D_0[1 - \alpha n]$$

and if the concentration γ in moles per liter is also substituted in this equation, it assumes the form:

$$D = D_0[1 - A\gamma] \quad (16)$$

where:

$$A = \frac{N}{1000}\alpha$$

Thus, finally, taking all of this into account:

$$\frac{\delta}{\bar{\Delta}} = A \frac{\gamma\nu\gamma'}{\gamma+\nu\gamma'} \frac{\Sigma \nu_i z_i^2}{\nu} \frac{\varepsilon^2}{2D_0 akT} \quad (17)$$

Because of the form of this equation, the ratio of δ/Δ to $\gamma\nu\gamma'/(\gamma + \nu\gamma')$ (in which ν is equal to 2, since the molecule KCl decomposes into $\nu = 2$ ions), is recorded in the sixth column of the table.

Right now, our problem is to deduce from the numbers in the sixth column the limit to which they tend in the limit $\gamma = \gamma' = 0$. However, the values vary relatively slightly and there is a certain arbitrariness in drawing a curve through them when they are plotted as a function of the concentration or the square root of the concentration of the salt. A value of 0.22 for this limit cannot be far wrong in the case of KCl. Now, if the numerical values $\varepsilon = 4.77 \times 10^{-10}$, $D_0 = 88.2$, $k = 1.37 \times 10^{-16}$, and $T = 273$ are substituted in eq. (17), then:

$$\frac{\delta/\bar{\Delta}}{\gamma\nu\gamma'/(\gamma+\nu\gamma')} = A\,\frac{\Sigma\nu_i z_i^2}{\nu}\,\frac{3.45\cdot 10^{-8}}{a} \qquad (18)$$

In the present case of KCl, the right hand side is 0.22 according to the data of Rivett just discussed. Further, $\nu = 2$ and $\Sigma\nu_i z_i^2/\nu = 1$. Moreover, there exist observations on the dielectric constants of sugar solutions by Harrington.[12] His results can be represented well by the empirical formula:

$$D = D_0[1 - 0.079\,\gamma]$$

Therefore, according to eq. (16), A is equal to 0.079. Finally, the average radius of the K and Cl ions can be determined from eq. (18) and follows from:

$$0.22 = 0.079\,\frac{3.45 \times 10^{-8}}{a}$$

which yields:*

$$a = 1.23 \times 10^{-8}\ \text{cm.}$$

This radius is actually of the expected magnitude. It seems somewhat too large; perhaps this is connected with the assumption (in no case rigorously correct) that the medium is homogeneous, with a fixed dielectric constant, even when ions are present.

Just as in the case in which KCl is the electrolytic addition, we have examined the other observations of Rivett on the effect of K_2SO_4, $Mg(NO_3)_2$, and $CuSO_4$ on a sucrose solution. Table II contains a collection of the limiting values of the ratio $\delta/\bar{\Delta}$ to $\nu\gamma\gamma'/(\gamma + \nu\gamma')$ deduced from the experimental results and the resulting average ionic radii for the four different salts.

* The average radius a is defined by equation (9), which in the special case of KCl is:

$$\frac{1}{a} = \frac{1}{2}\left(\frac{1}{a_k} + \frac{1}{a_{Cl}}\right)$$

Table II

Solution	$\delta/\bar{\Delta}$ $\nu\gamma\gamma'/(\gamma + \nu\gamma')$	a
Sugar + KCl	0.22	1.2×10^{-8}
+ K_2SO_4	0.22	2.5×10^{-8}
+ $Mg(NO_3)_2$	0.30	1.8×10^{-8}
+ $CuSO_4$	0.20	5.0×10^{-8}

It must be pointed out that the observations do not permit a satisfactory extrapolation for the ratio in the second column. There is no systematic variation with sugar concentration for a given electrolytic concentration or vice versa, just as there was none in Table I. Consequently, the numbers recorded are surely in considerable error. In spite of this, it seems evident that the agreement of the a values with their expected magnitude is an indication that an essentially physical theory of the salt effects, like that developed here, is probably correct in its main features. The following observations indicate the same thing. Harrington remarks in the same paper from which we took the formulae for the dielectric constant of the sugar solutions that he has found a substance, namely, urea, which when added to a sugar solution causes an increase in the dielectric constant. He gives numbers which agree roughly at small concentrations with the formula:

$$D = D_0 [1 + 0.038 \gamma]$$

According to the above analyses, it must be expected that consequently a mixture of urea and a strong electrolyte would show a smaller lowering of the freezing point than the sum of the individual lowerings, in contrast to Rivett's solutions. McAulay has carried out a few preliminary measurements which confirm this conjecture.

If, as in the discussion of the freezing point lowering, the mole concentration is substituted in equation (14) for the logarithm of the activity, we arrive at:

$$\log f = A \nu \gamma' \frac{\Sigma \nu_i z_i^2}{\nu} \frac{3.45 \cdot 10^{-8}}{a} \qquad (19)$$

in which the logarithm is the natural logarithm. The increase of the activity of the sugar subsequent to addition of 0.4 mole KCl per liter can be found by inserting the numerical data found above for the effect of KCl into eq. (19); the result of this is:*

$$\log f = 0.44\ \gamma' = 0.176$$

This means that for an 0.4 N solution of KCl:

$$f = 1.19$$

so that the activity of the sugar is increased by about 19%. Arrhenius reports that this solution occasions a 15% increase in the inversion velocity. Accordingly, Bronsted's hypotheses that the inversion rate is not proportional to the concentration but to the activity of the sugar seems substantially confirmed. It must not be forgotten, however,

* Observe that the factor of $\nu\gamma'$ in (19) is the ratio given in eq. (18).

in considering this agreement that it establishes nothing definitely either for or against the particular theory given here. It is always possible to calculate the activity from the freezing point measurements.[13] Only when we ask why the activity increases by 19% in the case under discussion and relate this increase to the size of the ionic sphere, a size which determines the heat of hydration, do we encounter the special nature of the present theory.

Bibliography

1. A thorough, now somewhat outdated, summary is given in the monograph of V. Rothmund: "Löslichkeit und Löslichkeitsbeeinflussung," in *Handbuch der angewandten physikalischen Chemie*, Vol. 7, G. Bredig, Leipzig, 1907.
2. *Z. phys. Chem.*, *9*, 108 (1892).
3. *Z. phys. Chem.*, *11*, 259 (1893).
4. F. Spohr, *Z. phys. Chem.*, *2*, 194 (1888).
5. S. Arrhenius, *Z. phys. Chem.*, *1*, 110 (1887); *4*, 240 (1889).
6. A complete explanation of the conditions can be found, for example, in the work of S. Arrhenius *Conférences sur quelques problèmes actuels de la chimie physique et cosmique*, Paris, 1923, p. 52 ff.
7. J. N. Brönsted, *Z. phys. Chem.*, *102*, 169 (1922).
8. N. Bjerrum, *Z. phys. Chem.*, *108*, 82 (1924).
9. P. Debye, *Physik. Z.*, *25*, 97 (1924).
10. This same picture of spherical ions has already been used by M. Born, *Z. Physik*, *1*, 45 (1920), in determining the heat of hydration as a function of the size of the ions.
11. A. C. D. Rivett, *Med. Vetenskapsakad. Nobelinst.*, *2*, No. 9.
12. E. A. Harrington, *Phys. Rev.*, *8*, 581 (1916).
13. See the earlier assumptions of H. Jahn, *Z. phys. Chem.*, *37*, 490 (1901); *38*, 125 (1901); *41*, 257 (1902); *50*, 129 (1905), and W. Nernst, *ibid*, *38*, 484 (1901).

THE ELECTRIC FIELD OF IONS AND NEUTRAL SALTING OUT
(Das elektrische Ionenfeld und das Aussalzen)

P. Debye *

Translated from
Zeitschrift fur physikalische Chemie, Cohen Festband, 1927, pages 56-64

(1) Quite some time ago, I tried to understand the effect of ions on the solubility of nonelectrolytes as a separation brought about by the strong electric fields surrounding the ions.[1] In fact, it is easy to see that a mixture of two dielectric liquids in an inhomogeneous electric field cannot have a mixing ratio which is constant throughout its volume. Rather, at equilibrium the mixing ratio varies from point to point in such a way that the dielectric constant is larger at places of high field intensity. In the earlier publication, this effect was calculated in first approximation. However, it is possible to sketch a clear picture of what goes on in the neighborhood of an ion if we start with the assumption that the electrical forces are the only important ones. This will be assumed here, the result being not only a better description of the phenomenon but also an improvement of the formulae. To avoid misunderstanding, it should be remarked here that this assumption does not imply that other forces or chemical combinations in the ordinary sense do not exist. However, it is implied that the effect described here does constitute an important part of the salting-out effect and that individual effects of the ions can be important only in the remainder.

(2) Let us assume that an ion is a sphere of radius a and charge e. Now imagine it surrounded by a mixture of N_1 molecules of a substance 1 and N_2 molecules of a substance 2. If n_1 and n_2 are the numbers of molecules per unit volume and $c_1 = n_1/(n_1 + n_2)$ and $c_2 = n_2/(n_1+n_2)$ the concentrations, then for an ideal mixture:

$$n_1 v_1 + n_2 v_2 = 1 \tag{1}$$

* Received May 23, 1927.

in which volumetric constants v_1 and v_2 have been introduced for the two types of molecules. The free energy of a volume element dS is:*

$$[n_1(\varphi_1 + kT\log c_1) + n_2(\varphi_2 + kT\log c_2)]\,dS \tag{2}$$

This expression does not include the energy of the electric field surrounding the ion, which for a volume element dS a distance r from the ion is:

$$\frac{e^2}{8\pi r^4}\frac{1}{\varepsilon}dS \tag{2'}$$

where ε is the dielectric constant of the mixture. This expression is the square of the dielectric displacement (which at a distance r from the ion is e/r^2) divided by the dielectric constant ε and multiplied by $1/8\,\pi$ (since we express all quantities in electrostatic units), and finally multiplied by the volume element dS.

In order to find the free energy of dS including the electrical energy, we must add (2) and (2').

Thus the free energy of the whole system becomes:

$$\Phi = \int dS\left[n_1(\varphi_1 + kT\log c_1) + n_2(\varphi_2 + kT\log c_2) + \frac{e^2}{8\pi r^4}\frac{1}{\varepsilon}\right] \tag{3}$$

and obviously depends on the spatial distribution of components 1 and 2 around the ion. It is well known that the equilibrium distribution is that for which Φ is a minimum subject to the subsidiary conditions that in the variation the numbers of molecules

$$N_1 = \int n_1\,dS \quad \text{und} \quad N_2 = \int n_2\,dS \tag{4}$$

must remain constant, and in addition that at each point n_1 and n_2 must be chosen in such a way that eq. (1) is satisfied. If n_1 is varied by δn_1, and n_2 by δn_2, then, since the dielectric constant ε depends on the mixing ratio

$$\delta\Phi = \int dS\left[\delta n_1\left(\varphi_1 + kT\log c_1 - \frac{e^2}{8\pi r^4}\frac{1}{\varepsilon^2}\frac{\partial\varepsilon}{\partial n_1}\right) + \delta n_2\left(\varphi_2 + kT\log c_2 - \frac{e^2}{8\pi r^4}\frac{1}{\varepsilon^2}\frac{\partial\varepsilon}{\partial n_2}\right)\right]$$

However, eq. (1) requires that $v_1\delta n_1 + v_2\delta n_2 = 0$. In order to satisfy this equation we set:

$$\delta n_1 = v_2\,\delta f$$

and

$$\delta n_2 = -\,v_1\,\delta f \tag{5}$$

in which δf is an arbitrary variation. As a result:

* When the mixture is not ideal, the above is still valid if activities instead of concentrations are inserted, and v_1 and v_2 are no longer considered constant.

$$\delta\Phi = \int dS\delta f\left[v_2\left(\varphi_1 + kT\log c_1 - \frac{e^2}{8\pi r^4}\frac{1}{\varepsilon^2}\frac{\partial\varepsilon}{\partial n_1}\right)\right. \\ \left. - v_1\left(\varphi_2 + kT\log c_2 - \frac{e^2}{8\pi r^4}\frac{1}{\varepsilon^2}\frac{\partial\varepsilon}{\partial n_2}\right)\right] \tag{6}$$

Because of (5), the variations of the subsidiary conditions (4) both reduce to a single equation:

$$0 = \int dS\delta f \tag{6'}$$

Since at equilibrium $\delta\Phi = 0$, the solution of the variational problem can be found in the usual way by multiplying eq. (6') by an arbitrary factor, adding it to eq. (6), and setting the factor of δf under the integral equal to zero. Thus, the mixing ratio must be such that for each distance r from the ion:

$$v_2\left(\varphi_1 + kT\log c_1 - \frac{e^2}{8\pi r^4}\frac{1}{\varepsilon^2}\frac{\partial\varepsilon}{\partial n_1}\right) \\ - v_1\left(\varphi_2 + kT\log c_2 - \frac{e^2}{8\pi r^4}\frac{1}{\varepsilon^2}\frac{\partial\varepsilon}{\partial n_2}\right) = \text{const.} \tag{7}$$

Naturally, at considerable distances from the ion, the concentrations are unaffected by the electric field; we will call their values there c_1^0 and c_2^0. Consequently, the constant in eq. (7) has the value

$$\text{const.} = v_2(\varphi_1 + kT\log c_1^0) - v_1(\varphi_2 + kT\log c_2^0) \tag{7'}$$

If this is inserted in eq. (7), the final result for the distribution equation becomes:

$$v_2\log\frac{c_1}{c_1^0} - v_1\log\frac{c_2}{c_2^0} = \frac{e^2}{8\pi kT}\frac{1}{\varepsilon^2}\left(v_2\frac{\partial\varepsilon}{\partial n_1} - v_1\frac{\partial\varepsilon}{\partial n_2}\right)\frac{1}{r^4} \tag{8}$$

When the dielectric constant ε is given as a function of the mixing ratio, eq. (8) gives us a means of calculating the concentrations of the two components at any distance r from the ion.* The numerical details of such a calculation illustrate in an interesting manner how, for example, traces of water accumulate in the neighborhood of an ion.

(3) In general, calculations must be based on eq. (8) and the total effect of an ion calculated from the resulting curves (which, for example, represent c_1 as a function of r) by integrating over the whole volume

* It should be emphasized that ε is treated in a way which implies the assumption that the phenomenon of electrical saturation in the vicinity of the ion can be neglected.

surrounding the ion. If one of the components, for example 2, is present only in small quantities, it is useful to replace eq. (8) by an approximate solution. For convenience, we set the small concentration $c_2 = c$ for any distance r and in particular c^0 for large distance, so that $c_1 = 1 - c$ and $c_1^0 = 1 - c^0$. Then, the right-hand side of eq. (8) becomes:

$$v_2 \log \frac{1-c}{1-c^0} - v_1 \log \frac{c}{c^0} = v_2 (c^0 - c) - v_1 \log \frac{c}{c^0}$$

and if substance 2 is salted out, we will assume not only that $c/c^0 < 1$ but also that c^0 is small. Consequently, the first term merely plays the part of a correction term and the right-hand side is approximately $-v_1 \log c/c_0$. Now eq. (8) reduces to:

$$\log \frac{c}{c^0} = -\frac{\frac{e^2}{v_1}}{8\pi kT} \frac{1}{\varepsilon^2} \left(v_2 \frac{\partial \varepsilon}{\partial n_1} - v_1 \frac{\partial \varepsilon}{\partial n_2} \right) \frac{1}{r^4}$$

Obviously, the factor $1/r^4$ on the right-hand side has the dimensions of the fourth power of a characteristic length, and so we define such a length R by:

$$R^4 = \frac{\frac{e^2}{v_1}}{8\pi kT} \frac{1}{\varepsilon^2} \left(v_2 \frac{\partial \varepsilon}{\partial n_1} - v_1 \frac{\partial \varepsilon}{\partial n_2} \right) \tag{9}$$

In this way we obtain for the concentration of component 2 at any distance r the formula:

$$c = c^0 e^{-\frac{R^4}{r^4}} \tag{10}$$

which shows how the salting out effect decreases with increasing r. Now the problem is to calculate the characteristic length R. With the use of some obvious approximations, equation (9), from which this length must be calculated, can be simplified. We will assume that at the small concentrations with which we are dealing, the dielectric constant is a linear function of the concentrations, and put:

$$\varepsilon = \varepsilon_1 c_1 + \varepsilon_2 c_2 \tag{11}$$

Then we have exactly:

$$v_2 \frac{\partial \varepsilon}{\partial n_1} - v_1 \frac{\partial \varepsilon}{\partial n_2} = (\varepsilon_1 - \varepsilon_2) \frac{v_1 n_1 + v_2 n_2}{(n_1 + n_2)^2} = \frac{\varepsilon_1 - \varepsilon_2}{(n_1 + n_2)^2}$$

Referring to eq. (1), we see that for small concentrations of the second component, $1/(n_1 + n_2) = 1/n_1 = v_1$; moreover, according to eq. (11), the dielectric constant ε_1 of the first component can be substituted for ε, so that

$$R^4 = \frac{e^2 v_1}{8\pi kT} \frac{\varepsilon_1 - \varepsilon_2}{\varepsilon_1^2} \tag{12}$$

or multiplying numerator and denominator by the Loschmidt number

$$R^4 = \frac{e^2 V_1}{8\pi BT} \frac{\varepsilon_1 - \varepsilon_2}{\varepsilon_1^2} \tag{12'}$$

in which the volume of one mole of the first component and the universal gas constant (8.31×10^7 ergs) are designated by V_1 and B respectively.

(4) If all of the above simplifications are accepted, we have first to determine the dielectric constants of mixtures of 1 and 2 and to verify that the linear expression (11) is correct. Since

$$c_1 = 1 - c_2 = 1 - c$$

eq. (11) has the form

$$\varepsilon = \varepsilon_1 - (\varepsilon_1 - \varepsilon_2)c$$

If the constant $\varepsilon_1 - \varepsilon_2$ is positive, that is, if the dielectric constant decreases when the second component is added, R^4 is positive and equation (10) shows that the concentration of this component is less in the vicinity of an ion. In other words, the second component is salted out. It remains only to calculate the exact total amount which is separated from the solution by an ion's electrical effect. Clearly, this can be done as follows:

In general the numbers of molecules present can be expressed in terms of the concentrations as follows:

$$n_1 = \frac{c_1}{v_1 c_1 + v_2 c_2}, \qquad n_2 = \frac{c_2}{v_1 c_1 + v_2 c_2}$$

Thus, for small concentrations of the second component:

$$n_2 = \frac{c_2}{v_1} = \frac{c}{v_1} \tag{13}$$

On the other hand, eq. (10) gives the concentration c in terms of the distance. According to it, when we have a saturated solution of component 2 in component 1 in which there are p ions per unit volume, the number of molecules of component 2 surrounding an ion is on the average:

$$\int n_2 dS = \frac{c^0}{v_1}\int e^{-\frac{R^4}{r^4}} dS = \frac{c^0}{v_1}\left[\int dS - \int\left(1 - e^{-\frac{R^4}{r^4}}\right) dS\right] \tag{14}$$

in which the integration extends over the average volume available to an ion.

If we make the obvious assumption that for large distances from the ion the concentration has the value it would have were there no ions present, then c^0 is the concentration at every point of a mixture to which no ions have been added. Before the salt is added

$$\frac{c^0}{v_1}\int dS \tag{14'}$$

molecules of 2 are present in the volume available to an ion after the salt is added. Accordingly, if Z_2 is the number of molecules of 2 in the solution before and Z_2' the corresponding number after ions are added, the ratio Z_2'/Z_2 can be obtained by dividing eq. (14) by (14'), so that:

$$\frac{Z_2'}{Z_2} = 1 - \frac{\int\left(1 - e^{-\frac{R^4}{r^4}}\right)dS}{\int dS} \tag{15}$$

Since the integration is to be done over only the volume available to an ion, $\int dS = 1/p$, if p is the number of ions per c.c. The integration of the numerator of equation (15) can be extended to ∞, since the integrand disappears sufficiently rapidly at large distances. Consequently, if we define a "salt factor" by:

$$\sigma = \int_{r=a}^{r=\infty}\left(1 - e^{-\frac{R^4}{r^4}}\right)4\pi r^2 dr \tag{16}$$

then a measure of the effect can be obtained from the relation:

$$Z_2' = Z_2[1 - \sigma p] \tag{17}$$

Since in the above we have calculated the effect of added ions obviously only for small salt concentration, the total effect will be proportional to the number of ions p per c.c. Moreover, I have convinced myself that this same result can be obtained by inserting eq. (10) in the expression for the free energy and in this way calculating the increase of the activity of the second component brought about by the addition of ions. This calculation will be omitted here for the sake of brevity.

(5) The integral (16) which defines the salt factor cannot be expressed in finite form in terms of simple functions. However, if we set:

$$\sigma = \frac{4\pi}{3}R^3 s \tag{18}$$

then s is:

$$s = s\left(\frac{R}{a}\right) = 3\int_{\varrho=\frac{R}{a}}^{\varrho=\infty}\left(1 - e^{-\frac{1}{\varrho^4}}\right)\varrho^2 d\varrho \tag{19}$$

that is, s is a function of the ratio R/a. This function can be expanded in the following series:

$$\left.\begin{aligned} s &= 3{\cdot}626 - \left(\frac{a}{R}\right)^3 + \left(\frac{a}{R}\right)^3 e^{-\left(\frac{R}{a}\right)^4}\left[\frac{3}{4}\left(\frac{a}{R}\right)^4 - \frac{3}{4}\frac{7}{4}\left(\frac{a}{R}\right)^8 + \cdots\right] \\ s &= 3\frac{R}{a}\left[1 - \frac{1}{2!}\frac{1}{5}\left(\frac{R}{a}\right)^4 + \frac{1}{3!}\frac{1}{9}\left(\frac{R}{a}\right)^8 - \frac{1}{4!}\frac{1}{13}\left(\frac{R}{a}\right)^{12} + \cdots\right] \end{aligned}\right\} \tag{20}$$

of which the first is suitable for "large" values of R/a and the second for "small" values of R/a. Numerical values of the function s are collected in Table I following. Earlier analyses (*loc. cit.*) were based on the approximation obtained by setting $s = 3R/a$ in the second equation of (20). Here it is shown that for large values of the ratio, s approaches a limiting value, namely 3.63. This could have been anticipated, since an infinitesimally small ion (a = 0) can produce only a finite effect.

TABLE I

R/a	s
0	0
0.5	1.49
1.0	2.75
1.5	3.33
2.00	3.50
∞	3.63

In order to see what part of the observed effect is described by this effect, I requested Dr. J. W. Williams of the University at Madison to measure the dielectric constant of a saturated solution of ethyl ether in water. He found at a temperature of 23°C for water ε = 80.3 and for the saturated ether solution ε = 74.5. If the linear formula (11) is correct,* then ε_1 = 80.3 and ε_2 = -295, since according to measurements by Y. Osaka the saturated solution (at 23°C) contains 6.07 grams ethyl ether and 93.93 grams of water, so that c_1 = 1 - 0.0154 and c_2 = 0.0154. Adding 1 mol of KCl to 1 liter of mixture introduces $p = 12.1 \times 10^{20}$ ions per c.c. (neglecting the volume change which occurs). For the limiting case of small ions, the salt factor σ is:

$$\sigma = 3{\cdot}63 \frac{4\pi}{3} R^3$$

First, we must calculate the characteristic radius R from (12'). If in this equation are substituted $e = 4.77 \times 10^{-10}$, V_1 = 18, $B = 8.31 \times 10^7$, t = 296, and from Williams' measurements $\varepsilon_1 - \varepsilon_2$ = 375 and ε_1 = 80.3, there results $R = 2.49 \times 10^{-8}$ cm. Consequently, $p\sigma$ = 0.285. Were all our assumptions valid for such concentrated solutions, we would expect that according to eq. (17), in the case considered 28% of the ethyl ether is salted out (by sufficiently small ions). According to Euler, 33%, 42%, and 40% of ethyl ether are salted out, respectively, by 1 mol per liter of LiCl, NaCl, and KCl.

Obviously, only a rough comparison is possible. Nevertheless, as was remarked at the beginning, the theoretical value is near enough to the experimental value, in spite of the many neglected factors and the extrapolation to large concentrations, to convey the impression that the electrical effect actually plays an important role in salting out. In the comparison of theory with experiment it should be borne in mind that the formulae used in the numerical computation contain no essential experimental constants besides the dielectric constants of pure water and of the satured ether solution. It is clear that only exact

* Earlier measurements of Sack carried out at the Zürich Institute of Physics confirm the linear formula.

measurements on the salting out effect, like those, for example, by Linderström-Lang, in combination with good measurements (at this time still practically nonexistent) of the dielectric constant of suitable mixtures, can finally decide the value of our suggestions.

Bibliography

1. P. Debye and J. McAulay, *Phys. Z.*, *26*, 22 (1925).

FREQUENCY DEPENDENCE OF THE CONDUCTIVITY AND DIELECTRIC CONSTANT OF A STRONG ELECTROLYTE *

(Dispersion von Leitfähigkeit und Dielektrizitätskonstante bei starken Elektrolyten)**

P. Debye and H. Falkenhagen

Translated from
Physikalische Zeitschrift, Vol. 29, 1928, pages 121-132

As will be shown in this paper, there exists a relaxation time for the motion of ions in strong electrolytes which is an important factor among those determining the frequency dependence of the conductivity and the dielectric constant. Here a treatment will be presented which, although not completely rigorous, will bring out the characteristics of the phenomena to be expected. First some fundamental points of the theory of electrolytes will be recalled.[1]

The most important forces between ions are Coulomb forces, particularly in dilute solutions. If we single out an ion of a certain kind, its heat motion - at first we imagine that there are no external forces on the ions - takes some complicated form. Imagine a radius vector r drawn from the ion, and let this radius vector have always the same length and direction. Imagine attached to the end of r a small volume element dS in which excess charges will be present from time to time. At different times the amount of charge in dS is different. At a certain time it will be positive; at another, negative. In the following, the time average of the electric charge in dS divided by dS, that is, the time average of the charge density in the volume element considered, will be referred to simply as the density of the ionic atmosphere.

In consequence of the Maxwell-Boltzmann principle, which is certainly valid in the present case of static equilibrium, the density of the ionic atmosphere of a positive ion, for example, is everywhere negative and, in addition, spherically symmetrical. Naturally, the total charge of this ionic atmosphere must be opposite to that of the central ion. This means that the potential around an ion, which is a Coulomb potential for infinite dilution, deviates from a Coulomb potential at finite concentrations. The thermodynamics of strong electrolytes is based on the energy directly associated with the existence of this

* Presented at the January 7 meeting of the Thüringen-Saxony-Silesia Section of the German Physical Society in Halle.

**Received January 30, 1928

ionic atmosphere. Going a step further and proceeding along the same lines, it has also been possible to derive laws for the conductivity of strong electrolytes in the limit of large dilution. If an ion moves under the influence of an electric field, its mobility depends on how many ions are present in its vicinity. First, the Stokes friction is changed, a phenomenon referred to as cataphoretic action. Second, the ionic atmosphere loses its spherical symmetry. Both effects diminish the mobility.

In this note no effort will be made to take the cataphoretic action into account; we will investigate only how the dissymmetry of the ionic cloud changes with frequency. Because of this our results can be only qualitative. In a later paper we will make the necessary improvements.

If an ion moves through a liquid, a braking force appears because of the dissymmetry mentioned. This force is present because the density distribution of the ionic cloud requires a small but finite time for its creation or annihilation. Thus it is to be anticipated that a noticeable frequency dependence occurs when the period of the oscillations used in the measurements is comparable with the relaxation time. Onsager has shown that the Brownian motion which the ions execute around their straight-line paths affects the magnitude of the relaxation time. His improvement of the formula for the conductivity is based on this remark. We neglect also this refinement in this paper, but we will take it into account in the future paper.

Suppose that on the (time) average, $n_1 dS \ldots n_p dS \ldots n_s dS$ ions of the kinds $1 \ldots p \ldots s$ whose charges are $e_1 \ldots e_p \ldots e_s$ are found in the volume element dS. Further let D be the dielectric constant of the solute, T the absolute temperature of the solution, k Boltzmann's constant (1.372×10^{-16} erg/degree), and κ^2 the characteristic quantity for the theory of strong electrolytes:

$$\kappa^2 = \frac{4\pi}{DkT} \sum \bar{n}_p e_p^2 \tag{1}$$

Further let v_p be the constant velocity of ions of the kind p and $\rho_p v_p$ Stokes' frictional force. If the abbreviations

$$\omega_p = \bar{\varrho} \frac{v_p}{kT}$$
$$\bar{\varrho} = \frac{\sum \bar{n}_p e_p^2 \varrho_p}{\sum \bar{n}_p e_p^2} \tag{2}$$

are introduced, this additional force is:

$$-e_p^2 \frac{\omega_p}{6D} \kappa \tag{3}$$

In general it is small compared with Stokes' force. The quantity $1/\kappa$ has the dimension of a length which characterizes the charge distribution of the ionic cloud. In the following it will be frequently referred to as the thickness of the ionic cloud (its order of magnitude in cm.: 10^{-8} divided by the square root of the concentration in mol/liter).

I. The Role of the Relaxation Time in Nonstationary Cases (Example of a Vanishing Equilibrium State)

First we will turn to some simple questions which one of us has already briefly mentioned previously.[2] We imagine a charge E of small dimensions at a certain position in the electrolyte. Let the symmetrical ionic atmosphere, which corresponds to the distribution of the potential around the static ion, have achieved its equilibrium state, which is determined by the Maxwell-Boltzmann principle. Imagine the central charge E removed at a certain moment $(t = 0)$. We ask how the ionic cloud will disappear under these circumstances. Now the Maxwell-Boltzmann distribution function is no longer useful. Rather, we must return to the basic equation developed in the paper mentioned previously,[1] which is based on the general equation for the Brownian motion. The variation in time of the number of ions of the kind p in some element of volume has two causes: first, it changes because of heat motion, and second, it changes because a force $\mathfrak{K}$ is acting on the particles. The mathematical expression which describes this nonstationary process is:

$$\frac{\partial n_p}{\partial t} = \operatorname{div}\left(\frac{kT}{\varrho_p}\operatorname{grad} n_p - \frac{n_p \mathfrak{K}}{\varrho_p}\right) \qquad (4)$$

We will now assume that we are dealing with a very dilute solution. Then in the expression containing $\mathfrak{K}$ we can substitute for n_p the value $\overline{n}_p$, which it takes on at large distances from the central ion. With this simplification eq. (4) reduces to:

$$\frac{\partial n_p}{\partial t} = \operatorname{div}\left(\frac{kT}{\varrho_p}\operatorname{grad} n_p - \frac{n_p \mathfrak{K}}{\varrho_p}\right) \qquad (5)$$

In the present case, there are no external forces on the ions. Therefore we will assume only an internal force derivable from a potential Φ:

$$\mathfrak{K} = -e_p \operatorname{grad} \Phi$$

Further we put:

$$n_p = \overline{n}_p + \nu_p \qquad (6)$$

and thus obtain for the deviation ν_p of the number of ions of kind p from the constant value $\overline{n}_p$ the following differential equation:

$$\frac{\partial \nu_p}{\partial t} = \operatorname{div}\left(\frac{kT}{\varrho_p}\operatorname{grad} \nu_p + \overline{n}_p \frac{e_p}{\varrho_p}\operatorname{grad} \Phi\right) \qquad (7)$$

To this must be added Poisson's equation:

$$\operatorname{div}\operatorname{grad} \Phi = -\frac{4\pi}{D}\sum e_p \nu_p \qquad (8)$$

In determining how the equilibrium distribution vanishes, we will simplify the problem by singling out the case of uni-univalent KCl, the ions of which possess equal mobilities. Consequently, in this special case:

$$\bar{n}_1 = \bar{n}_2 = n; \quad e_1 = E; \quad e_2 = -E$$
$$\nu_1 = f; \quad \nu_2 = -f; \quad \varrho_1 = \varrho_2 = \varrho \tag{9}$$

For this example, the basic differential equations are:

$$\frac{\varrho}{kT}\frac{\partial f}{\partial t} = \Delta f + \frac{nE}{kT}\Delta \Phi \tag{10}$$

$$\Delta \Phi = -\frac{8\pi E}{D} f \tag{11}$$

If we introduce the characteristic quantity (1), which for this example is:

$$\varkappa^2 = \frac{8\pi n E^2}{DkT}$$

then, substituting $\Delta \Phi$ from eq. (11), we find for the equation which determines the time variation of f:

$$\frac{\varrho}{kT}\frac{\partial f}{\partial t} = \Delta f - \varkappa^2 f \tag{12}$$

The form of equation (12) is similar to that of the equation which describes the process of heat conduction. Next we will write this equation in a dimensionless form by introducing a time Θ which is important for all nonstationary processes and which we will call the relaxation time. Thus we define:

$$\frac{\varrho}{\varkappa^2 kT} = \Theta \tag{13}$$

$$\varkappa r = s; \frac{t}{\Theta} = \tau \tag{14}$$

With this introduction of dimensionless quantities, we obtain the following differential equation for f:

$$\frac{1}{s^2}\frac{\partial}{\partial s} s^2 \frac{\partial f}{\partial s} - f = \frac{\partial f}{\partial \tau} \tag{15}$$

At time $t = 0$, $f = f_0$ is known; it is:

$$f_0 = -\frac{D}{8\pi E}\varkappa^2 \Phi_0 = -\frac{\varkappa^3}{8\pi}\frac{e^{-s}}{s} \tag{16}$$

An expression for f which satisfies all the conditions is the following:

$$f = -\frac{\varkappa^3}{8\pi}\frac{e^{-s}}{s}\frac{1}{\sqrt{\pi}}\int_{\sqrt{\tau}-\frac{s}{2\sqrt{\tau}}}^{\infty} e^{-x^2} dx \tag{17}$$

Now this expression for f must be discussed. Notice that since:

$$\int_{-\infty}^{+\infty} e^{-x^2}\,dx = \sqrt{\pi} \qquad (18)$$

f reduces to f_0 for $t = 0$. For any other time t the integral:

$$\frac{1}{\sqrt{\pi}} \cdot \int_{\sqrt{\tau}-\frac{s}{2\sqrt{\tau}}}^{\infty} e^{-x^2}\,dx = J(\tau, s) \qquad (19)$$

can be evaluated easily. If J is known, f is known for different times t and for different values of s. The following Table I gives a summary of some values of J as a function of s for some different times t ($\tau = t/\theta$).

TABLE I

Values of $J(\tau, s)$

s	$\tau = 0.25$	$\tau = 1$	$\tau = 4$	se^{-s}
0.2	0.336	0.102	---	0.164
0.4	0.444	0.129	---	0.268
0.5	0.500	0.144	---	0.303
0.6	0.556	0.161	---	0.329
0.8	0.664	0.198	---	0.360
1.0	0.760	0.240	0.0066	0.368
1.2	0.839	0.286	---	0.361
1.4	0.898	0.336	---	0.345
1.6	0.940	0.389	---	0.323
1.8	0.967	0.444	---	0.298
2.0	0.983	0.500	0.017	0.271
2.4	0.996	0.611	---	0.218
2.8	0.999	0.714	---	0.170
3.0	0.999	0.760	0.039	0.149
4.0	0.999	0.921	0.078	0.073
6.0	0.999	0.998	0.240	0.0149
8.0	1	1	0.500	0.0027
10.0	1	1	0.760	0.00045
12	1	1	0.921	

If we ask when $J = \frac{1}{2}$, the answer is: $J = \frac{1}{2}$ if:

$$\sqrt{\tau} - \frac{s}{2\sqrt{\tau}} = 0$$

or:

$$s = 2\tau$$

or:

$$\varkappa r = \frac{2t}{\Theta}$$

or:

$$t = \frac{\varkappa r}{2}\Theta$$

Table I demonstrates these simple rules. We have also recorded se^{-s} in the last column of Table I for the following reason: The way in which the equilibrium distribution vanishes can be visualized best by considering the amount of electricity dQ in a spherical shell of thickness dr at different times. This is:

$$dQ = 2Ef\,4\pi r^2 dr = -E\varkappa s e^{-s} J dr$$
$$= -yE\varkappa dr \qquad (20)$$

where:

$$y = se^{-s}J(s, \tau) \qquad (21)$$

y is a measure of the density of the ionic cloud in a spherical shell of thickness dr at different distances from the central ion and at different times. In Fig. 1 we have plotted y for $\tau = 0$, $\tau = 0.25$, and $\tau = 1$. y possesses a maximum at each moment which shifts to

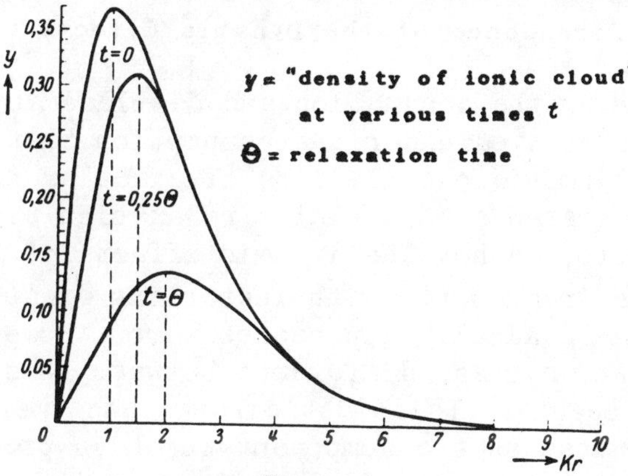

Fig. 1.

higher κr values as the time increases, while the maximum itself decreases simultaneously. Thus the maximum of the density y lies, as we will say simply from now on at time $t = 0$ at $\kappa r = 1$, at $t = 0.5\theta$ at $\kappa r = 1.5$, and finally at time $t = \theta$ at $\kappa r = 2.0$. At the same times the maximum itself decreases in the ratios 0.37:0.31:0.14. Further, Fig. 1 shows that the ionic cloud's density at $t = \theta$ is relatively small compared to its original density at time $t = 0$. At 4 times the relaxation time θ, the density has already nearly vanished. Were we to plot the curve corresponding to $\tau = 4$, it would hardly be distinguishable from the abscissa. Its maximum value amounts to 0.006 and lies at about $\kappa r = 3.5$.

According to eq. (13), in the special case of KCl, the relaxation time is:

$$\Theta = \frac{0.553 \cdot 10^{-10}}{\gamma} \qquad (22)$$

that is, for a concentration γ = 0.001 mol per liter, for example, θ is of the order of magnitude 10^{-7} sec. It depends strongly on temperature, the same way as viscosity does. Besides this, it is a function of the characteristic quantity $1/\kappa$ and, through this, the concentration γ of the solution. The relaxation time which plays a role in an entirely different connection in the theory of anomalous dispersion by dipolar liquids is in ordinary cases 1000 times smaller than the time referred to here. Therefore we predict a frequency dependence of conductivity and dielectric constant in a relatively accessible wave-length region. The conclusions which we draw from the theory in the case of small field strengths which we treat here have not yet been verified experimentally. However, it can be surmised that the observation of M. Wien[3] on the deviations from Ohm's law at high field strengths makes our effect probable, especially since Joos and Blumentritt[4] have interpreted the Wien effect, at least in its chief features, on the basis of the modern theory of electrolytes. These authors treat the stationary case taking into account higher powers of the field strength. In our paper a frequency dependence for small field strengths is being investigated. The general problem, still unsolved, corresponds to a superposition of both effects.

II. Frequency Dependence of the Dynamic Effects Caused by the Ionic Clouds

In the following the conduction and displacement currents in the electrolyte are calculated under the assumption that the ions of the solution execute harmonic oscillations of circular frequency ω. On the basis of the existence of a finite relaxation time of the ionic cloud, it is easy to see how the dynamic effects of the ionic cloud will depend on the frequency. With increasing frequency the absolute value of this force gradually approaches zero. However, because of the relaxation time, a phase difference between this force and the motion of the ion occurs. This last circumstance means that the current density also possesses a component which is proportional to the time variation of the electric field and which is thus 90° out of phase with the electric field strength. This means that besides the conduction current a displacement current exists. This displacement current is directly related to the dielectric constant whereas the conduction current determines the conductivity. We proceed in the following way. We take the differential equations (4) as our starting point. The differentiation in this equation refers to a system of coordinates X, Y, Z at rest. Now since the ions carry out harmonic oscillations and we are essentially interested in the force on an ion resulting from the existence of the ionic atmospheres, which in conjunction with the exterior electric field parallel to X and the Stokes frictional force determines the displacement and conduction currents, it is convenient to refer the partial differential equation to a coordinate system which moves with the ion and which we call x, y, z. Integration of this partial differential equation gives the potential and with it the force on the ion. Thus the chief question is: How do the dynamic effects on a vibrating ion of charge e_p differ from those on a uniformly moving ion? The partial differential equation which governs the nonstationary case gives for the number of ions of the kind p per c.c.:

$$\frac{\partial n_p}{\partial t} = \mathrm{Div}\left(\frac{kT}{\varrho_p}\,\mathrm{Grad}\, n_p - \frac{n_p \mathfrak{K}}{\varrho_p}\right) \quad (23)$$

This is referred to the rest coordinate system X, Y, Z. To indicate this we have written Grad and Div in eq. (23). It is clear that the potential Φ which we are seeking for the nonstationary case is not stationary even in the coordinate system moving with the ion parallel to X. Let the coordinates of the ion be $\xi, 0, 0$. Then:

$$\xi = a e^{i\omega t}; \qquad \dot{\xi} = a i\omega e^{i\omega t} \quad (24)$$

Now we introduce a coordinate system x, y, z whose origin is the moving ion and which moves with it:

$$x = X - \xi; \quad y = Y; \quad z = Z; \quad t = t' \quad (25)$$

Consequently:

$$\left.\begin{array}{ll} \dfrac{\partial}{\partial t} = -\dot{\xi}\dfrac{\partial}{\partial x} + \dfrac{\partial}{\partial t'}; & \dfrac{\partial}{\partial X} = \dfrac{\partial}{\partial x} \\ \dfrac{\partial}{\partial Y} = \dfrac{\partial}{\partial y};\ \dfrac{\partial}{\partial Z} = \dfrac{\partial}{\partial z}; & \begin{array}{l}\mathrm{div} = \mathrm{Div} \\ \mathrm{grad} = \mathrm{Grad}\end{array} \end{array}\right| \quad (26)$$

Relative to this coordinate system our differential equation is (omitting the primes on t):

$$\frac{\partial n_p}{\partial t} - \dot{\xi}\frac{\partial n_p}{\partial x} = \mathrm{div}\left\{\frac{kT}{\varrho_p}\,\mathrm{grad}\, n_p - \frac{n_p \mathfrak{K}}{\varrho_p}\right\} \quad (27)$$

For the force $\mathfrak{K}$ we can write:

$$\mathfrak{K} = e_p \mathfrak{E} - e_p\, \mathrm{grad}\, \Phi \quad (28)$$

Consequently, the system of equations (27) reduces to:

$$\left.\begin{array}{l} \dfrac{\partial n_p}{\partial t} - \dot{\xi}(t)\dfrac{\partial n_p}{\partial x} \\ = \mathrm{div}\left\{\dfrac{kT}{\varrho_p}\,\mathrm{grad}\, n_p - \dfrac{n_p e_p \mathfrak{E}}{\varrho_p} + \dfrac{n_p e_p\, \mathrm{grad}\,\Phi}{\varrho_p}\right\} \\ \cdots\cdots\cdots\cdots\cdots\cdots \\ \mathrm{div}\,\mathrm{grad}\,\Phi = -\dfrac{4\pi}{D}\sum n_p e_p \end{array}\right| \quad (29)$$

For sufficiently dilute solutions we can again substitute the constant value $\bar{n}_p$ for n_p in the last two terms of the bracket on the right-hand side. We call the deviation of the number of ions of the kind p per c.c.

from this constant value ν_p. Thus we put:

$$n_p = \bar{n}_p + \nu_p$$

If we substitute this in eq. (29) we get the system (div $\mathfrak{E} = 0$):

$$\left.\begin{aligned} &\frac{\partial \nu_p}{\partial t} - \xi \frac{\partial \nu_p}{\partial x} = \operatorname{div}\left\{\frac{kT}{\varrho_p}\operatorname{grad}\nu_p + \frac{\bar{n}_p e_p}{\varrho_p}\operatorname{grad}\Phi.\right\} \\ &\cdots\cdots\cdots\cdots\cdots\cdots \\ &\operatorname{div}\operatorname{grad}\Phi = -\frac{4\pi}{D}\Sigma e_p \nu_p \end{aligned}\right\} \quad (30)$$

In the special case $\xi = 0$, this gives the Maxwell-Boltzmann distribution:

$$\begin{aligned} \nu_p = \bar{\nu}_p &= -\frac{\bar{n}_p e_p}{kT}\Phi_0 \\ \Delta\Phi_0 &= \varkappa^2 \Phi_0 \\ \Phi_0 &= \frac{e_p}{D}\frac{e^{-\varkappa r}}{r} \end{aligned} \quad (31)$$

In the general case we introduce the deviation of the number of ions per c.c. from the number in the stationary case and denote it by μ_p; further we call the deviation of the potential Φ from its stationary value φ. Thus:

$$\nu_p = \bar{\nu}_p + \mu_p; \quad \Phi = \Phi_0 + \varphi \quad (32)$$

Now we try to solve the system (30) using eq. (32), where $\Phi_0 = e_p e^{-\kappa r}/Dr$. Further, we put:

$$\mu_p = f_p e^{i\omega t} \quad \varphi = v e^{i\omega t} \quad (33)$$

Then, for f_p and $\mathfrak{v}$ the differential equations are:

$$\left.\begin{aligned} &i\omega f_p - i\omega a \frac{\partial \bar{\nu}_p}{\partial x} \\ &\quad = \operatorname{div}\left\{\frac{kT}{\varrho_p}\operatorname{grad} f_p + \frac{\bar{n}_p e_p}{\varrho_p}\operatorname{grad} v\right\} \\ &\cdots\cdots\cdots\cdots\cdots\cdots \\ &\Delta v = -\frac{4\pi}{D}\Sigma e_p f_p. \end{aligned}\right\} \quad (34)$$

We can give at once three solutions, (35), (36), and (37), of these differential equations, which are of the form:

$$\left.\begin{aligned} v &= a\frac{\partial \Phi_0}{\partial x} \\ f_p &= -\frac{\bar{n}_p e_p}{kT} a \frac{\partial \Phi_0}{\partial x} \end{aligned}\right\} \quad (35)$$

$$\left.\begin{aligned} v &= A\frac{\partial}{\partial x}\frac{e^{-Kr}}{r} \\ f_p &= -\frac{n_p e_p}{kT}\frac{K^2}{K^2 - \frac{i\omega\varrho_p}{kT}}v \end{aligned}\right| \quad (36)$$

$$\left.\begin{aligned} v &= B\frac{\partial}{\partial x}\left(\frac{1}{r}\right) \\ f_p &= 0 \end{aligned}\right| \quad (37)$$

K is given by the relation:

$$\frac{DkT}{4\pi} = \sum \frac{\overline{n}_p e_p^2}{K^2 - \frac{i\omega\varrho_p}{kT}} \quad (38)$$

The correctness of these solutions can be easily shown by substituting them in eq. (34). We can find a general solution of our problem by making the potential Φ equal to a sum of the three solutions (35), (36), and (37):

$$\Phi = \frac{e_p}{D}\left|\frac{e^{-\varkappa r}}{r} + a e^{i\omega t}\frac{\partial}{\partial x}\frac{e^{-\varkappa r}}{r} + A e^{i\omega t}\frac{\partial}{\partial x}\frac{e^{-Kr}}{r} + B e^{i\omega t}\frac{\partial}{\partial x}\left(\frac{1}{r}\right)\right| \quad (39)$$

In order to determine the constants A and B, we expand the expressions

$$\frac{\partial}{\partial x}\frac{e^{-\varkappa r}}{r} \quad \text{and} \quad \frac{\partial}{\partial x}\frac{e^{-Kr}}{r}$$

in series:

$$\frac{\partial}{\partial x}\frac{e^{-\varkappa r}}{r} = \left(-\frac{1}{r^2} + \frac{\varkappa^2}{2} - \frac{\varkappa^3 r}{3} + \frac{\varkappa^4 r^2}{1\cdot 2\cdot 4} - \frac{\varkappa^5 r^3}{1\cdot 2\cdot 3\cdot 5} + \cdots\right)\cos\vartheta$$

$$\frac{\partial}{\partial x}\frac{e^{-Kr}}{r} = \left(-\frac{1}{r^2} + \frac{K^2}{2} - \frac{K^3 r}{3} + \cdots\right)\cos\vartheta$$

The additional potential cannot be infinite in the neighborhood of an ion; therefore the terms in eq. (39) proportional to $1/r^2$ must vanish. Furthermore, the charge density resulting from the additional potential must not be infinite in the neighborhood of an ion. All in all, the following conditions must be satisfied:

$$a + A + B = 0$$
$$\varkappa^2 a + K^2 A = 0$$

that is:

$$A = -\frac{\varkappa^2}{K^2}a; \quad B = \left(\frac{\varkappa^2}{K^2} - 1\right)a \quad (40)$$

Thus a definite expression for the potential * Φ is found.

* For the special case of KCl, eq. (41) can be solved in the following way also. If we proceed in a way entirely similar to the way Debye and Hückel proceeded in their work, only with the generalization that the velocity of the ions is now written as:

$$\bar{v}\, e^{i\omega t}$$

then we are lead to the following partial equation for v:

$$\Delta(\Delta v - K^2 v) = -\varkappa^2 \bar{\omega} \frac{\partial \Phi_0}{\partial x}$$

The method of variation of constants gives the following particular solution of this differential equation:

$$-\frac{\partial \Phi_0}{\partial x} \frac{\bar{\omega}}{\varkappa^2 - K^2} e^{i\omega t} = \frac{e}{D} \bar{\omega} \frac{e^{-\varkappa r}\left(\frac{\varkappa}{r} + \frac{1}{r^2}\right)}{\varkappa^2 - K^2} \cos\vartheta\, e^{i\omega t}$$

The particular integrals of the homogeneous equation

$$\Delta(\Delta v - K^2 v) = 0$$

are:

$$\frac{d}{dr}\frac{e^{-Kr}}{r}\cos\vartheta;\quad \frac{d}{dr}\frac{e^{+Kr}}{r}\cos\vartheta;\quad r^2\cos\vartheta;\quad \frac{\cos\vartheta}{r^2}$$

Consequently the general solution is:

$$\Phi = \Phi_0 + \left| \frac{e}{D}\omega \frac{e^{-\varkappa r}\left(\frac{\varkappa}{r} + \frac{1}{r^2}\right)}{\varkappa^2 - K^2} + A' \frac{d\frac{e^{-Kr}}{r}}{dr} \right.$$

$$\left. + A'' \frac{d\frac{e^{-Kr}}{r}}{dr} + \frac{B'}{r^2} + B'' r^2 \right| e^{i\omega t} \cos\vartheta$$

Since the potential cannot be infinite at $r = \infty$, $A'' = B'' = 0$. The finiteness of the additional potential and the additional density in the neighborhood of the ion requires the constants A' and B' to have the values:

$$A' = \frac{e}{D} \frac{\bar{\omega}}{\varkappa^2 - K^2} \frac{\varkappa^2}{K^2}$$

$$B' = \frac{e\bar{\omega}}{D(K^2 - \varkappa^2)} + \frac{e\,\omega\,\varkappa^2}{D(\varkappa^2 - K^2)K^2}$$

If we compare these constants with those in (40) above, then we must have:

$$A' = \frac{e}{D} \frac{\bar{\omega}}{\varkappa^2 - K^2} \frac{\varkappa^2}{K^2} = -\frac{\varkappa^2}{K^2} \frac{ae}{D}$$

(footnote continued)

If this is developed in powers of r:

$$\Phi = \frac{e_p}{D}\left[\frac{e^{-\varkappa r}}{r} + \frac{\varkappa^2}{3}(K-\varkappa)\, a\, e^{i\omega t}\, r\cos\vartheta + \cdots\right] \qquad (41)$$

Thus the additional field strength to which the ion is subject is:

$$-\frac{\partial\Phi}{\partial x} = -\frac{e_p}{D}\frac{\varkappa^2}{3}(K-\varkappa)\, a\, e^{i\omega t} \qquad (42)$$

If the velocity $\dot{\xi}$ of the ion, which is:

$$\dot{\xi} = i\omega a e^{i\omega t}$$

is substituted in eq. (42), the force to which the vibrating ion is subject can also be written:

$$-e_p\frac{\partial\Phi}{\partial x} = -\frac{e_p^2}{3D}\frac{\varkappa^2\dot{\xi}}{i\omega}(K-\varkappa) \qquad (42')$$

Footnote to previous page continued

that is:

$$a = -\frac{\overline{\omega}}{\varkappa^2 - K^2}$$

That this condition is fulfilled can be shown simply in the following way. Since:

$$\dot{\xi} = a i\omega e^{i\omega t} \text{ and } \overline{\omega} = \frac{\varrho\overline{v}}{kT}$$

there must hold:

$$a = -\frac{\varrho}{kT}\frac{ai\omega}{\varkappa^2 - K^2}$$

or:

$$1 = -\frac{\varrho}{kT}\frac{i\omega}{\varkappa^2 - K^2}$$

From this it follows that:

$$K^2 = \varkappa^2 + \frac{i\omega\varrho}{kT}$$

This relation is identical with the relation (43), and thus the proof of the correctness of our assertion is complete.

From eq. (38) the quantity K can be calculated for any electrolyte and from this the force on the vibrating ion. The calculations are particularly simple in the special case of KCl solution treated above. For this case:

$$e_1 = -e_2 = e,\ \varrho_1 = \varrho_2 = \varrho,\ \overline{n}_1 = \overline{n}_2 = n,\ \dot{\xi} = \overline{v} e^{i\omega t}$$

and K^2 becomes:

$$K^2 = \varkappa^2 \left(1 + i \frac{\omega \varrho}{kT} \frac{1}{\varkappa^2}\right) = \varkappa^2 (1 + i\omega\Theta) \qquad (43)$$

For this example we write:

$$\varkappa + K = \varkappa + \varkappa \sqrt{1 + i\omega\Theta} = \varkappa(a + ib) \qquad (44)$$

According to eq. (13) we now have for the force acting on an ion the complex form

$$-\frac{\varkappa e^2}{3D} \frac{\varrho \overline{v} e^{i\omega t}}{kT(a+ib)} = -\frac{\varkappa e^2}{3D} \frac{\varrho \overline{v}(a-ib) e^{i\omega t}}{kT(a^2+b^2)} \qquad (45)$$

We put this force equal to:

$$-K^* e^{i(\omega t + \Delta)} = K^* e^{i(\omega t + \pi + \Delta)} \qquad (46)$$

Consequently:

$$K^* = \frac{\varkappa e^2}{3DkT} \frac{\varrho \overline{v}}{\sqrt{a^2+b^2}} \qquad (47)$$

$$\mathrm{tg}\,\Delta = -\frac{b}{a} \qquad (48)$$

If we put:

$$a + ib = 1 + \alpha \cos\beta + i\alpha \sin\beta$$

so that:

$$\beta = \frac{1}{2} \operatorname{arctg} \omega\Theta$$

$$\alpha = \sqrt{\frac{\omega\Theta}{\sin 2\beta}}$$

then it is easy to calculate K^* and Δ for each value of $\omega\theta$.

As can be seen from eq. (47), the force to which a vibrating ion is subject differs in absolute value from that to which it is subject in the stationary case. As is immediately evident, the absolute value of this force reduces to the Debye and Hückel value (3) when $\omega = 0$. For frequencies ω which are very small compared with the quantity $1/\theta$, the absolute value of this force differs from the stationary value only by terms of the second order in ω. Consequently, in this case the absolute value of this force is only slightly less than the absolute value of the force on ions moving with constant velocity. As the frequency approaches $\omega = 1/\theta$, the absolute value of this force differs more and more from the stationary value, and for large values of ω compared to θ it approaches zero as $1/\omega$.

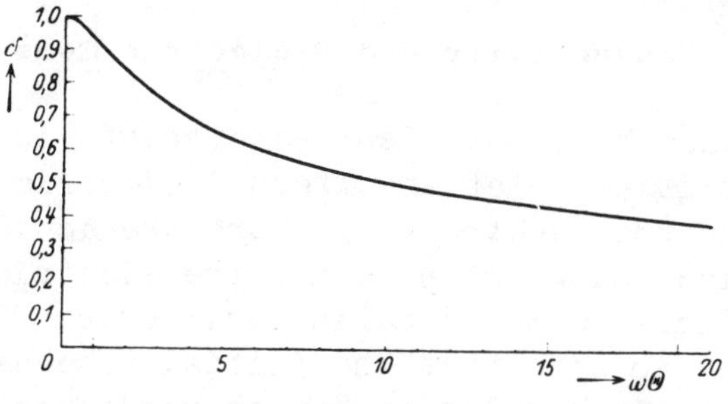

$\delta = \dfrac{\text{force on a vibrating ion}}{\text{force on a uniformly moving ion}}$

(Θ = relaxation time, ω = frequency)

Fig. 2.

In Fig. 2 the ratio of the absolute value of the force on a vibrating ion to the force on a uniformly moving ion is plotted as a function of $\omega\theta$. This ratio is denoted by δ in Fig. 2.

Now that we know the absolute value of the force as a function of the quantity $\omega\theta$ and thus the frequency ω, we turn to the phase difference $\pi + \Delta$ between the force on and the velocity of a vibrating ion. The formula (48) gives immediately the phase difference $\pi + \Delta$ between the force and the velocity, and of course phase Δ is negative. For very small values of ω the negative phase Δ increases linearly with ω. At $\omega = 1/\theta$ it is about -12°. Finally, as the frequency ω becomes very large compared with the frequency $1/\theta$, it approaches the value -45°. In Fig. 3 the phase difference Δ is plotted as a function of $\omega\theta$. $\Delta + \pi$ is the phase between the force on and the velocity of a vibrating ion.

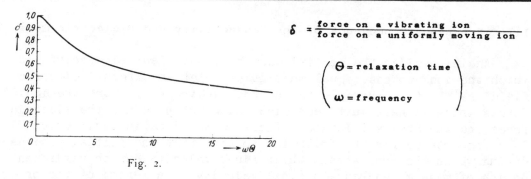

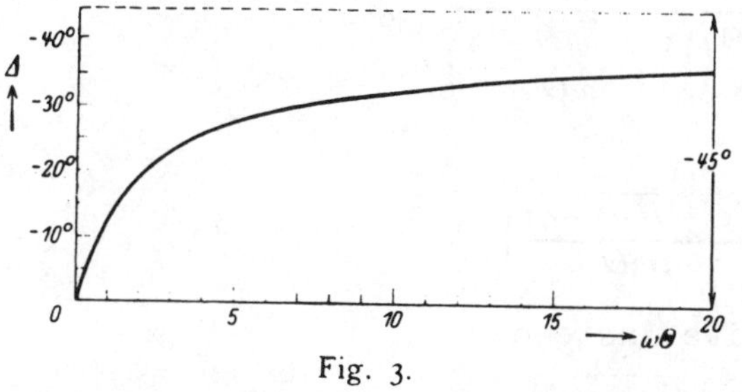

$\pi + \Delta$ = phase difference between force and velocity

(Θ = relaxation time, ω = frequency)

Fig. 3.

Since the expression for the force on a vibrating ion is complex, both the conductivity and the dielectric constant of a strong electrolyte are a function of frequency.

Summarizing (compare also Table II), we can remark that for very small ω the force on a vibrating ion and the Stokes frictional force are in phase, while on the contrary for large frequencies ω these two forces are no longer in phase. Furthermore, the absolute value of the force on an ion decreases with increasing frequency and approaches zero for large frequencies. During this decrease we see that finally with increasing frequency only Stokes' force remains. Consequently we expect that at high frequencies the molecular conductivity of a strong electrolyte will approach its value for infinite dilution, at least if cataphoretic action is being neglected. We now come to formulae for the conductivity and dielectric constant of a solution as function of ω.

III. Frequency Dependence of the Conductivity and Dielectric Constant

Again consider the special case of a univalent electrolyte in which the ions possess equal mobilities. Let an external electric field* $Ee^{i\omega t}$ of frequency ω act on the electrolyte. Were the ionic forces which we have just been discussing not present, the electric force and the Stokes frictional force $\rho\bar{v}$ would be in equilibrium. It is of importance for the following to note that the initial processes during which the ions attain their final velocity can be neglected. An ion of mass m reaches its final velocity in a period of the order of magnitude $m/\rho \sim 10^{-13}$ sec. This time is entirely negligible compared to those which are of importance in our problem. Therefore, we can assume that the ions follow the electric field instantaneously.

In our case the exterior electric field, Stokes frictional force, and braking ionic force must be in equilibrium. The equation which expresses this condition is:

$$eEe^{i\omega t} - \varrho i\omega a e^{i\omega t} - \frac{e^2\varkappa^2}{3D}\varkappa \cdot a e^{i\omega t}\left(\sqrt{1 + i\omega\Theta} - 1\right) = 0 \tag{49}$$

From this the velocity $\dot{\xi}$ can be calculated and is:

$$\dot{\xi} = i\omega a e^{i\omega t} = \frac{eEe^{i\omega t}}{\varrho}\,\frac{1}{1 + \dfrac{e^2\varkappa^2}{3D}\dfrac{\varkappa\Theta}{\varrho}\left(\dfrac{\sqrt{1+i\omega\Theta}-1}{i\omega\Theta}\right)} \tag{50}$$

Now since the quantity

$$\frac{e^2\varkappa^2}{3D}\,\frac{\varkappa}{\varrho/\Theta}\left(\frac{\sqrt{1+i\omega\Theta}-1}{i\omega\Theta}\right)$$

is small compared to 1, we can write instead:

$$\dot{\xi} = \frac{eEe^{i\omega t}}{\varrho}\left[1 - \frac{e^2\varkappa}{3DkT}\,\frac{\sqrt{1+i\omega\Theta}-1}{i\omega\Theta}\right] \tag{51}$$

In terms of the velocity, the electric current density is $i = 2ne\dot{\xi}$. If the value of the velocity $\dot{\xi}$ given by eq. (51) is substituted, the current density becomes:

$$i = \frac{2ne^2}{\varrho}Ee^{i\omega t}\left[1 - \frac{e^2\varkappa}{3DkT}\,\frac{\sqrt{1+i\omega\Theta}-1}{i\omega\Theta}\right] \tag{52}$$

Consequently the sum of the conduction and displacement current densities must be:

* In section III E is the absolute value of the electric field strength, while in section I it is the central charge of the ions.

$$J = \frac{i\omega D}{4\pi} E e^{i\omega t} + \frac{2ne^2}{\varrho} E e^{i\omega t}\left[1 - \frac{e^2\varkappa}{3DkT}\,\frac{\sqrt{1+i\omega\Theta}-1}{i\omega\Theta}\right] \qquad (53)$$

Now according to the phenomenological Maxwell theory, the total current density J is equal to the sum of the displacement and conduction currents. Therefore we have the equation:

$$J = \lambda\mathfrak{E} + \frac{D'}{4\pi}\dot{\mathfrak{E}} = \lambda E e^{i\omega t} + \frac{i\omega D'}{4\pi} E e^{i\omega t} \qquad (54)$$

The real part of the right-hand side of equation (54) gives the electric conductivity; the imaginary part is related to the dielectric constant of the solution. In detail, the conductivity λ and the increase in the dielectric constant D' of the solution over that of the solvent D are

$$\lambda = \frac{2ne^2}{\varrho}\cdot \text{Real part of}\left[1 - \frac{e^2\varkappa}{3DkT}\,\frac{\sqrt{1+i\omega\Theta}-1}{i\omega\Theta}\right] \qquad (55)$$

$$D' - D = \frac{4\pi}{\omega}\frac{2ne^2}{\varrho}\ \text{Imaginary part of}\left[1 - \frac{e^2\varkappa}{3DkT}\,\frac{\sqrt{1+i\omega\Theta}-1}{i\omega\Theta}\right] \qquad (56)$$

In forming the real and imaginary parts, we can either apply the trigonometriç formula on p. 386 or write:

$$\frac{\sqrt{1+i\omega\Theta}-1}{i\omega\Theta} = \sqrt{\frac{(1+\omega^2\Theta^2)^{1/2}-1}{2\omega^2\Theta^2}} - i\left[\sqrt{\frac{(1+\omega^2\Theta^2)^{1/2}+1}{2\omega^2\Theta^2}} - \frac{1}{\omega\Theta}\right] \qquad (57)$$

In this way we find for the formulae for the conductivity and the increase in dielectric constant:

$$\lambda = \frac{2ne^2}{\varrho}\left[1 - \frac{e^2\varkappa}{3DkT}\sqrt{\frac{(1+\omega^2\Theta^2)^{1/2}-1}{2\omega^2\Theta^2}}\right] \qquad (58)$$

$$D' - D = 4\pi\frac{2ne^2}{\varrho}\,\frac{e^2\varkappa}{3DkT}\,\Theta\left[\frac{\sqrt{\frac{(1+\omega^2\Theta^2)^{1/2}+1}{2}}-1}{\omega^2\Theta^2}\right] \qquad (59)$$

These determine the quantities λ and $D'-D$ for each value of $\omega\theta$. If we introduce the value of the conductivity at ∞ dilution λ_∞:

$$\lambda_\infty = \frac{2ne^2}{\varrho} \qquad (60)$$

and put:

$$\frac{e^2\varkappa}{6DkT}\,2\sqrt{\frac{(1+\omega^2\Theta^2)^{1/2}-1}{2\omega^2\Theta^2}} = \frac{e^2\varkappa}{6DkT}x_1(\omega\Theta) \qquad (61)$$

we can rewrite the formula (58) for the conductivity:

$$\frac{\lambda_\infty - \lambda_\omega}{\lambda_\infty} = \frac{e^2\varkappa}{6DkT}x_1(\omega\Theta) \qquad (62)$$

As $\omega\theta$ approaches 0, x_1 approaches 1. If we denote the conductivity in the stationary case by λ_0, then we find:

$$\frac{\lambda_\infty - \lambda_0}{\lambda_\infty} = \frac{e^2 \varkappa}{6DkT} \qquad (63)$$

Now if we form the quotient of eqs. (62) and (63) we find the relation:

$$\frac{\lambda_\infty - \lambda_\omega}{\lambda_\infty - \lambda_0} = x_1(\omega\Theta) \qquad (64)$$

We will now carry through a similar analysis for the increase in the dielectric constant. Formula (59) can be written:

$$D'_\omega - D = 4\pi \frac{2ne^2}{\varrho} \frac{e^2\varkappa}{3DkT} \Theta x_2(\omega\Theta) \qquad (63')$$

with the abbreviation:

$$x_2(\omega\Theta) = \left[\frac{\sqrt{\dfrac{(1+\omega^2\Theta^2)^{1/2}+1}{2}} - 1}{\omega^2\Theta^2} \right] \qquad (65)$$

As $\omega\theta$ approaches zero, x_2 approaches 1/8 (compare Table II). Consequently, if we denote the increase in the dielectric constant at 0 frequency by $D_0' - D$:

$$D'_0 - D = 4\pi \frac{2ne^2}{\varrho} \frac{e^2\varkappa}{3DkT} \Theta \frac{1}{8} \qquad (66)$$

Summarizing, we remark that with the abbreviations (64) and (65) the formulae for the conductivity defect and the increase in the dielectric constant can be put in the form:

$$\lambda_\infty - \lambda_\omega = \lambda_\infty \frac{e^2\varkappa}{6DkT} x_1(\omega\Theta)$$

$$D'_\omega - D = 4\pi \frac{2ne^2}{\varrho} \frac{e^2\varkappa}{3DkT} \Theta x_2(\omega\Theta)$$

In calculating the conductivity defect or the dielectric constant excess, it is best to start from the equations (62) and (63'). For a graphical representation, on the other hand, it is better to plot the quantities

$$\frac{\lambda_\infty - \lambda_\omega}{\lambda_\infty - \lambda_0} \quad \text{and} \quad \frac{D'_\omega - D}{D'_0 - D}$$

as functions of $\omega\theta$. In Figs. 4 and 5 we have plotted these quantities, which are also given by $x_1(\omega\theta)$ and $8x_2(\omega\theta)$ respectively, as functions of $\omega\theta$. Fig. 4 contains the quotient of the conductivity defect at frequency ω and the conductivity defect at 0 frequency plotted as a function of $\omega\theta$. That a frequency dependence of the conductivity exists

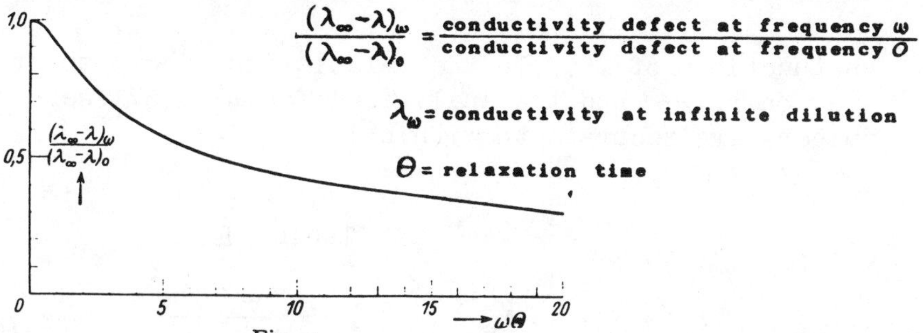

Fig. 4.

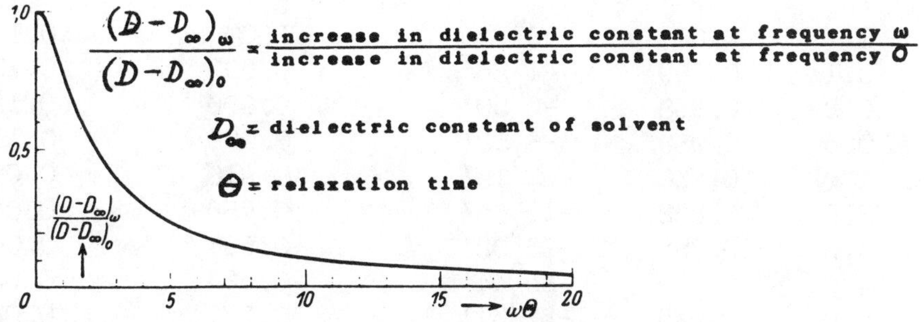

Fig. 5.

for strong electrolytes is shown clearly by Fig. 4. For frequencies large compared to $1/\theta$ the conductivity λ approaches its value at infinite dilution*. For very small values of ω compared to $1/\theta$,

$$\frac{\lambda_\infty - \lambda_\omega}{\lambda_\infty - \lambda_0}$$

deviates from 1 again only in the second order in $\omega\theta$. Therefore, for very small values of $\omega\theta$ the curve $x_1(\omega\theta)$ is parabolic in shape and possesses a horizontal tangent at $\omega = 0$. In Fig. 5 the quotient of the increase in the dielectric constant of the solution at the frequency ω and the increase in the dielectric constant at frequency 0 is plotted as a function of $\omega\theta$.

For very small values of ω the curve has again the form of a parabola with a horizontal tangent at $\omega = 0$, while for large values of ω the dielectric constant of the solution approaches the dielectric constant of the solvent. An increase in the dielectric constant over the normal value for the solvent does not prove that this behavior is caused by dipole effects. Our result means that this same effect can occur even in a completely dissociated solution in consequence of the reaction of the ionic clouds.

In the following Table II we have recorded the absolute value of the force on a vibrating ion, the phase, and the quantities

$$\frac{\lambda_\infty - \lambda_\omega}{\lambda_\infty - \lambda} \quad \text{and} \quad \frac{D'_\omega - D}{D'_0 - D}$$

* As has been emphasized several times already, this result needs correcting because of the cataphoretic action.

as functions of $\omega\theta$. In the calculations, both the trigonometric formulae on p. 386 and the analytical formula (57) were used as checks. The numbers are accurate to within 1%.

Table II

$\omega\theta$	δ	Δ	$x_1 = \dfrac{\lambda_\infty - \lambda_\omega}{\lambda_\infty - \lambda_0}$	$x_2(\omega\theta)$	$\dfrac{D_\omega' - D}{D_0' - D} = 8x_2$
0	1.00	0°	1.00	0.125	1.000
0.1	0.999	$-1^\circ 26'$	0.999	0.125	0.999
0.2	0.998	$-2^\circ 50'$	0.996	0.124	0.996
0.5	0.980	$-6^\circ 50'$	0.972	0.116	0.928
1.0	0.930	$-12^\circ 15'$	0.910	0.098	0.784
2.0	0.832	$-19^\circ 05'$	0.786	0.068	0.544
3.0	0.755	$-23^\circ 02'$	0.694	0.0493	0.3944
4.0	0.690	$-25^\circ 44'$	0.624	0.0375	0.3000
5.0	0.645	$-27^\circ 31'$	0.572	0.0298	0.2384
6.0	0.608	$-28^\circ 54'$	0.532	0.0245	0.1960
7.0	0.573	$-30^\circ 07$	0.498	0.0206	0.1648
8.0	0.547	$-30^\circ 55'$	0.470	0.0176	0.1408
9.0	0.524	$-31^\circ 40'$	0.446	0.0153	0.1224
10.0	0.507	$-32^\circ 19'$	0.426	0.0135	0.1080
12.0	0.469	$-33^\circ 30'$	0.392	0.0108	0.0864
15.0	0.431	$-34^\circ 41'$	0.354	0.00813	0.0650
18.0	0.398	$-35^\circ 25'$	0.324	0.00644	0.0515
20.0	0.380	$-35^\circ 57'$	0.308	0.00506	0.0405
∞	0	-45°	0	0	0

In conclusion some numerical details will be given for the example of KCl.

Since the molecular conductivity of KCl is 65 + 65 = 130, we have the relation:

$$\frac{\gamma}{1000}\, 135\, E\, 3 \times 10^9 = \frac{2ne^2 E}{\varrho \cdot 300} = \frac{2\gamma N e^2}{1000 \cdot \varrho} \frac{E}{300}$$

In this equation the electric field strength is in volts/cm. Since N, the Loschmidt number per mol, has the value 6.06×10^{23}, the mobility of the KCl ions is

$$\varrho = 0{,}236 \cdot 10^{-8} \text{ dyn sec cm}^{-1}$$

From this, the quantities:

$$\lambda_x, \frac{\lambda_x - \lambda_0}{\lambda_x}, D_0' - D$$

are found to be:

$$\lambda_\infty = 1.171 \cdot 10^{11} \cdot \gamma$$

$$\frac{\lambda_\infty - \lambda_0}{\lambda_\infty} = 0.386 \cdot \sqrt{\gamma}$$

$$D'_0 - D = 2.42 \cdot \pi \sqrt{\gamma}$$

These results are for frequency $\omega = 0$. The expressions for frequency ω can be found easily from eqs. (62) and (63'). In doing this, according to eq. (22), we must use for θ:

$$\Theta = \frac{0.553 \cdot 10^{-10}}{\gamma}$$

which implies that $T = 291.2$, at which temperature $D = 81.3$.

If we choose for the concentration $\gamma = 1/1000$, the increase in the dielectric constant for this concentration and frequency 0, for which it takes its largest value, amounts to 0.24. If the wave length at which this increase has decreased to half of this value is calculated, it is found that:

$$\lambda_{im} = \frac{4.74 \cdot 10^{-2}}{\gamma}$$

where λ is measured in meters.

If, as above, the concentration is $\gamma = 1/1000$, the wave length is:

$$\lambda_{im} = 47.4 \cdot m$$

On the other hand, the conductivity defect is half of its maximum value at a wave length $\lambda'_{im} = \frac{1.49 \cdot 10^{-2}}{\gamma}$, that is for $\gamma = 1/1000$ at $\lambda'_{im} = 14.9$ m.

In conclusion we can make the following remarks. A large number of measurements of the dependence of the dielectric constant of strong electrolytes on concentration exist. Up to now it has not been possible to bring these measurements into harmony with one another*. Many authors find a variation; others, none. The measurements of Walden and of Sack, which are theoretically intelligible, show a decrease in the dielectric constant proportional to the concentration. On the other hand, our theory predicts an increase proportional to $\sqrt{\gamma}$. However, it is easy to see that our effect is in no way contradictory to these experiments. According to Sack, the cause of the increase in the dielectric constant is completely different from the effect discussed here. Actually, both effects are superimposed, and at extremely small concentrations, unaccessible to experiment, the dielectric constant possesses a maximum. It

* Compare the monograph which will appear shortly: P. Debye, *Polar Molecules*, Chem. Catalogue Co., New York. Published in Germany by S. Hirzel, Leipzig.

increases over the dielectric constant of the solvent at very low concentrations and finally decreases with increasing concentration. This latter region is the one accessible to experiment at the present.

In conclusion it is emphasized that this paper contains essentially only a survey of the phenomena to be expected. As has been mentioned in the text many times, their quantitative explanation is not yet complete. The necessary additions will be contained in a paper to appear shortly.

One of us wishes to express here his sincere thanks to the International Education Board for a grant which made possible his stay in Leipzig.

Bibliography

1. See P. Debye and E. Huckel, *Physik. Z.*, *24*, 185, 305 (1923); and the discussion of strong electrolytes, *Trans. Faraday Soc.*, 334-554 (April 1927).
2. P. Debye, *Trans. Faraday Soc.*, 334 (April 1927).
3. M. Wien, *Ann. Physik*, *83*, 327-361 (1927); *Physik. Z.*, *28*, 834-836 (1927).
4. G. Joos and M. Blumentritt, *Physik. Z.*, 836-838 (1927).

FREQUENCY DEPENDENCE OF THE CONDUCTIVITY AND THE DIELECTRIC CONSTANT OF STRONG ELECTROLYTES*

(Dispersion der Leitfähigkeit und der Dielektrizitätskonstante starker Elektrolyte)

P. Debye and H. Falkenhagen**

Translated from
Physikalische Zeitschrift, Vol. 29, 1928, pages 401-426

Recently[1] we have suggested that experiment*** should reveal a frequency dependence of the conductivity and dielectric constant of a strong electrolyte. Theoretically, these phenomena are the result of the existence of a finite relaxation time of the ionic atmosphere. In our first paper, we pointed out that two essential refinements, omitted there for simplicity's sake, must be added to our calculations in order to solve in a satisfactory quantitative way the problem of the dependence of the conductivity and dielectric constant on the frequency of an alternating electric field. First, the ions in the solution execute Brownian motions around their average straight-line paths parallel to the external electric field, and second, the electrophoretic effect, which is as important in determining the conductivity as the relaxation effect, has to be taken into account. In this paper we will treat the necessary additions in detail. First, by way of survey, we give an outline of our treatment.

*Presented at the June 2 meeting at Dresden of the Thüringen-Saxony-Silesia Section of the German Physical Society.

**Received May 15, 1928.

***M. Wien (*Ann. Physik, 83,* 340/41 (1927)) found indications of an effect of the collision time on the magnitude of the conductivity which is closely related to our effect.

Summary of Contents

A.I. The General Basic Equations of Nonstationary Process in Strong Electrolytes

We begin by discussing how the Brownian motion of the ions affects the frequency effect. If, for example, an external electric field acts on an electrolyte, the ions do not move parallel to the field but execute fluctuations which depend on the heat motion of the molecules of the solvent. We must take account of these random motions of the ions and so will present first general basic equations.*

*For the stationary case, Onsager has calculated a formula for the conductivity (compare L. Onsager, *Physik. Z.*, *28*, 277 (1927)). This section contains in part only a superficially altered presentation of his method.

Let ions of the kinds $1 \ldots i \ldots s$ with the charges $e_1 \ldots e_i \ldots e_s$, respectively, be present. Let $n_1 \ldots n_i \ldots n_s$ be the time averages of the number of corresponding ions per c.c. We imagine (compare Figure 1) two volume elements fixed in space at the points P and Q, which we

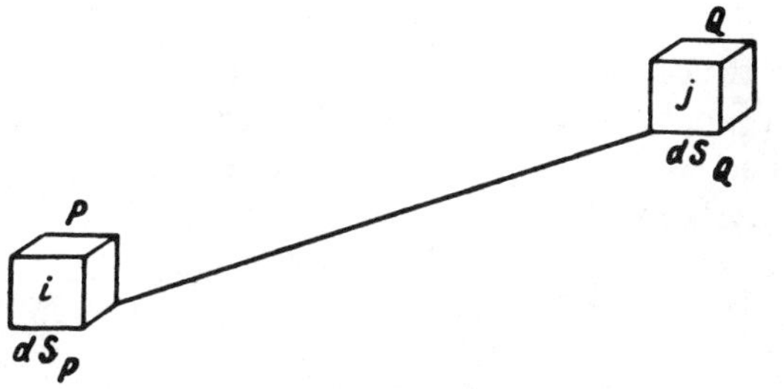

Figure 1

will call dS_P and dS_Q, respectively. Let the coordinates of P be x_P, y_P, z_P; those of Q, x_Q, y_Q, z_Q. Now we will observe these two volume elements for a time τ which is long compared to one of the elementary displacements of which the Brownian motion of an ion is made up. During the time τ we note the time intervals during which an ion of the kind i is in dS_P or an ion of the kind j in dS_Q. Naturally, occasionally several ions of the kind i will be found in dS_P or several of the sort j in dS_Q. If we take the sizes of these volume elements to be small, this occurs relatively infrequently, so that we can neglect these cases. We call the sums of these occasional time intervals t_i and t_j. Therefore, since τ is sufficiently large:

$$\frac{t_i}{\tau} = n_i dS_P \qquad \frac{t_j}{\tau} = n_j dS_Q \tag{1}$$

Further, we form the sums of all time intervals during which simultaneously an ion of the kind i is in dS_P and an ion of the kind j in dS_Q. We call the sum of such intervals within the time interval τ: t_{iP}^{jQ}. If the time average of the number of ions of the kind j per c.c. which lies in the neighborhood of an i ion is denoted by n_{iP}^{jQ} and the concentration (number of ions of the kind in question per c.c.) of the i ions in the neighborhood of a j ion by n_{jQ}^{iP}, then:

$$\frac{t_{iP}^{jQ}}{t_i} = n_{iP}^{jQ} dS_Q \qquad \frac{t_{jQ}^{iP}}{t_j} = n_{jQ}^{iP} dS_P \tag{2}$$

Now the times t_{iP}^{jQ} and t_{jQ}^{iP} must be equal for a fixed position of the volume elements dS_P and dS_Q, so that:

$$t_{iP}^{jQ} = t_{jQ}^{iP} \tag{2'}$$

Equations (2,2') in conjunction with (1) give the relation:

$$t_{iP}^{jQ} = n_i n_{iP}^{jQ} \cdot \tau dS_P dS_Q = n_j n_{jQ}^{iP} \tau dS_P dS_Q$$

that is:

$$n_i n_{iP}^{jQ} = n_j n_{jQ}^{iP} \tag{3}$$

Now for the quantity $n_i n_{iP}^{jQ}$ we introduce the abbreviation:

$$n_i n_{iP}^{jQ} = n_j n_{jQ}^{iP} = w_{iP}^{jQ} = w_{jQ}^{iP} \tag{3'}$$

w_{iP}^{jQ} is the distribution function for the frequency of those configurations in which an ion of the kind i is in dS_P and simultaneously an ion of the kind j in dS_Q. If the probability function w_{iP}^{jQ} is known, the time interval (t_{iP}^{jQ}) during the time of observation τ in which an i ion is in dS_P and simultaneously a j ion in dS_Q is known. Imagine a large number of pairs of volume elements dS_P and dS_Q. We will denote this number by N.* If we note the time interval during which an ion of the kind i is in dS_P and simultaneously an ion of the kind j is in dS_Q - this time was called t_{iP}^{jQ} - and note the corresponding number N of pairs of elements dS_P, dS_Q, then:

$$\frac{N_{iP}^{jQ}}{N} = \frac{t_{iP}^{jQ}}{\tau} = w_{iP}^{jQ} dS_P dS_Q \tag{4}$$

The observations just described might be carried out during a small time τ following the moment t. Now we make the same observations during a time interval τ following the moment $t + \Delta t$. Now instead of N_{iP}^{jQ} combinations we find $N_{iP}^{jQ} + \Delta N_{iP}^{jQ}$. Thus in all nonstationary processes the distribution function w_{iP}^{jQ} varies with time. This variation ΔN_{iP}^{jQ} can be calculated in the following way. While an i ion is in dS_P, a j ion can wander into dS_Q. Let this mode of change occur A_{io}^{ij} times during time τ. Or while a j ion is in dS_Q, an i ion can wander into dS_P. Let this second mode of change occur A_{oj}^{ij} times during time τ. We put:

$$A_{oj}^{ij} + A_{io}^{ij} = Z_+$$

Or, however, while an i ion is in dS_P, a j ion can wander out of dS_Q, or while a j ion is in dS_Q, an i ion can wander out of dS_P. Let the first happen A_{ij}^{iO} times during time τ; the second, A_{ij}^{Oj} times. We put:

$$A_{ij}^{oj} + A_{ij}^{io} = Z_-$$

Consequently:

$$\frac{\Delta N_{iP}^{jQ}}{N} = \frac{Z_+ - Z_-}{\tau} \Delta t$$

or if we put:

$$Z_+ - Z_- = Z$$

then:

$$\frac{\Delta N_{iP}^{jQ}}{N} = \frac{Z}{\tau} \Delta t$$

*The N used in Part A has nothing to do with the N used in Part D; in Part D, N is the Loschmidt number per mol.

Thus, according to equation (4):

$$\frac{\partial w_{iP}^{jQ}}{\partial t} dS_P dS_Q = \frac{Z}{\tau} \tag{5}$$

Now we must calculate Z. For this purpose we introduce the average velocity of the j ions in the neighborhood of an i ion. We call this velocity $\mathfrak{v}_{iP}^{jQ}$. The number of j ions wandering into dS_Q per unit time minus the number wandering out of it per unit time is:

$$-\operatorname{div}_Q (n_{iP}^{jQ} \mathfrak{v}_{iP}^{jQ}) dS_Q$$

The index Q means that the partial derivatives in the divergence are to be taken with repect to the coordinates x_Q, y_Q, z_Q.

If we multiply this quantity by the sum of all time intervals t_i which fall in the time interval τ and during which an ion is actually in dS_P, we get:

$$-t_i \operatorname{div}_Q (n_{iP}^{jQ} \mathfrak{v}_{iP}^{jQ}) dS_Q$$

However, this excess of j ions in the neighborhood of an i ion during the time τ must be equal to $A_{io}^{ij} - A_{ij}^{io}$. Thus:

$$A_{io}^{ij} - A_{ij}^{io} = -t_i dS_Q \operatorname{div}_Q (n_{iP}^{jQ} \mathfrak{v}_{iP}^{jQ})$$

In the same way, the increase in the number of i ions in dS_P while simultaneously a j ion is in dS_Q during the time τ is:

$$A_{oj}^{ij} - A_{ij}^{oj} = -t_j \operatorname{div}_P (n_{jQ}^{iP} \mathfrak{v}_{jQ}^{iP}) dS_P$$

Consequently:

$$Z = -t_i \operatorname{div}_Q (n_{iP}^{jQ} \mathfrak{v}_{iP}^{jQ}) dS_Q - t_j \operatorname{div}_P (n_{jQ}^{iP} \mathfrak{v}_{jQ}^{iP}) dS_P$$

or:

$$\frac{Z}{\tau} = -\left[n_i \operatorname{div}_Q (n_{iP}^{jQ} \mathfrak{v}_{iP}^{jQ}) + n_j \operatorname{div}_Q (n_{jQ}^{iP} \mathfrak{v}_{jQ}^{iP})\right] dS_P dS_Q$$

This combined with equation (5) gives:

$$\begin{aligned} \frac{\partial w_{iP}^{jQ}}{\partial t} &= -\left[n_i \operatorname{div}_Q (n_{iP}^{jQ} \mathfrak{v}_{iP}^{jQ}) + n_j \operatorname{div}_P (n_{jQ}^{iP} \mathfrak{v}_{jQ}^{iP})\right] \\ &= -\left[\operatorname{div}_Q (w_{iP}^{jQ} \mathfrak{v}_{iP}^{jQ}) + \operatorname{div}_P (w_{jQ}^{iP} \mathfrak{v}_{jQ}^{iP})\right] \end{aligned} \tag{6}$$

The indices P and Q on div indicate that the corresponding derivatives are to be taken with respect to x_P, y_P, z_P and x_Q, y_Q, z_Q. According to equation (3'), $w_{iP}^{jQ} = w_{jQ}^{iP}$. Now we will apply equation (6), which governs all nonstationary processes, to the conductivity problem. For this purpose we need only remember that the causes of the velocity of the ions are (1) the electric field and (2) heat motion. With Debye and Hückel, we can neglect the currents in the liquid in the case of very dilute solutions. If the external force on a j ion in dS_Q when it is in the neighborhood of an i ion is called $\mathfrak{K}_{iP}^{jQ}$, the Boltzmann constant k, the absolute temperature T, and the frictional constant of ions of the kind j ρ_j, the average velocity of the j ions due to heat motion is $(-kT/\rho_j\, w_{iP}^{jQ})\,(\mathrm{grad}_Q w_{iP}^{jQ})$, and the velocity of the j ions due to the force is $\mathfrak{K}_{iP}^{jQ}/\rho_j$. Therefore the total velocity of the j ions is:

$$\mathfrak{v}_{iP}^{jQ} = \left[\mathfrak{K}_{iP}^{jQ} - \frac{kT}{w_{iP}^{jQ}}\,\mathrm{grad}_Q\, w_{iP}^{jQ}\right]\frac{1}{\varrho_j} \tag{7}$$

In the same way:

$$\mathfrak{v}_{jQ}^{iP} = \left[\mathfrak{K}_{jQ}^{iP} - \frac{kT}{w_{jQ}^{iP}}\,\mathrm{grad}_P\, w_{jQ}^{iP}\right]\frac{1}{\varrho_i} \tag{7'}$$

Thus our equation (6) becomes:

$$\begin{aligned}\frac{\partial w_{iP}^{jQ}}{\partial t} = &-\frac{1}{\varrho_j}\,\mathrm{div}_Q\left[w_{iP}^{jQ}\mathfrak{K}_{iP}^{jQ} - kT\,\mathrm{grad}_Q\, w_{iP}^{jQ}\right]\\ &-\frac{1}{\varrho_i}\,\mathrm{div}_P\left[w_{jQ}^{iP}\mathfrak{K}_{jQ}^{iP} - kT\,\mathrm{grad}_P\, w_{jQ}^{iP}\right]\end{aligned} \tag{8}$$

For the conductivity problem, if $\mathfrak{E}$ denotes the external electric field:

$$\mathfrak{K}_{iP}^{jQ} = -e_j\,\mathrm{grad}_Q\,\psi_{iP}^{Q} + e_j\mathfrak{E} \qquad \mathfrak{K}_{jQ}^{iP} = -e_i\,\mathrm{grad}_P\,\psi_{jQ}^{P} + e_i\mathfrak{E} \tag{9}$$

Here ψ_{iP}^{Q} is the potential resulting from the ionic atmospheres of the i ions at the position of a j ion. Likewise, ψ_{iQ}^{P} is the potential resulting from the ionic atmospheres of j ions at the position of an i ion. These two potentials must satisfy Poisson's equation:

$$\mathrm{div}_Q\,\mathrm{grad}_Q\,\psi_{iP}^{Q} = -\frac{4\pi}{D}\sum_j n_{iP}^{jQ} e_j = -\frac{4\pi}{n_i D}\sum_j w_{iP}^{jQ} e_j \tag{10}$$

$$\mathrm{div}_P\,\mathrm{grad}_P\,\psi_{jQ}^{P} = -\frac{4\pi}{D}\sum_i n_{jQ}^{iP} e_i = -\frac{4\pi}{n_j D}\sum_i w_{jQ}^{iP} e_i \tag{10'}$$

If we consider the case of equilibrium, that is, if we consider a homogeneous electrolyte in which no current is flowing, then according to (7):

$$-w_{iP}^{jQ} e_j\,\mathrm{grad}_Q\,\psi_{iP}^{Q} - kT\,\mathrm{grad}_Q\, w_{iP}^{jQ} = c$$

That is, in this case the distribution function* becomes:

$$w_{iP}^{jQ} = n_i n_j e^{-\frac{e_j \psi_{iP}^{Q}}{kT}} \tag{11}$$

Likewise there follows:

$$w_{jQ}^{iP} = n_i n_j e^{-\frac{e_i \psi_{jQ}^{P}}{kT}} \tag{11'}$$

*The multiplicative constant $n_i n_j$ is adjusted so that at large distances from the central ions the above distribution reduces to that corresponding to random distribution of the ions.

If we put:

$$\psi_{iP}^{Q} = e_i g(r) \qquad (11'')$$

$$\psi_{jQ}^{P} = e_j g(r) \qquad (11''')$$

where $g(r)$ depends only on the distance r of the i ion from the j ion; then, if as in the original method of Debye and Hückel we replace the exponential function by its expansion, we get:

$$\Delta g(r) = g(r)\frac{4\pi}{DkT}\sum n_j e_j^2 = \varkappa^2 g(r) \qquad (11'''')$$

Thus, in this stationary case, even when the Brownian motion is taken into account, the differential equation which Debye and Hückel have used for the case of static ions is obtained. It holds only for very dilute solutions. Equations (11-11''') also fulfill equation (3') in this case. If equation (9) is substituted into (8) the result is:

$$\begin{aligned}\frac{\partial w_{iP}^{jQ}}{\partial t} &= \frac{1}{\varrho_j}\operatorname{div}_Q\left[e_j w_{iP}^{jQ}\operatorname{grad}_Q \psi_{iP}^{Q} + kT\operatorname{grad}_Q w_{iP}^{jQ} - e_j \mathfrak{E} w_{iP}^{jQ}\right] \\ &+ \frac{1}{\varrho_i}\operatorname{div}_P\left[e_i w_{jQ}^{iP}\operatorname{grad}_P \psi_{jQ}^{P} + kT\operatorname{grad}_P w_{jQ}^{iP} - e_i \mathfrak{E} w_{jQ}^{iP}\right]\end{aligned} \qquad (12)$$

Now we will examine closely the deviation from the stationary distribution characterized by the relations (11-11''''). Since (compare equation 3):

$$w_{iP}^{jQ} = w_{jQ}^{iP}$$

we simplify the writing by setting:

$$w_{iP}^{jQ} = w_{jQ}^{iP} = w_{ij} \qquad (13)$$

Further, let f_{ij} denote the deviation from complete random distribution of the ions; that is, we write:

$$w_{ij} = n_i n_j + f_{ij} \qquad (14)$$

Thus the differential equation for f_{ij} becomes:

$$\begin{aligned}\frac{\partial f_{ij}}{\partial t} &= \frac{1}{\varrho_j}\left|-\frac{n_i n_j e_j}{n_i D}4\pi\sum_p f_{ip}e_p + kT\Delta_Q f_{ij} - e_j \mathfrak{E}\operatorname{div}_Q f_{ij}\right| \\ &+ \frac{1}{\varrho_i}\left|-\frac{n_i n_j e_i}{n_j D}4\pi\sum_p f_{pj}e_p + kT\Delta_P f_{ij} - e_i \mathfrak{E}\operatorname{div}_P f_{ij}\right|\end{aligned} \qquad (15)$$

or:

$$\begin{aligned}\frac{\partial f_{ij}}{\partial t} &= \frac{1}{\varrho_j}\left|kT\Delta_Q f_{ij} - \frac{4\pi n_j e_j}{D}\sum_p f_{ip}e_p - e_j \mathfrak{E}\operatorname{div}_Q f_{ij}\right| \\ &+ \frac{1}{\varrho_i}\left|kT\Delta_P f_{ij} - \frac{4\pi n_i e_i}{D}\sum_p f_{pj}e_p - e_i \mathfrak{E}\operatorname{div}_P f_{ij}\right|\end{aligned} \qquad (15')$$

Further, we substitute the relative coordinates x, y, z in the distribution function f_{ij} in which we are interested:

$$f_{ij} = F_{ij}(x, y, z) \tag{16}$$

Here:

$$x = x_Q - x_P, \quad y = y_Q - y_P, \quad z = z_Q - z_P \tag{17}$$

so that:

$$\frac{\partial f_{ij}}{\partial x_Q} = \frac{\partial F_{ij}}{\partial x}, \quad \frac{\partial f_{ij}}{\partial x_P} = -\frac{\partial F_{ij}}{\partial x} \cdots$$
$$\frac{\partial^2 f_{ij}}{\partial x_Q^2} = \frac{\partial^2 F_{ij}}{\partial x^2}. \quad \frac{\partial^2 f_{ij}}{\partial x_P^2} = \frac{\partial^2 F_{ij}}{\partial x^2} \cdots \tag{18}$$

Further, let the external field act parallel to x and denote $\mathfrak{E}_x$ by X. Then the differential equation for F_{ij} is:

$$\frac{\partial F_{ij}}{\partial t} = kT\left(\frac{1}{\varrho_i} + \frac{1}{\varrho_j}\right)\Delta F_{ij} + \left(\frac{e_i}{\varrho_i} - \frac{e_j}{\varrho_j}\right)X\frac{\partial F_{ij}}{\partial x}$$
$$-\frac{4\pi}{D}\left(\frac{n_i e_i}{\varrho_i}\sum_p e_p F_{pj} + \frac{n_j e_j}{\varrho_j}\sum_p e_p F_{ip}\right) \tag{19}$$

The equations (19) are the general basic equations on which the conductivity problem is founded in a nonstationary case. Now we will simplify the equations (19) by limiting ourselves to the special case in which only two kinds of ions are present. The following discussion will always refer to this case. When we speak of simple electrolytes in the following, we mean those which contain two kinds of ions—for example, KCl, $MgCl_2$, and so on.

B. Frequency Dependence of Dynamic Effects for Strong Electrolytes II. The Basic Equations for the Case of Two Kinds of Ions

Let the electrolyte contain two kinds of ions. Then four functions, F_{11}, F_{22}, F_{12}, and F_{21}, appear. These four functions must satisfy the four equations:

$$\frac{\partial F_{11}}{\partial t} = \frac{2kT}{\varrho_1}\Delta F_{11} - \frac{4\pi}{D}\frac{n_1 e_1}{\varrho_1}(e_1 F_{11} + e_2 F_{21}) - \frac{4\pi}{D}\frac{n_1 e_1}{\varrho_1}(e_1 F_{11} + e_2 F_{12})$$
$$\frac{\partial F_{22}}{\partial t} = \frac{2kT}{\varrho_2}\Delta F_{22} - \frac{4\pi}{D}\frac{n_2 e_2}{\varrho_2}(e_1 F_{12} + e_2 F_{22}) - \frac{4\pi}{D}\frac{n_2 e_2}{\varrho_2}(e_1 F_{21} + e_2 F_{22})$$
$$\frac{\partial F_{12}}{\partial t} = kT\left(\frac{1}{\varrho_1} + \frac{1}{\varrho_2}\right)\Delta F_{12} + \left(\frac{e_1}{\varrho_1} - \frac{e_2}{\varrho_2}\right)X\frac{\partial F_{12}}{\partial x}$$
$$-\frac{4\pi}{D}\left[\frac{n_1 e_1}{\varrho_1}(e_1 F_{12} + e_2 F_{22}) + \frac{n_2 e_2}{\varrho_2}(e_1 F_{11} + e_2 F_{12})\right] \tag{20}$$
$$\frac{\partial F_{21}}{\partial t} = kT\left(\frac{1}{\varrho_1} + \frac{1}{\varrho_2}\right)\Delta F_{21} + \left(\frac{e_2}{\varrho_2} - \frac{e_1}{\varrho_1}\right)X\frac{\partial F_{21}}{\partial x}$$
$$-\frac{4\pi}{D}\left[\frac{n_2 e_2}{\varrho_2}(e_1 F_{11} + e_2 F_{21}) + \frac{n_1 e_1}{\varrho_1}(e_1 F_{21} + e_2 F_{22})\right]$$

We will now assume that X is very small and oscillates with frequency ω. Thus:

$$X = Ee^{i\omega t}$$
$$F_{ij} = F_{ij}^0 + g_{ij} \tag{21}$$

where g_{ij} is very small compared to F_{ij}^{o}, an assumption which is correct in the region of validity of Ohm's law. According to equations (11,11', ...) F_{ij}^{o} is:

$$F_{ij}^{0} = n_i n_j \left(1 - \frac{e_i e_j}{DkT} \frac{e^{-\varkappa r}}{r}\right) \tag{22}$$

The deviations g_{ij} of the distribution functions are also harmonic oscillations with the same circular frequency ω. Thus we set:

$$g_{ij} = G_{ij} e^{i\omega t}$$

Substituting these expressions in equation (20) gives the equations:

$$\frac{2}{\varrho_1} kT\Delta G_{11} - i\omega G_{11} - \frac{4\pi n_1 e_1}{D\varrho_1}(e_1 G_{11} + e_2 G_{21}) - \frac{4\pi n_1 e_1}{D\varrho_1}(e_1 G_{11} + e_2 G_{12}) = 0$$

$$\frac{2}{\varrho_2} kT\Delta G_{22} - i\omega G_{22} - \frac{4\pi n_2 e_2}{D\varrho_2}(e_1 G_{12} + e_2 G_{22}) - \frac{4\pi n_2 e_2}{D\varrho_2}(e_1 G_{21} + e_2 G_{22}) = 0 \tag{23}$$

$$\left(\frac{1}{\varrho_1} + \frac{1}{\varrho_2}\right) kT\Delta G_{12} - i\omega G_{12} - \frac{4\pi n_1 e_1}{D\varrho_1}(e_1 G_{12} + e_2 G_{22}) - \frac{4\pi n_2 e_2}{D\varrho_2}(e_1 G_{11} + e_2 G_{12}) = -\left(\frac{e_1}{\varrho_1} - \frac{e_2}{\varrho_2}\right) E \frac{\partial F_{12}^{0}}{\partial x}$$

$$\left(\frac{1}{\varrho_1} + \frac{1}{\varrho_2}\right) kT\Delta G_{21} - i\omega G_{21} - \frac{4\pi n_2 e_2}{D\varrho_2}(e_1 G_{11} + e_2 G_{21}) - \frac{4\pi n_1 e_1}{D\varrho_1}(e_1 G_{21} + e_2 G_{22}) = \left(\frac{e_1}{\varrho_1} - \frac{e_2}{\varrho_2}\right) E \frac{\partial F_{21}^{0}}{\partial x}$$

Now the problem is to find solutions of these equations. So long as the field strength is not too high, say considerably under 100,000 volts/cm., Ohm's law is valid. In this case:*

$$G_{12} = -G_{21} \quad \text{from which follows } G_{11} = G_{22} = 0 \tag{24}$$

These relations (24) between the distribution functions have already been used by Onsager in the problem of conductivity in the stationary case. Physically, they mean nothing but that the ionic atmospheres of the two kinds of ions are changed in a dissymmetric way by the fields, as was shown by Debye and Hückel. Therefore our differential equations reduce to a single equation which governs the problem of the frequency dependence of the conductivity and the dielectric constant of a simple electrolyte:

$$\left(\frac{1}{\varrho_1} + \frac{1}{\varrho_2}\right) kT\Delta G_{12} - i\omega G_{12} - \frac{4\pi}{D}\left(\frac{n_1 e_1^2}{\varrho_1} + \frac{n_2 e_2^2}{\varrho_2}\right) G_{12} = -\left(\frac{e_1}{\varrho_1} - \frac{e_2}{\varrho_2}\right) E \frac{\partial F_{12}^{0}}{\partial x} \tag{25}$$

For the further treatment of this differential equation (25) we introduce the relaxation time Θ of the ionic atmosphere around ion 1 or ion 2:

$$\Theta = \frac{\varrho_1 \varrho_2}{\varrho_1 + \varrho_2} \frac{1}{kT\varkappa^{*2}} \tag{26}$$

where:

$$\varkappa^{*2} = q\varkappa^2 \tag{27}$$

*It is easy to see by examining the entire complex of solutions that the special relations (24) are the appropriate solution in our case.

and:

$$q = \frac{\frac{e_1}{\varrho_1} - \frac{e_2}{\varrho_2}}{(e_1 - e_2)\left(\frac{1}{\varrho_1} + \frac{1}{\varrho_2}\right)} \qquad (28)$$

Further, we introduce a frequency function K by:*

$$K^2 = \varkappa^{*2}(1 + i\omega\Theta) \qquad (29)$$

On substitution of equation (29) in the differential equation (25), it reduces to:

$$\Delta G_{12} - K^2 G_{12} = -\frac{e_1\varrho_2 - e_2\varrho_1}{\varrho_1 + \varrho_2}\frac{E}{kT}\frac{\partial F_{12}^0}{\partial x} \qquad (30)$$

It is useful to introduce the additional potential ψ'_1 around ion 1 which exists because the original potential has been disturbed by the superimposed field. According to equations (10, 11, 11',..., 21, and 22):

$$G_{12}e^{i\omega t} = -\frac{D}{4\pi}\frac{n_1}{e_2}\Delta\psi'_1 \qquad (31)$$

According to equations (30) and (31), this potential ψ'_1 must satisfy the differential equation:

$$\Delta\Delta\psi'_1 - K^2\Delta\psi'_1 = \Omega\varkappa^{*2}\frac{\partial}{\partial x}\frac{e^{-\varkappa r}}{r}e^{i\omega t} \qquad (32)$$

Here we use the abbreviation:

$$\Omega = E\frac{e_1 e_2}{DkT} \qquad (33)$$

Furthermore, the equation (22):

$$F_{12}^0 = n_1 n_2 - \frac{n_1 n_2}{DkT}\frac{e_1 e_2 e^{-\varkappa r}}{r}$$

has been used in deriving equation (32). Our differential equation (32) is of the same general form as that on which Onsager based his work on the conductivity of strong electrolytes. The difference is that K^2 now is complex, which leads to frequency dependence.

III. Calculation of the Asymmetric Potentials of Oscillating Ions

Now we must find the additional potential at the position of an ion. First we remark that, since according to (24') and (31):

$$n_1 e_1 \psi'_1 + n_2 e_2 \psi'_2 = 0$$

we have:

$$\psi'_1 = \psi'_2 \qquad (34)$$

*It is to be kept in mind that for the simple electrolytes treated here

$$n_1 e_1 + n_2 e_2 = 0$$

(since $n_1 e_1 = -n_2 e_2$). That is, the additional potential which results from a change in the symmetric ionic atmosphere of one ion is the same as the additional potential resulting from a change in the ionic cloud of the other ion when both are averaged in time. As a particular solution of equation (32) we can give immediately:

$$\frac{q\Omega}{\varkappa^2 - K^2} \frac{\partial}{\partial x} \frac{e^{-\varkappa r}}{r} e^{i\omega t}$$

Now we must add an arbitrary solution of the homogeneous equation:

$$\Delta(\Delta\psi' - K^2\psi') = 0$$

to this particular solution. A general solution of equation (32) is:

$$\psi_1' = \left[\frac{q\Omega}{\varkappa^2 - K^2} \frac{d}{dr}\left(\frac{e^{-\varkappa r}}{r}\right) + A' \frac{d\left(\frac{e^{-Kr}}{r}\right)}{dr} + \frac{B'}{r^2}\right] e^{i\omega t} \cos\vartheta \tag{35}$$

ϑ is the angle between the direction of the field and the radius vector $\mathfrak{r}$. Now the potential of the oscillating ions must not be infinite at the position of an ion. Likewise, the charge density there must not be infinite. If we expand the following quantities in the way indicated:

$$\frac{d}{dr}\left(\frac{e^{-\varkappa r}}{r}\right) = -\frac{1}{r^2} + \frac{\varkappa^2}{2} - \frac{\varkappa^3 r}{3} + \cdots$$

$$\frac{d}{dr}\left(\frac{e^{-Kr}}{r}\right) = -\frac{1}{r^2} + \frac{K^2}{2} - \frac{K^3 r}{3} + \cdots$$

we see that the constants A' and B' must satisfy the equations:

$$-\frac{q\Omega}{\varkappa^2 - K^2} - A' + B' = 0 \qquad \frac{q\Omega}{\varkappa^2 - K^2}\varkappa^2 + A'K^2 = 0$$

From these the constants A' and B' are found to be:

$$A' = -\frac{q\Omega}{\varkappa^2 - K^2}\frac{\varkappa^2}{K^2} \qquad B' = A' + \frac{q\Omega}{\varkappa^2 - K^2} \tag{36}$$

Consequently, in the neighborhood of the ion the additional potential is:

$$\psi_1' = \left(-\frac{q\Omega}{\varkappa^2 - K^2}\frac{\varkappa^3 r}{3} - \frac{A'K^3}{3} r \cdots\cdots\right) e^{i\omega t}\cos\vartheta = \frac{-\Omega q \varkappa^2}{3(\varkappa + K)} r e^{i\omega t}\cos\vartheta \tag{37}$$

IV. Calculation of the Dynamic Effect on a Vibrating Ion (Frequency Dependence of the "Absolute Value of the Force" and "Phase")

According to equation (37) the field strength at the origin, that is, at the position of ion 1, is:

$$\frac{\Omega q \varkappa^2}{3(\varkappa + K)} \cdot e^{i\omega t}$$

If according to equations (27) and (29) we put:

$$K = \varkappa\sqrt{q}\sqrt{1 + i\omega\Theta} \tag{38}$$

the field strength at the position of the ion can be written:

$$\frac{e_1 e_2}{3DkT}\frac{q\varkappa E e^{i\omega t}}{1 + \sqrt{q}\sqrt{1 + i\omega\Theta}} \tag{39}$$

Now in the limit:

$$\varrho_1 v_1 = e_1 E \tag{40}$$

Therefore to a sufficient degree of approximation the force on ion 1 is:

$$\frac{e_1 e_2}{3DkT} \frac{\varkappa \varrho_1 v_1 q e^{i\omega t}}{1 + \sqrt{q}\sqrt{1 + i\omega\Theta}}$$

or:

$$\frac{e_1 e_2}{3DkT} \sqrt{q}\, \varrho_1 v_1 \frac{a - ib}{a^2 + b^2} e^{i\omega t} \tag{41}$$

where:

$$\frac{1}{\sqrt{q}} + \sqrt{1 + i\omega\Theta} = a + ib \tag{42}$$

The force which we wanted to calculate is represented by:

$$-K^* e^{i(\omega t + \Delta)} = K^* e^{i(\omega t + \Delta + \pi)} \tag{43}$$

An expression in which:

$$K^* = \frac{|e_1 e_2|}{3DkT} \frac{\varkappa \varrho_1 v_1 \sqrt{q}}{\sqrt{a^2 + b^2}} \tag{44}$$

and the phase Δ is given by:

$$\operatorname{tg} \Delta = -\frac{b}{a} \tag{45}$$

$\pi + \Delta$ is the phase difference between the force on an oscillating ion and its velocity. For any particular simple electrolyte, the force K^* and the phase Δ can be calculated as a function of $\omega\theta$. For this purpose it is only necessary to calculate the quantities q, $\varkappa$, and θ (compare, for example, Table III on p. 414); then equations (44) and (45) give K^* and Δ as functions of the frequency ω or the wave length of the alternating electric field.[1] In the stationary case, the absolute value of the force is:

$$\frac{|e_1 e_2|}{3DkT} \varkappa \varrho_1 v_1 \frac{q}{1 + \sqrt{q}} \tag{46}$$

a value which agrees with Onsager's.

Now as in our earlier work we will investigate quantitatively the case of KCl and see just how large the changes from the qualitative treatment (compare DFI) are.

V. Example of KCl ("Absolute Value of the Force" and "Phase" as Functions of $\omega\theta$; θ = Relaxation Time, ω = Circular Frequency)

KCl is a binary electrolyte ($e_1 = -e_2 = e$). According to equation (28), for all binary electrolytes:

$$q = 0.5$$

[1] We may remark that for complicated electrolytes for which $q \neq 0.5$ the curves for δ and Δ do not deviate markedly from the curves of Figures 2 and 3.

The relaxation time can be calculated from equation (26) and is, since $\rho_1 = \rho_2 = \rho$

$$\Theta = \frac{\varrho}{kT\varkappa^2} = \frac{0.553}{\gamma} \cdot 10^{-10}\,\mathrm{sec} \qquad (\gamma = \text{concentration in mol per liter})$$

which is exactly the same as the value upon which we based our earlier report (compare equation (2) in DFI). This refers to a temperature of 18°C if the solvent is water. In the stationary case ($\omega = 0$) and for binary electrolytes in general, the absolute value of the force is:

$$\frac{e^2\varkappa}{6DkT}(2-\sqrt{2})\,E \qquad (46')$$

Now we take the ratio of the absolute value of the force on a vibrating ion and the absolute value of that on a uniformly moving ion (46'), which we will call δ. We also calculate the phase Δ as a function of $\omega\theta$. In the following Table I, δ and Δ are given for a sequence of values of $\omega\theta$. The numbers in the third and fifth columns result when the Brownian motion is neglected; we have calculated them earlier (DFI, Table II) and have put them in parentheses in Table I. It is seen that the values of δ and Δ differ little from those calculated earlier. The drop in δ is somewhat less rapid than before. Again, for very small values of $\omega\theta$, δ is hardly distinguishable from 1. For very small values of $\omega\theta$ the curve is of parabolic shape and has a horizontal tangent at $\omega\theta = 0$, and for large values of $\omega\theta$ it decreases as $1/\omega\theta$. In Figure 2 the solid curve represents δ as a function of $\omega\theta$ as it follows from the quantitative theory, while the dashed line

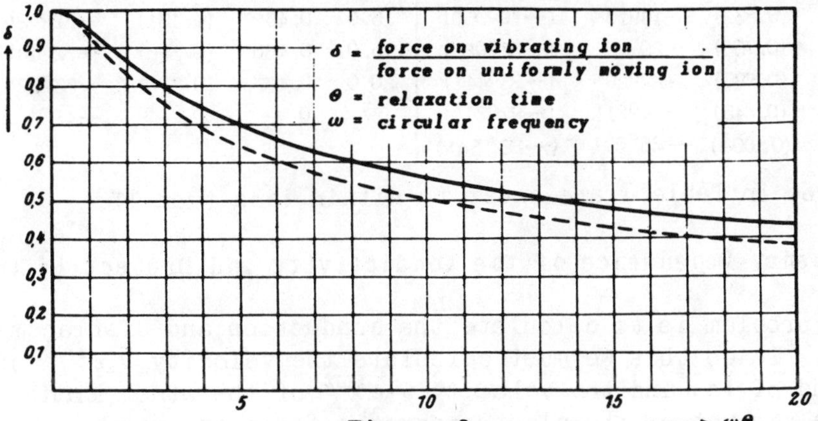

Figure 2

represents the function which results when the Brownian motion is neglected. In Figure 3 the phase Δ is plotted as a function of $\omega\theta$ (solid curve). Again, the dashed curve refers to the values which result when the Brownian motion is not taken into account and which are designated by (Δ). For very small values of $\omega\theta$ the negative phase Δ increases linearly; for large values of $\omega\theta$ it approaches the value -45°. According to equation (46'), for the stationary case, the absolute value of the force turns out less than that resulting when the Brownian motion is not accounted for. For

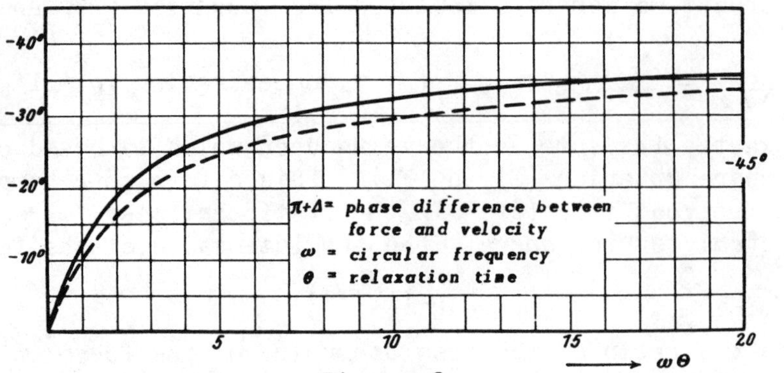

Figure 3

binary electrolytes, the ratio of the two is:

$$2 - \sqrt{2} = 0.586$$

Table I

$\omega\theta$	δ	(δ)	Δ	(Δ)	$\omega\theta$	δ	(δ)	Δ	(Δ)
0	1.00	(1.00)	0°	(0°)	7.0	0.628	(0.573)	—27° 4'	(—30°07')
0.1	1.00	(0.999)	—1°11'	(— 1°26')	8.0	0.601	(0.547)	—28°	(—30°55')
0.2	0.998	(0.998)	—2°22'	(— 2°50')	9.0	0.579	(0.524)	—28°46'	(—31°40')
0.5	0.983	(0.980)	—5°41'	(— 6°50')	10.0	0.558	(0.507)	—29°30'	(—32°19')
1.0	0.945	(0.930)	-10°16'	(—12°15')	12.0	0.524	(0.469)	—30°42'	(—33°30')
2.0	0.863	(0.832)	-16°18'	(—19°05')	15.0	0.482	(0.431)	—32° 2'	(—34°41')
3.0	0.793	(0.755)	-20°	(—23°02')	18.0	0.450	(0.398)	—32°49'	(—35°25')
4.0	0.741	(0.690)	-22°29'	(—25°44')	20.0	0.433	(0.380)	—33°32'	(—35°57')
5.0	0.696	(0.645)	-24°21'	(—27°31')	∞	0	0	—45°	(—45°)
6.0	0.658	(0.608)	—25°50'	(—28°54')					

The figures in Table I are exact to within less than 1%.

C. Frequency Dependence of the Conductivity and Dielectric Constant

Our problem is to calculate the conduction and displacement currents. To do this we must calculate the velocity $v_1e^{i\omega t}$ of the first kind of ion and the velocity $v_2e^{i\omega t}$ of the other kind. We will take account of the electrophoretic force and assume that it is independent of frequency*. It is caused by changes in the Stokes frictional force which occur because of the presence of ions. Since this electrophoretic force depends on the potential around the static ion, we can take for it the value given by Debye and Hückel, a procedure justifiable for small velocity. The force due to the

*It is not certain that this assumption is valid. However, we believe that it is probably correct and hope that further investigation will clarify this point. If this assumption is not correct, the dispersion effect will only be magnified.

external field, the Stokes force, the electrophoretic force, and the relaxation force must be in equilibrium. The following relation expresses this condition (η = coefficient of internal friction):*

$$
\begin{aligned}
& e_1 E - \varrho_1 v_1 + \frac{e_1 e_2}{3DkT} \frac{\varrho_1 v_1 \sqrt{q}}{\frac{1}{\sqrt{q}} + \sqrt{1 + i\omega\Theta}} - \frac{e_1 E \varkappa \varrho_1}{6\pi\eta} = 0 \\
& e_2 E - \varrho_2 v_2 + \frac{e_1 e_2}{3DkT} \frac{\varrho_2 v_2 \sqrt{q}}{\frac{1}{\sqrt{q}} + \sqrt{1 + i\omega\Theta}} - \frac{e_2 E \varkappa \varrho_2}{6\pi\eta} = 0
\end{aligned}
\tag{47}
$$

From this the velocities $v_1 e^{i\omega t}$ and $v_2 e^{i\omega t}$ can be calculated and are

$$
v_j e^{i\omega t} = E\left| e_j \omega_j + \frac{e_1 e_2}{3DkT} \varkappa e_j \omega_j \chi(q, \omega\Theta) - \frac{e_j \varkappa}{6\pi\eta} \right| e^{i\omega t} \tag{48}
$$

where j = 1 or 2 according to the kind of ion, and where:

$$
\chi(q, \omega\Theta) = \frac{\sqrt{q}\left| \sqrt{1 + i\omega\Theta} - \frac{1}{\sqrt{q}} \right|}{1 + i\omega\Theta - \frac{1}{q}} \tag{49}
$$

$$
\omega_j = \frac{1}{\varrho_j} \tag{49'}
$$

Thus χ is related to the frequency-dependent part of the conductivity and depends only on q and the product $\omega\theta$. From equation (48) we obtain for the current density $\mathfrak{i}$:

$$
\mathfrak{i} = \sum n_j e_j v_j e^{i\omega t} = E e^{i\omega t} \left| \sum_j n_j e_j^2 \omega_j + \frac{e_1 e_2}{3DkT} \varkappa \sum_j n_j e_j^2 \omega_j \chi(q, \omega\Theta) - \sum_j \frac{n_j e_j^2 \varkappa}{6\pi\eta} \right| \tag{50}
$$

According to Maxwell's phenomenological theory, the total current density J is equal to the sum of the conduction and displacement currents. Consequently:

$$
J = \lambda \mathfrak{E} + \frac{D'}{4\pi} \dot{\mathfrak{E}} = \lambda E e^{i\omega t} + \frac{i\omega}{4\pi} D' E e^{i\omega t} \tag{51}
$$

and:

$$
J = \frac{D}{4\pi} i\omega E e^{i\omega t} + \left| \sum n_j e_j^2 \omega_j + \frac{e_1 e_2}{3DkT} \varkappa \sum n_j e_j^2 \omega_j \chi - \sum \frac{n_j e_j^2 \varkappa}{6\pi\eta} \right| E e^{i\omega t} \tag{51'}
$$

D is the dielectric constant of the solvent. The real part of the right-hand side of equation (51') gives the conductivity λ, and is:

$$
\lambda = \sum n_j e_j^2 \omega_j + \frac{e_1 e_2}{3DkT} \varkappa \sum n_j e_j^2 \omega_j \times \textbf{real part of } \chi - \sum \frac{n_j e_j^2 \varkappa}{6\pi\eta} \tag{52}
$$

while the imaginary part of equation (51') is related to the increase in the dielectric constant in the following way:

*The value of the electrophoretic force is taken from the work of Debye and Hückel (*Physik. Z.*, *24*, 305 ff. (1923)).

$$D' - D = \frac{4\pi}{\omega}\frac{e_1 e_2}{3DkT}\varkappa \sum n_j e_j^2 \omega_j \times \textbf{imaginary part of } \chi \tag{53}$$

The conductivity and the dielectric constant depend on the frequency through χ. Now we must find the general formulae for the conductivity and the increase in the dielectric constant by taking the real and imaginary parts, respectively, of χ.

VI. General Formula for the Conductivity

Because of equation (52), the conductivity λ can be written:

$$\lambda = \lambda_\infty + \frac{e_1 e_2}{3DkT}\varkappa\lambda_\infty \times \textbf{real part of } \chi - \sum \frac{n_j e_j^2}{6\pi\eta}\varkappa \tag{54}$$

where:

$$\sum n_j e_j^2 \omega_j = \lambda_\infty \tag{55}$$

Now, for the contribution to the conductivity resulting from the relaxation force of the ionic atmospheres, we introduce the abbreviation:

$$\frac{|e_1 e_2|}{3DkT}\varkappa\lambda_\infty \times \textbf{real part of } \chi = \lambda_{\text{I}} \tag{56}$$

and for the electrophoretic contribution, the abbreviation:

$$\sum \frac{n_j e_j^2 \varkappa}{6\pi\eta} = \lambda_{\text{II}} \tag{57}$$

Consequently:

$$\lambda = \lambda_\infty - \lambda_{\text{I}} - \lambda_{\text{II}} \tag{58}$$

Now we consider the ratio of the decreases in the conductivity resulting from the relaxation force at the frequencies ω and 0. That is, we form the quotient:

$$\frac{\lambda_{\text{I}\omega}}{\lambda_{\text{I}0}} = \frac{(\textbf{real part of } \chi)_\omega}{(\textbf{real part of } \chi)_{\omega=0}} \tag{59}$$

According to equation (49):

$$(\textbf{real part of } \chi)_{\omega=0} = \frac{q}{1+\sqrt{q}} \tag{60}$$

If we put:

$$\sqrt{1 + i\omega\Theta} = \overline{R} + i\,\overline{Q} \tag{61}$$

so that:

$$\overline{R} = \frac{1}{\sqrt{2}}\sqrt{\sqrt{1+\omega^2\Theta^2}+1} \qquad \overline{Q} = \frac{1}{\sqrt{2}}\sqrt{\sqrt{1+\omega^2\Theta^2}-1} \tag{61'}$$

then:

$$\textbf{real part of } \chi = \frac{\sqrt{q}}{\left(1-\frac{1}{q}\right)^2 + \omega^2\Theta^2}\left[\left(1-\frac{1}{q}\right)\left(\overline{R}-\frac{1}{\sqrt{q}}\right) + \omega\Theta\overline{Q}\right] \tag{62}$$

Formula (54) in conjunction with (62) determines the conductivity λ as a function of $\omega\theta$. We will now introduce the molar conductivity Λ defined by:

$$\Lambda = \frac{1000}{\gamma}\lambda \tag{63}$$

Here γ is the molar concentration of the dissolved salt. The molar conductivity at infinite dilution is:

$$\Lambda_\infty = \frac{1000}{\gamma}\lambda_\infty \tag{64}$$

Thus we have to write:

$$\Lambda = \Lambda_\infty - \Lambda_{I} - \Lambda_{II} \tag{65}$$

in which:

$$\Lambda_{I} = \frac{|e_1 e_2|}{3DkT}\varkappa\Lambda_\infty \times \text{real part of } \chi \tag{66}$$

$$\Lambda_{II} = \sum \frac{n_j e_j^2 \varkappa}{6\pi\eta}\frac{1000}{\gamma} \tag{67}$$

In practice the molar conductivity is expressed in practical units (ohm^{-1} cm.^{2}); that is, if Λ is the molar conductivity in these units, then:

$$\overline{\Lambda} = \frac{\Lambda}{9 \cdot 10^{11}} \cdots \tag{68}$$

In practical units our formula for the conductivity is:

$$\overline{\Lambda} = \overline{\Lambda}_\infty - \overline{\Lambda}_{I} - \overline{\Lambda}_{II} \tag{69}$$

where $\overline{\Lambda}_I$ and $\overline{\Lambda}_{II}$ are:

$$\overline{\Lambda}_{I} = \frac{|e_1 e_2|}{3DkT}\varkappa\overline{\Lambda}_\infty \times \text{real part of } \chi \tag{70}$$

$$\overline{\Lambda}_{II} = \sum \frac{n_j e_j^2 \varkappa}{6\pi\eta}\frac{1000}{\gamma}\frac{1}{9\ 10^{11}} \tag{71}$$

We base the discussion of particular examples in part D on these formulae (70,71).

VII. General Formula for the Increase in the Dielectric Constant

For the increase in the dielectric constant we found (compare equation 53):

$$D' - D = \frac{4\pi}{\omega}\frac{e_1 e_2 \varkappa}{3DkT}\sum n_j e_j^2 \omega_j \times \text{imaginary part of } \chi \tag{72}$$

or:

$$D' - D = \frac{4\pi}{\omega}\frac{e_1 e_2 \varkappa}{3DkT}\lambda_\infty \times \text{imaginary part of } \chi \tag{73}$$

The imaginary part of χ is:

$$\text{imaginary part of } \chi = \frac{\sqrt{q}}{\left(1-\frac{1}{q}\right)^2 + \omega^2\Theta^2}\left[\overline{Q}\left(1-\frac{1}{q}\right) - \omega\Theta\left(\overline{R}-\frac{1}{\sqrt{q}}\right)\right] \tag{74}$$

Thus the formula which gives the increase in the dielectric constant reduces to:

$$D' - D = \frac{4\pi e_1 e_2}{3DkT}\frac{\Theta\sqrt{q}}{\omega\Theta\left|\left(1-\frac{1}{q}\right)^2 + \omega^2\Theta^2\right|}\left|\overline{Q}\left(1-\frac{1}{q}\right) - \omega\Theta\left(\overline{R}-\frac{1}{\sqrt{q}}\right)\right| \tag{75}$$

For any particular example $D' - D$ can be calculated easily from equation (75) as a function of $\omega\theta$. Curves are obtained which are similar to those given in our first paper (compare DFI, Fig. 5). However (compare the concluding remarks in DFI), since the nature of this phenomenon has not yet been cleared up experimentally, we will avoid a special discussion of the formula (75) here. Now we turn to a discussion of the way in which the conductivity depends on the frequency in special examples and will investigate how the different factors affect our results.

D. Discussion of Particular Examples

VIII. q Values, κ Values, Relaxation Times θ, and Mobilities for the Special Examples Considered in Part D

In this last part of our paper we choose as examples the electrolytes HCl, KCl, $MgCl_2$, $CdSO_4$, $LaCl_3$, and $K_4Fe(CN)_6$. First we give the mobilities which will be used here.* If a molecule decomposes into ions numbered $\nu_1 \ldots \nu_s$ with valencies $z_1 \ldots z_s$ and frictional constants $\rho_1 \ldots \rho_s$, the molar conductivity at infinite dilution can be written

$$\overline{\Lambda}_\infty = \sum \nu_i \overline{L}_i \tag{75'}$$

where:

$$\overline{L}_i = \frac{N\varepsilon^2 \; z_i^2}{9\cdot 10^{11}\,\varrho_i} \tag{76}$$

$$(N = 6.06\cdot 10^{23}$$
$$\varepsilon = 4.774\cdot 10^{-10}\text{ e. s. e.})$$

We call the quantities L_i the mobilities. Putting in the values of N and ε gives:

$$\overline{L}_i = \frac{15.34}{\varrho_i} z_i^2 \cdot 10^{-8} \tag{77}$$

Since the mobilities are known from experiment, it is useful to express the quantities appearing in our formula for the conductivity in terms of them. Doing this, we find the formulae:

*The general custom is tc call L_i/z_i the mobility.

$$\rho_i = \frac{15.34 \cdot z_i^2}{\overline{L}_i} 10^{-8} \tag{78}$$

$$q = \frac{\overline{L}_1 z_2 + \overline{L}_2 z_1}{(z_1 + z_2)(\overline{L}_1 z_2^2 + \overline{L}_2 z_1^2)} z_1 \cdot z_2 \tag{79}$$

$$\Theta = \frac{z_1^2 z_2^2}{(z_2^2 \overline{L}_1 + z_1^2 \overline{L}_2)} \frac{15.34 \cdot 10^{-8}}{kTq\varkappa^2} \tag{80}$$

The quantities ρ_i and q depend only on the valencies and the mobilities; the latter are very temperature-dependent because the internal friction of the solvent depends strongly on temperature. On the other hand, the relaxation time contains besides the valencies and the mobilities the important quantity κ^2. This latter quantity depends on the dielectric constant, the concentration, the valencies, the kind of electrolyte, and the temperature. It depends only slightly on temperature. We can write κ^2 in the form:

$$\varkappa^2 = \frac{4\pi}{DkT} \frac{\varepsilon^2 N\gamma}{1000} \sum \nu_i z_i^2 \tag{81}$$

If water is chosen as the solvent, then

$$\varkappa^2 = 0.05342 \cdot 10^{16} \cdot \gamma \sum \nu_i z_i^2 \text{ (at } t = 18^0\text{C)}$$

$$\varkappa^2 = 0.05385 \cdot 10^{16} \gamma \sum \nu_i z_i^2 \text{ (at } t = 25^0\text{C)}$$

If the kind of electrolyte is known, and further the solvent and temperature, then for some definite concentration γ of the electrolyte, the quantities q, θ, and κ^2 which determine the frequency dependence of the conductivity can be calculated easily from equations (79), (80), and (81). The values of the valencies and the mobilities (at the given temperature of the solvent) are listed in Table II.

Table II

	z_i	$\overline{L}_i$	t
H	1	315	18 °C
K	1	65	18 °C
Cl	1	65	18 °C
Mg	2	92	18 °C
SO_4	2	136.6	18 °C
Cd	2	92	18 °C
La	3	50	18 °C
$Fe(CN)_6$	4	380	25 °C

The following remarks concerning Table II are in order. We use values of the mobilities which agree with those usually given in the literature to within about 1%. Since the dispersion effect causès a change in the molar conductivity which is relatively small at infinite dilution, the mobilities used are sufficiently exact for the dispersion phenomenon. The measurement of the value of the molar conductivity for $K_4Fe(CN)_6$ is based on measurements of Burrows[2] made at a temperature of 25°C. He found 680. If we recall that the mobility of K at t = 25° C is 75, we find for the mobility of $Fe(CN)_6$ the value given in Table II. In Table III we have given the characteristic quantities κ^2, q, and $\theta\gamma$. The values in the first column are given in order to illustrate the thickness of the ionic cloud.

Table III

	$\frac{I}{\kappa}\cdot\sqrt{\gamma}\times10^8$	κ^2	q	$\theta\gamma$
HCl	3.06	$0.107 \times 10^{16}\cdot\gamma$	0.5	0.189×10^{-10}
KCl	3.06	$0.107 \times 10^{16}\cdot\gamma$	0.5	0.553×10^{-10}
$MgCl_2$	1.77	$0.321 \times 10^{16}\cdot\gamma$	0.421	0.324×10^{-10}
$CdSO_4$	1.53	$0.428 \times 10^{16}\cdot\gamma$	0.5	0.315×10^{-10}
$LaCl_3$	1.25	$0.642 \times 10^{16}\cdot\gamma$	0.454	0.162×10^{-10}
$K_4Fe(CN)_6$	0.96	$1.08 \times 10^{16}\cdot\gamma$	0.344	0.102×10^{-10}

Table III shows that the q values are subject to slight variations. The thickness of the ionic cloud $1/\kappa$ in cm. or the relaxation time θ in seconds can be found easily from Table III for any concentration γ. Since the thicknesses of the ionic clouds or the relaxation times for a fixed concentration and for the electrolytes of Table III are rather different, the dispersion effects must vary considerably. Now we will investigate the different effects of different factors on the dispersion phenomenon. In this we regard $\overline{\Lambda}_{I\omega}/\overline{\Lambda}_{Io}$, which is equal to $\underline{\Lambda}_{I\omega}/\overline{\Lambda}_{Io}$, as the essential quantity. We shall return to the value of $\overline{\Lambda}_{Io}$ in section XII.

IX. Influence of the Concentration on the Frequency Dependence of the Conductivity

We choose KCl as an example. Let the solvent be water, which has a dielectric constant D = 81.3 at the temperature t = 18°C. As concentrations we choose γ = 0.01, γ = 0.001, and γ = 0.0001. We take the corresponding relaxation times from Table III; they are $\theta = 0.553 \times 10^{-8}$ sec., $\theta = 0.553 \times 10^{-7}$ sec.; and $\theta = 0.553 \times 10^{-6}$ sec. For the corresponding κ^2 values, Table III gives:

$$\kappa^2 = 0.107 \times 10^{-14} \quad \kappa^2 = 0.107 \times 10^{-13} \quad \kappa^2 = 0.107 \times 10^{-12}$$

Furthermore, since we are dealing with a symmetrical electrolyte, q = 0.5. Thus everything necessary to calculate the quantity $\Lambda_{I\omega}/\Lambda_{Io}$ as a function of $\omega\theta$ or the wavelength l (in meters) is known. The following Table IV gives an overall view of the dispersion effect. Here $\Lambda_{I\omega}/\Lambda_{Io}$ is given for a series of wave lengths (within 2%).

At the wave length 0.2 m. the contribution of the relaxation force to the decrease in the conductivity is already relatively slight. At this wave length the conductivity of the electrolyte must be larger than its stationary value. Figure 4 illustrates the

Table IV

l in m.	γ=0.01	γ=0.001	γ=0.0001	l in m.	γ=0.01	γ=0.001	γ=0.0001
0	0	0	0	40	0.995	0.774	0.320
0.2	0.230	0.075	0.023	60	1	0.850	0.385
0.4	0.320	0.105	0.033	80	1	0.897	0.435
0.6	0.385	0.130	0.040	100	1	0.925	0.470
0.8	0.435	0.147	0.048	200	1	0.977	0.625
1.0	0.470	0.164	0.053	400	1	0.995	0.774
2	0.625	0.230	0.075	600	1	1	0.850
4	0.774	0.320	0.105	800	1	1	0.897
6	0.850	0.385	0.130	1000	1	1	0.925
8	0.897	0.435	0.147	2000	1	1	0.977
10	0.925	0.470	0.164	4000	1	1	0.995
20	0.977	0.625	0.230	6000	1	1	1

effect of the concentration on the dispersion effect. $\Lambda_{I\omega}/\Lambda_{Io}$ is plotted as ordinate and the wave length l in meters as abscissa on a logarithmic scale. From Figure 4 it can be deduced, for example, that the

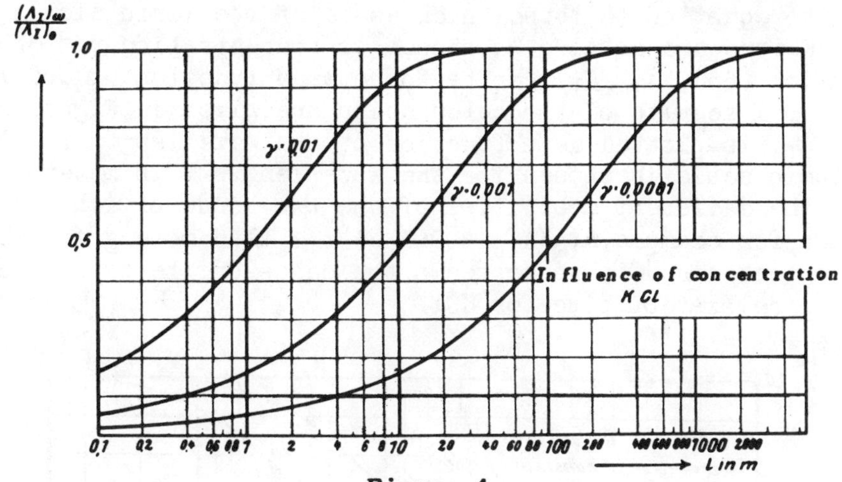

Figure 4

wave lengths at which:

$$\Lambda_{I\omega} / \Lambda_{Io} = 0.5$$

are:

$$l = 1.16 \text{ m. or } 11.6 \text{ m. or } 116 \text{ m.}$$

corresponding to the concentrations:

$$\gamma = 0.01 \quad \gamma = 0.001 \quad \gamma = 0.0001$$

At this wave length and for the corresponding concentrations the dispersion effect has reached its half-way value. If the concentration is reduced by a factor $1/U$, the same value of $\Lambda_{I\omega}/\Lambda_{Io}$ is reached at a wave length $l' = lU$. This is true since the relaxation time is proportionally larger. Had we plotted $\Lambda_{I\omega}/\Lambda_{I0}$ as a function of $\omega\theta$, for small values of $\omega\theta$ the curve would have been parabolic with a horizontal tangent at $\omega\theta = 0$. For very large $\omega\theta$ it approaches zero like $1/\omega\theta$ as equations (62) and (66) show immediately (compare also DFI, p.130).

X. Effect of the Mobility

In investigating the effect of different mobilities $\bar{L}_i$ or the related ρ_i values (eq.78), we have chosen HCl and KCl as examples. Again the temperature is taken to be 18°C and water is chosen as the solvent; consequently the dielectric constant is $D = 81.3$. The mobilities are taken from Table II. With these data it is found that the molar conductivities are:

$$\bar{\Lambda}_{HCl} = 380; \qquad \bar{\Lambda}_{KCl} = 130$$

From Table III it is found that the associated relaxation times are:

$$\Theta_{HCl} = \frac{0.189 \cdot 10^{-10}}{\gamma}; \qquad \Theta_{KCl} = \frac{0.553 \quad 10^{-10}}{\gamma}$$

According to equation (81) the thicknesses of the ionic clouds around the H and K ions are equal. We choose as concentration $\gamma = 0.001$. From equation (66), $\Lambda_{I\omega}/\Lambda_{I0}$ can be found as a function of the wave length l. The results of the calculation are given in Figure 5, in which $\Lambda_{I\omega}/\Lambda_{I0}$ is plotted as a function of the wave length l. Again a logarithmic scale is chosen for the wave length l in meters. The effect of the different mobilities of the two kinds of ions is apparent. Thus, for example, $\Lambda_{I\omega}/\Lambda_{I0} = 0.5$ at $l = 11.6$ meters for KCl and at $l = 4$ meters for HCL, where again the cause of the difference is the smaller relaxation time for HCl.

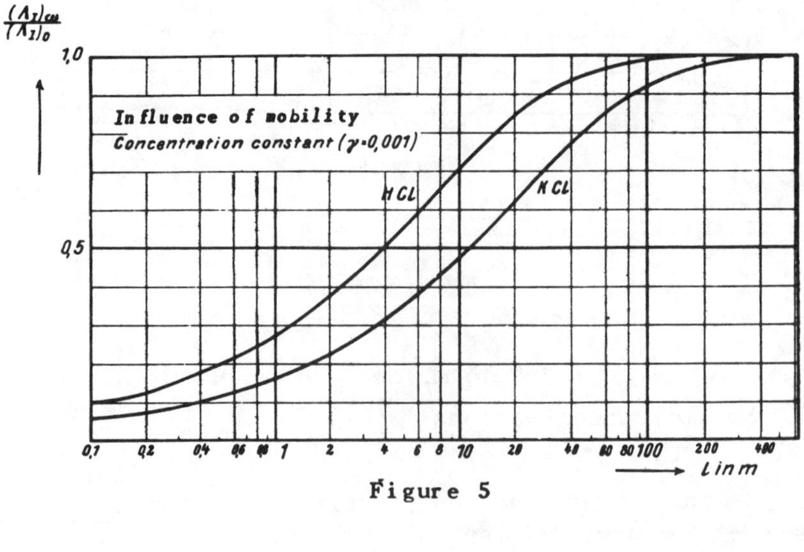

Figure 5

XI. Effect of the Dielectric Constant of the Solvent and the Temperature

As an example we take KCl in water and methyl alcohol. Let the temperature be 25°C this time. The corresponding dielectric constants are:

$$D_{H_2O} = 78.8 \qquad D_{\text{methyl alcohol}} = 30$$

In these cases the molecular conductivities at infinite dilution are:

$$\bar{\Lambda}\infty(H_2O) = 1.50 \qquad \bar{\Lambda}\infty(\text{methyl alcohol}) = 108$$

According to equation (78), the frictional constants of the K ions are:

$$\rho(\text{in } H_2O) = 0.205 \times 10^{-8} \qquad \rho(\text{in methyl alcohol}) = 0.284 \times 10^{-8}$$

The values of the mobilities in methyl alcohol are taken from the investigations of Hartley[1] and Raikes. With these data, we find for the corresponding κ^2 values and relaxation times tne following:

$$\varkappa^2 \text{ (in } H_2O) = 0.1077 \cdot \gamma \cdot 10^{16}$$

$$\varkappa^2 \text{ (in } CH_3OH) = 0.2828 \cdot \gamma \cdot 10^{16}$$

$$\Theta \text{ (in } H_2O) = \frac{0.466}{\gamma} \cdot 10^{-10} \text{ sec}$$

$$\Theta \text{ (in } CH_3OH) = \frac{0.245}{\gamma} \cdot 10^{-10} \text{ sec}$$

That a changed dielectric constant changes our effect is to be ascribed to the resulting change in tne θ value again. $\Lambda_{I\omega}/\Lambda_{Io}$ is again plotted as a function of the wave length in Figure 6. The concentration is taken to be $\gamma = 0.001$. In the case of water, $\Lambda_{I\omega}/\Lambda_{Io}$ = 0.5 when the wave length l = 9.8 m.; and in the case of methyl alcohol, when l = 5.1 m.

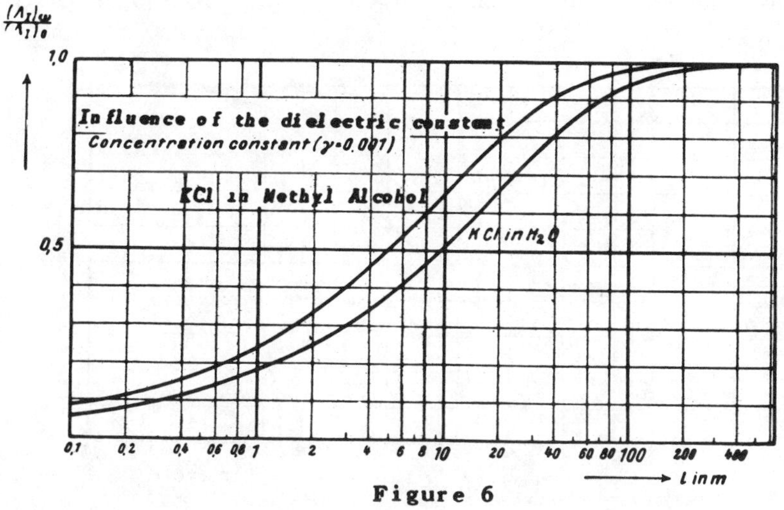

Figure 6

Further, the effect of the temperature shows up, since the mobilities at 25°C are already fairly larger than at 18°C. At higher temperatures the relaxation time (compare equations 80 and 81) is smaller, if the other quantities which determine the relaxation time are constant. κ^2 varies only slightly with temperature. Thus, for example, in the case of KCl at a temperature of 18°C, $\Lambda_{I\omega}/\Lambda_{Io}$ = 0.5 at a wavelength l = 11.6 m., while at a temperature of 25°C, $\Lambda_{I\omega}/\Lambda_{Io}$ = 0.5 at l = 9.8 m. Thus, in spite of the slight temperature change, there is a noticeable change in the dispersion curve. The effect of the valency is quite important and will be discussed now.

XII. Effect of the Valency, and Conductivity Defect at $\omega = 0$ for the Examples Chosen

While up to now we have always considered symmetrical electrolytes, we will now consider more complicated electrolytes and see how the valency affects the frequency dependence of the conductivity. We investigate examples of electrolytes with valency combinations one-two, two-two, one-three, and one-four. Water is taken to be the solvent. First note that for nonsymmetrical electrolytes, q = 0.5 (according to eq. 79); it takes values which can lie between 0 and 1; they are usually less than 0.5 (compare Table III on p. 414). Let the concentration be the same in each example; we choose as concentration γ = 0.0001. The relaxation times for our examples are rather different (compare Table III).

The temperatures of Table III will be used here. The difference of the relaxation times causes a large dependence of the dispersion effect on the valency. Of less importance is the difference in the q values. In Figure 7 $\Lambda_{I\omega}/\Lambda_{Io}$ is plotted as a function of the wave length for the electrolytes KCl (one-one valence combination), $MgSO_4$ (one-two)*, $LaCl_3$ (one-three), and $K_4Fe(CN)_6$ (one-four). Again the

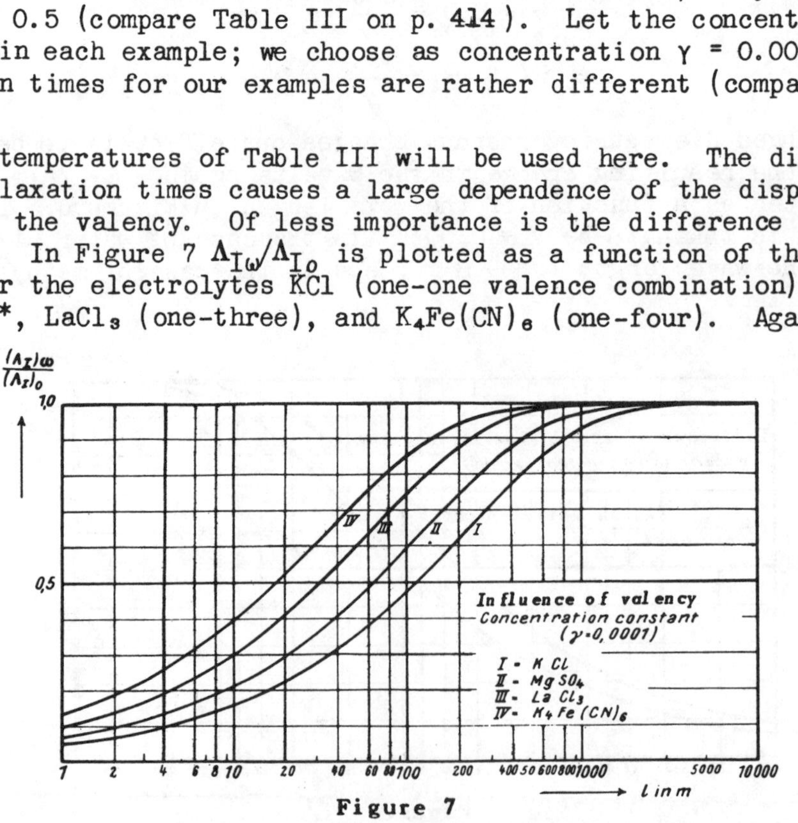

Figure 7

*The dispersion curve for $MgCl_2$ coincides with that for $MgSO_4$, and so we have not included it in Figure 7.

wave length l is plotted on a logarithmic scale and expressed in meters. With increasing valency the curves move to the left and thus the dispersion effect becomes noticeable at shorter and shorter wave lengths. From Figure 7 it can be gathered that for the sequence chosen above, KCl..., $K_4Fe(CN)_6$, $\Lambda_{I\omega}/\Lambda_{Io}$ = 0.5 at the wave lengths: 116 m., 66m., 30.5m., and 17.6m., respectively.

What occurs at other concentrations can be found immediately by referring to section IX. If we want to know, for example, at what wave length $\Lambda_{I\omega}/\Lambda_{Io}$ = 0.5 at a concentration γ = 0.001, we find corresponding to the above sequence the values 11.6m., 6.6m., 3.05m., and 1.76m. The following Table V gives a summary of some values of $\bar{\Lambda}_{I\omega}/\bar{\Lambda}_{Io}$ for differing wave lengths and the electrolytes selected above. Since $\bar{\Lambda}_{Io}$ is known for these electrolytes (compare Table VI), we can find how the conductivity varies with the wave length immediately. Now we will examine more closely the stationary values of the conductivity for our examples.

Table V

Values of $\Lambda_{I\omega}/\Lambda_{Io}$ for different wave lengths l

l in m.	KCl	$MgSO_4$	$LaCl_3$	$K_4Fe(CN)_6$
1	0.055	0.070	0.10	0.138
2	0.073	0.098	0.14	0.190
5	0.115	0.154	0.22	0.295
10	0.163	0.216	0.305	0.40
20	0.23	0.30	0.42	0.53
50	0.355	0.455	0.60	0.73
100	0.48	0.60	0.75	0.86
200	0.63	0.75	0.88	0.95
500	0.82	0.91	0.97	0.99
1000	0.93	0.97	0.997	0.997
2000	0.98	0.994	1	1
5000	1	1	1	1
10000	1	1	1	1

The values in Table V are exact to within less than 2%.

In the following Table VI we have given the values $\bar{\Lambda}_{Io}$ for the examples chosen when water is the solvent. Furthermore, the values of the molar conductivities $\bar{\Lambda}_{\infty}$ and the electrophoretic contributions $\bar{\Lambda}_{II}$ are also given in the table, so that in the stationary case the conductivity can be calculated for any concentration from the formula:

$$\bar{\Lambda} = \bar{\Lambda}_{\infty} - \bar{\Lambda}_{Io} - \bar{\Lambda}_{II}$$

Table VI

	$\bar{\Lambda}_\infty$	$\bar{\Lambda}_{II}$	$\bar{\Lambda}_{Io}$
KCl	130	$50.5\sqrt{\gamma}$	$0.224\ \bar{\Lambda}_\infty\sqrt{\gamma} = 29.1\sqrt{\gamma}$
$MgSO_4$	229	$404\ \sqrt{\gamma}$	$1.79\ \bar{\Lambda}_\infty\sqrt{\gamma} = 410\ \sqrt{\gamma}$
$LaCl_3$	245	$742.6\sqrt{\gamma}$	$1.52\ \bar{\Lambda}_\infty\sqrt{\gamma} = 373\ \sqrt{\gamma}$
$K_4Fe(CN)_6$	680	$1890\ \sqrt{\gamma}$	$2.12\ \bar{\Lambda}_\infty\sqrt{\gamma} = 1430\ \sqrt{\gamma}$

Again note that the values given for potassium ferrocyanide refer to a temperature of 25°C. They are calculated with the help of the values:

$$\varkappa = 0.3282 \cdot 10^8 \sqrt{\gamma}\sqrt{10}$$
$$\eta = 0.00895$$

For a concentration $\gamma = 0.0001$ the values of $\bar{\Lambda}_{II}$ and $\bar{\Lambda}_{Io}$ are those given in Table VII. In conjunction with equation (66) this gives the complete behavior of the electrolytes mentioned relative to the dependence of their conductivities on wave length. If, as an example, we take the wave length to be 20 m., the concentration $\gamma = 0.0001$, the value of $\Lambda_{I\omega}/\Lambda_{Io}$ can be taken from Figure 6 (or Table V) and the value of $\bar{\Lambda}_{Io}$ from Table VI. The values of $\bar{\Lambda}_{I\omega}/\bar{\Lambda}_{Io}$

Table VII

	∞	II	Io
KCl	130	0.505	0.29
$MgSO_4$	229	4.04	4.1
$LaCl_3$	245	7.43	3.73
$K_4Fe(CN)_6$	680	18.9	14.3

and $\bar{\Lambda}_{I\omega}$ found in this way for the four electrolytes selected are given in the following Table VIII and refer to a wave length l = 20 m. and a concentration $\gamma = 0.0001$.

Table VIII

	$\bar{\Lambda}_{I\omega}/\bar{\Lambda}_{Io}$	$\bar{\Lambda}_{I\omega}$
KCl	0.23	$0.00515\ \bar{\Lambda}_\infty = 0.067$
$MgSO_4$	0.30	$0.00537\ \bar{\Lambda}_\infty = 1.23$
$LaCl_3$	0.45	$0.00685\ \bar{\Lambda}_\infty = 1.68$
$K_4Fe(CN)_6$	0.53	$0.0111\ \bar{\Lambda}_\infty = 7.54$

The corresponding increases of the molecular conductivities over the stationary values are given for our examples in Table IX in per cent of the molecular conductivity at infinite dilution.

Table IX

KCl	0.17 %
$MgSO_4$	1.25 %
$LaCl_3$	0.84 %
$K_4Fe(CN)_6$	1 %

If the concentration $\gamma = 0.001$, then the value of the wave length corresponding to the changes given in Table VIII shifts to 2 m. The resulting changes in the molecular conductivity are given in Table X in per cent of $\overline{\Lambda}_\infty$.

Table X

KCl	0.54 %
$MgSO_4$	3.95 %
$LaCl_3$	2.65 %
$K_4Fe(CN)_6$	3.16 %

From the experimental viewpoint it is of interest that for any fixed wave length a value of the concentration γ exists for which the corresponding increase in the molecular conductivity is greatest. Equations (66) and (62) show this immediately. For example, for potassium ferrocyanide it lies between $\gamma = 0.0001$ and 0.01.

The magnitude of the dispersion effect is most simply measured by the difference of the molecular conductivity $\overline{\Lambda}\omega$ over its value for infinite dilution expressed in per cent of the same difference measured for direct current (compare Figures 8 and 9). For these the concentration is chosen $\gamma = 0.001$. We consider first the example of $CdSO_4$, to which Figure 8 applies. From Table VI the molecular conductivity is $\overline{\Lambda}_\infty = 229$. $\overline{\Lambda}_{II}$ amounts to 12.8 (in per cent of $\overline{\Lambda}_\infty$, 5.6%), $\overline{\Lambda}_{Io}$ is 12.9 (in per cent of $\overline{\Lambda}_\infty$, 5.7%). Consequently the total decrease of the conductivity amounts to 11.3%. $\overline{\Lambda}_{I\omega}$ results from the relaxation forces and decreases with decreasing wave length. Thus, for example, $\overline{\Lambda}_{I\omega}$ amounts to only half of this at $l = 6.5$ m. Expressed in per cent of $\overline{\Lambda}_\infty$, the molecular conductivity is larger by 2.83% at this frequency.

Now we discuss briefly the example of potassium ferrocyanide, to which Figure 9 applies. According to Table II the molar conductivity at infinite dilution at $t = 25°C$ is $\overline{\Lambda}_\infty = 680$. Furthermore

$$\overline{\Lambda}_{Io} = 2.12\sqrt{\gamma}\,\overline{\Lambda}_\infty = 47.5 \text{ (in per cent of } \overline{\Lambda}_\infty : 7\,\%)$$

$$\overline{\Lambda}_{II} = 1890\sqrt{\gamma} = 59.8 \text{ (in per cent of } \overline{\Lambda}_\infty : 8.8\,\%)$$

The total decrease of the molecular conductivity again in per cent of $\overline{\Lambda}_\infty$ is 15.8%. At the wave length $l = 1.63$ m., at which $\overline{\Lambda}_{I\omega}/\overline{\Lambda}_{Io}$ has decreased to $\frac{1}{2}$, the molecular conductivity is about 3.5% larger than this.

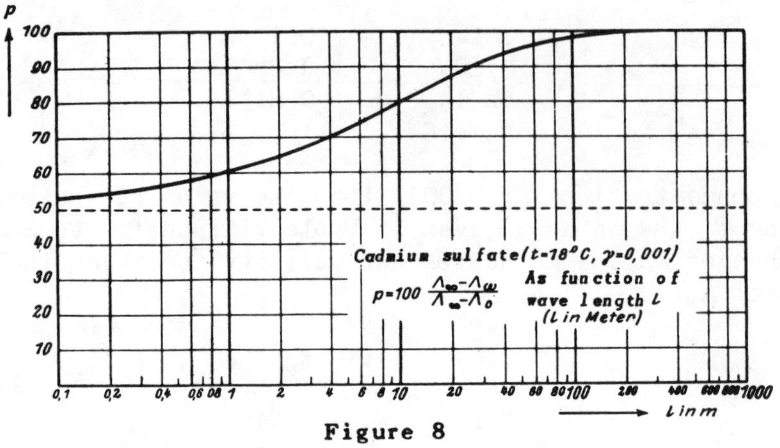

Figure 8

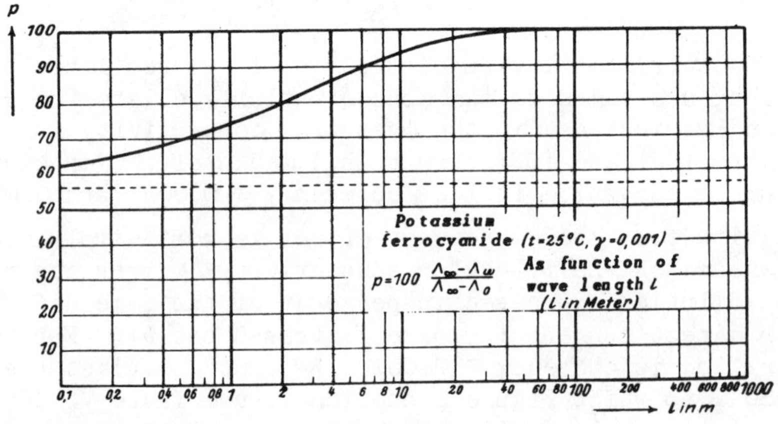

Figure 9

XIII. The General Formulae and the Interpolation Table for Calculating the Conductivity of Any Simple Electrolyte

In conclusion we will give a simple method for quickly determining the frequency dependence of the conductivity for any electrolyte which contains two kinds of ions. The solvent, temperature, electrolyte, and concentration can be chosen in any way. From equations (79), (80), and (81), q, $\varkappa^2$, and θ can be determined since the mobilities are known from experiment. Here they are summarized once more for convenience:

$$q = \frac{\overline{L}_1 z_2 + \overline{L}_2 z_1}{(z_1 + z_2)(\overline{L}_1 z_2^2 + \overline{L}_2 z_1^2)} z_1 \cdot z_2$$

$$\varkappa^2 = \frac{4\pi}{DkT} \frac{\varepsilon^2 N \gamma}{1000} \sum \nu_i z_i^2 \left\{ = 0.05342 \cdot 10^{16} \cdot \gamma \sum \nu_i z_i^2 \text{ for water at } 18^0 \text{ C} \right\}$$

$$\Theta = \frac{z_1^2 z_2^2}{z_2^2 \overline{L}_1 + z_1^2 \overline{L}_2} \frac{15.34 \cdot 10^{-8}}{kTq\varkappa^2}$$

Now according to equation (62), $\overline{\Lambda}_{I\omega}/\overline{\Lambda}_{Io}$ depends only on q and $\omega\theta$. If we prepare an interpolation Table XI for different values of q, which coordinates values of $\omega\theta$ with those of $\overline{\Lambda}_{I\omega}/\overline{\Lambda}_{Io}$, then by means of equation (82) for the wave length:

$$l = \frac{18.85 \cdot 10^{10}}{\omega \Theta} \Theta \qquad \text{(this formula gives } l \text{ in meters)} \qquad (82)$$

we can find $\overline{\Lambda}_{I\omega}/\overline{\Lambda}_{Io}$ as a function of the wave length for any example. If, for example, $q = 0.33$, it is only necessary to interpolate between columns 7 and 8 of Table XI and to calculate the wave length correspond-

Table XI

$\omega\theta$	q=1.0	q=0.75	q=0.5	q=0.45	q=0.40	q=0.35	q=0.30
0	1.00	1.00	1.00	1.00	1.00	1.00	1.00
0.1	0.998	0.998	0.999	0.999	0.999	0.999	0.999
0.2	0.995	0.996	0.997	0.997	0.997	0.997	0.997
0.35	0.985	0.987	0.989	0.990	0.990	0.990	0.991
0.5	0.972	0.976	0.980	0.981	0.981	0.981	0.982
0.75	0.943	0.948	0.956	0.959	0.962	0.965	0.968
1	0.910	0.919	0.930	0.934	0.938	0.942	0.946
1.25	0.876	0.888	0.904	0.909	0.913	0.918	0.922
1.5	0.844	0.859	0.876	0.882	0.888	0.894	0.900
2	0.785	0.804	0.826	0.833	0.841	0.848	0.856
2.5	0.736	0.755	0.783	0.791	0.799	0.808	0.817
3	0.692	0.715	0.745	0.753	0.763	0.773	0.785
4	0.624	0.649	0.682	0.691	0.701	0.713	0.727
6	0.531	0.557	0.593	0.603	0.614	0.628	0.643
8	0.469	0.494	0.531	0.540	0.553	0.565	0.582
10	0.425	0.449	0.486	0.496	0.507	0.520	0.537
15	0.353	0.374	0.409	0.417	0.428	0.442	0.458
20	0.308	0.328	0.360	0.368	0.378	0.390	0.406
25	0.277	0.296	0.326	0.334	0.344	0.354	0.369
30	0.254	0.272	0.299	0.307	0.317	0.327	0.342
35	0.235	0.252	0.279	0.286	0.295	0.305	0.318
40	0.221	0.237	0.261	0.267	0.277	0.287	0.299
45	0.204	0.223	0.247	0.254	0.263	0.272	0.288
50	0.198	0.213	0.235	0.242	0.250	0.259	0.271
75	0.162	0.174	0.193	0.199	0.206	0.214	0.223
100	0.1405	0.151	0.168	0.173	0.179	0.186	0.195
150	0.1148	0.1236	0.138	0.142	0.147	0.153	0.160
200	0.0996	0.1072	0.1195	0.1232	0.1278	0.1328	0.1394
300	0.1815	0.0877	0.0983	0.1011	0.1046	0.1088	0.1145
500	0.0631	0.0680	0.0759	0.0783	0.0812	0.0844	0.0888
700	0.0534	0.0574	0.0643	0.0663	0.0688	0.0717	0.0751
1000	0.0446	0.0481	0.0538	0.0555	0.0576	0.0599	0.0629
5000	0.0200	0.0215	0.0241	0.0249	0.0258	0.0269	0.0282
10000	0.0141_4	0.0152_3	0.01706	0.01755	0.01824	0.0190	0.0199

The values in Table XI are exact to within less than 1%.

ing to the value of $\omega\theta$ by means of equation (82). Equations (60) and (70) give the decrease in the molecular conductivity which results from the relaxation force; (71) the decrease $\bar{\Lambda}_{I_0}$ which results from the electrophoretic effect for the stationary case. Thus all that is necessary to determine the dispersion effect is completely known.

Summary

As a result of the Coulomb force acting between ions, a certain regularity occurs in the distribution of the ions; ionic atmospheres exist. This idea has become important not only in the theory of thermodynamic properties of ionic solutions but also in the theory of the irreversible processes of electric conduction. In this paper the behavior of these ionic clouds in the presence of an external alternating electric field is investigated taking account of the Brownian motion of the ions in order to obtain a quantitative answer to the question as to how the conductivity of strong electrolytes depends on the frequency of the electric field. This investigation reveals that two quantities related to the ionic clouds are important: first, their thickness $1/\kappa$, and second, their relaxation time θ necessary for the density distribution of the clouds to appear or disappear. The thickness of an ionic atmosphere is of the order of magnitude $10^{-8}/\sqrt{\gamma}$ cm. (compare Table III), where γ is the concentration in mols per liter; if an ion is suddenly removed, a finite relaxation time is required for the transition of the regularity in the ionic distribution around the ion into random order. This time is of the order of magnitude $10^{-10}/\gamma$ sec. (compare Table III). The thickness and the relaxation time determine the difference of the molecular conductivity at small concentration from its value at infinite dilution. Even assuming complete dissociation, a braking force on the ions occurs because of their electrostatic interaction which is effective over rather large distances. For this force two reasons exist; that is why it is divided in two parts: the relaxation force and the electrophoretic force. The relaxation force results from the existence of a relaxation time. If this time did not exist, the relaxation force would also vanish. If an ion moves under the influence of an electric field, a defect in the charge density in front of the ion appears and an excess behind. This dissymmetric charge distribution causes a braking force, "the relaxation force." It is proportional to the ionic velocity for small velocities. The electrophoretic force results from an additional motion of the solvent, which causes an increase in the Stokes frictional force. Both forces are inversely proportional to the thickness of the ionic cloud, which causes a decrease in the molar conductivity in agreement with Kohlrausch's law. Investigations of M. Wien on the effect of very high field strengths (order of magnitude 100,000 volt/cm.) on the conductivity, at which velocities of the order of 1 m./sec. occur and at which deviations from Ohm's law are observed, can be understood on the basis of the picture which has been sketched above.

At high field strengths the ionic cloud can no longer appear; thus, the conductivity must increase* to the value which corresponds to infinite dilution, assuming that the dissociation is complete. Examination of the Wien effect or also the dispersion effect discussed in this paper discloses the possibility of eliminating experimentally** the effect of Coulomb interaction. Thus we have a new method of obtaining information about the degree of electrolytic dissociation. In a later paper we hope to discuss this more thoroughly for the case of KCl for which the increase of the molar conductivity can be understood theoretically as a function of the field strength up to very high field strengths. Even although the hydrodynamic aspect of the process is unproportionally complicated and still by no means fully understood, it is possible to represent Wien's experimental curve over the entire range.

Wien's experiments lead us to expect the existence of another as yet unobserved effect, which we have closely examined in this paper. If an electric alternating field acts on the ions of a solution, the ionic clouds show at every instant a dissymmetry of the charge distribution which corresponds to the instantaneous velocity of the ions. However, if the period of a vibration of the ions is comparable to or even less than the relaxation time, the dissymmetric charge distribution in the ionic clouds can no longer form completely. Consequently, the relaxation force disappears for very high frequencies, and the molar conductivity must increase. In this paper, the general theory of the frequency dependence of the conductivity has been investigated for the case of dilute solutions, taking the Brownian motion of the ions into account. The effects of the concentration, the mobility of the ions, the dielectric constant, the temperature, and the valencies are also investigated in detail. The increase in the molar conductivity which occurs corresponds to the contribution resulting from the relaxation force and can be appreciable. When this was written the dispersion effect on the conductivity of strong electrolytes had not yet been demonstrated experimentally. M. Wien[4] found indications of it. Recently, Sack in this laboratory has been able to show that two solutions (KCl and $MgSO_4$) which have the same conductivity at low frequency show different resistances if measured with high frequency (wave length 18 m.). The conductivity difference is in the direction predicted and of the expected order of magnitude. In the near future Sack himself will report on these investigations, which are still in progress.

*M. Wien has kindly informed us that his most recent experiments appear to confirm this conjecture which he himself announced as probable in 1927.

**It seems instructive to us to compare the degree of dissociation implied by the above with that derived from the law of mass action as found by W. Nernst. (Compare W. Nernst, *Z. Elektrochemie*, 428 (1927); Orthmann, *Ergeb. Naturwiss.*, 6 (1927); Naudé, *Z. Elektrochemie*, 532 (1927); W. Nernst, *Berlin Akad. Ber.* 2, 4 (1928).

Bibliography

1. P. Debye and H. Falkenhagen, *Physik. Z.*, *29*, 121-132 (1928) (referred to as DFI in the text).
2. G. J. Burrows, *J. Chem. Soc.*, *123*, 2026 (1923).
3. H. Hartley and H. R. Raikes, *Trans. Faraday Soc.*, 393-396 (April, 1927).
4. M. Wien, *Ann. Physik*, *38*, 340/41 (1927).

DISPERSION OF CONDUCTIVITY OF STRONG ELECTROLYTES
(Dispersion der Leitfähigkeit starker Elektrolyte)

P. Debye and H. Falkenhagen*

Translated from
Zeitschrift für Elektrochemie, Vol. 34, 1928, pages 562-565

The latest development of electrolytic theory[1] is characterized by the fact that the special significance of the Coulomb force which ions exert on one another due to their charges is recognized. As a result of this force a completely random distribution of ions in an electrolytic solution is impossible. It is more probable that two oppositely charged ions are near each other than that two ions of like charge are found at the same distance apart. In order to describe the deviation from random distribution, the concept of an "ion cloud or ion atmosphere" has been introduced. If one follows any ion and imagines a volume element rigidly connected with it at a certain distance, upon averaging over the time, we can speak of an average charge density existing in this volume element. The smaller the distance the greater this average density. The regularity present in an electrolytic solution can thus be represented by means of a system of ion clouds surrounding each separate ion with a charge density, which, integrated over all space, represents a total charge equal in magnitude to the charge of the central ion, but of opposite sign.

In the classical theory it has been assumed that two ions are completely free at all distances from each other as long as they have not united by appropriate contact to a neutral molecule. Now we realize that even when such a chemical process has not yet occurred, a certain mutual binding is already present which extends up to rather large distances and decreases only slowly with increasing separation.

Just as in the classical theory, where the magnitude of the

*International Education Board Fellow.

ionization energy determines the degree of dissociation, here the potential of an ion relative to its surroundings will determine the strength of binding to all surrounding ions and will therefore be important with **respect** to the thermodynamic behavior of the solution. From this point of view, it becomes necessary to investigate the characteristic properties of the ion cloud.

Its essential properties are *(a)* its thickness and *(b)* its relaxation time. The investigation of these properties can be carried out theoretically as well as experimentally. Experience proves that the efficiency of such an effort is much greater if both methods are employed simultaneously. In the present case it is difficult for the theory to advance beyond the range of diluted solutions in the same way as it is difficult to derive an equation of state for a substance of high density. Therefore, theory will be used in order to derive quantitative limiting laws which are the more precise the more dilute the solutions are, but which, even from purely theoretical considerations, must deviate from experiment for higher concentrations. In the process of the derivation of such laws it will become evident which are the conspicuous properties of ionic solutions most strongly at variance with the expected behavior according to hitherto accepted rules. Such properties once spotted will then have to be investigated experimentally up to concentrations beyond the reach of the theory.

Let us consider *(a)*, the thickness d of the ionic atmosphere, which is a dimension characterizing the decrease of the charge density in the ionic cloud with increasing distance from the central ion. The theory developed for dilute solutions shows that d is represented by an expression which depends only on the concentration, on the ionic charges, on the dielectric constant of the solvent, and on the temperature. The order of magnitude of d in cm. is:

$$d = 10^{-8}/\sqrt{\gamma}$$

where γ is the concentration in moles per liter. Table I gives values of the density d for $\gamma = 0.001$ in water at 18°C.

Table I

Type of electrolyte	d
Uni-uni valent	96.6×10^{-8}
Uni-bi valent	55.9
Bi-bi valent	48.3
Uni-tri valent	39.5
Uni-tetra valent	30.7

The characteristic dependence of this thickness on the concentration (it is proportional to the reciprocal square root) is the reason for the characteristic concentration dependence of the activity which defies explanation by means of the classical concept of partial dissociation. The detailed formula shows that the thickness is dependent on the ionic strength,* a weighted concentration built up simultaneously of concentration and valency. It is the ionic strength and not the concentration itself which determines the thermodynamic behavior of strong electrolytes. Valency and dielectric constant as essential and determinative factors are characteristic of the practical importance of the Coulomb interaction. In this paper, however, we want to concern ourselves chiefly with the second characteristic property of ionic clouds which was introduced above under (*b*) as "relaxation time." Its influence is decisive in investigating, instead of equilibrium conditions, the irreversible process of current conduction. The ionic clouds can neither disappear nor be built up infinitely quickly. Suppose, for example, that some ion is suddenly removed from the solution. The charge distribution present in its surroundings now will no longer have any reason for its existence, since its presence was due only to the electric action of the central ion. The subsequent transformation to random orientation relative to the point where the central ion previously was situated will occur only gradually. A time θ called the relaxation time determines this process. This relaxation time is, as calculation indicates, determined not only by the quantities on which the thickness d depends but contains also the ionic mobilities. Its order of magnitude in seconds is:

$$\theta = \frac{10^{-10}}{\gamma}$$

in which γ again is the concentration in moles per liter. Table 2 gives values at $\gamma = 0.001$ in water as a solvent. The temperature of the solution is assumed to be 18°C. with the exception of potassium ferrocyanide for which it is 25°C.

Table II

Electrolyte	Relaxation time, seconds	Corresponding wave length, meters
KCl	0.553×10^{-7}	16.6
HCl	0.189×10^{-7}	5.67
$MgCl_2$	0.324×10^{-7}	9.72
$CdSO_4$	0.315×10^{-7}	9.45
$LaCl_3$	0.162×10^{-7}	4.86
$K_4Fe(CN)_6$	0.102×10^{-7}	3.06

It is to be noted that the relaxation time is very sensitive to temperature in consequence of the strong dependence of mobility on temperature. Since we have selected for potassium ferrocyanide the recent value found by Burrows[2] for the mobility of the $Fe(CN)_6$ ion, referring to a temperature of 25°C, the relaxation time for potassium ferrocyanide refers to that temperature. If we should assume that the corresponding mobilities vary inversely as the viscosity, the relaxation time for potassium ferrocyanide at 18°C would amount to $\theta = 0.120 \times 10^{-7}$ seconds. The two values d, as

*Ionic strength = sum of products of ion concentration times square of valency.

well as θ, determine the difference between the molar conductivity in small concentrations and its value for infinite dilution Λ_∞ in contrast to the classical explanation, according to which these differences would be attributed to a changing degree of dissociation. As a result of the actual electrostatic binding of ions to each other which is active over rather large distances, the ionic movement under the influence of a driving electric force is hindered, and we observe a decrease in mobility. Two reasons exist for this decrease of mobility with increasing concentration. As we saw, each ion is surrounded by a charge density of opposite sign.[2]

Under the influence of the electric field which generates the current, the solvent is therefore induced into additional motion, an effect commonly known as electrophoresis. This additional movement causes an increased frictional force which we call the "electrophoretic force".

A second force which likewise appears as an increase of friction is called the "relaxation force." Since in a solution which carries a current each ion moves in a positive or negative direction, its ion cloud must be continuously built up around new centers, whereas it disappears around the centers previously occupied. Consequently, the average charge density distribution of the ionic cloud can no longer have radial symmetry, since for its generation or its disappearance it requires a finite relaxation time. We will find a charge density too small in front of the ion and too large behind, and, of course, the greater the average velocity of the ion the larger will be this dissymmetry. As result of this dissymmetry a relaxation force appears which for small velocities will be proportional to the velocity and therefore will be observed as a decrease in mobility.

Both forces are inversely proportional to the thickness of the ionic cloud and since the latter is inversely proportional to the square root of the concentration, we will observe with increasing concentration a decrease of the molar conductivity which is proportional to its square root. This is the law found experimentally by Kohlrausch.

However, the calculation of the dissymmetry of the ionic cloud for dilute solutions is only an approximation good for small ionic velocities. In all ordinary cases this approximation is fully sufficient. But this is no longer true if, as was accomplished by Wien, abnormally large ionic velocities are developed by means of very strong electric fields. The velocities attained in Wien's experiments are of the order of magnitude of 1 m./sec. while the normal values are around 0.01 mm./sec. Consider as an **example** a 0.001 normal solution of potassium ferrocyanide for which, according to Table II, the relaxation time is roughly 10^{-8} seconds. In this time ions moving with a velocity of 1 m./sec. cover a distance of 10^{-6} cm. On the other hand, according to Table I, the thickness of the ion clouds for uni-tetra-valent electrolytes is about 0.3×10^{-6} cm. Ions of the velocity obtained by Wien would therefore cover approximately three times the thickness of their atmosphere during the relaxation time, *i.e.*, in the time which is necessary for building up the atmosphere. It is clear that under such conditions the ion cloud cannot appear at all. It was to the existence of the ionic cloud that the decrease in mobility had to be attributed. Accordingly, we expect that in strong fields the corresponding additional frictional force can not occur, *i.e.*, we expect the conductivity to increase in strong fields. This deviation from Ohm's law is the effect discovered by Wien.[3] The calculations of Joos[4] and Blumentritt represent a quantitative discussion of the same fundamental ideas.

If the experiments of Wien are to be interpreted in the manner just stated the existence of another effect, as yet unobserved, is to be anticipated. Suppose that **no d.c.** but a.c. **measurements** are performed. Each ion now will execute a vibrational motion. As long as the frequency is small enough, the ion cloud will at each moment, show a dissymmetry of its charge distribution corresponding to the instantaneous velocity of the ion. However, if the frequency is so large that the duration of one period becomes comparable with or even **smaller** than the relaxation time, the dissymmetry must disappear. Consequently, the corresponding frictional force exists no more and the ion mobility has increased. We expect, therefore, an increase of the conductivity for wavelengths of the order of magnitude of the product: velocity of light times the relaxation time (compare Table 2). Thus the Wien effect is characterized by a total disappearance of the ionic cloud, whereas we believe that there will also exist a dispersion effect, which is characterized by the disappearance of the dissymmetry in the ionic cloud.

We have now worked out[5] the general theory of this effect for dilute solutions and with due consideration of the Brownian motion fluctuations of the ions.

The influence of concentration, mobility of ions, temperature, dielectric constant, and valency have been discussed. Here we wish only to illustrate the expected order of magnitude of the new effect. A $CdSO_4$ solution at a molecular concentration of $\gamma = 0.001$ may serve as an example. From the mobilities of **Cd:92 and SO_4:136.6** the molar conductivity at infinite dilution was taken to be $\Lambda_\infty = 228.6$; the electrophoretic conductivity decrease Λ_E amounts to 12.8 (in per cent of Λ_∞: 5.6%) and the conductivity decrease as a result of the relaxation force for the stationary case (frequency zero) amounts to $\Lambda_R = 12.9$ (in per cent of Λ_∞: 5.7%). Therefore the total decrease in our example is 11.3% of the molar conductivity at infinite dilution. In order to illustrate the expected dispersion effect we have plotted the logarithm of the wavelength l in meters as abscissa in Figure 1. The difference of the molecular conductivity from its value for infinite dilution, expressed in per cent of its value for frequency zero is the ordinate. It is assumed that the part of this difference which is due to "cataphoretic action" is frequency-independent. The other part, which is due to relaxation and which at long wave lengths appears in full, becomes smaller with decreasing wave length and is, for example, reduced by half at $\lambda = 6.5$ meters. It amounts to 2.8% of the molecular conductivity. Meanwhile Sack was able to show in this laboratory and in agreement with the theory that two solutions of KCl and $MgSO_4$, which had the same conductivities at lower frequencies, conducted quite differently at high frequencies. Indications of such an effect have already been noted by Wien, *Ann. Physik, 38*, 340-341 (1927).

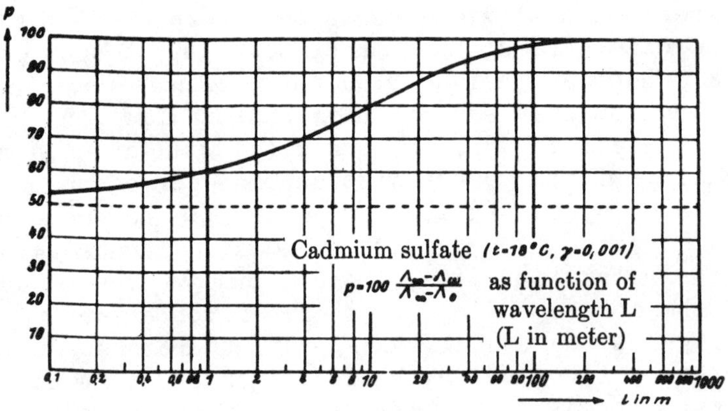

Figure 1

Another aspect of the dispersion effect is of some significance. An ion of constant velocity **experiences** in experiments of the usual kind (field strength not too high) a relaxation force which is proportional to its velocity. However, if the ion is oscillating periodically then the instantaneous value of the relaxation force corresponds no longer to the speed existent at the same instant. This can be **expressed** by saying that a phase difference exists between the field strength and the speed produced. This, however, means that in an a.c. field we will observe two current components, one in phase with the field and the other in phase with the time derivative of the field strength. A current of the latter type is what has been known as a displacement current since its discovery by Maxwell; its magnitude is formally measured by the dielectric constant. Because of this phase difference, besides the dispersion of conductivity, an increase of the dielectric constant is to be expected which, of course, itself will again be frequency-dependent. The possibility of such an effect is understood most simply perhaps by means of the following consideration. Suppose that an ion is suddenly moved over a small distance. The ion cloud, in consequence of its finite relaxation time, cannot follow instantaneously but is still distributed around the original position of the ion. So a kind of quasi-elastic force must exist which pulls the ion toward its earlier position. Although no real binding between ions has been assumed, the electrostatic interaction produces a kind of quasi-elastic interconnection which manifests itself as an increase of the dielectric constant. Increases of the dielectric constant have often been interpreted as due to the orientation effect of undissociated molecules acting as electrical dipoles. We see now that not only polar undissociated molecules can effect an increase but that an effect of the same sign will occur even in the case of complete dissociation.

In the present state of the theory a quantitative explanation of the different effects is rigorously possible only for dilute solutions. The general concepts upon which the theory is based cover, of course, a wider range. It is of special interest if those concepts lead us to the prediction of new effects for their experimental verification, which, even if only qualitative, is sufficient to underline the practical importance of ionic interaction. In this paper we have tried to feature this qualitative aspect by avoiding the introduction of any kind of mathematical formula. Following the effects under discussion up to higher concentrations will most probably lead to definite conclusions concerning the question of whether, under such circumstances, the degree of dissociation is still 100 per cent or less. A first attempt in this direction in which the law of mass action is combined with the interaction theory has recently been published by Nernst (*Ber.*, *II*, 4 (1928)).

Bibliography

1. Compare P. Debye and E. Hückel, *Physik. Z.*, *24*, 185, 305 (1923); also the summarizing article of E. Hückel, *Ergeb. exakt. Naturw.*, *3*, 195 ff. (1924); W. Orthmann, *ibid.*, *6*, 155 ff. (1927); E. Baars, in Geiger-Scheel, *Handbuch der Physik*, Vol. 13, 1928, pp. 397-516; or discussion about strong electrolytes in *Trans. Faraday Soc.*, 334-544 (1927).
2. G. I. Burrows, *J. Chem. Soc.*, *123*, 2026 (1923).
3. M. Wien, *Ann. Physik*, *83*, 327-361 (1927); *Physik, Z.*, *28*, 834-836 (1927).
4. G. Joos and M. Blumentritt, *Physik. Z.*, *28*, 836-838 (1927).
5. P. Debye and H. Falkenhagen, *Physik. Z.*, *29*, 121-132 (1928). See also a second paper on the same subject to appear shortly in this journal.

TRANSIENT PROCESSES IN ELECTROLYTIC SOLUTIONS

(Die zeitlichen Vorgänge in Elektrolytlösungen)

P. Debye

Translated from
Sommerfeld Festschrift, 1923, pages 52-57.

(1) In a completely dissociated electrolytic solution the ions are not distributed entirely uniformly; in the neighborhood of every ion the opposite charge predominates on the average. Quantitatively, this property can be expressed by the introduction of a series of distribution functions f_{ij} which are defined as follows.[1] Consider a volume element dS_P at a point P and a volume element dS_Q at a point Q at a distance r from P. Then the probability of finding an ion of type j in dS_Q if at the same time an ion of type i is in dS_P is:

$$f_{ij} dS_P dS_Q$$

If the ions were completely independent of one another we would have:

$$f_{ij} = n_i n_j$$

where n_i, etc., denote the average number of ions per unit volume. If the electrolyte is in its natural state, the distribution function is spherically symmetrical; that is, it depends only on r. However, a current passing through the solution, say in the x-direction, will cause a dissymmetry with the direction of the current as its axis, and this dissymmetry will cause a decrease in the mobility.

For the determination of the distribution functions when a time-dependent field in the x-direction is present, the following differential equation has been derived:[2]

$$\frac{\partial f_{ij}}{\partial t} = kT\left(\frac{1}{\varrho_i}+\frac{1}{\varrho_j}\right)\Delta f_{ij} + \left(\frac{e_i}{\varrho_i}-\frac{e_j}{\varrho_j}\right)X\frac{\partial f_{ij}}{\partial x} - \frac{4\pi}{D}\left(\frac{n_i e_i}{\varrho_i}\sum_p e_p f_{pj} + \frac{n_j e_j}{\varrho_j}\sum_p e_p f_{ip}\right) \tag{1}$$

which is valid for dilute solutions and in which e_i etc. denote the charges and ρ_i etc. the frictional constants of the ions, while D = dielectric constant of the solvent, k = Boltzmann constant, T = absolute temperature.

Corresponding to the space-time variables of this equation, two quantities can be defined, (*a*) a characteristic length and (*b*) a characteristic time. The length in question is a measure of the size of the region over which an ion exerts its ordering influence, while the characteristic time provides a measure for the time which elapses before the equilibrium distribution is reattained after a spontaneously produced variation in the distribution. If a field is used whose period is comparable with the characteristic time, the conductivity is changed.[3] If, as M. Wien* did, a strong field is used and its intensity is increased to such a point that the velocity of the ions is sufficient to move them over a distance which is comparable with the characteristic length in the characteristic time, a change in conductivity will again be the result. Although the effects described here - frequency dependence of the conductivity and deviations from Ohm's law - are the proper subjects for experimentation, the fundamental aspects naturally do not appear too clearly in the corresponding detailed theoretical calculations. I would like, therefore, to describe in the following, with the help of the basic equations, how an electrolyte whose ionic distribution has been disordered in some way reestablishes its natural order. In order to eliminate unnecessary complications, I will restrict myself to the case in which there are only two kinds of ions present with charges $+\varepsilon$ and $-\varepsilon$ having the same frictional constant ρ and the same average density n. This case is realized by an electrolyte like KCl.

(2) Four distribution functions, f_{11}, f_{22}, f_{12}, and f_{21}, describe the present case; according to eq. (1) they must satisfy the following system of differential equations:

$$\left.\begin{aligned}
\frac{\Theta\varkappa^2}{2}\frac{\partial f_{11}}{\partial t} &= \Delta f_{11} - \frac{\varkappa^2}{4}(2f_{11}-f_{12}-f_{21})\\
\frac{\Theta\varkappa^2}{2}\frac{\partial f_{22}}{\partial t} &= \Delta f_{22} - \frac{\varkappa^2}{4}(2f_{22}-f_{12}-f_{21})\\
\frac{\Theta\varkappa^2}{2}\frac{\partial f_{12}}{\partial t} &= \Delta f_{12} - \frac{\varkappa^2}{4}(2f_{12}-f_{11}-f_{22})\\
\frac{\Theta\varkappa^2}{2}\frac{\partial f_{21}}{\partial t} &= \Delta f_{21} - \frac{\varkappa^2}{4}(2f_{21}-f_{11}-f_{22})
\end{aligned}\right\} \tag{2}$$

where the quantity κ^2 (dimension = square of a reciprocal length) is defined by the relation:

$$\kappa^2 = \frac{8\pi}{DkT} n\varepsilon^2 \tag{2'}$$

and the relaxation time Θ is given by the equation:

$$\Theta = \frac{\varrho}{\kappa^2 kT} \tag{2''}$$

In the case of complete disorder we would have:

$$f_{11} = f_{22} = f_{12} = f_{21} = n^2 \tag{3}$$

On the other hand, in previous publications it has been shown that the distribution functions for the actual distribution are given by the formulae:

$$\left.\begin{aligned} f_{11} = f_{22} &= n^2\left(1 - \frac{\varepsilon^2}{DkT}\frac{e^{-\kappa r}}{r}\right) \\ f_{12} = f_{21} &= n^2\left(1 + \frac{\varepsilon^2}{DkT}\frac{e^{-\kappa r}}{r}\right) \end{aligned}\right\} \tag{3'}$$

The factor multiplying the function $e^{-\kappa r}/r$ has been determined from the equation for the electrostatic potential in order to correspond to a central charge of $\pm\varepsilon$. The problem now is to obtain time-dependent solutions of eq. (2) which reduce to eq. (3) for $t = 0$ and to eq. (3') for $t = \infty$. This then describes how the actual distribution arises from the totally disordered distribution which is supposed to prevail at the start.

(3) It is possible to retain throughout the equalities:

$$f_{11} = f_{22} \text{ and } f_{12} = f_{21} \tag{4}$$

so that only the two equations:

$$\left.\begin{aligned} \frac{\Theta\kappa^2}{2}\frac{\partial f_{11}}{\partial t} &= \Delta f_{11} - \frac{\kappa^2}{2}(f_{11} - f_{12}) \\ \frac{\Theta\kappa^2}{2}\frac{\partial f_{12}}{\partial t} &= \Delta f_{12} - \frac{\kappa^2}{2}(f_{12} - f_{11}) \end{aligned}\right\} \tag{5}$$

must be solved with the boundary conditions (3').

In preparation for the final solution, suppose:

$$\left.\begin{aligned} f_{11} &= A_{11}\, e^{-\beta\frac{t}{\Theta}} \frac{e^{i\lambda r}}{r} \\ f_{12} &= A_{12}\, e^{-\beta\frac{t}{\Theta}} \frac{e^{i\lambda r}}{r} \end{aligned}\right\} \tag{6}$$

where A_{11} and A_{12} are constants and β and λ parameters. Since

$$\Delta\left(\frac{e^{i\lambda r}}{r}\right) = -\lambda^2\left(\frac{e^{i\lambda r}}{r}\right)$$

A_{11} and A_{12} must satisfy the following equations:

$$\left.\begin{aligned} &\left(\beta - 1 - 2\frac{\lambda^2}{\varkappa^2}\right) A_{11} + A_{12} = 0 \\ &A_{11} + \left(\beta - 1 - 2\frac{\lambda^2}{\varkappa^2}\right) A_{12} = 0 \end{aligned}\right\} \tag{7}$$

These are compatible with either:

$$A_{11} = B, \qquad A_{12} = -B, \qquad \beta = 2\left(1 + \frac{\lambda^2}{\varkappa^2}\right) \tag{8}$$

or:

$$A_{11} = C, \qquad A_{12} = C, \qquad \beta = 2\frac{\lambda^2}{\varkappa^2} \tag{8'}$$

As a possible particular solution we have then:

$$\left.\begin{aligned} f_{11} &= \left(Be^{-2\frac{t}{\Theta}} + C\right) e^{-2\frac{\lambda^2}{\varkappa^2}\frac{t}{\Theta}} \frac{e^{i\lambda r}}{r} \\ f_{12} &= \left(-Be^{-2\frac{t}{\Theta}} + C\right) e^{-2\frac{\lambda^2}{\varkappa^2}\frac{t}{\Theta}} \frac{e^{i\lambda r}}{r} \end{aligned}\right\} \tag{9}$$

We seek now another particular solution which for $t = 0$ reduces to:

$$f_{11} = \frac{n^2\varepsilon^2}{DkT}\frac{e^{-\varkappa r}}{r}, \qquad f_{12} = -\frac{n^2\varepsilon^2}{DkT}\frac{e^{-\varkappa r}}{r} \tag{10}$$

and which vanishes for $t = \infty$. If we add this solution to the stationary values (3'), the solution of our problem is completed.

(4) We remark that for $0 < r < \infty$ the relation exists:

$$e^{-\varkappa r} = \frac{1}{2\pi}\int_{\infty}^{+\infty} e^{i\lambda r}\frac{d\lambda}{\varkappa + i\lambda} \tag{11}$$

which can be proved easily by application of the Fourier integral theorem. Now if we consider B in eq. (9) a function of λ and integrate over λ, we see immediately that the above-defined special problem is solved if we suppose that:

$$B(\lambda) = \frac{1}{2\pi}\frac{n^2\varepsilon^2}{DkT}\frac{1}{\varkappa + i\lambda}, \qquad C = 0$$

Thus the particular result is of the form:

$$\left.\begin{aligned} f_{11} &= \frac{1}{2\pi}\frac{n^2\varepsilon^2}{DkT} e^{-2\frac{t}{\Theta}} \int_{-\infty}^{+\infty} e^{-2\frac{\lambda^2}{\varkappa^2}\frac{t}{\Theta}} \frac{e^{i\lambda r}}{r}\frac{d\lambda}{\varkappa + i\lambda} \\ f_{12} &= -f_{11} \end{aligned}\right\} \tag{12}$$

By adding the expressions (12) to the formulae (3') we obtain the final solution of the problem in the form:

$$\left.\begin{aligned} f_{11} = f_{22} &= n^2\left[1 - \frac{\varepsilon^2}{DkT}\left(\frac{e^{-\varkappa r}}{r} - \frac{e^{-2\frac{t}{\Theta}}}{2\pi}\int_{-\infty}^{+\infty} e^{-2\frac{\lambda^2}{\varkappa^2}\frac{t}{\Theta}}\frac{e^{i\lambda r}}{r}\frac{d\lambda}{\varkappa + i\lambda}\right)\right] \\ f_{12} = f_{21} &= n^2\left[1 + \frac{\varepsilon^2}{DkT}\left(\frac{e^{-\varkappa r}}{r} - \frac{e^{-2\frac{t}{\Theta}}}{2\pi}\int_{-\infty}^{+\infty} e^{-2\frac{\lambda^2}{\varkappa^2}\frac{t}{\Theta}}\frac{e^{i\lambda r}}{r}\frac{d\lambda}{\varkappa + i\lambda}\right)\right] \end{aligned}\right\} \quad (13)$$

(5) These formulae can be put in another form which makes it easier to see the space-time course of the processes. The factor $1/(\kappa+i\lambda)$ under the integral is identically equal to:

$$\int_0^\infty e^{-\mu(\varkappa+i\lambda)}d\mu$$

and can be replaced by it. By reversing the order of integration we obtain:

$$\int_{-\infty}^{+\infty} e^{-2\frac{\lambda^2}{\varkappa^2}\frac{t}{\Theta}}\frac{e^{i\lambda r}}{r}\frac{d\lambda}{\varkappa+i\lambda} = \frac{1}{r}\int_0^\infty d\mu\, e^{-\mu\varkappa}\int_{-\infty}^{+\infty} d\lambda\, e^{-2\frac{\lambda^2}{\varkappa^2}\frac{t}{\Theta}} e^{-i\lambda(\mu-r)}$$

Evaluating the inner integral (by completing the square in the exponent) gives immediately:

$$\int_{-\infty}^{+\infty} e^{-2\frac{\lambda^2}{\varkappa^2}\frac{t}{\Theta}}\frac{e^{i\lambda r}}{r}\frac{d\lambda}{\varkappa+i\lambda} = \sqrt{\frac{\pi\varkappa^2\Theta}{2t}}\int_0^\infty d\mu\frac{e^{-\mu\varkappa}}{r}e^{-(\mu-r)^2\frac{\varkappa^2\Theta}{8t}}$$

If μ is replaced by a new variable:

$$\xi = \sqrt{\frac{\Theta}{8t}}\left[\varkappa(\mu - r) + \frac{4t}{\Theta}\right]$$

the simple form:

$$\int_{-\infty}^{+\infty} e^{-2\frac{\lambda^2}{\varkappa^2}\frac{t}{\Theta}}\frac{e^{i\lambda r}}{r}\frac{d\lambda}{\varkappa+i\lambda} = 2\sqrt{\pi}\, e^{2\frac{t}{\Theta}}\frac{e^{-\varkappa r}}{r}\int_{\xi_0}^\infty e^{-\xi^2}d\xi \quad (14)$$

is obtained, in which:

$$\xi_0 = \sqrt{\frac{\Theta}{2t}}\left(\frac{2t}{\Theta} - \frac{\varkappa r}{2}\right) \quad (14')$$

Thus with the help of the error integral:

$$\Phi(\xi_0) = \frac{1}{\sqrt{\pi}}\int_{\xi_0}^\infty e^{-\xi^2}d\xi \quad (15)$$

the solutions can be expressed in the following form:

$$\left.\begin{aligned} f_{11} = f_{22} = n^2\left[1 - \frac{\varepsilon^2}{DkT}\frac{e^{-\varkappa r}}{r}(1 - \Phi(\xi_0))\right] \\ f_{12} = f_{21} = n^2\left[1 + \frac{\varepsilon^2}{DkT}\frac{e^{-\varkappa r}}{r}(1 - \Phi(\xi_0))\right] \end{aligned}\right\} \quad (16)$$

According to eq. (14'), when $t = 0$, $\xi_0 = -\infty$ no matter what the value of r, so that according to eq. (15) $\Phi = 1$ and all the distribution functions in eq. (16) reduce to n^2, corresponding to complete randomness. On the other hand, when $t = \infty$, $\xi_0 = +\infty$, again independent of r, so that $\Phi = 0$ and (16) reduces to the prescribed stationary distribution. For $\xi_0 = 0$ we have $\Phi = \frac{1}{2}$, and according to eq. (16) when this condition is fulfilled half of the stationary ordering is realized. However, according to eq. (14') $\xi_0 = 0$ when:

$$\varkappa r = 4\frac{t}{\Theta}$$

The greater the distance from the center of the ion, the longer time is required before ordering appears. At the characteristic distance $r = 1/\kappa$ half of the ordering is achieved in time $t = \Theta/4$. It is worth while noting that for a solution of concentration γ mol per liter $1/\kappa$ and Θ are in order of magnitude:

$$\frac{1}{\varkappa} = \frac{10^{-8}}{\sqrt{\gamma}}\,\text{cm} \quad \text{and} \quad \Theta = \frac{10^{-10}}{\gamma}\,\text{sec}$$

Finally notice that for large positive values of ξ_0 we have the asymptotic equation:

$$\Phi_{(\xi_0)} = \frac{1}{2\sqrt{\pi}}\frac{e^{-\xi_0^2}}{\xi_0}$$

For sufficiently large times (which must be larger the larger the distance r) the distribution functions are approximately equal to:

$$\left.\begin{aligned} f_{11} = f_{22} = n^2\left[1 - \frac{\varepsilon^2}{DkT}\frac{e^{-\varkappa r}}{r}\left(1 - \frac{1}{2\sqrt{\pi}}\frac{e^{-\frac{\Theta}{2t}\left(\frac{2t}{\Theta}-\frac{\varkappa r}{2}\right)^2}}{\sqrt{\frac{\Theta}{2t}}\left(\frac{2t}{\Theta}-\frac{\varkappa r}{2}\right)}\right)\right] \\ f_{12} = f_{21} = n^2\left[1 + \frac{\varepsilon^2}{DkT}\frac{e^{-\varkappa r}}{r}\left(1 - \frac{1}{2\sqrt{\pi}}\frac{e^{-\frac{\Theta}{2t}\left(\frac{2t}{\Theta}-\frac{\varkappa r}{2}\right)^2}}{\sqrt{\frac{\Theta}{2t}}\left(\frac{2t}{\Theta}-\frac{\varkappa r}{2}\right)}\right)\right] \end{aligned}\right\} \quad (17)$$

(6) In the preceding part we discussed how each ion forms its stationary ionic atmosphere. Similarly, it can be shown how the dissymmetry of the ionic atmosphere induced by the passage of a current disappears when, for example, the current-producing field is suddenly switched off. It is immediately evident from the form of the basic equations that the relaxation time for this process is the same as for the process discussed here. From the experimental point of view the essence of these considera-

tions is that the Wien deviation from Ohm's law on the one hand and the frequency dependence of the conductivity on the other are interrelated. The relation between the two effects can be evaluated as follows. Appreciable frequency dependence is to be expected when the period of the a.c. field used for determining the conductivity is equal to Θ. Under the influence of a field strength X a univalent ion reaches the velocity $\varepsilon X/\rho$ and thus moves a distance $\varepsilon X\Theta/\rho$ in time Θ. As soon as this distance is equal to the characteristic thickness $1/\kappa$ of the ionic atmosphere, a well-developed Wien effect is to be expected. Thus for the necessary field strength the equation:

$$X = \frac{\varrho}{\Theta}\frac{1}{\varkappa\varepsilon} = \frac{\varkappa k T}{\varepsilon}$$

is obtained, the second part of which follows using the definition of θ according to eq. (2"). At ordinary temperatures kT/ε is of the order of magnitude 10^{-4}, and since for concentration γ mol per liter $1/\kappa$ is of the order of magnitude $10^{-8}\sqrt{\gamma}$ as stated above, we get:

$$X = 10^4\sqrt{\gamma}\, e\cdot s\cdot E = 3\cdot 10^6\sqrt{\gamma}\ \text{Volt/cm.}$$

Thus the order of magnitude as well as the characteristic dependence on the concentration is correctly predicted by this relation. Therefore it seems very probable that the key to the understanding of the new effects detected in electrolytic solutions will be furnished by a proper consideration of the Coulomb interaction.

Bibliography

1. P. Debye and E. Hückel, *Physik. Z.*, *24*, 185, 305 (1923); L. Onsager, *Physik. Z.*, *28*, 277 (1927).
2. P. Debye and H. Falkenhagen, *Physik. Z.*, *29*, 401-26 (1928), equation (19).
3. P. Debye and H. Falkenhagen, *Physik. Z.*, *29*, 121-32 (1928).
4. M. Wien, *Ann. Physik* , *83*, 327 (1927); *85*, 795 (1928). For the theoretical interpretation of the Wien effect, see G. Joos and M. Blumentritt, *Physik. Z.*, *28*, 836 (1927); M. Blumentritt, *Ann. Physik*, *85*, 812 (1928); and P. Debye and H. Falkenhagen, *Z. Elektrochemie* (1928).

LIGHT SCATTERING

THERMAL DIFFUSION OF POLYMER SOLUTIONS

P. DEBYE AND A. M. BUECHE

Department of Chemistry, Cornell University, Ithaca, New York.

I. Introduction

Thermal diffusion was reported in 1856 by Ludwig.[1] In 1879, apparently unaware of Ludwig's work, Soret[2] reported the effect again. Both Ludwig and Soret observed that in a vertical tube filled with an originally homogeneous salt solution, heated at the upper end and cooled at the lower end, a concentration difference developed between the two ends. The temperature gradient had to be maintained several weeks before this concentration difference reached its maximum value. Their experiments showed that the cold end always contained the more concentrated solution, that is, under the influence of a temperature gradient, some of the particles in solution were forced to the cooler regions. Their observations were confirmed by the experiments of Arrhenius,[3] Wereide,[4] and others.[5-14]

In 1916, S. Chapman,[15] using the kinetic theory, calculated that a mixture of two gases should show a transport effect which he named thermal diffusion. He derived an expression for the dependence of a thermal diffusion constant on the masses of molecules and their law of interaction. Experiments with mixtures of gases were carried out by Chapman and many others, which confirmed his expectations. Most of this work on gases is summarized by Ibbs,[16] and Fleischmann and Jensen.[17] Previously, D. Enskog[18] had mentioned such an effect in his kinetic theory studies. In following years, he developed his theory very thoroughly and some of his results[19] have been used to correct minor errors in Chapman's papers.

Chapman,[20] in 1919, suggested gaseous thermal diffusion as a

*Reprinted from *High Polymer Physics*, H. A. Robinson, ed., Chemical Publishing Company, Brooklyn, 1948, pp. 497-527.

method of separating isotopes. R. S. Mulliken[21] made a critical survey of methods for this purpose and came to the conclusion that thermal diffusion could not be utilized as effectively as several other procedures. However, a revision of this statement was made necessary in 1938 when K. Clusius and G. Dickel[22] reported a combination of thermal diffusion and a special type of thermal circulation which proved to be very effective in gases and led to a practical separation of the chlorine isotopes. Experimental work of the same kind was done in the United States by Brewer and Bramley.[23]

Korsching and Wirtz[24-25] suggested in 1939 that the Clusius method be applied to the original thermal diffusion in solution as a method of separating isotopes. They performed many experiments and verified a phenomenological theory for their instrument which had been proposed by P. Debye.[26] Other similar investigations were undertaken and to date, many liquid systems have been examined.[27-29] A good summary of this work and some new experiments are given in a monograph by S. de Groot.[29]

All this work was on low molecular weight substances. Whether or not polymers would show a marked thermal diffusion effect cannot be predicted, since no adequate theoretical treatment of solutions has been reported. Many attempts[30-42] have been made both from the thermodynamic and kinetic points of view but they all fail to predict what a given liquid mixture will do. The problem, then, is an experimental one guided only by the instrumental constants which can be expected to be important for the process.

The authors have been able to show that a large thermal diffusion effect does exist for polymers and to learn some of the more important aspects involved. The following treatment includes a discussion of the effect itself and the instrumental constants, followed by a description of pertinent experimental results. Most of this work was done during 1945 and 1946 in connection with the program carried out at Cornell University for the Rubber Reserve Company.

II. Molecular Interpretation of Thermal Diffusion

A molecular description of thermal diffusion must contain a description of both thermal and ordinary diffusion. The combined effects in a dilute solution containing n solute particles per unit volume can be described easily in a formal way. Suppose at a certain moment, t, the concentration $n(x)$ is a function of only one coordinate, x, and that a temperature gradient:

$$\theta = \frac{\partial T}{\partial x}$$

is maintained. Through a unit surface perpendicular to x, an excess of particles will move in one direction. This excess can be measured by the specific current, j, defined as the net number of particles passing through 1 sq. cm. per second. For this current, we can assume an expression of the form:

$$j = -D_0 \frac{\partial n}{\partial x} - D_1 \theta n \tag{1}$$

where D_0 is the ordinary diffusion constant, and D_1 is the thermal diffusion constant. Both are measured in square centimeters per second. The first term gives the contribution of ordinary diffusion and the second that of thermal diffusion. If D_1 is positive, the particles will be driven to the cooler regions of the solution under the influence of the temperature gradient.

Einstein[43] has shown that the ordinary diffusion is a result of the Brownian motion of the solute particles. If ξ designates the x component of an irregular displacement, during time τ, of a particle in Brownian motion, the ordinary diffusion constant is given by the expression

$$D_0 = \frac{\overline{\xi^2}}{2\tau} \tag{2}$$

where $\overline{\xi^2}$ is the average square of a number of such displacements.

Einstein considered the isothermal case, where $\theta = 0$. In this special case, calling $W(\xi)d\xi$ the probability that a particle will be found at a distance between ξ and $\xi + d\xi$ from its original position after time τ, the probability function, $W(\xi)$, will be symmetrical around $\xi = 0$ and $\bar{\xi}$ will be zero. However, if a temperature gradient, θ, exists, it is expected that $W(\xi)$ will be unsymmetrical which will lead to a finite value of $\bar{\xi}$. At the same time, the width of the probability curve, measured by $\overline{\xi^2}$, will vary with the original position, x, of the particle; the higher the temperature at the position x, the larger will be $\overline{\xi^2}$. Then $W(\xi)$ will also be a function of x.

Consider a solution with the concentration and temperature varying in the x direction. The dependence of concentration on x will be indicated by $n(x)$. Choose a point x_0 in the center of an infinitesimally small division of x denoted by Δ. The number of particles leaving this small region Δ during time τ is:

$$\Delta n(x_0) \tag{3}$$

The number of particles entering during time τ is:

$$\Delta \int_{-\infty}^{x_0-\frac{\Delta}{2}} n(x)W(x, \xi)dx + \Delta \int_{x_0+\frac{\Delta}{2}}^{+\infty} n(x)W(x, \xi)dx = $$

$$= \Delta \int_{-\infty}^{+\infty} n(x)W(x, \xi)dx \tag{4}$$

Subtracting equation (3) from equation (4) to obtain the change in the number of particles in division Δ during time τ gives:

$$\tau\Delta \frac{\partial n(x_0)}{\partial t} = \Delta\left[\int_{-\infty}^{+\infty} n(x)W(x, \xi)dx - n(x_0)\right] \tag{5}$$

making a change in variables such that $\xi = x_0 - x$. It can be seen that both of the functions under the integral may be expanded around x_0 in terms of ξ. On expanding and integrating (since the integral over $W(x, \xi)$ must equal unity), we arrive at a continuity-relation, which can be written in the usual form

$$\frac{\partial n}{\partial t} + \frac{\partial j}{\partial x} = 0 \tag{6}$$

provided we represent the specific current, j, by the expression:

$$j = -\frac{\overline{\xi^2}}{2\tau}\frac{\partial n}{\partial x} - \left[\frac{\partial}{\partial T}\left(\frac{\overline{\xi^2}}{2\tau}\right) - \frac{1}{\theta}\frac{\overline{\xi}}{\tau}\right]\theta n \tag{7}$$

This confirms the formal relation according to equation (1). Comparison with equation (2) shows that the thermal diffusion constant is made up of two parts; that due to the temperature derivative of the ordinary diffusion constant and that due to the unsymmetrical character of the probability curve $W(\xi)$

$$D_1 = \frac{\partial D_0}{\partial T} - \frac{1}{\theta}\frac{\bar{\xi}}{\tau} \qquad (8)$$

The first term in equation (8) is known to be only about 2.8 percent of D_0 in aqueous solutions. If the particle in question is driven to the cooler regions, $\bar{\xi}$ will be negative. This transport effect has been found for polymer solutions to be much larger than the effect due to the temperature dependence of D_0. Even if the transport-effect measured by $\bar{\xi}$ would be zero, the ordinary Ludwig-Soret arrangement would still yield a concentration-gradient, the equilibrium condition being in this case nD_0 = const.

To extend this treatment so that one could predict the value of the thermal diffusion constant would require an exact knowledge of how the characteristics of a particle affect its Brownian motion. As was previously stated, this knowledge does not seem to be available.

III. Theory of the Separation Column

Two main types of separation columns employing the thermal circulation of Clusius have been in use. The first type consists of two vertical metal plates, one heated and one cooled, held a short distance apart by gaskets at the edges. The slit between the plates and the small reservoirs at the top and bottom, connected with the slit, are filled with the solution. The other type, which was used in the experiments discussed in this paper, consists of two vertical concentric metal tubes of length h forming a cylindrical slit of width $2a$. The inner tube is cooled by running water; the outer tube is heated electrically. If the temperature of the inner tube is T_1, and that of the outer tube T_2, the temperature gradient is given by

$$\theta = \frac{T_2 - T_1}{2a}$$

At each end of the tube there is a cylindrical reservoir connected with the slit. The reservoirs and the interspace are filled with the solution to be investigated and the heating current and the cooling water adjusted.

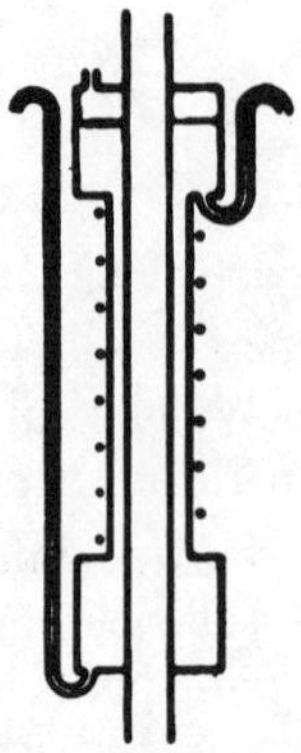

FIG. 1. Schematic diagram of the thermal-diffusion separation column.

Near the warm tube, the solution expands and becomes less dense; near the cool tube, the solution is more dense. As a result of this density difference, thermal circulation occurs; near the warm tube the liquid goes up, near the cool tube it goes down. These streams empty into the reservoirs. It is expected that thermal diffusion will take place during the time the solution is in the slit and under the influence of a relatively high temperature gradient. If this is so, some of the particles in solution will be transported to the cool stream and after a time the concentration in the upper reservoir will be less than that in the lower one. The next few paragraphs will be devoted to a discussion of the methods for obtaining an expression for this concentration difference.

Considering the cylinders as two flat plates a distance $2a$ apart, the solution velocity between them can be obtained by equating the forces on a small element of liquid volume and integrating the resulting expression over the entire slit. The value for this velocity, assuming the temperature gradient, θ, and the solution viscosity, η, to be constant, is then found to be

$$v(x) = \frac{\alpha\rho g}{6\eta} a^3\theta \frac{x}{a}\left(1 - \frac{x^2}{a^2}\right) \qquad (9)$$

where α = cubical expansion coefficient
ρ = solution density
g = acceleration of gravity

η = solution viscosity (midway between the plates)
$2a$ = distance between the plates
x = distance from a point midway between the plates

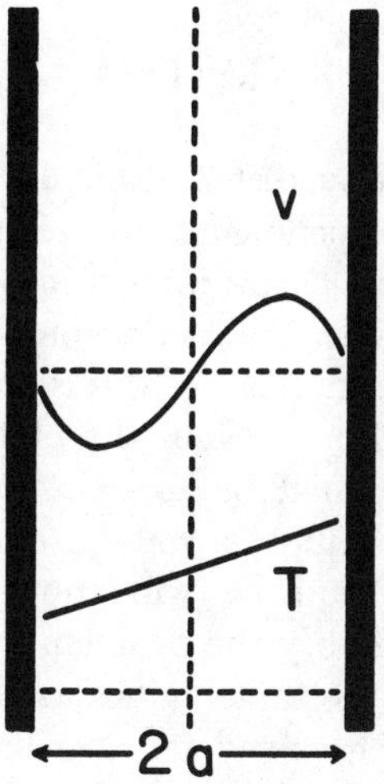

FIG. 2. Thermal circulation in the separation column.

Figure 2 is a plot of this velocity distribution and the temperature gradient in the slit. The velocity is zero at the plates and at a point midway between the plates. The error introduced by the approximation that the viscosity throughout the interspace is that at $x = 0$ was investigated by de Groot.[29] He found that even when it was twice as large at the cold plate as at the warm one, the errors in the final results were very small. The physical influence of this viscosity variation is to increase the maximum velocity in the upward stream and decrease it in the cool stream. Since the amount of solution going up must be equal to that going down, this requires that the velocity is zero at a point slightly in the positive x direction from $x = 0$ instead of at $x = 0$.

The average velocity, v_m, in one half the slit can be shown to be

$$v_m = \frac{\alpha \rho g}{24\eta} a^3 \theta \tag{10}$$

In the experiments to be described here, v_m was only a few millimeters per second.

The continuity equation which will be fundamental in the consideration of the thermal diffusion column is

$$\frac{\partial}{\partial x}\left[D_0 \frac{\partial n}{\partial x} + D_1 \theta n\right] - v(x)\frac{\partial n}{\partial z} = 0 \tag{11}$$

where $v(x)$ is the velocity distribution described above and z is the vertical direction perpendicular to x. The equation states that in case more solute particles are carried into an element of volume by the liquid motion than are carried out, the gain will be compensated by ordinary and thermal diffusion. The diffusion in the z-direction is being neglected. Since the plates are impermeable to solution, the equation must be solved subject to the boundary conditions that the specific current, j, equation (1), is zero at $x = \pm a$. Actually D_1 and D_0 will show some dependence on temperature, but to simplify the problem, it is assumed that they are constant and have the same values as they have at $x = 0$.

Equation (11) can be solved for two limiting cases. The first occurs when $D_1 \gg D_0$. Experimental evidence indicates that this equation will apply to solutions of most of the large molecules or high polymers here investigated. The second, an adaptation of the theory of Furry, Jones, and Onsager[44, 45] liquids, assumes that $D_1 \ll D_0$. The approximations made in this case will be best satisfied by solutions of small molecules whose diffusion constants are of the order of $D_0 = 1 \times 10^{-5}$.

(a) *Thermal Diffusion Large Compared with Ordinary Diffusion*

When thermal diffusion is large compared with ordinary diffusion, $D_1 \gg D_0$, and the D_0 terms in equation (11) may be omitted leaving·

$$D_1\theta \frac{\partial n}{\partial x} - v(x)\frac{\partial n}{\partial z} = 0 \tag{12}$$

Introducing a new variable, u, for x such that

$$u = \frac{1}{D_1\theta}\int_0^x v(x)dx = \frac{av_m}{D_1\theta}\xi^2(2 - \xi^2) \tag{13}$$

in which $\xi = x/a$, the differential equation becomes:

$$\frac{\partial n}{\partial u} - \frac{\partial n}{\partial z} = 0 \tag{14}$$

The general solution is:

$$n = F(u + z) \tag{15}$$

in which F is an arbitrary function of the argument $(u + z)$. Suppose now for $z = 0$, where the bottom of the slit joins the lower reservoir, the concentration is n_1 independent of x between $x = -a$ and $x = +a$. At this value of z, the variable ξ goes from -1 to 1 and u goes from $-(av_m/D_1\theta)$ to $av_m/D_1\theta$. In this range $n = n_1$. Since F was an arbitrary function a new one $\Phi = F/n_1$ may be defined such that

$$\frac{n}{n_1} = \Phi(u + z) = \Phi\left[z + \frac{av_m}{D_1\theta}\xi^2(2 - \xi^2)\right] \tag{16}$$

Then Φ is a function which is unity as long as the interval in which the variable $(u + z)$ moves is between $-(av_m/D_1\theta)$ and $av_m/D_1\theta$ and is zero for all other values of the argument.

On a streamline of a particle in the slit, the argument $(u + z)$ will be constant, so the equation of a streamline which begins at $z = 0$ and $\xi = \xi_1$ will be

$$z + \frac{av_m}{D_1\theta}\xi^2(2 - \xi^2) = \frac{av_m}{D_1\theta}\xi_1^2(2 - \xi_1^2) \tag{17}$$

According to this formula, ξ will be equal to zero for $z = z_1$, i.e., where:

$$z_1 = \frac{av_m}{D_1\theta}\xi_1^2(2 - \xi_1^2) \tag{18}$$

This is the maximum value of z which a particle can obtain if it begins at $\xi = \xi_1$, and $z = 0$. The maximum value of z obtained in a given instrument by a particle having a thermal diffusion coefficient D_1 will be that achieved by a particle starting its upward motion at $\xi_1 = 1$. Its streamline will be given by:

$$z + \frac{av_m}{D_1\theta}\xi^2(2 - \xi^2) = \frac{av_m}{D_1\theta} \tag{19}$$

Substituting the value $\xi = 0$ to find the value of z attained just as

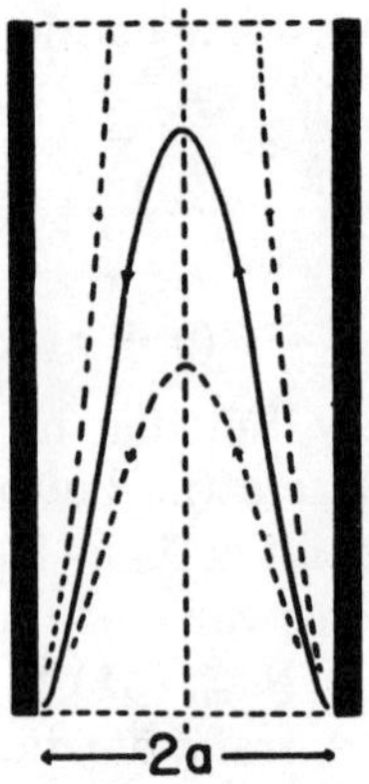

FIG. 3. Thermal-diffusion stream lines.

the particle starts to go down in the cool stream, shows the maximum height to be:

$$H = \frac{av_m}{D_1\theta} = \frac{\alpha\rho g}{24\eta}\frac{a^4}{D_1} \tag{20}$$

Figure 3 is a plot of streamlines leading to this maximum height, H, for particles having three different thermal diffusion constants. Two of the particles are able to reach their maximum heights in the instrument represented. The third has a smaller thermal diffusion coefficient. Actually, the streamlines will not be as sharp as pictured due to the ordinary diffusion of the particles.

It is convenient to define an "instrumental diffusion constant," Δ, such that:

$$\Delta = \frac{v_m}{\theta} \tag{21}$$

Then the expression for H becomes:

$$\frac{H}{a} = \frac{\Delta}{D_1} \tag{22}$$

The preceding discussion, which neglects ordinary diffusion, shows that by building a column whose height, h, is larger than the maximum height, H, for a particle, one can eliminate all of these particles from the upper reservoir. This is possible because after a time, all of the solution will have been in the lower reservoir and in the upward stream due to the thermal circulation, but the solute particles can never again reach the upper reservoir. A question as to what happens to the particles which move to the cold plate in the downward stream arises at this point. These particles will tend to be held at the cold plate due to the thermal gradient. It is expected that these will form a very concentrated layer at this plate, diffuse into the downward stream, and be carried to the lower reservoir. Experiments have been conducted to determine what fraction of the polymer particles remains there. It was found to be so small as to be negligible in ordinary work.

Suppose, now, the column ends at $z = h$ smaller than the maximum height, H, which is necessary to revert all of the entering particles to the downward stream. At what value of ξ (denoted by ξ_2) at the level $z = 0$ would a particle have to start so that it just reaches $\xi = 0$ at $z = h$? From equations (18) and (20), this can be seen to be:

$$h = H\xi_2^2(2 - \xi_2^2) \tag{23}$$

To obtain expressions for the concentrations in the top and bottom reservoirs in the steady state, it is necessary to know the fraction, f, of all the particles entering with the upward stream at $z = 0$ which will reach the downward stream and in the end return to the lower reservoir. The fraction $1 - f$ will then enter the upper reservoir. Call the width of the column, or the circumference of the tube, b. The total number of particles, ν, entering per second is

$$\nu = n_1 b \int_0^a v(x)dx = n_1 a b v_m \tag{24}$$

where n_1 is the concentration in the lower reservoir. The number of particles returned to the lower reservoir per second is

$$f\nu = n_1 b \int_0^{a\xi_2} v(x)dx = n_1 a b v_m \xi_2^2(2 - \xi_2^2) \tag{25}$$

From these and equation (23) then

$$f = \frac{f\nu}{\nu} = \frac{h}{H} \tag{26}$$

At equilibrium for this case in which ordinary diffusion is very small compared with thermal diffusion, the concentration of solute in the upper reservoir (n_T) divided by that in the lower one (n_B) is:

$$\left(\frac{n_T}{n_B}\right)_{eq.^*} = \begin{cases} 1 - \dfrac{h}{H} & \text{for} \quad h < H \\ 0 & \text{for} \quad h \geq H \end{cases} \tag{27}$$

This line of reasoning may be easily extended to find the course of the approach to the steady state in terms of time. By considering the change in the number of particles in the upper and lower reservoirs per unit time and combining these with the expression for the conservation of mass it can be shown that for reservoirs having equal volumes:

$$n_T = \frac{1 - (h/H)}{1 - (h/2H)} n_0 + \frac{(h/2H)}{1 - (h/2H)} n_0 e^{-t/t_c} \tag{28a}$$

$$n_B = \frac{1}{1 - (h/2H)} n_0 - \frac{(h/2H)}{1 - (h/2H)} n_0 e^{-t/t_c} \tag{28b}$$

where n_0 is the initial concentration of the solution, e is the base of the natural logarithms, t is the time since the beginning of the experiment and t_c is a characteristic time:

$$t_c = \frac{V}{2abv_m} \frac{1}{1 - (h/2H)} \tag{29}$$

where V is the volume of one of the reservoirs. Equations (28a, 28b) lead to (27) for large values of t compared with t_c and also show that:

$$n_B - n_T = \frac{h/H}{1 - h/2H} (1 - e^{-t/t_c}) n_0 \tag{30}$$

The preceding expressions for the rate were derived assuming the volume of the reservoirs to be equal. If this is not the case, the volumes will enter the expressions explicitly in such a way that they will cancel when the formula for $(n_T/n_B)_{eq.}$ is obtained but will appear in $n_B - n_T$.

* eq. = equilibrium.

(b) *Thermal Diffusion Small Compared with Ordinary Diffusion*

The second limiting case to be considered is that in which thermal diffusion is small compared to ordinary diffusion, i.e., $D_1 \ll D_0$. To find the total current of particles in the z direction, multiplying equation (11) by dx and integrating from $x = -a$ to $x = +a$ yields

$$\frac{\partial}{\partial z}\int_{-a}^{+a} nv(x)dx = 0 \qquad (31)$$

since the sum of the thermal and ordinary diffusion currents vanishes at the walls of the instrument ($x = \pm a$). If b is again the width of the column, and I the total current in the z direction, the integral in equation (31) is I/b. The steady state will be characterized by the fact that $I = 0$, that is,

$$\frac{I}{b} = \int_{-a}^{+a} nv(x)dx = 0 \qquad (32)$$

Assuming now that n takes the form

$$n = \Phi(x)e^{-\gamma z} \qquad (33)$$

where $\Phi(x)$ and γ are to be determined, substitution in (11) yields the following equation for Φ

$$\frac{\partial}{\partial x}\left[D_0\frac{\partial\Phi}{\partial x} + D_1\theta\Phi\right] + \gamma v(x)\Phi = 0 \qquad (34)$$

To solve this equation, it is possible to use an approximation procedure which corresponds to the development of γ in powers of D_1. The first approximation is $D_1 = 0$, $\gamma = 0$ simultaneously, and, therefore, $\Phi = \Phi_0$, a constant. This corresponds to the beginning of the experiment where the concentration is constant throughout the instrument. In a second approximation, we can now substitute $\Phi = \Phi_0$ in the last term of our equation (34). Integrating over x this leads to

$$D_0\frac{\partial\Phi}{\partial x} + D_1\theta\Phi = -\gamma\Phi_0\int_{x=-a}^{x=x} v(x)dx = -\gamma\Phi_0 F(x) \qquad (35)$$

where $F(x)$ stands for the integral of $v(x)$. By selecting the lower limit of the integration $x = -a$, a choice was made of the initially arbitrary integration constant. This choice satisfies the boundary conditions since $F(x) = 0$ at $x = \pm a$, which makes the left side of equation (35) (actually the specific current, j) zero at these points. Continuing the approximation procedure by setting $\Phi = \Phi_0$ in the second term of (35) and integrating over x gives:

$$\Phi = \Phi_0 - \frac{\Phi_0}{D_0}\int_0^x [\gamma F(x) + D_1\theta]\, dx \tag{36}$$

The addition of the constant Φ_0 and the choice of zero for the lower limit of integration again makes a definite choice of an arbitrary integration constant such that $\Phi = \Phi_0$, for $x = 0$, that is, a line midway between the plates. Actually this choice is immaterial for what follows, but perhaps makes the procedure a little easier to visualize.

We now know (in the proper approximation) the concentration-distribution $n(x, z)$ according to equation (33) and can calculate the total current I. The condition that $I = 0$ will determine the steady state and enable us to find the relation between γ and D_1. This condition is:

$$\int_{-a}^{+a} \{v(x) \int_0^x [\gamma\Phi(x) + D_1\theta] dx\}\, dx = 0 \tag{37}$$

or, expressed differently:

$$\frac{\dot{\gamma}}{\theta D_1} = -\frac{\int_{-a}^{+a} xv(x)dx}{\int_{-a}^{+a}\left\{v(x)\int_0^x \Phi(x)dx\right\}dx} \tag{38}$$

Upon integrating it is found that

$$\gamma = \frac{21}{16}\frac{D_1\theta}{av_m} \tag{39}$$

Remembering the value for the maximum height, H, according to equation (20), this can be expressed as

$$\gamma = \frac{21}{16}\left(\frac{1}{H}\right) \tag{40}$$

This means the concentration diminishes along the column according to the expression:

$$e^{-\frac{21}{16}\frac{z}{H}} \tag{41}$$

For a column of height, h, the concentration of the solution in the top reservoir divided by that in the bottom at equilibrium is

$$\left(\frac{n_T}{n_B}\right)_{\text{eq.}} = e^{-\frac{21}{16}\frac{h}{H}} \tag{42}$$

The two equations (27) and (42) are compared graphically in figure 4. It is obvious that there is little difference until the height, h, of the column approaches the maximum height of the particles, H. Figure 4 shows that if ordinary diffusion is small compared with

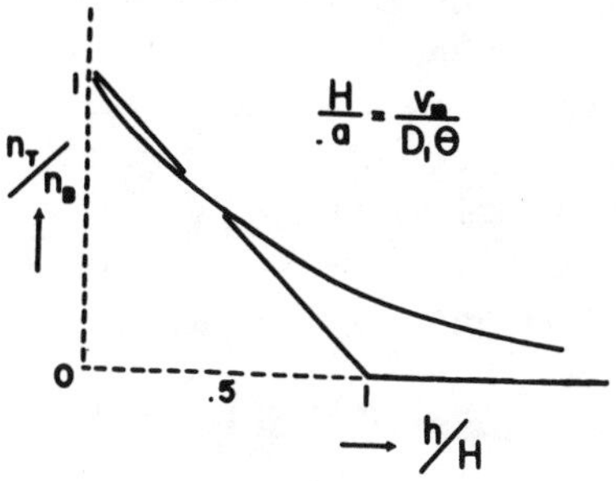

FIG. 4. Comparison of two approximations used in solving the diffusion equation. *Curve* is thermal diffusion small when compared with ordinary diffusion. *Line segments* are thermal diffusion large when compared with ordinary diffusion.

thermal diffusion, it is possible to remove nearly all of the solute from the upper reservoir, while if the opposite is true, it is possible to come as close to complete removal as desired only by building a column with the proper dimensions, i.e., one whose height, h, is larger than the maximum height, H, for a particle. By comparison of equations (27) and (42) with (20), it may be seen the $(n_T/n_B)_{\text{eq}}$ is independent of θ.

In addition to the assumptions of constant viscosity and thermal diffusion coefficient discussed previously, there are two other implicit assumptions which were made in the derivation of equations (27) and (42). These were the use of a constant value for the expansion coefficient and the neglect of the effect of concentration on the density of the solution. In general, the expansion coefficient does not change significantly over the ranges of concentration and temperature used so that the assumption of its constancy is not serious. However, the density of a solution could be changed appreciably during the thermal diffusion process because of changes in concentration at different x-values in the solution. This marked dependence on concentration could give erroneous results as to the thermal diffusion properties of the mixture. This does not appear to be true when working with polymer solutions. A 1.5 percent solution of polystyrene in xylene has a density which is different from that of pure xylene by only about 0.4 percent. However, the temperature difference of 50°C between the plates be which is usually employed in our experiments corresponds to a change in density of 7 percent. In other systems, this change may much larger as discussed by de Groot.[29] A striking example of this is the electrophoresis-convection method for separating proteins[46] where convection currents of the type described here are established by the concentration gradient alone.

IV. Experimental Results with Polymers

The separation columns used in the experiments with polymers were of the cylindrical type described in section III. They were made of brass tubes about two cm in diameter joined together through the reservoirs so as to be as well aligned as possible. The temperature gradient was established by running cool water through the inner tube and heating the outer tube electrically. The upper reservoir on each column had two openings to the exterior, one to fill the instrument and one to which a metal capillary was attached so that the solution in the reservoir could be removed without disturbing that below. The lower reservoir was also fitted with a capillary for removing the remaining solution.

(a) *Time of Separation*

The time dependence of the separation is given by equations (28a, 28b). The characteristic time, according to equation (29),

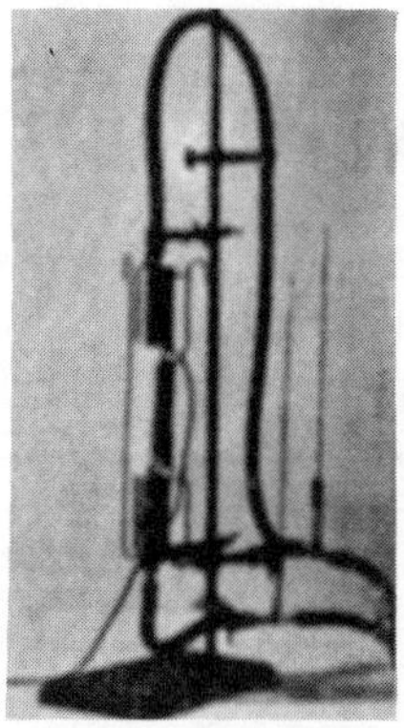

FIG. 5. The cylindrical thermal-diffusion separation column.

depends on the temperature gradient, the characteristics of the solution, and the dimensions of the column including the volume of the reservoirs. Korsching and Wirtz[24] have thoroughly investigated the effects of these variables and have shown their experiments to be in accord with the expressions of Debye[26] for the operation of their instrument.

Experiments were conducted to find the approximate times necessary to reach a steady state for the columns used in this study. The columns were filled with a polymer solution, the temperature gradients established, and the solutions in the upper reservoirs were investigated at intervals. This was done by removing 5 cc samples, measuring their flow times in an Ostwald viscometer and returning them. Figure 6 is a semi-logarithmetic plot of such flow times, t, divided by the flow times of the solvent t_0 against the time of the experiment. The solution was of a polystyrene, of an average molecular weight of 410,000, in toluene and had an initial viscosity of twice that of the solvent. The concentration was 0.5 percent; the temperature gradient was $\theta = 537°C$ per cm as measured by thermocouples at the warm and cool tubes. The volume of each reservoir was about 50 cc. This column is slow but very effective. The shape of the curve is representative of all the others except that most of the columns were constructed so as

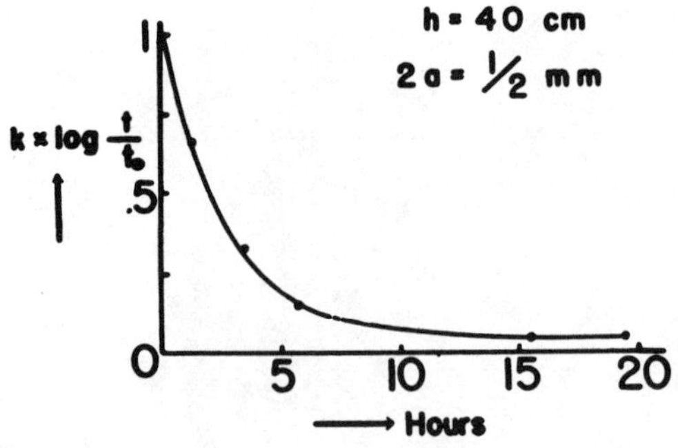

FIG. 6. Variation of viscosity with time.

to be less effective and, therefore, faster and the process was almost completed after 2 or 3 hours.

(b) *Dependence of the Concentration Quotient on the Column*

Equations (27) and (42) show that the efficiency of a column is dependent on its height, h, and on the fourth power of the distance, $2a$, between the tubes. Experiments were performed which help substantiate these expressions. Three instruments were constructed, two to find the dependence on h and the third to find the effect of a variation of a. Two types of solutions were investigated, a solution of a small molecule, naphthalene in xylene, and a solution of a polystyrene of molecular weight of 410,000 in toluene. Table I is a tabulation of the results for naphthalene. The temperature difference was about 50°C. in all cases and was maintained until the concentration in the upper reservoir became practically constant. The concentrations were measured by refractive index methods using an instrument designed by P. P. Debye[47] The

TABLE I. Thermal diffusion of naphthalene in xylene

h (cm)	$2a$ (mm)	n_T/n_B	n_T/n_B calculated	H experimental
40	1	0.93	0.94	722
20	1	0.97	0.97	861
15	0.5	0.74	0.70	65

third and fourth columns of data afford a comparison between the concentration of naphthalene in the top reservoir divided by that in the bottom, choosing an average value of the maximum particle height, $H = 880$ for a 1 mm slit-width. The fifth column shows the H-values as derived from the observed concentration-quotients, calculated according to equation (42).

Table II gives the same type of data for a toluene solution of a polystyrene having a molecular weight of 410,000 except that the calculated values were obtained from equation (27). The viscosity of the original solution was 1.59 times that of the solvent.

TABLE II. Thermal diffusion of polystyrene in toluene

h (cm)	*2a (mm)*	n_T/n_B	n_T/n_B *calculated*	*H* *experimental*
40	1	0.90	0.93	417
20	1	0.96	0.97	500
15	0.5	0.70	0.58	49

A great increase in efficiency is again noted when going from $2a = 1$ mm, to $2a = 0.5$ mm. The values in the fourth column are calculated with an average value of $H = 570$ for a 1-mm slit-width. The values of the maximum particle height, H, in the last column have been calculated from the observed concentration-quotients according to equation (27).

These experiments verify that high polymers as well as small molecules are subject to thermal diffusion. The theory predicts that a column can be made twice as effective by doubling the length and 16 times as effective by halving the width of the interspace. Although the experiments are not very accurate, they give an indication of this dependence on the specifications of the column. Changes in the slit-width are shown to have a particularly strong influence on the thermal diffusion.

(c) *Dependence of Thermal Diffusion on Molecular-Weight*

Since none of the existing theories of thermal diffusion is adequate to predict its dependence on molecular weight in liquid solution, the problem is of a purely experimental nature. Three samples of polystyrene having molecular weights* of 69,000, 450,000, and 1,050,000 were dissolved in toluene. Their concentrations of 2.9, 0.45 and 0.10 percent were such that they all had about the same relative flow times, t/t_0, in the viscometer. The solutions were placed in a column having a height, $h = 15$ cm and a slit-width, $2a = 0.5$ mm. The temperature gradient of 960°C per cm, as estimated from the rate of flow of the cooling water, its temperature increase, and the thermal conductivities, was maintained for 12 to 13 hours. At the end of this time, the concentrations of the solutions in the top and bottom reservoirs were measured by refraction. Table III gives the values of the experimentally determined concentration quotients. In the fourth column, average values of the maximum particle height, H, calculated from equation (27) for the limiting case $D_1 \gg D_0$ are tabulated. In the fifth column, values of H which were calculated from equation (42) for the limiting case $D_1 \ll D_0$ are given.

TABLE III. Thermal diffusion of polystyrene of various molecular weights in toluene

Weight Average Mol. Wt.	(t/t_0) initial	$\frac{n_T}{n_B}$	H_{cm} *Calc. from eq. (27)*	H_{cm} *Calc. from eq. (42)*	$D_1 \times 10^7$
69,000	2.15	0.70	52	58	0.40
69,000	1.96	0.72			
450,000	2.07	0.59	38	39	0.54
450,000	2.04	0.62			
450,000	2.07	0.61			
450,000	2.07	0.59			
1,050,000	2.21	0.37	24	20	0.86

Approximate values for D_1 are given. They were obtained using the physical constants of benzene at 50°C. The viscosity value used was twice that of benzene. These data indicate that the

* All the molecular weights mentioned are weight average molecular weights determined by J. R. McCartney using light scattering methods.[48]

thermal diffusion increases with increasing molecular weight. Actually, the increase is somewhat greater than that shown because of the slow depolymerization of the highest molecular weight polymer under the conditions of the experiment. It was found that by using a smaller thermal gradient and waiting longer for the establishment of the steady state, the decomposition could be slowed down relative to the thermal diffusion process. This procedure resulted in a still lower value of $n_T/n_B = 0.22$. The other polymers showed little or no depolymerization under the conditions of the experiment. Although no accurate measurements of D_0 are available for polymers, it is expected that for a molecular weight of 58,000 it will not be larger than 1×10^{-7}. The values for D_1 show that it is about equal to and much greater than the temperature gradient of D_0.

(d) *Fractionation of Polymers by Thermal Diffusion*

The polymers investigated in the previous section were unfractionated samples and thus consisted of a distribution of molecular weights. The experiments with that column gave little or no fractionation into samples of different molecular weights.

A column having a slit-width, $2a = 0.5$ mm and a height, $h = 40$ cm was constructed. Since the polymer with an average molecular weight of 450,000 has a maximum particle height, H, equal to 38 cm when $2a = 0.5$ mm, it was expected that it would be fractionated by this instrument. A toluene solution of the polymer having twice the viscosity of the solvent, was put into the instrument and the temperature gradient established. Figure 6 shows the variation with time of the viscosity of the solution in the upper reservoir. At the end of the experiment, the solution was removed from this reservoir. Then glass capillaries were dipped into the cylindrical interspace between the tubes and samples taken at depths of 10, 20, 30 and 40 cm. The remainder of the solution was removed from the bottom reservoir. The concentration of the solution from the top reservoir was so small that it could not be measured accurately by its refractive index. Obviously another more sensitive method must be used.

A Beckman Quartz Spectrophotometer was employed to obtain

the absorption curves for toluene solutions of the 3 polystyrenes. At each wave length, the optical density of toluene was taken as zero. The absorption coefficients, α, were found from the expression

$$\frac{I}{I_0} = e^{-\alpha d}$$

where I_0 and I are the incident and transmitted light intensities and d is the distance light travels in the solution. This distance was one cm in our matched quartz cells.

Since Beer's Law was found to hold within narrow limits for each polymer at the wave lengths used, curves of α vs λ could be normalized at 294$m\mu$. When this is done, it can be seen (figure 7, solid lines) that there are differences between the shapes of the curves. The higher the molecular weight, the flatter the curve. The concentrations of the polymer having a molecular weight of 69,000 at which the measurements were made and the concentrations the other solutions would have to have so that the absorption coefficients at 294$m\mu$ would all be the same are indicated below the molecular weight values.

The three solid curves form a scale with which we can compare the absorption data for the solutions from the top and bottom reservoirs of the column. If a fractionation does take place, one would expect the data for the bottom sample to resemble that for the original more closely than that for the top, since the bottom constitutes more than one half of the total solution. Figure 7a shows the results (dotted lines) for the polystyrene having a molecular weight of 450,000; figure 7b is for the one with a molecular weight of 69,000.

The spectral region investigated in which this specific effect of the molecular weight occurs is at the long wave length edge of a strong absorption band. When the samples are compared on this scale, it seems evident that some kind of fractionation occurs.

A good explanation as to why these absorption curves show the reported differences does not seem to be available. Repeated purifications of the polymers by precipitation in methanol change the curves slightly but apparently do not eliminate the differences. Until a satisfactory explanation is found, the scale must remain an empirical one.

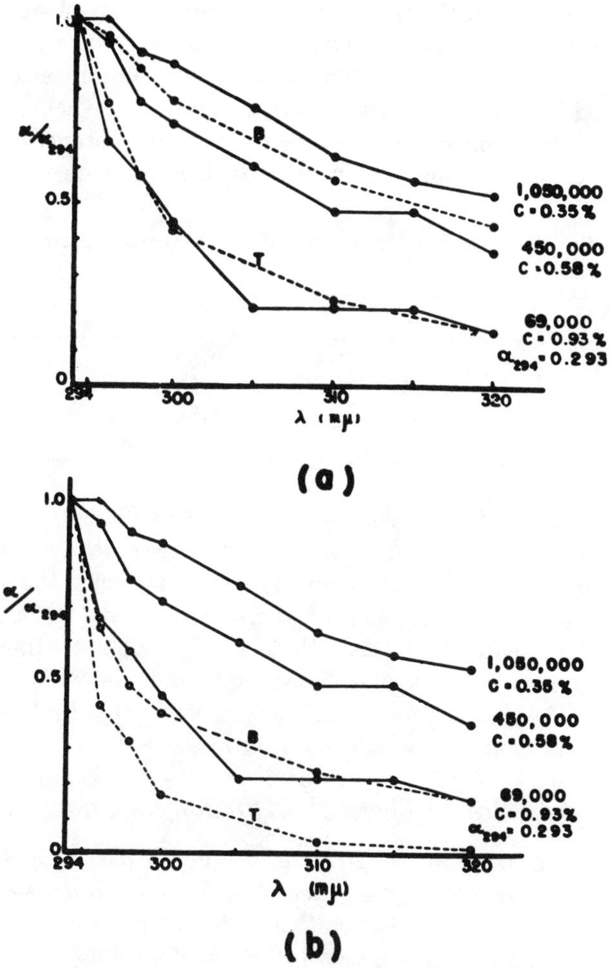

FIG. 7. Fractionation of Polystyrene.

The intrinsic viscosities, $[\eta]$, of the samples taken from different levels in the column by the capillaries were measured. The intrinsic viscosity, $[\eta]$, here is defined by

$$\ln \frac{t}{t_0} = [\eta]c - Bc^2$$

where t and t_0 are the flow times of solution and solvent in the viscometer, c is the concentration in grams per 100 cc and B is a constant for a given polymer and solvent. The concentrations were measured by weighing after evaporation in a vacuum oven. Table IV gives the values of $[\eta]$ for the average distances, $\bar{d}$, from the top of the column from which the samples were taken. The

TABLE IV. Variation of intrinsic viscosity with average distance from top of column

$\bar{d}$	$[\eta]$
5	0.50
15	0.92
25	1.62
35	1.60

value of $[\eta]$ for the original solution was 1.53. Since the intrinsic viscosity increases with molecular weight these data support the picture of the phenomenon (figure 3) and substantiate the previous findings on the dependence of thermal diffusion on molecular weight. It seems clear then, that it is possible to fractionate polymers into samples of different intrinsic viscosities by this method. Thermal diffusion might also be valuable for the determination of molecular weight distributions.

(e) *Dependence of Thermal Diffusion on Concentration*

According to equation (20), the maximum particle height, H, should decrease and a given thermal diffusion column should become more effective as the solution viscosity increases if the other factors remain constant. For solutions of high polymers, the viscosity is much higher than that of the solvent but one would expect the expansion coefficient and density to vary but little, at least for dilute solutions. If this is true, H can be written as

$$H = k/D_1\eta$$

Knowing the viscosity of the solution and determining the value of H experimentally, one can find the dependence of the thermal diffusion constant, D_1, on viscosity and thus on concentration.

Much experimental work has been done in the past to find the dependence of D_1 on concentration for ordinary compounds.[7,12,13] The results are somewhat conflicting, but they show that D_1 is not a true constant in a great number of cases. The deviation from constancy seems to depend on the nature of the compound and upon the solvent used.

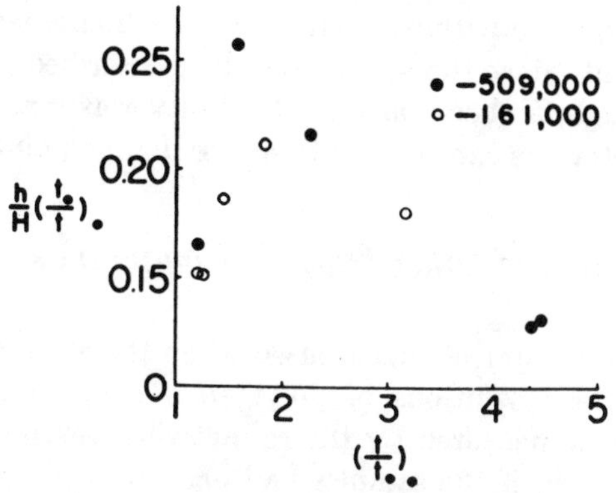

FIG. 8. Thermal diffusion *vs.* viscosity.

To determine the dependence on viscosity for polymer solutions of different molecular weights, a sample of Styron* was fractionated into several molecular weight groups by methanol precipitation. Two of these were chosen (161,000 and 509,000) and used to prepare toluene solutions of different viscosities. The solutions were introduced into a column having $h = 15$ cm and $2a = 0.5$ mm and a temperature gradient of about 960°C per cm was established. This was maintained for about 48 hours to insure the attainment of the steady state. At the end of this time, samples were withdrawn from the top and bottom reservoirs and their concentrations measured by viscosity. Figure 8 is a plot of the experimentally determined values of h/H (see equation (27)) multiplied by the reciprocal of the original relative flow-times in the viscometer *vs.* these relative flow times. The values on the ordinate may then be considered proportional to the thermal diffusion constant, D_1. It shows that D_1 is not a constant for

* A polystyrene (M.W. 410,000) supplied by The Dow Chemical Company.

polymers; there might even be some reversal of the molecular weight dependence at higher concentrations. Furthermore, there seems to be an optimum value of the viscosity at which to carry out a fractionation by thermal diffusion. It does not seem likely that the decrease of D_1 either for high or low viscosities can be due to insufficient time for the experiment to reach the steady state because the viscosity of the solution in the upper reservoir was always constant to within 1 percent over 4 hours near the end of an experiment. Since the whole procedure described in this section is rather rough, a check on reproducibility was made. The result was satisfactory as can be seen from the double points in figure 8.

(f) *Separation of Other Organic Compounds*

Considerable interest was aroused when the authors were working with xylene solutions of polymers. It was found that the concentrations measured by the refractive index method were in appreciable error if the samples had been treated by the thermal diffusion process. Since the xylene used was a mixture of the isomers, it was suspected that they too were being separated. As a result, Dr. A. L. Jones performed some preliminary experiments on these isomers and other compounds. The column used had a height, $h = 40$ cm and a slit-width, $2a = 0.5$ mm. The temperature gradient was about 900°C per cm. The compounds were chosen so as to find the effects of shape and molecular weight on the separation achieved. If possible, concentrations were determined by refractive index measurements. The results are tabulated in table V. Some of these results were checked by viscosity-measurements. Previously water had been separated from alcohol even at the constant boiling mixture.

Experiments 1, 2, and 5 show that shape is important in determining the thermal diffusion properties of a compound. Perhaps the most striking example is mixture number 5. Here the molecular weights are exactly the same, the difference being only that between a branched and a straight chain. Experiment number 3 gives some indication as to the dependenoe on molecular weight for somewhat similarly shaped compounds. These preliminary results

TABLE V. Thermal diffusion of various organic substances

Compounds	Initial Vol.%	Final Vol.% (Top)
1. *m*-Xylene	80	85
o-Xylene	20	15
2. *p*-Xylene	20	27.5
o-Xylene	80	72 5
3. Toluene	90	100
Chlorobenzene	10	0
4. Heptane	20	36
Toluene	80	64
5. *n*-Heptane	80	100
Triptane	20	0
6. *n*-Heptane	75	91.5
Isooctane	25	8.5
7. Benzene	25	26
Cyclohexane	75	74

show the applicability of thermal diffusion to the problem of purification of organic liquids and the separation of isomers. In a more thorough investigation, attention should also be given to the dependence of the separation on the relative concentrations.

(g) *A Column of Different Design*

The construction of a column such as those previously described presents some mechanical difficulties. With such small slit-widths ($2a$), between the tubes, they must be of a high grade and the machine work must be very precise. In an attempt to eliminate these difficulties, a different type was constructed.

The efficiency of a column for a given molecule depends on the quotient of two velocities, that in the vertical direction due to the thermal circulation and that in the horizontal direction due to thermal diffusion. If the former is decreased relative to the latter, the entering solute molecule will have a longer time in which to reach the downward stream and the column will be more effective. This can be accomplished by packing the cylindrical space between the warm and cool tubes with some material which hinders the thermal circulation but does not greatly affect the thermal diffusion velocity.

To accomplish this, a column having a height, $h = 10$ cm, and a slit-width, $2a = 4$ mm but of the same general design as those previously described was constructed. Such a column would be about 16,000 times less effective than one having $h = 40$ cm and $2a = 0.5$ mm, and would give an extremely small and undetectable decrease in concentration in the upper reservoir for a polystyrene of molecular weight 450,000. In this case, however, the effect was enhanced by packing the space between the tubes with glass wool. The column was filled with a toluene solution of the polystyrene and the temperature gradient established. At intervals, 5 cc. samples were taken from the top reservoir, their flow times measured in an Ostwald viscometer and then returned. Figure 9

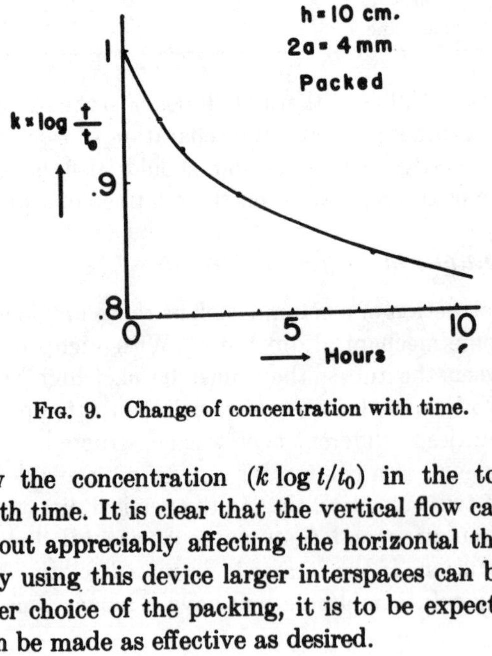

FIG. 9. Change of concentration with time.

shows how the concentration ($k \log t/t_0$) in the top reservoir changes with time. It is clear that the vertical flow can be slowed down without appreciably affecting the horizontal thermal diffusion and by using this device larger interspaces can be tolerated. By a proper choice of the packing, it is to be expected that the column can be made as effective as desired.

V. Summary

The history and the molecular interpretation of thermal diffusion in solutions in terms of the Brownian motion of the particles is treated. A thermal-diffusion-column based on the thermal circulation principle of Clusius and adapted to liquids is described. Its theory is developed. It is applied especially to polymer solutions and it is shown that they have a rather great thermal-diffusion effect. The dependence of the thermal-diffusion separation on the molecular weight of the polymer suggests its use as a new fractionating device. Experiments on mixtures of low-molecular-weight substances indicate that the same method can be applied to the purification of organic compounds and the separation of isomers. A new design of the column shows that its efficiency can be increased by slowing down the circulation, for the case discussed in this paper, by the use of an appropriate packing material. This publication is a condensation of some of the reports submitted to the Office of Rubber Reserve, Reconstruction Finance Corporation, which has been supporting this work in connection with the Government's synthetic rubber program.

VI. Bibliography

1. C. Ludwig, *Sitzber. Acad. Wiss Wien* **20**, 539 (1856).
2. Ch. Soret, *Arch. sci. phys. nat., Genève,* **3**, 48 (1879). *Ann. Chim. Phys.* **22**, 293 (1881).
3. S. Arrhenius, *Öfversigt Vetensk. Akrd. Förh.* **51**, 61 (1894); Z. phys. Chem. **26**, 187 (1898).
4. T. Wereide, *Ann. phys.* **2**, 55, 67 (1914).
5. R. Abegg, *Z. physik. Chem.*, **26**, 161 (1898).
6. O. Scarpa, *Attiaccad. Lincei, R.C. Cl. Sci.*, 5 **17**I, 362 (1908).
7. A. Eilert, *Z. anorg. Chem.* **88**, 1 (1914).
8. H. Wessels, *Z. physik. Chem.* **87**, 215 (1914).
9. N. L. Bowen, *J. Geol.* **29**, 295 (1921).
10. C. Benedicks, *Trans. Am. Inst. Mining Met. Engrs.* **71**, 597 (1925).
11. D. Hanson and W. Rosenhain, *Bull. Inst. Mining Met.* **259** (1926); *Trans. Inst. Min. Met.* **35**, 301 (1926); *J. Inst. Metals* **37**, 241 (1927).

12. J. Chipman, *J. Am. Chem. Soc.* **48**, 2577 (1926).
13. H. R. Bruins, *Z. physik. Chem.* **130**, 601 (1927).
14. C. C. Tanner, *Trans. Faraday Soc.* **23**, 75 (1927).
15. S. Chapman, *Phil. Trans.* **217A**, 115 (1916).
16. T. L. Ibbs, *Physica* **4**, 1133 (1937).
17. H. Fleischmann and H. Jensen, *Ergeb. exakt. Naturw.* **20**, 121 (1942).
18. D. Enskog, *Physik. Z.* **12**, 56, 533 (1911); Doctor's Thesis, Upsala (1917) (Almquist and Wiksels).
19. D. Enskog, *Arkiv. Mat. Astron. Fysik* **16**, No. 16, 1 (1921).
20. S. Chapman, *Phil. Mag.* **38**, 182 (1919).
21. R. S. Mulliken, *J. Am. Chem. Soc.*, **44**, 1033 (1922).
22. K. Clusius and G. Dickel, *Naturwissenschaften* **26**, 546 (1938); **27**, 148 (1939); *Z. physik. Chem.* **B44**, 397 (1939).
23. A. K. Brewer and A. Bramley, *Phys. Rev.* **55**, 590 (1939).
24. H. Korsching and K. Wirtz, *Naturwissenschaften* **27**, 110, 367 (1939).
25. K. Wirtz, *Naturwissenschaften* **27**, 369 (1939).
26. P. Debye, *Ann. Physik* **36**, 284 (1939).
27. N. Riehl, *Z. Electrochem.* **49**, 306 (1943).
28. K. Wirtz, *Naturwissenschaften* **31**, 416 (1943).
29. S. R. de Groot, L'Effet Soret (1943) N. V. Noord-Hollandsche Uitgevers Maatschappij, Amsterdam.
30. E. D. Eastman, *J. Am. Chem. Soc.* **48**, 1482 (1926); **49**, 794 (1927); **50**, 283, 292 (1928).
31. C. Wagner, *Ann. Physik.* **3**, 629 (1929); **6**, 370 (1930).
32. J. H. Van't Hoff, *Z. physik. Chem.* **1**, 481 (1887).
33. W. Nernst, *Z. physik. Chem.* **4**, 129 (1889).
34. W. Duane, *Ann. Phys.* **2** [9], 67 (1914).
35. G. S. Hartley, *Trans. Faraday Soc.* **27**, I, 1 (1931).
36. Th. Wereide, *Ann. phys.* **2**, 67 (1914).
37. S. Chapman, *Proc. Roy. Soc. (London)* **A119**, 34, 55 (1928); *Phil. Mag.* [7] **5**, 630 (1928); [7] **7**, 1 (1929).
38. A. W. Porter, *Trans. Faraday Soc.* **23**, 314 (1927).
39. L. J. Gillespie, *J. Chem. Phys.* **7**, 438, 530 (1939).
40. J. W. Westhaver and A. K. Brewer, *J. Chem. Phys.* **8**, 314 (1940).
41. R. Fürth, *Proc. Roy. Soc. (London)* **A179**, 461 (1942).

42. K. Wirtz and J. W. Higby, *Physik. Z.* **44**, 369 (1943).
43. A. Einstein, *Ann. Physik.* **17**, 4, 549 (1905).
44. W. H. Furry, R. C. Jones, and L. Onsanger, *Phys. Rev.* **55**, 1083 (1939).
45. W. H. Furry and R. C. Jones, *Phys. Rev.* **69**, 459 (1946).
46. L. E. Nielson and J. G. Kirkwood, *J. Am. Chem. Soc.* **68**, 181 (1946).
47. P. P. Debye, *J. Applied Phys.*, **17**, 392 (1946).
48. P. Debye, *J. Applied Phys.* **15**, 338 (1944).

REACTION RATES IN IONIC SOLUTIONS* **

By P. Debye†

ABSTRACT

Smoluchowski's method of evaluating the fundamental frequency factor for the rate of a reaction like the coagulation of colloidal suspension by employing a purely diffusional treatment is extended to include the electrostatic effects arising from the presence of net charges. The introduction of the concept of a diffuse ionic cloud and the potential calculated therefrom leads to the following results: (a) The well-established Bronsted-Debye primary salt effect formula is derived. (b) The so-called "solvent" term of the Christiansen-Scatchard equation arising from the self potential of the ions is also derived but appears as the linear approximation of an exponential expression. The conspicuous absence of quantum theory in evaluating the absolute rate of a kinetic reaction merits attention.

V. K. LaM.

The customary way followed in order to answer the question how reaction rates are influenced by electric charges which may be carried by the reacting particles is that proposed by G. Scatchard[1] following a derivation of the Bronsted[2]-Bjerrum[3] activity relationship given by J. A. Christiansen.[4] On the other hand the method followed by v. Smoluchowski[5] for the calculation of the rate of coagulation of colloidal suspensions (within the range of rapid coagulation) can certainly be considered as a calculation of reaction rates in a special case. The fundamental question answered by Smoluchowski's treatment is: How many times per second will one of a number, n, of particles per cc. suspended in a liquid be hit by one of the others as a result of their Brownian motion? Smoluchowski obtains the desired number of collisions ν per sec. by stating that it can be calculated by solving a diffusion problem in which particles are supposed to diffuse steadily into a hole surrounding the particle in question. At the boundary of this hole the concentration is supposed to be kept 0 and at large dis-

* Manuscript received July 20, 1942.

† Professor of Physical Chemistry, Cornell University, Ithaca, N. Y.

[1] G. Scatchard, J. Am. Chem. Soc. **52**, 52 (1930); Chem. Rev. **10**, 229 (1931).

[2] J. N. Bronsted, Z. physik. Chem. **102**, 109 (1922); **115**, 337 (1925).

[3] N. Bjerrum, Z. physik. Chem. **108**, 82 (1924).

[4] J. A. Christiansen, Z. physik. Chem. **113**, 35 (1924).

[5] Smoluchowski, Physik. Z. **17**, 557, 585 (1916); Z. physik. Chem. **92**, 129 (1917).

**Reprinted from *Transactions of the Electrochemical Society*, 82, 265-272 (1942).

tances from this boundary the concentration is considered to be equal to the normal concentration n per cc. In order to take care of the Brownian motion of the central particle as well as that of all the others, the diffusion constant involved has to be taken twice as large as the normal diffusion constant δ of the particles, all considered as spheres of equal size with the diameter R. Under these circumstances the hole itself is also a sphere with a radius equal to R.

The result obtained in this way is

$$\nu = 8\,\pi\,\delta\,R\,n \tag{1}$$

If the mobility of the particles is calculated, by applying Stokes' law for the motion of a sphere in a viscous liquid of viscosity η, then their friction coefficient ρ, defined as the quotient of the applied force divided by the velocity obtained, becomes

$$\rho = 3\,\pi\,\eta\,R \tag{2}$$

At the same time Einstein's relations for the intensity of the Brownian motion expresses the diffusion constant δ in terms of ρ in the form

$$\delta = \frac{k\,T}{\rho} \tag{3}$$

in which k is Boltzmann's constant and T the absolute temperature.

Combining Eq. (2) and (3) with the fundamental relation (1) leads to the equivalent relations

$$\nu = \frac{8\,k\,T}{3\,\eta}\,n \tag{4}$$

The new form Eq. (4) shows that the number of collisions per second is independent of the size of the particles, a peculiar but well known result in colloid chemistry.

In many instances, *e.g.*, in calculating the mobility of ions or in calculating the relaxation time of dipolar molecules, it has been shown that the application of the relations of Stokes for a sphere in linear or even rotational motion to particles of molecular size leads to an estimate of the right order of magnitude. Therefore, in order to estimate the frequency factor of reaction rates in cases in which the reactions are carried out in the liquid phase, it seems advisable to use the relation, Eq. (4), instead of the customary collision theory which only applies to molecules in the gaseous state.

The independence of particle size holds only for collisions of spherical particles of equal size. If collisions between spherical particles of different sizes, say diameters R_1 and R_2 are considered, Smoluchowski's reasoning leads to the relation

$$\nu = \frac{1}{4}\left(2 + \frac{R_1}{R_2} + \frac{R_2}{R_1}\right)\frac{8kT}{3\eta} n \qquad (4')$$

However, the influence of particle size is not very important. For instance, for $R_1/R_2 = 3$, the new factor introduced in Eq. (4′) is only 4/3 as compared with 1 in relation Eq. (4).

A practical example in which the validity of relation Eq. (4) can be tested is the quenching of fluorescence, *e.g.*, of fluoresceine-solutions by iodine ions. The result of such a comparison is very satisfactory. In this paper we are concerned with the influence of intermolecular forces on reaction rates, especially with the case in which the long-range Coulomb forces due to charges carried by the particles are proponderant. If we adopt Smoluchowski's statement about the equivalence of the diffusion problem with the collision problem, his method can easily be generalized to include the effect of a potential energy U depending on the distance between the centers of spherical particles. For particles with diameters R_1 and R_2 the final result* is

$$\nu = \frac{(R_1 + R_2)/2}{\int_{(R_1+R_2)/2}^{\infty} e^{U/kT} \frac{dr}{r^2}} \frac{8kT}{3\eta} n \qquad (5)$$

In the special case $U = 0$ Eq. (5) reduces to the relation Eq. (4′).

Let us now suppose that the only force to be considered is the Coulomb force between two particles of charges $z_1 \epsilon$ and $z_2 \epsilon$ in a medium of the dielectric constant D. In this case

$$U = \frac{z_1 z_2 \epsilon^2}{D} \frac{1}{r}$$

the integral occurring in Eq. (5) can be evaluated and the result becomes

$$\nu = f \frac{1}{4}\left(2 + \frac{R_1}{R_2} + \frac{R_2}{R_1}\right)\frac{8kT}{3\eta} n \qquad (6)$$

This is the old result Eq. (4′) except for a correction factor f represented by the formula

$$f = \frac{z_1 z_2 \epsilon^2}{DkT} \frac{2}{R_1 + R_2} \Big/ \left(e^{\frac{z_1 z_2 \epsilon^2}{DkT} \frac{2}{R_1+R_2}} - 1 \right) \qquad (6')$$

The combination $z_1 z_2 \epsilon^2/DkT$ has the dimension of a length and its

* See Appendix for details of this calculation.

absolute value represents the distance at which two charges of positive valencies z_1 and z_2 have a potential energy equal to the thermal energy $k\,T$. Consequently the characteristic quantity $\dfrac{z_1 z_2\,\epsilon^2}{D\,k\,T}\;\dfrac{2}{R_1+R_2}$ represents the quotient of this length divided by the distance of the centers of the particles at contact. For two univalent ions in water of room temperature the characteristic equilibrium distance $z_1 z_2\,\epsilon^2/D\,k\,T$ is 7 Ångström units. For molecules of average size the effect, therefore, is important and corresponds to a decrease in the number of collisions if z_1 and z_2 are both positive or both negative, that is, if the particles repel each other.

The assumption that the Coulomb force between the two particles considered is the only action, which has to be taken into account, will hold only in the limiting case of very dilute solutions. As soon as the concentration increases, the screening effect, which is fundamental for the theory of strong electrolytes, will set in. It will decrease the range of action of the forces and so counteract their influence on the frequency factor of the reaction rates.

Again the calculation can be based on Eq. (5); it is only necessary to decide upon an expression of the potential energy U as a function of the distance r. Defining the reciprocal thickness $\varkappa$ of the ionic layer according to the theory of strong electrolytes by the formula

$$\varkappa^2 = \frac{4\,\pi\,\epsilon^2}{D\,k\,T}\,\Sigma\, n_i z_i^2 \tag{7}$$

this potential energy can be expressed in the form

$$U = \frac{1}{2}\left\{\frac{e^{\frac{\varkappa R_2}{2}}}{1+\varkappa\frac{R_1}{2}} + \frac{e^{\frac{\varkappa R_1}{2}}}{1+\varkappa\frac{R_2}{2}}\right\}\frac{z_1 z_2\,\epsilon^2}{D\,k\,T}\,\frac{e^{-\varkappa r}}{r} \tag{8}$$

at least for large distances r. Since the main effect of the concentration occurs at larger distances where the familiar decrease proportional to $1/r$ is materially changed by the occurrence of the exponential factor $e^{-\varkappa r}$, it can be expected that the introduction of the approximate expression Eq. (8) for the energy will be adequate for the calculation of the concentration effect at least for small concentrations which are according to Eq. (7) related to small values of $\varkappa$.

The integral appearing in Eq. (5) now is more complicated and cannot be expressed by a simple function. A similar integral appears in the theory of the influence of strong electrolytes on the surface tension. For our immediate purpose it may suffice to indicate the result in the limit for small values of $\varkappa$.

In this case the number of collisions ν per second can again be expressed by an equation similar to Eq. (6), it is only necessary to replace the correction factor f by another factor f.* If we introduce the equilibrium distance l by putting

$$\frac{z_1 z_2 \epsilon^2}{D k T} = l \tag{9}$$

and furthermore the thickness of the ionic layer λ as

$$\lambda = \frac{1}{\kappa} \tag{10}$$

the former correction factor f as expressed by Eq. (6′) is

$$f = \frac{2l}{R_1 + R_2} \Big/ \left(e^{\frac{2l}{R_1+R_2}} - 1 \right) \tag{11}$$

The new correction factor f^* becomes

$$f^* = \gamma \frac{2l}{R_1 + R_2} e^{\gamma \frac{l}{\lambda}} \Big/ \left(e^{\gamma \frac{2l}{R_1+R_2}} - 1 \right) \tag{12}$$

with

$$\gamma = \frac{1}{2} \left\{ \frac{e^{R_1/2\lambda}}{1+\frac{R_1}{2\lambda}} + \frac{e^{R_1/2\lambda}}{1+\frac{R_2}{2\lambda}} \right\} \tag{12′}$$

In the limit for very small concentrations (λ being infinite) γ becomes equal to unity and f^* becomes identical with f.

Finally f^* can be developed in powers of κ or $\frac{1}{\lambda}$. If this is carried out and if only the term proportional to the first power of κ is retained, the result can be simply expressed by stating that

$$\frac{f^*}{f} = 1 + \kappa l = 1 + \frac{l}{\lambda} \tag{13}$$

since the difference of γ from unity is of the second order in κ. The results here obtained are very similar to those published in the literature, a comparison can be most easily made by referring to the book of Glasstone, Laidler and Eyring on the Theory of Rate Processes, page 423 and following pages.

In detail, significant differences exist, for instance, with respect to the absolute value of the rate constants. The main advantage seems to be that the proposed treatment translates the physical picture of the processes involved into the necessary formulae without the helo of any intermediary.

APPENDIX

If ρ_1 and ρ_2 are the friction constants for the two kinds of particles of diameters R_1 and R_2, which have to be considered, the diffusion equation takes the form

$$\frac{I}{4\pi r^2} = -(\delta_1 + \delta_2)\frac{dn}{dr} - \left(\frac{1}{\rho_1} + \frac{1}{\rho_2}\right)\frac{dU}{dr}\,n$$

δ_1 and δ_2 are the diffusion constants and I is the specific current, that is, the number of particles passing through one square centimeter per second under the influence of the combined effect of an established concentration gradient and the forces between the particles. According to Einstein's theorem δ_1 and δ_2 can be replaced by the expressions kT/ρ_1 and kT/ρ_2. A general solution of the equation now is

$$n = -\frac{I}{4\pi k T}\,\frac{e^{-U/kT}}{\frac{1}{\rho_1} + \frac{1}{\rho_2}}\int^{r} e^{U/kT}\frac{dr}{r^2}$$

In order that $n = 0$ at the boundary of the central sphere with a radius equal to $(R_1 + R_2)/2$, the lower limit of the integral has to be taken equal to $r = (R_1 + R_2)/2$. On the other hand in order that for $r = \infty$ the concentration becomes equal to the overall concentration, which for our immediate purpose shall be called n_0, the condition

$$n_0 = -\frac{I}{4\pi k T}\,\frac{1}{\frac{1}{\rho_1} + \frac{1}{2\rho}}\int_{(R_1+R_2)/2}^{\infty} e^{U/kT}\frac{dr}{r^2}$$

has to be satisfied, since the energy-function U has been normalized to become zero at infinite distance. Applying Stokes' law means replacing ρ_1 and ρ_2 by $3\pi\eta R_1$ and $3\pi\eta R_2$ respectively, and since the total current, I, is identical with the number of impacts ν as defined in the text, our result is identical with Eq. (5).

MOLECULAR-WEIGHT DETERMINATION BY LIGHT SCATTERING[1] *

P. DEBYE

Department of Chemistry, Cornell University, Ithaca, New York

Received August 8, 1946

I. INTRODUCTION

Non-absorbing gases or liquids are not perfectly transparent but scatter light. The main part of this light has not changed its wave length, the fraction of the intensity corresponding to the displaced spectral lines of the Raman effect is small, and the change in frequency is unimportant for the following considerations. The scattering is due to the non-homogeneous molecular structure, and we would expect that if a solvent is made more inhomogeneous by adding a solute the scattered intensity would increase. The question to be answered is how this increase in scattered intensity can be used in order to count the number of solute particles and how in appropriate cases conclusions about the structure of such particles can be drawn from observations on the angular distribution of the scattered light.

The problem can be approached either by making a detailed calculation of the

[1] Presented at the Twentieth National Colloid Symposium, which was held at Madison, Wisconsin, May 28–29, 1946.

*Reprinted from *The Journal of Physical & Colloid Chemistry,* 51, 18-32 (1947).

electromagnetic field surrounding a particle in order to derive the loss of primary light energy due to its radiation (14) or by treating the effect of the molecular inhomogeneities on the light in a second approximation as due to spontaneous fluctuations of the density and the concentration in a medium which in a first approximation is considered to be perfectly homogeneous (7).

Each method has advantages and disadvantages. In the application of the direct method it is commonly considered unavoidable to start with a particle of definite shape (i.e., a sphere). It will be shown in the following that this restriction is unnecessary. This offers the advantage of drawing conclusions about the special effects related to the particle structure which will begin to occur as soon as some dimension of the particle becomes comparable to the wave length of the light. On the other hand, it becomes very difficult to extend the calculations to the case of higher concentrations, when the particles begin to interact with each other. This is not so if the method of fluctuations is applied. By this method the intimate connection between light scattering and osmotic pressure is revealed, and the effects of concentration on osmotic pressure and on light scattering can be related to each other. However, now it becomes difficult to handle larger particles.

II. SMALL PARTICLES; HIGH DILUTION

Making the decision that we will be interested only in the additional scattering due to a small amount of solute in a much larger amount of solvent, we are permitted to consider the solvent as perfectly homogeneous. The dielectric constant of this medium will be called ϵ_0 and its index of refraction (for the particular wave length which is used in a primary beam) μ_0. The relation $\epsilon_0 = \mu_0^2$ holds, and can be considered as a definition of the dielectric constant for the frequency in question.

After the addition of the solute, we shall find regions in the solution at large mutual distances in which the dielectric (optical) properties have been changed in a way not known in detail. Each such region we call a particle.

If the solution is subjected to the influence of a homogeneous electric field of intensity F, every cubic centimeter of the solvent will acquire an electric moment

$$(\epsilon_0 - 1)\frac{F}{4\pi}$$

The homogeneous field will be disturbed in the region of the particles in a complicated way, but if we observe this effect in a point at a larger distance from the particle, the lines of force superimposed on the homogeneous field will be those of a dipole m, which can be either positive or negative, and which by its strength and direction defines the particle in all the details we need for our purpose. In general, m and F will not have the same direction, with the ultimate result that the scattered light observed under an angle of 90° with the direction of the primary beam will not be plane polarized. At the end of this paragraph we shall indicate the correction which takes care of this depolarization effect, but since in all the practical cases which have come to my attention so far this correction is negligible, we shall from now on assume that m and F point in the same direction.

If now a volume V of solution containing n particles per cubic centimeter in high dilution is subjected to the field F, the total electric moment in the direction of F will be

$$V\left[(\epsilon_0 - 1)\frac{F}{4\pi} + nm\right]$$

which means that the dielectric constant ϵ which we observe as characteristic for the solution follows from the relation:

$$\epsilon - \epsilon_0 = 4\pi n\frac{m}{F} \qquad (1)$$

Let us now suppose that the extension of the particle is so small that even at the distance where its whole disturbing effect can be described as the field of the dipole m this distance is still small compared with the wave length of the light (measured in the medium).

In order to find the radiation field surrounding the particle at larger distances we have merely to adjust it as a solution of Maxwell's equations in such a way that at small distances it equals the electrostatic field of our dipole m vibrating with the frequency of the light. This is a familiar calculation. It is found that at a large distance r (large compared with the wave length) the electric intensity E and the magnetic intensity H are

$$\begin{cases} E = m\kappa^2 \dfrac{\sin\vartheta}{r}\cos(\omega t - \mu_0\kappa r) \\ H = \mu_0 E \end{cases}$$

in which:

$$\kappa = \frac{\omega}{v} = \frac{2\pi}{\lambda}$$

where λ = wave length in a vacuum, v = velocity of light in a vacuum, and ω = frequency of the light vibrations.

On a large sphere surrounding the dipole, with its center at the position of the dipole and its north pole at the place where the vector representing the dipole cuts through this sphere, E is in the south–north direction everywhere, and the lines of force of the magnetic field are circles of constant latitude. The angle ϑ measures the latitudes angle from the north pole, and $E = H = 0$ at both the south and the north pole.

The energy radiated per second through 1 cm.² of the sphere is represented by the time average of Poynting's vector and amounts to

$$\frac{v}{8\pi}\mu_0 m^2\kappa^4\frac{\sin^2\vartheta}{r^2}$$

making the total energy loss per second and per particle equal to

$$\frac{1}{3}\frac{v}{\mu_0}\mu_0^2 m^2 \kappa^4$$

For a (polarized) primary light beam its intensity I equal to the energy carried through 1 cm.[2] per second, is

$$I = \frac{v}{\mu_0}\frac{\mu_0^2 F^2}{8\pi}$$

If such a beam goes through our solution, it will in the direction of propagation lose intensity according to the relation

$$-\frac{dI}{dx} = n\frac{1}{3}\frac{v}{\mu_0}\mu_0^2 m^2 \kappa^4 = \left(\frac{8\pi}{3} n\kappa^4 \frac{m^2}{F^2}\right) I$$

The quantity in brackets is what is generally called the "turbidity", to be indicated by τ. We therefore come to the conclusion that

$$\tau = \frac{8\pi}{3} n\kappa^4 \frac{m^2}{F^2} \tag{2}$$

According to equation 1 the difference in dielectric constant between the solvent and the solution is proportional to n, the number of particles per cubic centimeter, and to m/F, the electric moment characterizing the particle in a field of unit intensity. The turbidity according to equation 2 is also proportional to n, but, unlike the dielectric constant, proportional to the square of m/F. This corresponds to the fact that the change in the velocity of the light (measured by the dielectric constant) is essentially an interference effect, which involves only the amplitudes of the field, whereas the turbidity measures energy losses and therefore intensities, which are proportional to the square of the amplitude.

Like the osmotic pressure, the turbidity is proportional to the number of particles per cubic centimeter. Therefore either can be used to determine the molecular weight. However, in the first case the proportionality constant is universally the same and equal to kT (k = Boltzmann's constant; T = absolute temperature). In the second case the proportionality constant depends on the optical properties of the particle, but it can be determined unambiguously from a measurement of the change in refraction connected with a shift from the solvent to the solution. In order to express this interconnection between the two experiments which have to be carried out if the turbidity method is used, the unknown quantity m/F can be eliminated between equations 1 and 2 and we arrive at the relation (4):

$$\tau = \frac{32\pi^3}{3}\frac{\mu_0^2(\mu - \mu_0)^2}{\lambda^4}\frac{1}{n} \tag{3}$$

where, instead of the dielectric constants, the indices of refraction, μ for the solution and μ_0 for the solvent, are introduced. For practical purposes it is customary to express n in the concentration c (in grams per cubic centimeter), the molecular weight M, and Avogadro's number N ($N = (6.0228 \pm 0.0011) \times 10^{23}$) (1).

Relation 3 can then be written in the form

$$\frac{\tau}{c} = HM \qquad (3')$$

in which the proportionality constant H must be derived from refraction measurements and is defined by the relation

$$H = \frac{32\pi^3}{3} \frac{\mu_0^2}{N\lambda^4} \left(\frac{\mu - \mu_0}{c}\right)^2 \qquad (3'')$$

It has the dimension cm.^{2}/g.^{2}

The relations 3, 3′, and 3″ hold for the case in which the light scattered at an angle of 90° with the direction of the primary beam is plane polarized. It can be expected that this condition is not quite satisfied, although so far a correction for depolarization has not been found necessary. If the primary light is unpolarized, the depolarization ρ of the scattered light can be measured by building the quotient of the intensity of the light which (scattered at an angle of 90°) passes a nicol with its plane of polarization coinciding with the plane of primary and secondary ray divided by the intensity passed by this nicol turned 90°. The correction factor for the molecular weight is then

$$\frac{6 - 7\rho}{6 + 3\rho}$$

and is known as Cabannes' factor (2). For $\rho = 1$ per cent the actual molecular weight is 1.56 per cent smaller than that derived from relation 3′.

III. SMALL PARTICLES; MORE CONCENTRATED SOLUTIONS

In Einstein's fluctuation theory the additional scattering from a solution as compared to that of the solvent is considered due to spontaneous fluctuations in concentration. The final result expresses that these fluctuations are the smaller in magnitude the larger the osmotic work which has to be performed in order to bring about a change in concentration. At the same time it relates the intensity of scattering also to the change in index of refraction connected with a change in concentration. The relation can be written in the form:

$$\tau = \frac{32\pi^3}{3} \frac{\mu_0^2}{N\lambda^4} \frac{(c\partial\mu/\partial c)^2}{c\dfrac{\partial}{\partial c}\left(\dfrac{P}{RT}\right)} \qquad (4)$$

in which P is the osmotic pressure and $R = Nk$ is the gas constant. In most practical cases it is found that $\mu - \mu_0$ is proportional to the concentration with a high degree of precision, within the concentration range generally considered for polymer solutions. So instead of $\partial\mu/\partial c$ we can substitute $(\mu - \mu_0)/c$. The occurrence of the concentration gradient of the osmotic pressure in the denominator suggests that the reciprocal of the turbidity be considered instead of the turbidity itself in all cases where the effect of concentration on the turbidity is to be compared with that on the osmotic pressure. So we are led to transform relation 4 into

$$H\frac{c}{\tau} = \frac{\partial}{\partial c}\left(\frac{P}{RT}\right) \tag{4'}$$

in which H is the same refraction constant as defined by relation 3″. Since, according to van't Hoff's law, P/RT is equal to c/M for dilute solutions, relations 3′ and 4′ are identical in this case.

Since it is well known that, although the deviations of van't Hoff's law for high-polymer solutions are important even at high dilutions, the osmotic pressure can be expressed adequately by a two-term expression of the form

$$\frac{P}{RT} = \frac{c}{M} + B\,c^2$$

we have to expect according to equation 4′ that for the reciprocal specific turbidity the linear relationship

$$H\frac{c}{\tau} = \frac{1}{M} + 2Bc \tag{5}$$

will be a good approximation.

The constant B depends on the solvent. In a good solvent B is large and positive; in a poor solvent B can be zero and even negative (for the theory see references 9, 10, 12). In all practical applications of light scattering to the determination of molecular weights it has now become the custom to determine H by refraction measurements and τ for a series of concentrations, and then plot Hc/τ as a function of c. A straight line through the observed points now cuts the vertical axis of the figure at an ordinate which is equal to $1/M$ (6).[2] The straight-line relationship holds also for solutions containing polymers with a distribution of molecular weights. It can easily be shown from relation 4′ that in this case the intercept is the reciprocal of the weight-average molecular weight (15). As an illustration figures 1 and 2 represent measurements of McCartney on polystyrene dissolved in different mixtures of a good solvent (benzene) and a precipitant (methanol). Figure 1 represents the excess turbidity (normalized in such a way that the effect of addition of methanol on the specific refractive index is eliminated) of polystyrene in solvent mixtures containing 0, 7.5, 10.0, 12.5, 15, and 22.5 volume per cent methanol. Whereas in a good solvent the

[2] The first application of light scattering to molecular-weight determinations in protein solutions, based on Rayleigh's theory of the scattering of a medium in which spheres are suspended, seems to be that of P. Putzeys and J. Brosteaux (Trans. Faraday Soc. **31,** 1314 (1935)). Recently a summary of a paper by H. Staudinger and I. Henel-Immendörfer (J. Makromol. **1,** 185 (1944)), entitled "Determination of molecular weights of glycogenes by the use of Rayleigh's law," came to my attention. The authors do not yet realize that the constant H can be determined experimentally without making assumptions about the particles or their so-called optical constants. See also C. V. Raman (Indian J. Phys. **2,** 1 (1927)).

turbidity increases slowly with increasing concentration, and indeed becomes essentially constant at higher concentrations, in a poor solvent the increase is more rapid and in the mixture containing the highest per cent of methanol the turbidity increases proportionally to the concentration. If these curves are converted into another set, plotting this time Hc/τ *vs.* c, a series of straight lines is obtained (figure 2). Thus benzene gives the steepest slope, and as the solvent gets poorer, the slope decreases until for the poorest (22.5 per cent methanol) the line is practically horizontal.

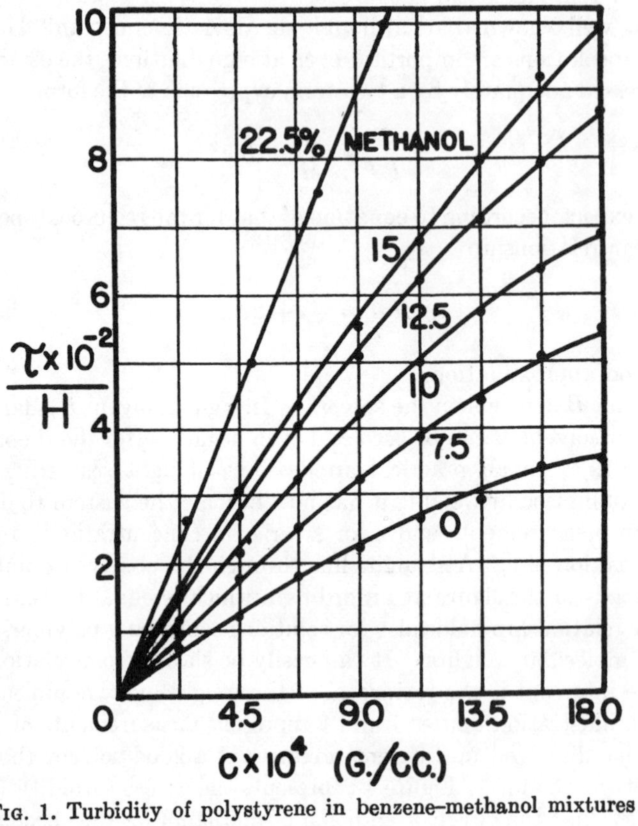

FIG. 1. Turbidity of polystyrene in benzene–methanol mixtures.

A peculiar feature of the straight lines in our case is that they have not the same intercept. Considered superficially, this seems to indicate that the molecular weight increases with increasing content of non-solvent in the mixture. It can, however, be shown that the effect is due to preferential adsorption of the benzene on the polymer particle (8) and its interpretation furnishes a quantitative measurement of this preference. In simple solvents no such effect occurs, except for possible real agglomeration of the polymer particles.

IV. LARGER PARTICLES; HIGH DILUTION

The preceding has been built up on the assumption that the radiation of the particle can be represented as that of a single dipole and this implies, as we have seen, that the dimensions of the particles are small compared with the wave length. In the case of a larger particle different parts of it will not be submitted to the same exciting field intensity either in magnitude or in phase, and at a distant point the radiation field will be made up of a superposition of waves coming from the different parts of the particle, interfering with each other. If we want to calculate the resulting radiation field according to this picture, we have not only to know what the phase differences are between the different elementary waves at the point of observation, but also what the field intensities

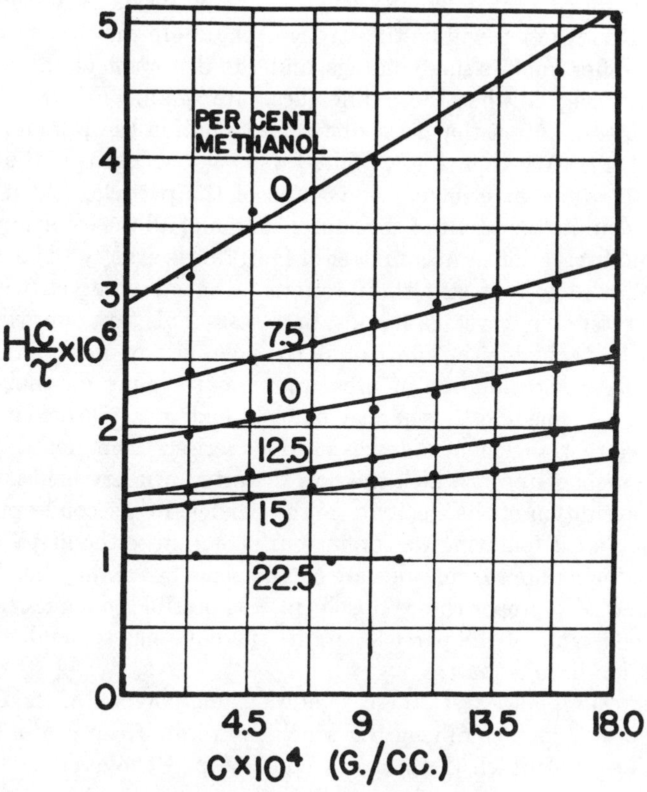

FIG. 2. Reciprocal specific turbidity of polystyrene in benzene–methanol mixtures

are in the elements of volume from which each of these waves emerges, since this determines their amplitude. In all its generality this problem can only be solved by finding an exact solution of Maxwell's equations with the proper boundary conditions. The mathematical problem is so difficult that the only case of any importance to our subject which could be treated is that of a homogeneous sphere (3, 11).[3] Such a restriction to a very special form of the particle is obviously not what we want in connection with our problem.

[3] For applications of these calculations see the recent articles of V. K. La Mer and Marion D. Barnes on colloidal sulfur (J. Colloid Sci. **1**, 71, 79 (1946)).

Under these circumstances we propose to attack it in a different way, which eliminates the restrictions as to shape or structure of the particle from the beginning. In Mie's treatment the total radiation field is eventually represented as a superposition of fields of di-, quadru-, and higher poles all emanating from the center of the sphere. Their amplitudes and phases follow from the exact solution of Maxwell's equations. The consecutive terms of this series represent in their sequence the parts which have to be taken into account to push the representation of reality to higher and higher orders of approximation. Now the main difficulty in determining the strength of these poles lies in the fact that the original field which excites the particle is distorted by its own electromagnetic reaction. This reaction will be the stronger the more the optical constants of the particle differ from its surroundings, and the distortion of the primary field will become negligible when those differences are small. For our purpose this suggests the assumption that to a first approximation the primary field is not distorted at all. We then know right away the strength of the elementary dipoles in the different elements of volume of the particle, and it is a simple matter to calculate the result of the superposition of all the elementary waves in the point of observation. As a matter of fact, this is exactly the method of calculation followed to represent the interference effects observed with x-rays or electrons scattered by crystals, liquids, or gases. This first approximation contains from the beginning the effect of all the poles of any order, but at the same time, as a first approximation, it uses approximate values for their amplitudes and phases. Applied to a spherical particle and in a mathematical sense it amounts to a rearrangement of terms in Mie's series. The first step can be followed by a second step in which the now distorted primary field is used to calculate the excitation of the elements of the particle in a second approximation, and so on. In the following we confine our attention to the first step, not only because the higher approximations are cumbersome but mainly because in many instances and by a proper choice of solvent it is possible to make the differences in optical constants of the particle and its surroundings so small that our first approximation is adequate.

As a first example we treat the case of a flexible polymer of the familiar kind consisting of n links each of length a which can rotate freely and make an angle with each other of which the cosine is equal to p. The average square of the distance from beginning to end of such a chain is

$$R^2 = na^2 \frac{1+p}{1-p} \tag{6}$$

and the usual formula (which neglects any kind of interaction of parts of the chain on each other) representing the probability that the end point is found in a shell of radii between r and $r + dr$ with the beginning of the chain in its center is

$$w(r)4\pi r^2\, dr = \left(\frac{3}{2\pi R^2}\right)^{3/2} e^{-(3/2)(r^2/R^2)}\, 4\pi r^2\, dr \tag{6'}$$

If we visualize the chain as a series of emitters, each located at the intersection of two bonds and all of the same strength, it is well known from the procedure followed in the case of x-rays that the average intensity of scattering which is to be observed in any direction is proportional to

$$I = \sum_{\mu} \sum_{\nu} \frac{\sin\ [\kappa s r_{\mu\nu}]}{\kappa s r_{\mu\nu}} \tag{7}$$

Here s stands for $2 \sin \frac{\vartheta}{2}$, in which ϑ is the angle between the secondary and the primary beam, $\kappa = 2\ \pi/\lambda$, in which λ is the wave length as measured in the liquid surrounding the particle, and $r_{\mu\nu}$ is the distance between the emitter μ and another emitter ν. The summation goes twice over all emitters. Attention should be drawn to the fact that in equation 7 the effect of polarization has been omitted. Here and in the following the term "intensity" refers therefore to the observed intensity after correction for the influence of the polarization effect.

Relation 7 holds for a molecule of definite and unvariable shape in the average over all orientations in space which it occupies without discrimination. The only thing left to do in our case is to average relation 7 a second time over all the forms the flexible molecule can acquire with probabilities according to relations 6 and 6′. The final result is

$$I_{\text{average}} = \frac{2}{x^2}\left[e^{-x} - (1 - x)\right] \tag{8}$$

with:

$$x = \frac{\kappa^2 s^2 R^2}{6} \tag{8'}$$

if we normalize the average intensity so as to make it equal to unity for $x = 0$, which corresponds to $\vartheta = 0$, that is, the direction of the primary beam. Since, according to equation 8, the intensity decreases steadily with increasing x, we have to expect that in cases in which R/λ is large enough the forward scattering will exceed the backward, and experimental evidence of such an angular dissymmetry will enable us to draw conclusions about the size of the polymer molecule. For $R = 0.1\ \lambda$ it follows, for instance, that the backward scattering is 8.2 per cent smaller than that in the forward direction. Our own experience, as well as that of other observers (13), seems to indicate that in many cases the dissymmetry actually observed is more pronounced than we would expect if R is calculated from the chemical structure by equation 6. This then means that the polymer chain is stiffer than free rotation combined with the assumption of negligible interaction of chain parts would lead us to expect. This is, of course, no surprise; we see the value of dissymmetry measurements in the fact that in this way we learn experimentally about the actual stiffness of the polymer chain, without having recourse to doubtful preconceived notions.

We prefer to visualize other types of polymers, like proteins, as rigid particles of definite shape. The summation of equation 7 must then be replaced by an integration over the volume of the particle. Applied to a spherical shape (radius a) the intensity–distribution function is

$$I_{\text{average}} = \left[\frac{3}{x^3}(\sin x - x\cos x)\right]^2 \tag{9}$$

with :

$$x = \kappa s a \tag{9'}$$

and again normalized to $I = 1$ for $x = 0$. This relation has been used with good results in the optical analysis of rubber lattices.[4]

Formulae for other forms can easily be derived. However, from a more general point of view it seems worthwhile to present the argument in the following way: if dissymetry measurements have been made with a rigid polymer, it will be possible to represent the (normalized) intensity as a function of s in the form

$$I = 1 - \alpha_1 s^2 + \alpha_2 s^4 - \tag{10}$$

Knowing then the coefficients α, we can ask what they tell us about the shape of the particle. If only α_1 is known or in case the two-term expression $1 - \alpha_1 s^2$ is sufficient to represent the intensity from $\vartheta = 0°$ to $\vartheta = 180°$, it can easily be seen that we have learned only about the average size of the particle in the following sense. It turns out that

$$\alpha_1 = \frac{\kappa^2}{3}\frac{1}{V}\int r^2\, dS \tag{10'}$$

in which V is the volume of the particle and r is the distance of any point in the interior of the particle from its center of gravity, whereas the integration extends over all the elements of volume dS in the interior of the particle. What we have learned then by determining α_1 is the value of the average square of all the distances within the particle from its center of gravity divided by the square of the wave length. Only if the particle is large enough, such that the two-term expression is not sufficient to represent the intensity distribution, can we learn more about the shape by light scattering. The next term tells about the average fourth power of distances within the particle and in connection with the preceding term would enable us to approximate its shape by an ellipsoid and so on.

One final remark has to be made concerning the use of formulae like equation 5 for molecular-weight determinations in cases in which a dissymmetry effect can be observed. The turbidity τ appearing in the relations of paragraph 2 has been calculated under the assumption that the particles are very small com-

[4] McCartney has compared equation 9 with the numerical calculations of H. Blumer (Z. Physik **32,** 119 (1925)) based on Mie's formula and has found that equation 9 gives values of I_{180}/I_0 in error by less than 10 per cent for particles as large as $2a/\lambda = 1/3$ for refractive-index ratios of 1.5 and barely perceptible errors for particles even larger in the refractive-index range normally encountered in polymer solutions.

pared with the wave length. This makes the scattered intensity proportionl to $\frac{1}{2}(1 + \cos^2\vartheta)$, and τ of the formula can be and has been obtained by integrating over all directions in space. It is obvious that if a dissymmetry effect exists, the actual total turbidity will be diminished by the interference effect which makes the backward scattering smaller than the forward scattering. The turbidity τ appearing in the formula represents the value which would be obtained if the interference effect did not exist. Practically, the correction can easily be evaluated by comparison of the actual angular distribution with that represented by $\frac{1}{2}(1 + \cos^2\vartheta)$.

V. EXPERIMENTAL ARRANGEMENTS

In the application of the turbidity method to the determination of molecular weights two measurements must be made.

Firstly, the difference between the index of refraction of the solution μ and that of the solvent μ_0 has to be measured. One single measurement at a given concentration c, say of 1 per cent, is usually sufficient, since the difference $\mu - \mu_0$ is generally very accurately proportional to c. Usually this difference is of the order 0.001 for a 1 per cent solution. The fifth decimal must be known in order to ensure an accuracy of the order of 1 per cent in the quotient $(\mu - \mu_0)/c$. In our laboratory we use a differential refractometer, consisting of a hollow large-angle prism (140° angle) which contains the solution and which is immersed in a rectangular cell containing the solvent. The cell is interposed between two lenses (focal length 70 cm.). In the focal plane of the first lens is a slit illuminated with the same monochromatic light as that used for the scattering measurements. In the focal plane of the second lens the image of the slit is observed in an eyepiece equipped with a filar micrometer. As a monochromatic light source a mercury arc, AH-4, with the appropriate light filter is used. The instrument is calibrated with sucrose solutions and water. The displacement of the slit image is proportional to $\mu - \mu_0$, and a difference of 0.001 corresponds to a displacement of 3 mm., which is amply sufficient to ensure the required accuracy.

Secondly, the turbidity must be measured. The instrument used for this purpose is so constructed that a slightly convergent beam of monochromatic light, again furnished by a mercury arc, and the appropriate filter is focused in the interior of a test tube containing the liquid, entering through the bottom of the tube. The light scattered in directions near 90° with respect to the direction of the primary beam falls on a photocell. The resulting photocurrent is amplified in a D.C. amplifier (supplied by the Photovolt Corporation, New York City) and read on a microammeter. Another photocell receives a small part of the primary light and serves as a check on the constancy of the light source. With current regulation of the mercury arc by means of a General Electric B-47 ballast lamp circuit, this arrangement gives stable, reproducible meter readings. The excess scattering of the solution as compared with the solvent is measured as the difference in reading of the microammeter for the test tube filled with solution and filled with solvent. For the most sensitive setting of the amplifier, one scale division of the microammeter corresponds to a turbidity of 5×10^{-6} cm.$^{-1}$, a result which means practically that reliable measurements can be made with solutions which scatter as little as 20 per cent more than the pure solvent. As an illustration of the high sensitivity of this method, McCartney determined the molecular weight of sucrose by 90° scattering. Figure 3 shows the close agreement between the experimental values of the turbidity and the turbidity calculated from equation 4, using osmotic-pressure data. The molecular weight calculated from experiment is 380, 10 per cent different from the osmotic-pressure molecular weight (342).

The absolute value of the turbidity is determined with the help of a standard, a solution of polystyrene in toluene of known turbidity in a sealed test tube. This standard has been calibrated with a special instrument. A monochromatic parallel beam passes first through a small rectangular cell and the light scattered under 90° by the solutions to be investigated falls on a photographic plate.

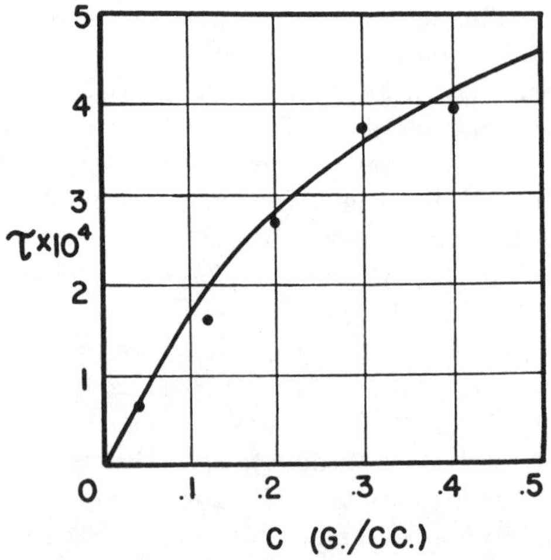

FIG. 3. Turbidity of aqueous sucrose solutions. Curve calculated from osmotic-pressure data. Points determined by 90° scattering.

After the primary light has passed this first cell it enters a second cubical cell filled with water in which a hollow cube is immersed. Three vertical sides of this cube are glass plates; the fourth vertical side is open. The light passes first the open side, is then three times reflected on the three glass sides, and eventually leaves the cube and the second cell in a direction perpendicular to the primary beam, after which it is also intercepted by the photographic plate. The reflection coefficient of each glass plate immersed in water can be calculated from the refractive indices of glass and water according to Fresnel's formula. Since this coefficient is small and since it has to be raised to the third power to account for the three reflectors, the final reflected beam has a known intensity comparable to that of the scattered beam emitted by the solution to be investigated. The intensities of the two spots on the photographic plate are finally compared with a microphotometer.

In order to indicate the order of magnitude of the effects we may mention that if a 1 per cent solution has $\mu - \mu_0 = 10^{-3}$, the refractive constant H of equation 3″ is of the order 10^{-6} cm.²/g.² This corresponds roughly, according to equation 3′, to a turbidity $\tau = 10^{-3}$ reciprocal cm. for a 1 per cent solution of a polymer with the molecular weight $M = 10^5$ g.

Our dissymmetry measurements have all been made with the photographic method, although measurements with a photocell and amplifier are perfectly feasible. The second procedure now seems more advisable.

In one instrument a parallel beam of light passes through a cell and is received in a light trap. At the side of the cell is a flat glass plate. The scattered light leaves the solution through this plate and is restricted by a small circular hole in a metal plate. It is intercepted by a film which is bent to a half-circle around this opening. The blackening of the film is measured as a function of the scattering angle with a microphotometer and reverted to light intensity by the usual method. If this cell is used, corrections must be made for the effect of the refraction of the scattered light in leaving the cell in order to construct the angular distribution curve.

In another instrument a cell of octagonal cross section is used. The primary beam enters one side and leaves by the parallel side. The three other sides to the right or to the left are provided with screens, each having a small circular hole followed by a tube acting as collimator.

The scattered light leaving those tubes is again intercepted by a photographic film. In this case only the light scattered in three directions, under 45°, 90°, and 135° against the direction of the primary beam, is measured, and since it passes through the glass plates normal to their plane no corrections for refraction are indicated.

Most of the instruments have been designed by P. P. Debye (5) and constructed by our mechanician, H. Bush.

This paper is a summary, with additions and omissions, of reports submitted during the war to the Office of Rubber Reserve, Reconstruction Finance Corporation, which supports our program of research at Cornell University. Collaborators in this research were P. P. Debye, F. W. Billmeyer, Jr., J. R. McCartney, and A. M. Bueche.

REFERENCES

(1) Birge, R. T.: Rev. Modern Phys. **13**, 233 (1941).
(2) Cabannes, P.: La diffusion moleculaire de la lumière. Presses Universitaires de France, Paris (1929).
(3) Debye, P.: Ann. Physik [4] **30**, 57 (1909).
(4) Debye, P.: J. Applied Phys. **15**, 338 (1944).
(5) Debye, P. P.: J. Applied Phys. **17**, 392 (1946) (description of instruments of improved design).
(6) Doty, P., Zimm, B. H., and Mark, H.: J. Chem. Phys. **12**, 144 (1944).
(7) Einstein, A.: Ann. Physik **33**, 1275–98 (1910).
(8) Ewart, R. H., Roe, C. P., Debye, P., and McCartney, J. R.: J. Chem. Phys. **14**, 687 (1946).

(9) Flory, P. J.: J. Chem. Phys. **10,** 51 (1942).
(10) Huggins, M. L.: J. Phys. Chem. **46,** 151 (1942).
(11) Mie, G.: Ann. Physik [4] **25,** 377 (1908).
(12) Miller, A. R.: Proc. Cambridge Phil. Soc. **38,** 109 (1942); **39,** 54, 131 (1943).
(13) Stein, R. S., and Doty, P.: J. Am. Chem. Soc. **68,** 159 (1946).
(14) Strutt, J. W. (Lord Rayleigh): Phil. Mag. **41,** 107–20, 274–9, 447–54 (1871).
(15) Zimm, B. H., and Doty, P. M.: J. Chem. Phys. **12,** 203 (1944).

Light Scattering in Solutions *

By P. Debye
Cornell University, Ithaca, New York
(Received November 6, 1943)

It is a well-known fact that the scattering of light in solutions becomes more prominent the smaller the number of ultimate particles is in which a definite amount of solute has been broken up in the course of the solution process. Since with increasing intensity experimental procedures for the observation of the scattering become easier to handle the application of methods of optical analysis to solutions of polymers seems appropriate. This address deals with some of the conclusions which can be drawn from the results of such measurements.

Our reagent, the primary light, is characterized by its wave-length. Therefore we will have to subdivide our discussion into two parts, the first dealing with particles which are small, the other with particles of a size comparable with or larger than the wave-length. The dividing line can be drawn at a particle size of $\frac{1}{20}$ to $\frac{1}{10}$ of the wave-length, which will correspond to a molecular weight somewhere between 10 and 100 million.

Considering small particles first, it can be said that the scattered light in this case has two main characteristics. If the scattering is due to a parallel primary beam we observe (a) as much backward as forward scattering and (b) the intensity of scattering is essentially proportional to the reciprocal fourth power of the wave-length (not taking account of an additional small effect due to dispersion). The main thing to be considered, therefore, is the absolute intensity of scattering and not its angular distribution or its dependence on the color of the primary light.

The simplest case, already treated by Lord Rayleigh, is that of an ideal gas of low density. If this gas contains n molecules per cc and the polarizability of a molecule is α, then the turbidity τ of the gas due to scattering can be calculated by the formula

$$\tau=\frac{8\pi}{3}\left(\frac{2\pi}{\lambda}\right)^4 n\alpha^2. \qquad (1)$$

The turbidity τ is defined as the fractional decrease of the primary intensity over unit distance. λ indicates the wave-length. The polarizability α is the electric moment induced in the molecule by an electric field of unit strength.

Since we know from the beginning that the scattered intensity will be proportional to the first power of n (dimension cm^{-3}) and the second power of α (dimension cm^3) it is seen without any further calculation that a factor containing the 4th power of $1/\lambda$ has to be added in order to calculate the turbidity with the dimension cm^{-1}.

Equation (1) has sometimes been used to determine the number of molecules contained per cc. In order to do this we must know the polarizability α. This quantity is connected with the index of refraction μ by the formula

$$\mu-1=2\pi n\alpha. \qquad (2)$$

We can now eliminate α between the two equations, (1) and (2), and arrive at the relation

$$\tau=\frac{32\pi^3}{3}\frac{(\mu-1)^2}{\lambda^4}\frac{1}{n}, \qquad (3)$$

which leads to a calculation of n from the results of 2 experiments, the one measuring the refraction, the other the turbidity.

Next we consider the case of a liquid. It now becomes apparent at once that the method followed by Lord Rayleigh carries with it some complications. We are supposed to develop not only a molecular theory of scattering starting with the optical properties of a single molecule, but also a second theory which is to enable us to express the refraction of the medium by the same fundamental constants. Since now the molecules are so near each other both these theories are much more difficult than previously. In the end we would again have to eliminate the molecular constants between the two theories in order to obtain a formula equivalent to Eq. (3). A. Einstein[1] has shown how these complications can

*Reprinted from *Journal of Applied Physics*, 15, 338-342 (1944).

be avoided. He considers the scattering as due to local thermal fluctuations in the density of the liquid, which make the medium optically inhomogeneous. Their magnitude he derives from a comparison of the thermal energy kT (k = Boltzmann's constant, T = absolute temperature) with the work which has to be supplied in order to accomplish a change in density by outside pressure, relating these two quantities with the help of Boltzmann's probability concept of the entropy. In this way elaborate molecular theories become unnecessary and his result for the turbidity of a liquid can be put in the form

$$\tau = \frac{32\pi^3}{3}\frac{1}{\lambda^4}\frac{kT}{\kappa}\left(\mu\frac{\partial\mu}{\partial p}\right)^2 \tag{4}$$

in which p is the hydrostatic pressure and κ the compressibility. Applied to ideal gases Eq. (4) coincides of course with Eq. (3).

Our next step brings us to solutions. Here another reason for irregular thermal changes of the refractive index becomes apparent. We now have also to consider local differences due to fluctuations of the concentration. In the final part of his paper A. Einstein has already considered fluctuations of this kind in mixtures. They can, however, be treated in a different way, analogous to the treatment of the density fluctuations by introducing the osmotic pressure, which seems more appropriate to our immediate problem. Instead of the volume v and its changes the concentration c and its changes enter and the work which has to be compared with the thermal energy is now the work to be supplied from the outside (for instance, with the help of an imaginary semipermeable piston) in order to make reversible changes in the concentration. As an intermediary on the way to Eq. (4), the scattered intensity is found to be proportional to

$$\frac{kT}{\lambda^4}\frac{\left(v\dfrac{\partial\epsilon}{\partial v}\right)^2}{-v\dfrac{\partial p}{\partial v}}$$

in which ϵ, the dielectric constant, is ultimately to be replaced by μ^2, the square of the refractive index, and p is the hydrostatic pressure. It can be shown that the fluctuations in concentration lead to a scattered intensity proportional to

$$\frac{kT}{\lambda^4}\frac{\left(c\dfrac{\partial\epsilon}{\partial c}\right)^2}{c\dfrac{\partial P}{\partial c}}$$

in which P is the osmotic pressure.

In all the cases which have to be considered in dealing with solutions for our purpose the difference in refractive index of the solution and the solvent is proportional to the concentration. Taking this into account the final result for the turbidity as due to fluctuations in concentration can be given the form:

$$\tau = \frac{32\pi^3}{3}\frac{\mu_0^2(\mu-\mu_0)^2}{\lambda^4}\frac{1}{c\dfrac{\partial}{\partial c}\left(\dfrac{P}{kT}\right)}, \tag{5}$$

in which μ_0 is the refractive index of the solvent and μ the refractive index of the solution.

From Eq. (5) it follows immediately that $\partial P/\partial c$ is proportional to c/i if i is the intensity of the scattered light. This is equivalent to saying that actual measurements of the osmotic pressure as a function of the concentration can be replaced by measurements of the scattered intensity. It is well known that osmotic pressure-concentration curves show large deviations from the limiting straight line relation in the case of solutions of polymers, especially if the polymers are of the rubber type with free rotating bonds. This connection can be illustrated by applying our relation to scattering measurements performed on rubber solutions by S. D. Gehman and J. E. Field[2] or by those made with protein solutions by P. Putzeys and J. Brosteaux.[3] In the limit for small enough concentrations van't Hoff's classical relation

$$P = nkT$$

holds in every case (n = number of independent particles per cc). Therefore the turbidity of diluted solutions due to fluctuations in concentration can according to Eq. (5) be expressed in the form

$$\tau = \frac{32\pi^3}{3}\frac{\mu_0^2(\mu-\mu_0)^2}{\lambda^4}\frac{1}{n}, \tag{5'}$$

which should be compared with Eq. (3) referring to gases. From the number of particles per cc and the concentration the mass of one particle and its molecular weight follows at once. Equation (5′) can therefore be interpreted as showing how by the combination of two measurements, the first of the turbidity, the second of the difference in refraction of solution and solvent, the molecular weight of the substance in solution can be evaluated, without introducing any kind of empirical constants.

In their paper on protein solutions P. Putzeys and J. Brosteaux[3] report measurements which show that the scattered intensity is proportional to the molecular weight for 4 proteins with molecular weights ranging from $3\cdot10^4$ to $5\cdot10^6$. Their theoretical background is a formula due to Rayleigh which represents the scattering of a medium in which spherical particles are suspended. The same result follows from Eq. (5′).

For the method to be of practical value it is necessary that the scattering due to fluctuations in concentration is large compared with that due to density fluctuations. We will expect this to be the case for high enough molecular weights. In Table I nitrogen gas of 1-atmos. pressure and 0°C, pure water at room temperature, and a 1 percent solution of ovalbumin with a molecular weight of 34,500 are compared.

TABLE I.

Substance	τ (cm^{-1})	D
Nitrogen	$8\cdot9\cdot10^{-8}$	100 km
Water	$1\cdot0\cdot10^{-5}$	1 km
1 percent ovalbumin solution	$3\cdot1\cdot10^{-3}$	3 m

The second column contains the values of the turbidity calculated from Eqs. (3), (4), or (5′); in the third column an approximate value of the traveling distance D is given which will reduce the intensity of the primary beam to $1/e$ of its original value as a result of scattering. Values for the compressibility κ and for $\partial\mu/\partial p$ were taken from tables; the data for ovalbumin were derived from the experimental results of P. Putzeys and J. Brosteaux.[3]

I believe the conclusion is in favor of the scattering method. Attention should perhaps be drawn to the fact that the solvent is preferably chosen so as to make $\mu-\mu_0$ not too small. In most cases which have come to my attention it can easily be arranged to have $\mu-\mu_0$ of the order 10^{-3} for a 1 percent solution, which is quite sufficient. In experiments which are being carried out at Cornell a hollow large angle prism filled with the solvent combined with two compensating half-prisms filled with the solution, has been used for the direct determination of $\mu-\mu_0$. Scattering has been measured with a photo-tube connected to a Vance amplifier.

The discussion presented so far is not complete. One correction has to be added, which in most cases is not very important with regard to the molecular weight, but is interesting for its own sake.

In Einstein's theory one omission is made, which makes the theoretical result deviate from the experimental evidence. The theory predicts that, for instance, in the case of a liquid with only one component the scattered light observed at an angle of 90° against the primary beam will be totally polarized. In most cases the experiment shows to the contrary the existence of a depolarization effect which can be measured by the quotient of the intensity of the light with electrical vibrations in the plane of the primary beam and the direction of observation to that of the light vibrating perpendicular to this plane. The theory predicts $\rho=0$.

This deficiency is due to the fact that every element of volume is considered to remain optically isotropic notwithstanding its thermal fluctuations. Instead we have to expect that even in an ordinary liquid the fluctuations will change the optical properties of an element of volume momentarily from those of an isotropic to those of a crystalline medium. This can be taken into account and the improved calculation shows that under these circumstances a depolarization effect exists. The amount of this depolarization can be tied up with the work performed in exciting double refraction with the help of a strong electric or magnetic field (Kerr effect). In this way the constants characteristic for these effects are seen to be related to the depolarization constant ρ. For solutions this connection is not very practical for obvious reasons, but if we confine our attention to diluted solutions an approach similar to Lord Rayleigh's theory for gases gives all the information which is needed. If the particles of the solute are such that they must be characterized by an ellipsoid of polarization corresponding to 3 main polarizabilities α_1, α_2, α_3 in the direction of 3 mutually perpendicular axes, we will have to consider on top of the fluctuations in concentration fluctuations in orientation of the particles. A theory along these lines shows that the turbidity is increased and to the right-hand side of Eq. (5′) the factor

$$\frac{6+3\rho}{6-7\rho}$$

has to be added. A factor of this form has already previously been indicated by Cabannes. In order to be exact an additional measurement of the depolarization constant ρ should be made and the molecular weight calculated from Eq. (5′) has to be corrected by applying a correction factor

$$f=\frac{6-7\rho}{6+3\rho}. \tag{6}$$

In most practical cases ρ is of the order of only a few percent. The highest imaginable value of ρ is $\frac{1}{2}$ which would correspond to infinitely thin particles of needle-shape ($\alpha_1=\alpha_1$, $\alpha_2=0$, $\alpha_3=0$). In this case the correction factor f would be $\frac{1}{3}$. Although in general only of minor importance for molecular weight determinations the observation of the depolarization is interesting since it furnishes information about the optical qualities of the single particle. If two optical constants A and B are defined by the equations

$$A^2=\left(\frac{\alpha_1+\alpha_2+\alpha_3}{3}\right)^2,$$
$$B^2=\frac{1}{15}[(\alpha_1-\alpha_2)^2+(\alpha_2-\alpha_3)^2+(\alpha_3-\alpha_1)^2], \tag{7}$$

it can be shown that

$$\frac{B^2}{A^2}=\frac{6\rho}{6-7\rho}. \tag{7'}$$

Up to this point the discussion applies to particles which are small enough if compared with the wave-length. A few remarks concerning the case of larger particles will now be added.

If the size of the particle becomes comparable with the wave-length the electric currents which are excited in the particles by the electric force of the primary light wave will be out of phase with each other and this the more so the larger the particle is. The scattering intensity in a definite direction with respect to the primary beam will now be a result of a superposition of elementary waves coming from different points inside the particle which reach the observation point with phase differences depending on the position of those points and on the direction of observation. Only for small scattering angles those phase differences will be small. They increase with increasing scattering angle and are zero in the direction of the primary beam. This accounts for the fact that with increasing particle size the scattered light will become more concentrated in the forward direction and it is easy to see that experiments on this dissymmetry in angular distribution of scattered intensity can be used to obtain information about particle size. At the same time the scattered intensity will no longer remain proportional to $1/\lambda^4$ and so another possibility for determining such sizes proposed by W. Heller consists in following up the dependence of the intensity of scattering in a definite direction on the color of the primary light. The two methods are equivalent.

In a general way the theoretical work involved is rather complicated. However, in the special field of polymers advantage can be taken from the fact that it is possible in many cases to adjust the solvent in such a way that the difference in refractivity between the solvent and the solute can be considered as small. Under these circumstances the distortion of the primary wave inside the particle can be neglected and then it is much easier to relate particle form and particle size to the observed scattering.

In direct connection with the first part of this address is the following remark. In directions near the direction of the primary beam the phase differences of the wavelets coming from different parts of the particle tend to zero. In these directions the particle acts therefore as if its size is negligible. The net result is that for these special directions in calculating the intensity of scattering the reasoning followed in the first part of this paper can still be applied even if the particle is not small. So in this case observations of small angle scattering can be used for determining the number of particles or what is equivalent to their "molecular weight." I believe it more practical, however, to observe the actual angular distribution of scattering and by comparing this with the distribution to be expected for infinitely small particles to determine an experimental value for a reduction coefficient which introduced on the right-hand side of equations such as (5′) will yield a value for n from the observed turbidity. The intentions of these final remarks on larger particles are to indicate possible methods for intermediate sizes only, too small to be investigated directly by the microscope and at the same time not quite small enough to be treated as molecules of negligible dimensions.

BIBLIOGRAPHY

(1) A. Einstein, Ann. d. Physik **33**, 1275 (1910).

(2) S. D. Gehman and J. E. Field, Ind. Eng. Chem. **29**, 793 (1937).

(3) P. Putzeys and J. Brosteaux, Trans. Faraday Soc. **31**, 1314 (1935).

ANGULAR DISSYMMETRY OF SCATTERING AND SHAPE OF PARTICLES *

P. Debye

I. Introduction

It now seems well established that a measurement of the intensity of the scattered light emitted by a polymer solution as compared to that of the primary light or any other measurement which can yield a numerical value for the turbidity τ of the solution is a practical method for the determination of the molecular weight M of the polymer. **

Briefly the procedure is as follows:

Prepare a solution of concentration c expressed in grams per cubic centimeter and determine the *refraction difference* of this solution against the solvent. If μ is the index of refraction of the solution and μ_0 that of the solvent it turns out that $\mu-\mu_0$ is within very narrow limits proportional to the concentration. (In most cases $\mu-\mu_0$ is of the order of 1 unit in the third decimal for a 1% solution.)

Next calculate from the observed value of $(\mu-\mu_0)/c$ a constant K which characterizes the solution as to its scattering power from the relation:

$$K = \frac{32\pi^3}{3} \frac{\mu_0^2}{N\lambda^4} \left(\frac{\mu-\mu_0}{c}\right)^2$$

in which N is Avogadro's number ($N=6.061\times10^{23}$) and λ is the wavelength in centimeters of the primary light measured in vacuum. (The usual values of K are of the order 10^{-6} cm.^{2}/g.^{2}).

Now determine the *turbidity* τ in cm.$^{-1}$ for different concentrations c (ranging practically from a few tenth of 1% to a few per cent). Plotting finally $K\frac{c}{\tau}$ as ordinates against c as abscissas a straight line is found with an intercept for which is equal to $1/M$, in which M is the molecular weight.

*Technical Report No. 637 to Rubber Reserve Company, April 9, 1945.

** The turbidity τ is defined as the fractional decrease of the primary intensity over unit distance. In the following the letter τ always refers to the additional turbidity of the solution due to the presence of the polymer. It represents the difference in turbidity of solution and pure solvent.

For polymers of the same kind but with different degrees of polymerization all measured in the same solvent the straight lines are nearly parallel to each other and differ only in their intercepts. The better the solvent the steeper is the slope of these lines.

If the polymer is a mixture of particles with different degrees of polymerization the average molecular weight measured by the intercept is the *weight-average* molecular weight.

The statement as it is presented here implies that the particles in the solution are small compared with the wavelength of the light measured in the solution. If we are dealing with coiling polymers it seems at a first glance that this condition is amply verified up to rather high molecular weights. Let us suppose as an example that we have to deal with a straight polystyrene chain consisting of 10,000 linked monomers. This would correspond to a molecular weight $M = 104 \times 10^6$. The chain would contain $N =$ 20,000 carbon-to-carbon links. According to the usual calculation the average distance R from beginning to end of the chain (defined as the square root of the average value of the square of this distance) is:

$$R = \sqrt{2Na^2} = 308 \text{ A.U.}$$

if we take a, the C-C distance, equal to 1.54 A.U. This would still be at least 10 times smaller than the wavelength.*

At this stage of the argument it can be questioned whether we are right to judge the size of the polymer by such an average distance R, since we know that the molecule is continuously changing its shape and will pass through forms with a much larger distance from beginning to end. Our subsequent calculation will show, however, that this objection does not carry weight. In the averaging process involved in the calculation of the scattering of an assembly of all possible kinds of shapes it is again the same average distance R as before which is important.

The first observable effect to be expected as soon as the size becomes comparable with the wavelength would be that the angular distribution of the scattered intensity loses its symmetry around the 90° direction, the scattering becoming more pronounced in the forward direction. As well our own experience as that from other laboratories, accumulated during the last year, indicates that such an effect can be observed for polymers for which it would not be expected, judging from the value of the average distance R as calculated from the customary formula. Although from an experimental point of view there can be no question as to the existence of such a dissymmetry effect, questions can be raised as to its interpretation.

*For the formula compare for instance H. Mark, "*Physical Chemistry of High Polymeric Systems*," Interscience, New York, 1940, p. 72.

The most straightforward interpretation is that in reality the average distance R in a polymer molecule of the coiling type is larger than the value calculated from the customary formula. This is rather to be expected if we consider the assumptions which are made in deriving the relation. It is assumed that the chain consists of links with perfectly free rotations around each other with no amount of hindering at all. It is further assumed that the monomeric parts of the chain do not occupy any space, this even to such an extent that in the theoretical picture of the molecule links can cut through each other without any hindrance. This leads us to believe that the chains in nature will be considerably "stiffer" than those of the picture and as a consequence we have to anticipate larger values of R.

However, there exist other less fundamental reasons why a solution can show angular dissymmetry of scattering. Apart from the obvious effect of large size impurities, the dissymmetry may be due to gel content. An example of this is contained in Report CR-578 of W. O. Baker, J. H. Heiss, Jr., and R. W. Walker. Two solutions of identical composition of a GR-S copolymer are compared (A) containing no gel, (B) containing 50% of microgel. Whereas with respect to the intensity of the 90° scattering A:B = 2.38:3.06, the total turbidities are much farther apart. For these, A:B = 4.8:33.1, demonstrating that in the solution with a high gel content a large portion of the light is scattered in directions approaching the directions of the primary beam. This part of the scattered light obviously is not observed in the 90° scattering, whereas on the other hand it has to be counted as contributing to the total turbidity.

What an investigation of the dissymmetry effect shows immediately is the existence or nonexistence of particles of relatively large size in the solution. If such particles are observed the question still remains as to the constitution of those particles. It is evident that in order to answer this last question additional evidence, preferably independent of experiments on light scattering alone, will be very important.

In the following the case of unbranched polymers of the coiling type is considered. The angular distribution of scattering is calculated and expressed with the help of the average distance R from beginning to end of the chain. It is further shown how abnormally large values of R can be explained by hindered rotation and a simple formula is found which connects the increase in R with the restriction of rotation for a special type of hindrance.

II. Light Scattering by an Unbranched Chain

As a model to start with, we assume a chain of $N + 1$ atomic groups, each with its center at the intersection of two links. Each link may have the length a and for the present these bonds are considered to make a valence angle with each other and to have perfect freedom of rotation.

The amplitude of the scattered light emitted by a molecule of this kind will be proportional to:

$$\sum_{n=1}^{n=N+1} e^{ik(\bar{s},\ \bar{r}_n)}$$

We define $k = 2\pi/\lambda$ with λ equal to the wavelength of the primary light, measured in the medium; $\bar{s}$ is the vectorial difference between two unit vectors, one drawn in the direction of the scattered beam, the other in the direction of the primary beam; $\bar{r}_n$ is the vectorial distance of the position of the center of the nth atomic group from a fixed reference point. In this formula the proportionality factor measuring the response of the monomeric atomic group is left out, since we will only be interested in the angular distribution of the scattered intensity and not in its absolute value.

A definite configuration of the chain can still have all possible orientations in space, without changing the shape of the molecule by bond rotations. So we will first obtain the intensity averaged with respect to all possible orientations by multiplying the expression for the amplitude with its conjugate and then averaging about all the orientations. The result of this procedure is:

$$I = \sum_m \sum_n \frac{\sin[ksr_{mn}]}{ksr_{mn}} \tag{1}$$

I is the intensity, s is the length of the vector $\bar{s}$, which is equal to 2 sin $\vartheta/2$ with ϑ equal to the angle between the primary and the scattered beam and r_{mn} is the distance from group m to group n. The averaging process used in obtaining this expression for I is the same as that used in calculating the scattering of x-rays or electrons from a gas molecule. The equality sign is to be interpreted as mentioned before. For polymers another averaging process is still to be carried out, since the shape of the molecules will vary continuously. This can be done if we know for every distance r_{mn} what the probability is that such a distance will occur in an interval between r_{mn} and $r_{mn} + dr_{mn}$. Now it is well-known that for a chain of the kind here considered this probability for the distance r of beginning to end of the chain is:

$$w(r)\ dr = 4\pi \left(\frac{3}{2\pi R^2}\right)^{3/2} e^{-3r^2/2R^2} r^2 dr \tag{2}$$

in which the factor is so adjusted that $\int w(r)\,dr = 1$ and in which R^2 is the average value of r^2. We know also that for free rotating bonds:

$$R^2 = za^2 \frac{1 + p}{1 - p} \tag{3}$$

if the chain consists of z links ($z + 1$ groups) and p is the cosine of the angle between two bonds. The relation for the probability of occurrence of the distance r is the customary expression, which from a mathematical point of view is only exact in the limit for an infinitely long chain. The same is true for the relation expressed by equation (3). We will accept the validity of both relations for any chain lengths, a procedure which can be shown not to introduce any appreciable errors in all practical cases to be considered.

Knowing the probability function for r we can now determine at once the average value of one of the terms contained in the double sum representing the intensity I. We find:

$$\text{Av}\ \frac{\sin[ksr]}{ksr} = e^{-k^2s^2R^2/6} \tag{4}$$

and substituting this in equation (1) with due regard to equation (3), which expresses R^2 in terms of the number of links, we obtain:

$$I = \sum_{mn}\sum \exp\left(-\frac{k^2s^2a^2}{6}\,\frac{1 + p}{1 - p}\,|m - n|\right) \tag{5}$$

Each summation index m, as well as n, goes from 1 to $N + 1$; of the difference $m - n$ the absolute value is to be taken, which is indicated by writing $|m - n|$. If this summation is performed, the result is:

$$I = N^2 \frac{2}{x^2}\left[e^{-x} - (1 - x)\right] \tag{6}$$

with:

$$x = \frac{1 + p}{1 - p}\,\frac{k^2s^2}{6}\,Na^2 = \frac{k^2s^2R^2}{6} \tag{7}$$

in which R^2 now indicates the average square of the distance from beginning to end of the whole chain.

Details of the Calculation

If one term of the sum in equation (5) is written as $t_z = e^{-zb}$ with:

$$b = \frac{k^2 s^2 a^2}{6} \frac{1 + p}{1 - p}$$

the whole double sum can be arranged in a quadratic scheme of the following form:

$$I = \begin{vmatrix} t_0 + t_1 + t_2 + \ldots\ldots\ldots + t_N \\ t_1 + t_0 + t_1 + t_2 + \ldots + t_{N-1} \\ t_2 + t_1 + t_0 + t_1 + \ldots + t_{N-2} \\ \ldots\ldots\ldots\ldots\ldots\ldots\ldots\ldots \\ t_N + t_{N-1} + t_{N-2} + \ldots + t_1 + t_0 \end{vmatrix}$$

Taking out the diagonal terms first and then observing that the part in the upper right is equal to the part in the lower left, the result is:

$$I = (N + 1) + 2 \left\{ \sum_{z=1}^{z=N} t_z + \sum_{z=1}^{z=N-1} t_z + \ldots + \sum_{z=1}^{z=N-(N-1)} t_z \right.$$

or:

$$I = (N + 1) + 2 \sum_{0}^{N-1} \sum_{1}^{N-1} t_z$$

Since t_z is an exponential function the series over which the summations have to be carried out are simple geometrical series and we find:

$$I = (N + 1) + 2 \frac{e^{-b}}{1 - e^{-b}} N - \frac{1 - e^{-Nb}}{e^{b} - 1}$$

Now in every case b has a very small numerical value and only Nb can be finite provided the chain is long enough. The way of interpreting the result therefore is to put $Nb = x$ and $b = x/N$, and to determine the value of I in the limit for finite x and increasingly large values of N. In this way we find:

$$I = (N + 1) + 2 \frac{N^2}{x} \left(1 - \frac{1}{2}\frac{x}{N} + \ldots\right) \quad 1 - \frac{1 - e^{-x}}{x} \left(1 - \frac{1}{2}\frac{x}{N} + \ldots\right)$$

Retaining only terms of the order N^2 this yields the final result:

$$I = N^2 \frac{2}{x}\left[1 - \frac{1 - e^{-x}}{x}\right] = N^2 \frac{2}{x^2}\left[e^{-x} - (1-x)\right]$$

Since we do not care about the strength of the scattering but are only concerned with the angular distribution, we will omit the factor N^2 and have in this way normalized the scattered intensity so as to make $I = 1$ for $s = 0$, which, since $s = 2 \sin \theta/2$, corresponds to $\theta = 0$. In the following discussion we therefore take:

$$I = \frac{2}{x^2}\left[e^{-x} - (1 - x)\right] \quad \ldots x = \frac{k^2R^2}{6} s^2 \qquad (8)$$

For small values of the angle θ we will have:

$$I = 1 \frac{k^2R^2}{18} s^2 + \frac{k^4R^4}{72} s^4 - + \ldots$$

If the size of the polymer is so large that for large angles our variable x itself becomes large the intensity can be represented by the relation:

$$I = \frac{12}{k^2R^2}\frac{1}{s^2} - \frac{72}{k^4R^4}\frac{1}{s^4}$$

In order to facilitate applications, Table I gives I as a function of x.

Table I

x	I	x	I	x	I
0	1	1.0	0.736	3.5	0.413
0.2	0.937	1.5	0.643	4.0	0.377
0.4	0.878	2.0	0.568	5.0	0.321
0.6	0.828	2.5	0.506	6.0	0.278
0.8	0.780	3.0	0.455	∞	0

The largest value $s = 2 \sin \vartheta/2$ can have in the angular range at our disposition is $s = 2$. Taking the polystyrene polymer mentioned in the introduction of the molecular weight $M = 1.04\times10^6$ as an example, we found $R = 308$ A.U. For $\vartheta = 180^\circ$ $(s = 2)$ and with a wavelength of 3000 A.U. (measured in the solvent) we find $x = 0.31$. According to (8) this means that only a slight dissymmetry effect should exist, which would decrease the intensity to be observed in the backward direction $(\vartheta = 180^\circ)$ to 90% of the intensity in the forward direction $(\vartheta = 0^\circ)$. The reason for the small effect is of course our small assumed value of R (308 A.U.), which is only 1/10 of the wavelength. This average distance was the result of the usual formula which follows from calculations assuming highly idealized conditions. According to our discussion in the introduction we have every reason to believe that in this calculation the distance is underestimated.

As can easily be seen the whole reasoning which leads to equation (8) as an expression describing the angular distribution is independent of the special way in which R^2 is supposed to follow from the atomic distances of the chain (equation 3). The only important thing is that R^2 be proportional to the number of links, which is characteristic for all molecules of the coiling type. The factor entering in the relation between the average square of the distance of the ends and the number of links will depend essentially on the "stiffness" of the chain and we can safely surmise that the value given to this factor in equation (3) is the lowest estimate possible. Realizing this situation, we suggest that relations (8) be taken as describing the angular distribution and that the parameter R^2 be determined not by calculation from a model of the molecule but be derived from the experimental evidence. In this way we learn experimentally something about the actual size of the polymer. This done, we can then speculate about the experimental value of R as compared to theoretical values derived from a model. Attempts of this kind are to be found in Fred W. Billmeyer's report CR622.

III. Influence of Hindrance of Rotation on the Average Chain Length

In order to estimate how much increase of the size of a polymer can result from hindered rotation the following special model is considered. We again have N links each of length a making the valence angle with each other, the projection of one bond on the next being pa. But now instead of assuming free rotation we restrict it in the following manner. Two consecutive bonds define instantaneously a plane, we assume that the third following bond can rotate freely only, within an angle 2α around its central position which is in the plane and parallel to the first bond of the three. The rotation is in this way restricted to an interval $-\alpha$ to $+\alpha$ and we would have free rotation only if we took $\alpha=\pi$. We want to know the average square of the distance from beginning to end of this chain.

Considering the bonds as vectors $\bar{a}_1 \ldots \bar{a}_N$, the vectorial distance from beginning to end is the sum of these vectors. In order to obtain the square of this distance we have to take the scalar product of this sum with itself. In this way we obtain N^2 terms each of which represents the scalar product of one vectorial bond with another. So if we are going to determine the average value R^2, we will have to know first what the average is of the scalar product of two vectorial bonds taken at random.

Now considering a continuous part of the chain containing the bonds $\bar{a}_{n+1}$ to $\bar{a}_{n+z}$ inclusive and supposing that we want to find the average value of $(\bar{a}_{n+1}\ \bar{a}_{n+z})$ the averaging process can be built up of consecutive steps. First we allow the bond $n + z$ to perform its motion and calculate the resulting average, then we do the same with the following bond $n + z - 1$ and so on. Now it can be shown that the result of the first step transforms the product:

$$(\bar{a}_{n+1}\ \bar{a}_{n+z})$$

into:

$$\varepsilon(\bar{a}_{n+1}\ \bar{a}_{n+z-1}) + \eta(\bar{a}_{n+1}\ \bar{a}_{n+z-2}) \tag{9}$$

with two constants ε and η defined by the relations:

$$\varepsilon = p(1 - \frac{\sin\alpha}{\alpha}) \tag{9'}$$

$$\eta = \frac{\sin\alpha}{\alpha}$$

Details of the Calculation

Let Figure 1 represent a picture of the situation with respect to the 3 last bonds and let us draw the vector $\bar{a}_{n+1}$ from the center of the circle over which the end of vector $\bar{a}_{n+z}$ can move. Let further $\bar{u}_0, \bar{u}_0', \bar{v}_0, \bar{v}$, and $\bar{w}$ represent vector components of the main vectors to be considered, as indicated in Figure 1.

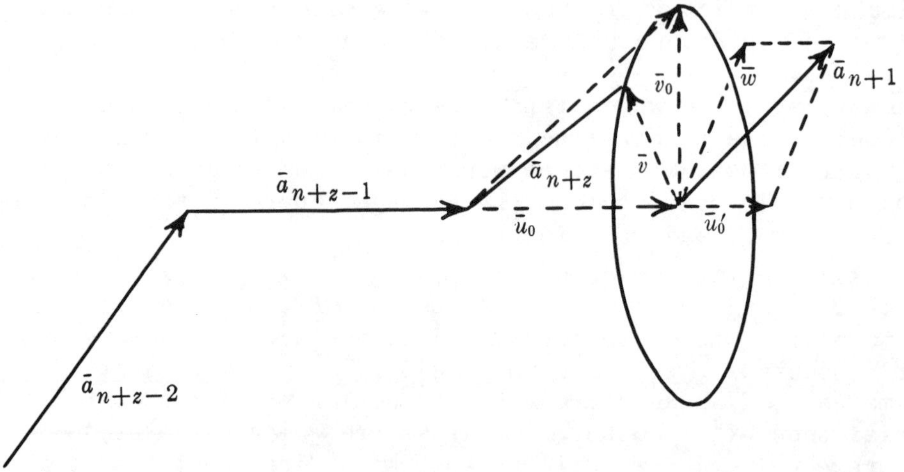

Figure 1

We then have;

$$(\bar{a}_{n+1}\ \bar{a}_{n+z}) = (\bar{a}_{n+1} . \bar{u}_0) + (\bar{a}_{n+1}\ \bar{v})$$

but:

$$(\bar{a}_{n+1}\ \bar{v}) = (\bar{w}\ \bar{v}) + (\bar{u}'_0\ \bar{v}) = (\bar{w}\ \bar{v})$$

since $\bar{u}'_0$ is perpendicular to $\bar{v}$. So we can write:

$$(\bar{a}_{n+1}\ \bar{a}_{n+z}) = (\bar{a}_{n+1}\ \bar{u}_0) + (\bar{w}\ \bar{v})$$

The first term is constant through the motion of bond $n + z$. If ψ is the angle between $\bar{w}$ and $\bar{v}$ and ψ_0 the angle between w and v_0 the motion of bond $n + z$ will be confined to the interval $\psi = \psi_0 + \alpha$ to $\psi = \psi_0 - \alpha$. For the average value of the second term we therefore find:

$$wv\ \frac{1}{2\alpha}\int_{\psi_0+\alpha}^{\psi_0-\alpha} \cos\psi d\psi = \frac{\sin\alpha}{\alpha}\ wv\ \cos\psi_0$$

However:

$$wv\ \cos\psi_0 = (\bar{w}\ \bar{v}_0) = (\bar{a}_{n+1}\ \bar{v}_0)$$

So we can say that the result of the first step in our averaging process of $\bar{a}_{n+1}\ \bar{a}_{n+z}$ is:

$$(\bar{a}_{n+1}\bar{u}_0) + \frac{\sin\alpha}{\alpha}\ (\bar{a}_{n+1}\ \bar{v}_0)$$

The two vectors $\bar{u}_0$ and $\bar{v}_0$ are both situated in the plane of the two vectors $\bar{a}_{n+z-1}$ and $\bar{a}_{n+z-2}$. Therefore the first two are linear combinations of the second two. It is easily seen that

$$\bar{u}_0 = p\ \bar{a}_{n+z-1},\ \bar{v}_0 = \bar{a}_{n+z-2} - p\ \bar{a}_{n+z-1}$$

Substituting this in our last result, relations (9) and (9') of the text are obtained.

Let us now call P_z the average value of the product $(\bar{a}_{n+1} a_{n+z})$ after all the steps of the averaging process are carried out. From (9) it then follows immediately that:

$$P_z = \varepsilon P_{z-1} + \eta P_{z-2} \qquad (10)$$

We further know that:

$$P_1 = a^2 \text{ and } P_2 = pa^2 \qquad (10')$$

and from these two relations (10) and 10') an expression for P_z can be calculated. The result is:

Define λ_1 and λ_1 as the two roots of the quadratic equation:

$$\lambda_2 - \varepsilon\lambda - \eta = 0 \qquad (11)$$

then:

$$\frac{P_z}{a^2} = \frac{p-\lambda_2}{\lambda_1-\lambda_2}\lambda_1^{z-1} - \frac{p-\lambda_1}{\lambda_1-\lambda_2}\lambda_2^{z-1} \qquad (12)$$

Details of the Calculations

In order to solve the recurrence equation (10) we put $P_z = \lambda^z$ and see immediately that this can be done provided λ is a root of the quadratic equation (11). We then assume $P_z/a^2 = A_1\lambda_1^z + A_2\lambda_2^z$ in which λ_1 and λ_2 are the 2 roots and A_1 and A_2 arbitrary constants. We can then determine these 2 constants by the 2 conditions that $P_1 = a^2$ and $P_2 = pa^2$, which leads to:

$$A_1\lambda_1 + A_2\lambda_2 = 1$$

$$A_1\lambda_1^2 + A_2\lambda_2^2 = p$$

or:

$$A_1 = \frac{1}{\lambda_1}\frac{p-\lambda_2}{\lambda_1-\lambda_2}. \quad -A_2 = \frac{1}{\lambda_2}\frac{p-\lambda_1}{\lambda_1-\lambda_2}$$

As mentioned before, the required average R^2 is the sum of the N^2 average values of all the products $(\bar{a}_m \bar{a}_n)$. In a quadratic arrangement this summation appears in the following form:

$$R^2 = \left| \begin{array}{l} P_1 + P_2 + P_3 + \ldots + P_N \\ P_2 + P_1 + P_2 + \ldots + P_{N-1} \\ \ldots\ldots\ldots\ldots\ldots\ldots \\ \ldots\ldots\ldots\ldots\ldots\ldots \\ P_N + P_{N-1} + P_{N-2} + \ldots + P_1 \end{array} \right| \tag{13}$$

This summation is of the same kind as that which has to be carried out in the case of free rotation. The final result (in the limit for large values of N) is:

$$\frac{R^2}{a^2} = \frac{1 + (\sin \alpha/\alpha)}{1 - (\sin \alpha/\alpha)} \; \frac{1 + p}{1 - p} \tag{14}$$

Details of the Calculation

If we substitute in (13) the values of P_z following from equation (12) we see that R^2 can be expressed by two sums Σ_1 and Σ_2 in the form:

$$\frac{R^2}{a^2} = \frac{p - \lambda_2}{\lambda_1 - \lambda_2} \Sigma_1 - \frac{p - \lambda_1}{\lambda_1 - \lambda_2} \Sigma_2$$

Each of these sums can be written in a quadratic scheme in the form:

$$\Sigma = \left| \begin{array}{l} 1 + \lambda + \lambda^2 + \ldots + \lambda^{N-1} \\ \lambda + 1 + \lambda + \ldots + \lambda^{N-2} \\ \\ \ldots\ldots\ldots\ldots\ldots \\ \ldots\ldots\ldots\ldots\ldots \\ \lambda^{N-1} + \lambda^{N-a} + \lambda^{N-3} + \ldots + 1 \end{array} \right|$$

For Σ_1 we have to substitute $\lambda = \lambda_1$ and for Σ_2 the other root $\lambda = \lambda_2$. Taking the diagonal terms first and observing that the upper right part is equal to the lower left part, it follows that:

$$\Sigma = N + 2 \sum_{\sigma=1}^{\sigma=N-1} \sum_{z=1}^{z=N-1} \lambda^{z}$$

In the limit for large values of N this reduces to:

$$\Sigma = N \left\{ 1 + \frac{2\lambda}{1-\lambda} \right\}$$

and we find:

$$\frac{R^2}{a^2} = N \left\{ 1 + 2 \frac{p - \lambda_2}{\lambda_1 - \lambda_2} \frac{\lambda_1}{1 - \lambda_1} - 2 \frac{p - \lambda_1}{\lambda_1 - \lambda_2} \frac{\lambda_2}{1 - \lambda_2} \right\}$$

or:

$$\frac{R^2}{a^2} = N \left\{ 1 + 2 \frac{p - \lambda_1 \lambda_2}{(1-\lambda_1)(1-\lambda_2)} \right\}$$

since λ_1 and λ_2 are the roots of equation (11) we have (considering also equation 9'):

$$\lambda_1 + \lambda_2 = \varepsilon = p(1-\eta), \quad \lambda_1 \lambda_2 = 1-\eta$$

Substituting gives:

$$\frac{R^2}{a^2} = N \frac{1 + \eta}{1 - \eta} \frac{1 + p}{1 - p}$$

Since according to (9') we have $\eta = \sin \alpha/\alpha$ this is the final result indicated in the text in equation (14).

The relation (14) for R^2 is the same as that of the current literature in the limit for free rotation, since in this case we have to substitute $\alpha=\pi$. In general for hindered rotation R^2 will be larger by a factor:

$$\frac{1 + (\sin\alpha/\alpha)}{1 - (\sin\alpha/\alpha)}$$

Since the quotient $\sin\alpha/\alpha$ approaches unity when the angle α becomes small the relation illustrates how R^2 increases with increasing hindrance.

As an illustration for the use which can be made of the relations presented in this report let us suppose the same linear polystyrene polymer, which we already discussed, of a molecular weight $M = 1.04\times10^6$ containing $N = 20,000$ C-C bonds, each of length $a = 1.54$ A.U. Formerly we calculated for free rotating bonds $R = 308$ A.U. and concluded that if this was the actual distance we could expect only a small dissymmetry effect, the backward scattering being 90% of the intensity in the forward direction if observed with a wavelength of 3000 A.U. (measured in the medium).

Let us now assume that the experiment yielded a larger effect, say 25% more scattering under an angle $\vartheta=60^\circ$ than under the angle $\vartheta=120^\circ$. (For actually observed values see Report CR622). The two values of $s = 2 \sin \vartheta/2$ corresponding to $\vartheta=60^\circ$ and $\vartheta=120^\circ$ are $s = 1$ and $s = 3$. According to equation (8) the two values of the parameter x will therefore be in the ratio 1:3. If x has a numerical value say β for $\vartheta = 60^\circ$, it will be 3β for $\vartheta=120^\circ$ and since we have observed that $I_\beta = 1.25\ I_{3\beta}$ we have to choose $x = \beta$ so that this relation holds. This can easily be done with the help of the Table I and we find $x = \beta = 0.37$. This means that we should have:

$$\frac{k^2R^2}{6} s^2 = \frac{\pi^2}{6} \frac{R^2}{\lambda^2} = 0.37$$

in order to explain the 25% difference in intensities. With $\lambda = 3000$ A.U. it follows then that $R = 711$ A.U. If finally we want to interpret this larger distance as due to hindering of rotation, we have to determine the angle α from the relation (see equation 14).

$$\frac{1 + (\sin\alpha/\alpha)}{1 - (\sin\alpha/\alpha)} = \left(\frac{711}{308}\right)^2$$

since we calculated $R = 308$ A.U. for the case of free rotation. The result is:

$$\alpha = 84^\circ$$

A restriction to a total angle 2α of 168° instead of the 360° in the case of free rotation would explain the 25% intensity difference assumed to be the result of the experiment.

Reprinted from pages 573-579 of

THE JOURNAL OF CHEMICAL PHYSICS VOLUME 16, NUMBER 6 JUNE, 1948

Intrinsic Viscosity, Diffusion, and Sedimentation Rate of Polymers in Solution

P. DEBYE AND A. M. BUECHE

(Received March 2, 1948)

Intrinsic viscosity, diffusion and sedimentation rate of polymers in solution is calculated by a generalization of Einstein's theory for impermeable spheres. For the coiled polymer molecule a sphere is substituted which hinders the liquid flow through its interior only to a degree depending on the average density in space of the polymer molecule in solution. The amount of shielding of the liquid flow which is introduced in this way determines the exponent in the customary exponential relation between intrinsic viscosity, diffusion, or sedimentation rate and molecular weight. This relation is shown to have only the merits of an interpolation formula. It is shown how the dimensions of the molecular coil can be derived from the experimental data on viscosity, and these dimensions are compared with those derived from interference measurements. The point is stressed that the relation between intrinsic viscosity and molecular weight is rather indirect and depends essentially on the type of polymer molecule under consideration. In most cases polymer molecules are decidedly stiffer and therefore cover a larger space in solution than would be expected from models with free rotation around bonds.

IN the preceding paper Kirkwood and Riseman refer to some theoretical calculations concerning intrinsic viscosity of Debye and similar calculations of Bueche concerning diffusion and sedimentation. Since these papers are not readily available* it seems appropriate, in order to facilitate a comparison of the different lines of attack, to present here the main features of the theoretical approach as conceived by one of the authors of this paper.**

If the mutual interaction of the beads of a

* The theory of viscosity was the subject of a report to the Office of Rubber Reserve presented in a meeting in Chicago, Oct. 31 and Nov. 1, 1946. A later report of A. M. Bueche dealt with sedimentation and diffusion. A paper containing the results concerning intrinsic viscosity as well as diffusion and sedimentation was submitted to Section 11 of the eleventh International Congress of Pure and Applied Chemistry (London: July 17th–24th, 1947). This paper has not yet been printed. Two short notes appeared, one in Phys. Rev. **71**, 486 (1947), the other in the Jan.–April issue of the Record of Chemical Progress, 1947.

** Calculations along the same line were made independently by H. C. Brinkman, Proc. Amsterdam Acad. **50**, No. 6 (1947); App. Sci. Res. **A1**, 27 (1947). Brinkman's results coincide with ours.

pearl string, which is substituted for a linear chain molecule with more or less free rotation, is neglected, the intrinsic viscosity is represented by the following relation:***

$$[\eta]=\frac{1}{36}\frac{f/\eta}{m}R^2. \qquad (1)$$

$[\eta]$ is the intrinsic viscosity expressed in cc/gr (which is 100 times the value expressed in customary units); f the friction factor of one bead, defined in such a way that in order to drag one single bead through the liquid with a velocity v a force fv is needed; η is the viscosity of the solvent, m the mass of one bead, and R^2 the average square of the distance between the two ends of the chain. Since R^2 is proportional to the degree of polymerization, the relation confirms Staudinger's rule.

I

The assumption that the interaction can be neglected obviously cannot be accepted. As a matter of fact, each individual bead will move in a velocity field which deviates from the original undisturbed field because of the superposition of the disturbances caused by all the other beads. The situation is analogous to a case in which a dielectric medium is placed in an electric field and in which we find that in a point in the dielectric the field is distorted because of the disturbances caused by the polarization of all the other parts of the medium. In an attempt to formulate this idea such that care is taken of the essential features and at the same time the mathematical complications are reduced to a minimum, Debye comes to the following scheme.

In a homogeneous liquid, considered incompressible and with a viscosity η, the velocity $\bar{v}$ can be determined from the fundamental equations

$$\eta \operatorname{curl}\operatorname{curl}\bar{v}+\operatorname{grad}p=0, \quad \operatorname{div}\bar{v}=0, \qquad (2)$$

in which p denotes the pressure. The first equation merely expresses that equilibrium exists between the frictional force on an element of volume and the force caused by local variations in pressure. As soon, however, as parts of the polymer molecule are situated in the element of volume under consideration, we have to take account also of the frictional force between the beads and the liquid. If we call the relative velocity of the liquid with respect to the molecule at the point in question, $\bar{v}_r$, the average value of this force acting on the liquid in an element of volume $d\tau$ will be

$$-\nu f\bar{v}_r d\tau,$$

provided the average number of beads contained in $d\tau$ is $\nu d\tau$. Adding this force to the forces already considered in (2) we arrive at a new set of equations, namely,

$$\eta \operatorname{curl}\operatorname{curl}\bar{v}+\nu f\bar{v}_r+\operatorname{grad}p=0, \quad \operatorname{div}\bar{v}=0. \qquad (3)$$

From the statistics of the chain molecule we know the bead density ν as a function of the distance from the center of gravity. For the relative velocity $\bar{v}_r$ we can substitute the actual velocity of the liquid $\bar{v}$ minus a velocity $\bar{v}_0$ corresponding to the proceeding and rotating molecule. In this way the problem of finding the velocity of the liquid everywhere, inside and outside of the space occupied in the average by the polymer molecule, is now reduced to finding the appropriate solution of the set of Eqs. (3). And, of course, as soon as we know this velocity distribution, the additional heat developed due to the immersion of one polymer molecule can be found, and the intrinsic viscosity can be calculated along familiar lines.

The main point to consider before proceeding with the solution of Eq. (3) is the average density in space of the beads. From a consideration of the statistics of the chain it is found that this density is a maximum around the center of gravity of the polymer molecule, and with increasing distance r of this point ν decreases gradually. The function $F(r)$ which represents this density has a shape which resembles a Gaussian probability curve. At this stage I have not considered it worth while to go ahead and solve our problem using the exact representation of the density. It is more than probable that the essential features of the final result will depend only in minor details on the exact form of $F(r)$. Instead I propose to substitute as a picture for the average space occupied by the polymer molecule a sphere of radius R_s in which the bead density ν is supposed to be constant throughout,

*** P. Debye, J. Chem. Phys. **14**, 636 (1946).

whereas outside of this sphere $\nu=0$. If N is the total number of beads of the string, we then have inside the sphere

$$\nu=(3/4\pi)(N/R_s^3). \tag{4}$$

In this formulation we have to find solutions of Eq. (3), one for the outside of the sphere of radius R_s with $\nu=0$, another for the inside with ν equal to the constant value given by Eq. (4). The constants appearing in these solutions have to be adjusted in such manner that at large distances we have the original undisturbed velocity distribution. At the surface of the sphere the velocities inside and outside must be the same as well as the stresses.

The mathematical method which leads to this goal is a generalization of a method derived by *G. Kirchhoff*, formerly used by *A. Einstein* in his well-known but simpler problem of calculating the effect of a number of immersed *impermeable* spheres on the over-all viscosity.†

The angular velocity of the whole molecule can be left indeterminate at the outset, but using the condition, as in the simpler case in which no interaction was considered, that the average total torque on the molecule must be zero, it turns out that just as before this angular velocity still is equal to half the gradient of the velocity of the liquid.

Accepting the form which has now been given to the problem, it is readily seen that apart from the number of beads, N, only two constants each of the dimension of a length enter into it. The first is, of course, the radius R_s of the substituted spherical space occupied in the average by the polymer molecule. The second is a length L defined by the relation

$$1/L^2=\nu(f/\eta), \tag{5}$$

in which ν is the bead density according to Eq. (4). I would like to call L the "*shielding length*" for the following reason. Consider a much simpler case in which a liquid is flowing free in the x direction with a velocity having a gradient α in the z direction everywhere above the plane $z=0$, whereas below this plane the flow is hindered by an accumulation of beads, distributed with constant density ν. Using Eq. (3), we find that in this latter part, that is for negative values of z, the velocity v_x decreases exponentially and is proportional to $\exp(z/L)$. The lower part of the medium is therefore shielded by the accumulated effect of the beads, and the distance in this part over which the velocity decreases by a factor $1/e$ is our shielding length L. I understand that the same formulation as expressed by Eq. (3) is used by engineers in discussing the flow of water through sand.

The advantage of using the approximation of constant bead density in our problem of the rotating sphere is that all the functions which have to be introduced as solutions of the equations can be expressed by elementary functions. By a queer coincidence the fundamental equation which has to be satisfied is of the same type as that describing the potential around an ion in a strong electrolyte.

Skipping now all further mathematical details, the final result for the intrinsic viscosity appears in the following form:

$$[\eta]=(\Omega_s/M)\phi(\sigma)\cdots\sigma=R_s/L. \tag{6}$$

$\Omega_s=(4\pi/3)R_s^3$ is the spherical volume occupied by the sphere substituted for the polymer molecule, M is its actual mass, and $\phi(\sigma)$ is a function of a number $\sigma=R_s/L$, which I would like to call the "shielding ratio."†† This function is represented by Eq. (7):

$$\phi(\sigma)=\frac{5}{2}\frac{(1+(3/\sigma^2)-(3/\sigma)\,\mathrm{cotgh}\sigma)}{1+(10/\sigma^2)(1+(3/\sigma^2)-(3/\sigma)\,\mathrm{cotgh}\sigma)}. \tag{7}$$

For small values of the shielding ratio σ we have

$$\phi(\sigma)=\sigma^2/10[1-(2/35)\sigma^2+\cdots]. \tag{8}$$

Substituting the first approximation of ϕ for small values of σ in Eq. (6) and remembering the definition of ν and L as expressed by Eqs. (4) and (5), we find in the limit for $\sigma=0$

$$[\eta]=1/10(f/\eta m)R_s^2. \tag{8'}$$

This relation should be compared with Eq. (1), which was derived directly (without using the approximation of the substituted sphere with constant bead density) for the case of infinitely small shielding effect. If we do this we see that in

† A. Einstein, Ann. d. Physik (4) **19**, 289 (1906); **34**, 591 (1911).

†† The intrinsic viscosity is again expressed in cc/g.

order to obtain coincidence we should take $R_s^2/10$ equal $R^2/36$, and this means that the diameter $2R_s$ of our spherical volume should be taken equal to $1.054R$. This diameter is therefore very nearly equivalent to R, the square root of the average square of the distance between ends of the chain.

For large values of the shielding ratio we have

$$\phi(\sigma)=5/2[1-(3/\sigma)-\cdots], \tag{9}$$

and

$$[\eta]=5/2(\Omega_s/M)=5/2(\Omega_s/Nm). \tag{9'}$$

This is Einstein's relation for rigid impenetrable spheres, as it should be.

It is of interest to observe that the important features which determine the intrinsic viscosity are *not* primarily concerned with the molecular weight. They are, as seen from (6): firstly, the specific volume of the material of the polymer molecule as distributed over the average space occupied by this molecule in the liquid; secondly, the shielding ratio σ. However, if we know the structure of the molecule we can connect both these quantities with the degree of polymerization N (or the molecular weight). So for linear chains we know that R or R_s is proportional to $N^{\frac{1}{2}}$, and we see immediately from (8′) and (9′) that in this case we can find anything between proportionality with N to proportionality with $N^{\frac{1}{2}}$ depending on the shielding ratio. However, if the chain is not linear, if, for instance, the molecule is highly cross-linked, a very different dependence on the molecular weight will prevail. Under such circumstances we can have very high molecular weights combined with relatively small intrinsic viscosities.

It remains to be seen how the customary formula for the intrinsic viscosity with an exponent of the molecular weight between 1 and $\frac{1}{2}$ has to be interpreted.

If we assume a linear chain ($R_s \sim N^{\frac{1}{2}}$) and draw according to Eq. (6) a curve for the intrinsic viscosity as a function of the degree of polymerization, we will find that it begins proportional to N and that its slope gradually decreases with increasing N until for very large N we end up with a proportionality to $N^{\frac{1}{2}}$. At any point of this curve for a given value $N=N_0$ we can approximate the curve by its tangent, representing it by the linear function

$$[\eta]=\alpha+\beta(N-N_0) \tag{10}$$

with two constants α and β. This would be the usual procedure. But of course this procedure is arbitrary inasmuch as we decided arbitrarily on the outset that a linear approximation was what we wanted to use. We can just as well decide that in the vicinity of $N=N_0$ we want to approximate the curve by an expression of the form

$$[\eta]=AN^{\epsilon} \tag{10'}$$

with two other constants A and ϵ. If we take the logarithm on both sides of (10′), we see that the meaning of this second form of approximation is to draw a curve for $\log[\eta]$ as a function of $\log N$ and to approximate this new curve by its tangent. Fundamentally and from a purely mathematical point of view, there is no difference between the two procedures, but in our special case the exponential approximation according to (10′) has a peculiar advantage. We know that our formula, as well as any other formula which might be proposed, makes $[\eta]=0$ for $N=0$. The exponential approximation formula (10′) does this also, automatically, whereas the linear approximation (10) does not. In this way it turns out that the exponential approximation has really 3 points in common with the actual curve, whereas the linear approximation has only 2. We will therefore expect that the exponential approximation will not usually deviate much from the actual curve over a large range of values N. This is actually what is observed. We have here before us a case in which a special formula fits the experimental data very well, and still it would be wrong to conclude that its special form has any fundamental importance.

From what has been said about the meaning of the exponential approximation, it follows immediately that *for a linear chain*

$$\epsilon=\frac{d\log[\eta]}{d\log N}=\frac{N}{[\eta]}\frac{d[\eta]}{dN}=\frac{1}{2}+\frac{1}{4}\frac{\sigma}{\phi}\frac{d\phi}{d\sigma}. \tag{11}$$

To every value N of the degree of polymerization belongs a definite exponent ϵ, which can be calculated from (11) as a function of the shielding ratio σ. The following table gives corresponding values of the shielding ratio σ, the function ϕ (which for reasons apparent from Eq. (6) is called the volume factor), and the exponent ϵ.

In order to illustrate the use of this table, we can take some measurements of R. H. Ewart on polystyrene in benzene. Ewart represents his results by the interpolation formula

$$[\eta]=7.54\cdot 10^{-3}\,(MW)^{0.783} \tag{12}$$

in which the intrinsic viscosity is measured in cc/g. and MW indicates the customary molecular weight. The measurements covered a range from $MW=50000$ to $MW=800000$. The average molecular weight of this interval (the average being taken in the logarithmic diagram) is $MW=200000$. We will assume that the exponent 0.783 corresponds exactly to this average MW. Now assuming that polystyrene is a linear polymer, we find by interpolation in Table I that $\epsilon=0.783$ corresponds to $\sigma=3.92$ and $\phi=0.836$. At the same time Ewart's relation (12) gives for $MW=200000$ the value $[\eta]=107$ cc/g. From Eq. (6) it follows that the specific volume of the polymer substance within the substituted sphere is $\Omega_s/M=128$ cc/g. Since we know the mass of the polymer molecule, we now can calculate the volume Ω_s and the diameter $2R_s$ of this sphere and find $2R_s=431$A. If we decide that the relation $2R_s=1.054R$, which holds for the limiting case $\sigma=0$, can be used, we will conclude that for a linear polystyrene molecule in benzene of $MW=200000$ the root mean square distance between ends $R=409$A, which is 3 times as large as the theoretical value of a C$-$C chain with absolutely free rotation.

II

A. M. Bueche has applied the same line of reasoning to the problem of diffusion and sedimentation. If a polymer molecule is dragged through a liquid with the velocity v, a force Fv will have to be applied. The friction constant F is the essential quantity in sedimentation experiments, the diffusion constant D follows from Einstein's relation

$$D=kT/F. \tag{13}$$

For the friction constant F Bueche finds

$$F=6\pi\eta R_s\psi(\sigma) \tag{14}$$

TABLE I. (Linear polymers) $[\eta]=(\Omega_s/M)\phi$.

Shielding ratio σ	Volume factor ϕ	Exponent ϵ
0	0	1.000
1	0.0947	0.973
2	0.327	0.910
3	0.600	0.839
4	0.857	0.778
5	1.07	0.731
6	1.25	0.693
7	1.40	0.664
8	1.52	0.642
9	1.62	0.625
10	1.70	0.611
15	1.96	0.569
20	2.10	0.549
∞	2.50	0.500

in which ψ is again a function of the shielding ratio σ analogous to ϕ. It is defined by the relation

$$\psi(\sigma)=\frac{(1-(1/\sigma)Tgh\sigma)}{1+(3/2\sigma^2)(1-(1/\sigma)Tgh\sigma)}. \tag{15}$$

For small σ we have

$$\psi=(2/9)\sigma^2[1-(4/15)\sigma^2+\cdots], \tag{16}$$

which leads in the limiting case $\sigma=0$ to

$$F=Nf, \tag{16'}$$

remembering the definition of the shielding length L by Eq. (5). This is exactly what has to be expected since, if we neglect all interaction, the force on N beads is N times as large as that on a single bead.

For large σ we have

$$\psi(\sigma)=1-(1/\sigma)-\cdots, \tag{17}$$

which leads to Stokes' formula

$$F=6\pi\eta R_s. \tag{17'}$$

TABLE II. (Linear polymer) $F=6\pi\eta R_s\psi$.

Shielding ratio σ	Volume factor ψ	Exponent ϵ
0	0.000	1.000
1	0.176	0.899
2	0.432	0.752
3	0.600	0.661
4	0.701	0.610
5	0.770	0.574
8	0.858	0.544
10	0.890	0.532
∞	1.000	0.500

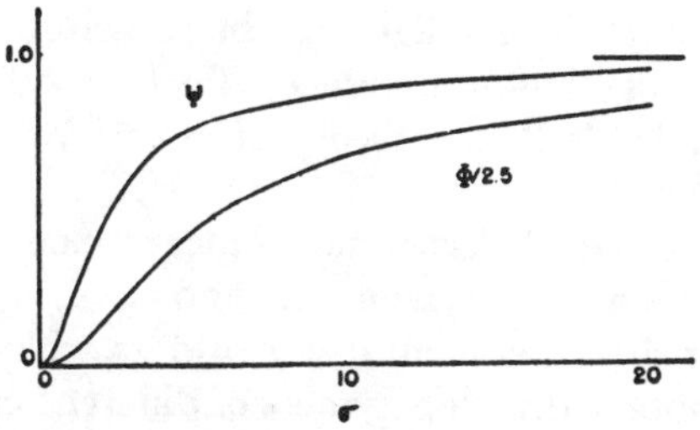

FIG. 1. Volume factor Φ and radius factor Ψ as functions of the shielding ratio σ.

Again as in the case of the intrinsic viscosity, the friction constant F can be represented by an exponential formula of the type

$$F = AN^{\epsilon}$$

over a limited range of molecular weights and again there will exist a one-to-one correspondence between ϵ and σ, according to the relation

$$\epsilon = \frac{1}{2} + \frac{1}{4}\frac{\sigma}{\psi}\frac{d\psi}{d\sigma}. \tag{18}$$

Table II contains corresponding values of the shielding ratio σ, the "radius factor" ψ, and the exponent ϵ for the case of linear polymers. In Fig. 1 the two functions $\phi/2.5$ and ψ are shown plotted as functions of the shielding ratio σ.

III

We now have two theoretical representations of the intrinsic viscosity and two of the friction constant, one type attributable to Kirkwood and Riseman, which we will indicate by the letters K.R., and one type discussed in this paper to be indicated by the letters D.B.

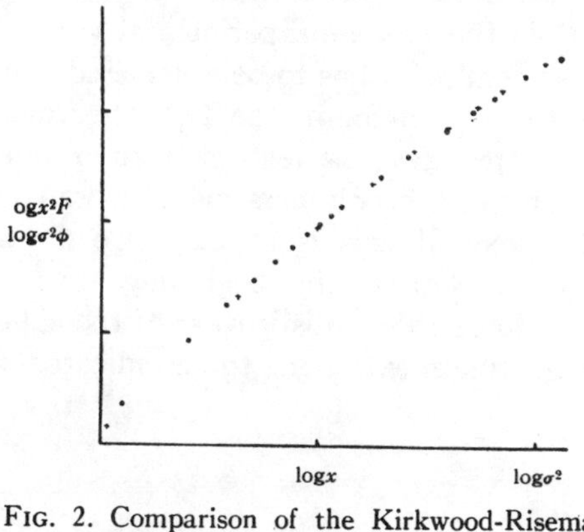

FIG. 2. Comparison of the Kirkwood-Riseman representation of intrinsic viscosity with that of Debye-Bueche. Circles for K.R., cross for D.B.

In making the calculations of the K.R. type a pearl string with small beads is substituted for the polymer and it is assumed that the links connecting the beads do not experience any friction. The very interesting mathematical treatment cannot be carried out quite rigorously; interchanges of averaging processes have to be introduced. In the calculations of the D.B. type a permeable sphere is substituted for the polymer, and its permeability is connected with the monomer number in the polymer. Once the problem has been formulated mathematically, it is solved rigorously.

The question now arises as to which of the two types of representation is better as compared with the experimental facts. Confirming our attention to the intrinsic viscosity alone, K.R. as well as D.B. are satisfied with the performance in the few examples for which sufficiently accurate data are available. It seems, therefore, that we should ask how much difference exists between the two types when each of them is used to represent a definite set of experimental values.

In the K.R. case the intrinsic viscosity is represented by a formula of the type

$$[\eta] = Ax^2F(x), \quad x = BN^{\frac{1}{2}}, \tag{19}$$

in which A and B are adjustable constants and N is the degree of polymerization. The function F and the variable x are those used in the K.R. paper. In the D.B. case the intrinsic viscosity is represented by a formula of the type

$$[\eta] = A^*\sigma^2\phi(\sigma), \quad \sigma^2 = B^*N^{\frac{1}{2}}, \tag{19'}$$

with two adjustable constants A^* and B^*. The function ϕ and the variable σ are those used in this paper.

A comparison of the two types can most easily be carried out, without any reference to specific experimental values, by plotting on one sheet of paper $\log(x^2F)$ as a function of $\log x$, and on another transparent sheet $\log(\sigma^2\phi)$ as a function of $\log\sigma^2$. After superimposing the two sheets, the transparent sheet is moved until the two curves are estimated to coincide as well as possible, always keeping the corresponding axes of the two sheets parallel. This has been done using the

F values from the K.R. table and the ϕ values from the D.B. table. The result of this superposition is represented in Fig. 2, the circles are K.R. points, the crosses D.B. points. The length of the decades is indicated on the axes. Since the horizontal axis covers 2 decades, which means a factor 10^4 in the degree of polymerization, and the vertical axis comprises 3 decades, which means a factor 10^3 in the intrinsic viscosity, it seems that only one conclusion can be drawn. Both types of representation show such exceedingly small deviations from each other that with proper adjustment of the constants, A and B in the one case, A^* and B^* in the other, no preference can be given to either. It is, of course, still possible that a set of experimental data may deviate, but if it does so, we have to expect that these deviations will be just as serious in the one case as in the other.

We now come to the interpretation of the constants in terms of the fundamental molecular quantities. One of these quantities is the friction factor of the monomer: ζ in the K.R. paper, f in the D.B. paper. The four authors agree that no theory exists which links this factor adequately with the structure of the monomer and that an attempt to represent it by a Stokes sphere turns out to be very unsatisfactory indeed. The other quantity appears in two forms, in the K.R. theory it is the root mean square distance R of the ends of the linear chain, in the D.B. theory it is the diameter $2R_s$ of the substituted sphere. In this last case a comparison with the R value of a linear chain can only be made for the limiting case of small shielding and then it is found that $2R_s = 1.054R$. It is evident that further progress can only be made if the R value can be determined by some other independent method. As such we now have the interference method based on the difference of the intensity of the light scattered by the solution in different directions,††† for which in appropriate cases also the dependence of the scattered intensity on the wave-length of the light can be substituted. The theory of the interference effect is decidedly more reliable than a theory of viscosity can ever expect to be.

Three sets of dissymmetry measurements performed on polystyrene are known to us. Reducing the experimental sets of data which have been obtained with polymers of different molecular weights to a common molecular weight of 10^6, the following values of R in A units have been found: Doty for polystyrene in toluene 1080, Ewart for polystyrene in carbon tetrachloride 1050, Bueche for polystyrene in benzene 1110. I believe that the very close agreement is accidental to some degree. In the K.R. paper R is found from viscosity measurements equal to 1125, in this paper we found as a result from our discussion of one set of measurements 914. It seems, therefore, that the experimental evidence from light scattering favors the K.R. representation. However, viscosity measurements of Bueche on the same samples on which he measured the light scattering, discussed according to the D.B. schedule, lead to $R = 1030$. Taking the average of all the values without any preference and disregarding the use of different solvents in the interference measurements, since all 3 liquids are good solvents, we arrive at $R = 1050$. As far as the experimental evidence goes we have to conclude that we do not yet have a clear-cut case.

There is general agreement about the one point that polystyrene is much more extended than we would calculate from the usual formula in which free rotation is assumed, since this formula leads to $R = 300$ which is 3.5 times smaller than the average experimental value.

Another final point has to be considered which happens to be in favor of the D.B. treatment. The K.R. treatment is restricted to a linear chain. Many of the polymers may be branched or cross-linked. If this is so or if we do not definitely know from the beginning that the polymer is linear, the substitution of the sphere of the D.B. treatment seems to be indicated.

††† P. Debye, J. Phys. a. Coll. Chem. 51, 18 (1947).

LIGHT SCATTERING IN SOAP SOLUTIONS[1] *

P. DEBYE

Department of Chemistry, Cornell University, Ithaca, New York

Received August 19, 1948

I. INTRODUCTION

The light scattering (Tyndall effect) in colloidal solutions depends primarily on an interference effect originating at the suspended particles. If the particles are small compared with the wave length of the incident light each acts as a point source, and the scattering depends largely upon a microscopic heterogeneity of the solution due to fluctuations in concentration. The extent of these spontaneous fluctuations in concentration is a measure of the osmotic work necessary to produce them, and it is found that a simple relation exists between the intensity of the scattered light and the osmotic pressure of the solution, both observed as functions of the concentration.

The fundamental relation on which the method depends may be written in the form:

$$\tau = \frac{32\pi^3 n_0^2 c(dn/dc)^2}{3\lambda^4 N \frac{\partial}{\partial c}(P/RT)} \tag{1}$$

The turbidity, τ, is the extinction coefficient in $cm.^{-1}$ due to the fluctuations in concentration. The meaning of the other symbols is more or less standard, with n as refractive index, c as concentration, dn/dc as refractive index increment, λ as the wave length of the incident light, N as Avogadro's number, and M_2 as the molecular weight of the solute (1). In most practical cases and in dilute solution $n - n_0$ is directly proportional to the concentration. Hence this equation may be written in the form

$$H\frac{c}{\tau} = \frac{\partial}{\partial c}\left(\frac{P}{RT}\right) \tag{2}$$

where

$$H = \frac{32\pi^3}{3}\frac{n_0^2}{N\lambda^4}\left(\frac{n - n_0}{c}\right)^2$$

Also, since

$$\frac{P}{RT} = \frac{c}{M_2} + Bc^2 \tag{3}$$

$$H\frac{c}{\tau} = \frac{1}{M_2} + 2Bc$$

[1] Presented at the Twenty-second National Colloid Symposium, which was held under the auspices of the Division of Colloid Chemistry of the American Chemical Society at Cambridge, Massachusetts, June 23–25, 1948.

*Reprinted from *The Journal of Physical & Colloid Chemistry*, 53, 1-8 (1949).

The constant B is familiar as a measure of the deviation from ideality in the van't Hoff osmotic pressure law. Thus, if Hc/τ is plotted as a function of c, a straight line is obtained with limiting ordinate equal to the reciprocal of the weight-average molecular weight of the solute. Two determinations are required,—those of turbidity and of refractive index as they change with concentrations.

II. EXPERIMENTAL

The method has been applied with marked success in the determination of the weight-average molecular weight of a number of organic high polymers in solution. It is now used with certain soap solutions, with and without added electrolyte, in order to determine the size of the micelles or, as we shall prefer to say, the

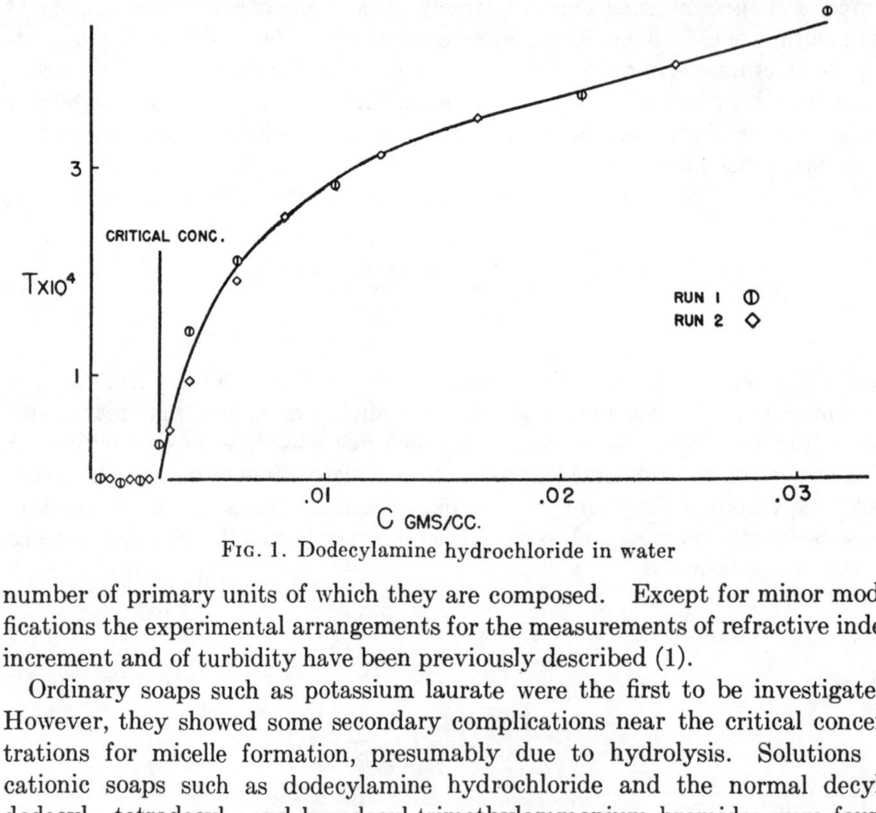

FIG. 1. Dodecylamine hydrochloride in water

number of primary units of which they are composed. Except for minor modifications the experimental arrangements for the measurements of refractive index increment and of turbidity have been previously described (1).

Ordinary soaps such as potassium laurate were the first to be investigated. However, they showed some secondary complications near the critical concentrations for micelle formation, presumably due to hydrolysis. Solutions of cationic soaps such as dodecylamine hydrochloride and the normal decyl-, dodecyl-, tetradecyl-, and hexadecyl-trimethylammonium bromides were found to be free of these complications and were subjected to detailed study. The excess turbidity of these solutions over that of water is nearly negligible up to a concentration which is identical with the critical concentration as determined by the usual methods. From this point on the turbidity increases rapidly and follows a curve similar to that which starts from zero concentration in the case of high-polymer solutions.

Typical data are shown in figures 1 to 3. In the first of these graphs the turbid-

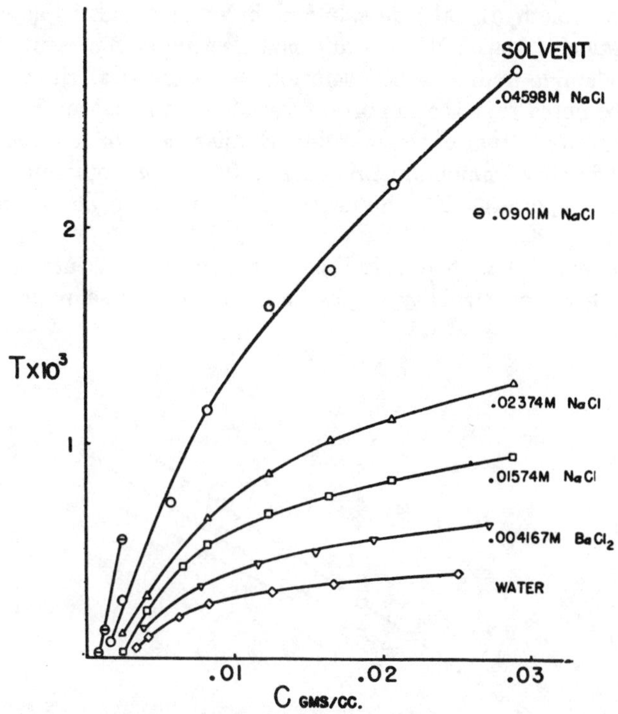

FIG. 2. Dodecylamine hydrochloride in water: effect of added electrolytes

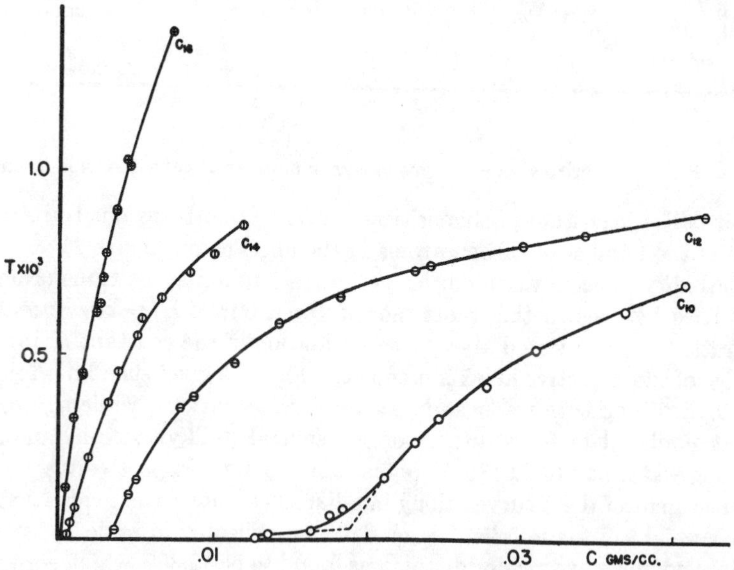

FIG. 3. *n*-Alkyltrimethylammonium bromides in 0.0130 *M* potassium bromide

ity of dodecylamine hydrochloride solutions in water is plotted as a function of the concentration of soap. When additional electrolyte is present the turbidity of the dodecylamine hydrochloride solutions is modified as shown in figure 2. Here it will be noted that the important variable is the concentration of added ion of sign opposite to that of the micelle. Similar data for solutions of the four normal alkyltrimethylammonium bromides in 0.0130 *M* potassium bromide are plotted to form figure 3. The influence of chain length on turbidity is here clearly evident.

It will be observed that there is in each case a critical concentration, which we shall call c_0, where the turbidity begins its abrupt rise as soap concentration is

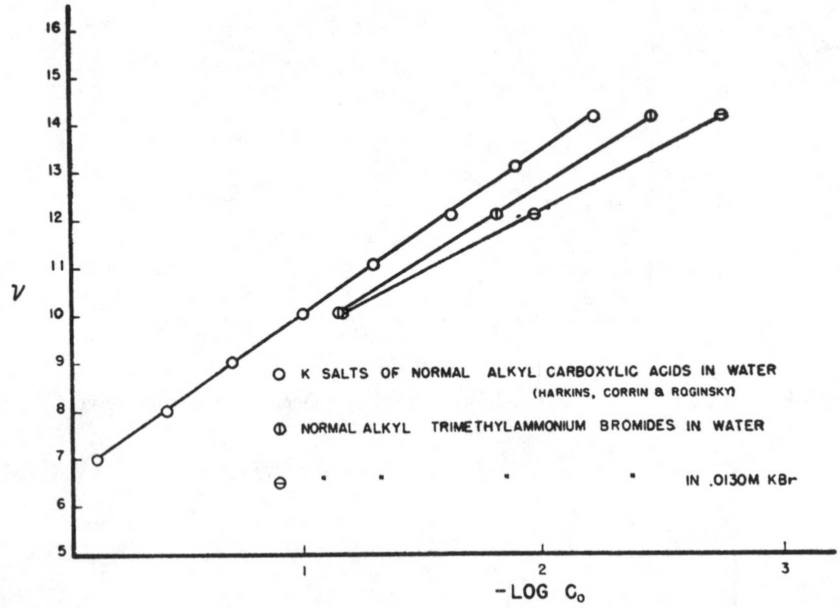

FIG. 4. Plot of log critical concentration *versus* number of carbon atoms in chain

increased. The logarithm of this critical concentration turns out to be a linear function of the number of carbon atoms in the alkyl chain (figure 4).

The turbidity–concentration curves of figures 1 to 3 may be transformed into straight lines by constructing plots, not of Hc/τ, but of $H(c-c_0)/\tau$ *versus* soap concentration. As indicated above, the evaluation of the constant H involves a knowledge of the refractive index increment. Figure 5 gives such lines for dodecylamine hydrochloride in water and in several salt solutions, while figures 6 and 7 present similar data for solutions of the several *n*-alkyltrimethylammonium bromides in water and in 0.0130 *M* potassium bromide, respectively.

The treatment of these curves along familiar lines leads to an evaluation of the molecular weight of the micelle. Such data are collected in table 1.

For dodecylamine hydrochloride this was found to be 12,300, which corresponds to a conglomerate of 55 single dodecylamine hydrochloride molecules. It is well

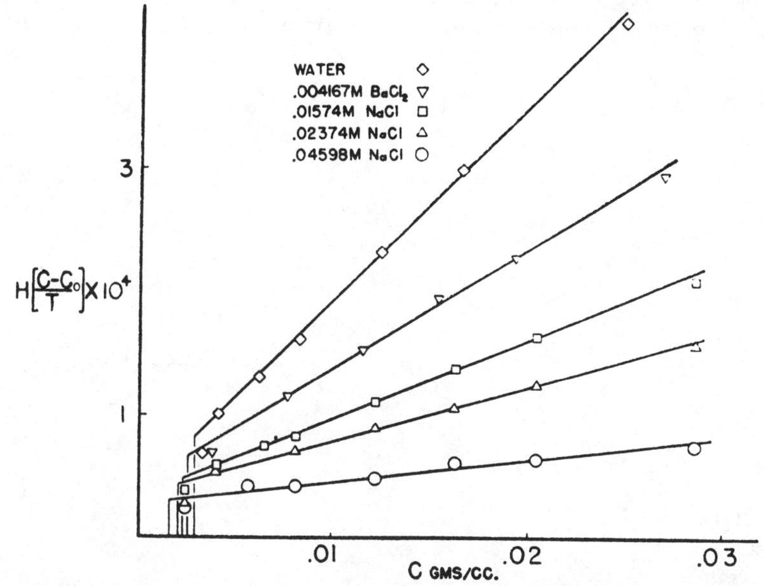

FIG. 5. Plot of $H(c - c_0)/\tau$ *versus* soap concentration for dodecylamine hydrochloride in water and in several salt solutions.

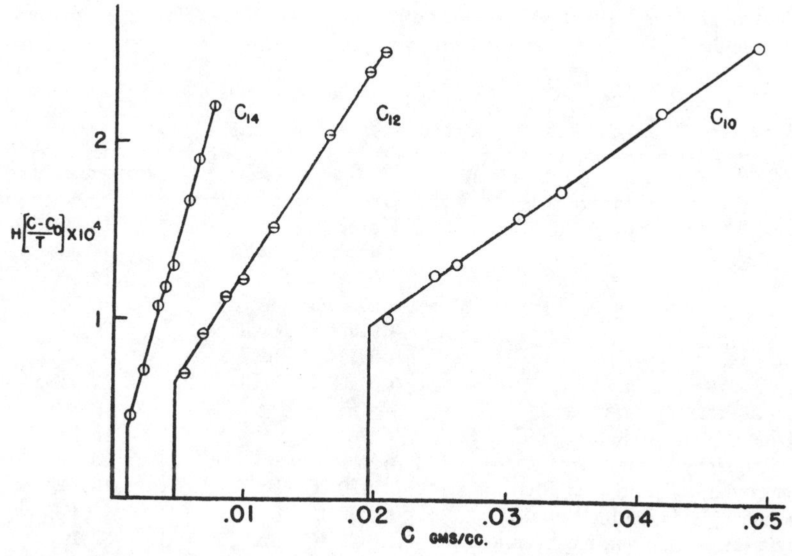

FIG. 6. Plot of $H(c - c_0)/\tau$ *versus* soap concentration for solutions of *n*-alkyltrimethylammonium bromides in water.

known that the critical concentration as determined by the usual methods decreases if salts like sodium chloride, barium chloride, or lanthanum chloride are added. It was found that the same is true for the critical concentration determined by light scattering. Moreover, it was found that the molecular weight is raised markedly by the addition of salt. In a 0.046 molar solution of sodium

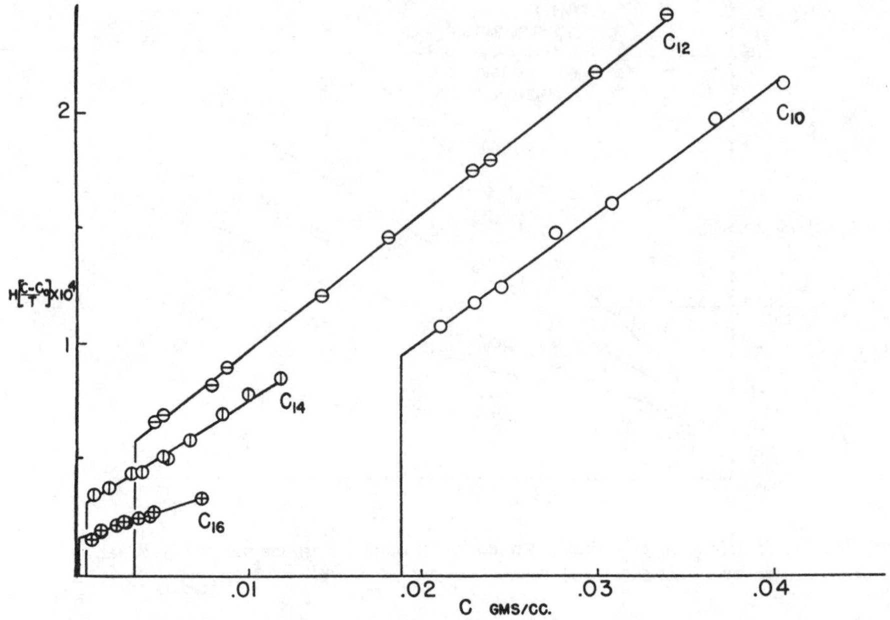

FIG. 7. Plot of $H(c - c_0)/\tau$ *versus* soap concentration for solutions of *n*-alkyltrimethylammonium bromides in 0.0130 *M* potassium bromide.

TABLE 1

Molecular weight data for several cationic soaps

SOAP	SOLVENT	MOLECULAR WEIGHT
Dodecylamine hydrochloride	Water	12,300
	0.00204 *M* $LaCl_3$	14,500
	0.00417 *M* $BaCl_2$	14,600
	0.00264 *M* $LaCl_3$	15,100
	0.01574 *M* NaCl	20,500
	0.02374 *M* NaCl	22,400
	0.04598 *M* NaCl	31,400
n-Decyltrimethylammonium bromide	Water	10,200
n-Dodecyltrimethylammonium bromide	Water	15,500
n-Tetradecyltrimethylammonium bromide	Water	25,300
n-Decyltrimethylammonium bromide	0.0130 *M* KBr	10,700
n-Dodecyltrimethylammonium bromide	0.0130 *M* KBr	17,400
n-Tetradecyltrimethylammonium bromide	0.0130 *M* KBr	32,100
n-Hexadecyltrimethylammonium bromide	0.0130 *M* KBr	61,700

chloride it is 31,400 instead of 12,300 as in pure water. For the molecular weight as well as for the critical concentration itself only the concentration of the chloride ions is important, whereas the nature or the valency of the compensating ion (which has the same sign of charge as the micelle) is immaterial. At the same time other negative univalent ions can be substituted for the chloride ions without changing materially either the critical concentration or the molecular weight.

Comparing dodecylamine hydrochloride with its homologues of varying lengths of hydrocarbon chains it was found that the decrease of the critical concentration with increasing number of CH_2 groups in the chain is coupled with an increase in the number of single molecules contained in the micelle.

III. THEORETICAL

In accordance with thermodynamic principles the gain of free energy associated with the formation of a micelle can be calculated from the critical concentration. The light-scattering measurements establish the number of monomers contained in the micelle. A tentative theory of the structure of the micelle is now proposed which is based upon the double-layer or "sandwich" model which has been recently adopted by Harkins and associates (2). Here it is assumed that the electrically charged "heads" of the molecules occupy the flat sides, whereas the interior is made up of the long hydrocarbon "tails." Work has to be done in order to bring the charged heads together; on the other hand, energy is gained when the hydrophobic tails leave the water and reach hydrocarbon surroundings. Thus the micelle structure represents an equilibrium between the repulsive long-range Coulomb forces due to the charges concentrated in the "heads" of the soap molecules and the short-range attractive van der Waals forces which come into play when the "tails" are brought from the surrounding water into the hydrocarbon part of the micelle structure. If N is the number of single molecules in the micelle it can be shown that the work W_e done against the electrical Coulomb forces in assembling the micelle is proportional to $N^{3/2}$:

$$W_e = N^{3/2} w_e$$

In this relation w_e is a fundamental electrical energy which can be expressed in the ionic charge of the soap molecule and the distance of the heads of the layer. At the same time molecular work is gained due to the short-range forces and this work is proportional to N.

$$W_m = -N w_m$$

in which w_m is a fundamental molecular energy. The curve for the total work $W = W_e + W_m$ has a minimum for a certain $N = N_0$, and this minimum is taken to represent the equilibrium structure. If W_0 is the total work done (a negative quantity) in assembling, it is found that

$$\frac{w_m}{w_e} = \frac{3}{2} N_0^{1/2} \quad \text{and} \quad w_m = -3 \frac{W_0}{N_0}$$

From the light-scattering experiments the number N_0 has been established. The energy W_0 can be calculated from the critical concentration according to classical thermodynamic principles. In the case of dodecylamine hydrochloride with a critical concentration of 0.0131 molar, we have $N_0 = 55$ and $W_0/N_0 = -8.3kT$. This leads (at 25°C.) to

$$w_m = 25kT \qquad \text{and} \qquad w_e = 2.2kT$$

The question is whether these values are reasonable. The heat of vaporization at 25°C. of nonane is 11,099 kg.-cal./mole and the increase for one CH_2 group is 1.18 kg.-cal./mole (3). Extrapolating to dodecane this corresponds to a heat of vaporization per dodecane molecule of $25kT$

According to Harkins the surface area occupied by one molecular head is 27 sq. Å. Taking for the charge the electronic charge and assuming for the effective dielectric constant the average of that of water and that of a hydrocarbon leads to $w_e = 2.8kT$.

From thermodynamic principles coupled with the assumption that w_m is a linear function of the number of CH_2 groups in the tail it follows that the log of the critical concentration should be a linear function of the number of these groups. This checks with the experimental facts (figure 4) and the observed slope of this straight line corresponds per CH_2 group to $w_m = 2.14kT$, whereas the increment in heat of vaporization of 1.18 kg.-cal./mole mentioned before corresponds to $2.0kT$.

If electrolyte is added to the soap solution we have to expect that, owing to the same shielding effect which is of fundamental importance in the interionic attraction theory of strong electrolytes, the electrical work W_e will diminish. This means that the minimum in the curve for the total work W will be displaced to larger values of N. So the micelle will increase in size as observed. So far, however, the quantitative aspect of this question has not yet been worked out in detail.

The independence of the micelle structure on the kind or valency of added ions with the same sign of charge seems peculiar at first sight, especially if one is confronted with the principle of ionic strength. However, it should not be forgotten that the formulation of this principle is linked essentially with the assumption that the electrical work performed by the charges in changing their position within the solution always is small compared to kT. This assumption is certainly not valid for ions which approach the micelle. A calculation according to Gouy reveals that the potential at the micelle should be of the order of 200 mv., and this means that at this point the concentration of ions of equal sign will be at least 3×10^{-4} times smaller than in the bulk of the solution. Keeping this in mind it seems probable that eventually also this point will fall in line.

This research was supported by the Office of Rubber Reserve. It was carried out experimentally by Mr. E. W. Anacker, who, in turn, was assisted in part by Mr. R. M. Hagen, holder of a Procter & Gamble Fellowship.

REFERENCES

(1) Debye, P.: J. Applied Phys. **15**, 338 (1944); J. Phys. Colloid Chem. **51**, 18 (1947).
(2) Mattoon, R. H., Stearns, R. S., and Harkins, W. D.: J. Chem. Phys. **15**, 209 (1947).
(3) U. S. Bureau of Standards, American Petroleum Institute Research Project No. 44.

LIGHT SCATTERING IN SOAP SOLUTIONS **

By P. Debye*
Baker Laboratory, Cornell University, Ithaca, New York

Introduction

Soap solutions exhibit even lower osmotic activity than would be predicted if one assumed that soap existed in solution as simple undissociated molecules. Soap solutions also conduct the electric current far better than would be expected from the observed osmotic effects. Attempting to explain these anomalies, McBain,[1] in 1913, suggested that the fatty soap ions aggregated in solution. Such colloidal aggregations of ions, which were termed micelles, would explain the low osmotic activity and relatively high conductivity of soap solutions.

Since 1913, investigators have shown considerable interest in the determination of the size and shape of the micelle. McBain[2] proposed two different micelle species, which he said could coexist in solution: one a small, spherical, hydrated, ionic micelle, and the other a large, lamellar, weakly conducting micelle. While agreeing with McBain that the behavior of soap solutions pointed to the existence of micelles, Hartley[3] took the view that only the small spherical micelle was feasible. On the basis of geometrical considerations, Hartley[4] calculated that the micelle of a 16 carbon soap consisted of approximately 50 cetyl chains. He and Runnicles[5] carried out diffusion experiments with cetyl pyridinium chloride and calculated from their results that the micelle of this soap contained about 70 paraffin chains.

Ultracentrifuge and diffusion measurements by Miller and Andersson[6] on Duponal (sodium salts of sulfated aliphatic alcohols of chain length C_8 to C_{18}) led to a molecular weight of 12,500 for the mixed micelle.

Hakala[7] has made diffusion measurements on sodium dodecyl sulfate solutions. If a spherical model for the micelle is assumed, the introduction of his results into the Stokes-Einstein equation gives a value of 23.6 Å for the radius. A molecular weight of about 25,000 (87 paraffin chains per micelle) is obtained from this value of the radius if a density equal to that of dodecane is taken.

Vetter[8] studied the sodium salt of sulfonated di (2-hexyl) succinate, known commercially as Aerosol MA, and from density, viscosity, and diffusion data calculated an aggregation number of 24 for the micelle.

Gonick and McBain[9] obtained cryoscopic evidence of micelle formation in aqueous solutions of several non-ionic detergents. Assuming ideal behavior, their data indicate that a micelle consists of no more than 7 detergent molecules.

* All the measurements quoted in this paper have been made by E. W. Anacker in the Chemistry Department of Cornell University. His work was supported by the Office of Rubber Reserve.

**Reprinted from *Annals of the New York Academy of Sciences*, 51, 575-592 (1949).

X-ray work of various German investigators has been cited by McBain[10] as proof of the presence of lamellar micelles in solution. The more recent X-ray work of Harkins[11] and co-workers at the University of Chicago tends to disprove the lamellar micelle theory. Their results may be interpreted as showing that the micelle consists of a double layer of soap molecules, roughly cylindrical in shape. According to this model, the hydrocarbon chains in each layer are in alignment and form the body of the cylinder. The polar groups make up the ends of the cylinder. The Chicago group has also calculated from their data the number of soap molecules per micelle. Their values range from 30 molecules per micelle for a 9 per cent by weight potassium laurate solution to 270 molecules per micelle for a 14.1 per cent solution of dodecylamine hydrochloride.

It should be stated, at this point, that the diffusion and ultracentrifuge investigations mentioned were carried out in aqueous salt solutions of the soaps. Hakala, Hartley and Runnicles, and Vetter found that, above a certain salt concentration, the diffusion coefficients were constant and independent of salt or soap concentration. According to Vetter,[8] "This behavior indicates constancy of micelle size, micelle shape, and degree of solvation. . .if these three factors do not change simultaneously in such a manner that they neutralize their separate effects. The latter possibility seems highly improbable." The work of the Chicago group indicates that the micelle increases in size with increasing soap concentration. Light scattering work seems to indicate that the size of the micelle is constant in a given solvent in at least a range of soap concentrations above the so-called "critical concentration" for micelle formation. It also appears that the size of the micelle is dependent upon the salt concentration of the solvent.

It would not be correct to say that the results from the various methods are in complete disagreement, since all of the investigations mentioned were conducted under different conditions and, for the most part, with different soaps. If the soaps used were the same, the concentrations most likely were not. Soap solutions of at least 9 per cent by weight were used in the X-ray work of the Chicago group, whereas diffusion experiments by Hakala were carried out with solutions whose soap concentrations were about 1.5 per cent or less.

Light scattering offers an independent method of determining the number of solute particles in a given solution and thus enables one to determine the molecular weight of the solute. The quantities which have to be measured for such a determination are turbidities of a series of solutions at different concentrations and the difference in refractive index between the solvent and one solution of known solute concentration. If the particles under investigation have a dimension larger than one-twentieth of the wave length of the light employed, an additional quantity, the dissymmetry coefficient, must be found. This is not necessary for the case in hand.

The turbidity τ is defined as the fractional decrease of the intensity I of the incident beam of light due to scattering per unit length l of its path

$$\tau = -\frac{1}{I}\frac{dI}{dl}. \qquad (1)$$

For small particle dimensions, such as are encountered in soap solutions, the angular distribution of the scattered light does not depend on the properties of the particle and a measurement of the intensity scattered at a 90° angle to the incident beam is adequate.

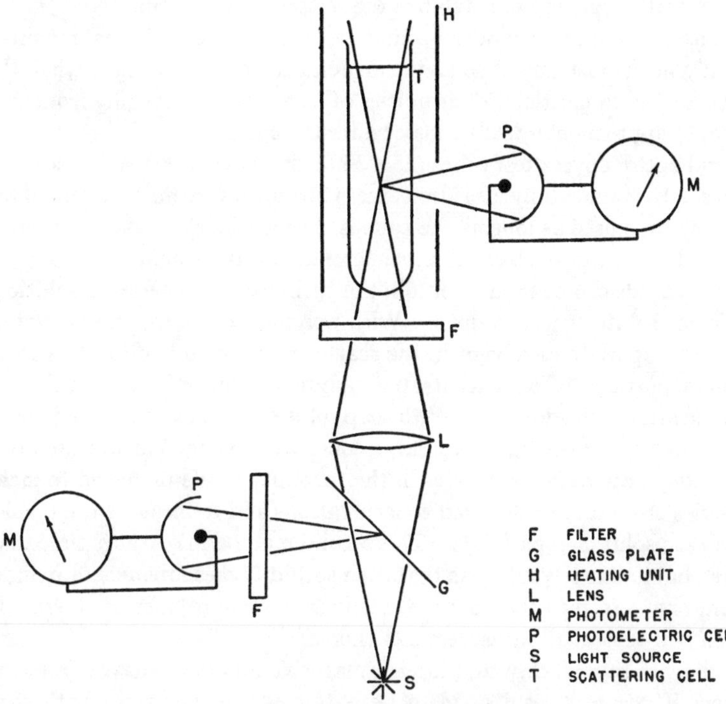

FIGURE 1. 90° scattering instrument.

We have used the simple fluorimeter type of instrument illustrated schematically in FIGURE 1.

The light source is a low pressure mercury arc (G.E. AH-4), which is located at the bottom of the instrument. The primary beam passes upward through a lens and monochromatizing filters and is focused in the center of the scattering cell, opposite the photo tube. The deflection of the needle of a current amplifier connected to the photo tube is proportional to the intensity of the scattered light and hence to the combined turbidity of the solution and scattering cell. The proportionality constant is obtained by noting the deflection of the needle of the current amplifier when the standard of known turbidity is placed in the instrument. The significant result is the excess turbidity of a solution over that of the solvent, which we obtain by subtracting the turbidity of the scattering cell and solution from the turbidity of the scattering cell and solvent.

In order to detect and correct for fluctuations in the primary intensity, a glass plate is set at an angle of 45° in the path of the primary beam. The light reflected by this glass plate is intercepted by a second photoelectric cell. Readings of an amplifier connected to this photo tube give the intensity of the primary beam. A voltage stabilizer and ballast lamp are included in the electrical circuit to help stabilize the output of the lamp.

An insulated brass cylinder wound with resistance wire is mounted vertically in the instrument so as to surround the scattering tube when the latter is placed in position. Turbidities at temperatures ranging from that of room to approximately 90°C. may be measured.

Round bottom pyrex test tubes (25 x 200 mm.) have been used as scattering cells. Experimentally measured excess turbidities are independent of the particular tube used as long as the tube is thoroughly cleaned and oriented in the instrument in the same manner for each measurement.

Two methods can be used to obtain the turbidities of a series of solutions of different solute concentrations. With polymer solutions, one may start with a 50 cc. portion of solvent in the scattering tube and add to this small measured portions of a concentrated polymer solution. Turbidities are measured after each addition. With soap solutions, it has been found necessary to alter this procedure since large concentration increments are often needed to get appreciable changes in the turbidity. We preferred to make up a series of solutions at desired concentrations and to measure the turbidities of each solution separately. This second method takes more time than the first, but has an advantage in that each turbidity measurement is an independent one. In the other method, the introduction of dust or an error in dilution will violate all measurements thereafter.

Needless to say, all dust and foreign material must be removed from the solutions if accurate results are to be obtained. In our work with soap solutions, we have found it convenient and satisfactory to remove dust by filtering the solutions through a fine sintered glass filter. In scattering measurements, it is a good precaution to distill solvents before use. The water used in our studies of soap solutions was freshly distilled.

In general, $\mu - \mu_0$ for 1 per cent polymer or soap solution is of the order 10^{-3}. Since this difference in refractive index between the solution and solvent occurs squared in the refraction constant H, its value must be determined with some degree of accuracy. Ordinary commercial refractometers are not adequate for this purpose. A suitable differential refractometer has been described by P. P. Debye.[12] We have used such an instrument with occasional minor variations of the cell construction.

Interpretation of Light Scattering

The theory of light scattering has been described in greater detail elsewhere.[13] For randomly distributed, small particles the result of the theory may be described by the equation

$$\tau = \frac{32\pi^3}{3} \frac{\mu_0^2(\mu - \mu_0)^2}{\lambda^4} \frac{1}{n}, \tag{2}$$

where τ denotes the excess turbidity as defined by EQUATION 1, μ is the refractive index of the solution and μ_0 that of the solvent. λ stands for the wave length (in vacuum) of the light and n for the number of particles per cubic centimeter.

Relation (2) may be rewritten in terms of the molecular weight M and the concentration c (in grams per cubic centimeter) as follows

$$\tau/c = HM. \tag{3}$$

Here, the constant of proportionality H must be derived from refraction measurements according to the equation

$$H = \frac{32\pi^3}{3} \frac{\mu_0^2}{N\lambda^4} \left(\frac{\mu - \mu_0}{c}\right)^2 \tag{4}$$

where N denotes Avogadro's number $(6.0228 \pm 0.0011) \times 10^{23}$.

According to EQUATION 2, the light scattering from a solution of colloidal particles should be proportional to the concentration. This relation, however, holds only exactly in the limit for infinite dilution. For most solutions of polymers, the scattering per particle tends to decrease with increasing concentrations. A typical case is illustrated in FIGURE 2, where it is seen that the effect is more pronounced the better the solvent. Now EQUATION 2 depends on two assumptions. One is that the refractive index is a linear function of the concentration. This hypothesis can be tested directly by refraction measurements and, in nearly all cases, the linear relation is found valid to a high degree of approximation. The second hypothesis is, that not only in calculating the refraction, but also in the discussion of the total scattering the particles can be considered to act independently of each other. This last assumption, however, although correct for high dilutions, begins already to fail at rather low concentrations, well within the range in which experiments are performed ordinarily.

In order to discuss this interaction effect, we can go back to a theory developed by Einstein[15] for the additional scattering from a solution as compared to the scattering from the solvent according to which the difference in question is due to the spontaneous fluctuations of the concentration. Their magnitude, in turn, depends on the osmotic properties of the solution. This also implies, of course, that scattering measurements can be used as a substitute for straightforward osmotic measurements. Einstein's result may be written in the form

$$\tau = \frac{32\pi^3}{3} \frac{\mu_0^2}{N\lambda^4} \frac{(c\partial\mu/\partial c)^2}{c\frac{\partial}{\partial c}\left(\frac{P}{RT}\right)} \tag{5}$$

where P denotes the osmotic pressure and $R = Nk$ is the gas constant. This formula is valid as long as the dimensions of the particles are reasonably small compared to λ. The variation of μ with c may be determined in an

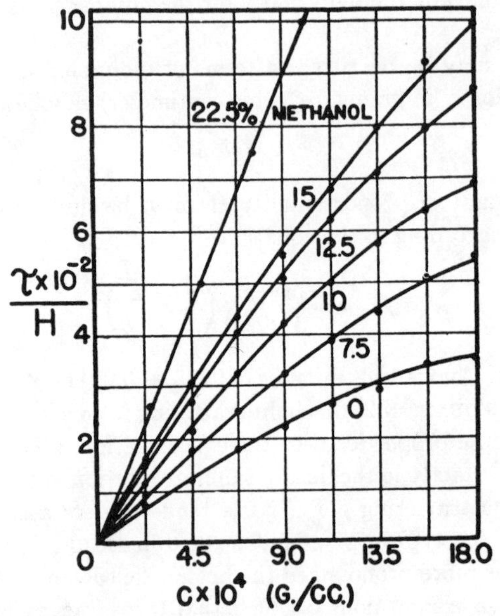

FIGURE 2. Turbidity of polystyrene in benzene-methanol mixtures (MCCARTNEY[14]).

auxiliary measurement; as mentioned above it is usually linear, so that $\frac{\partial\mu}{\partial c}$ may be replaced by $(\mu - \mu_0)/c$. Thus, we are led to transform EQUATION 5 into

$$H\frac{c}{\tau} = \frac{\partial}{\partial c}\left(\frac{P}{RT}\right) \tag{6}$$

in which H is the same refraction constant as defined by EQUATION 4.

For dilute solutions, according to van't Hoff's law, P/RT is equal to c/M and EQUATION 6 simplifies to EQUATION 3 in this case.

The light scattering measurements illustrated in FIGURE 2 confirm what has been inferred from other experiments: that solutions of high polymers commonly exhibit considerable deviations from van't Hoff's law even in rather dilute solutions. It has been found that a two-term expression of the form

$$\frac{P}{RT} = \frac{C}{M} + Bc^2 \tag{7}$$

is often valid over a wide range of concentrations. In such cases, we expect, instead of EQUATION 3, the relation

$$H\frac{c}{\tau} = \frac{1}{M} + 2\,Bc. \tag{8}$$

In FIGURE 3, in which the experimental results of FIGURE 2 have been used in order to plot Hc/τ as a function of the concentration, it is seen that now we obtain straight lines in accord with EQUATION 8.

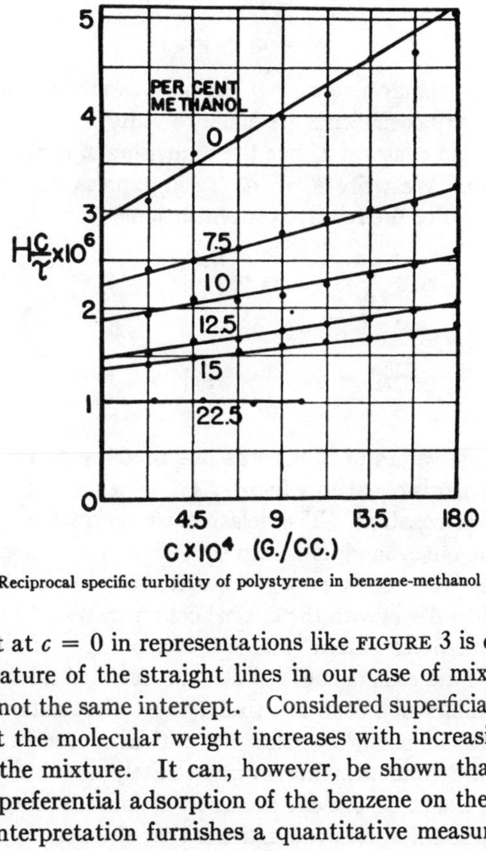

FIGURE 3. Reciprocal specific turbidity of polystyrene in benzene-methanol mixtures.

The intercept at $c = 0$ in representations like FIGURE 3 is equal to $1/M$.

A peculiar feature of the straight lines in our case of mixed solvents is that they have not the same intercept. Considered superficially, this seems to indicate that the molecular weight increases with increasing content of non-solvent in the mixture. It can, however, be shown that the effect is due instead to preferential adsorption of the benzene on the polymer particle[14] and its interpretation furnishes a quantitative measurement of this preference. In simple solvents, no such effect occurs, except for possible real agglomeration of the polymer particles.

Mass Action

The relations valid for high polymers are not directly applicable to soap solutions. However, we shall presently consider the mass action equilibrium between simple ions and micelles, and arrive at an extremely simple modification of these relations.

Consider the following idealized reaction between fatty ions A and micelles A_n, where n is the number of fatty ions per micelle:

$$nA \rightleftarrows A_n \tag{9}$$

If we let c_n be the concentration of micelles, c_1 the concentration of unaggregated paraffin chains, and c the total concentration of fatty ions the following relationships hold:

$$\frac{c_1^n}{c_n} = K \tag{10}$$

$$c = c_1 + nc_n \tag{11}$$

K is the equilibrium constant. It has been assumed that, for this simple treatment, activity coefficients are equal to unity.

The equilibrium constant K has the dimension of a concentration to the power $(n - 1)$. We write $K = c_0^{n-1}$ and express our concentrations as multiples of c_0. For the relative concentrations

$$\gamma_1 = \frac{c_1}{c_0}, \qquad \gamma_n = \frac{cn}{c_0}, \qquad \gamma = \frac{c}{c_0}, \tag{12}$$

the relations

$$\gamma_1^n = \gamma_n, \qquad \gamma_1 + n\gamma_n = \gamma \tag{13}$$

hold.

For very large values of n, it turns out that the relative concentration γ_1 of the monomer is equal to γ for $\gamma < 1$. From $\gamma = 1$ on the concentration γ_1 remains constant. The relative concentration of the polymeric particle, on the other hand, is 0 from $\gamma = 0$ to $\gamma = 1$ and equal to $\gamma - 1$ from there on. It is seen that $\gamma = 1$ corresponds to a critical point and we shall have to identify c_0 with the critical concentration.

For large but finite values of n, the sharp edges of the curves for γ_1 and γ_n at $\gamma = 1$ are rounded off. A straight-forward calculation for $n = 65$ leads to the values of TABLE 1. Under these circumstances, there would be some micelles immediately below the critical concentration (see the first part of the third column), whereas immediately above this concentration, more micelles would be present as expected for n infinite (see the fourth column). These considerations indicate that the soap is practically unaggregated up to the critical concentration c_0 and that, from here on, all but a negligible fraction of the soap which is added in excess will appear in the form of micelles. In FIGURE 4, excess turbidities of solutions of the 10, 12, 14, and 16 carbon *n*-alkyl trimethylammonium bromides above that of the solvent are plotted against total soap concentration *c*. Our qualitative expectations are confirmed. The fact that little or no turbidity difference

TABLE 1

γ	γ_1	$n\gamma_n$	$n\gamma_n - (\gamma - 1)$
0.8408	0.8400	0.0008	—
0.8636	0.8600	0.0036	—
0.8960	0.8800	0.0160	—
0.9234	0.8900	0.0334	—
0.9690	0.9000	0.0690	—
1.0514	0.9100	0.1414	0.0900
1.2079	0.9200	0.2879	0.0800
2.1051	0.9400	1.1651	0.0600
5.5360	0.9600	4.5760	0.0400
18.472	0.9800	17.592	0.0200
34.835	0.9900	33.845	0.0100
66.000	1.0000	65.000	0.0000

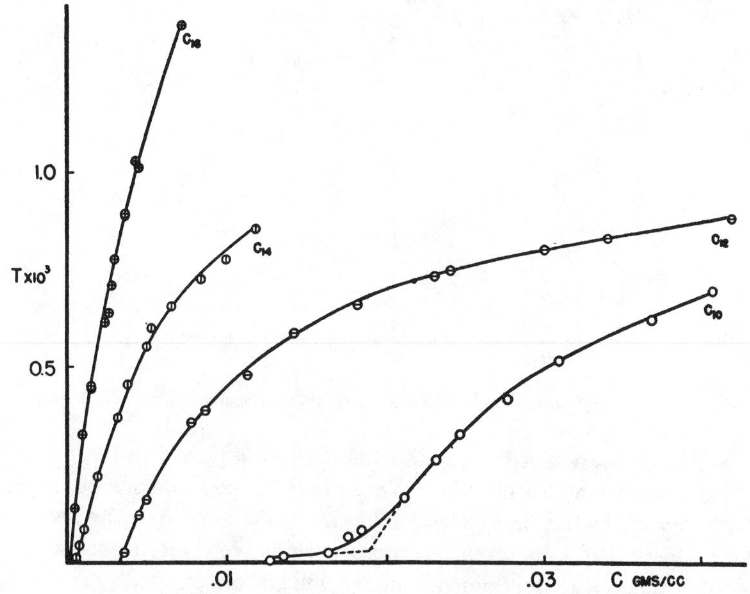

FIGURE 4. Turbidities of *n*-alkyl trimethylammonium bromides in 0.0130M KBr.

between solutions and solvent could be detected means that either no micelles exist at low concentrations or that their number is very small. The rapid rise of the curve for concentrations just above the critical concentration indicates that here the micelles are rapidly increasing in number. From the critical concentration on, the curves resemble typical τ vs. c plots for polymer solutions. This then means that, in treating the light scattering data for soap solutions, we should not plot Hc/τ but $H\frac{(c - c_0)}{\tau}$ as a function of the total soap concentration c; and that we should extrapolate to $c = c_0$ rather than to $c = 0$ in order to obtain the reciprocal of the molecular weight.

In FIGURE 5, data for aqueous solutions of the 10, 12, and 14 carbon trimethylammonium bromides are plotted in accordance with the preceding discussion. The heights of the vertical lines drawn at the critical concentrations represent the reciprocal molecular weights of the micelles. It is readily seen that the molecular weight increases as the hydrocarbon tail of the soap molecule is lengthened.

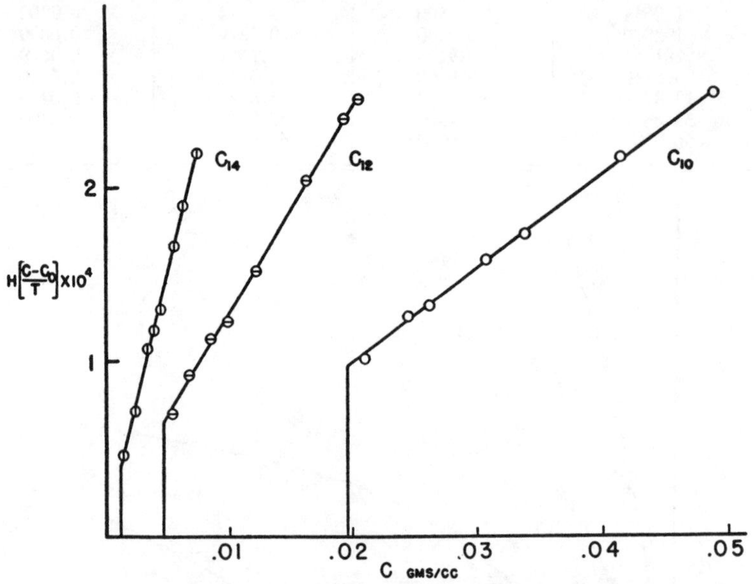

FIGURE 5. Reciprocal specific turbidities of *n*-alkyl trimethylammonium bromides in water.

In FIGURE 6, are plotted the observed data for aqueous and salt solutions of dodecylamine hydrochloride. We see that the addition of salt lowers the critical concentration and increases the size of the micelle. The decrease in slope shows that the solvent becomes progressively more poor as salt is added. If enough salt is added, the soap will be "salted out." The molecular weights (TABLE 2) are plotted against the added chloride ion concentration in FIGURE 7. Although the single run made with $BaCl_2$ solution as the solvent is not sufficient to draw definite conclusions, the results of that run indicate that the size of the micelle is dependent only upon the concentration of the ion opposite in charge to that on the fatty ions comprising the micelle.

The critical concentrations of dodecylamine hydrochloride in the various salt solutions were taken from a curve plotted from data given by Corrin

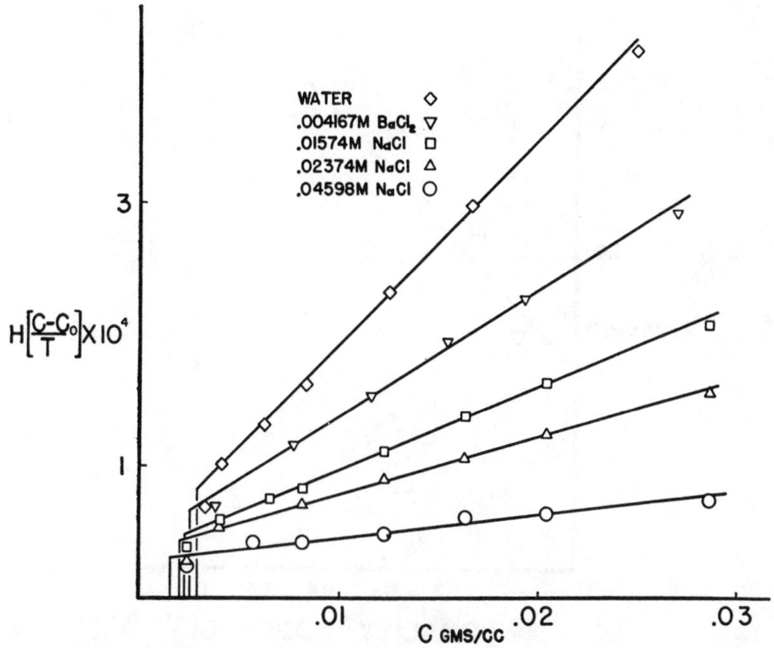

FIGURE 6. Reciprocal specific turbidity of dodecylamine hydrochloride in water and salt solutions.

TABLE 2
DODECYLAMINE HYDROCHLORIDE IN AQUEOUS AND SALT SOLUTIONS

Solvent	*Crit. conc.*	*M.W.*
Water	1.31×10^{-2} M	12,300
.004167 M $BaCl_2$	1.14×10^{-2} M	14,600
.01574 M NaCl	1.04×10^{-2} M	20,500
.02374 M NaCl	9.25×10^{-3} M	22,400
.04598 M NaCl	7.22×10^{-3} M	31,400

and Harkins.[16] All other critical concentrations given hereafter have been determined by a least squares method, which also gives at the same time the best linear plots of $H\frac{(c - c_0)}{\tau}$ vs. c.

In FIGURE 8 are $H\frac{(c - c_0)}{\tau}$ *vs.* c plots for the *n*-alkyl trimethyl bromides in 0.0130 *M*. KBr. From a comparison of this figure with FIGURE 5 and the results listed in TABLE 3, we see that the effect of salt is much more pronounced for the longer chain lengths than for the short. The *n*-decyl soap micelle is scarcely changed in size, while the micelle of the *n*-tetradecyl micelle increases in size by a comparatively large amount.

If we consider the change made in the "total" bromide ion concentration (TBIC) when bromide ion is added in the form of a salt, the resulting changes

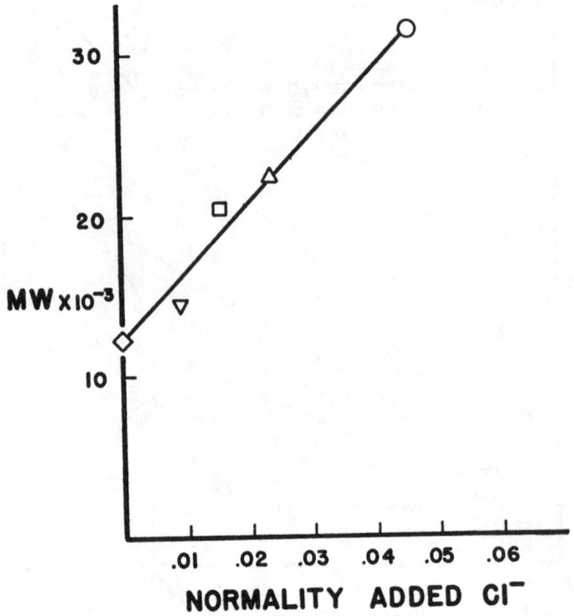

FIGURE 7. Molecular weight of dodecylamine hydrochloride as a function of added Cl^-.

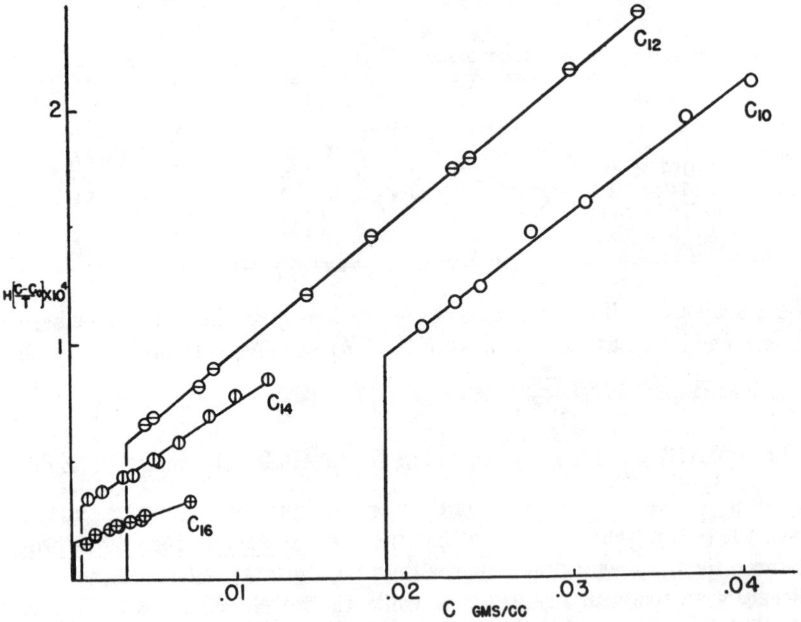

FIGURE 8. Reciprocal specific turbidities of *n*-alkyl trimethylammonium bromide in 0.0130M KBr.

in micelle size for the various soaps are not startling. By TBIC, we mean the sum of the critical concentration and the concentration of the added bromide ion. We are tacitly assuming that the charge of the micelle is zero. To emphasize this point, we might compare the results obtained for C_{10} and C_{14} in water and 0.0130*M* KBr. In the former case, we increase TBIC from 0.0700 to 0.0802 and *N* from 36 to 38. In the latter case, we increase

TABLE 3

n-ALKYL TRIMETHYLAMMONIUM BROMIDES IN WATER AND 0.0130M KBr

Solvent	*Soap*	C_0	*MW*	*Monomeric soap ions per micelle*
Water	C_{10}	0.0700M	10,200	36
	C_{12}	0.0151	15,500	50
	C_{14}	0.00341	25,300	75
0.0130M KBr	C_{10}	0.0672	10,700	38
	C_{12}	0.0105	17,400	56
	C_{14}	0.00176	32,100	95
	C_{16}		61,700	170

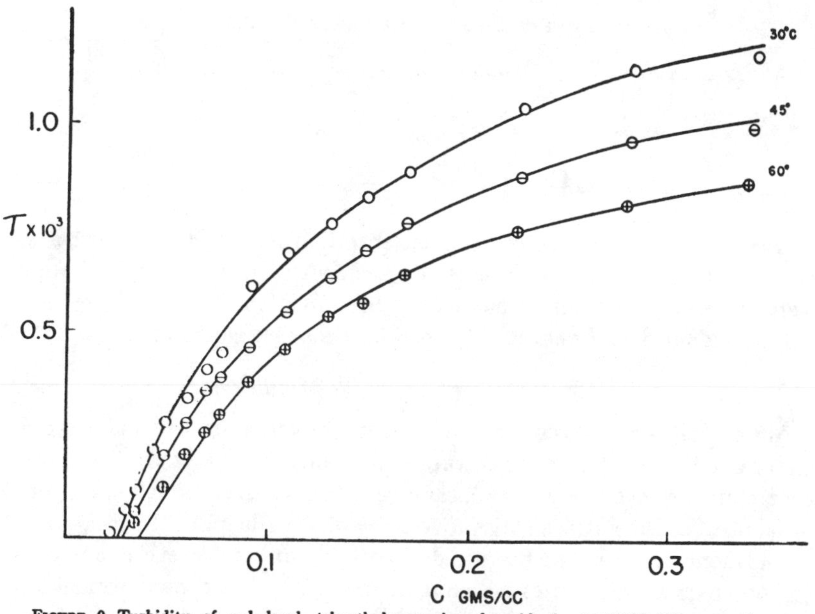

FIGURE 9. Turbidity of *n*-dodecyl trimethylammonium bromide in 0.03403M KBr as a function of temperature.

TBIC from 0.00341 to 0.01476 and *N* from 75 to 95. Thus, in one instance we increase the TBIC by 14.6 per cent while in the other we increase it by 333 per cent. It is therefore not surprising that *N* increases only by 5.5 per cent for the *n*-decyl soap and increases 26.7 per cent for the *n*-tetradecyl soap.

In FIGURE 9, are plotted the results of turbidity measurements carried out with solutions of *n*-dodecyl trimethylammonium bromide in 0.03403*M* KBr at different temperatures. Molecular weights have been calculated on the assumption that H is independent of temperature. Our differential refractometer, in its present form, cannot be used to determine experimentally at high temperatures the factor $\left(\frac{\mu - \mu_0}{c}\right)^2$ which occurs in H. A calculation employing the Lorenz-Lorentz formula and the density data of Scott and Tartar[17] was carried out to determine the approximate change in this factor with temperature. According to this computation, $\left(\frac{\mu - \mu_0}{c}\right)^2$ decreases by 2.5 per cent as the temperature goes from 30° to 60°C. Varying just the last figure of the density data by one changes the percentage decrease to 2.5 ± 3.3 per cent. This shows that the calculation is very uncertain, consequently we have not taken into account any temperature variation of H.

TABLE 4
n-DODECYL TRIMETHYLAMMONIUM BROMIDE IN 0.03403M KBr

Temperature	*Critical concentration*	*MW*
30°C.	0.00848M	20,200
45	0.00931	17,000
60	0.01201	16,400

As seen in TABLE 4, the critical concentration for micelle formation increases with temperature. This is in agreement with the findings of other workers employing different methods of investigation.[18,19]

It is evident that the effect of temperature changes is not large.

Sketch of a Theory of the Micelle

Although it is not possible, at this time, to present a finished quantitative picture of the micelle, its constitution and the meaning of the constants characterizing its size and its appearance, I believe that the following very incomplete considerations may prove to be of some help, in the meantime.

The monomeric ions we have to deal with in our case consist of a hydrocarbon chain which carries a charge at one end. If we want to make a micelle out of a larger number of such ions, which has the general form proposed by McBain and adopted by Harkins and co-workers as a result of their X-ray work, we see that two different kinds of energy will come into play. We shall gain energy because we remove a number of hydrocarbon tails from the surrounding water and bring them in contact with each other in the micelle. The molecular forces of importance in this process are all short range forces often indicated as van der Waal's forces. At the same time, however, we have to bring the charged ends of the monomer nearer to each other until they fill both the flat ends of the sandwich-micelle with a surface density ultimately determined by the space requirements of the monomer tail. This process requires energy and now the forces we have to overcome are long range electrical forces. The interplay between short range and long range forces seems to be responsible for the structure of the micelle, as can be seen from the following calculation.

Suppose a circular disk of radius a covered with a constant surface density of electricity σ. The total charge of this sheet will be proportional to σa^2 and the potential at the rim of the disk will be proportional to σa. If we add to this disk an additional ring of thickness da which has the surface $2\pi a da$ and the charge $2\pi\sigma a da$, we will do work against the coulomb forces which is proportional $\sigma^2 a^2 da$. The total electrical work involved in building up the disk is seen to be proportional to a^3 and, since the number of molecules involved in the micelle arrangement is proportional to a^2, we come to the following conclusion: in order to build a micelle containing N molecules, we have to supply an electrical energy

$$W_e = N^{3/2} w_e \tag{14}$$

in which w_e is a fundamental electrical energy left unspecified for the moment. The main point is that, due to the long range character of the electrical forces, the energy involved increases faster than the number of molecules. We can assume, on the other hand, that the energy gained by bringing N hydrocarbon tails in contact (which involves only short range molecular forces) can be represented as

$$W_m = -N w_m \tag{15}$$

with the introduction of another fundamental molecular energy w_m. If we draw the curve for the total energy $W = W_e + W_m$ as a function of the number of molecules N, it is seen that it has a minimum for a certain value $N = N_0$ and, at this point, the energy W_0 of the micelle is negative. This means that the micelle is more stable than N_0 separate molecules and that work is required in order to either increase or decrease the equilibrium number N_0. From the EQUATION $dW/dN = 0$ and the expression for W, it follows that

$$N_0^{1/2} = \frac{2}{3}\frac{w_m}{w_e}$$
$$W_0 = -N_0 \frac{w_m}{3}. \tag{16}$$

Accepting these relations, we see that we can determine the 2 fundamental constants w_m and w_e of the micelle, if we know N_0 and W_0. We can determine N_0 from light scattering. In the case of dodecylamine hydrochloride in water, N_0 is 55. From the first equation of (16) it follows immediately that

$$\frac{w_m}{w_e} = 11.1 . \tag{17}$$

If we accept the reasoning in the preceding paragraph concerning the application of the mass action law, we can also determine W_0. According to thermodynamical principles, the natural logarithm of the equilibrium constant K multiplied with kT is a measure for the free energy difference between the micelle and the equivalent number of free molecules. Since $K = c_0^{N_0-1}$ we come to the conclusion that

$$W_0 = -N_0 \frac{w_m}{3} = (N_0 - 1)kT \ln c_0 \tag{18}$$

(in which c_0 is of course to be expressed in mol-fractions). If we neglect the difference between N_0 and $N_0 - 1$, we find

$$w_m = -3\, kT \ln c_0. \tag{19}$$

In our case, c_0 is 1.3110^{-2} moles per liter, so we arrive at

$$w_m = 25\, kT \text{ and } w_e = 2{,}2\, kT. \tag{20}$$

We now can, at once, draw a conclusion about the question whether we shall have a broad distribution of micelle sizes or not.

In the vicinity of $N = N_0$, the micelle energy W can be developed in powers of $N - N_0$. Doing this, we find

$$W = W_0 \left[1 - \frac{3}{4}\left(\frac{N - N_0}{N_0}\right)^2\right]. \tag{21}$$

The average natural fluctuations in size will involve energy differences kT. Assuming $W - W_0 = kT$ and remembering that $W_0 = -N_0 \frac{w_m}{3}$, it follows from (21) that the average natural fluctuation of N_0 is

$$N - N_0 = \sqrt{4N_0 \frac{kT}{w_m}}.$$

With $N_0 = 55$ and $w_m/kT = 25$ we have

$$N - N_0 = 2.97$$

which means that the micelles containing each in the average 55 molecules, will have a narrow distribution which practically does not cover much more than the range from 52 to 58.

Next, we want to investigate whether the values found in EQUATION 20 for w_m and w_e are such that we can reasonably expect values like that.

First of all, with respect to w_m, we shall guess that it should not be very different from the heat of vaporization of a dodecane molecule from the liquid. According to the measurements of the National Bureau of Standards, in the American Petroleum Institute Research Project 44, the heat of vaporization, at 25°C., of nonane is 11.099 Kcal/mole and the increase for one CH_2-group is 1.18 Kcal/mole. Extrapolating to dodecane, this corresponds to a heat of vaporization of 14.63 Kcal/mole. Assuming for RT a value of 600 cal., this corresponds to 24.4 kT to be compared with 25 kT. On the other hand, w_e can be calculated for a disk covered with a constant density of electricity in a medium of dielectric constant D. It is found that

$$w_e = \frac{4}{3}\sqrt{\frac{2}{\pi}}\frac{\epsilon^2}{D\sqrt{w}} \tag{22}$$

in which ϵ is the electronic charge and ω the surface occupied by one monomer. Taking a value of 27 Å² for ω, the electronic charge for ϵ, and assuming for the effective dielectric constant the average of that of water and that of a hydrocarbon, we obtain $w_e = 2.8\ kT$. This is to be compared with 2.2 kT.

We can also apply relation (18) in order to make an estimate of the influence of the number of C-atoms in the chain on the critical concentration. Calling ν the number of C-atoms, we know that w_m can very adequately be represented by a linear function of ν. But, if this is so, EQUATION 18 predicts that c_0 will be proportional to an exponential function of Hν, in which the coefficient of ν is the energy gain per CH_2-group in going over from free monomer molecules in water to their associated form in the micelle, divided by 3 kT. Dr. Harkins and his co-workers have published a table of the critical concentration of fatty acid soaps covering the range from $\nu = 7$ to $\nu = 14$. The relation is indeed exponential and a good representation of the experimental values is given by the relation

$$\ln c_0 = 4.811 - 0.714\nu \tag{23}$$

(in which c_0 is taken in mols/liter). This means that the energy equivalent of one CH_2-group is 0.714. $3\ kT = 2.14\ kT$ or in molar quantities $2.14\ RT =$ 1280 cal. To be confronted with this value is the observation that the molar heat of vaporization of hydrocarbons increases 1180 cal. with every added CH_2-group.

It remains to be seen what an added electrolyte will do to the micelle. Since, in an electrolyte solution, every charge will be surrounded by an excess of ions of opposite sign, its electrical actions will be screened out for larger distances. In the theory of strong electrolytes, the characteristic distance is the thickness of the ionic layer which, in water and for a uni-univalent electrolyte, is 100 Å for a 0.001 molar and 10 Å for a 0.1 molar solution. We can say, at once, that an added electrolyte will screen the action of the charges on the micelle, will reduce the electrical work W_e, and therefore will increase the equilibrium size of the micelle as has been observed. The concentration of surface charges on the micelle (one electronic charge on a surface of 27 $Å^2$) is so high that the potential in the neighborhood of this surface will certainly exceed 25 millivolts. This value of 25 millivolts corresponds to kT divided by the charge of the electron and is a kind of fundamental unit for the potential. As long as the potentials which the ions encounter are small compared with 25 millivolts, the solution can be characterized by its ionic strength and the variations in number of positive and negative ions from their equilibrium values can be considered as small. As soon, however, as those potentials are a few times 25 millivolts there will be a huge difference between the concentrations of oppositely charged ions. Some considerations led me to a value of 150 to 200 millivolts, 6 to 8 times the fundamental potential for the potential at the surface of the micelle. This would mean that here the positive ions are so strongly repulsed that their concentration is $e^6 = 400$ to $e^8 = 3000$ times smaller than at larger distances. Perhaps this is the way to understand the fact that foreign ions of the same sign as the micellar ion are of very minor importance for its behavior and that the principle of "ionic strength" (which is essentially linked with the existence of potentials small as compared with 25 millivolts) does not hold.

References

1. McBain. 1913. Trans. Faraday Soc. **9:** 99.
2. McBain. 1944. Alexander's Colloid Chemistry. **5.** Reinhold. New York.
3. Hartley. 1936. Aqueous Solutions of Paraffin-chain salts. Hermann et Cie. Paris.
4. Hartley, Collie, & Somis. 1936. Trans. Faraday Soc. **32:** 795.
5. Hartley & Runnicles. 1938. Proc. Roy. Soc. London. **A168:** 420.
6. Miller & Andersson. 1942. J. Biol. Chem. **144:** 475.
7. Hakala. 1943. Dissertation. Univ. of Wisconsin. Quoted in Ann. N. Y. Acad. Sci. 1945. **46(5):** 326.
8. Vetter. 1947. J. Phy. & Colloid Chem. **51:** 262.
9. Gonick & McBain. 1947. J. Am. Chem. Soc. **69:** 334.
10. McBain. 1942. Advances in Colloid Science. **1:** 124. Interscience Publishers.
11. Harkins. 1948. J. Chem. Phys. **16:** 156.
12. Debye, P. 1948. J. App. Phy. **52:** 260.
13. Debye, P. 1947. J. Phys. & Colloid Chem. **51:** 18.
14. Ewart, Roe, Debye, & McCartney. 1946. J. Chem. Phys. **14:** 687.
15. Einstein. 1910. Ann. Physik. **33:** 1275.
16. Corrin & Harkins. 1947. J. Am. Chem. Sci. **69:** 683.
17. Scott & Tartar. 1943. J. Am. Chem. Soc. **65:** 692.
18. Klevens. 1948. J. Phys. Colloid Chem. **52:** 130.
19. Wright, Abbott, Sivertz, & Tartar. 1939. J. Am. Chem. Soc. **61:** 549.

MICELLE SHAPE FROM DISSYMMETRY MEASUREMENTS[1] *

P. DEBYE AND E. W. ANACKER[2]

Department of Chemistry, Cornell University, Ithaca, New York

Received April 27, 1950

INTRODUCTION

In work (2) previously carried out in this laboratory it was found that the addition of inorganic salts to aqueous solutions of various detergents causes the colloidal aggregates to increase in size. The molecular weights of these aggregates, called micelles, in the presence of low salt concentrations indicate that the micelles are still too small to cause dissymmetry in the intensity of the scattered light. It was one of the aims of the present investigation to check this point.

Solutions of the longer-chain detergents to which relatively large amounts of salt have been added are visibly turbid. The aggregates are undoubtedly large. A second aim of this investigation was to measure the dissymmetry of light scattered by these solutions and to gain thereby some information regarding the shape of the micelle.

THEORY

The question of the shape of the micelle has received considerable attention from many investigators. McBain (7) has proposed models which are spherical and lamellar. Hartley (6) believes that only the spherical model is tenable. Harkins (5) and coworkers have supported a lamellar model on the basis of x-ray measurements. Recently Corrin (1) has stated that the x-ray scattering of soap solutions can also be explained by the spherical micelle. Debye (3) has shown from energy considerations that in aqueous solutions of the detergents or in solutions of low salt concentration the lamellar model is feasible and consistent with light-scattering measurements.

Dissymmetry measurements are incapable of resolving the problem of the shape of the micelle in aqueous solutions or in solutions of low salt concentration. The micelles are too small to cause appreciable dissymmetry in the scattered light. However, in the presence of relatively high salt concentrations the micelles of the longer-chain detergents are large enough to cause measurable dissymmetry in the intensity of the scattered light. This dissymmetry may be used to calculate characteristic dimensions of various models. Assuming a reasonable density of the aggregate, one may calculate from these characteristic dimensions molecular weights for each of the models. These molecular weights in turn may be compared with the molecular weight determined by 90° scattering. The value which agrees best with the experimentally determined molecular weight indicates the model which approximates most closely the correct one.

[1] The work reported in this paper was done in connection with the Synthetic Rubber Program of the United States Government under contract with the Office of Rubber Reserve, Reconstruction Finance Corporation.

[2] Present address: Department of Chemistry, Montana State College, Bozeman, Montana.

*Reprinted from *The Journal of Physical & Colloid Chemistry*, *55*, 644-655 (1951).

The theoretical relationship which connects the intensity of the scattered light, the angle of observation, and a length characteristic of the given model is usually written as a function of the parameter X, which is defined as:

$$X = \frac{2\pi L}{\lambda} \sin \frac{\theta}{2} \tag{1}$$

λ is the wave length of the light in the scattering medium, L is the characteristic dimension of the model, and θ is the angle of observation. One may proceed in two ways to obtain the characteristic dimension from the experimental data. One may plot the intensity of the scattered light, I, as a function of θ for various chosen values of L. From these plots one obtains $I_\theta/(I_{180-\theta})$ for any given pair of angles symmetrical about 90° as a function of L and plots the results. This plot is then entered with the experimentally determined ratio $I_\theta/(I_{180-\theta})$ for the same pair of angles and a value of L is read off immediately. This method has advantages, in that it is relatively rapid and requires no correcting of the observed intensities for polarization and change in scattering volume with angle of observation. This assumes, of course, that the correction factors are symmetrical about 90°.

In the second method, one computes for various values of X corresponding values of I, using the theoretical scattering equation. Then a plot is made of log I *vs.* log X. One then plots the experimentally determined log I as a function of log sin $\theta/2$, using a separate transparent sheet of graph paper. The two plots are superimposed and moved relative to each other, keeping the corresponding coördinate axes parallel, until the best alignment of the curves is found. The separation of the origins on the log X and log sin $\theta/2$ axes is equal to log $2\pi L/\lambda$, from which L may be obtained. This method takes more time than the first to carry out, but it is more accurate, as the intensity values over a considerable range of angles are employed.

EXPERIMENTAL TECHNIQUES

All dissymmetry measurements were made in an instrument recently designed by A. M. Bueche. An AH-4 mercury arc serves as the light source and is housed outside the instrument. The primary light enters the instrument through a pin hole in one end, is collimated by a single lens, and then proceeds through an aperture direct to the scattering cell. The primary beam is trapped after passage through the solution under observation. The pin hole and aperture are adjustable with respect to size, and the aperture and lens with respect to position.

A mirror located outside the cell reflects the scattered light downwards through a monochromatic filter and onto a phototube (electron multiplier type). The intensity of the scattered light is recorded as a deflection of the galvanometer connected to the phototube. The mirror may be moved in a semicircle about the cell to the desired position. The phototube moves as a unit with the mirror.

The scattering cell is an upright glass cylinder, sealed off on the bottom and open at the top. It has two horizontal arms extending outward from opposite sides of the cell and coaxial with the primary beam. A piece of microscope slide glass is sealed to the end of the arm nearer the light source; it serves as the entrance window of the cell. The extremity of the other cell arm is drawn down and to a point to form a light trap. The arms and one side of the cell are painted black. Any light reflected from the cell entrance is not permitted by this design to reach the traveling mirror.

The *n*-tetradecyl- and *n*-hexadecyl-trimethylammonium bromides were of the same lot prepared for earlier work (2). In this investigation they were twice recrystallized from water for additional purification. Stock solutions were prepared from the wet crystals; other solutions of varying detergent concentration were prepared from the stock solutions by dilution. The concentration of detergent in the stock solutions was determined by refractive-index measurements. The potassium bromide was of reagent grade and not further purified. All water was slowly distilled from an alkaline potassium permanganate solution. It showed very little dissymmetry. Since the intensity of the light scattered from this water was little more than the sensitivity of the instrument as used, no values can be given. Except for low concentrations, the intensities of the light scattered from detergent solutions were considerably larger than that from water. It was observed in the course of the work that undistilled water could seemingly be cleaned up to show very little dissymmetry by the addition of small amounts of salt and subsequent filtration. This observation was not investigated thoroughly.

All solvents and solutions were filtered slowly through an ultra-fine Pyrex sintered-glass filter into receiving tubes from which they were poured into the scattering cell. Receiving tubes were cleaned in dichromate solution, rinsed with tap water, and then copiously rinsed with the carefully distilled water. Steaming out by the use of external heat concluded the treatment. The scattering cell was rinsed repeatedly with solvent until no further decrease in intensity of scattered light at 40° was observed. It was then filled with the detergent solution of lowest concentration. After all data for this solution had been obtained, the detergent solution was poured out of the cell and the solution of immediately higher concentration poured in. Some error in detergent concentration, due to dilution, occurs in this procedure, but contamination of the cell by dust is kept to a minimum. Since the cell takes a relatively large amount of solution (60 ml.) and the difference in concentration between successive solutions is taken small, it is felt that the dilution error is not of serious consequence. On several occasions two detergent solutions of the same concentration have been introduced successively into the scattering cell after a solution of lower detergent concentration had been in the cell. Little or no difference in the intensities of the light scattered from the solutions of identical detergent concentration was observed, indicating that the error of dilution is not serious.

Intensity measurements for both solutions and solvents were made at 10° intervals from $\theta = 40°$ to $\theta = 140°$, inclusive. θ is the angle of observation, as measured in the horizontal plane from the forward direction of the primary beam. Turbidities (as measured by 90° scattering) were obtained by comparing the 90° galvanometer deflections with the 90° galvanometer deflection produced by the light scattered from a Styron block placed in the primary beam. The

Styron block was calibrated with the aid of a solution of polystyrene in toluene. The difference in turbidity between the polystyrene solution and toluene was determined in the absolute intensity camera designed by P. P. Debye (4).

All measurements were made with the mercury blue line, $\lambda = 4358$ Å. Because unpolarized light was used and because the scattering volume is a function of angle, the observed intensities must be corrected. This may be accomplished in one of two ways. One may calculate experimental correction factors from the intensity *vs.* angle curve obtained from a solution which has no dissymmetry. Or one may correct for unpolarized light with the factor $1/(1 + \cos^2 \theta)$ and for different scattering volumes with the factor $\sin \theta$. Both methods are open to objections. Because the mirror accepts scattered light over a small range of angles instead of at a given angle, the experimentally determined correction factor will be somewhat dependent upon the refractive index of the solution used. The correction factor $\sin \theta/(1 + \cos^2 \theta)$ does not take into account the refractive-index difference between the solution and the glass of the cell and may not completely correct for scattering volume differences. Throughout this work the correction factors calculated from $\sin \theta/(1 + \cos^2 \theta)$ have been used.

DISCUSSION AND RESULTS

In table 1 are given the results of two runs with *n*-tetradecyltrimethylammonium bromide in 0.202 *M* potassium bromide. θ is the angle of observation, I is the galvanometer deflection in excess of that produced by the solvent alone, and $I_\theta/(I_{180-\theta})$ represents the dissymmetry for a pair of angles symmetrical about 90°. Dissymmetry measurements on solutions of this same detergent at lower detergent and lower potassium bromide concentrations were also made. The values of I_{40}/I_{140} for these runs are given in table 2.

Unless a scattering particle has some dimension greater than about 0.1 the wave length of the exciting radiation, no dissymmetry in the intensity of the light scattered by it will be detected. The observance of extremely little or no dissymmetry in the systems just mentioned enables us to say that, aside from the possibility of the existence of a minute fraction of large particles, the micelles have a maximum dimension no greater than 330 Å.

In earlier work (2) it was found that the effect of a given amount of added salt in increasing the molecular weight of the micelle was more pronounced the longer the hydrocarbon tail of the detergent molecule. It was found, for example, that *n*-decyltrimethylammonium bromide increased in molecular weight from 10,200 to 10,700 when the solvent was changed from water to 0.0130 *M* potassium bromide. For the same shift in solvents the molecular weight of the micelle of *n*-tetradecyltrimethylammonium bromide increased from 25,300 to 32,100. Because of this dependence of molecular weight on chain length, the sixteen-carbon-atom trimethylammonium bromide was picked for dissymmetry work at the high salt concentrations.

It was found necessary to conduct the experiments with *n*-hexadecyltrimethylammonium bromide at temperatures somewhat above 30°C., since precipitation readily occurred below this temperature. All measurements were made at

TABLE 1

n-Tetradecyltrimethylammonium bromide in 0.202 M potassium bromide

1.1 PER CENT C_{14}		θ	2.1 PER CENT C_{14}	
I	$\frac{I_\theta}{I_{180-\theta}}$		I	$\frac{I_\theta}{I_{180-\theta}}$
5.64	1.027	40	11.61	1.017
4.26	1.012	50	8.69	0.992
3.33	1.006	60	6.97	1.007
2.73	1.000	70	5.72	1.004
2.41	1.004	80	5.03	1.000
2.28	1.000	90	4.81	1.000
2.40		100	5.03	
2.73		110	5.70	
3.31		120	6.92	
4.21		130	8.76	
5.49		140	11.41	

TABLE 2

n-Tetradecyltrimethylammonium bromide

SYSTEM	$\frac{I_{40}}{I_{140}}$
0.13% C_{14} in 0.083 *M* KBr....................................	1.005
0.16% C_{14} in 0.083 *M* KBr....................................	1.024
0.18% C_{14} in 0.083 *M* KBr....................................	1.028

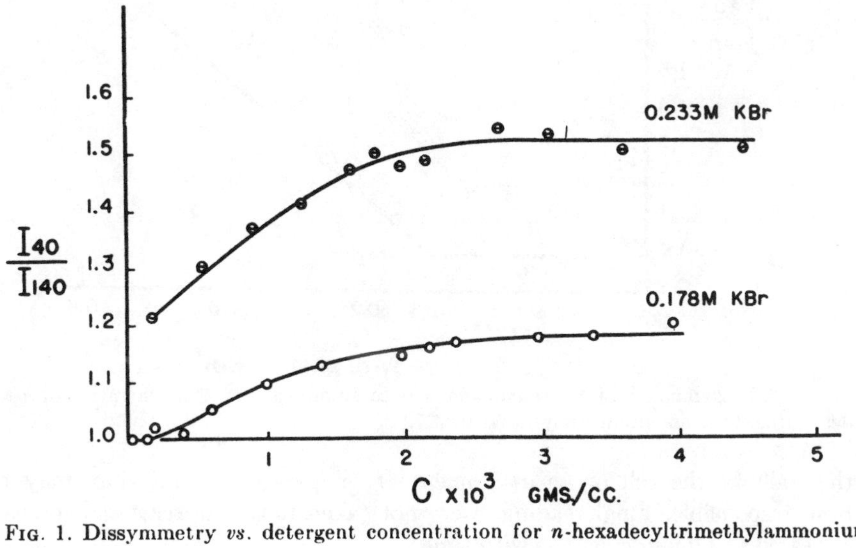

FIG. 1. Dissymmetry *vs*. detergent concentration for *n*-hexadecyltrimethylammonium bromide.

34°C. ± 1°. At the higher soap concentrations it was found that the intensity of the scattered light decreased for a period of time after the solutions were placed in the scattering cell. Mixing with the previous solution still clinging to

TABLE 3

n-Hexadecyltrimethylammonium bromide

SOLVENT: 0.178 *M* KBr			SOLVENT: 0.233 *M* KBr		
C_{16} CONCENTRATION	$\tau \times 10^3$	$\frac{I_{40}}{I_{140}}$	C_{16} CONCENTRATION	$\tau \times 10^3$	$\frac{I_{40}}{I_{140}}$
g./ml.	*cm.*$^{-1}$		*g./ml.*	*cm.*$^{-1}$	
0.000039	0.074	1.00	0.000179	0.389	1.215
0.000138	0.179	1.00	0.000537	2.15	1.304
0.000198	0.205	1.02	0.000896	4.08	1.372
0.000395	0.606	1.01	0.001254	6.27	1.414
0.000593	1.29	1.05	0.001612	9.15	1.474
0.000988	2.42	1.098	0.001791	10.0	1.503
0.001384	3.79	1.129	0.001970	10.7	1.480
0.001977	6.48	1.147	0.002149	11.0	1.489
0.002175	7.49	1.161	0.002687	15.1	1.545
0.002372	8.28	1.170	0.003045	17.3	1.535
0.002966	10.4	1.178	0.003582	18.0	1.508
0.003361	12.1	1.181	0.004478	21.4	1.511
0.003954	14.5	1.202			

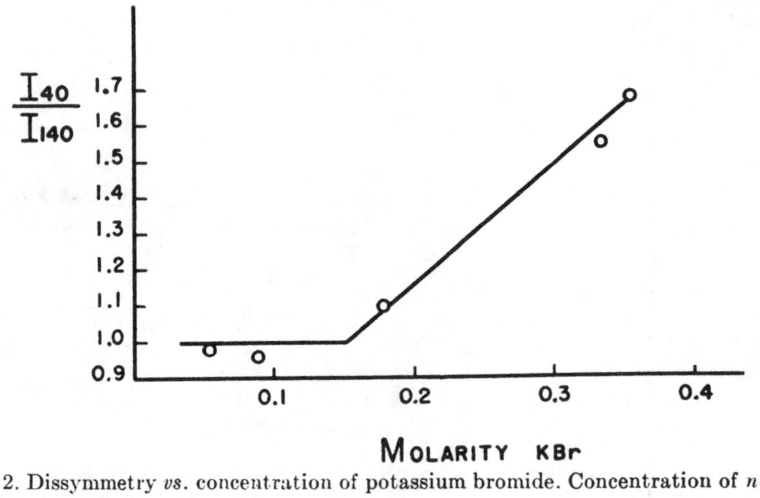

FIG. 2. Dissymmetry *vs.* concentration of potassium bromide. Concentration of *n*-hexadecyltrimethylammonium bromide = 0.00272 *M*.

the walls of the cell or an attainment of temperature equilibrium may have been responsible. Final readings were not taken until the scattered intensities appeared to remain constant with time.

In figure 1 are plotted experimentally determined ratios of I_{40}/I_{140} as a function of detergent concentration for two separate runs at different potassium bromide concentrations. The striking feature of these plots is the rise in dissymmetry as the detergent concentration increases. An initial drop in the dissymmetry curve before the levelling-off process takes place is the usual behavior. The limiting value of the dissymmetry is theoretically characteristic of the scattering particle. In the case at hand the micelles evidently increase rapidly

TABLE 4
n-Hexadecyltrimethylammonium bromide
Concentration = 0.00272 *M*

KBr CONCENTRATION	$\frac{I_{40}}{I_{140}}$
M	
0.053	0.98
0.089	0.96
0.178	1.10
0.335	1.55
0.356	1.68

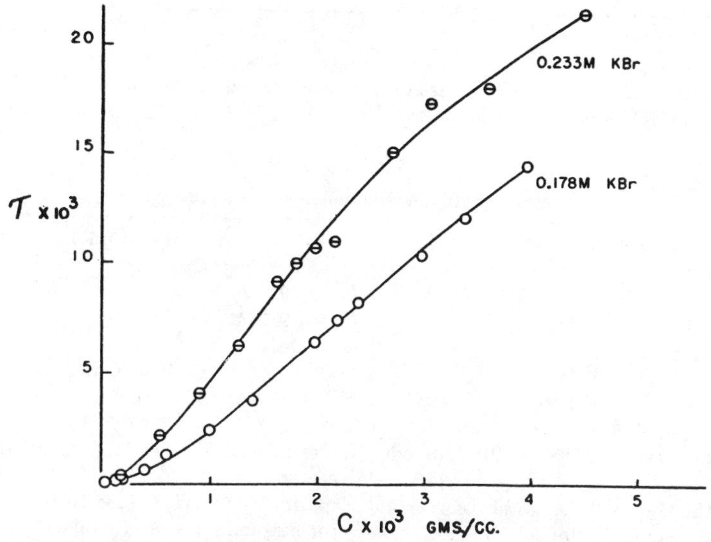

FIG. 3. Turbidity, τ, from 90° scattering measurements *vs.* concentration of detergent (*n*-hexadecyltrimethylammonium bromide).

in size in a concentration range starting roughly at the critical concentration. Presumably only after a certain detergent concentration is reached do the micelles attain constant size. In this work intensity curves obtained at the highest detergent concentrations on the relatively level portions of the curves have been used in the determinations of the characteristic dimensions. The experimental data used to plot figure 1 are given in table 3.

In figure 2 are plotted the results of an experiment (data in table 4) in which the detergent concentration was maintained constant and the concentration of the added potassium bromide varied. The increase in dissymmetry with increase in salt concentration is independent proof that salt increases the size of the micelle.

Turbidity *vs.* concentration plots as obtained from 90° scattering (data in table 3) make up figure 3. Considerable rounding off in the region of the critical concentration is evident. In figure 4 are the corresponding $H(C - C_0)/\tau$ plots. H is the well-known refraction constant (2). C_0, the critical concentration, was

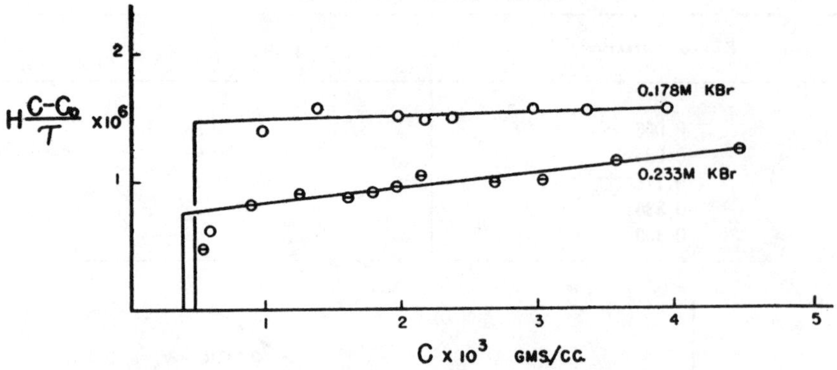

FIG. 4. Plots of $H(C - C_0)/\tau$ *vs.* C used for determination of the molecular weight of *n*-hexadecyltrimethylammonium bromide.

TABLE 5

n-Hexadecyltrimethylammonium bromide

SOLVENT: 0.178 *M* KBr	SOLVENT: 0.233 *M* KBr
L_d = 683 Å.	L_d = 1020 Å.
L_s = 615 Å.	L_s = 920 Å.
L_r = 870 Å.	L_r = 1312 Å.
MW (disk) = 10.6×10^6	MW (disk) = 23.6×10^6
MW (sphere) = 73.4×10^6	MW (sphere) = 246.0×10^6
MW (rod) = 0.948×10^6	MW (rod) = 1.44×10^6
Experimental MW = 0.690×10^6 (not corrected)	Experimental MW = 1.34×10^6 (not corrected)
Experimental MW = 0.795×10^6 (corrected for dissymmetry; rod model)	Experimental MW = 1.86×10^6 (corrected for dissymmetry; rod model)

obtained for each run by a least-squares method, which gives also the best plot of $H(C - C_0)/\tau$. C is the detergent concentration and τ is the turbidity of the detergent solution in excess of that of the solvent. The molecular weight of the solute is equal to the reciprocal of the intercept of the $H(C - C_0)/\tau$ plot at the critical concentration. When dissymmetry is present these molecular weights must be corrected accordingly. Values are given in table 5.

Since the two most frequently mentioned models for the micelle have been the cylinder and the sphere, it was decided to apply the experimental data to them first. In the cylindrical lamellar model of the micelle the polar heads form the ends of the cylinder, while the hydrocarbon tails aligned in two adjacent layers make up the body. Such a micelle would approach the shape of a disk if large enough to produce dissymmetry of a measurable value. Its thickness would be approximately twice the length of the detergent molecule. A disk of negligible thickness (negligible compared with the wave length) was calculated to scatter light (incident light polarized perpendicular to the plane of observation) according to the relation:

$$I = 1 - \frac{X^2}{6} + \frac{X^4}{72} - \frac{X^6}{1440} + \frac{X^8}{43200} - \cdots \tag{2}$$

This expression may be written

$$I = \frac{X^0}{1} - \frac{X^2}{6} + \sum_{n=3}^{n=\infty} \frac{X^{2(n-1)}}{\left(2n + \frac{b_{n-1}}{b_{n-2}}\right) b_{n-1}} \cos (n+1)\pi \tag{3}$$

in which b_n represents the denominator of the nth term. $b_1 = 1$ and $b_2 = 6$. $X = 2\pi L_d/\lambda \sin \theta/2$. L_d is the diameter of the disk. The intensity of the light scattered in the forward direction is taken as unity.

Debye has calculated that a sphere will scatter light according to the relation

$$I = \left[\frac{3}{X^3} (\sin X - X \cos X)\right]^2 \tag{4}$$

where

$$X = \frac{2\pi L_s}{\lambda} \sin \frac{\theta}{2}$$

L_s is the diameter of the sphere.

With the aid of relationships 2 and 4 and the second method of determining characteristic dimensions described under the heading of theory, disk and sphere diameters for two detergent–potassium bromide systems were calculated. Micelle molecular weights were computed from these diameters, assuming a density of unity[3] and a length for the detergent molecule of 24 Å. The results are

[3] In our calculations we have assumed the density of the micelle to be unity. Although this value is not likely to be exact, it is a good enough approximation for our purpose. We have estimated micelle densities (d) from the results of viscosity measurements made in this laboratory and Einstein's formula for the rigid sphere:

$$[\eta] = \frac{2.5}{d}$$

In the case of *n*-dodecyltrimethylammonium bromide in 0.0235 *M* potassium bromide a density of 0.612 g./ml. was calculated for the aggregates.

The application of Einstein's equation to the micelle is, of course, open to criticism as to details. However, the deduction that in solution soap micelles have a very high density as compared with the usual polymers is reliable, whereas the accuracy of the calculated value of the density must be poor.

listed in table 5; there is no agreement between the calculated and observed molecular weights. Therefore it was decided to repeat the procedure, using the rod as a model for the micelle.

TABLE 6

Intensity of light scattered by rigid rod of length L

$$I = \frac{1}{X}\int_0^{2X} \frac{\sin u}{u}\, du - \left(\frac{\sin X}{X}\right)^2$$

$$X = \frac{2\pi L}{\lambda} \sin \frac{\theta}{2}$$

X	I	X	I
0.10	0.9990	1.50	0.7902
0.20	0.9956	1.75	0.7313
0.30	0.9901	2.00	0.6724
0.40	0.9824	2.25	0.6156
0.50	0.9728	2.50	0.5627
0.60	0.9611	3.00	0.4727
0.70	0.9476	3.50	0.4056
0.80	0.9324	4.00	0.3578
0.90	0.9156	4.50	0.3228
1.00	0.8973	5.00	0.2949
1.25	0.8465		

According to Debye a rod of negligible diameter will scatter light in accordance with the relation

$$I = \frac{1}{X}\int_0^{2X} \frac{\sin u}{u}\, du - \left(\frac{\sin X}{X}\right)^2 \tag{5}$$

where

$$X = \frac{2\pi L_r}{\lambda} \sin \frac{\theta}{2}$$

L_r is the rod length. In table 6 the values of this function are given for a series of values of the parameter X.

In figure 5 the plot of log I *vs.* log sin $\theta/2$ for one of the runs (solvent 0.233 M potassium bromide) has been projected upon the theoretical log I *vs.* log X plot for the rod. The separation of the origins for the horizontal axes is 0.405 and equal to log $2\pi L_r/\lambda$. L_r is computed to be 1312 Å. The molecular weight of a rod of this length, a diameter twice the length of the *n*-hexadecyltrimethylammonium bromide molecule (a value of 48 Å. was taken), and density of unity is calculated to be 1.44×10^6. The experimentally determined molecular weight is 1.86×10^6.

For the second run (solvent 0.178 M potassium bromide), L_r is found to be 870 Å. From this L_r a molecular weight of 9.48×10^5 is calculated. The experimentally determined molecular weight for this run is 7.95 × 10^5. With the rod model the agreement is as good as can reasonably be expected.[4]

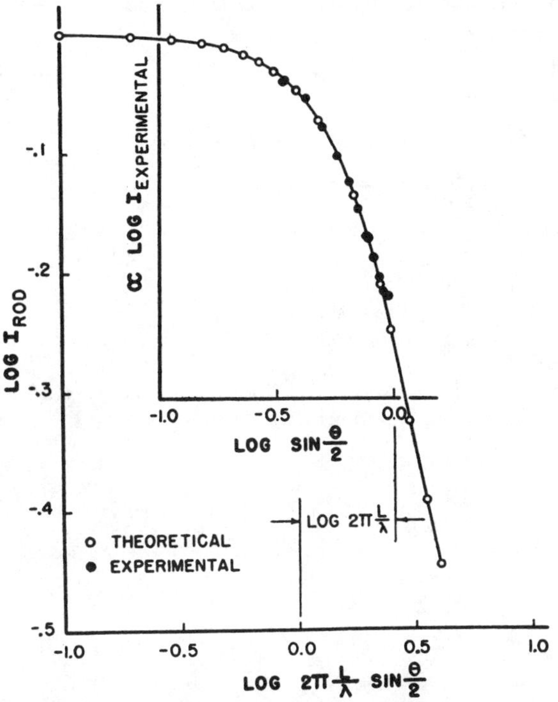

FIG. 5. Superposition of experimental data and theoretical curve, assuming a rod model for the micelle (*n*-hexadecyltrimethylammonium bromide).

CONCLUSIONS AND SUMMARY

1. No real dissymmetry was detected in solutions of *n*-tetradecyltrimethylammonium bromide in 0.083 *M* and 0.202 *M* potassium bromide. The detergent micelles in these solutions cannot have a dimension greater than roughly 300 Å.

2. The micelles of *n*-hexadecyltrimethylammonium bromide in 0.178 *M* and 0.233 *M* potassium bromide are large enough to produce measurable dissymmetry in the scattered light. Dissymmetry measurements showed conclusively that these micelles are not spherical or disk-like in shape; analysis of the data indicates that the micelles are rod-like. The cross section of such a rod would

[4] The objection might be raised that an experimental molecular weight derived for concentrations near the critical is being compared with molecular weights calculated for higher concentrations at which the dissymmetry of scattering is well developed. However, figure 4 shows only a small slope for the straight lines representing $H(C - C_0)/\tau$ *versus* C and this means that an unsophisticated calculation of the molecular weight directly from the observed turbidity values at higher concentrations will not differ materially (at least for our purpose) from the value given in the table. As a matter of fact, this admittedly rough calculation yields values ranging from 0.808 × 10^6 to 1.01 × 10^6 for the four highest concentrations in the case of 0.233 *M* potassium bromide (before correction for dissymmetry).

be circular, with the polar heads of the detergent lying on the periphery and the hydrocarbon tails filling the interior. The ends of such a rod would most certainly have to be rounded off with polar heads. This rod would differ from the cylindrical model, in that the hydrocarbon tails would be perpendicular to the axis of the symmetry rather than parallel to it.

3. The angular distribution of scattering for disks of negligible thickness has been calculated.

REFERENCES

(1) Corrin, M. L.: J. Chem. Phys. **16**, 844 (1948).
(2) Debye, P.: J. Phys. & Colloid Chem. **53**, 1 (1949).
(3) Debye, P.: Ann. N. Y. Acad. Sci. **51**, 575 (1949).
(4) Debye, P. P.: See Doctoral Dissertation of F. W. Billmeyer, Cornell University, Ithaca, New York, 1945.
(5) Harkins, W. D.: J. Chem. Phys. **16**, 156 (1948).
(6) Hartley, G. S.: *Aqueous Solutions of Paraffin-Chain Salts.* Hermann et Cie, Paris (1936).
(7) McBain, J. W.: In *Colloid Chemistry,* edited by Jerome Alexander, Vol. 5. Reinhold Publishing Corporation, New York (1944).

LIGHT-SCATTERING INVESTIGATIONS OF CAREFULLY FILTERED SODIUM SILICATE SOLUTIONS[1, 2] *

R. V. NAUMAN AND P. DEBYE

Department of Chemistry, Cornell University, Ithaca, New York

Received August 10, 1950

Many properties of sodium silicate solutions have been explained by assuming that the solutions contain complex aggregates, micelles, or colloidal material. Harman (10) concluded from conductivity, diffusion, and freezing-point data that sodium silicate solutions of $SiO_2:Na_2O$ ratio greater than 2:1 contain some colloidal silica, with the amount increasing with both increasing $SiO_2:Na_2O$ ratio and increasing concentration. This colloidal silica was described as being present in the form of complex aggregates and ionic micelles. Cann and Gilmore (3) interpreted the low boiling-point elevations found in concentrated solutions of the 3.75:1 ratio as being due to micelle formation. The high viscosities which the sodium silicate solutions have at high concentrations seemed to be further evidence for the existence of large particles.

The application of light-scattering theory and techniques seemed to be in order. Crude light-scattering experiments had been done before (2, 8), and these experiments had indicated that there was a large increase in the intensity of the scattered light with ratio in sodium silicate solutions of $SiO_2:Na_2O$ ratio greater than 3.0. There was much more scattering from these silicate solutions than from ordinary monomolecular solutions, and again these experiments were interpreted as being evidence for the presence of colloidal particles or micelles.

With the refined light-scattering techniques we have investigated sodium silicate solutions of $SiO_2:Na_2O$ ratio 0.5:1 to 3.75:1 in an attempt to measure the molecular weights of the particles present in these solutions. The first results, which have been reported previously (5), showed that there is no aggregation in relatively dilute sodium metasilicate solutions. Since then the turbidities of other ratios have been measured at low concentrations, and the turbidities of several ratios have been measured up to very high concentrations. In no case have we found evidence for the existence of large particles. It should be emphasized that these silicate solutions are specially prepared, very pure solutions that have been very carefully filtered. A few experiments with commercial solutions have given turbidities characteristic of large-particle solutions. The viscosity data have been reconsidered and have been shown to be interpretable in terms of the small-particle concept.

EXPERIMENTAL

The 90° turbidimeter used to measure the scattering intensity was built in this laboratory and has been described by Debye (4). The angular scattering instru-

[1] Presented at the Twenty-Fourth National Colloid Symposium, which was held under the auspices of the Division of Colloid Chemistry of the American Chemical Society at St. Louis, Missouri, June 15–17, 1950.

[2] This work was supported by the Sodium Silicate Manufacturers Institute.

*Reprinted from *The Journal of Physical & Colloid Chemistry*, 55, 1–9 (1951).

ment is similar in design to that described by P. P. Debye (6) but has been improved by Dr. A. M. Bueche in order to eliminate stray light. The viscometer used for the determination of intrinsic viscosity is an Ostwald-type viscometer, in which the capillary is bent into a coil in order to give a long flow time with an instrument of convenient size. Differences in refractive index were measured with a differential refractometer (6). The pH's of the systems were measured with a Beckman Model G pH meter having a type E electrode.

The sodium silicate solutions were specially prepared samples of very high purity, made by fusing crystal quartz powder with either C.P. sodium hydroxide or C.P. sodium carbonate. The specimens of 0.5:1 and 2:1 ratios were prepared by the Diamond Alkali Company; that of 3.32:1 ratio by the Philadelphia Quartz Company; those of 3.75:1, 2.85:1, and 2.50:1 ratios by the Grasselli Chemicals Department of E. I. du Pont de Nemours and Company; and that of 1:1 ratio in this laboratory. Except for the 0.5:1 and 1:1 ratios the glasses formed by the fusion were dissolved in steam under pressure to give concentrated stock solutions. The 0.5:1 ratio sample was dissolved directly in water. The 1:1 metasilicate was dissolved in water and then recrystallized several times in the form of the nonahydrate; solutions were made from the wet undried crystals. The stock solutions or solids of all ratios were analyzed to determine the SiO_2:Na_2O ratio and the carbon dioxide content and then analyzed spectrographically for metallic ion impurities. These analyses showed that in all cases the silicates were at least 99.7 per cent pure. Less concentrated solutions were made from the stock solutions by dilution with freshly redistilled carbon dioxide-free water.

Storage and all operations of preparation and measurement were carried out under an atmosphere of nitrogen whenever possible.

Cleaning the solutions for turbidity measurements has proved to be a formidable task. Sintered-glass filters were first used but proved to be unsatisfactory; they frequently contaminated the solutions and caused the turbidity to increase in no regular manner. Reproducible results were not obtainable when glass filters were used. Sintered stainless-steel filters were useful for investigating dilute solutions after they had been carefully cleaned with soap, sodium hydroxide, and sodium silicate solutions. In the presence of halides or in the case of con centrated solutions the steel filters were useless, because the filter contaminated the solution with large colored particles. Filter crucibles of sintered platinum have been used satisfactorily for all the high-concentration work and are now used for all the filtrations. The pore sizes of the platinum filter are unknown, but comparative filtration tests indicate that the platinum filter is as efficient a cleaning agent as either the stainless-steel or the glass filters (maximum pore size of 5–10 microns) for 90° turbidity measurements. It is sometimes found necessary to filter the silicate solutions many times before a reproducible turbidity is obtained from the solution. Repeated filtering does not remove any appreciable amount of material from the solution. No satisfactory filtering device has been found for preparation of the aqueous silicate solutions for low-angle scattering measurements.

TURBIDITY MEASUREMENTS

In dilute solutions of small particles the turbidity, τ, can be represented by the expression (4)

$$\tau = HMc \tag{1}$$

in which H is a refraction constant defined by

$$H = \frac{32}{3}\pi^3 \frac{\mu_0^2}{N\lambda^4}\left(\frac{\mu - \mu_0}{c}\right)^2 \tag{2}$$

M is the weight-average molecular weight, and c is the concentration (g./cc.). In the second equation $\mu - \mu_0$ is the refractive-index difference between solution and solvent, N is Avogadro's number, and λ is the wave length of the light measured *in vacuo*.

TABLE 1

Molecular weights of sodium silicates

RATIO	$(\mu - \mu_0)/c$	MOLECULAR WEIGHT
0.48:1	0.260	60
1.01:1	0.242	70
2.03:1	0.188	150
2.51:1	0.177	175
2.85:1	0.164	200
3.32:1	0.156	325
3.75:1	0.144	400
Sucrose	0.144	342
Polystyrene	0.111	248,000

It can be seen that from two series of measurements—turbidity and difference in index of refraction as functions of concentration—molecular weights can be determined. These two measurements are made in a 90° turbidimeter by comparison with solutions of known turbidity and in the differential refractometer standardized with a solution of known refraction.

From turbidity measurements on dilute solutions having concentrations up to 0.1 g./cc. molecular weights have been crudely determined. In this concentration range the turbidity can be represented within experimental error as a linear function of concentration. The turbidities are very low for all ratios in this concentration range, and the relative error in any one turbidity measurement may be very large, but from a large number of measurements it is possible to obtain a linear turbidity–concentration relationship that is reproducible within 10 to 20 per cent. Table 1 summarizes the molecular weights found for the various ratio sodium silicates at low concentrations along with the $(\mu - \mu_0)/c$ values measured and used to calculate the molecular weights. It must be emphasized that the molecular weights given represent only the order of magnitude and are not to be interpreted as precise determinations.

It is thus apparent that, at low concentrations at least, there are small particles present in all ratios up to 3.75:1. Some investigators have concluded that there are some large particles even in dilute solutions of the higher ratios. Many of those experiments which indicated the presence of large particles were done with commercial samples, and a few experiments which we have made with commercial samples have indicated that these samples do indeed contain large particles (molecular weights were not measured), because they have huge turbidities. These large particles and the resulting large turbidities are, we believe, due to impurities in the commercial samples. Other experiments have shown that contamination with carbon dioxide, for example, brings about the formation of

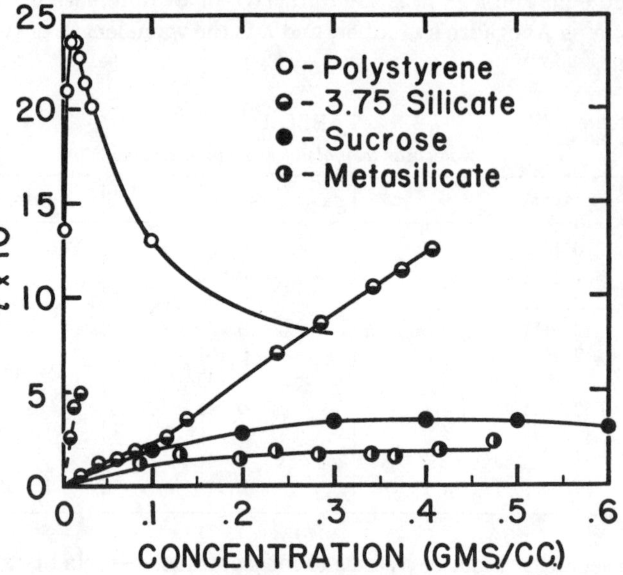

FIG. 1. Comparison of the turbidities of high-molecular-weight solutions (polystyrene, MW = 248,000) and low-molecular-weight solutions (sucrose, MW = 342) with the turbidities of solutions of 1:1 and 3.75:1 SiO_2:Na_2O ratio sodium silicates.

large particles. If a solution with a low turbidity is allowed to stand in air, the turbidity as measured by our turbidimeter increases after a few days and continues to increase; after a month or more there is frequently a visible turbidity. Similar solutions stored under a nitrogen atmosphere show no turbidity change. In the case of the 1:1 ratio unreproducible molecular weights of 700–1800 were found when dried crystals were used to make the solutions, although analysis showed that these crystals contained only 0.01 per cent carbon dioxide. When wet freshly crystallized crystals were used to make the solutions, the reproducible molecular weight of 70 was found.

Thus far we have considered only the dilute solutions. Turbidity measurements have been extended to very high concentrations in the cases of the 1:1, 2.0:1, 2.85:1, and 3.75:1 ratios. These data are shown in figures 1 and 2 along with data for a 248,000 molecular weight polystyrene (1) and sucrose (9). In the case of the sucrose data Halwer (9) has shown that the experimental turbidity and that calculated from osmotic pressure data are in agreement even at the high concentration of 0.6 g./cc. It can be observed that qualitatively the turbidity curves of the sodium silicates of 1:1 and 2:1 ratio are similar to that of sucrose even up to very high concentrations. The turbidities of the sodium silicates of 1:1 ratio seem to increase at about 0.5 g./cc., but there is some doubt concerning the validity of the results for this 1:1 ratio above 0.4 g./cc. because of the filtration difficulties and the fact that at the very high concentrations the work was not repeated. These silicate turbidity curves which are similar to that for the monomolecular sucrose indicate by analogy that there are small particles in the 1:1 and 2:1 ratio sodium silicates even in very highly concentrated solutions. These silicate and sucrose turbidity curves can be compared with more confidence when one considers (see table 1) that the $(\mu - \mu_0)/c$ values for the silicates and the sucrose are of the same order of magnitude.

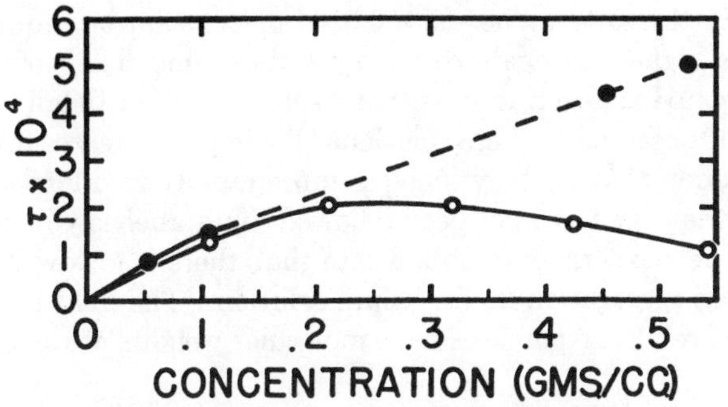

Fig. 2. Comparison of the turbidities of 2.85 ratio sodium silicate (●) with those of the 2.0 ratio sodium silicate (○).

The turbidity–concentration curves for the sodium silicates of 3.75:1 and 2.85:1 ratios have a different shape from those of sucrose and of the silicates of lower ratios. There appears to be a break in the curve of the sodium silicate of higher ratio. At low concentrations these curves are like those of the silicates of lower ratios, but in the region of 0.1 g./cc. for the silicate of 3.75:1 ratio the curve appears to bend upward. This bend may be illusory, since all the data from low to very high concentrations could be represented by a straight line without assuming an inconceivable experimental error. If the linear relationship is valid for the whole concentration range, the data could indicate that the molecular weight measured at low concentrations is valid at all concentrations and that water is a poor solvent (4) for sodium silicate of this ratio. If, on the other hand, the turbidity–concentration curve does actually bend upwards, the upward curvature might be due to the occurrence of aggregation as the concentration increases. There is no theory available to measure the degree of such an aggregation, but one has the intuitive feeling that such an aggregation, if it occurs, is small, because in all other cases where there is a large amount of aggregation there is a rapid, large increase in turbidity. Thus we have the feeling that, although there may be some aggregation in concentrated solutions of the sodium silicates of higher ratios, there are no really large particles.

The turbidity investigations at high concentrations have not been completed in the case of the silicate of 2.85 ratio, but the data obtained show conclusively that this silicate has much higher turbidities at the very high concentrations than do the silicates of 2.0 or lower ratios. Experiments are being done at intermediate concentrations in order to see if a break in the turbidity *versus* concentration plot similar to that of the sodium silicate of 3.75 ratio can be detected.

CHANGE IN TURBIDITY OF THE 3.75 RATIO SODIUM SILICATE WITH TIME

An increase in the turbidity of dilute solutions of the sodium silicate of 3.75 ratio has been observed in solutions of concentration less than 0.04 g./cc. The turbidity increases more rapidly the more dilute the solution; however, the more concentrated solutions attain the higher turbidity in the end. A solution of 0.012 g./cc. concentration required only 4 days to attain a turbidity which changed only very slowly thereafter, whereas a 0.0204 g./cc. solution required about 12 days to attain a higher but again relatively stable value. The turbidity increase and the accompanying reaction have not been studied in detail, but increases in pH have been observed in these solutions. These pH increases follow the same pattern of behavior, that is, they increase more rapidly in dilute solutions. The turbidities and the pH's have not been followed simultaneously in the same solutions, but these few experiments do indicate that there is a slow hydrolysis and an accompanying aggregation in the dilute solutions. The final turbidity values indicate that the resulting particles have molecular weights of the order of 5000–10,000.

In solutions of concentration greater than 0.05 g./cc. neither turbidity nor pH increases have been observed in a month's time.

There is meagre evidence for a similar turbidity increase in the sodium silicate of 3.3:1 ratio. This reaction seems to occur at lower concentrations and with slower rate than that of the sodium silicate of 3.75:1 ratio.

VISCOSITY

The intrinsic viscosity of the silicate of 1:1 ratio (sodium metasilicate) was determined from the flow times, t, of solutions of concentration, c, and that of water, t_0, by plotting $(t - t_0)/t_0c$ against c, extrapolating to infinite dilution, and taking the intercept as the intrinsic viscosity $[\eta]$. The intrinsic viscosity at 30°C. was found to be 3.89 cc./g. Solutions made from either a commercial sample or our specially prepared silicate gave essentially the same result. From the intrinsic viscosity the density, ρ, of the particles in the solution can be calculated from the equation

$$[\eta] = \frac{1}{\rho}\Phi \tag{3}$$

by assuming that the particle is dense enough not to allow the flow to penetrate it; in this case the volume factor Φ has the extreme Einstein value 2.50, from which is calculated $\rho = 0.64$. The intrinsic viscosities of sucrose and the sodium silicate of 3.75:1 ratio were not measured, but from the data which are plotted in figure 3 the intrinsic viscosities have been estimated. Using these estimated values and equation 3, densities of 0.43 and 0.96 have been calculated for the 3.75:1 ratio sodium silicate particles and the sucrose particles, respectively. A comparison of these densities with one of the order of 0.001 found for high polymers indicates that the particles in the silicates and sucrose solutions are relatively dense. These high densities indicate that the particles are tightly packed and not extended.

In figure 3 are plotted relative viscosity *vs.* concentration data for a polystyrene of molecular weight 259,000 (7), for 3.80:1 and 1:1 ratio sodium silicates (12), and for sucrose (11).

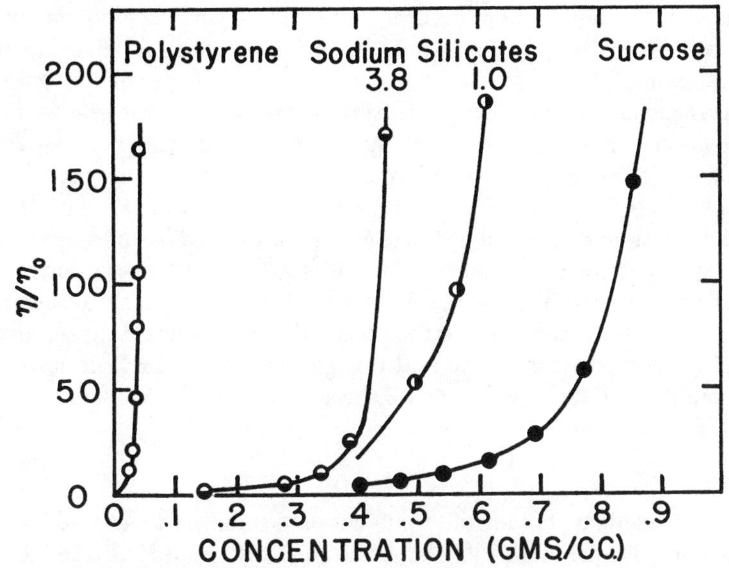

Fig. 3. Comparison of the relative viscosities of polystyrene, 3.8 ratio sodium silicate, 1.0 ratio sodium silicate, and sucrose as functions of concentration.

It is observed that the large rapid increase in viscosity occurs at very low concentrations in polystyrene, which is known to contain large particles, and at very high concentrations in monomolecular sucrose solutions. The silicates occupy an intermediate position but seem to be more nearly like the sucrose than the polystyrene; this conclusion becomes more apparent when we crudely calculate the distance between molecules at the concentration at which the viscosity begins to increase rapidly. From this concentration and the molecular weight calculated from low-concentration turbidity data we find a distance of about 300 Å. between polystyrene molecules, 12.5 Å. between particles in the 3.75:1 ratio silicate, 6.5 Å. distance in the metasilicate, and about 9 Å. between sucrose molecules. All these distances apply of course only at the concentration at which the viscosity begins to increase rapidly. From these calculations it is evident that the conception of small particles in the silicate solutions is not incompatible with the observed viscosities because, for example, when the 3.75:1 ratio silicate solutions begin to become very viscous at a concentration of 0.35 g./cc. particles of molecular weight 400, as found from low-concentration light-scattering data, would be only about 12.5 Å. apart and are thus approaching the point where there is little room for particles with their water atmosphere to move without touching one another; thus big particles are not necessary to explain the viscosity. Larger particles than molecular weight 400 may exist at the 0.35 g./cc. concentration, but the viscosity may be understood even if there is no increase in aggregation with concentration.

ANGULAR DEPENDENCE OF SCATTERING INTENSITY

If all the particles in these silicate solutions are truly small the intensity of the scattered light should show no dissymmetrical dependence on the angle of observation; it should show an angle *vs.* scattering intensity relationship that is symmetrical around 90°. This has not been found; all experiments completed thus far have given a dissymmetrical scattering pattern with the greater intensities at the lower angles, a result which indicates the presence of large particles. We assume the large particles to be impurities which a more thorough cleaning would eliminate. Some evidence for this assumption is found in the fact that salt solutions known to contain small particles cannot be cleaned by the same methods to give a symmetrical scattering pattern. Before the 90° measurements and their interpretations can be accepted with complete confidence the dissymmetrical scattering must be proved to be the result of the experimental procedure, or a relationship between symmetrical scattering intensity and angle of observation must be obtained, or the angular and 90° scattering results must be shown to be compatible.

SUMMARY

Turbidity measurements indicate that there are small particles in dilute solutions of sodium silicates with SiO_2:Na_2O ratio as high as 3.75:1. At high concentrations the turbidities of the sodium silicates of ratio 2.0:1 or lower are similar to those of sucrose and give no evidence for any aggregation. Sodium silicates with SiO_2:Na_2O ratio greater than 2.0 give turbidities that continue to increase with concentration, and these turbidities are much greater than those of sucrose at high concentrations. These higher turbidities may indicate either that water is a poor solvent for these sodium silicates of higher ratios or that there is a small amount of aggregation in the sodium silicates of higher ratios at high concentrations, but in any case there is no evidence for the existence of very large particles. All of these silicates have been very carefully filtered.

In the 3.75:1 ratio sodium silicate the turbidities and the pH's of the dilute solutions are found to increase with time. Particles of 5000–10,000 molecular weight are present when the changes stop.

The viscosities of the sodium silicates are considered and are shown to be quite low in dilute solution, indicating the presence of small, tightly packed particles. At high concentrations the viscosities increase rapidly at a concentration at which the molecules are very close to one another, showing that it is not necessary to assume aggregation to explain and understand the high viscosities at high concentrations.

The angular dependence of the intensity of the scattered light has not been satisfactorily investigated, but there is an indication that with an improvement in experimental technique the angular dependence too will be consistent with the idea that sodium silicate solutions contain small particles.

REFERENCES

(1) Bueche, A. M.: Private communication.
(2) Burgess, L., and Krishnamurti, K.: Trans. Faraday Soc. **26**, 574 (1930).
(3) Cann, J. Y., and Gilmore, K. E.: J. Phys. Chem. **32**, 72 (1928).
(4) Debye, P.: J. Phys. & Colloid Chem. **51**, 18 (1947).
(5) Debye, P., and Nauman, R. V.: J. Chem. Phys. **17**, 664 (1949).
(6) Debye, P. P.: J. Applied Phys. **17**, 392 (1946).
(7) Fox, T. G., Jr., and Flory, P. J.: J. Am. Chem. Soc. **70**, 2384 (1948).
(8) Ganguly, P. B.: J. Phys. Chem. **30**, 706 (1926).
(9) Halwer, M.: J. Am. Chem. Soc. **70**, 3985 (1948).
(10) Harman, R. W.: J. Phys. Chem. **32**, 44 (1928).
(11) National Bureau of Standards Circular No. 440.
(12) Vail, J. G.: *Soluble Silicates in Industry*. The Chemical Catalog Company, Inc., New York (1928).

Reprinted from pages 1423–1425 of

THE JOURNAL OF CHEMICAL PHYSICS VOLUME 18, NUMBER 11 NOVEMBER, 1950

Light Scattering by Concentrated Polymer Solutions*

P. DEBYE AND A. M. BUECHE**
Department of Chemistry, Cornell University, Ithaca, New York
(Received April 24, 1950)

The light scattered by benzene solutions of polystyrene up to a volume fraction of polymer of 0.62 has been measured. The results are compared with what was to be expected on the basis of recent thermodynamic theories for polymer solutions.

This comparison has led to the recognition of a new and rapid method for estimating molecular weights. The method also seems to be a convenient one for obtaining thermodynamic information.

WE have recently measured the 90° light scattering of polystyrene solutions in benzene up to a volume fraction of polymer $v_2=0.62$. The data are compared with the curves to be expected on the basis of current thermodynamic theories. A new method for the determination of molecular weights is indicated. The method is general and appears to be an easy and rapid one for obtaining thermodynamic information about concentrated as well as dilute solutions.

EXPERIMENTAL

A sample of Dow Styron, having an intrinsic viscosity at 35°C in benzene of 1.0, was carefully washed with freshly distilled methyl alcohol to remove dust and was dried in vacuum. Portions of this sample were weighed into test tubes to be used in our light scattering apparatus, different amounts of benzene were added and the tubes sealed off. They were then heated to about 100°C for several days until the polymer and solvent were homogeneously mixed. After cooling to room temperature the 90° scattering, for $\lambda=4358$A, was measured in an apparatus[1] described elsewhere. Since the dissymmetry of scattering for this sample is small, it was neglected and the 90° values were used to calculate the turbidities, τ_0, of the solutions. The data are shown in Fig. 1. The value at $v_2=0$ is that obtained with the tubes filled with benzene and represents the scattering of benzene plus a small amount of stray light in the instrument.

The curve obtained shows the usual rapid increase of τ_0 in the low concentration region. This initial slope is proportional to the weight average molecular weight of the polymer. Data in this region have been used many times in the past for molecular weight determinations.[2] The curve then goes through a rather sharp maximum and finally remains almost constant after about $v_2=0.4$. This maximum is of considerable importance. Its use in the determination of molecular weights will be discussed in the next section.

DISCUSSION

It has been shown[3] that the total turbidity, τ_0, of a solution of a single substance in a solvent is given by

$$\tau_0=\tau'+HRT\bar{V}_0C/[\partial(P\bar{V}_0)/\partial C], \quad (1)$$

where τ' is the background scattering due to bulk density fluctuations given by the Smoluchowski-Einstein calculation, R, T, $\bar{V}_0$, C, and P are the gas constant, absolute temperature, partial molar volume of solvent, concentration in g/cc, and osmotic pressure respectively. H is the usual light scattering constant given by

$$H=(32\pi^3/3)(\bar{\mu}^2/N\lambda^4)(\partial\bar{\mu}/\partial C)^2,*$$

* The work reported in this paper was done in connection with the Government Research Program on Synthetic Rubber under contract with the Office of Rubber Reserve, Reconstruction Finance Corporation.

** Present address: General Electric Research Laboratory, Schenectady, New York.

[1] P. Debye, J. Phys. Colloid Chem. 51, 18 (1947).

[2] G. Oster, Chem. Rev. 43, 319 (1948).

[3] A. Einstein, Ann. d. Physik 33, 1275 (1910).

* Instead of the quantity $\bar{\mu}^2(\partial\bar{\mu}/\partial C)^2$ the formula should in its exact form contain $(\frac{1}{4})(\partial\epsilon/\partial C)^2$ where ϵ is the dielectric constant of the solution. The quantity ϵ at 1.5 Mc has been measured in this laboratory up to concentrations of polymer of over 25 percent for this system and found to be linear and extrapolates to the value for the pure polymer. (See P. Debye and F. J. Bueche,

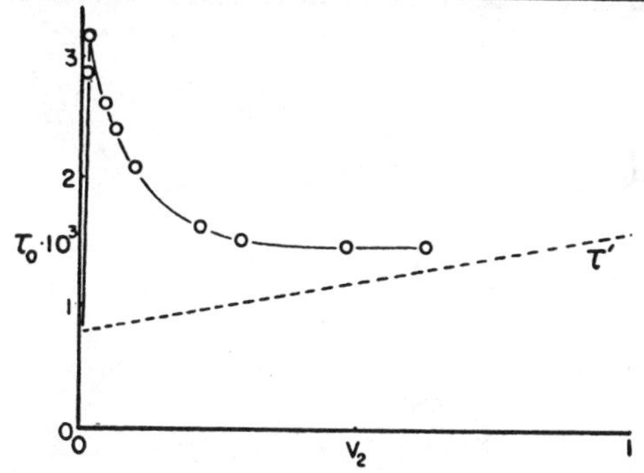

FIG. 1. Turbidity of polystyrene in benzene (τ_0) *vs.* volume fraction of polymer (circles).

where $\bar{\mu}$ is the refractive index of the solution, λ the wave-length of the light in vacuum and N is Avogadro's number. Writing C in terms of the volume fraction of polymer as $C=v_2d_2$ (d_2 was found to be 1.09 for this system at 30°C from measurements in the dilute solution range) and using Flory's[4] expression for the osmotic pressure assuming both $\bar{V}_0$ and d_2 to be independent of C the excess turbidity due to concentration fluctuations can be written as

$$\tau=\tau_0-\tau'=HMC\bigg/\left[1-nv_2\left(\frac{1}{1-v_2}-2\mu\right)\right]$$

or

$$\tau=\frac{HMd_2}{n}\bigg/\left(\frac{1+(n-1)v_2}{nv_2(1-v_2)}-2\mu\right),$$

where μ is the constant in the van Laar expression for the heat of mixing, n is the ratio of the molecular volumes of polymer and solvent, or the number of links in the polymer chain in the Flory-Huggins sense, and

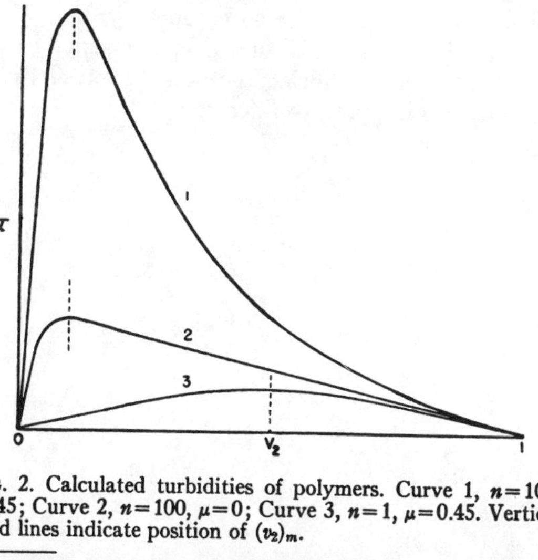

FIG. 2. Calculated turbidities of polymers. Curve 1, $n=100$, $\mu=0.45$; Curve 2, $n=100$, $\mu=0$; Curve 3, $n=1$, $\mu=0.45$. Vertical dashed lines indicate position of $(v_2)_m$.

J. Phys. Colloid Chem., to be published.) Since we know of no theoretical reason why it should not be linear we assume that the optical dielectric constant behaves similarly.

[4] P. Flory, J. Chem. Phys. **10**, 51 (1942).

M is the molecular weight of the polymer molecule. In the case of a molecular weight distribution of the polymer, H. C. Brinkman and J. J. Hermans[5] have shown by means of a formula developed in 1915 by Zernike that n must be replaced by the quotient formed by dividing the weight average molecular volume of the polymer by the molecular volume of the solvent. The quantity M is then the weight average molecular weight of the polymer. Thus the turbidity curve depends on only the weight average molecular weight rather than some combination of averages. This will always be true provided the free energy has the general form given in Eq. (18) of the article by Brinkman and Hermans. The Flory-Huggins formula is of this form.

From Eq. (2) it is evident that a maximum of the turbidity will occur where the term preceding 2μ in the denominator goes through a minimum. This means that the position of the maximum on the concentration axis will be *independent* of the interaction constant μ.

By differentiation of Eq. (2) with respect to v_2 it can be shown that the turbidity τ should have a maximum when

$$v_2=(v_2)_m=1/(1+n^{\frac{1}{2}}).$$

In the case of polymers of high molecular weight this reduces to $(v_2)_m=n^{-\frac{1}{2}}$.

It is interesting to notice that, according to the theories, $(v_2)_m$ is independent of the value of μ. The position of the maximum, then, allows one to determine the quotient of the molecular volumes for polymer and solvent independent of solvent type. For a given system it is to be expected from the theories that this quotient will be proportional to the molecular weight of the polymer. Always assuming the theory to be absolutely correct we have determined the absolute value of the actual volume displaced by a polymer molecule in the solution. After calibration for a given polymer-solvent system finding the position of the maximum may afford a rapid means of estimating molecular weights in the absence of a scattering standard and with a minimum of computation.

If the position of the maximum is to be used for this purpose one would certainly like the maximum to be as high and as sharp as possible. These conditions can be met by choosing a solvent in which the refractive index increment is large and for which the μ value is as close to 0.5 as possible. If $\mu=0.5$ the theories predict the polymer will precipitate when $v_2=(v_2)_m$.

In Fig. 2 we have plotted several theoretical curves using Eq. (2). It can be seen that while $(v_2)_m$ is independent of μ (curves 1 and 2) the height of the maximum and the shape of the curve will depend on μ. The slopes of the curves near $v_2=1$ correspond to a solution of solvent in pure polymer and are proportional to the molecular weight of the solvent. A comparison of the

[5] H. Brinkman and J. Hermans, J. Chem. Phys. **17**, 574 (1949); see also J. Kirkwood and R. Goldberg, *ibid.* **18**, 54 (1950) and W. Stockmayer, *ibid.* **18**, 58 (1950).

experimental data with the theoretical curves should enable one to decide whether a single value of μ will represent the data over the range studied and thus check the current theories of polymer solutions.

To compare our data with the calculated curves it is necessary to know how τ' depends on v_2. It is given by[3]

$$\tau' = (8\pi^3/27)(RT/N\lambda^4)\beta[(\bar{\mu}^2-1)(\bar{\mu}^2+2)]^2,$$

where β is the isothermal compressibility. However, β is unknown for these solutions. We can estimate τ' in the following manner. In the region near $v_2 = 1$ the slope of the experimental curve must correspond to that which represents the excess scattering produced by the solution of benzene in pure polystyrene. One can then choose an approximate value of μ and adjust the theoretical curve to the data such that they fit at some point in the region of the maximum. The slope that the calculated curve makes with the axis of $\tau = 0$ at high concentrations will not be far different from the slope the data should make with $\tau = 0$ after subtraction of τ'. In the present work we have, for the time being, followed this procedure, assumed τ' to be linear with v_2 and have represented it by the dotted line in Fig. 1. We hesitate, however, to say that the great variation in τ' is caused by changes in compressibility alone. Any dust added with the polymer would have the same effect.

The adjusted data are compared in Fig. 3 with curves calculated from Eq. (2) using $\mu = 0.45$ and 0.47 and $n = 5.77\times 10^3$. The data presented here are not sufficient to allow a calculation of a precise value of n. The value used is estimated from the position of the highest point. The molecular weight of the polymer is about 2.18×10^5 so that this value of n would lead to the conclusion that the effective volume of one styrene unit in the polymer chain is 2.76 times the volume of a benzene molecule. As will be seen the maxima fall approximately in the same place but neither curve fits the data exactly in the intermediate concentration range. The small discrepancies may be the result of the rough approximation for τ' or reflect the inadequacy of the thermodynamic theories for polymer solutions. A more complete investigation in which more data are obtained at low concentrations and in the region around the maximum would be interesting. Measurements at different temperatures may be carried out quite easily with the experimental arrangement used and would lead to estimates of the heats of solution enabling one to compute the entropies of mixing. Further work along these lines is in progress.

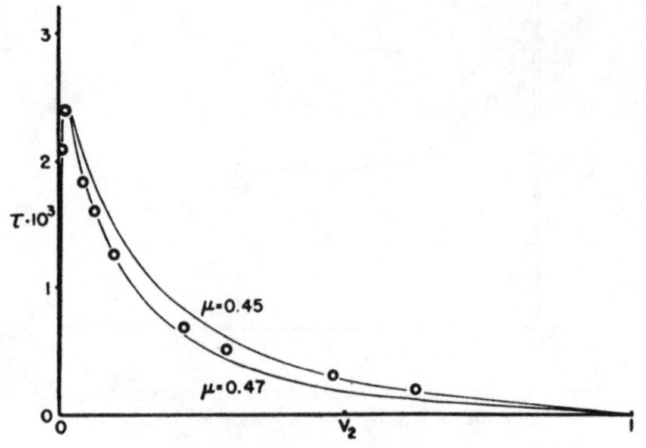

Fig. 3. Comparison of corrected turbidities for polystyrene in benzene with theory.

Originally this work was undertaken in order to approach the turbidity of the solid polymer from the side of its solution. Solutions of very high polymer concentrations are difficult to handle; however, the curve in Fig. 1 which goes to $v_2 = 0.62$ seems to indicate that a final value for the turbidity of an ideal solid polymer is already nearly reached at this point. This value still is definitely smaller than that found experimentally for any solid polystyrene so far investigated. Whether or not this difference is due to inhomogeneities which might be frozen in during the preparation of the solid polymer has not yet been decided.

Reprinted from pages 687–695 of
THE JOURNAL OF CHEMICAL PHYSICS VOLUME 14, NUMBER 11 NOVEMBER, 1946

The Determination of Polymeric Molecular Weights by Light Scattering in Solvent-Precipitant Systems

R. H. EWART AND C. P. ROE
General Laboratories, United States Rubber Company, Passaic, New Jersey

AND

P. DEBYE AND J. R. MCCARTNEY
*Department of Chemistry, Cornell University, Ithaca, New York**

(Received July 12, 1946)

The theory of scattering by an inhomogeneous dielectric medium has been extended so as to account for the turbidimetric behavior of polymer solutions in solvent-precipitant mixtures. It is predicted by this theory and verified by experiment that correct values of molecular weight are obtained by the usual interpretation of turbidity measurements if and only if the solvent and precipitant have the same refractive index. The practical utility of turbidimetry in high polymer solutions is shown to be greatly increased by the proper use of solvent-precipitant mixtures. If the solvent and precipitant have different refractive indices, scattering measurements give information about the extent of selective absorption of solvent by the polymer.

INTRODUCTION

WHEN the reciprocal of the specific turbidity, namely, c/τ, of a dilute high polymer solution in a pure solvent is multiplied by a constant H, characteristic of the scattering system, and the product $H(c/\tau)$ is plotted against the polymer concentration, c, in the solution, a linear relationship is obtained. The slope of this line varies with the solvent for a given polymer, and is in general greater for the so-called "good" solvents than for "poor" solvents. However, the intercept with the axis of $H(c/\tau)$ at $c=0$ is independent of the solvent in pure solvents. These results are in harmony with the currently accepted theory[1–6] of the scattering of monochromatic light by solutions. This theory gives a simple relation between the intercept and the molecular weight of the solute, and thus yields a means of estimating polymer molecular weights.

It has been observed by previous investigators[7] that, as one adds a precipitant to a high polymer solution, the turbidity progressively increases at concentrations of precipitant well below the threshold necessary to produce macroscopic separation into two phases. This fact suggests the possibility that scattering data capable of yielding accurate molecular weights may be obtained more easily from solutions in solvent-precipitant mixtures than in pure solvents, especially in cases where the turbidities in pure solvents are so small that they can be measured accurately only with very sensitive apparatus, and also only by taking extreme care to clarify the solutions.

Experiments have shown that in some solvent-precipitant systems correct values of molecular weight are obtained, whereas in others considerable departures from the correct values are found. Thus, the data shown in Fig. 2, on polystyrene in benzene-methanol mixtures, are the results of a careful investigation at concentrations as low as 0.000225 gram of polymer per milliliter of solution. The extrapolation of these results at very low concentrations proves definitely that the value of the intercept at infinite dilution is not constant for this system and depends on the concentration of methanol used. All of the known facts can be correlated by a proper consideration of the refractive indices of

* The portion of the work here reported which was done at Cornell University was carried out in connection with the Government Research Program on Synthetic Rubber under contract with the Office of Rubber Reserve, Reconstruction Finance Corporation.

[1] S. Bhagavantam, *Scattering of Light* (Chemical Publishing Company, Brooklyn, New York, 1942).

[2] J. Cabannes, *La Diffusion Moleculaire de la Lumiere* (University of France Press, Paris, 1929).

[3] M. Born, *Optik* (Julius Springer, Berlin, 1933).

[4] P. Debye, J. App. Phys. **15**, 338 (1944).

[5] B. Zimm, R. S. Stein, and P. Doty, Polymer Bull. **1**, 90 (1945).

[6] P. Doty, B. Zimm, and H. Mark, J. Chem. Phys. **13**, 159 (1945).

[7] Gehman and Field, Ind. and Eng. Chem. **30**, 1031 (1938).

the solvent and precipitant, coupled with the assumption that the solute selectively absorbs solvent from the solvent-precipitant mixture.

THEORETICAL

It was shown in 1897 by H. Kneebone Tompkins[8] that when rubber is soaked in a mixture of carbon disulfide and acetone, the solvent medium becomes poorer in carbon disulfide. A closely related phenomenon is observed in the precipitation of a polymeric solute from homogeneous solution by the addition of a precipitant. The addition of precipitant beyond a certain threshold causes the separation of two phases, one of which is richer than the other in polymer. In general the polymer-rich phase contains both solvent and precipitant, but the ratio of solvent to precipitant is higher than in the other phase. Prior to reaching the threshold necessary to cause incipient precipitation the progressive addition of precipitant causes steadily increasing turbidity. It is assumed for this discussion that this increase in turbidity is caused by two factors, namely, (1) shrinkage of the swollen polymer molecules in solution, and (2) selective absorption of solvent by the polymer. If we confine our attention to the problem of determining molecular weights by means of scattering, then it is necessary to consider only the effects of these two factors on the quantity $(1/H)(\tau/c)$ at infinite dilution, or at $c=0$. In order to do this, it is helpful to discuss the behavior of solutions in which the solute concentration is small but finite and subsequently to pass to the limit. Thus we shall show that shrinkage alone should produce augmented turbidity at finite concentrations, but should not affect the limiting value of $(1/H)(\tau/c)$. Also we shall show that both at finite concentrations and at infinite dilution selective absorption should produce changes in $(1/H)(\tau/c)$, and that in this case the changes may be either positive or negative. In terms of the properties of the conventional scattering curves, the characteristic effect of shrinkage is to decrease the slope but not to alter the intercept, whereas the effect of selective absorption is to change both the intercept and slope.

[8] H. Kneebone Tompkins, *The Physics and Chemistry of Colloids* (H. M. Stationery Office, London, 1921), p. 163.

Consider the properties of a suspension of spheres, the linear dimensions of which are small by comparison with the wave-length of light. Let us define the following symbols:

τ = excess turbidity of the suspension over that of the suspending medium in absence of any spheres.
v = volume of a sphere in the suspension.
c = weight concentration of solute per unit volume.
λ = wave-length of the exciting radiation.
n = refractive index.
N = number of spheres per unit volume.
N_0 = Avogadro's number.
M = molecular weight.
ϵ = homogeneous dielectric constant of the suspension.
ϵ_0 = dielectric constant of the suspending medium in absence of any spheres.
ϵ_1 = dielectric constant of the medium surrounding the spheres.
ϵ_2 = dielectric constant of a sphere.
ψ = volume fraction of solvent in a solvent-precipitant mixture, exclusive of the volume occupied by the solute.
ψ_0 = original value of ψ in absence of any spheres.
ψ_1 = value of ψ in the medium surrounding the spheres.
ψ_2 = value of ψ within a sphere.

If the suspension is very dilute, then from electromagnetic theory, values of the turbidity and dielectric constant may be calculated. The resulting expressions are

$$\tau=\frac{24\pi^3Nv^2}{\lambda^4}\epsilon_1{}^2\left(\frac{\epsilon_2-\epsilon_1}{\epsilon_2+2\epsilon_1}\right)^2, \tag{1}$$

$$\epsilon-\epsilon_1=3Nv\epsilon_1\left(\frac{\epsilon_2-\epsilon_1}{\epsilon_2+2\epsilon_1}\right). \tag{2}$$

Now the quantity $v\epsilon_1[(\epsilon_2-\epsilon_1)/(\epsilon_2+2\epsilon_1)]$ may be eliminated between (1) and (2) to yield

$$\tau=\frac{8\pi^3}{3N\lambda^4}(\epsilon-\epsilon_1)^2. \tag{3}$$

But since $N/N_0=c/M$,

$$\tau=M\frac{8\pi^3c}{3\lambda^4N_0}\left(\frac{\epsilon-\epsilon_1}{c}\right)^2. \tag{3a}$$

The usual discussion of this equation for the case in which the solvent medium is a pure compound assumes that ϵ_1 may be identified with ϵ_0, the dielectric constant for the solvent in absence of any dispersed material. This assumption is well justified by experiment, and we find that in the limit as $c \to 0$

$$\lim_{c\to 0}\left(\frac{\tau}{c}\right)=M\frac{8\pi^3}{3\lambda^4 N_0}\left(\frac{\partial\epsilon}{\partial c}\right)^2=M\frac{32\pi^3 n^2}{3\lambda^4 N_0}\left(\frac{\partial n}{\partial c}\right)^2. \quad (3b)$$

However, if we are dealing with a solvent medium having two components, one of which may be selectively absorbed by the solute, then it is no longer justifiable to set $\epsilon_1=\epsilon_0$, and Eq. (3a) must be rewritten as follows:

$$\tau=M\frac{8\pi^3}{3\lambda^4 N_0}c\left[\frac{\epsilon-\epsilon_0}{c}+\frac{\epsilon_0-\epsilon_1}{c}\right]^2. \quad (4)$$

The term $(\epsilon_0-\epsilon_1)/c$ alone requires discussion. Let us consider the behavior of ϵ in the neighborhood of a given value of ψ, namely ψ_0, corresponding to ϵ_0.

$$\epsilon=\epsilon_0+(\psi-\psi_0)(\partial\epsilon/\partial\psi). \quad (5)$$

If we assume that the factor $(\psi-\psi_0)$ owes its existence to selective absorption, we may, for sufficiently small values of c, rewrite Eq. (5)

$$\epsilon_1=\epsilon_0+\frac{\partial\epsilon_0}{\partial\psi}\frac{\partial\psi_1}{\partial c}c. \quad (5a)$$

Then we get at once

$$\frac{\epsilon-\epsilon_1}{c}=\frac{\epsilon-\epsilon_0}{c}-\frac{\partial\epsilon_0}{\partial\psi}\frac{\partial\psi_1}{\partial c} \quad (5b)$$

and

$$\lim_{c\to 0}\left(\frac{\epsilon-\epsilon_1}{c}\right)=\frac{\partial\epsilon}{\partial c}-\frac{\partial\epsilon_0}{\partial\psi}\frac{\partial\psi_1}{\partial c}. \quad (5c)$$

If we now set $\partial\psi_1/\partial c=-\alpha$,

$$\tau=M\frac{8\pi^3 c}{3\lambda^4 N_0}\left(\frac{\partial\epsilon}{\partial c}+\alpha\frac{\partial\epsilon}{\partial\psi}\right)^2. \quad (6)$$

Or if we assume $\epsilon=n^2$ (n = refractive index),

$$\lim_{c\to 0}\left(\frac{\tau}{c}\right)=\left(\frac{\tau}{c}\right)_0=M\frac{32\pi^3 n^2}{3\lambda^4 N_0}\left(\frac{\partial n}{\partial c}+\alpha\frac{\partial n}{\partial\psi}\right)^2. \quad (6a)$$

Equation (6a) contains all that is needed in order to give a satisfactory semiquantitative interpretation of the observed facts. If the solvent is a pure compound, $\alpha=0$, and Eq. (6a) reduces to (3b), i.e., to the conventional form of the molecular weight equation in turbidimetry. If we are dealing with a solvent-precipitant mixture in which the two components have the same refractive index, $\partial n/\partial\psi=0$, and again we arrive at Eq. (3b). If the two components have different refractive indices, then in general at values of $\psi<1$, the term $\alpha(\partial n/\partial\psi)\neq 0$, but may have either positive or negative values. If α is assumed to be positive, if $\partial n/\partial c>0$ and $\partial n/\partial\psi>0$, $(\tau/c)_0$ will be greater than in the pure solvent and the value of M obtained from Eq. (3b) will be higher than the true molecular weight. If $\alpha\partial n/\partial\psi>0$ and $\partial n/\partial c<0$, the value of M may be too small or too large, depending on the numerical magnitudes of these terms. If $\alpha\partial n/\partial\psi<0$ and $\partial n/\partial c>0$, the value of M in (3b) may again deviate either positively or negatively from the true molecular weight. If $\partial n/\partial c=0$ in a system such that $\alpha\partial n/\partial\psi\neq 0$, the theory predicts a finite value of $(\tau/c)_0$ which is traceable solely to selective absorption.

The unpredictable feature of the theory is the presence of the factor α introduced in Eq. (6). This factor is not a constant for any one system, nor does it appear to behave in the same way for different systems. It seems reasonable to suppose on general grounds that it may have positive values less than or of the order of unity in solvent-precipitant mixtures, and that it vanishes in pure solvents. It appears from experiment that it may sometimes have negative values for very small precipitant concentrations. Fortunately there has appeared to be no need to make more precise assumptions about this factor, since those mentioned above have enabled us to discuss the equations of the theory in all interesting cases.

The experiments presented here may be considered as a method for the determination of α or $(\partial\psi_1/\partial c)$. The experimental significance of this factor α is clear and unambiguous. It measures the change in composition of the solvent-precipitant mixture surrounding the particles due to selective absorption. A complete theoretical account of the factor α will have to await future developments.

The preceding paragraphs deal exclusively with very dilute solutions. It therefore seems appropriate to include a few remarks about the slopes of scattering curves at finite concentrations. The slope of the light scattering curve for a given system is predicted by the usual theory to be twice the slope of the corresponding osmotic pressure curve. Now it is well known that the introduction of a precipitant has the effect of reducing the slope of the osmotic pressure curve, and that this reduction in slope may be traced to the heat of mixing of the polymer and solvent medium. Since as c approaches zero, the ratio of the partial molal heat of mixing to the concentration also approaches zero, the lowering of osmotic pressure at finite concentrations does not lead to any change in the intercept at infinite dilution. Hence the use of solvent-precipitant mixtures, regardless of selective absorption, has no effect on the apparent molecular weight as derived from osmotic measurements. On the other hand, selective absorption produces a purely optical effect on the turbidity—an effect which is additive and proportional to the polymer concentration, and which therefore persists in the limiting value, $(c/\tau)_0$.

EXPERIMENTAL PART

Turbidity Measurements

All turbidities were measured by 90° scattering of monochromatic light. Absolute values were obtained by comparison with primary standards whose turbidities had been determined by transmission measurements, upon the assumption that Beer's law is valid. Correction for the turbidity of the solvent was made in all cases. No corrections for dissymmetry were introduced. Recent measurements have shown that although the dissymmetry of scattering by the polymers here studied was sufficient to necessitate an appreciable correction (about 12 percent) to the calculated molecular weights, there is very little variation of this correction with the composition of a solvent-precipitant mixture. Hence, the omission of such corrections does not affect the comparison of molecular weight values obtained on any one polymer sample in different mixtures.

Refractive Index

Refractive index measurements were made with an Abbe refractometer which was equipped with compensating Amici prisms adjusted so as to yield correct values of refractive index for $\lambda = 5890$A when dispersion was eliminated. Refractive indices at other wave-lengths were calculated with the aid of a dispersion chart furnished with the instrument.

Refractive Index Gradient with Polymer Concentration

All measurements of this quantity reported here have been made by means of an instrument built substantially according to a design obtained from Dr. P. P. Debye.[9] This is a differential instrument and has a sensitivity of approximately 5×10^{-6}. The instrument enables one to measure the angular deviation of light which passes through a hollow prism immersed in solvent, when the prism is first filled with the solvent and then with a dilute solution in the same solvent.

Materials

Most of the measurements on polystyrene were made with a fractionated sample which was derived from a thermally polymerized material. A few measurements were also made with a commercial unfractionated polystyrene. The polymethylmethacrylate used was also a fractionated material polymerized with the aid of benzoyl peroxide.

Solvents and precipitants were usually of c.p. grade. Where it was necessary to use commercial materials, these were suitably purified.

RESULTS AND DISCUSSION

This research was initiated in an attempt to extend the turbidimetric method of determining polymeric molecular weights to systems in which the dispersion medium, instead of being a pure solvent, is a mixture of a solvent and a precipitant. The experimental results will be presented by means of the conventional type of scattering curves, in conformity with the equation of Debye

$$H\frac{c}{\tau}=\frac{1}{M}+2Bc, \qquad (7)$$

where

c = concentration of polymer in grams per ml of solution.
τ = turbidity.
M = molecular weight.
B = a constant characteristic of the solvent-solute system.
$H=\frac{32\pi^3}{3\lambda^4 N_0}n^2\left(\frac{\partial n}{\partial c}\right)^2$.
N_0 = Avogadro's number.
λ = wave-length of exciting radiation.
n = refractive index of the solvent.
$\frac{\partial n}{\partial c}$ = refractive index gradient with respect to polymer concentration.

[9] P. P. Debye, Bull. Am. Phys. Soc. 21, 16 (1946).

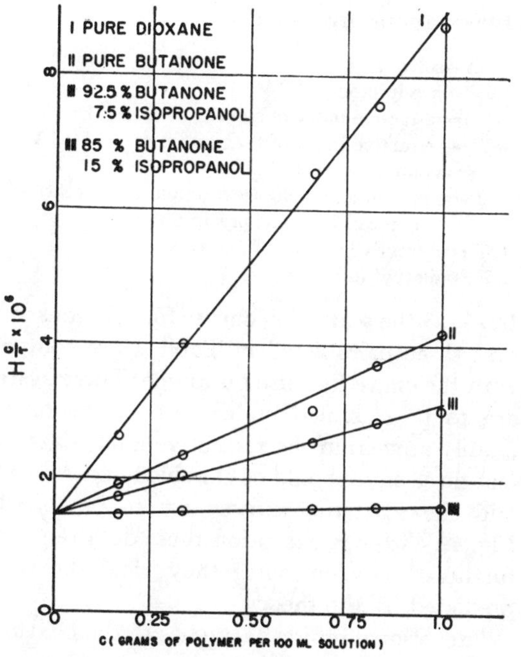

FIG. 1. Scattering curves: fractionated polystyrene.

If the linear relationship (7) be extrapolated to $c=0$, the result may be expressed

$$H(c/\tau)_0 = 1/M. \tag{7a}$$

This is equivalent to Eq. (3b), which was shown in a previous section to be valid for pure solvents.

If the solvent is not a pure compound, but is instead a mixture of a solvent and a precipitant, then the quantity M is in general not the true molecular weight but differs from the true value by an amount which depends on the composition of the entire system. The discrepancy occurs because of failure in the calculation of H to take account of selective absorption of solvent by the polymer. It is, nevertheless, convenient to represent the results of experiments with solvent-precipitant systems in terms of the Debye equation, retaining the definition of H which was given following Eq. (7). If this is done, then visual inspection of the scattering curves shows immediately which mixtures give polymeric molecular weights in agreement with those found in pure solvents.

Much of the experimental work was done prior to development of the theory. However the results will be considered under three cases, each of which is regarded as a test of the validity of Eq. (6a).

Case (1) $\alpha\partial n/\partial\psi$ Is Negligible with Respect to $\partial n/\partial c$

This condition is fulfilled if either α or $\partial n/\partial\psi$ is sufficiently small. Presumably α vanishes in a pure solvent but assumes appreciable values in solvent-precipitant mixtures. Hence all pure solvents fall under this case. α is probably very small for mixtures which are sufficiently poor in precipitant, but these are of slight practical interest and have not been studied. $\partial n/\partial\psi$ may be made small by the choice of a solvent and precipitant which have nearly equal refractive indices and a sufficiently small volume change on mixing.

Preliminary turbidimetric data are presented in Fig. 1. The polymer used was a polystyrene fraction having an intrinsic viscosity of 2.06 and an osmotic molecular weight of 0.8×10^6. Inspection of curves I and II in Fig. 1 makes it clear that the data in pure solvents may be represented by straight lines having the same intercept at $c=0$. The light scattering molecular weight is 0.69×10^6, from Eq. (7a). The discrepancy between this and the osmotic value is probably because of the inaccuracy of the latter at high values of M.

Choice of the solvent-precipitant pair, butanone-isopropanol, is a very fortunate one, since $\partial n/\partial\psi = 0.0004$, whereas $\partial n/\partial c = 0.219$ according to our measurements. Inspection of curves III and IV in Fig. 1 shows at once that extrapolation to $c=0$ leads to precisely the same intercept as that found with pure solvents. Thus all the data of Fig. 1 are in accord with theory.

Case (2) $\alpha\partial n/\partial\psi$ and $\partial n/\partial c$ Both Finite and of the Same Order

The solvent-precipitant pair, benzene-methanol, is used for the discussion of this case. The polymer was an unfractionated polystyrene. In mixtures of benzene and methanol it is found that

$$\frac{\partial n}{\partial c} = 0.108 + 0.1447(1-\psi),$$

$$\frac{\partial n}{\partial\psi} = 0.16, \text{ and is practically independent of } \psi.$$

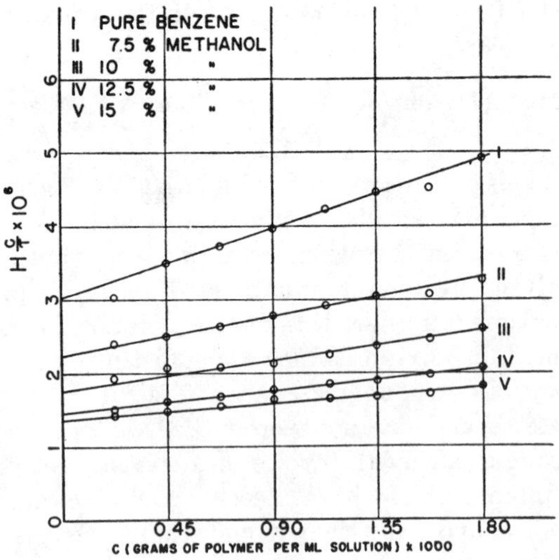

FIG. 2. Unfractionated polystyrene in $C_6H_6-CH_3OH$ mixtures.

Data on scattering are presented in Fig. 2. In order to explain these curves in terms of Eq. (6a), we must assume that α is positive and of the order of unity for $\psi<1$, also that α increases with increasing methanol concentration. Each of these assumptions appears physically reasonable. If they are granted, then theory predicts that the term $\alpha\partial n/\partial\psi$ will make a finite contribution to $(\tau/c)_0$ and that this contribution will increase as the methanol concentration increases. The steady decrease of $H(c/\tau)_0$ with increasing methanol content bears this out.

Further Examples Illustrating Cases (1) and (2)

A somewhat broader study of solvent-precipitant mixtures matched with respect to refractive index was suggested by the data presented in Figs. 1 and 2. Hence the turbidities of polystyrene solutions in the mixtures listed in Table I have been investigated. In this table, the following abbreviations are used:

S = solvent.
P = precipitant.
$n(S)$ = refractive index of solvent at $\lambda=5461$A.
$n(P)$ = refractive index of precipitant at $\lambda=5461$A.
ψ = volume fraction of solvent.
ψ_P = approximate volume fraction of solvent at the point of incipient precipitation.
$n(SP)$ = refractive index of mixture.
DEF = diethyl fumarate.

In Fig. 3 the scattering curves for solutions in the first three pairs listed in Table I are compared with the curve for pure butanone. These systems are to be classified under case (1) as may be readily seen from the refractive index data. The last three pairs should be classified under case (2), and the scattering curves are to be found in Fig. 4. With one exception these data require no further discussion, since they yield the results predicted by the theory.

Exceptional interest attaches to the mixture of butanone and 1,3 dichloro-isopropanol, since in this case $\partial n/\partial\psi<0$ and $\partial n/\partial c>0$. Thus application of the simple theory should lead to a value of M less than the true molecular weight. Inspection of curve II in Fig. 4 shows this to be the case, since the extrapolated curve has too high an intercept.

Case (3) $\partial n/\partial c$ Is Negligible with Respect to $\alpha\partial n/\partial\psi$

The system selected for study of this case consisted of a fractionated polymethylmethacrylate dissolved in mixtures of bromobenzene and methanol. It has been found that

$$\frac{\partial n}{\partial c}=-0.054+(1.568-n_0)\times1.06,$$

$$\frac{\partial n}{\partial\psi}=0.232,$$

TABLE I.

S	P	ψ	ψ_P	$n(S)$	$n(P)$	$n(SP)$	$\frac{\partial n}{\partial c}$	$\frac{\partial n}{\partial\psi}$
$C_2H_5COCH_3$	$(CH_3)_2CHOH$	0.85	0.80	1.3740	1.3736	1.3732	0.219	0.0004
DEF	CH_2ClCH_2OH	0.80	0.74	1.4372	1.4385	1.436	0.152	−0.0013
C_6H_6	$(CH_2Cl)_2CHOH$	0.33	0.28	1.495	1.476	1.483	0.112	0.019
$C_2H_5COCH_3$	$(CH_2Cl)_2CHOH$	0.46	0.33	1.3740	1.476	1.433	0.165	−0.102
C_6H_6	$(CH_3)_2CHOH$	0.65	0.59	1.4954	1.3736	1.450	0.154	0.122
C_6H_6	CH_3OH	0.80	0.75	1.4954	1.33	1.460	0.137	0.16

where n_0 = refractive index of mixture in question at $\lambda = 4358$A.

Data of the usual sort were obtained in several mixtures near the point where $\partial n/\partial c$ vanishes. Values of $(\tau/c)_0$ were obtained by extrapolation and were compared with the corresponding values of $\partial n/\partial c$ by referring both to n_0 as a parameter. This is shown graphically in Fig. 5. It is of interest to note that at the point where $\partial n/\partial c = 0$, $(\tau/c)_0$ has a finite value which is solely caused by selective absorption of solvent by dissolved polymer. This is, of course, in accord with our present theory.

The Magnitude of Selective Absorption

We have already noted the experimental significance of the factor α. It gives us at once the magnitude of the change in composition of the solvent-precipitant mixture surrounding the particles due to removal of solvent by selective absorption. If we now wish to calculate the composition of the solvent-precipitant mixture within the particles, we must introduce an assumption concerning the physical size of the particles in the solution. A reasonable and convenient assumption for this purpose is furnished

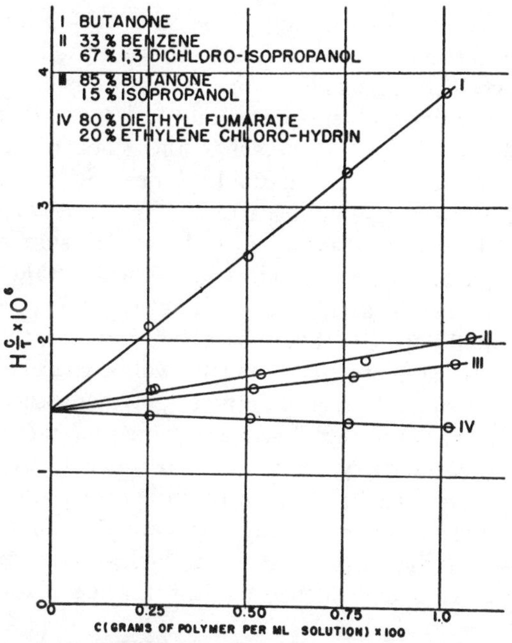

FIG. 3. Fractionated polystyrene in solvent-precipitant pairs matched with respect to refractive index.

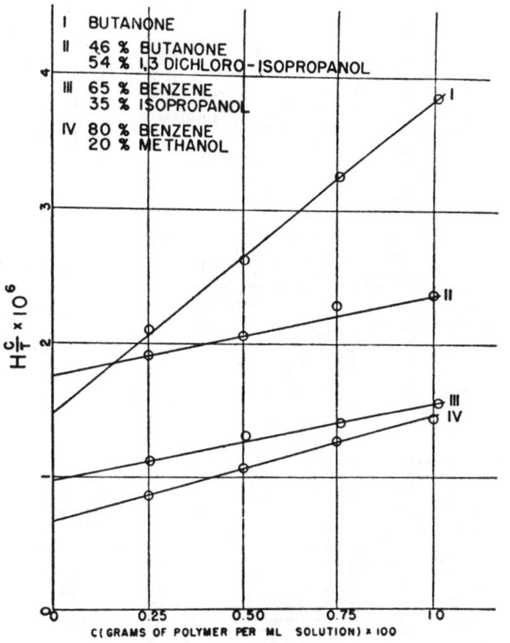

FIG. 4. Fractionated polystyrene in solvent-precipitant pairs unmatched with respect to refractive index.

by consideration of the intrinsic viscosity of the solution.

If we give the correct dimensions to the absolute intrinsic viscosity, $\{\eta\}$, (with concentration measured in grams per ml.) then $\{\eta\}$ is expressed in ml per gram, and Einstein's theory of viscosity leads to

$$\{\eta\} = 2.50\frac{v}{m}$$

in which v is the volume of the particle and m is its mass. In these units the absolute intrinsic viscosity $\{\eta\}$ is 100 times the conventional intrinsic viscosity, usually denoted by $[\eta]$. If, for instance, $[\eta] = 1.5$, this means that $\{\eta\} = 150$ ml per gram, and the polymer behaves in solution as if it possessed a specific volume of 60 ml per gram.

Now if we adopt this picture, it follows that

$$\alpha = \frac{\partial\psi}{\partial c} = \left(\frac{v}{m} - \bar{v}\right)(\psi_2 - \psi_1),$$

where $\bar{v}$ is the partial specific volume of the solute ($\bar{v} = 0.92$ for polystyrene in benzene). If we calculate $\partial\psi/\partial c$ from light scattering and v/m from viscosity data, we can obtain $(\psi_2 - \psi_1)$ at once. $(\psi_2 - \psi_1)$ is, of course, the difference in the concen-

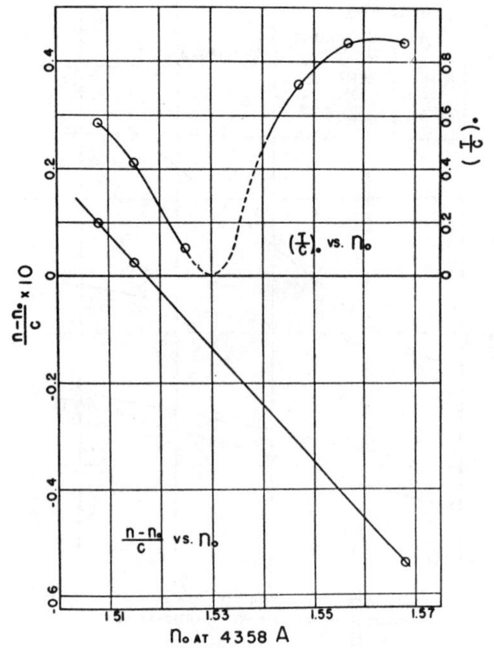

FIG. 5. Polymethylmethacrylate in mixtures of $C_6H_5Br-CH_3OH$.

trations of solvent in the solvent-precipitant mixture inside and outside the particles. The most convenient means of calculating α is to note that

$$\left(1+\alpha\frac{\frac{\partial n}{\partial \psi}}{\frac{\partial n}{\partial c}}\right)^2 = \frac{\left[H\left(\frac{c}{\tau}\right)_0\right]\text{pure solvent,}}{\left[H\left(\frac{c}{\tau}\right)_0\right]\text{solvent-precipitant mixture.}}$$

The results shown in the last two columns of Table II are derived from these considerations, on the basis of the data represented in Fig. 2. A similar calculation at $\psi=0.80$ from the data of Fig. 4 shows (since in this case $\{\eta\}=110$ and $v/m=44$) that $\alpha=0.42$, and $\psi_2-\psi_1=0.97\times10^{-2}$. This appears to be a remarkably good fit, since the two items of data were obtained by independent experimenters on different polymer samples in different laboratories.

GENERAL DISCUSSION

It might be thought that the phenomenon of increased turbidity with the addition of a precipitant to a polymer solution is caused by preflocculation or Van der Waals association of the solute. However, such an assumption is not only unnecessary to account for the observed results presented in this paper, but is indeed incompatible with these results. Thus, for example, the association hypothesis would require that the scattering curves bend upward to the correct value of $1/M$ at $c=0$, since at infinite dilution, association should vanish. This was not found to be the case from the data plotted in Fig. 2. Also in the cases of curves III and IV in Fig. 1, curve II in Fig. 4, and the picture presented by Fig. 5, the association hypothesis fails completely. It is shown furthermore to be untenable from consideration of the data on ψ_P in Table I, since the curves of Fig. 3, all of which lead to correct molecular weight values, represent systems in which the solvent compositions are much closer to the precipitation points than those of the curves in Fig. 2, all of which lead to values of M higher than the true molecular weight.

Measurements of molecular weight in mixed solvents by osmotic pressure also indicate that the displacement of the intercepts of the conventional $H(c/\tau)$ curves is an optical phenomenon and not caused by association. A. Dobry[10] studied the osmotic pressures of cellulose acetate in chloroform-alcohol mixtures and found that the conventional curves (pressure/c *versus* c) exhibited decreasing slopes as the precipitant concentration was increased. However, all the curves had the same intercept, regardless of the composition solvent-precipitant mixture. It is well known, of course, that the slope of the osmotic pressure curve should be one-half that of the conventional light scattering curve, also that the intercept at $c=0$ is a direct measure of molecular weight. Geoffrey Gee[11] has obtained similar results and has pointed out the usefulness of small slopes. It should be noted, however, that osmotic pressure measurements will be less sensitive to small degrees of polymer-polymer association than will light scattering measurements, since osmometry leads to a number average molecular weight whereas turbidimetry leads to a weight average. It thus appears that in the cases studied light scattering data present the most decisive evidence against the association

[10] A. Dobry, J. chim. phys. **36**, 102 (1939).
[11] Geoffrey Gee, Trans. Faraday Soc. **36**, 1171–8 (1940); **40**, 463–8 (1944).

hypothesis and in favor of selective absorption. This does not excluse the possibility that in some systems polymer-polymer association may occur prior to incipient precipitation. However, light scattering data should be interpreted as indicating polymer-polymer association in a solvent-precipitant mixture only after a suitable correction for the effect of selective absorption of the solvent has been made.

The increase of turbidity with added precipitant in high polymer solutions is a very great advantage in the determination of molecular weights by turbidimetry. It not only raises the relative accuracy obtainable in the measurement of absolute turbidity, but at the same time greatly reduces the slope of the conventional scattering curves. The latter circumstance enables one safely to make the linear extrapolation to infinite dilution from a smaller amount of data than would be necessary in case the data had to be relied on to determine both a high slope and the intercept. It is probable that in most cases conditions can be found which reduce the slope

TABLE II.

ψ	$[\eta]$	$\frac{v}{m}$	α	$(\psi_2-\psi_1)\times 100$
1	131.5	52.6	0	0
0.925	122.0	48.8	0.12	0.25
0.90	118.7	47.5	0.24	0.51
0.875	110.6	44.2	0.34	0.79
0.85	102.6	41.0	0.40	1.00

practically to zero, so that τ/c may be set equal to $(\tau/c)_0$. The sole condition which must be satisfied is that $\partial n/\partial\psi$ is negligible with respect to $\partial n/\partial c$ in the system considered.

The theory presented by the writers gives a satisfactory correlation of all the observed facts. In particular it accounts for the apparently anomolous molecular weights of high polymers as obtained by the application of simple scattering theory to measurements in solvent-precipitant systems. This theory also enables us to estimate the magnitude of selective absorption in these systems.

MISCELLANEOUS

APPROXIMATE FORMULAE FOR THE CYLINDER FUNCTIONS FOR LARGE VALUES OF THE ARGUMENT AND ANY VALUE OF THE INDEX.
(Näherungsformeln für die Zylinderfunktionen für grosse Werte des Arguments und unbeschränkt veränderliche Werte des Index)

P. Debye

Translated from
Mathematische Annalen, pages 535-558, 1910.

I. Introductory Remarks

In the extensive literature on the cylinder functions two kinds of series developments are found almost exclusively, both of which are, by the way, directly adapted to the nature of their differential equation:

$$\frac{d^2u}{dx^2}+\frac{1}{x}\frac{du}{dx}+\left(1-\frac{\alpha^2}{x^2}\right)u=0 \text{ —}$$

Series of the one kind are ordinary power series in x, expanded about the singular point $x = 0$; they converge in the entire plane Those of the other kind, investigated most thoroughly by Hankel, are expanded about $x = \infty$; they are power series in $1/x$, and are semi-convergent according to Poincaré's definition. While those of the first kind are useful in practical calculations only for small values of x, Hankel's developments are useful for large values of the argument. However, it can be seen immediately from the formulae for the latter that they are valid only when the index α is small compared to the argument x, a circumstance which limits their usefulness in those optical problems in which light encounters a body large compared with the wavelength of the light. Similar difficulties are met quite generally at small wavelengths, not only in the case of spherical or cylindrical symmetry, and are to be overcome in an analogous way. For example, in a calculation of the electromagnetic field resulting from an incident plane wave falling on a cylinder, one obtains easily for any component of the electric field an expansion in cylinder functions (or more complicated complexes of such functions) whose argument is proportional to a large number (cylinder radius divided by the wavelength) and whose index takes all values from 0 to ∞. Terms in which α is comparable to x, in which case the Hankel approximation fails, make an appreciable contribution to the value of this series. In the following, it will be our purpose to find expansions which approach the cylinder functions asymptotically for large values of x and for any value of the index. In this paper we will limit ourselves to the simplest case, that in which α and x are real. I intend to give in another place the extension to complex values of α and x; this can be done by the same method used here. Meanwhile, in my Munich dissertation, I have shown how the formulae are to be used in a calculation of light pressure on a sphere, while their application to the theory of diffraction was indicated in an address to the Scientific Congress at Cologne.[2] It should be noted, however, that Lord Rayleigh[2] had already pointed out the necessity of knowing approximate representations of the cylinder functions of the type defined above in order to solve optical problems dealing with spheres and cylinders. In the following we will show how such representations follow naturally if use is made of contour integral representations of the cylinder functions. The manner of derivation is not new in itself; the principle of the method, as I noticed later, is clearly put forth in a fragment left by Riemann[3] and completed by H. A. Schwarz: "Sullo svolgimento del quoziente di due serie ipergeometriche in frazione continua infinita." The method is used in this work of Riemann's to determine the limit of the hypergeometric series:

$$H(a+n,\ b+n,\ c+2n,\ x)$$

for very large values of n. By means of the same method, Graf and Gubler[4] in their book determined the limit:

$$\lim_{a=\infty} J_a(ax)$$

of the Bessel function for real values of α and x. Nicholson[5] recently attacked the problem treated in the following with another, more formal, method. Basing his work on that of Lorenz (Oeuvres Scientifiques I, p. 405), he treats only the special case in which the lower index is an odd multiple of $\frac{1}{2}$ and less than the argument which is assumed large. In a later note[6] the case is also treated in which the argument and index are nearly equal. However, only the constant terms of the expansion coefficients B_n (compare the summary in section VII) are obtained so that joining with the cases $\alpha < x$ or $\alpha > x$ is not achieved. The original formula of Lorenz agrees with the first terms of the series (54) and (55) of section VII, as can easily be shown by setting $\alpha = n + \frac{1}{2}$.

II. The Riemann Method and Its Application to the Cylinder Functions

As a point of departure we choose Nielsen's definition of the two Hankel functions $H_1^{\alpha}(x)$ and $H_2^{\alpha}(x)$ which, according to the Work of Sommerfeld[7] on the "Mathematical Theory of Diffraction" can be written in the form:

$$(1)\qquad \begin{aligned} H_1^{\alpha}(x) &= -\frac{1}{\pi}\int_{(1)} e^{-ix\sin\tau} e^{i\alpha\tau}\, d\tau \\ H_2^{\alpha}(x) &= -\frac{1}{\pi}\int_{(2)} e^{-ix\sin\tau} e^{i\alpha\tau}\, d\tau \end{aligned}$$

Because we limit ourselves to the case of real values of x, the path (1) over which the first integral is to be done begins between $\pi/2 - i\infty$ and $-\pi/2 - i\infty$ and ends between $-\pi/2 + i\infty$ and $-3\pi/2 + i\infty$, while the contour (2) of the second integral begins

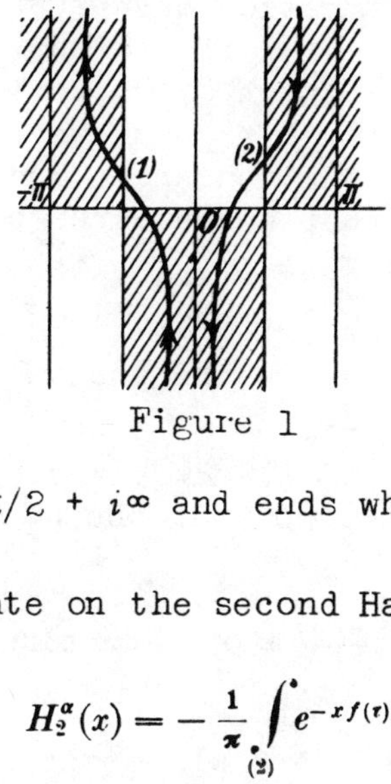

Figure 1

between $3\pi/2 + i\infty$ and $\pi/2 + i\infty$ and ends where the contour (1) begins.

First we concentrate on the second Hankel function and write it in the form:

$$(2)\qquad H_2^{\alpha}(x) = -\frac{1}{\pi}\int_{(2)} e^{-xf(\tau)}\, d\tau$$

where $f(\tau)$ is an abbreviation for the function:

$$(3)\qquad f(\tau) = i(\sin\tau - \xi\tau); \quad \xi = \frac{\alpha}{x}$$

If $f(\tau)$ is resolved into its real and imaginary parts:

$$(4)\qquad f(\tau) = R + iJ$$

then the absolute value of the integrand is e^{-xR}. For large values of x, in first approximation only the vicinities of points along the contour where R assumes its smallest value make an appreciable contribution to the integral. We put $\tau = a+ib$ and imagine R to be the ordinate of a surface spanning the a, b plane and seek to specialize the contour so that conditions for evaluation of the integral are most favorable.

First we note the following. Let the integration be carried out up to a certain point; of all the possible directions the path of integration can take at this point, we choose that in which R increases most rapidly. If s is this preferred direction and n perpendicular to it, then $\partial R/\partial n = 0$, or since $\partial R/\partial n = \partial J/\partial s$ the path which we wish to follow is given by the equation:

$$J = \text{const.}$$

That such a curve having the end points prescribed above exists is not certain and in general not the case. However, it is clear that if it exists, conditions along the curve J = const. are most favorable for evaluation of the integral for large x. Therefore, we try to find out whether or not such a curve with the prescribed end points actually exists. If we observe that (on convergence grounds) the location of each end point is chosen so that R is positively infinite there, it is clear that no matter what other properties the path of integration may have, R must have at least one minimum on this curve. Now, if the curve is the proposed J = const., along which by definition $\partial R/\partial n$ vanishes everywhere, it must pass at least once through a point where the derivative of R with respect to all directions vanishes. Such a point is necessarily a saddle point, so that we have the result:

Only those curves J = const. which pass through a saddle point can have their end points in the prescribed regions.

Above all, we must find how the position of the saddle point depends on the ratio $\alpha/x = \xi$.

Since, at the saddle point, $\partial R/\partial s$ vanishes for any direction s, the same holds for $\partial J/\partial s$, so that the position of the saddle point is determined by:

$$\frac{df(\tau)}{d\tau} = 0, \quad i.e. \quad \cos\tau = \xi \tag{5}$$

As is well known, (5) maps the ξ plane cut from 1 to ∞ and -1 to $-\infty$ on the strip $0 \ldots \pi$ of the τ plane, so that for any complex value of ξ an associated saddle point always lies interior to the strip $0 \ldots \pi$, for example, at $\tau = \tau_0$. Such points lie also at $\tau = \tau_0 + 2n\pi$ and $\tau = -\tau_0 + 2n\pi$, where:

$$n = 0, 1, 2, 3, \cdots, -1, -2, -3, \cdots$$

In our case, in which α and x are both assumed real, ξ is always real and can take all values from 0 to ∞; simultaneously, the associated saddle point $\tau = \tau_0$ takes values from $\pi/2$ to 0 on the real τ axis and from 0 to $0 - i\infty$ on the imaginary τ axis in such a way that the values $\tau = \pi/2, 0, -i\infty$ correspond to the values

$\xi = 0, 1, \infty$. At the same time, the second saddle point $\tau = -\tau_0$ takes values from $-\pi/2$ to 0 to $+i\infty$.

Now we will investigate the curves J = const. which pass through our saddle point $\tau = \tau_0$. Here we distinguish two cases, namely:

(1) $0 < \xi < 1$, that is, $0 < \alpha < x$

Saddle point on the real axis.

(2) $1 < \xi < \infty$, that is, $x < \alpha < \infty$

Saddle point on the imaginary axis.

If we put $\tau_0 = a_0 + ib_0$, we obtain in case (1) (in which $b_0 = 0$) for a curve J = const. passing through the saddle point (compare (3)) the formula:

$$\mathfrak{I}\,[i\,(\sin\tau - \tau\cos\tau_0)] = \mathfrak{I}\,[i\,(\sin\tau_0 - \tau_0\cos\tau_0)]^{*)} \tag{6}$$

that is:

$$\sin a \operatorname{Cos} b - a\cos a_0 = \sin a_0 - a_0\cos a_0 \tag{6'}$$

or:

$$\operatorname{Cos} b = \frac{\sin a_0 + (a - a_0)\cos a_0}{\sin a} \tag{6''}$$

If $\sin a$ is plotted against a and it is observed that the numerator of the right-hand side is the tangent to this curve at $a = a_0$, it is seen that while a varies from 0 to π the right-hand side varies from ∞ to 1, attaining the latter value at $a = a_0$, from which point on the right-hand side increases until it attains the value $+\infty$ at $a = \pi$. Therefore, when $\xi < 1$, the curves J = const. always have the form shown in Figure 2 and lie interior to the strip $0 \ldots \pi$. The

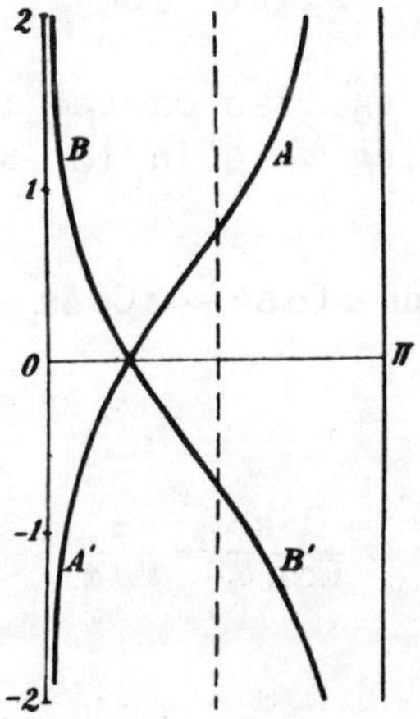

Figure 2

*$\mathfrak{I}$ = Imaginary part.

curve branches B and A', which are mirror images of each other, correspond to values of a such that $0 < a < a_0$; the branches A and B', also mirror images, to values of a such that $a_0 < a < \pi$.

The angles formed by the curves where they meet at the saddle point are all equal to $\pi/2$, as is easy to see, so that A goes continuously into A' and B into B'. On one of the two curves (BB'), R decreases on either side of the saddle point as rapidly as possible; on the other (AA'), R increases as rapidly as possible. Thus the latter satisfies our condition and besides has the prescribed end points, and therefore it will be used as the contour of integration.

In Figure 3a the way in which R varies along this curve is shown schematically.

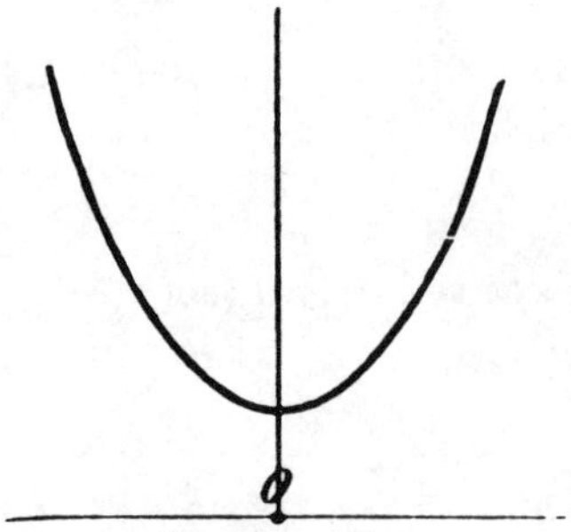

Figure 3a

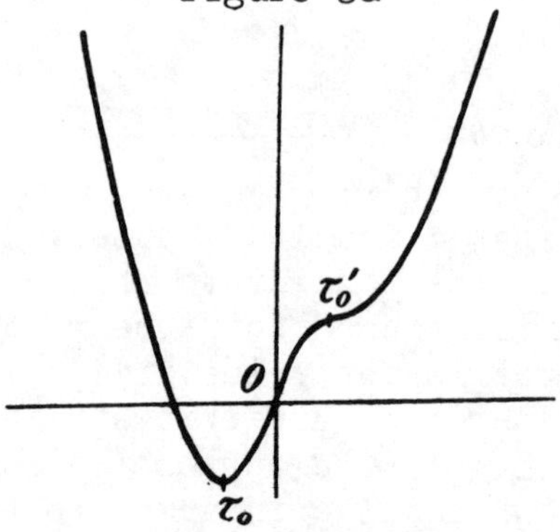

Figure 3b

In case (2) $(\xi > 1)$ τ_0 lies on the imaginary τ axis so that $a_0 = 0$, and by substituting this in (6) we obtain for the equation of a curve J = const.:

$$\sin a \operatorname{Cos} b - a \operatorname{Cos} b_0 = 0 \tag{7}$$

or:

$$\frac{\operatorname{Cos} b}{\operatorname{Cos} b_0} = \frac{a}{\sin a} \tag{7'}$$

According to (7) or (7') one of the curves passing the point $\tau = \tau_0$ is identical with the imaginary τ axis; the other goes from $-\pi - i\infty$ through the saddle point τ_0 to $+\pi - i\infty$.

The curves J = const. passing through the second saddle point (compare Figure 4) $\tau = \tau_0' = -\tau_0$ are obtained by reflecting the previous curves in the real τ axis; thus these are first the imaginary τ axis and second a curve which begins at $-\pi + i\infty$, passes through the saddle point $\tau = \tau_0$, and ends at $+\pi + i\infty$.

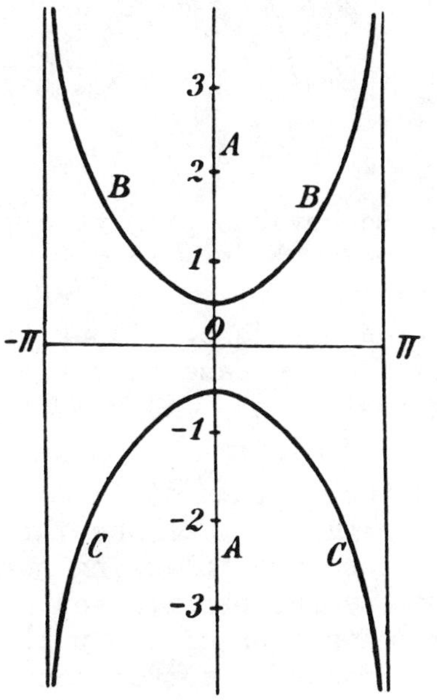

Figure 4

A curve J = const. joining the prescribed end points must now pass through both saddle points $\tau = -\tau_0$ and $\tau = +\tau_0$; as is obvious from the figure, such a curve is curve B from $\pi + i\infty$ to $\tau = -\tau_0$ and "curve A" (imaginary axis) from $\tau = -\tau_0$ to $\tau = \tau_0$ to $\tau = -i\infty$. On the part from $\tau = \pi + i\infty$ to $\tau = \tau_0' = -\tau_0$ to $\tau = +\tau_0$, R decreases steadily in a way shown in Figure 3b and from the last point on increases as rapidly as possible.

The limiting case $\xi = 1$ is a special one in which the two saddle points τ_0 and $-\tau_0$ coincide at the point $\tau = 0$, so that $f(\tau)$ behaves like τ^3 in the neighborhood of this point (compare equation 3). Correspondingly, three curves J = const. pass through this point at an angle of 60° to one another: the imaginary τ axis A and two transcendental curves B, C which have the equation:

$$\text{Cos}\, b = \frac{a}{\sin a} \tag{8}$$

and which are shown in Figure 5.

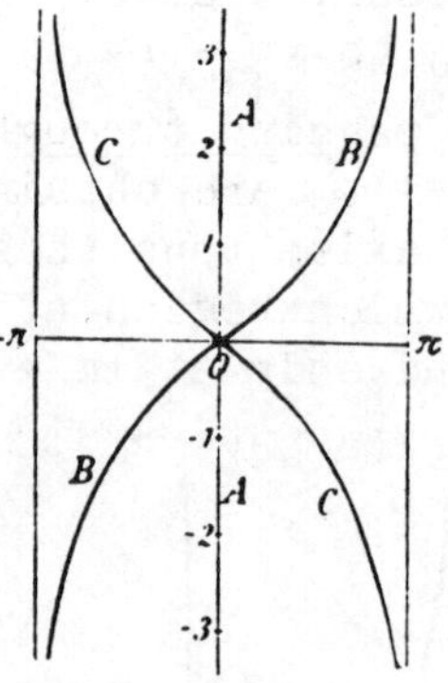

Figure 5

The following is the obvious path to choose as contour of integration: Curve B from $\tau = \pi + i\infty$ to $\tau = 0$ and the imaginary axis A from $\tau = 0$ to $\tau = 0 - i\infty$, on both of which from $\tau = 0$ on, R increases as rapidly as possible. Along this curve, R behaves qualitatively as it did in the case illustrated in Figure 3a, the only difference being that at $\tau = -\tau_0 = \tau_0'$ it touches the horizontal axis.

Having determined the contour of integration, we must now turn our attention to the carrying out of the integration. In order to fix the ideas, we will base the following analysis on Figure 2; that is, we assume $\xi < 1$. The other cases can be treated in exactly the same way. The definition of the path of integration by the formula $J = J_0 = \text{const.}$ makes it convenient not to retain the variable τ for the actual carrying out of the integral, but to introduce $f(\tau)$ or $f(\tau)$ - const. as a new variable. If we choose for the constant the value $f(\tau_0)$ taken on by $f(\tau)$ at the saddle point, then in the t plane defined by:

$$t = f(\tau) - f(\tau_0) \tag{9}$$

the integration is to be done along the real axis. Integration in the τ plane along path A from $\tau = \pi + i\infty$ to $\tau = \tau_0$ corresponds to integration from $t = +\infty$ to $t = 0$ in the t plane. Now, if it is observed that according to the definition of a saddle point, in the neighborhood of the point $\tau = \tau_0$ the function:

$$t = f(\tau) - f(\tau_0)$$

behaves like $(\tau - \tau_0)^2$, it is seen that the rotation through the angle π occurring on passage from a point of the upper part of the path A to a point of the lower part corresponds to a rotation through the angle 2π in the t plane. Thus, integration along the lower half of the curve A corresponds to integration from 0 to ∞

in the t plane. Upon introduction of our new variable of integration, the integral representation of our function $H_2^\alpha(x)$ takes the form:

$$H_2^\alpha(x) = \frac{e^{-xf(\tau_0)}}{\pi}\left(\int_0^\infty e^{-xt}\Phi(t)\,dt - \int_0^\infty e^{-xt}\Phi_1(t)\,dt\right) \tag{10}$$

where $\phi(t)$ and $\phi_1(t)$ are given as a function of t by (9) through the formulae:

$$\Phi(t), \quad \text{resp.} \quad \Phi_1(t) = \frac{1}{f'(\tau)} \tag{11}$$

These are distinguished only by the definition of the root $t^{\frac{1}{2}}$ in powers of which they are to be expanded about the origin of the t plane. In a first approximation $\phi(t)$ and $\phi_1(t)$ are given by const. $t^{-\frac{1}{2}}$ so that they become infinite in a permissible way. Accordingly, an asymptotic expression for H^α can be obtained from this approximation. If, on the other hand, several terms of the power series (in $t^{\frac{1}{2}}$) for $\phi(t)$ and $\phi_1(t)$ are retained, this gives a series for $H_2^\alpha(x)$ with the same number of terms which for constant ratio α/x is a power series in $x^{-1/2}$. Of course, the series so obtained does not converge; however, it is semiconvergent in the sense of Poincaré* and therefore excellently suited to practical calculations. The one problem which remains is thus of elementary nature and consists of a determination of the development coefficients of a quantity expressed only implicitly as a function of t.

III. The semi-convergent series in the first case ($0<\alpha<x$)

According to the discussion of section II, the function $H_2^\alpha(x)$ is defined for our purposes by equation (10), so the problem is to develop the function $\phi(t)$ given by (11) in powers of $t^{\frac{1}{2}}$. Here it is of course possible to proceed in different ways. The following procedure makes the rulé for forming the coefficients relatively clear. Temporarily, set:

$$f(\tau) - f(\tau_0) = t = T^2 \tag{12}$$

*The proof of the semi-convergence is easy if one notes that there exists an inequality series in t for $\phi(t)$ for the remainder of the broken off power similar to that used by E. B. Borel, "Leçons sur les séries divergentes," Paris, 1901, p. 25 in the Cauchy proof of the semi-convergence of the Stirling series.

where $f(\tau)$ is the function of τ defined by equation (3), so that according to (11) and (28):

$$(13) \qquad \Phi(t) = \frac{1}{f'(\tau)} = \frac{1}{2T}\frac{d\tau}{dT}$$

For $\tau = \tau_0$, or, what is the same thing, $T = 0$, $\frac{d\tau}{dT}$ remains finite so that a development in positive powers of T is possible. Thus if we set:

$$(14) \qquad \frac{d\tau}{dT} = \sum_0^\infty a_n T^n$$

then according to Cauchy:

$$(15) \qquad a_n = \frac{1}{2\pi i}\int \frac{dT}{T^{n+1}}\frac{d\tau}{dT}$$

where the integral is to be taken around the origin of the T plane, or if we go back to the τ plane by means of (12):

$$(16) \qquad a_n = \frac{1}{2\pi i}\int \{f(\tau) - f(\tau_0)\}^{-\frac{n+1}{2}} d\tau$$

where the path of integration is transformed into a closed path around $\tau = \tau_0$ in the τ plane. Equation (16) shows that the coefficients in question can all be obtained by developing:

$$\{f(\tau) - f(\tau_0)\}^{-\frac{n+1}{2}}$$

in powers of $\tau - \tau_0$. Namely, let:

$$(17) \qquad f(\tau) - f(\tau_0) = (\tau - \tau_0)^2[c_0 + c_1(\tau - \tau_0) + c_2(\tau - \tau_0^2) + \cdots]$$

and raise this to a power:

$$(18) \qquad \{f(\tau) - f(\tau_0)\}^{-\frac{n+1}{2}}$$
$$= (\tau - \tau_0)^{-n-1}[a_0(n) + a_1(n)(\tau - \tau_0) + a_2(n)(\tau - \tau_0)^2 + \cdots]$$

where the coefficients a can be calculated from a finite number of the coefficients c and the number n. Then the Cauchy residue for this series is $a_0(0)$ for $n = 0$, $a_1(1)$ for $n = 1$, $a_2(2)$ for $n = 2$, and so on. According to equation (16), $a_n(n)$ is identical with the required quantity a_n. Calculation gives:

$$
(19)\qquad \begin{cases} c_0 = -i\,\dfrac{\sin\tau_0}{2!}, & c_1 = -i\,\dfrac{\cos\tau_0}{3!}, \quad c_2 = i\,\dfrac{\sin\tau_0}{4!} \\ c_3 = \quad i\,\dfrac{\cos\tau_0}{5!}, & c_4 = -i\,\dfrac{\sin\tau_0}{6!}, \ldots \end{cases}
$$

while in general it is found that:

$$
(20)\qquad \begin{cases}
a_0(n) = c_0^{-\frac{n+1}{2}} \\
a_1(n) = c_0^{-\frac{n+1}{2}}\left[-\dfrac{n+1}{1!\,2}\dfrac{c_1}{c_0}\right] \\
a_2(n) = c_0^{-\frac{n+1}{2}}\left[-\dfrac{n+1}{1!\,2}\dfrac{c_2}{c_0} + \dfrac{(n+1)(n+3)}{2!\,2^2}\dfrac{c_1^{\,2}}{c_0^{\,2}}\right] \\
a_3(n) = c_0^{-\frac{n+1}{2}}\left[-\dfrac{n+1}{1!\,2}\dfrac{c_3}{c_0} + \dfrac{(n+1)(n+3)}{2!\,2^2}\,2\,\dfrac{c_1c_2}{c_0^{\,2}}\right. \\
\qquad\qquad\qquad \left. - \dfrac{(n+1)(n+3)(n+5)}{3!\,2^3}\dfrac{c_1^{\,3}}{c_0^{\,3}}\right] \\
a_4(n) = c_0^{-\frac{n+1}{2}}\left[-\dfrac{n+1}{1!\,2}\dfrac{c_4}{c_0} + \dfrac{(n+1)(n+3)}{2!\,2^2}\left(2\dfrac{c_1c_3}{c_0^{\,2}} + \dfrac{c_2^{\,2}}{c_0^{\,2}}\right)\right. \\
\qquad\qquad\qquad - \dfrac{(n+1)(n+3)(n+5)}{3!\,2^3}\dfrac{c_1^{\,2}c_2}{c_0^{\,3}} \\
\qquad\qquad\qquad \left. + \dfrac{(n+1)(n+3)(n+5)(n+7)}{4!\,2^4}\dfrac{c_1^{\,4}}{c_0^{\,4}}\right] \\
\ldots\ldots\ldots\ldots\ldots\ldots\ldots\ldots
\end{cases}
$$

Substituting the values of c_0, c_1, c_2 gives finally:

$$
(21)\qquad \begin{cases}
a_0 = a_0(0) = +\left(-i\,\dfrac{\sin\tau_0}{2}\right)^{-\frac{1}{2}} \\
a_1 = a_1(1) = -\left(-i\,\dfrac{\sin\tau_0}{2}\right)^{-1}\left[\dfrac{1}{3}\operatorname{cotg}\tau_0\right] \\
a_2 = a_2(2) = +\left(-i\,\dfrac{\sin\tau_0}{2}\right)^{-\frac{3}{2}}\left[\dfrac{1}{8} + \dfrac{5}{24}\operatorname{cotg}^2\tau_0\right] \\
a_3 = a_3(3) = -\left(-i\,\dfrac{\sin\tau_0}{2}\right)^{-2}\left[\dfrac{2}{15}\operatorname{cotg}\tau_0 + \dfrac{4}{27}\operatorname{cotg}^3\tau_0\right] \\
a_4 = a_4(4) = +\left(-i\,\dfrac{\sin\tau_0}{2}\right)^{-\frac{5}{2}}\left[\dfrac{3}{128} + \dfrac{7}{576}\operatorname{cotg}^2\tau_0 + \dfrac{385}{3456}\operatorname{cotg}^4\tau_0\right] \\
\ldots\ldots\ldots\ldots\ldots\ldots\ldots\ldots
\end{cases}
$$

If for brevity we put:

$$(22) \qquad u = \left(\frac{2it}{\sin\tau_0}\right)^{\frac{1}{2}}$$

then:

$$(23) \qquad \Phi(t), \text{ resp. } \Phi_1(t) = \frac{i}{\sin\tau_0}\frac{1}{u}\Big[1 - \frac{1}{3}\operatorname{cotg}\tau_0\, u + \left(\frac{1}{8} + \frac{5}{24}\operatorname{cotg}^2\tau_0\right)u^2$$
$$- \left(\frac{2}{15}\operatorname{cotg}\tau_0 + \frac{4}{27}\operatorname{cotg}^3\tau_0\right)u^3$$
$$+ \left(\frac{3}{128} + \frac{7}{576}\operatorname{cotg}^2\tau_0 + \frac{385}{3456}\operatorname{cotg}^4\tau_0\right)u^4 - \cdots\Big]$$

We still have to determine u unequivocally. Now, in the region of the point $\tau = \tau_0$:

$$(24) \qquad \frac{1}{f'(\tau)} = \frac{1}{i(\cos\tau - \cos\tau_0)} = \frac{i}{\sin\tau_0}\frac{1}{\tau - \tau_0}$$

and since the path of integration in the τ plane (compare Figure 2) cuts the real axis at an angle of 45°, for the first part of this path, which corresponds to the first integral in (10):

$$(25) \qquad \tau - \tau_0 = |\tau - \tau_0|\, e^{i\frac{\pi}{4}}$$

A comparison of (24) with (23) shows that $\phi(t)$ is obtained by substituting for u:

$$(26) \qquad u = \left(\frac{2t}{\sin\tau_0}\right)^{\frac{1}{2}} e^{i\frac{\pi}{4}}$$

and $\phi_1(t)$ by substituting:

$$(26') \qquad u = -\left(\frac{2t}{\sin\tau_0}\right)^{\frac{1}{2}} e^{i\frac{\pi}{4}}$$

if $(2t/\sin\tau_0)^{\frac{1}{2}}$ is chosen real and positive.

With these remarks, (10) gives immediately for $H_2^{\alpha}(x)$

$$(27) \qquad H_2^{\alpha}(x) = \frac{1}{\pi}\, e^{-ix(\sin\tau_0 - \tau_0\cos\tau_0)}\left[\frac{e^{i\frac{\pi}{4}}\Gamma\left(\frac{1}{2}\right)}{\left(\frac{x}{2}\sin\tau_0\right)^{\frac{1}{2}}} + \left(\frac{1}{8} + \frac{5}{24}\operatorname{cotg}^2\tau_0\right)\frac{e^{i\frac{3\pi}{4}}\Gamma\left(\frac{3}{2}\right)}{\left(\frac{x}{2}\sin\tau_0\right)^{\frac{3}{2}}}\right.$$
$$\left. + \left(\frac{3}{128} + \frac{7}{576}\operatorname{cotg}^2\tau_0 + \frac{385}{3456}\operatorname{cotg}^4\tau_0\right)\frac{e^{i\frac{5\pi}{4}}\Gamma\left(\frac{5}{2}\right)}{\left(\frac{x}{2}\sin\tau_0\right)^{\frac{5}{2}}} + \cdots\right]$$

in which the integrals with even powers of $t^{\frac{1}{2}}$ have canceled in the difference.*

We recapitulate:

If the index α is smaller than the argument x which is assumed large compared to 1, and a real angle τ_0 is defined by the formula:

$$\cos \tau_0 = \frac{\alpha}{x}$$

then $H_2^{\alpha}(x)$ is represented in a semi-convergent way by the series (27).

Notice the peculiar relation between α and x brought about by the presence of τ_0. It is easy to show that for small values of α the expression (27) reduces to a well known approximation. In this case, since $\cos \tau_0 = \alpha/x$, we have approximately:

$$\tau_0 = \frac{\pi}{2}, \qquad \sin \tau_0 = 1$$

so that the first term of (27) becomes:

$$H_2^{\alpha}(x) = \sqrt{\frac{2}{\pi x}} e^{-ix} e^{i\frac{\pi}{2}\alpha} e^{i\frac{\pi}{4}} \tag{28}$$

the same as Nielsen, for example, obtained as asymptotic values for large values of x and, as we must add, finite values of α.

In order to obtain the complete series obtained by Hankel[8] for $H_2^{\alpha}(x)$ under the same assumptions (α small compared to x, x large compared to 1) it is more convenient not to start from (10) but to take into account from the beginning in the integral (1) that α is finite by writing:

$$H_2^{\alpha}(x) = -\frac{1}{\pi} \int_{(2)} e^{-ix\sin\tau} e^{i\alpha\tau} d\tau$$

and introducing only the exponent $i \sin \tau$ as the function $f(\tau)$. Then the same train of thought as before leads to introducing $i (\sin \tau - 1)$ as a new variable t and then expanding the function:

$$\frac{e^{i\alpha\tau}}{i \cos \tau}$$

in powers of t. In this way, the series results precisely in the Hankel form.

A disadvantage of expression (27) is that as soon as the ratio α/x is near 1, since then τ_0 is very small, the $\cot \tau_0$ appearing in (27) becomes very large. Near $\alpha/x = 1$, the expression behaves like a nonuniformly convergent series, since, as ξ approaches 1,

*Expression (27) is normalized in such a way that all fractional powers are to be taken real and positive.

x must become larger and larger in order that the 1st term, for example, is a sufficiently good approximation. But it is clear that this state of affairs could have been anticipated from the beginning. Namely, as the saddle point $\tau = \tau_0$ approaches zero, the saddle point $\tau = -\tau_0$ does also. Now, since the images of the saddle points in the t plane draw closer at the same time and are branch points of the function $\phi(t)$, as α approaches x the radius of convergence of the series for $\phi(t)$ becomes smaller and smaller, which means that (27) becomes less and less useful.

For this reason, in the next section another representation will be derived for the case in which α/x is close to 1.

IV. Semi-Convergent Series for $H_2^\alpha(x)$ in the Limiting Case $\alpha \cong x$

We put:

$$\frac{\alpha}{x} = 1 - \varepsilon \tag{29}$$

where according to the assumptions of this paragraph ε is a very small number (compared to 1). If ε vanishes, according to (1) $H_2^\alpha(x)$ becomes:

$$H_2^\alpha(x) = -\frac{1}{\pi}\int e^{-ix(\sin\tau - \tau)}\, d\tau \tag{30}$$

where the integral has to be taken over the path discussed in detail at the end of section II. In case ε is not zero, but still very small, we will use (29) and write the equation defining our cylinder functions in the form:

$$H_2^\alpha(x) = -\frac{1}{\pi}\int e^{-ix(\sin\tau - \tau)}\, e^{-i\varepsilon x\tau}\, d\tau \tag{31}$$

and consider the first factor as the most important. The train of thought is then exactly the same as that suggested but not carried through in section III for the Hankel series. Thus again first put:

$$t = i(\sin\tau - \tau) \tag{32}$$

The integration is to be performed over the real axis of the t plane, first from ∞ to 0, then from 0 to ∞.

The result is:

$$H_2^\alpha(x) = \frac{1}{\pi}\left\{\int_0^\infty e^{-xt}\Phi(t)\, dt - \int_0^\infty e^{-xt}\Phi_1(t)\, dt\right\} \tag{33}$$

where by means of (32) $\phi(t)$ is given as a function of t through the relation:

$$\Phi(t) = \frac{e^{-i\varepsilon x\tau}}{i(\cos\tau - 1)} \tag{34}$$

$\phi_1(t)$ differs from $\phi(t)$ only in the way the fractional power of t is defined. Evidently the T which was introduced in equation (12) and which was appropriate to the simple saddle point of Figure 2 should be replaced by a new auxiliary variable $t = T^3$, when the two saddle points of Figure 5 coincide. Thus, developments in powers of $t^{1/3}$ are now to be considered. If we write:

$$\Phi(t) = t^{-\frac{2}{3}} \sum_0^\infty a_n t^{\frac{n}{3}} \tag{35}$$

we obtain, just as in section III, for the coefficients a_n the expression:

$$a_n = \frac{1}{2\pi i} \frac{e^{-i\frac{n+1}{3}\frac{\pi}{2}}}{3} \int \frac{e^{-i\varepsilon x\tau}}{(\sin\tau - \tau)^{\frac{n+1}{3}}} dt \tag{36}$$

where the integral is to be taken around the origin of the τ plane. If as in the foregoing paragraph we define the fractional powers occurring in $\phi(t)$ and $\phi_1(t)$ in the correct way following from the analysis given there, calculation gives:

$$\begin{aligned}
\Phi(t) = \left(\frac{2}{9}\right)^{\frac{1}{3}} e^{i\frac{\pi}{6}} t^{-\frac{2}{3}} \Big[1 &- i\varepsilon x 6^{\frac{1}{3}} e^{i\frac{\pi}{6}} t^{\frac{1}{3}} - \left(\frac{\varepsilon^2 x^2}{2} - \frac{1}{20}\right) 6^{\frac{2}{3}} e^{i\frac{2\pi}{6}} t^{\frac{2}{3}} \\
&+ i\left(\frac{\varepsilon^3 x^3}{6} - \frac{\varepsilon x}{15}\right) 6^{\frac{3}{3}} e^{i\frac{3\pi}{3}} t^{\frac{3}{3}} + \left(\frac{\varepsilon^4 x^4}{24} - \frac{\varepsilon^2 x^2}{24} + \frac{1}{280}\right) 6^{\frac{4}{3}} e^{i\frac{4\pi}{3}} t^{\frac{4}{3}} \cdots \Big] \\
\Phi_1(t) = \left(\frac{2}{9}\right)^{\frac{1}{3}} e^{-i\frac{\pi}{2}} t^{-\frac{2}{3}} \Big[1 &- i\varepsilon x 6^{\frac{1}{3}} e^{-i\frac{\pi}{2}} t^{\frac{1}{3}} - \left(\frac{\varepsilon^2 x^2}{2} - \frac{1}{20}\right) 6^{\frac{2}{3}} e^{-i\pi} t^{\frac{2}{3}} \\
&+ i\left(\frac{\varepsilon^3 x^3}{6} - \frac{\varepsilon x}{15}\right) 6^{\frac{3}{3}} e^{-i\frac{3\pi}{2}} t^{\frac{3}{3}} \left(\frac{\varepsilon^4 x^4}{24} - \frac{\varepsilon^2 x^2}{24} + \frac{1}{280}\right) 6^{\frac{4}{3}} e^{-i2\pi} t^{\frac{4}{3}} \cdots \Big]
\end{aligned} \tag{37}$$

Substituting this in (33) and integrating term by term gives for $H_2^\alpha(x)$ the representation:

$$\begin{aligned}
H_2^\alpha(x) = \frac{2i}{3\pi} \Bigg[& e^{-i\frac{\pi}{6}} 6^{\frac{1}{3}} \sin\frac{\pi}{3} \frac{\Gamma\left(\frac{1}{3}\right)}{x^{\frac{1}{3}}} - i(\varepsilon x) e^{-i\frac{2\pi}{6}} 6^{\frac{2}{3}} \sin\frac{2\pi}{3} \frac{\Gamma\left(\frac{2}{3}\right)}{x^{\frac{2}{3}}} \\
& - \left(\frac{\varepsilon^2 x^2}{2} - \frac{1}{20}\right) e^{-i\frac{3\pi}{6}} 6^{\frac{3}{3}} \sin\frac{3\pi}{3} \frac{\Gamma\left(\frac{3}{3}\right)}{x^{\frac{3}{3}}} + i\left(\frac{\varepsilon^3 x^3}{6} - \frac{\varepsilon x}{15}\right) e^{-i\frac{4\pi}{6}} 6^{\frac{4}{3}} \sin\frac{4\pi}{3} \frac{\Gamma\left(\frac{4}{3}\right)}{x^{\frac{4}{3}}} \\
& + \left(\frac{\varepsilon^4 x^4}{24} - \frac{\varepsilon^2 x^2}{24} + \frac{1}{280}\right) e^{-i\frac{5\pi}{6}} 6^{\frac{5}{3}} \sin\frac{5\pi}{3} \frac{\Gamma\left(\frac{5}{3}\right)}{x^{\frac{5}{3}}} + \cdots \Bigg]
\end{aligned} \tag{38}$$

This expression is normalized in such a way that all fractional powers are to be taken real and positive.

In the special case $\varepsilon = 0$, where α is not approximately but exactly equal to x, (38) reduces to:*

$$(39)\quad H_2^{\alpha}(x)=\frac{2i}{3\pi}\left[e^{-i\frac{\pi}{6}}6^{\frac{1}{3}}\sin\frac{\pi}{3}\frac{\Gamma\left(\frac{1}{3}\right)}{x^{\frac{1}{3}}}+\frac{1}{20}e^{-i\frac{3\pi}{6}}6^{\frac{3}{3}}\sin\frac{3\pi}{3}\frac{\Gamma\left(\frac{3}{3}\right)}{x^{\frac{3}{3}}}+\frac{1}{280}e^{-i\frac{5\pi}{6}}6^{\frac{5}{3}}\sin\frac{5\pi}{3}\frac{\Gamma\left(\frac{5}{3}\right)}{x^{\frac{5}{3}}}+\cdots\right]$$

V. Semi-Convergent Series for $H_2^{\alpha}(x)$ in the Second Case ($\alpha > x$)

In this case, as emphasized in section II, a new situation arises in that the path used in integrating passes through two saddle points. If we again take the exponent as variable of integration, the integrand becomes infinite at the two places on the path of integration corresponding to the two saddle points. However, only the neighborhood of one saddle point need be taken into account, since the contribution from the neighborhood of the other saddle point vanishes exponentially in comparison. In the present case, as is clear from Figure 4, the neighborhood of the lower saddle point denoted earlier by τ_0 (Fig. 3b) makes the determining contribution. We set then, as earlier in section III:

$$(40)\qquad t=f(\tau)-f(\tau_0)=i\{(\sin\tau-\tau\cos\tau_0)-(\sin\tau_0-\tau_0\cos\tau_0)\}$$

and have first to expand the function:

$$(41)\qquad \frac{1}{f'(\tau)}=\frac{1}{i(\cos\tau-\cos\tau_0)}$$

in powers of t. In the integration over the upper part of the path A (Fig. 3), that is, over the imaginary axis of the τ plane above the saddle point $\tau = \tau_0$, or in the t plane over the real axis from ∞ to 0, one obtains:

$$(42)\quad \Phi(t)=\frac{i}{\sin\tau_0}\frac{1}{u}\Big[1-\frac{1}{3}\operatorname{cotg}\tau_0\,u+\left(\frac{1}{8}+\frac{5}{24}\operatorname{cotg}^2\tau_0\right)u^2 -\left(\frac{4}{15}\operatorname{cotg}\tau_0+\frac{2}{27}\operatorname{cotg}^3\tau_0\right)u^3+\left(\frac{3}{128}+\frac{7}{576}\operatorname{cotg}^2\tau_0+\frac{385}{3456}\operatorname{cotg}^4\tau_0\right)u^4+\cdots\Big]$$

*In (38) the third term of the bracket vanishes because $\sin 3\pi/3 = 0$; in (39), the second for the same reason.

with the abbreviation:

$$u = \left(\frac{2t}{i\sin\tau_0}\right)^{\frac{1}{2}} e^{i\frac{\pi}{2}} \tag{43}$$

which agrees with the earlier equation (23), except it is written differently. On the lower part of path A in the τ plane, after the saddle point $t = \tau_0$ has been passed, or in the t plane on the real axis as we move again away from the origin, the expansion of $\phi(t) = \phi_1(t)$ is again (42), but now u is defined by:

$$u = -\left(\frac{2t}{i\sin\tau_0}\right)^{\frac{1}{2}} e^{i\frac{\pi}{2}} \tag{43'}$$

By substituting in (10) and integrating to ∞,* we get for $H_2^{\alpha}(x)$ the expression:

$$H_2^{\alpha}(x) = \frac{i}{\pi} e^{-ix(\sin\tau_0 - \tau_0\cos\tau_0)} \cdot \left[\frac{\Gamma\left(\frac{1}{2}\right)}{\left(i\frac{x}{2}\sin\tau_0\right)^{\frac{1}{2}}} - \left(\frac{1}{8} + \frac{5}{24}\operatorname{cotg}^2\tau_0\right) \frac{\Gamma\left(\frac{3}{2}\right)}{\left(i\frac{x}{2}\sin\tau_0\right)^{\frac{3}{2}}} \right.$$
$$\left. + \left(\frac{3}{128} + \frac{7}{576}\operatorname{cotg}^2\tau_0 + \frac{385}{3456}\operatorname{cotg}^4\tau_0\right) \frac{\Gamma\left(\frac{5}{2}\right)}{\left(i\frac{x}{2}\sin\tau_0\right)^{\frac{5}{2}}} + \cdots \right] \tag{44}$$

where again the fractional powers are to be taken real and positive. (Note that the denominators of this expression seem imaginary, while actually $ix/2 \sin\tau_0$ is a real positive number.) Recapitulating and comparing this last expression with equation (27), we see:

If the index α is larger than the argument x which itself is large compared to 1 and an imaginary angle τ_0 is defined by the negative imaginary roots of the equation:

$$\cos\tau_0 = \frac{\alpha}{x} \tag{45}$$

then the series (44) semi-converges to $H_2^{\alpha}(x)$ and differs from (27) only in that the definition of the roots of unity which appear is different.

Now from (44) we can derive the limiting value of the Hankel function for the case that α is large compared to x. If we write $\tau = -ib$, then (since $\cos\tau_0 = \alpha/x$) in first approximation for large values of b:

*In order to see that this can be done without impairing the semi-convergence of the series, remember that the neglected part vanishes exponentially compared to the right hand side of (44).

$$\frac{e^b}{2} = \frac{\alpha}{x} \quad \text{or} \quad b = \log \frac{2\alpha}{x}$$

so that we have:

$$\tau_0 = -i \log 2 \frac{\alpha}{x}$$

$$\cos \tau_0 = \frac{\alpha}{x}; \qquad \sin \tau_0 = -i \frac{\alpha}{x}$$

and by substitution in (44):

$$H_2^\alpha(x) = \frac{i}{\pi} \frac{\Gamma\left(\frac{1}{2}\right)}{\left(\frac{\alpha}{2}\right)^{\frac{1}{2}}} e^{-\alpha + \alpha \log 2 \frac{\alpha}{x}}$$

or:

$$(46) \qquad H_2^\alpha(x) = \frac{i}{\pi} \sqrt{\frac{2\pi}{\alpha}} \left(\frac{2}{x}\right)^\alpha \alpha^\alpha e^{-\alpha}$$

That this expression agrees with that already known is easy to see. According to formula 1, (p. 16) and 8 (p. 11) in Nielsen's[9] book, $H_2^\alpha(x)$ is asymptotically equal to:

$$H_2^\alpha(x) = J_\alpha(x) - i Y_\alpha(x) = \frac{i}{\pi} \left(\frac{2}{x}\right)^\alpha \Gamma(\alpha)$$

and if the Stirling asymptotic formula* for $\Gamma(\alpha)$ is used:

$$\Gamma(\alpha) = \sqrt{\frac{2\pi}{\alpha}} \alpha^\alpha e^{-\alpha}$$

this reduces to our formula (46).

*Compare for example Nielsen, *Handbuch der Theorie der Gammafunktionen*, Leipzig, 1906, p. 96. We may remark that this formula can be derived immediately from Hankel's complex integral definition of $\Gamma(\alpha)$ by the method used here.

VI. Expressions for the Hankel Functions of the First Kind and for the Bessel Functions

In the preceding paragraphs only the second Hankel function was discussed in detail; however, if it is recalled that for real values of α and x the two Hankel functions are complex conjugates, corresponding series for the first Hankel function $H_1^{\alpha}(x)$ can be obtained from (27), (38), and (44) without further calculation and are written out in the summary of the following paragraph.

The following well known formulae which connect the Bessel and Hankel functions* can be used in the case $\alpha < x$ to obtain expressions for the ordinary Bessel functions $J_{\alpha}(x)$ and $J_{-\alpha}(x)$

$$(47)\qquad \begin{cases} J_{\alpha}(x) = \frac{1}{2}[H_1^{\alpha}(x) + H_2^{\alpha}(x)] \\ J_{-\alpha}(x) = \frac{1}{2}[e^{i\alpha\pi} H_1^{\alpha}(x) + e^{-i\alpha\pi} H_2^{\alpha}(x)] \end{cases}$$

When $\alpha>x$, however, the first equation (47) fails. In this case we would obtain:

$$J_{\alpha}(x) = 0$$

since here the two Hankel functions (52') and (53') are equal and opposite. However, all series in this part (for $H_1^{\alpha}(x)$ and $H_2^{\alpha}(x)$) are semi-convergent; that is, as follows from Poincaré's definition of semi-convergence, it can only be asserted that the difference between the series broken off at the nth term and the exact value of the function divided by the exact value approaches zero for large values of x like a negative power of x whose exponent always increases in absolute value as more terms of the series are taken into account. A deviation of the approximation from the actual value which for large values of x vanishes like an exponential, for example, can still occur and does not appear in the series. Thus when we find on the basis of the series (44) or (52') and (53') of the summary of section VII:

$$J_{\alpha}(x) = 0$$

this does not mean that $J_{\alpha}(x)$ actually vanishes, but only that in this case ($\alpha>x$) the Bessel function is so much smaller than the Hankel function that their ratio approaches zero faster than any power of x as x approaches ∞. Figures 1 and 4 immediately show the state of affairs. While $H_2^{\alpha}(x)$, for example, must be obtained by integrating the function $e^{-xf(\tau)}$ along a path which begins in

*The formulae (47) follow immediately from equations (3) and (4) of Nielsen (*loc. cit.*, section 17).

the positive imaginary part of the τ plane and ends in the negative imaginary part and must therefore pass through the two saddle points $\tau = \tau_0$ and $\tau = \tau_0'$ if a curve J = const. serves as path of integration, this is not the case for the Bessel function $J_\alpha(x)$. With reference to Figure 1, the first equation (47) shows that, $J_\alpha(x)$ can be defined by the integral:

$$J_\alpha(x) = -\frac{1}{\pi}\int_{(3)} e^{-x f(\tau)}\, d\tau \tag{48}$$

which is to be carried over path (3) which begins between $\pi/2+i\infty$ and $3\pi/2+i\infty$ and ends between $-\pi/2+i\infty$ and $-3\pi/2+i\infty$, and which can be constructed by the union of the paths of integration (1) and (2) Therefore the path J = const., which we chose as path of integration in our earlier analysis, is the curve B shown in Figure 4 which passes through only one saddle point τ_0'. Furthermore, it can be seen immediately that:

$$f(\tau_0') = i\,(\sin\tau_0' - \tau_0'\cos\tau_0')$$

has a positive value, while $f(\tau_0)$ is negative. Now, since for very large values of x only the immediate neighborhood of the saddle point determines the value of the function, it is clear that $J_\alpha(x)$ as well as the ratio $J_\alpha(x)/H_\alpha(x)$ vanishes exponentially as x approaches ∞. A semi-convergent series for $J_\alpha(x)$ can be derived from (48) in a way no different from the calculations carried out in sections III and V for the second Hankel function, so the final result can be given immediately.

If the index α is larger than the argument x which is assumed large compared to 1, and an angle τ_0 is defined as in section V as the negative imaginary root of the equation:

$$\cos\tau_0 = \frac{\alpha}{x} \tag{49}$$

then the series:

$$J_\alpha(x) = \frac{1}{\pi} e^{ix(\sin\tau_0 - \tau_0\cos\tau_0)} \left[\frac{\Gamma\left(\frac{1}{2}\right)}{\left(i\frac{x}{2}\sin\tau_0\right)^{\frac{1}{2}}} + \left(\frac{1}{8} + \frac{5}{24}\operatorname{cotg}^2\tau_0\right)\frac{\Gamma\left(\frac{3}{2}\right)}{\left(i\frac{x}{2}\sin\tau_0\right)^{\frac{3}{2}}} \right. \tag{50}$$

$$\left. + \left(\frac{3}{128} + \frac{7}{576}\operatorname{cotg}^2\tau_0 + \frac{385}{3456}\operatorname{cotg}^4\tau_0\right)\frac{\Gamma\left(\frac{5}{2}\right)}{\left(i\frac{x}{2}\sin\tau_0\right)^{\frac{5}{2}}} + \cdots \right]$$

semi-converges to $J_\alpha(x)$, when the fractional powers are to be taken real and positive.

Finally, we come to the function $J_{-\alpha}(x)$, for which in general an approximate representation follows from the second of the connecting formulae (47), even if $\alpha > x$. This fails only when at the same time α is an integer n, since the relation $H_1^\alpha(x) = -H_2^\alpha(x)$ which is a consequence of convergent series, leads, with the help of the formulae (47), again to $J_{-\alpha}(x) = 0$. In this case, however, it is not necessary to look further since $J_{-n}(x)$ and $J_n(x)$ do not form a linearly independent set of solutions of the differential equation of the cylinder function. We have:*

$$J_{-n}(x) = (-1)^n J_n(x)$$

so that this case is reduced to the previous one.

VII. Summary of Results

Finally, we give in this paragraph a summary of the approximate formulae for the different cylinder functions. We introduce first abbreviations A_n and B_n for the functions of the angles τ_0 appearing in the series, which in each of the three cases to be distinguished $\alpha < x$, $\alpha > x$, and $\alpha \cong x$, can be computed from the formulae given here from the ratio α/x of index to argument. The way in which the functions A_n originate is more closely explained in section III; the first three run:

$$(51)\quad \begin{cases} \mathsf{A}_0(\tau_0) = 1, \\ \mathsf{A}_1(\tau_0) = \frac{1}{8} + \frac{5}{24}\cot^2\tau_0, \\ \mathsf{A}_2(\tau_0) = \frac{3}{128} + \frac{7}{576}\cot^2\tau_0 + \frac{385}{3456}\cot^4\tau_0 \\ \dots\dots\dots\dots\dots \end{cases}$$

With these abbreviations the results can be summarized as follows:

(1) The Index α is Smaller than the Argument x

If the real angle τ_0 $(0 < \tau_0 < \pi/2)$ is defined by:

$$\cos\tau_0 = \frac{\alpha}{x}$$

*Compare, for example, N. Nielsen (*loc. cit.*, page 5).

then:

$$(52)\quad H_1^\alpha(x) = \frac{1}{\pi} e^{ix(\sin\tau_0 - \tau_0\cos\tau_0)} \sum_{n=0}^{n=n} A_n(\tau_0) \frac{e^{-i(2n+1)\frac{\pi}{4}}\Gamma\left(n+\frac{1}{2}\right)}{\left(\frac{x}{2}\sin\tau_0\right)^{n+\frac{1}{2}}}$$

$$(53)\quad H_2^\alpha(x) = \frac{1}{\pi} e^{-ix(\sin\tau_0 - \tau_0\cos\tau_0)} \sum_{n=0}^{n=n} A_n(\tau_0) \frac{e^{i(2n+1)\frac{\pi}{4}}\Gamma\left(n+\frac{1}{2}\right)}{\left(\frac{x}{2}\sin\tau_0\right)^{n+\frac{1}{2}}}$$

$$(54)\quad J_\alpha(x) = \frac{1}{\pi} \sum_{n=0}^{n=n} A_n(\tau_0) \frac{\Gamma\left(n+\frac{1}{2}\right)}{\left(\frac{x}{2}\sin\tau_0\right)^{n+\frac{1}{2}}} \cos\left\{x(\sin\tau_0 - \tau_0\cos\tau_0) - (2n+1)\frac{\pi}{4}\right\}$$

$$(55)\quad J_{-\alpha}(x) = \frac{1}{\pi} \sum_{n=0}^{n=n} A_n(\tau_0) \frac{\Gamma\left(n+\frac{1}{2}\right)}{\left(\frac{x}{2}\sin\tau_0\right)^{n+\frac{1}{2}}} \cos\left\{x(\sin\tau_0 - (\tau_0-\pi)\cos\tau_0) - (2n+1)\frac{\pi}{4}\right.$$

(2) The Index α is Larger than the Argument x

If a negative imaginary angle τ_0 is defined by:

$$\cos\tau_0 = \frac{\alpha}{x}$$

then:

$$(52')\quad H_1^\alpha(x) = -\frac{i}{\pi} e^{-ix(\sin\tau_0 - \tau_0\cos\tau_0)} \sum_{n=0}^{n=n} (-1)^n A_n(\tau_0) \frac{\Gamma\left(n+\frac{1}{2}\right)}{\left(i\frac{x}{2}\sin\tau_0\right)^{n+\frac{1}{2}}}$$

$$(53')\quad H_2^\alpha(x) = \frac{i}{\pi} e^{-ix(\sin\tau_0 - \tau_0\cos\tau_0)} \sum_{n=0}^{n=n} (-1)^n A_n(\tau_0) \frac{\Gamma\left(n+\frac{1}{2}\right)}{\left(i\frac{x}{2}\sin\tau_0\right)^{n+\frac{1}{2}}}$$

$$(54') \quad J_\alpha(x) = \frac{1}{\pi} e^{ix(\sin\tau_0 - \tau_0\cos\tau_0)} \sum_{n=0}^{n=n} A_n(\tau_0) \frac{\Gamma\left(n+\frac{1}{2}\right)}{\left(i\frac{x}{2}\sin\tau_0\right)^{n+\frac{1}{2}}}$$

$$(55') \quad J_{-\alpha}(x) = \frac{\sin\alpha\pi}{\pi} e^{-ix(\sin\tau_0 - \tau_0\cos\tau_0)} \sum_{n=0}^{n=n} (-1)^n A_n(\tau_0) \frac{\Gamma\left(n+\frac{1}{2}\right)}{\left(i\frac{x}{2}\sin\tau_0\right)^{n+\frac{1}{2}}}$$

(3) The Index α is about Equal to the Argument x.

If we write:

$$\frac{\alpha}{x} = 1 - \varepsilon$$

and define functions $B_n(\varepsilon x)$ by the formulae:

$$(56) \quad \begin{cases} B_0(\varepsilon x) = 1 \\ B_1(\varepsilon x) = \varepsilon x \\ B_2(\varepsilon x) = \frac{\varepsilon^2 x^2}{2} - \frac{1}{20} \\ B_3(\varepsilon x) = \frac{\varepsilon^3 x^3}{6} - \frac{\varepsilon x}{15} \\ B_4(\varepsilon x) = \frac{\varepsilon^4 x^4}{24} - \frac{\varepsilon^2 x^2}{24} + \frac{1}{280} \\ \cdots\cdots\cdots\cdots \end{cases}$$

then:

$$(52'') \quad H_1^\alpha(x) = -\frac{2}{3\pi} \sum_{n=0}^{n=n} B_n(\varepsilon x) 6^{\frac{n+1}{3}} e^{i(n+1)\frac{2\pi}{3}} \sin(n+1)\frac{\pi}{3} \frac{\Gamma\left(\frac{n+1}{3}\right)}{x^{\frac{n+1}{3}}}$$

$$(53'') \quad H_2^\alpha(x) = -\frac{2}{3\pi} \sum_{n=0}^{n=n} B_n(\varepsilon x) 6^{\frac{n+1}{3}} e^{-i(n+1)\frac{2\pi}{3}} \sin(n+1)\frac{\pi}{3} \frac{\Gamma\left(\frac{n+1}{3}\right)}{x^{\frac{n+1}{3}}}$$

$$(54'') \quad J_\alpha(x) = \frac{1}{3\pi} \sum_{n=0}^{n=n} B_n(\varepsilon x) 6^{\frac{n+1}{3}} \sin(n+1)\frac{\pi}{3} \frac{\Gamma\left(\frac{n+1}{3}\right)}{x^{\frac{n+1}{3}}}$$

$$(55'') \quad J_{-\alpha}(x) = \frac{1}{3\pi} \sum_{n=0}^{n=n} B_n(\varepsilon x) 6^{\frac{n+1}{3}} \left[\sin\left((n+1)\frac{\pi}{3} + \alpha\pi\right) - \sin(n+1+\alpha)\pi\right] \frac{\Gamma\left(\frac{n+1}{3}\right)}{x^{\frac{n+1}{3}}}$$

Remarks: That the formulae given above embrace all cases in which x is large and that for values of $x \cong \alpha$ there is no gap where neither the formulae (52) through (55) or (52') through (55') nor (52") through (55") are applicable can be made clear by the following analysis. According to (51), for small values of ε (compare (29)) the functions $A_n(\tau_0)$ become of order ε^{-n}, since τ_0 becomes small with ε; indeed, according to the definition (29):

$$\tau_0 = (2\varepsilon)^{\frac{1}{2}}$$

Now, if we consider two consecutive terms T_{n-1} and T_n, for example, of the series (52), we obtain for their ratio:

$$\frac{T_{n-1}}{T_n} = \frac{\text{const.}}{x\varepsilon^{\frac{3}{2}}}$$

if we replace $\sin \tau_0$ by τ_0 or $(2\varepsilon)^{\frac{1}{2}}$. This ratio is finite if $x\varepsilon^{3/2}$ is finite; that is, if ε approaches zero like $x^{-2/3}$ as x approaches ∞. Similarly, when (56) is taken into account, the ratio of two consecutive terms T_{n-1} and T_n of the series (52") for the case $\alpha \cong x$ is given by:

$$\frac{T_{n-1}}{T_n} = \text{const. } \varepsilon x^{\frac{2}{3}}$$

This also is finite if ε approaches zero like $x^{-2/3}$ as x approaches ∞. In cases (1) and (2), if ε approaches zero like $x^{-2/3+\eta}$, T_{n-1}/T_n vanishes like $x^{-3\eta/2}$; on the other hand, in case (3) if ε approaches zero like $x^{-2/3-\eta}$, T_{n-1}/T_n vanishes like $x^{-\eta}$. Therefore it is clear that the formulae (52) to (55) and (52') to (55') cease to be of practical use exactly where the region of applicability of (52") to (55") begins.

Bibliography

1. *Physik. Z.*, 9, 775 (1908) and *Verhandl deut. physik. Ges.*, 10, 741 (1908).
2. Note on Bessel's Functions, *Phil. Mag.*, 44, 337 (1872).
3. B. Riemann, "Ges. math. Werke und wissenschaftlicher Nachlass." Leipzig, 1876, p. 400.
4. J. H. Graf and E. Gubler, "Einleitung in die Theorie der Besselschen Funktionen." Bern, 1898, Part 1, p. 96.

5. J. W. Nicholson, *Phil. Mag.*, *14*, Ser. VI, 697 (1907).
6. *Phil. Mag.*, *16*, Ser. VI, 271 (1908).
7. A. Sommerfeld, *Math. Ann.*, *47*, 317 (1895).
8. *Math. Ann.*, [1], *1869*, 494.
9. N. Nielsen, "Handbuch der Theorie der Zylinderfunktionen." Leipzig, 1904.

PHYSICAL OPTICS-EXPERIMENTS ON THE DIFFRACTION OF LIGHT BY ULTRASONICS
(Optique Physique.-Experiences sur la Diffraction de la Lumière par des Ultrasons)*

P. Debye, H. Sack, and F. Coulon, communicated by M. Brillouin.

Translated from
Comptes rendus hebdomadaires des séances de l'académie des sciences, 198, 1934, pages 922-924.

The theory[1] of the diffraction of light by ultrasonic waves predicts[2] that the frequency ν of the diffracted light is slightly different from that of the incident light ν_0 and that the difference depends on the order into which the light is diffracted. If the frequency of the *ultrasonic* wave is ν', then the frequency of the light diffracted into order n will be $\nu_0 + n\nu'$.

This frequency difference is extremely small: for the green line of mercury (λ = 5460 Å.) and *ultrasonic* waves of frequency 10^7, the variation of wavelength $\Delta\lambda$ in the first order is about 10^{-4} Å. Thus this effect is very hard to detect. We wish to describe here a simple, although indirect, experimental method of proving the existence of this effect which is also a method of measuring the speed of *ultrasonic* waves.**

The spectra furnished by an ordinary grating are coherent, and the beams of order + 1 and − 1, for example, may be made to interfere. The light waves diffracted by an ultrasonic wave into order + 1 and − 1 have different frequencies and thus cannot interfere. Still, there remains a certain phase relation between these two beams, as between two rays having undergone opposite Doppler effects. Suppose that these two vibrations are superimposed at a point and that the optical path for the beam of order + 1 is x_1, for that of order **-1**, x_2. The resulting amplitude is of the form:

*Meeting of February 26, 1934 of the Académie des Sciences.

**P. Debye, at a conference of the Physical Society of Berlin (May 5, 1933) pointed out the possibility of these experiments for the first time; his investigations were continued by Sack and Coulon.

$$A = \cos 2\pi \left(\nu' t - \frac{x_1 - x_2}{2\lambda} \right)$$

where λ is the wavelength of the light wave. The intensity thus varies with a frequency $2\nu'$, twice the frequency of the ultrasonic wave. It can be said, also, that a system of fringes is obtained which moves with a speed of the order of that of the ultrasonic wave and which can be calculated from the formula.

The fringes will again be visible if illuminated stroboscopically with the frequency $2\nu'$. This is realized by inserting a Kerr cell excited by an alternating voltage of frequency ν' in the path of the light to be scattered (if a constant voltage is added to the alternating voltage, the Kerr cell interrupts the light with the frequency ν' and the interference between the beams of orders 0 and 1 can be observed). If the ultrasonic wave gives stationary waves, a superposition of the two frequencies $\nu_0 + \nu'$ is found in the order +1 and the same in the order −1, and the interference can thus be observed directly.

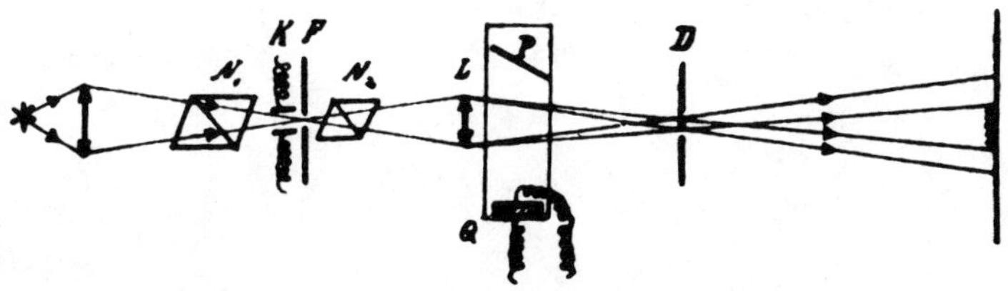

The figure shows the experimental arrangement: the light concentrated on a slit F traverses two Nicols N_1, N_2, a Kerr cell K (Nicols at 45° to the plane of the electrodes of the cell), and then a container having plane parallel faces and containing the quartz Q immersed in a liquid (in our experiment $C_2H_2Cl_4$). A lens L forms an image of the slit at D where a diaphragm having two openings passing the orders to be studied (such as orders +1 and −1, for example, or 0 and +1) is placed. The light converges behind the diaphram and interference is observed in the part common to the two beams. The Kerr cell K and the quartz Q are connected to the same induction coil coupled to a c. w. oscillator ($\nu' = 10^7$ approximately). In general, the ultrasonic wave gives standing waves in the tank and an interference pattern is seen immediately, even without putting the Kerr cell into operation. The stationary waves can be eliminated by putting an oblique plate P in the container; then the interference pattern disappears and only reappears if the cell is put into operation. The theory shows that this phenomenon of interference is independent of the color of the light so that an arc can be used. The wavelength of the ultrasonic wave (and thus its velocity) can be determined from the separation of the fringes.

These experiments can be varied in many ways, and different orders of diffraction can be made to interfere. The diaphragm *D* can even be dispensed with and all orders made to interfere; the fringes persist and give an "image" of the ultrasonic wave.* We will publish a complete study of these experiments shortly.

Bibliography

1. L. Brillouin, *Ann. Physik.*, *17*, 103 (1921); *Acta. Sci. et Ind.*, 59 (Hermann, Paris, 1933); P. Debye, *Physik. Z.*, *33*, 849 (1932).
2. Debye and Sears, *Proc. Natl. Acad.*, *18*, 409 (1932); *Sachs Akad.*, *84*, 125 (1932); Biquard and Lucas, *Compt. rend.*, *194*, 2132 (1932); *195*, 121 (1932); *J. Physik.*, *3*, 464 (1932).

*The experiment is analogous to that of Abbe on the resolving power of the microscope. Bachen, Hiedemann, and Asbach have described (*Nature*, January 1934) very briefly an experiment which seems analogous to ours; lacking detailed information, we cannot make an exact comparison.

A Method for the Determination of the Mass of Electrolytic Ions *

P. Debye, *University of Leipzig, Leipzig, Germany*

(Received November 18, 1932)

I.

THE question how many molecules of a solvent are intimately connected with the different ions in a solution is far from being solved. It occurred to me that we could perform some improvement if a method could be found enabling us to determine directly the masses of ions. Ordinarily the mass does not enter at all in our considerations and the reason is that the forces due to the friction of the ions in the solvent are so much larger than the dynamical reactions due to the masses. For an ion moving with a velocity v in a liquid, the frictional force may be put equal to ρv and the factor ρ is (for water as a solvent) of the order of magnitude 2.10^{-9}, as the friction is comparable to that of a sphere of radius 10^{-8} cm, according to Stokes' formula. On the other hand, the dynamical reaction due to acceleration is $m(dv/dt)$ if m is the mass of the ion. If therefore we consider an ion in a field of high frequency ω (number of vibrations in 2π sec.) the dynamical coefficient to be compared with ρ is ωm. In order to make this coefficient equal to ρ for an ion of the mass of an hydrogen atom ($1.64.10^{-24}$ g), the frequency would have to be equal to approximately $\omega = 10^{15}$ and for a mass a hundred times larger, we would still find $\omega = 10^{13}$. This means that mass effects such as we are considering here would not become important in comparison with friction effects except in the infrared region. However, in this region considering ordinary solutions in water as an example, it is doubtful if measurements of the absorption coefficient of infrared waves would give much information, as the proper absorption of the water itself will be important. That is why I tried to look for an effect, which is zero as long as the dynamical reactions of the ions are neglected and would therefore provide us with a direct measurement of these reactions alone. Suppose we introduce supersonic waves in a solution of an electrolyte. If then only frictional forces should exist between the ions and the solvent, there will be no difference between the velocity of the ions and the velocity of the surrounding liquid. The solution will behave like a homogeneous liquid. If, however, the dynamical reactions are taken in account, these reactions will be different for ions of different masses. As a result, the motion, say of the positive ions, may differ from that of the negative ions. But this means that periodically changing electric charge densities will accompany the sound waves. That is why we should expect potential differences of the period of the sound waves set up between different points of an electrolyte solution by the passing of sound waves. It is obvious that if we can detect and measure this effect, we will be able to draw definite conclusions as to the masses of the ions and in this way as to the number of solvent molecules connected with them.

II.

It remains first of all to be seen what the magnitude of the presumed effect will be. For this purpose the following calculation is presented. No account is taken of interionic attraction, but I think that we may safely wait for the experiments to be performed first as no fundamental change in the results is to be expected from this improvement, although the absolute value of the effect will be influenced especially for higher concentrations.

Suppose the solution contains per cc: $n_1, n_2, \cdots n_i, \cdots$ ions of charges $e_1, e_2, \cdots e_i, \cdots$ with the friction constants $\rho_1, \rho_2, \cdots \rho_i \cdots$ and the masses $m_1, m_2, \cdots m_i \cdots$. If at a certain moment, t, and in a certain place, which we will characterize by one single variable, x (only considering plane waves going in the direction x), the velocity of the solvent is v_0 and the velocity of the ions are $v_1, v_2, \cdots v_i, \cdots$ we have a first set of equations of the form

$$e_i X - \rho_i(v_i - v_0) = m_i(dv_i/dt). \qquad (1)$$

They are equal in number with the number of

*Reprinted from *The Journal of Chemical Physics*, 1, 13-16 (1933).

different ions contained in the solution and express that in every moment the total force acting on one ion has to be equal to its mass times its acceleration. The force on the left-hand side of (1) is made up of two parts, one part e_iX due to the electric component X of the forces, which in turn will be due to the electric charges in the solution, provided such charges appear. The other part, $-\rho_i(v_i-v_0)$ is the frictional force, which will result as soon as there is a difference between the actual velocities of the ion and the surrounding liquid.

A second set of as many equations as the number of different ions present is furnished by the equations of continuity, expressing that the gain in number of ions contained in an element of volume can only be due to ions entering or leaving through its surface.[1]

$$\frac{\partial n_i}{\partial t}+\frac{\partial}{\partial x}(n_iv_i)=0. \quad (2)$$

Calling the number of different ions z, we have now $2z$ equations for $(2z+1)$ variables, namely z numbers $n_1, n_2, \cdots n_i, \cdots$; z velocities $v_1, v_2, \cdots v_i \cdots$, and one additional variable, the field strength X. We make up for this deficiency in adding Poisson's equation, which has the form

$$D\frac{\partial X}{\partial x}=4\pi\sum_j n_je_j \quad (3)$$

if D is the dielectric constant of the solvent.

III.

In order to solve these equations for the case in consideration, we assume

$$v_0=a_0e^{i(\omega t-kx)}. \quad (4)$$

If g is the velocity of the sound waves, we have

$$k=\omega/g=2\pi/\Lambda \quad (4a)$$

denoting with Λ the wave-length of the sound waves in the liquid.

The number of ions per cc n_i, will always be very nearly equal to its equilibrium value $\bar{n}_i$. We put

$$n_i=\bar{n}_i+\nu_i \quad (5)$$

[1] Diffusion is not considered; its effect is altogether negligible for our problem.

and try the assumption

$$\nu_i=\alpha_ie^{i(\omega t-kx)} \quad (5a)$$

with the unknown values of the α_i.

At the same time we assume

$$X=Ae^{i(\omega t-kx)} \quad (6)$$

with a value of A which is also unknown and has to be determined.

Considering the two facts that ν_i is very small compared with $\bar{n}_i$ and that $\Sigma\bar{n}_je_j=0$, our equations take the form

$$\begin{cases} e_iX-\rho_i(v_i-v_0)=m_i(\partial v_i/\partial t). \\ \partial\nu_i/\partial t+\bar{n}_i(\partial v_i/\partial x)=0. \\ D(\partial X/\partial x)=4\pi\sum_j \nu_je_j. \end{cases} \quad (7)$$

Here moreover $\partial v_i/\partial t$ has been written instead of dv_i/dt, which is admissible as long as the velocities can be considered as small.

By introducing our special assumptions about the disturbance as due to a sound wave, the Eqs. (7) become

$$\left.\begin{array}{l} e_iA-\rho_i(a_i-a_0)=i\omega m_ia_i \\ i\omega\alpha_i-ik\bar{n}_ia_i=0 \\ -ikDA=4\pi\sum_j e_j\alpha_j \end{array}\right\}. \quad (8)$$

Combining the last equations of (8), we find

$$A=i(4\pi/D\omega)\sum_j \bar{n}_je_ja_j \quad (9)$$

and introducing this value of A into the equations of the first row, the result is

$$(\rho_i+i\omega m_i)a_i-i\frac{4\pi e_i}{D\omega}\sum_j \bar{n}_je_ja_j=\rho_ia_0 \quad (10)$$

representing z equations to determine the z values of $a_1, a_2, \cdots a_i \cdots$ as multiples of the value a_0 characteristic for the velocity of the solvent.

For all practical purposes it will be enough to give an approximate solution of these equations, keeping in mind that always ωm_i will be very small compared with ρ_i.

A zero order solution will be attained if we neglect ωm next to ρ altogether. Multiply every one of the Eqs. (10) with the corresponding product $\bar{n}_i\cdot e_i/\rho_i$ and take the sum of all the equations. The result will be

$$\sum_j \bar{n}_je_ja_j\left[1-i\frac{4\pi}{D\omega}\Sigma\frac{\bar{n}_ie_i^2}{\rho_i}\right]=a_0\sum n_ie_i=0,$$

showing that in a zero order approximation $\Sigma_j \bar{n}_j e_j a_j = 0$, which according to (9) would mean $A=0$ and therefore no potential differences at all. If they occur nevertheless we see that they will be exclusively due to the dynamical reactions of the ions measured by the products ωm_i. Now if as we found in the zero order approximation $A=0$, and if ωm is neglected, the first row of (8) shows that in this approximation $a_i = a_0$. In a next *first* order approximation we will assume

$$a_i = a_0 + \beta_i \tag{11}$$

and treat β_i as a small quantity, neglecting products of β_i with the small quantities $\omega m_i/\rho_i$. Divide every equation of (10) by the corresponding value of ρ_i and keep in mind the rule of the first order approximation. Instead of the Eqs. (10) we will then arrive at the equations

$$\beta_i - i\frac{4\pi}{D\omega}\frac{e_i}{\rho_i}\sum_j n_j e_j \beta_j = -i\frac{\omega m_i}{\rho_i}a_0. \tag{12}$$

This set of equations enables us at once to calculate the value of $\Sigma \bar{n}_j e_j \beta_j$; we find

$$\sum \bar{n}_j e_j \beta_j [1 - i(4\pi/D\omega)\sum(\bar{n}_i e_i^2/\rho_i)] = -i\omega \sum(\bar{n}_j e_j m_j/\rho_j) a_0. \tag{13}$$

Now according to (9) and (11),

$$A = i(4\pi/D\omega)\sum \bar{n}_j e_j a_j = i(4\pi/D\omega)\sum \bar{n}_j e_j \beta_j. \tag{13a}$$

In this approximation we will have an electric field and the field strength will be determined by the expression for A, which results from the combination of (13) and (13a), namely

$$\frac{A}{a_0} = \frac{4\pi}{D}\frac{\sum(\bar{n}_j e_j m_j/p_j)}{1 - i(4\pi/D\omega)\sum(\bar{n}_j e_j^2/\rho_j)}. \tag{14}$$

The field strength X itself is according to (6) represented by

$$X = A e^{i(\omega t - kx)}.$$

IV.

To discuss the magnitude of the effect we go back to the potential Φ of which the field strength X can be deduced by the well-known equation $X = -\partial\Phi/\partial x$. For this potential, the expression

$$\Phi = g a_0 \frac{(4\pi/D\omega)\sum(\bar{n}_j e_j m_j/\rho_i)}{i + (4\pi/D\omega)\sum(\bar{n}_j e_j^2/\rho_j)} e^{i(\omega t - kx)} \tag{15}$$

is easily found starting with the value for A indicated in (14). So we get the result that a sound wave travelling with the velocity g and in which the velocity of the medium oscillates between $\pm a_0$ will be accompanied by a potential wave travelling with the same velocity and in which the potential will oscillate between the values of $\pm\Phi_0$, this amplitude being represented by the formula

$$\Phi_0 = g a_0 \frac{(4\pi/D\omega)\sum(\bar{n}_j e_j m_j/\rho_j)}{[1 + ((4\pi/D\omega)\sum(\bar{n}_j e_j^2/\rho_j))^2]^{1/2}}. \tag{16}$$

At the same time the material wave and the potential wave will show a phase difference which, however, we need not consider here.

We observe that the expression $\Sigma(\bar{n}_j e_j^2/\rho_j)$ is nothing else than the conductivity of the solution expressed in electrostatic units, which we will denote by l. Introducing this notation a more convenient form for Φ_0 will be

$$\Phi_0 = g a_0 \frac{\sum(\bar{n}_j e_j m_j/\rho_j)}{\sum(\bar{n}_j e_j^2/\rho_j)} \cdot \frac{4\pi l/D\omega}{[1 + (4\pi l/D\omega)^2]^{1/2}}. \tag{16a}$$

If now we have a solution of one salt containing $\bar{n}$ molecules per cc, of which every one is split in $p_1, p_2, \cdots p_i \cdots$ different ions of valencies $\zeta_1, \zeta_2 \cdots \zeta_i$, we have $\bar{n}_j = p_j \bar{n}$; $e_j = \zeta_j \epsilon$, where ϵ is the electronic charge (4.77 10^{-10} e.s.u.) and the valency ζ_j is taken with its appropriate sign; $m_j = M_j m_H$, if M_j is the molecular weight of the ion, together with the bound solvent molecules and m_H is the mass of the hydrogen atom $1.64 \cdot 10^{-24}$ g. Introducing these notations we find

$$\Phi_0 = \frac{m_H}{\epsilon} g a_0 \frac{\sum(p_i \zeta_i M_i/\rho_i)}{\sum(p_i \zeta_i^2/\rho_i)} \cdot \frac{4\pi l/D\omega}{[1 + (4\pi l/D\omega)^2]^{1/2}}. \tag{16b}$$

Now, if as an example we are dealing with a 10^{-3} normal solution of a binary salt like KCl or LiBr, etc., the order of magnitude of l will be $l = 10^9$ (if we assume $\rho = 2.10^{-9}$) and therefore assuming $D = 80$ the quotient $4\pi l/D\omega$ will become equal to 1 for $\omega = 10^8$ approximately. This frequency would roughly correspond to radio waves of 30 meters wave-length and would make the last part of the expression (16b) equal to $2^{-1/2}$. Using a lower frequency (for the special solution considered) would mean bringing the value of that part of the expression nearer to 1. The same thing would occur if the concentration is increased. That is why in judging the magnitude of the effect we will take it that the expression $(4\pi l/D\omega)/[1+(4\pi l/D\omega)^2]^{1/2}$ can and has been made equal to 1. By adopting the usual values for m_H and ϵ and taking $g=1.4.10^5$ cm/sec., the first factor in (16b) becomes equal to $(m_H/\epsilon)ga_0=4.8\times10^{-10}a_0$. By using this factor the potential would be expressed in electrostatic units. Using volts instead, we find for the order of magnitude of Φ_0, the value

$$\Phi_0=1.4\times10^{-7}a_0\frac{\sum(p_i\zeta_iM_i/\rho_i)}{\sum(p_i\zeta_i^2/\rho_i)}\text{ volts.} \quad (17)$$

Now in our example, we would have $p_1=p_2=1$, $\zeta_1=+1$, $\zeta_2=-1$ and therefore instead of (17)

$$\Phi_0=1.4\times10^{-7}a_0\frac{M_1/\rho_1-M_2/\rho_2}{1/\rho_1+1/\rho_2}\text{ volts.} \quad (17a)$$

It is obvious that with $a_0=1$ cm/sec. and with appropriate salts, the oscillations of the potential should attain values of some 10^{-6} volts, which can easily be detected by ordinary amplification methods.

The question if $a_0=1$ cm/sec. is not too large a value to be assumed, remains to be answered. Now this velocity amplitude corresponds to a pressure-amplitude of 1/7 atmosphere and to an energy delivered per cm² and sec. of 1/200 Joule. It seems that this is not unreasonable at all thinking of a vibrating quartz-crystal, and that even more can be expected.

Experiments to detect the effect discussed here are being set up; no definite results are as yet available. As, however, the considerations put forward in this article seem to lead to an entirely new attack on the problem of solvation, I hope that I may be permitted to present them even in their present incomplete form in order to share in the congratulations at the birthday of this new journal.

[Contribution from the Gates Chemical Laboratory, California Institute of Technology, No. 66]

THE INTER-IONIC ATTRACTION THEORY OF IONIZED SOLUTES. IV. THE INFLUENCE OF VARIATION OF DIELECTRIC CONSTANT ON THE LIMITING LAW FOR SMALL CONCENTRATIONS *

By P. Debye and Linus Pauling

Received April 15, 1925   Published August 5, 1925

Introduction

There has recently been derived[1] with the aid of Poisson's equation and the Boltzmann principle an equation representing the mutual electrical effect of ions in solution and expressing the activities of ions and various properties of ionized solutes. The effect arises from the tendency of every ion to attract towards itself ions of unlike sign and repel those of like sign; as a consequence every ion is, owing to its ion-atmosphere, at an average potential P_0, of sign opposite to that of its charge.

It has been shown that for ions of any finite size in a solution of uniform dielectric constant κ, the expression for this quantity P_0 reduces for very dilute solutions to the limiting value

$$P_0 = -\frac{z\,e\,B}{\kappa} \qquad (1)$$

where

$$B^2 = \frac{4\pi\, e^2\, \Sigma(n_i z_i^2)}{\kappa\, kT} \qquad (1a)$$

[1] (a) Debye, "Hand. v. h. XIX Nederlansch Natuur en Geneeskundig Congres," April, 1923. (b) Debye and Hückel, *Physik. Z.*, **24**, 185 (1923). (c) Debye, *ibid.*, **25**, 97 (1924). For a re-presentation of this theory in somewhat different form see also Noyes, This Journal, **46**, 1080 (1924).

*Reprinted from *Journal of the American Chemical Society*, 47, 2129-2134 (1925).

in which z represents the valence (taken algebraically) of the ion under consideration, e the electronic charge, k Boltzmann's constant, T the absolute temperature, and n_i and z_i the concentration and valence of an ion of the ith sort, the summation of $n_i z_i^2$ being taken over all ion-sorts in the solution. The expression for the activation of the ion as derived from this potential P_0 is

$$-ln\ \alpha = \frac{z^2\ e^2\ B}{2\ \kappa\ kT} \quad (2)$$

It is known that the dielectric constant of a solution containing ions is not that of the pure solvent; furthermore, the very great electrical fields produced by an ion must cause in its immediate neighborhood variations in the dielectric constant. Question might therefore be raised as to whether the influence of these variations is such as to cause P_0 and $ln\ \alpha$ to be proportional not to the square root of the concentration, as required by Equations 1 and 2, but to some other power, or to cause the coëfficients to have values other than those obtained by introducing for κ the macroscopic value for the pure solvent.

It is the purpose of this paper to show that these variations of the dielectric constant do not change the limiting law that is approached as the concentration approaches zero, and that correct values of the coefficients are obtained by substituting for κ the value for the pure solvent.

Effect of Variation of the Dielectric Constant[2]

Let us consider why a variation of the dielectric constant in the immediate neighborhood of the ion is to be expected. Such variations may be produced in at least four ways. (1) In the presence of the very great electrical fields near the ion saturation of the dielectric, a phenomenon similar to magnetic saturation may occur, resulting in a decreased dielectric constant. (2) The dielectric constant would be expected to vary with the increase in density of the solvent produced by electrostriction. (3) The strong attraction by the ion of the permanent dipoles of the dielectric is not, in general, experienced by the other ions, which usually have no electric moment in the absence of a field, and become polarized only through deformation. The resultant electrostriction pressure of the medium is equivalent to superposing on the coulomb forces an effective repulsion of other ions, which can be expressed as due to a change in the dielectric constant near the ion. (4) At small distances the microscopic rather than the macroscopic point of view is required, in order to determine the influences of the sizes and shapes of the ions and of the molecules composing the dielectric medium. These influences have an effect only at small distances, and can accordingly be treated as equivalent to a change in the value of the dielectric constant near the ion.

[2] For approximate calculations of the effect of variation of the dielectric constant on the properties of concentrated solutions of strong electrolytes, see Hückel, *Physik. Z.*, **26**, 93 (1925).

Now let us consider the influence of the possible variations on the potential of the ion. In Fig. 1 is represented a particular ion whose valence is z and whose distance of closest approach to the centers of other ions is a. Each ion in the solution is considered as having some spherically symmetrical distribution of its total charge within a radius compatible with this distance of closest approach such, for example, as the location of the total charge at the center, or its uniform distribution throughout the volume or over the surface of a sphere. From the radius a to a certain greater radius r_1 the dielectric constant K is to be regarded as a function of r, and beyond r_1 it is to be considered constant and equal to K_1, that of the solution in mass.

The potential dP_0 produced at the ion by a given charge dQ distributed uniformly over a spherical shell of thickness dr and radius r greater than r_1 depends only on dQ, K and K_1, and not at all on the dielectric constant at points within the shell. This is evident when it is recalled that the energy change $-ze\ dP_0$ of removing the ion from the shell must equal that $-ze\ dQ/K_1 r$ of removing the shell to infinity, and this latter cannot involve the dielectric constant within the shell. Another evidence is given by the fact that in the interior of such a shell of constant surface-density no electric forces exist, and therefore the potential is constant throughout the whole volume. The contribution of the shell to the potential of the ion is, therefore, equal to the potential at the surface of the shell, and this can depend only on the dielectric properties of the medium beyond the radius of the shell.

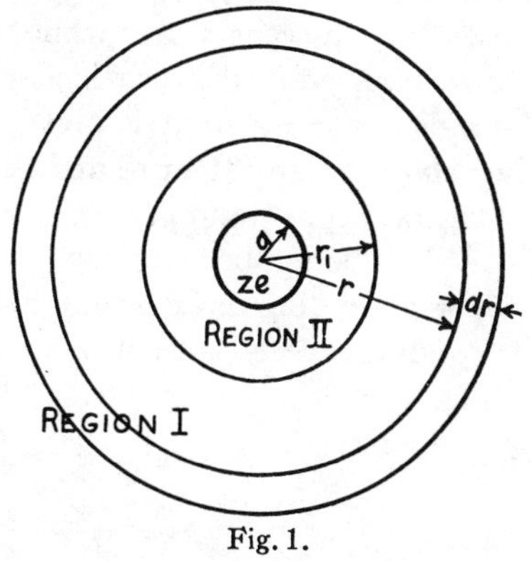

Fig. 1.

The charge dQ, however, is not arbitrary, but is induced by the charge of the ion and therefore depends on the values of the dielectric constant within the shell. The charge in a shell of radius r depends on the total potential at that radius produced by the central ion and its ion atmosphere, and the distribution and hence the potential of the ion atmosphere are influenced by variations in the dielectric constant. But as the solution is made more and more dilute the mutual electrical effect of ions becomes smaller and smaller; the charge of the ionic layer compensating the charge of the central ion will, therefore, be spread over a volume which increases beyond any limit as the concentration of the electrolyte tends to zero. For the case of a constant dielectric constant throughout the whole medium this corresponds to the fact that according to (2) the value $1/B$, which may be called the thickness of the corresponding ionic layer, is proportional to the reciprocal of the square root of the concentration. Now, the whole compensating charge of the layer is always equal (but opposite in sign) to the charge of the central ion. In the limit for low concentrations all but a negligible part of the charge will, therefore, be distributed in shells within the region of invariable dielectric constant κ_1. Therefore, the potential P_0 and the activity coefficient will be given by Equations 1 and 2 with κ replaced by κ_1; that is,

$$P_0 = -\frac{z\,e\,B_1}{\kappa_1} \tag{3}$$

and

$$-\ln \alpha = \frac{z^2\,e^2\,B_1}{2\kappa_1\,kT} \tag{4}$$

where

$$B_1^2 = \frac{4\pi\,e^2\,\Sigma(n_i z_i^2)}{\kappa_1\,kT} \tag{4a}$$

These arguments could easily be put into mathematical form, and constitute therefore a mathematical proof of the theorem that variations of the dielectric constant in the neighborhood of the ions do not affect the limiting law. Still it seems worth while to verify the arguments by a consideration of a simplified model for which explicit formulas can be derived. Suppose the dielectric constant is given the constant value κ_1 in the region I in which r is greater than r_1, and the constant value κ_2 in the region II where r lies between r_1 and a. Poisson's equation and the Boltzmann principle can then be applied here as in the case of a uniform dielectric constant. Upon expanding the exponential expressions and neglecting terms after the second in each expansion in the usual way, there are obtained the equations

$$\left.\begin{aligned} &\nabla^2 P_1 \equiv \frac{d^2P_1}{dr^2} + \frac{2}{r}\frac{dP_1}{dr} = B_1^2\,P_1, \text{ in region I} \\ \text{and}\quad &\nabla^2 P_2 \equiv \frac{d^2P_2}{dr^2} + \frac{2}{r}\frac{dP_2}{dr} = B_2^2\,P_2, \text{ in region II} \\ \text{with}\quad &B_1^2 = \frac{4\pi\,e^2\,\Sigma(n_i z_i^2)}{\kappa_1\,kT} \text{ and } B_2^2 = \frac{4\pi\,e^2\,\Sigma(n_i z_i^2)}{\kappa_2\,kT} \end{aligned}\right\} \tag{5}$$

The solutions of these equations are

$$\left.\begin{aligned} &P_1 = I_1\frac{e^{-B_1 r}}{r} + I_1'\frac{e^{B_1 r}}{r}, \text{ in region I} \\ \text{and}\quad &P_2 = I_2\frac{e^{-B_2 r}}{r} + I_2'\frac{e^{B_2 r}}{r}, \text{ in region II} \end{aligned}\right\} \tag{6}$$

Since the potential must tend to zero as r becomes infinite, I_1' is zero. Furthermore, at $r = r_1$ and at $r = a$ the potential and the induction must be continuous; that is,

$$P_1 = P_2, \text{ and } \kappa_1\frac{dP_1}{dr} = \kappa_2\frac{dP_2}{dr} \text{ at } r = r_1, \tag{7}$$

$$P_2 = \frac{ze}{\kappa_2 r} - \frac{ze}{\kappa_2 r_1} + \frac{ze}{\kappa_1 r_1} + P_0, \text{ and } \kappa_2\frac{dP_2}{dr} = -\frac{ze}{r^2} \text{ at } r = a \tag{8}$$

With these equations the three remaining constants of integration and the potential P_0 of the central ion due to its ion atmosphere may be evaluated.

The expression obtained for P_0 is

$$P_0 = \frac{ze}{K_2 r_1} - \frac{ze}{K_1 r_1} - \frac{ze}{K_2 a}$$

$$+ \frac{ze}{K_2 a} \cdot \frac{K_1(1 + B_1 r_1)\{e^{B_2(r_1-a)} - e^{-B_2(r_1-a)}\} - K_2\{(1 - B_2 r_1)\,^{B_2(r_1-a)} - (1 + B_2 r_1)e^{-B_2(r_1-a)}\}}{K_1(1 + B_1 r_1)\{(1 + B_2 a)\, e^{B_2(r_1-a)} - (1 - B_2 a)\, e^{-B_2(r_1-a)}\} - K_2\{(1 - B_2 r_1)(1 + B_2 a)e^{B_2(r_1-a)} - (1 + B_2 r_1)(1 - B_2 a)e^{-B_2(r_1-a)}\}} \quad (9)$$

On expanding this it is found that the first term in the expansion is that given in Equation 3; the remaining terms involve the first power and higher powers of the concentration, and so are negligible in very dilute solutions.

The problem now remains to determine the effect resulting from the fact that K_1, the dielectric constant of the solution in mass, is not in general equal to K, that of the pure solvent. Theoretical considerations as well as experimental evidence, show that for sufficiently dilute solutions the variation of the dielectric constant from that of the pure solvent is proportional to the concentration of the solute; that is, we can write $K_1 = K(1+\beta n)$. Upon substituting this in Equations 3 and 4, and expanding, they reduce to Equations 1 and 2 multiplied by the factor $1 + 3\,\beta n/2 + \ldots$; the correction terms introduced, involving as they do higher powers of the concentration, drop out in dilute solutions.

In obtaining an expression such as (2) for the activation of an ion it is necessary to take into consideration not only the energy change accompanying the transfer of the ion from the potential P_0 due to its ion atmosphere in the solution considered to that in an infinitely dilute one, but also that accompanying the removal of the ion from one dielectric medium to the other. The change in potential involved in the latter operation is $\frac{ze}{r_1}\left(\frac{1}{K} - \frac{1}{K_1}\right)$; replacing K_1 by $K(1 + \beta n + \ldots)$ this becomes $\frac{ze}{Kr_1}\beta n + \ldots$, which evidently is negligible in comparison with P_0 in sufficiently dilute solutions.[3]

Conclusions and Summary

The considerations that have been presented in this article lead to the conclusion that neither the variation of the dielectric constant in the immediate neighborhood of the ions nor the deviation of the dielectric constant of the solution in mass from that of the pure solvent has any effect on the limiting law for very dilute solutions of strong electrolytes (as expressed by Equations 1 and 2 above). It is further proved that for such solutions the value of the ordinary dielectric constant for the pure solvent in mass is to be substituted. The experiments of Brönsted and La Mer[4] at *very* low concentrations have completely confirmed this theoretical limiting law.

[3] This conclusion has been previously stated by Debye, Ref. 1 c, p. 99.

[4] Brönsted and La Mer, THIS JOURNAL, **46**, 555 (1924).

The fact that some other experimental results at fairly low concentrations have led to smaller values of the numerical coefficient than those given by these equations can be attributed to variations of the dielectric constant only if it be assumed that the solutions investigated were still too concentrated to make the limiting law strictly applicable.

PASADENA, CALIFORNIA

ON THE SCATTERING OF LIGHT BY SUPERSONIC WAVES

By P. Debye and F. W. Sears

Department of Physics, Massachusetts Institute of Technology

Communicated April 23, 1932

1. *Introduction.*—In a paper published in 1922 Leon Brillouin[1] treated the problem of light scattering. In accordance with the fact that for low temperatures Einstein's theory of specific heat has to be abandoned for Debye's theory, Brillouin attributes the thermal density fluctuations in the body, which, in his theory, as in a previous theory of Einstein's,[2] are responsible for the scattering to a superposition of sound waves. He tries to apply his theoretical results to the explanation of x-ray scattering. We know now that this application is far from correct, as for such short waves the electronic density changes due to the atomic or molecular structure are much more important than the thermal fluctuations. For light waves, however, with a wave-length much longer than molecular distances, Brillouin's analysis leads to some remarkable results. They can be stated in the following manner. Suppose the primary light travels in figure 1 in a direction characterized by a vector $\underset{\rightarrow}{S_0}$ of length unity in this direction. Let it be assumed that of the scattered light a part is observed traveling in another direction characterized by a unit vector $\underset{\rightarrow}{S}$. Then firstly, of the sound waves of all possible directions, only those are important for the scattering which are traveling in or opposite to the direction of the vector $\underset{\rightarrow}{s} = \underset{\rightarrow}{S} - \underset{\rightarrow}{S_0}$. This can also be expressed by saying that the planes of the sound waves have to be situated such that the scattered light can be considered as optically reflected by these planes. But there is a second limitation. Of all the sound waves of direction $\pm \underset{\rightarrow}{s}$, only those of a definite wave-length Λ are effective. This wave-length is $\Lambda = \lambda/s$, if λ is the wave-length of the light and s is the length of the vector $\underset{\rightarrow}{s}$, which is $2 \sin \Theta/2$, calling Θ the angle between the primary and the secondary ray. This last condition can be expressed by saying that the consecutive planes of maximum density in the sound wave must be

*Reprinted from *Proceedings of the National Academy of Sciences*, 18, 409-414 (1932).

separated by a distance Λ such that the well-known relation of Bragg holds. In this case the light rays reflected by the consecutive planes will have path differences of one wave-length (of light) each and therefore the reflections will be strong. So we are left with only two sound waves, traveling with the velocity of sound q, one in the direction $+\vec{s}$ and the other in the direction $-\vec{s}$. The frequency of the reflected light, according to Doppler's principle, will be changed, and instead of the primary frequency v, we should find in the scattered light the two frequencies:

$$v = v_0\,[1 \pm 2n\; q/c \sin \theta/2]$$

where c is the velocity of light in vacuo and n is the index of refraction of the medium.

Gross[3] reports that with an echelon he has been able to photograph several new components of a spectral line created by the scattering process. Recently, Meyer and Ramon[4] have published actual photographs of the effect. In contrast with Gross's results they were able to obtain two components only, one at either side of the primary line, which is in accordance with the theory. The shift is approximately 0.06 A. U. and checks satisfactorily with the value calculated from the known velocities of sound in the liquids used. Besides the two components, a central line of unchanged frequency v_0 appears which may partly be due to unavoidable traces of dust in the liquid, and partly to thermal fluctuations of the index of refraction as a result of molecular rotation, an effect which was not considered in the derivation of Brillouin's theory.

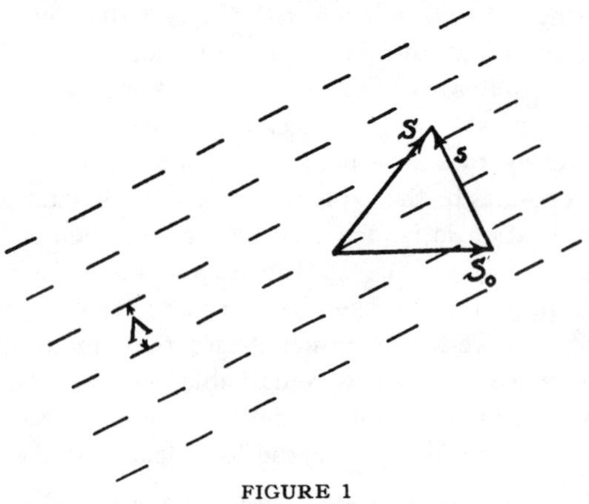

FIGURE 1

2. *Experiments.*—Brillouin[1] mentions the possibility of experimental verification of his calculations, making use of elastic waves set up in a liquid by a quartz crystal driven by a high-frequency oscillator, and it occurred to us that it would be interesting to try the scattering of light by these "artificial" sound waves. The experiment was set up in the following way. In a trough of rectangular cross-section, figure 2, filled with a liquid (benzene, carbon tetrachloride, etc.) a quartz crystal Q was immersed. The leads to the silvered faces of this crystal could be connected with a radio-frequency oscillator. Vibrations set up in the crystal

in this way excite supersonic waves traveling in the liquid in the direction $Q\ E$. Perpendicular to this direction a parallel beam of light from a slit S and a lens L_1 passes through the liquid. The parallel beams, scattered by the illuminated part of the liquid are focussed by a telescope lens L_2. As soon as the crystal is connected to the oscillator, spectra of different orders (like the spectra of an ordinary grating), appear at the left and right of the central image of the slit. The number of orders which can be observed depends on the intensity of the vibrations. Their spacing depends only on the frequency and increases if the frequency of the oscillator is increased.

As in ordinary grating spectra, the blue is the least deviated, and the red the most. With a mercury arc source, the different mercury lines can be observed in every order. Under favorable conditions as to the intensity of the vibration more than 10 spectra to the right and to the left have been obtained. In order to see if the supersonic waves are markedly transmitted by a solid, a glass block of cross-section equal to the trough was immersed in the liquid. Light passed through the liquid in front of and behind the block showed the spectra, but through the block itself produced only the central slit image. We may safely expect, however, that using a higher intensity the effect will also be seen in the solid. No marked decrease in the number of orders visible could be observed in passing the light through portions of the liquid at different distances from the crystal with liquids like benzene or carbon tetrachloride. With glycerine, however, this effect is very pronounced, no doubt due to its high viscosity.

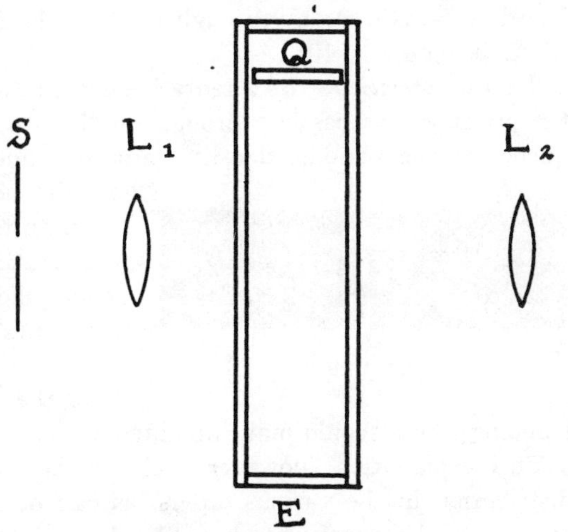

FIGURE 2

Fixing the attention on one of the spectra, preferably of higher order, one can observe that it attains its maximum intensity if the trough is turned through a small angle such that the primary rays are no longer parallel to the planes of the supersonic waves. Different settings are required to obtain highest intensities in different orders. If the trough is turned continuously in one direction, starting from a position which gave the highest intensity to one of the orders, the intensity decreases steadily, goes through zero, increases to a value much smaller than the first maximum, decreases to zero a second time and goes up and down again passing through a still smaller maximum. The same series of events occurs in turning the trough in the other direction. To make these observations it is necessary to watch the pattern carefully. One thing, however, which can be seen at first glance is the fact that the number of orders visible on the right and the left of the central image is different except for the case in which the primary rays are passing exactly parallel to the planes of the supersonic waves. Turning the trough continuously changes the number of orders to the right or the left in the same way as has been described for the intensity.

The whole effect is rather brilliant and can easily be projected on a screen so as to make it visible to an audience. Figure 3 shows a photograph of the effect obtained in passing monochromatic light of wavelength $\lambda = 5461$ A. U. through toluene, the frequency of the supersonic waves being 5.7×10^6.

3. *Interpretation.*—As a tentative explanation it was first thought that the primary rays passing through the liquid in sheets would, as a consequence of the periodic density variations due to the supersonic waves, acquire phase differences of periodical nature. The light emerging from the grating created in the liquid would show interference patterns of the same general kind as observed with an ordinary grating. The fact that the supersonic waves are not standing waves should make no difference.

FIGURE 3

This explanation, however, fails to explain the observed intensity distribution in the various orders, as can be shown by a more detailed analysis, and a theory has been developed based on the assumption of a volume scattering, in which every volume of the liquid contributes to the total scattering in accordance with Maxwell's equations. In this way, it seems at first, following Brillouin's theory, that one would expect only one reflection for a definite angle of incidence of the primary light and one other reflection on the other side for the same angle taken with negative sign. Moreover, light passing parallel to the planes of the supersonic wave should show no effect at all. These results are evidently at variance with the observations. Taking into account, however, that the dimensions of the illuminated volume of the liquid are finite it can easily be shown that in our case Bragg's reflection angle is not sharply defined and that reflection should occur over a rather appreciable angular range. If l is the length of the path of light in the liquid, Λ the wave-length of the supersonic waves and λ the wave-length of the light, then two quantities are of importance; namely, the quotients l/Λ and Λ/λ. Only if

l/Λ is large compared to Λ/λ does a sharp definition of Bragg's angle exist. Working with a frequency of 10^7 cycles, Λ is about 0.1 mm., l is of the order 10 mm. and λ is about 0.5×10^{-3} mm. In this case, therefore, $l/\Lambda = 100$ and $\Lambda/\lambda = 200$, the quotient of these two quantities is 1/2 and cannot be considered as large. A detailed analysis shows that in such a case reflection will occur over a range of angles left and right of the critical angle which follows from Bragg's relation. Moreover the intensity variations predicted by the theory in varying the angle continuously are just the same as described in relating the rather peculiar experimental results on this point.

We are, however, still left with another difficulty. The theory predicts only the first order spectrum to the right and the first order to the left. But in the theory so far it has been assumed that the variations of the index of refraction are of purely sinusoidal character. If they are not, then we can consider the disturbance as a superposition of variations of frequencies υ, 2υ, 3υ, etc., with the corresponding wave-lengths Λ, $\Lambda/2$, $\Lambda/3$, etc., provided a marked dispersion of the velocity of sound in the frequency region considered does not exist. A departure from sinusoidal character therefore accounts for the existence of the higher order spectra. This departure may be due to the non-sinusoidal cha acter of the crystal vibrations, although higher harmonics may be produced by the scattering itself, if the intensity of the supersonic waves is high enough. This is in agreement with the fact that the higher orders fade out if the intensity of the vibrations of the crystal is decreased.

Adopting the theory, the measurement of the angles θ of the different orders ρ with respect to the central image provides us with a measure of the wave-length of the supersonic wave in terms of the wave-length of the light which has been used. In fact, as in an ordinary grating

$$\sin \theta\rho = \rho\lambda/\Lambda.$$

The frequency can easily be determined with an ordinary wavemeter and so we get a very simple method for the determination of the velocity of sound.

The following table shows some preliminary measurements performed with a very simple spectrometer, together with values calculated from the density and the adiabatic compressibility.

	VELOCITY CALCULATED M./SEC.	FREQUENCY	VELOCITY OBSERVED M./SEC.	FREQUENCY	VELOCITY OBSERVED M./SEC.
Toluene	1290	1.7×10^6	1330	16.5×10^6	1310
Carbon tetrachloride	920	1.7×10^6	940	16.5×10^6	930

Up to now no indication of a change in velocity with frequency has been observed. If it exists, a more careful measurement of the angles for different orders should show it, as the measurement of say 10 orders simultaneously existing provides us with a frequency range of v to 10 v.

[1] L. Brillouin, *Ann. phys.*, **17,** 88 (1922).

[2] A. Einstein, *Ann. Phys.*, **33,** 1275 (1910).

[3] Gross, *Zeit. Phys.*, **63,** 685 (1930); *Naturwiss.*, **18,** 718 (1930); *Nature,* **126,** 201, 400, 603 (1930).

[4] Meyer and Ramon, *Phys. Zeitschr.*, **33,** 270 (1932).

SOUND WAVES AS OPTICAL GRATINGS

(Schallwellen als optische Gitter)

P. Debye

Translated from
Berichte über die Verhandlungen der sächsischen Akademie der Wissenschaften, Leipzig, Vol. 84, 1932, pages 125-127.

According to a theory of L. Brillouin[1] the scattering of light in a solid can be interpreted as Bragg reflections on thermal elastic waves. The general formulation of Brillouin's theory is not different from that of Einstein's[2] earlier theory. Here, as there, the density variations are made responsible for the light scattering. However, Brillouin takes account of the fact that the thermal motions of neighboring atoms are not independent of one another and thus describes the total motion of an atom (or molecule) as a superposition of sound waves in the same way as Debye did in calculating the specific heat.

The essential results of the Brillouin theory can be summarized briefly as follows:

(1) If a ray of scattered radiation and the primary ray producing it are observed, the scattered ray can be regarded as originating by optical reflection of the primary on sound waves of the proper direction. This means that of all possible sound waves which pass through the solid only those are important in the scattering which are in the proper direction; that is, are in such a direction that their wave fronts make the same angle with the primary ray and the scattered ray.

(2) Of all sound waves in this direction only those are important in the scattering for which the separation of planes of the same phase (that is their wavelength) is such that the rays reflected from these planes interfere constructively. Therefore the Bragg condition must be fulfilled by the light waves, where the wavelength of the sound waves plays the role of the lattice constant. The formula which expresses this condition is

$$2\,\Lambda \sin\frac{\vartheta}{2} = \lambda$$

where Λ = wavelength of the sound waves,

λ = wavelength of the light, and

ϑ = angle between the primary and the scattered rays.

In the case of ordinary light scattering, the interesting feature of this interpretation is that a Doppler effect must result from the reflection because of the finite velocity of propagation of the sound waves. I believe that this effect has been detected experimentally by Gross[3] and by Meyer and Ramm.[4] The interesting thing is that this Doppler effect occurs in liquids and thus demonstrates that here also the thermal motions of neighboring molecules are coupled.

While I was a visitor at the Massachusetts Institute of Technology in Cambridge, Mass., the idea occurred to me to try to detect the Bragg reflection on artificial sound waves. The experiment was carried out* with F. W. Sears of the Physics Department of M.I.T. A quartz crystal was provided with electrodes and a high frequency of some megacycles was applied. The crystal was immersed in a liquid (benzene, toluene, carbon tetrachloride, etc.) which was contained in a long trough with plane walls. The sound waves were of the order of a few tenths of a millimeter. A parallel bundle of light was passed through the trough perpendicular to the direction of propagation of the sound waves coming from a lens illuminated by light from a slit at focal distance. A second lens which refocused the light in its focal plane was placed behind the trough. The original purpose was to demonstrate Bragg reflection on sound waves by adjusting the reflection angle. However, another phenomenon appeared immediately. As long as the crystal did not vibrate, only the image of the slit was visible in the focal plane of the second lens. However, as soon as the quartz crystal was activated, diffraction images appeared to the left and right of the central image. They were completely analogous to the grating spectra of an ordinary grating. Each spectrum shows colors which appear in the usual order; the spectra are equidistant. The number of visible orders depends on the intensity of the sound vibrations. It was possible to make more than 20 orders visible to the left and right of the central image. The phenomenon is very brilliant and the images can be directly projected and demonstrated in a large auditorium. The angular separation of different orders depends on the wavelength of the sound waves, the wavelength playing the roll of the grating constant. Changing the capacity of the condensers and thus changing the high frequency produces corresponding changes in the wavelength of the sound waves and thus increases or decreases the separation of orders. With mercury light as the primary radiation, the mercury lines appear. The phenomenon therefore gives a very simple method for determining the speed of sound, since a simple measurement of angle yields the ratio of the wavelength of the light to the wavelength of the sound. This

* An account of this was submitted to the National Academy in Washington at the end of April.

is demonstrated, for example, by the different separations at constant frequency between orders in two liquids such as toluene and carbon tetrachloride.

The question remains how the phenomenon is related to Bragg reflection. The theory (which will be given elsewhere) as well as the experiment leads to the following opinion: Under the existing conditions, the spatial extension of the sound waves is not sufficient to bring about a Bragg reflection in a rigorous sense. Noticeable reflections over a finite angular interval occur. This goes hand in hand with the fact that different diffraction orders appear to the left and right of the central image. What remains of the rigorous Bragg reflection is the fact that the intensity distribution among the different orders is sensitive to the relative directions of the light and sound waves. The intensity distribution, for instance, is symmetrical only if the two are perpendicular to each other.

The phenomena described were photographed and the pictures presented to the Academy. These, as well as the complete report, will appear elsewhere.

Bibliography

1. L. Brillouin, *Ann. phys.*, *88* (1921).
2. A. Einstein, *Ann. Physik*, *33*, 1275 (1910).
3. Gross, *Z. Physik*, *63*, 685 (1930); *Naturwiss.*, *18*, 718 (1930); *Nature*, *126*, 201, 400, 603 (1930).
4. Meyer and Ramm, *Physik. Z.*, *33*, 270 (1932).

SOME REMARKS ON MAGNETIZATION AT LOW TEMPERATURES
(Einige Bemerkungen zur Magnetisierung bei tiefer Temperatur)

P. Debye

Translated from
Annalen der Physik, Vol. 81,. 1926, pages 1154–1160

It is a well known fact that the Langevin formula is capable of describing paramagnetic saturation phenomena. This was shown experimentally by **Kamerlingh-Onnes** in his experiments on gadolinium sulfate. That this is possible even though the assumptions which Langevin made in his derivation of the formula are not at all correct makes an exciting problem which until now has been treated only partially. In the following it will be shown that, on the one hand, the Langevin formula, in spite of its experimental verification, cannot be entirely correct, since it leads to conclusions which contradict the Nernst heat theorem. On the other hand, a quantitative **estimate** of the temperature change resulting from an adiabatic magnetic process will be calculated on the basis of this formula. This change appears to be fairly large and thus the question whether an effort should be made to use such a process in approaching absolute zero is raised.

1. Consider 1 gram of a magnetizable substance whose volume changes can be neglected and whose total energy is u. If an amount of heat dq is added to this body and its magnetic moment σ is changed by $d\sigma$ by changing the magnetic field H, the work done is $Hd\sigma$ and thus*

$$du = \delta q + H d\sigma = T ds + H d\sigma \tag{1}$$

where T is the temperature and s is the entropy of the body considered.

*Instead of to individual pieces of literature, reference may be made to my article, *Theorie der elektrischen und magnetischen Molekulareigenschaften* in *Handbuch der Radiologie*, Bd. 6, Leipzig, 1925, and to the recent excellent book of E. C. Stoner, *Magnetism and Atomic Structure*, London, 1926.

If σ and T are considered independent variables and the condition that ds be a total differential is written out, it follows that:

(2) $$\frac{\partial u}{\partial \sigma} = -T^2 \frac{\partial}{\partial T}\left(\frac{H}{T}\right)$$

a formula which is completely analogous to the well known relation:

$$\frac{\partial u}{\partial v} = T^2 \frac{\partial}{\partial T}\left(\frac{p}{T}\right)$$

which couples the caloric and thermal equations of state of a gas.

Now if we consider in particular the investigations on gadolinium **sulfate,** it seems certain from the experiments that σ is a function only of the ratio H/T, since with different fields and at different temperatures the experimental values of the magnetization all lie along a single curve if plotted as a function of this ratio. Conversely if this is so, H/T must be a function only of σ and according to (2):

$$\frac{\partial u}{\partial \sigma} = 0$$

so that u depends only on T.

With this assumption it follows from (1) that:

(3) $$ds = \frac{1}{T}\frac{du}{dT}dT - \frac{H}{T}d\sigma$$

or:

(3') $$s = s_1 + s_2$$

with a caloric contribution:

(4) $$s_1 = \int \frac{1}{T}\frac{du}{dT}dT$$

and a magnetic contribution:

(4') $$s_2 = -\int \frac{H}{T}d\sigma$$

If for brevity we put $H/T = x$ and $\sigma = f(x)$, then:

(5) $$s_2 = -\int x f'(x)\,dx = -x f(x) + \int f(x)\,dx$$

while s_1 is the ordinary entropy in zero field.

Up to now no use has been made of the form of the magnetization function. Now if the Langevin function, which certainly holds in the region of the experimental conditions of Kamerlingh-Onnes, is introduced for this, we have:

$$(6) \qquad \sigma = n\mu\left[\mathfrak{Cotg}\frac{\mu H}{kT} - \frac{1}{\mu H/kT}\right]$$

if n particles with magnetic moment μ are present in 1 gram and k is the Boltzmann constant. If for brevity we put $\frac{\mu H}{kT} = \xi$ and call the function in the brackets in (6) $L(\xi)$, then according to (5):

$$(5') \qquad s_2 = -nk\int \xi L'(\xi)\,d\xi = nk\left[\log\frac{\mathfrak{Sin}\,\xi}{\xi} + 1 - \xi\,\mathfrak{Cotg}\,\xi\right]$$

If now it is asked if this expression for s_2 is consistent with the Nernst heat theorem, (5') must be expanded for small values of T, that is, large values of ξ. It is found that:

$$(5'') \qquad s_2 = -nk[\log 2\xi - 1 + (2\xi + 1)e^{-2\xi} + \ldots]$$

Thus as absolute zero is approached ($\xi = \infty$) the magnetic contribution to the entropy approaches ∞ logarithmically. Thus (5") and with it the Langevin formula for the magnetization function are refuted by the following partial statement of the Nernst theorem: the entropy has a finite value at all finite temperatures.

If we reflect on the fundamental basis for the contradiction, we see without difficulty that it is to be sought in that the Langevin function represents a magnetization which approaches saturation too slowly with increasing field strength. In fact, for large values of ξ:

$$(7) \qquad L(\xi) = 1 - \frac{1}{\xi} + \ldots$$

and in first approximation $d\xi/\xi$ appears under the integral in (5') so that the logarithmic ratio appears. Obviously, any magnetization function which has an expansion similar to L's whose deviation from saturation is measured in first approximation by $1/\xi$ meets with the same difficulty.

With the help of the Weiss hypothesis of the inner field, the paramagnetic magnetization function can be used in order to describe the temperature dependence of the saturation magnetization of ferromagnetic bodies. The form of this curve at low temperatures is essentially determined by the form of the development of the paramagnetic magnetization function for large values of its argument. If this has a form similar to (7) then the saturation magnetization achieves its maximm value for $T = 0$ only relatively slowly. Now it is of special interest that the experimental curves at low temperatures approach the horizontal line which represents maximum saturation more rapidly than is consistent with the Langevin funtion. Thus there is experimental proof that the magnetization function does not have the form (7) so that the results deduced from the requirement that the magnetic contribution to the entropy remain finite are verified.

If the functions which are based on the existence of a finite number of discrete orientations as in quantum theory are tried, no difficulty is found. Thus the Lenz function:

$$\frac{\sigma}{n\mu} = \mathfrak{Tg}\,\xi$$

has on the one hand the expansion:

$$\mathfrak{Tg}\,\xi = 1 - 2e^{-\xi} - \ldots$$

On the other, it gives for the magnetic contribution to the entropy:

$$\begin{aligned} s_2 &= -nk[\xi\,\mathfrak{Tg}\,\xi - \log(\mathfrak{Cof}\,\xi)] \\ &= -nk[\log 2 - (2\xi + 1)e^{-2\xi} + \ldots] \end{aligned}$$

so that the requirement that this be finite for $T = 0$ is satisfied. Thus, in conclusion, it is probable that the Langevin function, in spite of its usefulness as an interpolation formula, is not the correct magnetization function.

2. However, the observations on gadolinium sulfate have shown that there is a region in which the Langevin function is valid as an interpolation formula. In this region correct conclusions can be drawn from it. For example, imagine an adiabatic process, characterized by the requirement that the entropy remain constant, in which the magnetic field is suddenly switched on or turned off. According to (4) and (5') the condition that the entropy remain constant is:

$$(8) \quad \left\{ \begin{aligned} s_1 + s_2 &= \int \frac{1}{T}\frac{du}{dT}\,dT \\ &\quad + nk\left[\log\frac{\mathfrak{Sin}\,\xi}{\xi} + 1 - \xi\,\mathfrak{Cotg}\,\xi\right] = \text{const.} \end{aligned} \right.$$

Now we consider the order of magnitude of the two contributions in a case which can be realized experimentally.

At 1.3° absolute and in a field of 22,000 gauss **Kamerlingh-Onnes has** reached about 84% of the saturation value. According to the Langevin formula this corresponds to a value of ξ which satisfies:

$$\mathfrak{Cotg}\,\xi - \frac{1}{\xi} = 0.84$$

and so is about $\xi = 6$. As ξ varies from 0 to 6, the square bracket varies from 0 to -1.48. If the process is adiabatic, (8) must be satisfied; that is, the caloric contribution to the entropy given by the first term must simultaneously increase about 1.48 nk. Unfortunately, the temperature change brought about by the adiabatic character of the process cannot be calculated exactly because there is no published data on the specific heat of gadolinium sulfate. In order to make an estimate of the change the following can be done. If 1 gram of the salt contains N atoms and at the low temperatures which are being considered the body is treated as a monatomic solid, the quantum theory of specific heats gives for s_1:

$$s_1 = \frac{4\pi^4}{5} N k \left(\frac{T}{\Theta}\right)^3$$

where θ is the characteristic temperature of gadolinium sulfate. Here N is the total number of atoms and n is the number of independent elementary magnets. Now if it is assumed that only the gadolinium atoms are provided with paramagnetic moments, the ratio of N to n is the ratio of the number of other kinds of atoms in the molecule to the number of gadolinium atoms. Since the molecule has the structure $Gd_2(SO_4)_3+8H_2O$ $N/n = 41/2$. Thus we can also write:

$$s_1 = \frac{4\pi^4}{5}\frac{41}{2}\left(\frac{T}{\Theta}\right)^3 nk = 1600\left(\frac{T}{\Theta}\right)^3 nk$$

Now if a very small value is taken for θ, for example, $\theta = 50$, and s_1 is calculated for $T = 2$, it is found that:

$$s_1 = 0.10\, nk$$

Above we agreed that ξ goes from 0 to 6 when a field is applied. We can imagine just as well that during an adiabatic switching off of the field ξ falls from 6 to 0. In this case the magnetic contribution to the entropy increases by 1.48 nk and the caloric contribution should compensate this by decreasing the same amount. However, our estimate that $s_1 = 0.10\, nk$ shows that under the assumed conditions even cooling to absolute zero would not suffice.

Naturally this does not mean that there is a possibility of reaching absolute zero with such an adiabatic magnetic process. Rather, it must be concluded that the Langevin formula is not correct. However, the facts that on the one hand the Langevin formula is verified by experiments at 1.3° absolute and on the other that a large difference was found above between the possible magnetic and caloric entropy changes make it very probable that cooling considerably beyond the region of validity of the formula can be achieved. Therefore, it may be of interest to make measurements at low temperatures on the adiabatic cooling of gadolinium sulfate which can be expected when the magnetic field is switched off suddenly.* We do not dare to venture a prediction of the magnitude of the cooling which can be realized; however, it appears not to be excluded that this can be large. Only experiments can decide, and the above analysis should stimulate the carrying out of these.

*In addition note that the specific heat measured in a constant magnetic field c_H should differ noticeably from the specific heat measured in 0 field c_0, for from (8) it follows immediately by differentiation that:

$$c_H - c_0 = n k \left[1 - \frac{\xi^2}{\mathfrak{Sin}^2 \xi}\right]$$

which at $\xi = 1$ is already $0.15\, nk$.

THE SCATTERING OF LIGHT BY SOUND WAVES
(Zerstreuung von Licht durch Schallwellen*)

P. Debye

Translated from
Physikalische Zeitschrift, Vol. 33, 1932, pages 849-856.

I. Introduction

Some time ago F. W. Sears and I were able to show that a fluid through which high-frequency sound waves are propagated strongly scatters light that passes through it. Moreover, interference phenomena appear that are completely analogous to those observed with ordinary ruled gratings and in which the wave length of the sound waves in the fluid plays the same part as the grating constant of the ruled grating.[1] In the paper with Sears it was already shown that a satisfactory theory of the phenomenon could be obtained only if the effect is considered as a volume effect in which every illuminated volume element of the fluid gives its own contribution. But even so all details could not be explained, at least as long as the disturbance of the light waves by the sound waves was treated as a small quantity of the first order. In fact, experiments show that even sound waves of relatively small amplitude produce a very great effect upon the propagation of light. Therefore it appears that a complete theory using calculations of the disturbances of the first order cannot be satisfactory but that better approximations have to be considered. A practical consequence of the magnitude of these interactions is the great light intensity in the interference phenomena which can easily be demonstrated before a large audience by projection.

In the following I shall bring together results which the theory of the phenomenon yields, in the first place in order to make possible the evaluation of the scattered intensity and simultaneously with this to calculate more rigorously the strength of the interaction and in the second place on account of the interesting questions of a more mathematical nature which arise through

*Received October 6, 1932.

the need of higher approximations.

For the detailed discussion of the interference, consider the following case. The sound waves travel in the x direction through a channel of rectangular cross section with dimensions b and l. A parallel beam of light of rectangular cross section whose dimensions are a and b is sent through perpendicular, or nearly perpendicular, to the direction of propagation of the sound. After

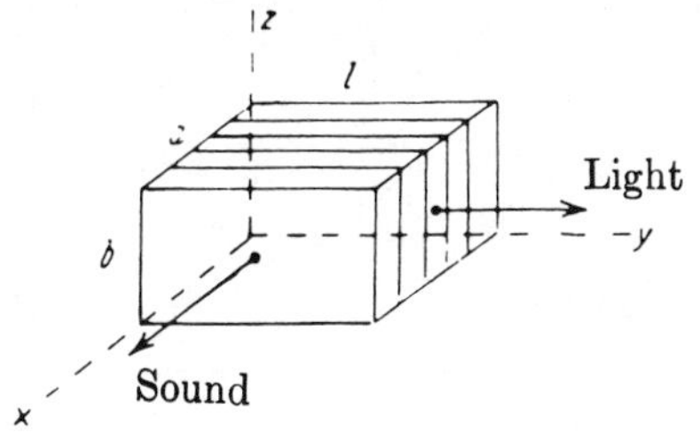

passing through the illuminated parallelepiped the parallel beam of light may be considered to have been collected by a lens and projected upon a screen. The distribution of light upon the screen is what interests us. We can discuss the light distribution in the usual way by placing ourselves at a very great distance R from the illuminated parallelepiped and considering the distribution of light upon the sphere of radius R.

II. General Principles for the Calculation of the Scattered Field

We assume that in a medium of dielectric constant ε there are within a bounded region of space charges of density ρ and currents of density $\mathfrak{J}$ which vary with time. We wish to know the electromagnetic fields $\mathfrak{E}$ and $\mathfrak{H}$ which are observed at a great distance R from this region in the medium.*

The Maxwell-Lorentz equations govern the field.

$$\left.\begin{aligned} \frac{1}{c}\dot{\mathfrak{H}} &= -\operatorname{rot}\mathfrak{E}, \quad \operatorname{div}\mathfrak{H} = 0 \\ \frac{4\pi}{c}\mathfrak{J} + \frac{\varepsilon}{c}\dot{\mathfrak{E}} &= \operatorname{rot}\mathfrak{H}, \quad \varepsilon \operatorname{div}\mathfrak{E} = 4\pi\rho \end{aligned}\right| \quad (1)$$

As is known the field produced by charges ρ and currents $\mathfrak{J}$ can be obtained most simply by introducing a scalar potential V and a vector potential $\mathfrak{A}$ from which the field $\mathfrak{E}$ and $\mathfrak{H}$ can be obtained by means of the formulas:

$$\mathfrak{H} = \operatorname{rot}\mathfrak{A}, \quad \mathfrak{E} = -\frac{\dot{\mathfrak{A}}}{c} - \operatorname{grad} V \qquad (2)$$

*In the experiment the scattered beam leaves the fluid and enters the air. In the discussion of the observed refraction angle of the interference bands, the diffraction occurring because of the departure of the beam from the fluid must, of course, be taken into account subsequently.

If the additional permissible assumption is made that:

$$\operatorname{div} \mathfrak{A} = -\frac{\varepsilon}{c} \dot{V}$$

then the well-known equations for the potentials follow.

$$\left.\begin{aligned} \Delta \mathfrak{A} - \frac{\varepsilon}{c^2}\ddot{\mathfrak{A}} &= -4\pi \frac{\mathfrak{J}}{c} \\ \varepsilon\left(\Delta V - \frac{\varepsilon}{c^2}\ddot{V}\right) &= -4\pi\varrho \end{aligned}\right| \quad (3)$$

We wish now to assume that all phenomena are periodic with frequency ω and we therefore set:

$$\varrho = \bar{\varrho} e^{i\omega t}, \quad \mathfrak{J} = \bar{\mathfrak{J}} e^{i\omega t}, \quad \mathfrak{A} = \bar{\mathfrak{A}} e^{i\omega t} \quad \textbf{etc.} \quad (4)$$

If dv is a volume element of the region in which currents and charges are present and D is the distance of this element from the point of observation, then, as is known, it follows from (3) that:

$$\bar{\mathfrak{A}} = \int \frac{\bar{\mathfrak{J}}}{c} \frac{e^{-ikD}}{D} dv \text{ and } \bar{V} = \int \frac{\bar{\varrho}}{\varepsilon} \frac{e^{-ikD}}{D} dv \quad (5)$$

We now assume that the point of observation is taken far away from that part of space in which the electromagnetic field is produced so that it no longer matters from which point of space we measure the distance and call this distance R. Moreover, let us characterize the direction of R (i.e., the direction of observation) by the unit vector $\mathfrak{S}$ in this direction. Under these conditions we can replace equations (5) for the potentials by the simpler formulas:

$$\left.\begin{aligned} \bar{\mathfrak{A}} &= \frac{e^{-ikR}}{R} \int \frac{\bar{\mathfrak{J}}}{c} e^{ik(\mathfrak{S}\mathfrak{r})} dv \\ \bar{V} &= \frac{e^{-ikR}}{R} \int \frac{\bar{\varrho}}{\varepsilon} e^{ik(\mathfrak{S}\mathfrak{r})} dv \end{aligned}\right| \quad (5')$$

In both (5) and (5') k is an abbreviation for:

$$k = \frac{\omega}{c} \sqrt{\varepsilon}$$

Further the scalar product of $\mathfrak{S}$ and $\mathfrak{r}$ is denoted by $(\mathfrak{S}\mathfrak{r})$ where $\mathfrak{r}$ is the vector which extends from the point of origin within the region of space which produces scattering to the volume dv which is being considered.

Proceeding from formulas (5') we can now calculate the field by means of the differentiations given in (2). The final result is:

$$\left.\begin{aligned}\overline{\mathfrak{E}} &= -i\frac{\omega}{c^2}\frac{e^{-ikR}}{R}\int\{\mathfrak{J}-\mathfrak{S}(\mathfrak{J}\mathfrak{S})\}\,e^{ik(\mathfrak{S}\mathfrak{r})}dv \\ \overline{\mathfrak{H}} &= -i\frac{\omega}{c^2}\sqrt{\varepsilon}\,\frac{e^{-ikR}}{R}\int[\mathfrak{S}\mathfrak{J}]\,e^{ik(\mathfrak{S}\mathfrak{r})}dv\end{aligned}\right| \qquad (6)$$

Therefore the field is determined by the spatial distribution of current densities alone and (as is seen during the course of the derivation) this is a consequence of the general equation of continuity:

$$\frac{\partial\varrho}{\partial t} + \operatorname{div}\mathfrak{J} = 0 \qquad (7)$$

Formulas (6) show that both $\mathfrak{E}$ and $\mathfrak{H}$ are perpendicular to R, that they are perpendicular to each other, and that their amplitudes in absolute value are in the ratio of 1 to $\sqrt{\varepsilon}$. Moreover, the polarization of the scattered beam is given by the fact that according to formula (6) only that component of $\mathfrak{J}$ perpendicular to $\mathfrak{S}$ (and hence to R) contributes to the field strengths.

III. The Scattered Field Produced by the Sound Waves

If sound waves pass through the fluid, variations in density are associated with them, and these variations of density produce in their turn variations of the dielectric constant (or of the index of refraction). The optical inhomogeneity of the medium produced in this way is considered responsible for the scattering effect. The same assumption is made by Einstein[2] in the calculation of the scattering of light by an undisturbed fluid. Only in this latter case the fluctuations are the result of the heat motion of the molecules. In this connection it may be remarked that in the original theory of Einstein it was not taken into account that the fluctuations in neighboring volume elements are not completely independent of one another. In fact the heat motion of a molecule is correlated with the motions of the surrounding molecules. In the case of a rigid body this was taken into account exactly by P. Debye by finding the complete motion of a molecule through the superposition of sound waves which pass through the entire solid. This feature was added to the Einstein theory by L. Brillouin.[3] The result was that the scattered light could no longer have exactly the same wave length as the incident light. If the incident light is concentrated in a single spectrum line, the scattered light should form a doublet.

More recent research by Gross[4] and Meyer and Ramm[5] have in fact verified this effect. The same causes which lead to the T^3 law for specific heats at low temperatures are here responsible for the splitting of the original spectrum line.

If the scattering takes place in a fluid rather than in a solid, then, as shown by experiment, a substantial part of the light is scattered with unchanged wave length. Therefore a triplet is formed. The splitting shows that in a fluid the molecular motion is very similar to that in a solid (a slow continuous motion with superimposed vibrations). The middle line of the triplet with unaltered wave length may arise in part from the difference in the motion of a bound molecule of a solid and the free molecule of a fluid. Another cause which I believe should be looked into is that optical inhomogeneities may occur without concurrent changes in density. Since, in general, every molecule must for its scattering effect be represented by a polarization ellipsoid, variations in orientation alone are sufficient to produce quasi-crystalline optical characteristics which vary from point to point in the fluid and which also cause the scattering.* This phenomenon is not taken into account in Brillouin's theory.

In our case we can completely neglect the thermal fluctuations and assume that the sole cause of the variations in density are artificially produced sound waves which outweigh all other causes.

Let the dielectric constant of the medium be given by:

$$\varepsilon = \varepsilon_0 + \varepsilon_1 \qquad (8)$$

*In this connection it may be remarked that the application of Brillouin's results to the scattering of x-rays attempted by him is not permissible. The variations of the optical characteristics which play an essential part are here principally determined by the electron structure for these short waves and are not appreciably changed by the molecular motion.

Where ε_1 gives a measure of the variations produced by the sound waves so that ε_1 is a small quantity which depends upon position and time. We can then separate the total electromagnetic field $\mathfrak{E}$ and $\mathfrak{H}$ into two components:

$$\mathfrak{E} = \mathfrak{E}_0 + \mathfrak{E}_1, \quad \mathfrak{H} = \mathfrak{H}_0 + \mathfrak{H}_1 \qquad (9)$$

We can also introduce the Maxwell-Lorentz equations in the form:

$$\left.\begin{aligned} \frac{1}{c}\dot{\mathfrak{H}} &= -\operatorname{rot}\mathfrak{E}, \quad \operatorname{div}\mathfrak{H} = 0 \\ & \qquad\qquad\qquad \mathfrak{D} = \varepsilon\mathfrak{E} \\ \frac{1}{c}\dot{\mathfrak{D}} &= \operatorname{rot}\mathfrak{H}, \quad \operatorname{div}\mathfrak{D} = 0 \end{aligned}\right| \qquad (10)$$

If we are satisfied with an approximation of the first order and take $\mathfrak{E}_0$ and $\mathfrak{H}_0$ as solutions of the equations for $\varepsilon = \varepsilon_0$, we obtain the following as equations which determine the scattering field.

$$\left.\begin{aligned} \frac{1}{c}\dot{\mathfrak{H}}_1 &= -\operatorname{rot}\mathfrak{E}_1 \\ \operatorname{div}\mathfrak{H}_1 &= 0 \\ \frac{1}{c}\frac{\partial}{\partial t}(\varepsilon_1\mathfrak{E}_0) + \frac{\varepsilon_0}{c}\dot{\mathfrak{E}}_1 &= \operatorname{rot}\mathfrak{H}_1 \\ \operatorname{div}(\varepsilon_1\mathfrak{E}_0) + \varepsilon_0 \operatorname{div}\mathfrak{E}_1 &= 0 \end{aligned}\right| \qquad (11)$$

By comparing equations (11) with the general equations (1) it is seen at once that the scattered field $\mathfrak{E}_1$ and $\mathfrak{H}_1$ is produced by a current $\mathfrak{J}$ and a density ρ of **magnitudes:**

$$\mathfrak{J} = \frac{1}{4\pi}\frac{\partial}{\partial t}(\varepsilon_1\mathfrak{E}_0), \quad \varrho = -\frac{1}{4\pi}\operatorname{div}(\varepsilon_1\mathfrak{E}_0) \qquad (12)$$

It is easily verified that $\mathfrak{J}$ and ρ satisfy the equation of continuity (7) as required.

We have seen in section II **above how the scattered field can** be calculated for any distribution of current. We therefore need apply these formulas only to the present case.

If the sound waves have the frequency Ω and are moving in a direction characterized by the unit vector $\mathfrak{S}^*$, ε_1 can be given in the form:

$$\varepsilon_1 = \delta e^{i\Omega t} e^{-iK(\mathfrak{S}^*\mathfrak{r})} \qquad (13)$$

Here:

$$K = \frac{\Omega}{q} \qquad (13')$$

where q designates the velocity of the sound waves.

The primary light waves whose amplitude is $\mathfrak{P}$ (taken vectorially), whose frequency is ω, and whose direction of propagation is given by the unit vector $\mathfrak{S}_0$ can be represented by:

$$\mathfrak{E}_0 = \mathfrak{P} e^{i\omega t} e^{-ik(\mathfrak{S}_0 \mathfrak{r})} \qquad (14)$$

where:

$$k = \frac{\omega}{c}\sqrt{\varepsilon} \qquad (14')$$

If, as in the usual manner, we regard ε_1 and $\mathfrak{E}_0$ given by the real parts of the right hand sides of equations (13) and (14), we obtain in the corresponding notation:

$$\varepsilon_1 \mathfrak{E}_0 = \frac{\delta \mathfrak{P}}{2}\left\{ e^{i(\omega+\Omega)t} e^{-i(k\mathfrak{S}_0 + K\mathfrak{S}^0, \mathfrak{r})} + e^{i(\omega-\Omega)t} e^{-i(k\mathfrak{S}_0 - K\mathfrak{S}^0, \mathfrak{r})} \right\} \qquad (15)$$

From (12) we obtain the current $\mathfrak{J}$ composed of two parts:

$$\mathfrak{J} = \mathfrak{J}' + \mathfrak{J}'' \qquad (16)$$

where:

$$\left.\begin{aligned} \mathfrak{J}' &= i\frac{\delta}{8\pi}(\omega+\Omega)\mathfrak{P} e^{i(\omega+\Omega)t} e^{-i(k\mathfrak{S}_0 + K\mathfrak{S}^0, \mathfrak{r})} \\ \mathfrak{J}'' &= i\frac{\delta}{8\pi}(\omega-\Omega)\mathfrak{P} e^{i(\omega-\Omega)t} e^{-i(k\mathfrak{S}_0 - K\mathfrak{S}^0, \mathfrak{r})} \end{aligned}\right| \qquad (16')$$

Therefore at this point there already appears the splitting of the the original frequency ω into a doublet $\omega \pm \Omega$.

If we correspondingly decompose the electrical field into two parts:

$$\mathfrak{E}_1 = \mathfrak{E}_1' + \mathfrak{E}_1'' \qquad (17)$$

and set:

$$\left.\begin{aligned} \mathfrak{E}_1' &= \overline{\mathfrak{E}_1'} e^{i(\omega+\Omega)t} \\ \mathfrak{E}_1'' &= \overline{\mathfrak{E}_1''} e^{i(\omega-\Omega)t} \end{aligned}\right| \qquad (17')$$

we find from (6):

$$\overline{\mathfrak{E}_1'} = \frac{\delta}{8\pi} \frac{\omega(\omega+\Omega)}{c^2} \frac{e^{-ikR}}{R} \left\{\mathfrak{P} - \mathfrak{S}(\mathfrak{P}\mathfrak{S})\right\} \int e^{i(k\mathfrak{s} - K\mathfrak{S}^*, \mathfrak{r})} dv$$
$$\overline{\mathfrak{E}_1''} = \frac{\delta}{8\pi} \frac{\omega(\omega-\Omega)}{c^2} \frac{e^{-ikR}}{R} \left\{\mathfrak{P} - \mathfrak{S}(\mathfrak{P}\mathfrak{S})\right\} \int e^{i(k\mathfrak{s} + K\mathfrak{S}^*, \mathfrak{r})} dv \quad (18)$$

The vector $\mathfrak{s}$ is given by:

$$\mathfrak{s} = \mathfrak{S} - \mathfrak{S}_0 \quad (19)$$

and represents the vectorial difference between the unit vectors $\mathfrak{S}_0$ and $\mathfrak{S}$ which are in the directions of the primary beam and the scattered beam, **respectively.** If the angle between these directions is ϑ, it is easily found that the absolute value s is given by:

$$|\mathfrak{s}| = s = 2\sin\frac{\vartheta}{2} \quad (19')$$

Therefore the scattered field is completely known.

In Brillouin's theory where scattering by thermal sound waves is considered, all possible directions $\mathfrak{S}^*$ of these waves occur and a large range of frequencies is possible. In this case it is only required to find the fields by summing over all such sound waves in (18). The volume over which the integration in (18) is to be taken can always in this case be assumed to be very great in comparison to the wave length Λ of the sound waves. A great simplification results from this assumption. If, for instance, the first component of the scattered field in (18) is considered, the only value of the field strength which has to be taken into account is that one for which the following condition is satisfied.

$$k\mathfrak{s} - K\mathfrak{S}^* = 0 \quad (19)$$

Only then is the exponential function which is to be integrated equal to 1 everywhere. In every other case the contributions of regions of space are alternately positive and negative and cancel each other when all contributions are considered. The geometrical meaning of the condition (19) is that the light scattered in a given direction ϑ can more properly be thought of as being reflected optically in the usual sense at the sound waves in the direction determined by the reflection law. This condition also gives the wave length Λ of these waves from the relation between the absolute values.

$$s = \frac{K}{k} = \frac{\lambda}{\Lambda} \quad (20)$$

or alternatively:

$$2\Lambda \sin\frac{\vartheta}{2} = \lambda \qquad (20')$$

This is exactly the Bragg condition when the reflecting planes are separated by the distance Λ. The only difference between this result and the usual formula is that integral multiples of λ do not appear in condition (20'). This means that only first order reflections are possible from the sound waves. This result is a consequence of the fact that the distribution in space of the density fluctuations, which cause the scattering, is completely periodic.

The scattered light waves being considered have the frequency $\omega + \Omega$ instead of ω, and from the definition of k and K together with formula (20):

$$\frac{\Omega}{\omega} = \frac{K}{k}\,\frac{q}{c/\sqrt{\varepsilon_0}} = 2\,\frac{q}{c/\sqrt{\varepsilon_0}}\sin\frac{\vartheta}{2}$$

This means that instead of the primary frequency ω we shall observe the altered frequency:

$$\omega\left(1 + 2\,\frac{q}{c/\sqrt{\varepsilon_0}}\sin\frac{\vartheta}{2}\right) \qquad (21)$$

Therefore the description of reflection at the sound waves is so successful that even the Doppler effect produced by the velocity of the sound waves is included. The second part of the scattered field would have the frequency:

$$\omega\left(1 - 2\,\frac{q}{c/\sqrt{\varepsilon_0}}\sin\frac{\vartheta}{2}\right) \qquad (21')$$

so that by means of the formulas (21) and (21') it is shown that ordinary scattering should lead to the splitting of the original line into a doublet as explained by Brillouin.

However, in the case which we wish to discuss here, the situation is different. We have to consider only a single, artificially produced train of sound waves. If we accepted Brillouin's theory as it stands, we should expect **noticeable** scattering only when the light fell upon the sound waves at exactly the proper angle calculated from (20'). But experiment shows that scattering appears even when this condition for the angle is not fulfilled, and further that the scattered light always appears within a finite angular region where both maxima and minima of intensity are present. We maintain that these facts do not contradict the basis of the previous calculations, but only that it is no longer correct to assume that the scattering volume is so large that the integrals in (18) can have appreciable values only for values of $\mathfrak{s}$ given by (19). In the next paragraph we wish to apply equations (18) for the scattered field in order to explain the observed interference phenomena from this point of view.

IV. Theory of the Observed Interference

We shall refer to the practical situation shown in the figure and limit our considerations at first to that half of the scattered field which has the frequency $\omega + \mathfrak{E}'$ and whose field is given by Ω in equation (18). If, as an abbreviation, we denote the vector $k\mathfrak{s} - K\mathfrak{S}^*$ by the vector $\mathfrak{v}$ with the three components v_1, v_2, v_3, the integral which appears in (18) takes on the form:

$$\int e^{i(k\mathfrak{s}-K\mathfrak{S}^*,\mathfrak{r})}\,dv = \int e^{i(v_1x+v_2y+v_3z)}\,dx\,dy\,dz \qquad (22).$$

Since the integration is to extend over the parallelepiped whose dimensions are a, l, b, the value of the integral is given by:

$$\int e^{i(v_1x+v_2y+v_3z)}\,dx\,dy\,dz = V\,\frac{\sin\frac{v_1a}{2}}{\frac{v_1a}{2}}\,\frac{\sin\frac{v_2l}{2}}{\frac{v_2l}{2}}\,\frac{\sin\frac{v_3b}{2}}{\frac{v_3b}{2}} \qquad (22')$$

where $V = alb$ denotes the illuminated volume.*

For the purpose of discussion we wish to consider now the special case in which the primary light is directed exactly parallel to the y axis. The direction vector $\mathfrak{S}_0$ associated with this case has the three components 0, 1, 0. The direction of observation $\mathfrak{S}$ is to be arbitrary and will be characterized by the three direction cosines α, β, γ. The vector $\mathfrak{s} = \mathfrak{S} - \mathfrak{S}_0$ will therefore have the three components α, $\beta - 1$, γ. Finally, the sound waves are to move in the direction of the x axis, so that the three components of $\mathfrak{S}^*$ equal 1,0,0. The vector $\mathfrak{v} = k\mathfrak{s} - K\mathfrak{S}^*$ which appears in the integral will then have the three components $2\pi\left(\frac{\alpha}{\lambda} - \frac{1}{\Lambda}\right)$, $2\pi\frac{\beta-1}{\lambda}$, $\frac{2\pi\gamma}{\lambda}$ since $k = 2\pi/\lambda$ and $K = 2\pi/\Lambda$ where λ and Λ denote the wavelengths of the light waves and sound waves, respectively. When all these results are used, the field $\mathfrak{E}_1'$ of one half of the scattered radiation, as found from (17') and (18), is given by the following expression:

$$\mathfrak{E}_1' = f'\{\mathfrak{P} - \mathfrak{S}(\mathfrak{P}\mathfrak{S})\}\,e^{i[(\omega+\Omega)t-kR]} \qquad (23)$$

*In carrying out the integration, the origin of the coordinate system is taken as the midpoint of the parallelepiped. The choice of origin has no influence upon the intensity distribution.

where:

$$f' = \frac{\delta}{8\pi} \frac{\omega(\omega+\Omega)}{c^2} \frac{V}{R} \frac{\sin\left[\pi a\left(\frac{\alpha}{\lambda} - \frac{1}{\Lambda}\right)\right]}{\left[\pi a\left(\frac{\alpha}{\lambda} - \frac{1}{\Lambda}\right)\right]} \frac{\sin\left[\pi l \frac{\beta - 1}{\lambda}\right]}{\left[\pi l \frac{\beta - 1}{\lambda}\right]} \frac{\sin\left[\pi b \frac{\gamma}{\lambda}\right]}{\left[\pi b \frac{\gamma}{\lambda}\right]} \tag{23'}$$

The direction $\mathfrak{S}$ of the scattered radiation, which we are considering, differs only slightly from the direction of propagation of the primary radiation $\mathfrak{S}_0$. Therefore the difference between the primary amplitude $\mathfrak{P}$ and the projection given by $\mathfrak{P} - \mathfrak{S}(\mathfrak{S}\mathfrak{P})$ will be negligible. Consequently we can say without any great error that the factor f' simply represents the ratio of the amplitude of the scattered field to that of the primary field.

The third factor in (23') given by:

$$\frac{\sin\left[\pi b \frac{\gamma}{\lambda}\right]}{\left[\pi b \frac{\gamma}{\lambda}\right]}$$

equals 1 for $\gamma = 0$, *i.e.*, for all points on the xy plane, and decreases very rapidly as we leave this plane because of the large value of the ratio b/λ. For this reason we wish to remain in this plane and set:

$$\alpha = \sin\vartheta, \quad \beta = \cos\vartheta \tag{24}$$

where ϑ denotes the angle between the secondary and primary beam. In the xy plane we can therefore write f' as:

$$f' = \frac{\delta}{8\pi} \frac{\omega(\omega+\Omega)}{c^2} \frac{V}{R} \frac{\sin\left[\pi \frac{a}{\lambda}\left(\sin\vartheta - \frac{\lambda}{\Lambda}\right)\right]}{\left[\pi \frac{a}{\lambda}\left(\sin\vartheta - \frac{\lambda}{\Lambda}\right)\right]} \frac{\sin\left[\pi \frac{l}{\lambda}(1 - \cos\vartheta)\right]}{\left[\pi \frac{l}{\lambda}(1 - \cos\vartheta)\right]} \tag{25}$$

The first factor in (25) takes on its maximum value 1 when:

$$\sin\vartheta = \frac{\lambda}{\Lambda} \qquad (26)$$

which is the direction for which an interference maximum, the only possible principal maximum, is observed. This maximum in its direction corresponds to a maximum of the first order of a ruled grating whose grating constant is Λ. In this direction we find that f' is given by:

$$f' = \frac{\delta}{8\pi}\,\frac{\omega(\omega+\Omega)}{c^2}\,\frac{V}{R}\,\frac{\sin\left|\frac{\pi}{2}\frac{l/\Lambda}{\Lambda/\lambda}\right|}{\left|\frac{\pi}{2}\frac{l/\Lambda}{\Lambda/\lambda}\right|} \qquad (27)$$

if we note that from (26) the angle ϑ is small and we may therefore replace sin ϑ by ϑ and 1 - cos ϑ by $\vartheta^2/2$. The important factor in this expression is the quantity:

$$\frac{\sin\left[\frac{\pi}{2}\frac{l/\Lambda}{\Lambda/\lambda}\right]}{\left[\frac{\pi}{2}\frac{l/\Lambda}{\Lambda/\gamma}\right]}$$

Here l is the length of the path of the light in the fluid, Λ the wavelength of the sound waves, and λ the wavelength of the light waves. If the ratio l/Λ were large compared to Λ/λ, the above expression would be very small, and the interference of first order when sin $\vartheta = \lambda/\Lambda$ would not be appreciable. As a matter of fact, however, in experiments with sound waves whose wavelength is of order of magnitude 0.1 mm. both ratios are nearly equal, each being about 100. Consequently, as confirmed by experiment, the intensity of the interference maximum is great. Despite this, it cannot be said that the theory reproduces all that is observed experimentally. Not only the first order is seen, but also higher orders for which sin $\vartheta = n\lambda/\Lambda$ (n an integer). We then have the question of how this comes about. It could be thought that the quartz crystal which produces the sound waves does not vibrate in a purely periodic way. Every overtone would in fact produce a superposed wave with a wavelength equal to a fraction of Λ. It is, however, much more probable that the interference maxima of higher order arise from the fact that the fluctuations in the index of refraction which are produced in the experiment by sound waves of very great intensity are so considerable that it is not sufficient to use the first approximation in the calculation of the scattered field as we have done here. The higher order approximations can be calculated, but it can be seen immediately without completely carrying out the calculations that with every new step in the approximation procedure new scattered fields appear with frequencies $\omega + 2\Omega$ $\omega + 3\Omega$, etc. whose interference maxima will produce the spectra of second, third, and higher orders. The only thing which should still be shown is the fact that the optical disturbances caused by the sound waves is indeed unexpectedly large, so that from the theoretical viewpoint all such higher approximations are shown to be necessary.

For this purpose it is not sufficient to evaluate the order of magnitude of f', for this amplitude factor only measures the ratio of amplitudes exactly at the position of the interference maximum. On the contrary, the scattered intensity over all directions must be summed in the vicinity of the maximum, and only then is there obtained a measure of the total energy concentrated at this maximum. I limit myself to a statement of the result of this calculation. If W denotes the total energy which is transported per unit time in the direction of the maximum of first order and its immediate neighborhood and if W_0 is the total light energy which goes into the parallelepiped per unit time, it is found that

$$\frac{W}{W_0} = \frac{\pi^2}{4} \frac{\delta^2}{\varepsilon_0^2} \frac{l^2}{\lambda^2} \frac{\sin^2\left[\frac{\pi}{2} \frac{l/\Lambda}{\Lambda/\lambda}\right]}{\left[\frac{\pi}{2} \frac{l/\Lambda}{\Lambda/\lambda}\right]^2} \tag{28}$$

The formula shows that W/W_0 is of order of magnitude 1, if δ/ε_0 is made equal λ/l. In this case the formula would predict that the total incident energy is to be found in the interference maximum. It is clear, then, the formula had been used far beyond its range of validity and that further approximations are necessary. But λ/l is of the order of magnitude of 10^{-4}; therefore the sound waves need only produce variations in the dielectric constant of order of magnitude of 10^{-4} in order to cause a strong optical disturbance. To make a relative change of 10^{-4} in the dielectric constant it is necessary in practice to have a pressure of approximately 1 atmosphere. Therefore sound waves with a pressure amplitude of 1 atmosphere are to be considered as very strong waves in the above sense. On the other hand, the energy carried by such waves is not particularly large, it is easily calculated that they carry about

1 watt/cm.² of energy per unit area and per unit time. The numerical results simply represent orders of magnitude and in practical cases, of course, should be made precise for the fluid used. They are sufficient to show, however, that the influence of sound waves upon the motion of light is certainly great, and they provide the explanation of why the interference phenomenon is so strong that it can easily be made visible by projection.

Up to this point we have discussed only one-half of the scattered field. The other half with the field strength $\mathfrak{E}_1''$ gives rise to an interference maximum on the other side of the primary beam with the angle:

$$\sin\vartheta = -\frac{\lambda}{\Lambda}$$

In the above, the frequency of the scattered radiation is $\omega + \Omega$ and is therefore different from the primary frequency ω. Up to this time no experiment has been undertaken to verify these small changes in frequency.

Bibliography

1. P. Debye and F. W. Sears, *Proc. Nat. Acad. Washington*, *18*, 409 (1932) (communicated April 23, 1932); P. Debye, *Sächs. Akad. Math. Kl.* *84*, 125 (1932). The same observation was made independently of us and almost simultaneously by R. Lucas and P. Biquard, *Compt. rend.*, *194*, 2132 (1932) (séance du 6 mai 1932, publiée 13 juin 1932).
2. A. Einstein, *Ann. Physik*, *33*, 1275 (1910).
3. L. Brillouin, *Ann. physique*, *17*, 88 (1922).
4. Gross, *Z. physik.*, *63*, 685 (1930); *Naturwiss.*, *18*, 718 (1930); *Nature*, *126*, 201,400,603 (1930); *129*, 722 (1932).
5. Meyer and Ramm, *Physik. Z.*, *33*, 270 (1932).

ON THE THEORY OF SPECIFIC HEATS
(Zur Theorie der spezifischen Wärmen*)

P. Debye

Translated from
Annalen der Physik, Vol. 39, 1912, pages 789-839.

The observations on the temperature dependence of specific heats carried out recently in the Nernst Laboratory have proven in a most convincing way that the theorem of equipartition of energy is incorrect for material bodies also. As is well known, Einstein[1] first called attention to the situation and obtained a formula giving the specific heat as a function of temperature using the quantum theory which Planck[2] had developed for radiation. While all measurements show a trend which agrees qualitatively with the Einstein formula, quantitative discrepancies between theory and experiment appear which are greater the lower the temperature. As an improvement, Nernst and Lindemann[3] altered the Einstein formula by introducing a second frequency $\nu/2$ in addition to ν. The introduction of this second frequency leads to a better formula than that of Einstein as far as practical needs are concerned; however, it has been impossible to find any sound basis for this particular value $\nu/2$. It will be seen in the following that there is no fundamental justification for it. Now even if the value $\nu/2$ itself has no theoretical significance, the necessity for the introduction of several frequencies can still be justified, as Einstein[4] has already suggested. The neighbors of a vibrating atom exert so strong an influence on it that its motion can have only a very slight resemblence to simple harmonic motion. The direct application of the Planck formula (with a single frequency) then naturally becomes questionable. If we consider the Fourier analysis of the motion of the atom, then the motion can be characterized by very many frequencies; thus, in agreement with Einstein, the necessity of several frequencies seems plausible.

While this suggestion is correct, the actual details of such a calculation are very laborious; the following method leads much

*Received July 24, 1912.

more directly to a sound formula for the specific heat. It is exactly analogous to Jeans' proof of the Rayleigh radiation formula.

If a solid is thought of as consisting of N atoms which can be treated as point masses, it represents a system of $3N$ degrees of freedom* which can carry out $3N$ different periodic motions with $3N$ different frequencies. The slowest of these vibrations are the readily observable sound vibrations. The most general motion which the body can carry out may be represented as a superposition of the above $3N$ forms of motion and can be described by a sum of these motions with the introduction of $6N$ constants.** Also for heat motion (a special case of general motion) this suggestion must also be valid.

Were the theorem of equipartition of energy correct, the "spectrum" of the solid would not have to be determined. From the **outset** to each eigenvibration would have to be ascribed the energy kT ($k = 1.47 \times 10^{-16}$ erg/degree, T = the absolute temperature) and the total energy of the solid would be $3NkT$, which, as is well known, corresponds to the Dulong-Petit law. Instead of this, however, Planck's formula states that the average energy at temperature T of a system with the frequency ν has the value

$$\frac{h\nu}{e^{\frac{h\nu}{kT}} - 1}$$

h = quantum of action = 7.10×10^{-27} erg sec.***

k = Boltzmann constant = 1.47×10^{-16} erg/degree.

We must first determine the eigen frequencies of the eigenvibrations of the solid, with each of these eigenvibrations we have to associate an energy corresponding to Planck's formula and finally we have to sum over the whole spectrum. Then the energy of the solid at temperature T is known and by differentiation also its specific heat C.

The first thing to be done is a determination of the "acoustic spectrum" of a solid of some arbitrary shape, for as is known the calculation of U does not depend on this shape. Now for this purpose the solid could be viewed as a point lattice consisting of $3N$ mass points and the $3N$ vibrational frequencies of the system determined from the associated system of $3N$ equations of motion.

*In terms of this definition a linear oscillator is a system with one degree of freedom.

**The number of constants is $6N$ because two data are required for the complete specification of one eigenvibration: amplitude and phase at some initial time.

***The numerical values are those given recently by Paschen-Gerlach, *Ann. Physik* 38, 41 (1912).

Then there would be no choice other than to make special **simple** hypotheses concerning the interactions of the atoms. There is no question that in so far as the hypotheses about the forces are correct this is completely correct. However, it is not said that a completely exact knowledge of the spectrum is necessary; on the **contrary**, it can be expected that from the practical point of view a sufficiently exact formula will be obtained in case only the characteristic properties of this spectrum be properly taken into account. We would like to show in the following that this is possible without considering in detail the mutual forces between atoms.

We proceed from the ordinary equations of elasticity. In the derivation of these the solid is treated as a continuum; doing this corresponds to ascribing infinitely many eigenvibrations to an elastic body. Now the latter is certainly false; the solid consisting of N atoms can have only $3N$ different eigenvibrations. Yet in another way a calculation based on these equations of elasticity will be of interest to us. Namely, it will give us a law for the number of spectral lines of the acoustic spectrum per frequency interval $d\nu$. In the following we will consider this law for the density of spectral lines entirely accurate. The discontinuous **structure** of the solid will be taken into account only in so far as it permits the following conclusion: the spectrum consists in its entirety only of $3N$ and not of infinitely many spectral lines. Accordingly, the spectrum calculated from the equations of elasticity will be cut off at the $3N$th spectral line. With **certainty** it can be asserted that the spectrum of the actual solid agrees completely with the calculated spectrum at low frequencies. With equal certainty, for high frequencies, at which the wavelength is comparable to the distance between two atoms, the so calculated distribution of spectral lines can be considered only as approximately correct. Yet we will neglect this error; the result supports us and shows that it is probably not of great importance. For, as will be shown in the following, this approximate calculation gives a formula for the specific heats, which agrees in every respect with the experimental facts.*

*A case which can easily be treated rigorously is that of a linear body of atomic structure (compare, for example, Lord Rayleigh, *Theory of Sound*, Vol. I, London, 1877, p. 129). By analogy with this case it can be concluded that the lines are more dense at the high frequency end than our formula indicates.

Since it has often been emphasized that a fundamental difference between the optical series and those otherwise known is to be found in that the first have a finite limit, I will not suppress the (**really** self-evident) remark that, in reality, besides the spectrum of the black body not a single one exists which does not have a finite limit. Of course, this limit is in no case an accumulation point in the mathematical sense, as must be expected according to the series formula. However, an accumulation of lines

(con'd)

In place of the simple frequency of Einstein there appears a complete spectrum, which, according to the previous discussion, is characterized by two properties: (a) by its cut-off; (b) by the number of lines per frequency interval $d\nu$. With these assumptions the cut-off frequency ν_m appears in these two data as the only essential parameter. This remark leads to the theorem:

If the temperature T is expressed as a multiple of a characteristic temperature θ of the substance concerned, the specific heat of all (monatomic) solids is represented by the same curve, in other words, the specific heat of a monatomic solid is a universal function of the ratio T/θ.

A second theorem which is not limited to the case of monatomic solids but claims a completely general validity is obtained if the density of line distribution is taken into account. It is found (compare section X) that in the frequency interval $d\nu$ there is a number of lines proportional to $\nu^2 d\nu$, that is, the same law which according to Jeans holds for black body radiation. This leads to the theorem:

At sufficiently low temperatures the specific heat of all solids is proportional to the third power of the absolute temperature.*

Then the energy content becomes proportional to T^4, analogous to the Stefan-Boltzmann law for black body radiation which holds at all temperatures. If the solid like the black body possessed an "acoustic spectrum" which had no finite limit, the theorem would also be correct at all temperatures. In reality it holds only below a temperature which depends on the solid. We will show later that it is correct within about 1% below temperatures equal to 0.1 of the temperature referred to above as the characteristic temperature.

As will be shown in section V, the cut-off frequency ν_m and with it the characteristic temperature θ can be represented as a

(con'd from preceding page)

always occurs in the neighborhood of this cut-off as far as I can see. Now, since it is possible to give arrangements of electrons whose spectra satisfy the Deslandre law of band spectra, it is perhaps to be expected that the series spectra can also be understood as eigenvibrations of electron systems of a finite number of degrees of freedom.

*After developing the two theorems in the Winter semester in my lectures on thermodynamics, I have reported on them later at the meeting of the Swiss Physical Society at Bern and published them in *Arch. de Genève*, March 1912, p. 256. Somewhat later, Born and von Karman developed independently of these communications a formula for the specific heat which agrees with the two theorems given, as can be verified easily.

function of the elastic constants of the solid. This relation corresponds in its structure with the well known formula of Einstein, which connects his single eigenfrequency with the compressibility. There is a difference in so far as, besides the compressibility, Poisson's ratio plays a roll. The appearane of these two quantities cannot be surprising if it is recalled that the properties of an elastic body are determined by two constants. Generally, it can be said that for crystals, for example, even these two quantities are not sufficient to determine θ but that all elastic constants are required.

For clarity the following is divided into a Part I in which a formula is developed for the specific heat on the basis described above and is compared with experimental results. In this part the formula for the constitution of the "acoustic spectrum" is used without proof. The proof of this is added in a Part II.

Part I

I. Energy Content and Specific Heat of Monatomic Solids

According to the theory of elasticity, a body of volume V possesses an unbounded spectrum. According to Part II, section X, the number of eigenvibrations z below a certain upper limit ν is:

$$(1) \qquad z = \nu^3 V F$$

F is a function of the elastic constants and the density ρ which, in terms of the compressibility κ and Poisson's ratio σ, for example, can be expressed in the form:

$$(2) \qquad F = \frac{4\pi}{3}\rho^{3/2}\kappa^{3/2}\left[2\left(\frac{2}{3}\right)^{3/2}\left(\frac{1+\sigma}{1-2\sigma}\right)^{3/2} + \frac{1}{3^{3/2}}\left(\frac{1+\sigma}{1-\sigma}\right)^{3/2}\right]$$

For the purposes of these paragraphs it suffices to know that F represents a quantity which can be calculated for **any body from its** density and its elastic constants.

Now the body whose energy content U will be calculated consists of N atoms and, as a system of $3N$ degrees of freedom, can carry out only $3N$ eigenvibrations. If we assume, as was explained in the introduction, that equation (1) can always be used with sufficient accuracy, the highest frequency ν_m can be calculated from the expression:

$$(3) \qquad 3N = \nu_m^3 V F$$

or:

$$(3') \qquad \nu_m = \left(\frac{3N}{VF}\right)^{1/3}$$

The density of spectral lines can be calculated immediately by differentiating (1). The number dz which lie in the frequency interval $d\nu$ is:

$$dz = 3\,VF\nu^2 d\nu$$

If, in this equation, VF is replaced by its value following from (3):

$$VF = \frac{3N}{\nu_m^3}$$

then:

$$(4) \qquad dz = 9N\frac{\nu^2 d\nu}{\nu_m^3}$$

On the other hand, according to the Planck formula, each eigenvibration with the frequency ν possesses the energy:

$$(5) \qquad \frac{h\nu}{e^{\frac{h\nu}{kT}} - 1}$$

if $h = 7.10 \times 10^{-27}$ erg sec. denotes the fundamental quantum of action, $k = 1.47 \times 10^{-16}$ erg/°C. the Boltzmann constant, and T the absolute temperature. By combining (4) and (5) we obtain for the energy U of the body the expression:

$$(6) \qquad U = \frac{9N}{\nu_m^3}\int_0^{\nu_m} \frac{h\nu}{e^{\frac{h\nu}{kT}} - 1}\nu^2 d\nu$$

where the integral has to be carried out over the entire acoustic spectrum from $\nu = 0$ to $\nu = \nu_m$.

If now a characteristic temperature θ of the body be defined by the expression:

$$(7) \qquad \theta = \frac{h\nu_m}{k}$$

and a new dimensionless variable:

$$\xi = \frac{h\nu}{kT} \tag{8}$$

is introduced in the integral, U can be written in the form:

$$U = 9NkT\left(\frac{kT}{h\nu_m}\right)^3 \int_0^{\frac{h\nu_m}{kT}} \frac{\xi^3 d\xi}{e^\xi - 1}$$

or if the definition (7) is taken into account:

$$U = 9NkT\left(\frac{T}{\Theta}\right)^3 \int_0^{\Theta/T} \frac{\xi^3 d\xi}{e^\xi - 1}. \tag{9}$$

As is well known (and in addition follows naturally from (9) for high temperatures):

$$U = 3NkT$$

corresponds to the law of Dulong and Petit. The relation expressed by (9) can be put in words as follows:

The energy of a body is obtained if the Dulong Petit value is multiplied by a factor which is a universal function of the ratio T/Θ, that is, temperature T divided by the characteristic temperature Θ.

If for brevity we put:

$$\frac{\Theta}{T} = x$$

then according to (9) this factor has the value:

$$\frac{3}{x^3}\int_0^x \frac{\xi^3 d\xi}{e^\xi - 1}$$

If N is understood to be the number of atoms per atomic weight, (9) represents the corresponding energy and by differentiation with respect to T the atomic heat at constant volume C_v is obtained, for which we write simply C without an index so long as there is no danger of confusion. Thus from (9):

$$C = 3Nk\left[\frac{12}{x^3}\int_0^x \frac{\xi^3 d\xi}{e^\xi - 1} - \frac{3x}{e^x - 1}\right] \tag{10}$$

if again x is the ratio Θ/T.

The quantity $3Nk$ has the value 5.955 cal.; if we denote it by C_∞ because it is reached in the limit $T = \infty$, (10) can be rewritten:

$$\frac{C}{C_\infty} = \frac{12}{x^3}\int_0^x \frac{\xi^3 d\xi}{e^\xi - 1} - \frac{3x}{e^x - 1} \tag{10'}$$

Thus the result stated in the introduction is verified:

The specific heat of a monatomic solid is a universal function of the ratio $T/\Theta = 1/x$.

II. Approximate Formula and Numerical Results

(a) Approximation for Small Values of x

If high temperatures at which the ratio Θ/T is small compared to 1 are considered first, the integrand in (10') can be replaced by a first approximation:

$$\frac{\xi^3}{e^\xi - 1} = \xi^2$$

Then the integral becomes $x^3/3$. On the other hand, for the limit $x = 0$:

$$\frac{3x}{e^x - 1} = \frac{3x}{x} = 3$$

which leads to C/C_∞:

$$\frac{C}{C_\infty} = \frac{12}{3} - 3 = 1$$

a value which corresponds to the law of Dulong and Petit.

For use in what follows some additional terms will be calculated. With the abbreviation $x = \Theta/T$, (9) can be rewritten:

$$\frac{U}{\Theta C_\infty} = \frac{3}{x^4}\int_0^x \frac{\xi^3 d\xi}{e^\xi - 1}$$

Since on the other hand:

$$\frac{d}{dT} = -\frac{\Theta}{T^2}\frac{d}{dx} = -\frac{1}{\Theta}x^2\frac{d}{dx}$$

we obtain for the ratio C/C_∞ the relation:

$$\frac{C}{C_\infty} = -3x^3 \frac{d}{dx}\left[\frac{1}{x^4}\int_0^x \frac{\xi^3 d\xi}{e^\xi - 1}\right] \tag{11}$$

The development of $1/e^\xi - 1$ in powers of ξ is well known; with the help of the Bernoulli numbers B it can be expressed in the form:*

$$\frac{1}{e^\xi - 1} = \frac{1}{\xi} - \frac{1}{2} + \xi \sum_{n=1}^{n=\infty} (-1)^{n-1} \frac{B_n}{(2n)!} \xi^{2n-2}$$

By means of this expansion and the operations indicated in (11), for C/C_∞ the series:

$$\frac{C}{C_\infty} = 1 - 3 \sum_{n=1}^{n=\infty} (-1)^{n-1} \frac{B_n}{(2n)!} \frac{2n-1}{2n+3} x^{2n} \tag{12}$$

is obtained. The first Bernoulli numbers have the values:

$$B_1 = \frac{1}{6}, \quad B_2 = \frac{1}{30}, \quad B_3 = \frac{1}{42}, \quad B_4 = \frac{1}{30}, \quad B_5 = \frac{5}{66},$$
$$B_6 = \frac{691}{2730}, \quad B_7 = \frac{7}{6}, \ldots$$

By use of these values (12) can also be written:

$$\frac{C}{C_\infty} = 1 - \frac{x^2}{20} + \frac{x^4}{560} - \frac{x^6}{18144} + \frac{x^8}{633600} - \frac{x^{10}}{23063040} + \cdots \tag{12'}$$

In the numerical calculations only the first four terms of this series were used. This results in less than 1% error in the interval $x = 0$ to $x = 2$. If the six terms written out are used, the interval can be extended with the same maximum error to beyond $x = 3$.

(b) Approximation for Large Values of x

In this case it is best to proceed directly from (10'). Again the value attained in the limit of low temperatures, that is, for very large values of x, is calculated first.

The term $3x/e^x - 1$ vanishes exponentially; on the other hand,

*Compare, for example, J. A. Serret, *Cours de calcul differentiel et integral*, Paris 1886, v. 2., p. 213. This series, like that for C/C_∞, converges only for values of the variable which are less than 2π.

the upper limit of the integral can be replaced by ∞ and thus a numerical can be obtained for the integral:

$$\int_0^\infty \frac{\xi^3 d\xi}{e^\xi - 1} = \int_0^\infty \xi^2 (e^{-\xi} + e^{-2\xi} + e^{-3\xi} + \dots) d\xi$$

$$= 6\left(1 + \frac{1}{2^4} + \frac{1}{3^4} + \frac{1}{4^4} + \dots\right)$$

The series in the parentheses appears in radiation theory[5]: it has the value (B_2 is again the second Bernoulli number):

$$\frac{8\pi^4}{4!} B_2 = \frac{\pi^4}{90} = 1.0823$$

By substituting in (10') we obtain for low temperatures:

$$\text{(13)} \qquad \frac{C}{C_\infty} = \frac{4}{5}\pi^4 \frac{1}{x^3} = \frac{4\pi^4}{5} \frac{T^3}{\Theta^3} = 77.938 \frac{T^3}{\Theta^3}$$

With this the second theorem emphasized in the introduction is verified:

For sufficiently low temperatures the specific heat is proportional to the third power of the absolute temperature.

Besides this the proportionality factor is found; according to (13) is has the value $77.938/\Theta^3$. Neither the Einstein nor the Nernst-Lindemann formula agrees with this law; both require that the specific heat vanish exponentially in the neighborhood of absolute zero. In the following (13) will be verified on the basis of the existing observations. It may, perhaps, be noted that (13) does not contradict the Nernst heat theorem, since the specific heat for low temperatures still vanishes sufficiently rapidly to give to the integral:

$$\int_0^T \frac{C}{T} dT$$

a finite value.

The proportionality to T^3 is a direct consequence of the circumstance that the number of vibrations per frequency interval $d\nu$ is proportional to the quantity $\nu^2 d\nu$. Now, on the one hand, at low temperatures chiefly slow vibrations come into consideration, while, on the other, the proportionality to $\nu^2 d\nu$ deduced from the elastic equations unquestionably can be applied to low frequencies. Therefore it can be concluded that the proportionality to T^3 follows not only for monatomic but for all bodies at sufficiently low temperatures.*

*Notice that according to (13) there is no temperature below which the radiation energy exceeds the energy of the atoms unless bodies exist with a characteristic temperature higher than 135,000°. This is in contrast to the Einstein and the Nernst-Lindemann formulae, for which such a temperature can always be found.

In order to test the exactness of the law (13) for higher temperatures and at the same time to be able to compute values of the formula (10') in such a region, we will also calculate here the series development of which (13) is the first term. If the integrand in (10') is developed in powers of $e^{-\xi}$:

$$\frac{\xi^3}{e^{\xi}-1} = \xi^3 e^{-\xi}(1+e^{-\xi}+e^{-2\xi}+\ldots) = \sum_{n=1}^{n=\infty} \xi^3 e^{-n\xi}$$

is obtained.

On the other hand there is found easily by integration by parts:

$$\int_0^x \xi^3 e^{-n\xi}\,d\xi = \int_0^\infty \xi^3 e^{-n\xi}\,d\xi - \int_x^\infty \xi^3 e^{-n\xi}\,d\xi$$

$$= \frac{6}{n^4} - x^4 e^{-nx}\left(\frac{1}{nx}+\frac{3}{n^2x^2}+\frac{6}{n^3x^3}+\frac{6}{n^4x^4}\right)$$

so that the integral appearing in (10') can be written:

$$\int_0^x \frac{\xi^3}{e^{\xi}-1}\,d\xi = 6\sum_{n=1}^{n=\infty}\frac{1}{n^4}$$

$$- x^4 \sum_{n=1}^{n=\infty} e^{-nx}\left(\frac{1}{nx}+\frac{3}{n^2x^2}+\frac{6}{n^3x^3}+\frac{6}{n^4x^4}\right)$$

For C/C_∞ (10') finally gives:

$$(14)\quad \begin{cases} \dfrac{C}{C_\infty} = \dfrac{72}{x^3}\displaystyle\sum_{n=1}^{n=\infty}\frac{1}{n^4} \\ \qquad - 12x\displaystyle\sum_{n=1}^{n=\infty} e^{-nx}\left(\frac{1}{nx}+\frac{3}{n^2x^2}+\frac{6}{n^3x^3}+\frac{6}{n^4x^4}\right) - \dfrac{3x}{e^x-1} \end{cases}$$

If $\Sigma 1/n^4$ is replaced by its value $\pi^4/90$ given earlier, (14) can be rewritten:

$$(14')\quad \begin{cases} \dfrac{C}{C_\infty} = \dfrac{4\pi^4}{5}\dfrac{1}{x^3} - \dfrac{3x}{e^x-1} \\ \qquad - 12x\displaystyle\sum_{n=1}^{n=\infty} e^{-nx}\left(\frac{1}{nx}+\frac{3}{n^2x^2}+\frac{6}{n^3x^3}+\frac{6}{n^4x^4}\right) \end{cases}$$

For the entire interval $x = \infty$ to $x = 2$ an error of less than .1 per cent is made if the first four terms of the above sum are used; for an error less than 1 per cent the first three terms at most suffice. From $x = 12$ on C/C_∞ is given with an error of less than 1 per cent by the expression $(4\pi^4/5)(1/x^3)$ alone; i.e., from $T = \theta/12$ to $T = 0$ the proportionality to T^3 holds with an error of less than 1 per cent.

Finally, by use of (12') and (14'), the values of C/C_∞ were calculated as a function of T/θ. They are recorded in Table I, in which the value of $1/x = T/\theta$ is also given.

Table I

$x = \frac{\theta}{T}$	$\frac{1}{x} = \frac{T}{\theta}$	$\frac{C}{C_\infty}$	$\frac{C}{C_\infty}$ Einstein	Deviation, %	$\frac{C}{C_\infty}$ Nernst-Lindemann	Deviation, %
0.250	4	0.997	0.994	- 0.3	0.997	0
0.333	3	0.994^5	0.989	- 0.5	0.994	0
0.500	2	0.988	0.979	- 0.9	0.988	0
0.667	1.5	0.978	0.959	- 1.9	0.977	- 0.1
1.000	1.0	0.952	0.920	- 3.3	0.951	- 0.1
1.111	0.9	0.941	0.902	- 4.1	0.939	- 0.2
1.250	0.8	0.926	0.878	- 5.2	0.924	- 0.2
1.429	0.7	0.904^5	0.841	- 7.0	0.904	0
1.667	0.6	0.872^5	0.799	- 8.3	0.873	0
2.000	0.5	0.825	0.724	-12.2	0.823	- 0.2
2.500	0.4	0.745	0.610	-18.1	0.744	- 0.1
3.333	0.3	0.607^5	0.428	-29.5	0.615	+ 1.2
4.000	0.25	0.503	0.304	-39.5	0.514	+ 2.2
4.021	0.2488	0.500	---	---	---	---
5.000	0.20	0.369	0.171	-51.1	0.390	+ 5.7
6.667	0.15	0.213	0.0598	-72.0	0.242	+ 13.6
10.00	0.10	0.0758	0.0045	-94.2	0.0875	+ 15.4
13.33	0.075	0.0328	---	---	0.0284	- 13.4
20.00	0.050	0.00974	---	---	0.00227	- 76.3
40.00	0.025	0.00122	---	---	0.000000412	-100
∞	0	0	0	-100	0	-100

In order to make possible a comparison with the two other formulae used up to now: that of Einstein:

$$\frac{C}{C_\infty} = \frac{x^2 e^x}{(e^x - 1)^2} \tag{15}$$

and that of Nernst-Lindemann:

$$\frac{C}{C_\infty} = \frac{1}{2}\left\{\frac{x^2 e^x}{(e^x - 1)^2} + \frac{\left(\frac{x}{2}\right)^2 e^{x/2}}{(e^{x/2} - 1)^2}\right\}$$

the values of these functions are reoorded in the 4th and 6th columns.*

The fifth column contains the deviation of the Einstein function from our function (columns 4 and 3) in per cent of the latter. It is seen that this per cent difference soon takes on relatively high values. The Nernst-Lindemann function behaves much better in this respect; down to T/θ = 0.4 the deviations are almost unnoticeable; they are, nevertheless, present and are negative.** In the table they do not appear because of our limitation to too few decimal places. At T/θ = approximately 0.4, the deviation vanishes. Then the per cent difference becomes larger again and at first is positive; however, it goes down again, becomes zero, and finally quickly attains high negative values. That the difference must be taken to be 100 per cent in the neighborhood of absolute zero is due to the fact that according to the formula of Einstein, or to that of Nernst-Lindemann, the specific heat must vanish exponentially, while according to our formula it vanishes only as T^3.

Finally, in Figure 1 the different curves for C/C_∞ are plotted as a function of $1/x = T/\theta$. The curve marked 1 corresponds to our formula (10'). The Nernst-Lindemann function is represented by curve 2, which is dashed so that it may be better distinguished from 1. The curve 3 represents the Einstein function. According to column 7 the Nernst-Lindemann function should run first below, then above, and again below the curve 1. Within the accuracy of the figure, however, the two curves fall together in the first interval. The Einstein function is smaller than (10') throughout. The ratio for low temperatures is very hard to represent on the scale adopted. However, in the same figure, the lower part of the three curves are plotted on a scale ten times enlarged with abscissae ten times enlarged. The numbers 1', 2', and 3' correspond to the numbers 1, 2, and 3 of the complete curves.

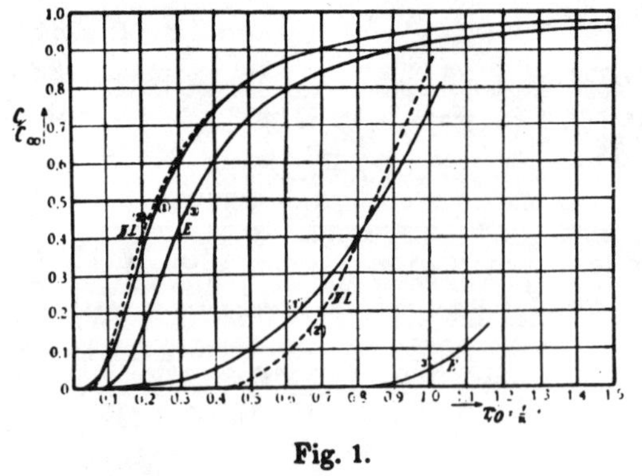

Fig. 1.

*The numerical values were not calculated anew, but taken from the tables of B. Pollitzer: Die Berechnung chemischer Affinitäten nach dem Nernstschen Wärmetheorem, Stuttgart 1912, p. 165 or interpolation between the values given there.

**Compare the remarks in section III.

III. Search for a Basis for the Nernst-Lindemann Form of the Function C/C_∞

The Nernst-Lindemann function has proven to be such a good representation of the observations in a large temperature range that the conclusion is inescapable that it must be connected with the theoretically founded formula in a simple if perhaps only superficial way. We will show in the following in what sense the Nernst-Lindemann function can be interpreted as an approximate representation of our function.

According to our analysis the formula for C/C_∞ is (10'):

$$\frac{C}{C_\infty} = \frac{12}{x^3}\int_0^x \frac{\xi^3 d\xi}{e^\xi - 1} - \frac{3x}{e^x - 1}$$

With the same notation ($x = \theta/T$) the Nernst-Lindemann formula is:

$$\frac{C}{C_\infty} = \frac{1}{2}\left\{\frac{x^2 e^x}{(e^x - 1)^2} + \frac{\left(\frac{x}{2}\right)^2 e^{x/2}}{(e^{x/2} - 1)^2}\right\}$$

We will now adhere to this introduction of $x/2$ besides x and pose the following question. If it is supposed that:

$$\frac{C}{C_\infty} = a\frac{x^2 e^x}{(e^x - 1)^2} + b\frac{\left(\frac{x}{2}\right)^2 e^{x/2}}{(e^{x/2} - 1)^2} \qquad (16)$$

how must the two constants a and b be chosen so that the integral of the difference of (10') and (16) taken over the temperature interval 0 to ∞, vanishes? Graphically, this means, referring to Figure 1, that the area enclosed between curves 1 and 2 should have in its entirety the content zero when sign is taken into account.*

*In the figure it appears as if the requirement is not satisfied. The reason for this lies in that on the scale used the negative differences which are actually present for high temperatures do not show up. The presence of these are best shown on the basis of the series developments for large values of T, that is, small values of x. According to (12'), our function is:

$$\frac{C}{C_\infty} = 1 - \frac{x^2}{20} + \frac{x^4}{560}$$

(Footnote continued)

The integral of (10') over the temperature is nothing more than the energy content, which has already been calculated in (9), divided by $C_\infty = 3Nk$. It has the value:

(17) $$\Theta \frac{3}{x^4}\int_0^x \frac{\xi^3 d\xi}{e^\xi - 1}$$

The corresponding integral of (16), which has a similar meaning, is:

(17') $$\Theta\left\{\frac{a}{e^x - 1} + \frac{b/2}{e^{x/2} - 1}\right\}$$

If our condition is to be satisfied, the difference of the two quantities (17) and (17') must vanish after the limits $T = 0$ and $T = \infty$ have been substituted, that is:

(18) $$\left[\frac{3}{x^4}\int_0^x \frac{\xi^3 d\xi}{e^\xi - 1} - \frac{a}{e^x - 1} - \frac{b/2}{e^{x/2} - 1}\right]_{T=0}^{T=\infty} = 0$$

Now when $T = 0$, $x = \infty$; however, for this value of x the bracket from (18) vanishes. On the other hand, when $T = \infty$, $x = 0$. For small values of x the quantities appearing in (18) can be expanded as follows:

*(continued from preceding page)

The Nernst-Lindemann function is obtained easily by using the development of $1/e^\xi - 1$ given in section II:

$$\frac{C}{C_\infty} = 1 - \frac{5}{96}x^2 + \frac{17}{7680}x^4$$

If the coefficients in our function are made to have the same denominators, our function is:

$$\frac{C}{C_\infty} = 1 - \frac{4.8}{96}x^2 + \frac{13.71}{7680}x^4$$

Since $5 > 4.8$, the value of the Nernst-Lindemann function, as asserted, is smaller than the value of our function for small x (large T). On the other hand, since $17 > 13.71$, the development indicates that for still smaller temperatures the ordinates become equal, as has been found earlier, when T/Θ is about equal to 0.4.

$$\frac{3}{x^4}\int_0^x \frac{\xi^3 d\xi}{e^\xi - 1} = \frac{3}{x^4}\int_0^x (\xi^2 - \tfrac{1}{2}\xi^3 + ..)d\xi = \frac{1}{x} - \frac{3}{8} + \ldots$$

$$\frac{a}{e^x - 1} = \frac{a}{x} - \frac{a}{2} + \cdots$$

$$\frac{b/2}{e^{x/2} - 1} = \frac{b}{x} - \frac{b}{4} + \cdot\cdot$$

where only terms have to be taken into account which are infinite for $x = 0$ or remain finite. The condition (18) can be written now:

$$(18') \quad \lim_{x=0}\left[\left(\frac{1}{x} - \frac{3}{8}\right) - a\left(\frac{1}{x} - \frac{1}{2}\right) - b\left(\frac{1}{x} - \frac{1}{4}\right)\right] = 0$$

that is, a and b must satisfy the two equations:

$$(19) \quad \begin{cases} a + b = 1 \\ \dfrac{a}{2} + \dfrac{b}{4} = \dfrac{3}{8} \end{cases}$$

From this follow:

$$(20) \quad a = \frac{1}{2} \text{ and } b = \frac{1}{2}$$

that is, just those values of a and b which Nernst and Lindemann used in their formula.

Therefore, it can be asserted that in so far as our formula is regarded as correct, the Nernst-Lindemann formula yields an approximation to it in the above sense.

IV. Comparison with Experiment

In these paragraphs the agreement of our formula with the observations on diamond, aluminum, copper, silver, and lead will be demonstrated with some graphs. In addition the correctness of the assertion that at low temperatures the specific heat is proportional to T^3 is shown. We take the observed values from the tables of Nernst-Lindemann.[6] These contain the specific heat at constant pressure C_p. This is converted to the specific heat $C_v = C$ at constant volume by means of the formula given there:

$$(21) \quad C_v = C_p - A C_p^2 T$$

in which A is a constant which for Al, Cu, Ag, and Pb is taken from the tables of Nernst-Lindemann. For diamond, the conversion is accomplished by the use of the other formula given there:

$$C_v = C_p - A_0 C_p^2 \frac{T}{T_s} \tag{21'}$$

in which A_0 = 0.0214 and T_s is the melting point for which we take T_s = 3600°.

In the following tables are compiled the temperature T, the specific heat at constant pressure C_p, the quantity ΔC_p from (21) or (21'), which must be subtracted from C_p to get C_v, the specific heat at constant volume C_v, and finally the ratio C_v/C_∞. For C_∞ the value C_∞ = 5.955 cal. is taken.

Table II

Diamond (T_s = 3600°).

T	C_p	ΔC_p	C_v	C_v/C_∞
30	0.00	---	0.00	0.000
42	0.00	---	0.00	0.000
88	0.03	---	0.03	0.005
92	0.03	---	0.03	0.005
205	0.62	---	0.62	0.104
209	0.66	---	0.66	0.111
220	0.72	---	0.72	0.121
222	0.76	---	0.76	0.128
232	0.86	---	0.86	0.145
243	0.95	---	0.95	0.160
262	1.14	---	1.14	0.191
284	1.35	---	1.35	0.227
306	1.58	---	1.58	0.266
331	1.84	0.01	1.83	0.308
358	2.12	0.01	2.11	0.354
413	2.66	0.02	2.64	0.444
1169	5.45	0.21	5.24	0.880

Table III

Aluminum ($A = 2.2 \cdot 10^{-5}$).

T	C_p	ΔC_p	C_v	C_v/C_∞
32.4	0.25	---	0.25	0.042
35.1	0.33	---	0.33	0.055
83.0	2.41	0.01	2.40	0.403
86.0	2.52	0.01	2.51	0.421
88.3	2.62	0.01	2.61	0.438
137	3.97	0.04	3.93	0.660
235	5.32	0.15	5.17	0.868
331	5.82	0.25	5.57	0.935
433	6.10	0.35	5.75	0.966
555	6.48	0.52	5.98	1.005

Table IV

Copper ($A = 1.3 \cdot 10^{-5}$).

T	C_p	ΔC_p	C_v	C_v/C_∞
23.5	0.22	---	0.22	0.037
27.7	0.32	---	0.32	0.054
33.4	0.54	---	0.54	0.092
87.0	3.33	0.01	3.32	0.558
88.0	3.38	0.01	3.37	0.566
137	4.57	0.04	4.53	0.761
234	5.59	0.09	5.50	0.924
290	5.79	0.13	5.66	0.951
323	5.90	0.15	5.75	0.966
450	6.09	0.22	5.87	0.985

Table V

Silver ($A = 2.5 \cdot 10^{-5}$).

T	C_p	ΔC_p	C_v	C_v/C_∞
35.0	1.58	---	1.58	0.266
39.1	1.90	---	1.90	0.319
42.9	2.26	---	2.26	0.380
45.5	2.47	0.01	2.46	0.413
51.4	2.81	0.01	2.80	0.470
53.8	2.90	0.01	2.89	0.502
77.0	4.07	0.03	4.04	0.678
100	4.86	0.06	4.80	0.806
200	5.78	0.17	5.61	0.944
273	6.00	0.25	5.75	0.965
331	6.01	0.30	5.71	0.960
535	6.46	0.56	5.90	0.990
589	6.64	0.65	5.99	1.006

Table VI

Lead ($A = 3.0 \cdot 10^{-5}$).*

T	C_p	ΔC_p	C_v	C_v/C_∞
23.0	2.96	0.01	2.95	0.495
28.3	3.92	0.01	3.91	0.656
36.8	4.40	0.02	4.38	0.735
38.1	4.45	0.02	4.43	0.744
85.5	5.65	0.08	5.57	0.935
90.2	5.71	0.09	5.62	0.945
200	6.13	0.22	5.91	0.993
273	6.31	0.34	5.97	1.002
290	6.33	0.35	5.98	1.004
332	6.41	0.41	6.00	1.007
409	6.61	0.54	6.07	1.02

*The constant A which is used in converting C_p to C_v seems somewhat too low. However, we have intentionally adhered to the values of Nernst-Lindemann.

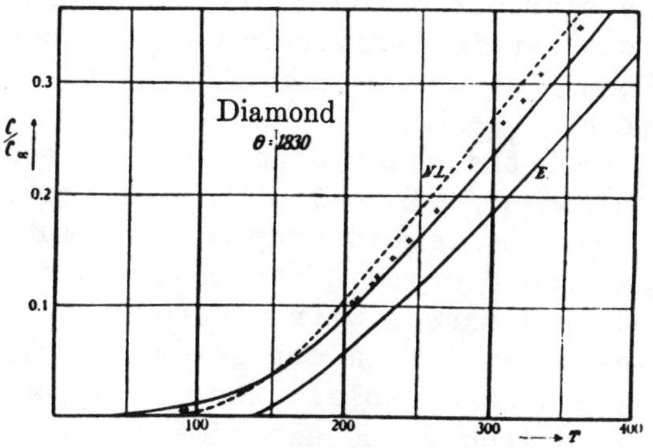

Figure 2

The ratio for diamond is plotted in Figure 2, and this case is used to test to a certain extend the accuracy of the different formulae. This is done as follows. The value of C/C_∞ = 0.880 at T = 1169° is taken to be exact and from this value alone the characteristic temperature θ is calculated, first assuming the Einstein formula is correct, second assuming the Nernst-Lindemann function is correct, and third on the basis of our formula.* It is found that to:

$$C/C_\infty = 0.880 \text{ at } T = 1169°$$

corresponds according to:

Einstein...............θ = 1450°
Nernst-Lindemann.......θ = 1884°
Our formula............θ = 1895°

Observe that in the last two cases numbers were found for θ which agree within about ½ per cent. This is the case throughout so that fortunately the values for θ calculated by Nernst-Lindemann remain essentially unchanged.

With these three values of θ the three curves of Figure 2 were plotted. The observed values given in Table II are designated by crosses. The curve marked E is the Einstein function, which, as can be seen, runs much too low. The dashed curve marked NL corresponds to the Nernst-Lindemann function; the full unmarked curve to

*It is no longer necessary to carry out the calculations but only to interpolate from the tables of section II. In order to be sure of the figures quoted, we have avoided interpolation in this case and calculated anew.

our theoretical law. The latter runs somewhat too high at first, finally too low; in consideration of the way θ was calculated, the agreement is to be considered good.

It must not be overlooked that first the value $C/C_\infty = 0.880$ at $T = 1169^\circ C$. is not particularly certain; the correction to C_v amounts to 4% and besides this is computed by the use of the hypothetical melting point $T_s = 3600^\circ$. Second, with another choice of θ better agreement with the observations can be achieved; $\theta = 1830^\circ$ satisfies this requirement. In order to avoid complicating the figure further, we have neglected to plot this latter curve.

Figures 3, 4, 5, and 6 contain the observed values for aluminum, copper, silver, and lead marked with crosses and the associated theoretical curve calculated from our formula or Table I. The figures are readily understandable. They show throughout a very good agreement between theory and experiment.

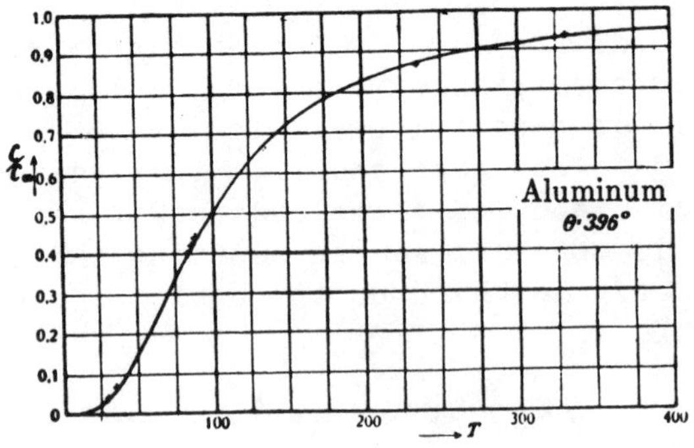

Figure 3

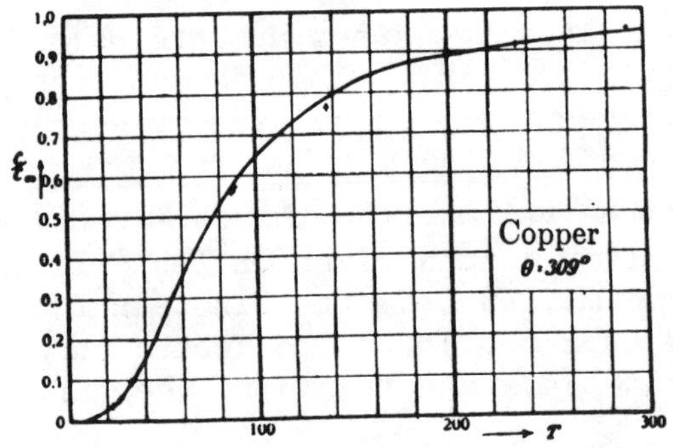

Figure 4

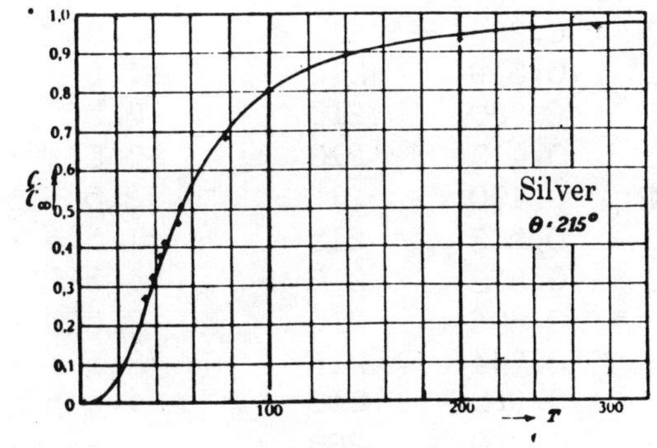

Figure 5

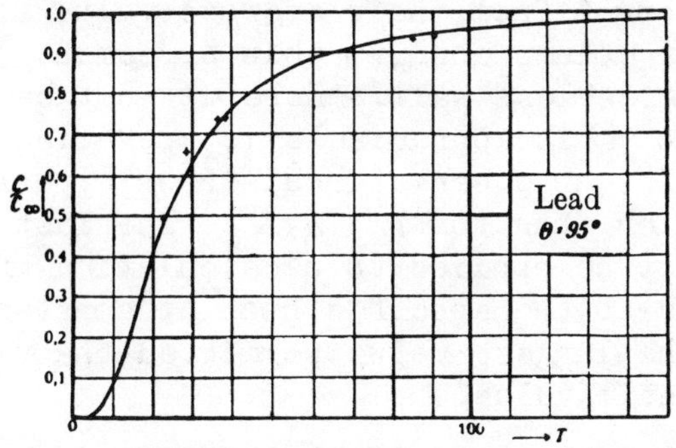

Figure 6

Finally, in order to do justice to the need of a discussion as exhaustive as possible, we have made the following test. The numbers given in Table I were used to calculate by linear interpolation the θ belonging to each observed value of C/C_∞ and T for silver. The result is contained in the following Table VII.

Table VII

T	C/C_∞	θ	Δ
35.0	0.266	210	-3.5
39.1	0.319	212	-1.5
42.9	0.380	210	-3.5
45.5	0.413	210	-3.5
51.4	0.470	216	+2.5
53.8	0.502	215	+1.5
77.0	0.678	218	+4.5
100	0.806	210	-3.5
200	0.944	216	+2.5
273	0.965	218	+4.5
331	0.960	288	---
535	0.990	236	---
589	1.006	---	---

The first ten values of θ show good agreement with one another. The deviations Δ from their average value 213.5 which are quoted in the last column confirm this agreement. The eleventh and twelfth numbers agree less well. Here it must be remembered that the correction to C_v is appreciable (compare Table V). Besides this, at $T = 331^\circ$, an observational error appears since C/C_∞ is given as smaller than for $T = 273^\circ$. The last number for C/C_∞ at $T = 579^\circ$ cannot be applied to a calculation of θ since C/C_∞ (after correction to the specific heat at constant volume) is 0.6% larger than the largest value permitted theoretically, that is, the Dulong-Petit value.

Finally, there is a question to be disposed of which is particularly important to our theory. Does the present numerical material show that our assertion that the specific heat should be proportional to T^3 at low temperatures is correct? This must be answered in the affirmative, as the Table VIII shows.

Table VIII

	T	C/C_∞	θ	θ
Diamond	205	0.104	1860	1830
Aluminum	32.4	0.0420	398	396
Copper	23.5	0.037	302	309

In column 3 is the observed value of C/C_∞ at the low temperature given in column 2 (compare the earlier table). Earlier in this limiting case C/C_∞ was found to be:

$$\frac{C}{C_\infty} = \frac{4\pi^4}{5}\frac{T^3}{\Theta^3} = 77{,}94\frac{T^3}{\Theta^3}$$

At sufficiently low temperatures θ, therefore, can be calculated from the formula:

$$\Theta = \left(77.94\frac{C_\infty}{C}\right)^{1/3} T$$

This has been done and the result is shown in column 4. For comparison, column 5 contains the value of θ which is based on the earlier graphs. It is seen that the agreement is very good.

Nevertheless, it may appear very desirable to test the validity of the proportionality to T^3 with a series of observations. Diamond would be most suitable for such research since the temperature region in which measurements would be done is relatively accessible. It extends from $T = 0$ up to about $T = 200$. Also, it should be pointed out that the expansion coefficient, like the specific heat, must be proportional to T^3 at low temperatures providing Gruneisen's statement that the two quantities are proportional is accepted.*

V. Calculation of the Characteristic Temperature θ from the Elastic Constants

In section I a quantity F is given which can be calculated from the density ρ, compressibility κ, and Poisson's ratio σ and which is connected in a simple way with the cut-off frequency ν_m of our elastic spectrum.

It was found that:

$$\nu_m = \left(\frac{3N}{VF}\right)^{1/3}$$

and:

$$F = \frac{4\pi}{3}\varrho^{3/2}\varkappa^{3/2}\left[2\left(\frac{2(1+\sigma)}{3(1-2\sigma)}\right)^{3/2} + \left(\frac{1+\sigma}{3(1-\sigma)}\right)^{3/2}\right]$$

For the proof of these formulae refer to Part II.

In equation (7) a quantity θ is fashioned from ν_m and the two universal constants h and k:

*It is permissible to ignore the difference between C_p and C_v at the temperatures considered here.

$$\Theta = \frac{h\nu_m}{k}$$

which plays the roll of a temperature characteristic of the substance and which alone determines the course of the specific heat as a function of temperature. The higher Θ, the higher the temperature T below which the substance begins to show deviation from the Dulong-Petit law.

If, in the expression for Θ, ν_m is replaced by its representation in N, V, and F, then:

$$\Theta = \frac{h}{k}\left(\frac{3N}{VF}\right)^{1/3} \tag{22}$$

If a piece whose mass is equal to the atomic weight is taken from the substance concerned, the number of atoms N in this piece is the same for all monatomic substances and has the value $N = 5.66 \times 10^{23}$.* On the other hand:

$$V = \frac{M}{\varrho}$$

so that (22) can be rewritten:

$$\Theta = \frac{h}{k} N^{1/3}\left(\frac{3\varrho}{MF}\right)^{1/3} = \frac{h}{k}\left(\frac{9N}{4\pi}\right)^{1/3} \frac{1}{M^{1/3}\varrho^{1/6}\varkappa^{1/2}} \frac{1}{f^{1/3}(\sigma)} \tag{23}$$

if $f(\sigma)$ denotes the expression:

$$f(\sigma) = 2\left(\frac{2}{3}\frac{1+\sigma}{1-2\sigma}\right)^{3/2} + \left(\frac{1+\sigma}{3(1-\sigma)}\right)^{3/2} \tag{24}$$

Finally, substitute:

$$N = 5.66\cdot 10^{23}, \quad h = 7.10\cdot 10^{-27}, \quad k = 1.47\cdot 10^{-16}$$

then:

$$\Theta = \frac{35.74\cdot 10^{-4}}{M^{1/3}\varrho^{1/6}\varkappa^{1/2}} \frac{1}{f^{1/3}(\sigma)} \tag{23}$$

As is well known, Einstein[7] first pointed out a connection between the "eigenfrequencies of the atom" and the elastic constants. If the characteristic temperature Θ is calculated from his eigenfrequency by multiplication by h/k, it is found that:

*We use (also for h and k) the value which was given recently by Paschen-Gerlach, *Ann. d. Phys.*, *38*, 41 (1912).

$$\theta = \frac{7.89 \cdot 10^{-4}}{M^{1/3} \varrho^{1/6} \varkappa^{1/2}} \tag{25}$$

Aside from the difference in the meaning of Einstein's simple frequency of the atoms and our cut-off frequency, the two formulae differ in two ways from one another. First, the numerical factors are different; second, the factor depending on Poisson's ratio σ is missing from Einstein's formula. Now, the two formulae are compared with the experimental material on the elastic constants and the specific heats. Table IX serves this purpose.

Table IX

	M	ρ	$\kappa \cdot 10^{12}$	$f(\sigma)$	θ (23')	θ (Specific heat)	θ (25)
Al	27.1	2.71	1.36	10.2	399	396	192
Cu	63.6	8.96	0.74	10.5	329	309	159
Ag	107.9	10.53	0.92	15.4	212	215	117
Au	197.2	19.21	0.60	24.7	166	---	---
Ni	58.7	8.81	0.57	7.38	435	---	---
Fe	55.9	7.85	0.62	5.86	467	---	---
Cd	112.4	8.63	2.4	7.89	168	---	---
Sn	119.0	7.28	1.9	8.50	185	---	---
Pb	207.1	11.32	2.0	61.0	72	95	63
Bi	208.0	9.78	3.2	8.98	111	---	---
Pd	106.7	11.96	0.57	18.8	204	---	---
Pt	195.0	21.39	0.40	17.1	226	---	---

In column 1 is the name of the material, column 2 contains the atomic weight M, column 3 the density ρ, column 4 and 5, respectively, the compressibility κ and the value of the function $f(\sigma)$ from equation (24). The values of κ and σ are taken from Gruneisen's observations.[8] In column 6 is the value of the characteristic temperature θ which is calculated from our formula (23'). For comparison, in column 7 are entered the values of θ which are adapted to the course of the specific heat curves (compare section IV). It is seen that the agreement is very good for Al and Ag, for Cu still good. It is less good for Pb, the error amounting to 20%. If it is remembered that the elastic constants and the specific heats are from different observations on different pieces of material, that the elastic properties of lead are not reliable, and besides that even for the other metals a relatively strong change of elastic behavior at different temperatures is to be expected, the agreement of the two series of numbers can be said good on the whole.

Column 8 serves for testing the Einstein formula (25). It is seen that the so calculated values give only coarsely the trend of the θ values. There is, however, no numerical agreement. The values are all considerably too small. This is not surprising, since the Einstein analysis can give only the order of magnitude of θ; nevertheless, Einstein's work removes any doubt of the connection of θ with the compressibility. Besides it must not be forgotten that the values of θ adjusted to the Einstein formula do not coincide with those corresponding to the most favorable choice for the Nernst-Lindemann or our formula.

Part II

VI. The Elastic Problem. Basic Equations. Introduction of Potentials

In the preceding parts the development has been based on the spectral law of elastic bodies. Up to section V only the theorem that the number of eigenfrequencies of such a body in the frequency interval $d\nu$ is proportional to $\nu^2 d\nu$ has been used. The calculation of the proportionality factor from the elastic constants first played a role in section V, where the question of how the course of the specific heats can be predicted on the basis of measurements of the elastic constants was treated. The proof of the correctness of the distribution law used must now be added.

We know already that for our purpose the shape of the elastic body upon which we base the calculations can have no effect on the final result. Actually, we have no choice, for at present a rigorous treatment is possible only for the sphere.[9] It can be shown that for the calculation of the distribution law it is not necessary to solve the boundary value problem rigorously; however, we will not discuss this possibility further here and must limit ourselves to the use of a spherical boundary for the body. The sphere should serve as a model of a heated body in thermal equilibrium. The boundary condition must be chosen so that the sphere does not lose or gain energy coming to it from the outside. Since no radiation is to be considered, this can be achieved if it is prescribed that the surface of the sphere is force free; or, what is just as good, the forces can be chosen so that no motion of the surface of the sphere is possible. Then the sphere forms a completely closed system. The latter condition, that the displacement on the surface of the sphere vanishes, is used in the following.

If we call the displacement of a point in the interior of the sphere $\mathfrak{B}$, the two elastic constants λ and μ, the density ρ, and the time t, the differential equation of elasticity runs, as is well known:

$$(26)\qquad (\lambda + 2\mu)\,\mathrm{grad}\,\mathrm{div}\,\mathfrak{B} - \mu\,\mathrm{rot}\,\mathrm{rot}\,\mathfrak{B} - \varrho\frac{\partial^2\mathfrak{B}}{\partial t^2} = 0$$

The **two** constants λ and μ are connected with the two used in practice, Young's modulus E and Poisson's ratio σ, through the equations:

$$(27)\qquad \lambda = E\frac{\sigma}{(1+\sigma)(1-2\sigma)}, \quad \mu = \frac{E}{2(1+\sigma)}$$

Recall that the compressibility κ can be computed from λ and μ or E and σ from the formula:*

$$(28)\qquad \varkappa = \frac{1}{\lambda + \frac{2}{3}\mu} = \frac{3(1-2\sigma)}{E}.$$

$\mathfrak{B}$ can be resolved into a vortex free part $\mathfrak{B}_1$ and a source free part $\mathfrak{B}_2$ as follows:

$$(29)\qquad \mathfrak{B} = \mathfrak{B}_1 + \mathfrak{B}_2 = \mathrm{grad}\,P + \mathrm{rot}\,\mathfrak{A}$$

Here P is a scalar potential; $\mathfrak{A}$, a vector potential.

To take account of (26), P is chosen so that it satisfies the equation:

$$(30)\qquad (\lambda + 2\mu)\Delta P - \varrho\frac{\partial^2 P}{\partial t^2} = 0$$

A similar equation holds for $\mathfrak{A}$. To obtain this it is convenient only to half carry out the substitution $\mathfrak{B}_2 = \mathrm{rot}\,\mathfrak{A}$ in (26); this is substituted only in the term differentiated with respect to time. Thus $\mathfrak{A}$ is defined by the two equations (instead of one):

$$(31)\qquad \begin{cases} -\dfrac{\varrho}{\mu}\dfrac{\partial^2\mathfrak{A}}{\partial t^2} = \mathrm{rot}\,\mathfrak{B}_2 \\ \mathfrak{B}_2 = \mathrm{rot}\,\mathfrak{A} \end{cases}$$

These equations have the advantage that they are now constructed similarly to the Maxwell equations for the electromagnetic field whose integration for spherical coordinates can be put in a very convenient form.

Since only states which are periodic in time will be investigated, P and $\mathfrak{A}$ are written:

$$(32)\qquad P = \Phi e^{i\omega t}, \quad \mathfrak{A} = \mathfrak{B} e^{i\omega t}$$

*Compare A. E. H. Love, p. 123 of ref. 9. Remember that the compression modulus defined there is equal to the reciprocal of the compressibility κ.

The quantity ω is the number of vibrations in 2π seconds and is equal to $2\pi\nu$.

For the same reason:

$$\mathfrak{V}_1 = \mathfrak{v}_1 e^{i\omega t}, \quad \mathfrak{V}_2 = \mathfrak{v}_2 e^{i\omega t} \tag{33}$$

so that the equations for the determination of $\mathfrak{v}_1$ are:

$$\begin{cases} \Delta \Phi + \dfrac{\varrho\,\omega^2}{\lambda + 2\mu}\,\Phi = 0 \\ \mathfrak{v}_1 = \operatorname{grad} \Phi \end{cases} \tag{34}$$

and for the determination of $\mathfrak{v}_2$:

$$\begin{cases} -\varrho\,\dfrac{\omega^2}{\mu}\,\mathfrak{B} = -\operatorname{rot}\mathfrak{v}_2 \\ \mathfrak{v}_2 = \operatorname{rot}\mathfrak{B} \end{cases} \tag{35}$$

The equations (34) are ready to use; so that (35) may be utilized a transformation like that which I[10] have used in the treatment of a similar electromagnetic problem is recommended. If $\mathfrak{B}$ is made to parallel the magnetic field strength H and $\mathfrak{v}_2$ the electric field E and the constants:

$$\frac{i\mu\omega}{c} \text{ bzw. } \left(\frac{i s\omega}{c} + \frac{\sigma}{c}\right)$$

are replaced by $-(\rho\omega^2/\mu)$ and 1, respectively, the following result is obtained. If spherical coordinates r, ϑ, φ are introduced and two scalar quantities Π and Ψ defined by:

$$\begin{cases} \Delta \Pi + \dfrac{\varrho\,\omega^2}{\mu}\,\Pi = 0 \\ \Delta \Psi + \dfrac{\varrho\,\omega^2}{\mu}\,\Psi = 0 \end{cases} \tag{36}$$

two independent expressions for the components of $\mathfrak{v}_2$ satisfying the basic equations are found by putting either:

$$\begin{cases} (\mathfrak{v}_2)_r = 0 \\ (\mathfrak{v}_2)_\vartheta = -\dfrac{1}{r\sin\vartheta}\dfrac{\partial}{\partial\varphi}\,r\,\Pi \\ (\mathfrak{v}_2)_\varphi = \dfrac{1}{r}\dfrac{\partial}{\partial\vartheta}\,r\,\Pi \end{cases} \qquad \left(\Delta\Pi + \frac{\varrho\,\omega^2}{\mu}\,\Pi = 0\right) \tag{37}$$

or:

$$(37')\quad \begin{cases} (\mathfrak{v}_2')_r = \dfrac{\partial^2}{\partial r^2} r\,\Psi + \dfrac{\varrho\,\omega^2}{\mu} r\,\Psi \\ (\mathfrak{v}_2')_\vartheta = \dfrac{1}{r}\,\dfrac{\partial^2}{\partial r\,\partial\vartheta} r\,\Psi \\ (\mathfrak{v}_2')_\varphi = \dfrac{1}{r\sin\vartheta}\,\dfrac{\partial^2}{\partial r\,\partial\varphi} r\,\Psi \end{cases} \qquad \left(\Delta\Psi + \frac{\varrho\,\omega^2}{\mu}\Psi = 0\right)$$

For convenient reference the differential equations for Π and Ψ are repeated; with the same intention, the definition of the third kind of motion is given below, but now in coordinate form:

$$(38)\quad \begin{cases} (\mathfrak{v}_1)_r = \dfrac{\partial\Phi}{\partial r} \\ (\mathfrak{v}_1)_\vartheta = \dfrac{1}{r}\,\dfrac{\partial\Phi}{\partial\vartheta} \\ (\mathfrak{v}_1)_\varphi = \dfrac{1}{r\sin\vartheta}\,\dfrac{\partial\Phi}{\partial\varphi} \end{cases} \qquad \left(\Delta\Phi + \frac{\varrho\,\omega^2}{\lambda+2\mu}\Phi = 0\right).$$

VII. The Boundary Conditions

As explained in section VI, the boundary condition is that the displacement vanish on the surface of the sphere $r = a$. If $\mathfrak{v} = \mathfrak{v}_2$ is computed from the potential Π according to (37), this boundary condition is fulfilled automatically if it is required that for $r = a$ the potential Π vanishes:

$$(39)\qquad [\Pi]_{r=a} = 0$$

On the other hand it is impossible to determine Ψ or Φ so that the boundary conditions are fulfilled individually for $\mathfrak{v}_2'$ and $\mathfrak{v}_1$. This can be done only for the sum $\mathfrak{v}_2' + \mathfrak{v}_1$. If the sum is taken, it can be concluded from the last two equations of (37') and the last two of (38) that for $r = a$, the expression:

$$\Phi + \frac{\partial}{\partial r} r\,\Psi$$

must vanish:

$$(40)\qquad \left[\Phi + \frac{\partial}{\partial r} r\,\Psi\right]_{r=a} = 0$$

This is clear if it is recalled that through the addition of the two expressions there results:

$$\frac{1}{r}\frac{\partial}{\partial\vartheta}\left(\frac{\partial}{\partial r} r \Psi + \Phi\right)$$

or:

$$\frac{1}{r \sin\vartheta}\frac{\partial}{\partial\varphi}\left(\frac{\partial}{\partial r} r \Psi + \Phi\right)$$

The remaining equations for $(\mathfrak{v}_2')_r$ and $(\mathfrak{v}_1)$ give the second boundary condition:

$$\text{(40')} \qquad \left[\frac{\partial\Phi}{\partial r} + \frac{\partial^2}{\partial r^2} r \Psi + \frac{\varrho\omega^2}{\mu} r \Psi\right]_{r=a} = 0$$

If Π, Φ, and Ψ have been determined in general and then the constants in Π determined so that (39) is satisfied and those in $\Phi + \Psi$ determined so that (40) and (40') are satisfied, the possible motions $\mathfrak{v}_2$ or $\mathfrak{v}_2' + \mathfrak{v}_1$ are found by application of the differentiation scheme of the equations (37), (37'), and (38). For our purposes we do not need to do this; what we need to know is only the position and number of the eigenvibrations, which are given already by (39), (40), and (40').

VIII. The Expressions for the Potentials and the Equations for the Eigenvibrations

For brevity, put:

$$\text{(41)} \qquad \alpha^2 = \frac{\varrho\omega^2}{\mu}, \quad \beta^2 = \frac{\varrho\omega^2}{\lambda + 2\mu}$$

Then for Π we have the equation:

$$\text{(42)} \qquad \Delta\Pi + \alpha^2\Pi = 0$$

A general solution in spherical coordinates is:*

$$\text{(43)} \qquad r\Pi = \sum_n A_n \psi_n(\alpha r) S_n(\vartheta, \varphi)$$

The A_n are real constants, $\psi_n(\alpha r)$ are up to a factor $(\pi\alpha r/2)^{\frac{1}{2}}$ ordinary Bessel functions of index $n + \frac{1}{2}$ and argument αr. $S_n(\vartheta, \varphi)$ is the nth surface harmonic. As is well known, the latter contains $2n - 1$ arbitrary constants if one of its constants is imagined combined with A_n; this follows directly from the Maxwell definition which

*Compare for the notation ψ_n and the associated series development P. Debye, *Ann. d. Phys.*, 30, 64 (1909).

uses the position of n points of intersection of n arbitrary axes to define S_n.[11] Now the boundary condition (39) requires that α be determined so that:

$$\psi_n(\alpha a) = 0 \tag{44}$$

If a root of this equation belonging to the index n has been determined, an eigenfrequency follows from (41). The associated motion can take several different forms, however, since the potential belonging to this eigenfrequency contains $2n$ arbitrary constants, as explained above.

As will be seen in the next paragraph, for fixed n (44) has not one, but infinitely many roots. We will denote the pth value of α belonging to the index n and satisfying (44) by α_n^p; then $r\Pi$ can be expressed as the following double sum:

$$r\,\Pi = \sum_n \sum_p A_n^p\, \psi_n(\alpha_n^p r)\, S_n^p(\vartheta, \varphi) \tag{43'}$$

which is to be summed over all roots which belong to the index n and then over n itself. The index p has been attached to the symbol S_n for the surface harmonics to signify that its $2n - 1$ constants can be chosen arbitrarily for each value of n and p. (43') is the most general expression for Π which satisfies the basic equations and the boundary conditions.

Now the **corresponding** analysis for the second kind of motion follows. Again, first without regard to the boundary conditions $r\Phi$ and $r\Psi$ can be written in the general form:

$$\begin{cases} r\,\Psi = \sum B_n\, \psi_n(\alpha r)\, S_n(\vartheta, \varphi) \\ r\,\Phi = \sum C_n\, \psi_n(\beta r)\, S_n(\vartheta, \varphi) \end{cases} \tag{45}$$

where B_n and C_n are arbitrary constants. Again, the S_n are surface harmonics and are identical in the two sums.

Now the first boundary condition (40) requires that:

$$C_n \frac{\psi_n(\beta a)}{a} + B_n \frac{\partial}{\partial a} \psi_n(\alpha a) = 0 \tag{46}$$

From the second condition (40') it follows that:

$$C_n \frac{\partial}{\partial a} \frac{\psi_n(\beta a)}{a} + B_n \left\{\frac{\partial^2}{\partial a^2} \psi_n(\alpha a) + \alpha^2 \psi_n(\alpha a)\right\} = 0 \tag{46'}$$

The curly bracket can be simplified somewhat; the differential equation for:

$$\frac{\partial^2}{\partial a^2}\psi_n(\alpha a) + \left(\alpha^2 - \frac{n(n+1)}{a^2}\right)\psi_n(\alpha a) = 0$$

makes it possible to replace (46') by:

$$(46'') \qquad C_n \frac{\partial}{\partial a}\frac{\psi_n(\beta a)}{a} + B_n n(n+1)\frac{\psi_n(\alpha a)}{a^2} = 0$$

The two equations (46) and (46") can be satisfied by finite values of B_n and C_n only if the determinant of the system of equations vanishes. If this is the case, the ratio of B_n to C_n is determined by the equations so that only one of them actually is arbitrary. After some easy reshuffling the equation which expresses the vanishing of the determinant can be put in the form:

$$(47) \qquad \frac{\psi_n'(\alpha a)}{\psi_n(\alpha a)/\alpha a} \cdot \frac{\beta a \dfrac{d}{d(\beta a)}\dfrac{\psi_n(\beta a)}{\beta a}}{\psi_n(\beta a)/\beta a} = n(n+1)$$

The actual variable in this equation is the frequency ω; according to (41), when the variables αa and βb are retained it must be remembered that:

$$(48) \qquad \alpha a\sqrt{\mu} = \beta a\sqrt{\lambda + 2\mu}$$

If two values of α and β which satisfy (47) and (48) have been found, one eigenfrequency follows from this. The associated potential and the associated state of motion contain $2n$ arbitrary constants, just as in the case of the potential Π, since, for definite values of α and β the ratio B_n/C_n is fixed and the surface harmonics in Φ and Ψ are identical.

IX. Asymptotic Position and Number of the Eigenfrequencies

After the frequency equations (44), (47), and (48) have been found, we must calculate the roots explicitly. In this a simplification is permissible which will be used from now on. The eigenfrequencies which must be taken into account in the calculation of specific heats are present in very large numbers. The highest eigenfrequencies are by far the most important, and these belong to large values of n, since according to Maxwell's definition n is a measure of the number of nodes in the radial direction interior to the sphere. Therefore, the scope of our formula can be limited to the case of large n, for which simple formulae for ψ_n exist. However, it is essential not to assume that the arguments of the functions are themselves large compared to n, which, as we will see, is certainly not the case. Therefore, we must not use the Hankel approximation, which depends on such assumptions, but formulae which permit the argument to take arbitrary values. Such formulae are:*

$$(49) \qquad \begin{cases} \psi_n(x) = \dfrac{1}{(\sin\tau)^{1/2}} \cos\left[x(\sin\tau - \tau\cos\tau) - \dfrac{\pi}{4}\right] \\ \cos\tau = \dfrac{n+\frac{1}{2}}{x} \end{cases}$$

as long as $x > n + \frac{1}{2}$, ψ_n vanishes exponentially. It can be concluded from this that the equation $\psi_n(x) = 0$ possesses no root smaller than the subscript n.

Under (a) our first frequency equation (44) is treated and under (b) the two equations (47) and (48).

(a) The frequency equation is:

$$(44) \qquad \psi_n(\alpha a) = 0$$

If αa is denoted by x in the following, then according to (49) the frequency equation can be rewritten:

$$(50) \qquad \begin{cases} \cos\left[x(\sin\tau - \tau\cos\tau) - \dfrac{\pi}{4}\right] = 0 \\ \cos\tau = \dfrac{n+\frac{1}{2}}{x} \end{cases}$$

from which roots belonging to a value n can be calculated. The number of roots which lie below a fixed, though arbitrary, frequency is more important than the exact values of the roots. This fixed frequency will be defined by an upper limit $x = x_0$ for our roots $x = \alpha a$ Then our question runs: For given n, how many roots lie below a certain upper limit $x = x_0$?

If x is replaced by its expression in n and τ in the first equation (50), there results:

$$(51) \qquad \cos\left[(n+\tfrac{1}{2})(\mathrm{tg}\,\tau - \tau) - \frac{\pi}{4}\right] = 0$$

*Compare P. Debye, *Math. Ann.*, 67, 535 (1909). The formulae are also collected in E. Jahnke and F. Emde, *Funktionentafeln*, Leipzig, 1909, p. 102. Calulations similar to ours are found in J. Reudler, *Dissertation*, Leiden, 1912, in which the electromagnetic field in a sphere is investigated in connection with Jeans' derivation of the radiation law.

According to the above it is necessary to consider only those cases in which $x > n + \frac{1}{2}$. The variable τ will start with a value which satisfies:

$$\cos \tau = 1$$

that is with:

$$\tau = 0 \tag{52}$$

Associated with the upper limit $x = x_0$ is a second value of $\tau = \tau_0$ determined by the equation:

$$\cos \tau_0 = \frac{n + \frac{1}{2}}{x_0} \quad \text{or} \quad \tau_0 = \text{arc cos} \frac{n + \frac{1}{2}}{x_0} \tag{52'}$$

where naturally τ_0 lies between 0 and $\pi/2$. Our question is now identical with the following: How many roots τ of the equation (51) lie between $\tau = 0$ and $\tau = \tau_0$?

Now if τ goes from 0 to τ_0, the argument of the cos in (51) goes from:

$$-\frac{\pi}{4} \quad \text{bis} \quad (n + \tfrac{1}{2})(\text{tg}\, \tau_0 - \tau_0) - \frac{\pi}{4}$$

The function:

$$(n + \tfrac{1}{2})(\text{tg}\, \tau - \tau) - \frac{\pi}{4}$$

is plotted against τ in Figure 7. Further, imagine horizontal straight lines drawn at heights $\pi/2$, $3\pi/2$, $5\pi/2$ and so on. Then each intersection of such a line with the curve has an abscissa τ which is a root of the equation in question.

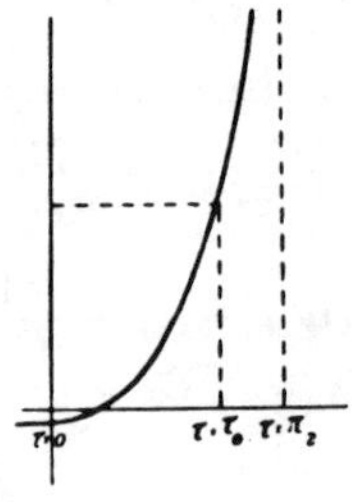

Figure 7

The total height which must be run through amounts to:

$$(n+\tfrac{1}{2})(\operatorname{tg}\tau_0-\tau_0)-\frac{\pi}{4}$$

The ordinates of the successive horizontal lines differ by π so that the number of roots which belong to the index n and which we denote by z_n is:

$$z_n=\frac{n+\frac{1}{2}}{\pi}(\operatorname{tg}\tau_0-\tau_0) \tag{53}$$

If $n+\frac{1}{2}$ is replaced by its expression in τ_0, this can be re-written:

$$z_n=\frac{x_0}{\pi}(\sin\tau_0-\tau_0\cos\tau_0) \tag{53'}$$

This is the expression for the number of eigenfrequencies which lie below a certain upper limit x_0 for given n.

(b) We treat now the same question for the two equations (47) and (48).

If, as above, αa is denoted by x and also βa by y, (48) becomes:

$$\frac{\psi_n'(x)}{\psi_n(x)/x}\;\frac{y\dfrac{d}{dy}\dfrac{\psi_n(y)}{y}}{\psi_n(y)/y}=n(n+1) \tag{54}$$

If again:

$$\cos\tau=\frac{n+\frac{1}{2}}{x}$$

then according to (49):

$$\psi_n(x)=\frac{1}{\sin^{1/2}\tau}\cos\left[(n+\tfrac{1}{2})(\operatorname{tg}\tau-\tau)-\frac{\pi}{4}\right]$$

From this follows easily by differentiation and to the same approximation:

$$\psi_n'(x)=-\sin^{1/2}\tau\sin\left[(n+\tfrac{1}{2})(\operatorname{tg}\tau-\tau)-\frac{\pi}{4}\right]$$

Besides y an angle t is introduced which is defined by:

$$\cos t=\frac{n+\frac{1}{2}}{y} \tag{55}$$

so that:

$$\psi_n(y) = \frac{1}{\sin^{1/2} t} \cos\left[(n + \tfrac{1}{2})(\operatorname{tg} t - t) - \frac{\pi}{4}\right]$$

and by carrying out the differentiation:

$$y \frac{d}{dy} \frac{\psi_n(y)}{y} = -\sin^{1/2} t \sin\left[(n + \tfrac{1}{2})(\operatorname{tg} t - t) - \frac{\pi}{4}\right]$$

If this is substituted in equation (54), the relation:

$$(56) \quad \begin{cases} \operatorname{tg}\left[(n + \tfrac{1}{2})(\operatorname{tg} \tau - \tau) - \frac{\pi}{4}\right] \operatorname{tg}\left[(n + \tfrac{1}{2})(\operatorname{tg} t - t) - \frac{\pi}{4}\right] \\ \qquad = \frac{n(n+1)}{(n + \frac{1}{2})^2} \frac{1}{\operatorname{tg} \tau \operatorname{tg} t}. \end{cases}$$

between τ and t is obtained. For brevity, the two square brackets will be denoted by $[\tau]$ and $[t]$. Observe that for large values of n:

$$\frac{n(n+1)}{(n + \frac{1}{2})^2} = 1$$

so that the frequency equation finally becomes:

$$(56') \qquad \operatorname{tg}[\tau] \operatorname{tg}[t] = \frac{1}{\operatorname{tg} \tau \operatorname{tg} t}$$

Besides this, it must be taken into account that according to (48):

$$\alpha a \sqrt{\mu} = \beta a \sqrt{\lambda + 2\mu}$$

or expresed in τ and t:

$$(57) \qquad \sqrt{\mu} \cos t = \sqrt{\lambda + 2\mu} \cos \tau$$

If two values of τ and t have been found which satisfy (56') and (57) simultaneously, then these define one eigenfrequency.

Just as in (a) the exact location of the roots is not important; only the number which lie below an upper limit determined by a fixed but arbitrary eigenfrequency is of interest to us. Let the upper limit on x be denoted by x_0; that on y, by y_0, where, of course, according to (57) or (48):

$$x_0 \sqrt{\mu} = y_0 \sqrt{\lambda + 2\mu}$$

With these values x_0 and y_0 may be associated the values τ_0 and t_0. Then our question is:

How many pairs of values of τ and t are there which satisfy (56') and (57) and for which τ lies between 0 and τ_0 and simultaneously t lies between 0 and t_0?

Imagine that $[\tau]$ is plotted along a horizontal axis, $[t]$ along the vertical axis, and:

$$\operatorname{tg}[\tau]\cdot\operatorname{tg}[t]$$

plotted perpendicular to this plane as the **ordinates of a surface** Then we know of this surface that it **cuts** the $|[\tau]\,[t]$**-plane along** vertical and horizontal lines (shown in Figure 8), the first passing through the abscissae o, π, 2π, 3π, etc., the others passing through the ordinates o, π, 2π, 3π.

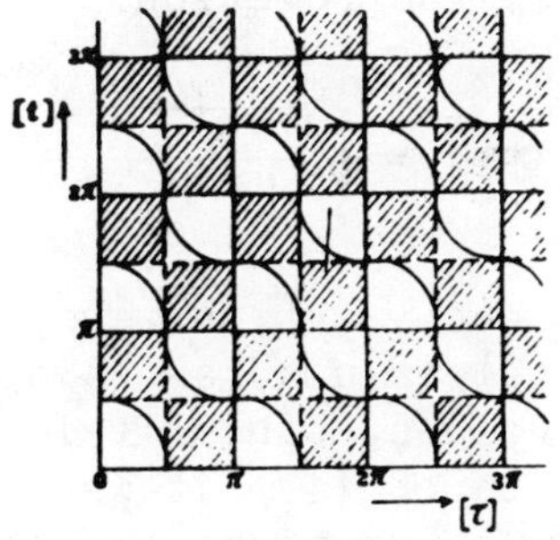

Figure 8

A similar lattice of dotted lines arising from the above through a horizontal displacement of $\pi/2$ and a vertical displacement through the same interval indicates the points of our plane at which the surface has infinitely large ordinates. Now the ordinates $\operatorname{tg}[\tau]$, $\operatorname{tg}[t]$ are alternately positive and negative. To make this clear, those squares on which the surface has negative ordinates are hatched. Thus Figure 8 takes on a checkerboard pattern in which on unhatched squares the ordinates take all positive values between 0 and ∞ and in such a way that the 0 values lie on the full lines and the infinite values above **the** dashed lines.

The right member of equation (56'):

$$\frac{1}{\operatorname{tg}\tau\operatorname{tg}t}$$

always has positive values in the whole region to be examined, since t, like τ, is less than $\pi/2$. Now equation (56') defines a series of curves in the $[\tau][t]$-plane which, since the right term is positive, pass through the unhatched squares; these curves are shown (schematically) in Figure 8. On any one of these curves, the first equation (56') is fulfilled.

Besides equation (56'), equation (57) must be fulfilled. The totality of points at which the latter equation is satisfied also defines a curve in the $[\tau][t]$-plane. Then the intersections of this curve with the curves defined by equation (56') give the values of $[\tau]$ and $[t]$ which satisfy (56') and (57) simultaneously and thus belong to an eigenfrequency.

If $\tau = 0$, then $[\tau] = -\pi/4$, just as when $t = 0$, $[t] = -\pi/4$. Since, however, in $[\tau]$ the quantity $(\mathrm{tg}\tau - \tau)$ is multiplied by the very large quantity $n + \frac{1}{2}$, the value of τ for which $[\tau] = 0$ is exceedingly small. The same holds for $[t]$ and t, so that without noticeable error it can be assumed that $\tau = 0$ and $[\tau] = 0$ simultaneously and that $t = 0$ and $[t] = 0$ simultaneously. According to (57), there exists a value of τ, which will be denoted by τ_1, for which $t = 0$; it is given by the relation:

$$\cos \tau_1 = \sqrt{\frac{\mu}{\lambda + 2\mu}}$$

Thus if t and consequently $[t]$ is zero, $[\tau]$ has the finite value $[\tau_1]$. As long as (49) is used as an approximation, τ cannot be allowed to take small values, since it is assumed that τ and t are real. In the derivation of (49) it was expressly assumed that the arguments of the two ψ functions are larger than the index, and according to the definition of τ and t this is covered by the assumption that τ and t are both real.

The purpose of Figure 9 is to illustrate the curve defined by (57).

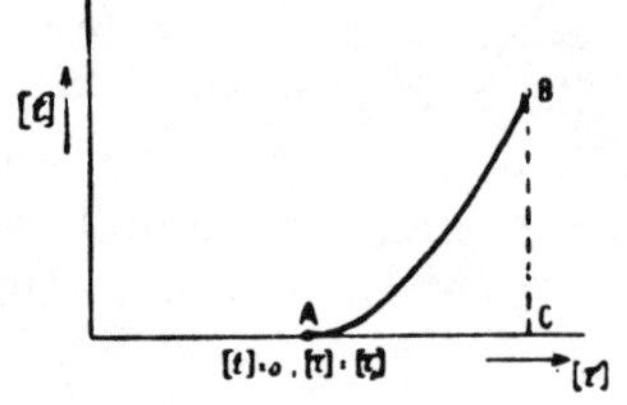

Figure 9

As was just explained, the **curve** begins at a point A with the coordinates 0 and $[\tau_1]$ and ends at a point B with the coordinates $[\tau_0]$ and $[t_0]$. This curve cannot be plotted in Figure 8 since these coordinates are very large because of the assumed magnitude of n. On the other hand, if the checkerboard structure of Figure 8 is imagined superimposed on Figure 9, the scale of the latter makes the dimensions of the squares extremely small. Imagine that this transformation has been carried out and also that the curves of Figure 8 have been marked in. Then the number of roots is equal to the number of intersections of the curve AB with those running through the unhatched squares. Now since twice the breadth of a panel of the checkerboard is π, the number of roots is obtained if the lengths AC and CB of Figure 9 are added and divided by π. Call this number of roots belong to the index n z_n'; then:

$$(58) \qquad z_n' = \frac{AC + CB}{\pi} = \frac{[t_0] + [\tau_0] - [\tau_1]}{\pi}$$

However, z_n' is only the number of roots of the approximate equation (56'), not the number of roots of the equation derived originally. We know that our functions ψ_n vanish exponentially (without oscillation) if the argument is less than the index. However, (54) covers the whole remaining region for the factor:

$$\frac{y \dfrac{d}{dy} \dfrac{\psi_n(y)}{y}}{\psi_n(y)/y}$$

since t starts from 0. For the other factor:

$$\frac{\psi_n'(x)}{\psi_n(x)}$$

there remains an uninvestigated region which belongs to the interval from $\tau = 0$ to $\tau = \tau_1$ and in which roots are actually present.

For this region $y < n$; on the other hand, x is always $> n$. In replacing (54) by an approximate equation suitable for this case, we can keep our earlier approximation for $\psi_n(x)$:

$$(49) \qquad \begin{cases} \psi_n(x) = \dfrac{1}{\sin^{1/2}\tau} \cos\left[(n + \frac{1}{2})(\operatorname{tg}\tau - \tau) - \dfrac{\pi}{4}\right] \\ \cos\tau = \dfrac{n + \frac{1}{2}}{x} \end{cases}$$

However, because $y < n$, $\psi_n(y)$ is approximated by:[12]

$$(59)\qquad \begin{cases} \psi_n(y) = \dfrac{1}{\mathfrak{Sin}^{1/2}\, t}\, e^{-(n+1/2)(\mathfrak{Tg}\, t - t)} \\ \mathfrak{Cos}\, t = \dfrac{n+\frac{1}{2}}{y} \end{cases}$$

in which Cos and Tg mean hyperbolic functions and t is positive.

From this it is found easily by differentiation that:

$$\frac{y \dfrac{d}{dy} \dfrac{\psi_n(y)}{y}}{\psi_n(y)/y} = -(n+\tfrac{1}{2})\,\mathfrak{Tg}\, t$$

and finally, that equation (54) takes the form:

$$(60)\qquad \operatorname{tg}\left[(n+\tfrac{1}{2})(\operatorname{tg}\tau - \tau) - \frac{\pi}{4}\right] = \frac{n(n+1)}{(n+\frac{1}{2})^2}\,\frac{1}{\operatorname{tg}\tau\,\mathfrak{Tg}\, t}$$

If again, as earlier, the square bracket is written simply $[\tau]$ and $n(n+1)/(n+\frac{1}{2})^2$ is replaced by 1, the frequency equation is:

$$(60')\qquad \operatorname{tg}[\tau] = \frac{1}{\operatorname{tg}\tau\,\mathfrak{Tg}\, t}$$

In each interval of length π, $\operatorname{tg}[\tau]$ runs through all values from $\tau = 0$ to $\tau = \tau_1$; in such an interval the function takes the value prescribed by the right side once and only once. Thus the number of roots which (60') possesses is obtained if the length of the interval of $[\tau]$ values is divided by π. If this number is called z_n'', then:

$$(61)\qquad z_n'' = \frac{[\tau_1]}{\pi}$$

With this all possibilities are disposed of and for case (b) the total number of roots is found from (58) and (61) and is equal to:

$$(62)\qquad z_n' + z_n'' = \frac{[t_0] + [\tau_0]}{\pi}$$

Summary of the Results of These Paragraphs

A state of vibration whose dependence on the angles ϑ and φ is given by a surface harmonic of the nth order $S_n(\vartheta, \varphi)$ possesses two kinds of motion, (a) and (b).

We now fix an upper limit for the frequency-say ω_0-and define two angles τ_0 and t_0 by means of this frequency using the equations:

$$(63)\qquad \begin{cases} \cos\tau_0 = \dfrac{n+\frac{1}{2}}{x_0} = \dfrac{n+\frac{1}{2}}{\alpha_0 a} = \dfrac{n+\frac{1}{2}}{a\,\omega_0}\sqrt{\dfrac{\mu}{\varrho}}, \\ \cos t_0 = \dfrac{n+\frac{1}{2}}{y_0} = \dfrac{n+\frac{1}{2}}{\beta_0 a} = \dfrac{n+\frac{1}{2}}{a\,\omega_0}\sqrt{\dfrac{\lambda+2\mu}{\varrho}} \end{cases}$$

where it is assumed that ω_0 is so large that the angles τ_0 and t_0 are real. Besides this, it will be assumed that they lie in the interval 0 to $\pi/2$.

Then the number of eigenfrequencies of the state (a) is:

$$(64)\qquad z_n = \frac{[\tau_0]}{\pi}$$

and of states (b):

$$(64')\qquad z_n' + z_n'' = \frac{[t_0] + [\tau_0]}{\pi}$$

where $[\tau_0]$ is an abbreviation for:

$$(65)\qquad [\tau_0] = (n+\tfrac{1}{2})(\operatorname{tg}\tau_0 - \tau_0) - \frac{\pi}{4}$$

where because of the size of n:

$$(65')\qquad [\tau_0] = (n+\tfrac{1}{2})(\operatorname{tg}\tau_0 - \tau_0)$$

is just as good as (65). The quantity $[t_0]$ is defined in a similar way.

X. Number of Degrees of Freedom which Belong to a Frequency Interval

An upper limit ν_0 is placed on the frequencies to be investigated, so that:

$$\omega_0 = 2\pi\nu_0$$

and answer the question: How large is the number of constants which must be given in order to completely define an arbitrary state of vibration of our sphere in which the individual frequencies do not exceed the limit ν_0?

We can call half of this number the number of degrees of freedom Z of the sphere below the frequency ν_0. By this definition, a simple linear dipole is a system of one degree of freedom.

Again consider the states of motion (a) and (b) and the numbers Z belonging to them, Z_a and Z_b.

(a) It was found above (compare equation 69) that to one term of the series belongs:

$$z_n = \frac{[\tau_0]}{\pi}$$

eigenfrequencies. However, if one of the eigenfrequencies is fixed, each term contains $2n$ arbitrary constants. Thus, to a complete definition of a state given by $S_n(\vartheta, \varphi)$ belong:

$$2n z_n = \frac{2n}{\pi}[\tau_0] \tag{66}$$

constants in all; however, only in so far as the potential can be put proportional to cos ωt. However, if this is done, the phase of the motion has been determined arbitrarily. To avoid this it is necessary to add a similar term proportional to sin ωt whereby the number of constants is doubled. Thus according to our definition of a degree of freedom (66) gives their number directly, in so far as they "belong to a surface harmonic $S_n(\vartheta, \varphi)$."

Then the total number of degrees of freedom of the states (a) is

$$Z_a = \sum 2n z_n = \sum \frac{2n}{\pi}[\tau_0] \tag{67}$$

The sum is to be taken over all values of n for which τ_0 belongs to the interval $0 < \tau_0 < \pi/2$. According to (63):

$$\cos \tau_0 = \frac{n+\frac{1}{2}}{a\,\omega_0}\sqrt{\frac{\mu}{\varrho}} = \frac{n+\frac{1}{2}}{2\pi a \nu_0}\sqrt{\frac{\mu}{\varrho}}$$

or since n is large:

$$\cos \tau_0 = \frac{n}{2\pi a \nu_0}\sqrt{\frac{\mu}{\varrho}} \tag{68}$$

that is:

$$n = 2\pi a \nu_0 \sqrt{\frac{\varrho}{\mu}} \cos \tau_0 \tag{68'}$$

With these values for n, Z_a is equal to:

$$Z_a = 4 a \nu_0 \sqrt{\frac{\varrho}{\mu}} \sum \cos \tau_0 \cdot [\tau_0] \tag{67'}$$

On the other hand:

$$[\tau_0] = (n + \tfrac{1}{2})(\operatorname{tg} \tau_0 - \tau_0)$$

If again $n + \frac{1}{2}$ is replaced by n and n by its expression in τ_0, then:

$$Z_a = 8\pi a^2 \nu_0{}^2 \frac{\varrho}{\mu} \sum \cos^2 \tau_0 (\operatorname{tg} \tau_0 - \tau_0) \tag{67''}$$

The sum is to be understood this way: in it the values of τ_0 are substituted successively which belong to successive integral values of n. In going from one term of the series to the next, τ_0 changes very little. Therefore, according to (68), the corresponding interval in n, which is of the magnitude 1, can be written:

$$1 = \Delta n = -2\pi a \nu_0 \sqrt{\frac{\varrho}{\mu}} \sin \tau_0 \Delta \tau_0$$

Substitution of these values changes our sum into an integral:

$$Z_a = -16\pi^2 a^3 \nu_0{}^3 \left(\frac{\varrho}{\mu}\right)^{3/2} \int_{\pi/2}^{0} \sin \tau_0 \cos^2 \tau_0 (\operatorname{tg} \tau_0 - \tau_0) \, d\tau_0 \tag{69}$$

At the lower limit of the integral $n = 0$, so that $\cos \tau_0 = 0$ there, and $\tau_0 = \pi/2$; the upper limit corresponds to the other end of the interval where:

$$\frac{n}{2\pi a \nu_0} \sqrt{\frac{\mu}{\varrho}}$$

has the value 1.

The integral can be evaluated easily; it is found that:

$$\int_0^{\pi/2} \sin \tau_0 \cos^2 \tau_0 (\operatorname{tg} \tau_0 - \tau_0) \, d\tau_0 = \frac{1}{9}$$

and thus finally:

$$Z_a = \frac{16\pi^2}{9} a^3 \nu_0{}^3 \left(\frac{\varrho}{\mu}\right)^{3/2} \tag{70}$$

Calling the volume of the sphere $V = 4/3\, \pi\, a^3$, (70) can be rewritten:

$$(70') \qquad Z_a = \frac{4\pi}{3} V \nu_0^3 \left(\frac{\varrho}{\mu}\right)^{3/2}$$

that is, the number of degrees of freedom is proportional to the volume of the sphere and the third power of the fixed maximum frequency ν_0.

(b) Just as above, according to (64''):

$$2n(z_n' + z_n'') = \frac{2n}{\pi}([t_0] + [\tau_0])$$

degrees of freedom belong to a surface harmonic $S_n(\vartheta, \varphi)$. The total number of degrees of freedom is:

$$(71) \quad Z_b = \sum \frac{2n}{\pi}([t_0] + [\tau_0]) = \sum \frac{2n}{\pi}[t_0] + \sum \frac{2n}{\pi}[\tau_0]$$

The second sum is calculated above; it was found that:

$$(72) \qquad \sum \frac{2n}{\pi}[t_0] = Z_a = \frac{4\pi}{3} V \nu_0^3 \left(\frac{\varrho}{\mu}\right)^{3/2}$$

The first sum can be calculated exactly in the same way since it differs from the earlier case only in that μ is replaced by $\lambda + 2\mu$. It is:

$$(73) \qquad \sum \frac{2n}{\pi}[t_0] = \frac{4\pi}{3} V \nu_0^3 \left(\frac{\varrho}{\lambda + 2\mu}\right)^{3/2}$$

Adding (70'), (72), and (73), we obtain our final result:

The number of degrees of freedom of an elastic body of volume V whose properties can be characterized by two elastic constants λ and μ is:

$$(74) \qquad Z = \frac{4\pi}{3} V \nu^3 \left[2\left(\frac{\varrho}{\mu}\right)^{3/2} + \left(\frac{\varrho}{\lambda + 2\mu}\right)^{3/2}\right]$$

where only eigenvibrations are taken into account whose frequencies are smaller than an arbitrary fixed frequency ν.

In order to connect this directly with the formula of the first part, we can replace λ and μ by the constants used in practice, the compressibility κ and Poisson's ratio σ.

In section VI, it was shown that if E is Young's modulus:

$$\lambda = E \frac{\sigma}{(1+\sigma)(1-2\sigma)} \quad \text{and} \quad \mu = \frac{E}{2(1+\sigma)}$$

while:

$$\varkappa = \frac{3(1-2\sigma)}{E}$$

Thus we can write:

$$\lambda = \frac{1}{\varkappa}\frac{3\sigma}{1+\sigma}, \quad \mu = \frac{1}{\varkappa}\frac{3(1-2\sigma)}{2(1+\sigma)}$$

and:

$$\lambda + 2\mu = \frac{1}{\varkappa}\frac{3(1-\sigma)}{1+\sigma}$$

When these are substituted in (74) it is found that:

$$(74') \quad Z = \frac{4\pi}{3} V \nu^3 \varrho^{3/2} \varkappa^{3/2} \left[2\left(\frac{2(1+\sigma)}{3(1-2\sigma)}\right)^{3/2} + \left(\frac{1+\sigma}{3(1-\sigma)}\right)^{3/2} \right]$$

and this is the result used in Part I. Earlier, the square bracket was denoted by $f(\sigma)$ for brevity.

Summary

(1) For a solid, there exists no simple frequency of the atoms as originally supposed by Einstein in calculating the specific heat. The solid can be characterized only by a complete spectrum of eigenfrequencies.

(2) The spectrum possesses a finite number of lines (equal to three times the number of atoms). The lines of least frequency are the ordinary acoustic vibrations.

(3) The spectrum can be characterized by the density of spectral lines per frequency interval $d\nu$. We find that this is proportional to $\nu^2 d\nu$. The proportionality factor can be calculated from the elastic constants of the material.

(4) By use of (3), formulae for the energy content and specific heats can be derived if to each degree of freedom the energy:

$$\frac{h\nu}{e^{\frac{h\nu}{kT}} - 1}$$

is ascribed, as in quantum theory.

(5) This procedure gives an expression for the specific heat of a monatomic solid which depends only on the ratio θ/T, where θ is a temperature characteristic of the body in question. The specific heat of a monatomic solid is therefore a universal function of the ratio θ/T.

(6) For low temperatures, as can be concluded from the resulting formulae (or also directly from 3), the specific heat is proportional to T^3 for all substances. The energy content is then proportional to T^4, just as it is for radiation according to the Stefan-Boltzmann law for all temperatures. This limiting law shows especially simply the difference between our formula and the Einstein or Nernst-Lindemann formula. According to the latter two, the specific heat vanishes exponentially at low temperatures.

(7) Comparison of our formula with the observations on diamond, aluminum, copper, silver, and lead shows a very good agreement between experiment and theory.

(8) In a certain sense, the formula of Nernst and Lindemann is an approximation to our function (compare section III). Thus this explains why in earlier investigations this formula showed such a good agreement with experiment.

(9) The characteristic temperature θ can be calculated from the elastic constants and the results are good. Aside from a numerical factor, our formula differs from that given earlier by Einstein since it contains not only the compressibility but Poisson's ratio as well.

Bibliography

1. A. Einstein, *Ann. Physik*, *22*, 180 (1907).
2. M. Planck, *Wärmestrahlung*, Leipzig, 1906, p. 157.
3. W. Nernst and Lindemann, *Z. Elektrochemie*, *1911*, 817; *Berl. Ber.* *1910*, 26.
4. A. Einstein, *Ann. Physik*, *35*, 679 (1911).
5. Compare M. Planck, *Wärmestrahlung*, Leipzig, 1906, p. 161.
6. W. Nernst-Lindemann, *Z. Elektrochem.* *1911*, 817.
7. A. Einstein, *Ann. d. Phys.*, *34*, 170 (1911).
8. E. Gruneisen, *Ann. d. Phys.*, *22*, 838 (1907); *25*, 845 (1908).
9. Compare A. E. H. Love, *Lehrbuch der Elektrizität.* Translated by Timpe, Leipzig, 1907, p. 320, and the literature cited there.
10. Compare P. Debye, *Ann. d. Phys.*, *30*, 61 ff. (1909).
11. J. C. Maxwell, *Treatise on Electricity and Magnetism*, Oxford, 1881, Vol. 1, p. 179.
12. Compare P. Debye, *Math. Ann.*, *67*, 535 (1909).

PROPOSAL OF A NEW METHOD FOR DETERMINING MOLECULAR WEIGHTS OF POLYMERS* **

P. Debye and P.P. Debye, Jr.

The method to be proposed is an attempt to make proper use of the high degree of accuracy with which high frequencies can be measured nowadays.

When a solution containing a solute, which has a higher polarizability than the solvent, is exposed to the influence of a static and inhomogeneous electric field the solute molecules tend to move to those places where the field is strongest. In this tendency they are hindered by their thermal motion. In the course of time, the length of which depends essentially on the diffusion constant of the solute, the concentration distribution reaches an equilibrium. After this is established the solution is no longer homogeneous with respect to concentration and as a consequence will have a dielectric constant varying with the position in the solution, being highest at those places where the static field is strongest.

The theory of these effects predicts the following relations:

(a) Suppose the dielectric constant of the solvent to be ε_0 and that of the solution ε_1, and suppose we consider a point in the solution where the static field has the strength E (in electrostatic units). Under these circumstances the actual dielectric constant ε in this point will be:

$$\varepsilon = \varepsilon_1 \left[1 + \frac{E^2}{F^2}\right] \tag{1}$$

The change in dielectric constant at the point considered is proportional to the square of the field-strength E. In order to obtain a numerical value for the effect on the dielectric constant

*Cornell University, February, 1953.

**The work discussed herein was performed as part of the research project sponsored by the Reconstruction Finance Corporation, Office of Synthetic Rubber, in connection with the Government Synthetic Rubber Program.

we have to compare E with a characteristic field strength F, which is determined by the properties of solute and solvent.

How F depends on these properties can be derived from the relation:

$$F^2 = \frac{\varepsilon_1}{\varepsilon_1 - \varepsilon_0} \frac{2kT}{\alpha} \qquad (2)$$

where: ε_0 is the dielectric constant of the solvent;
ε_1 is the dielectric constant of the solution;
kT measures the thermal energy (T is absolute temperature and k is Boltzmann's constant, $k = 1.37 \times 10^{-16}$ ergs);
α is the polarizability of a solute molecule as it appears in the solution surrounded by solvent molecules. Its value can be derived from the observed difference between ε_1 and ε_0 (difference between solution and solvent) according to the relation:

$$\varepsilon_1 - \varepsilon_0 = 4\pi n\alpha \qquad (3)$$

in which n is the number of solute molecules per c.c.

It has been assumed moreover that the application of the field E has not yet distorted the molecule so much that its original polarizability has changed appreciably. This assumption is open to discussion and must be weighed experimentally.

For a polymer solution of a given concentration (in gr. per c.c.) the difference $\varepsilon_1 - \varepsilon_0$ does not depend on the molecular weight, kT is a constant at a given temperature, finally α, the polarizability of the solute molecule, is proportional to the degree of polymerization or the molecular weight. Relation (2) therefore shows that F^2 is proportional to the reciprocal of the molecular weight, which makes the change produced in the dielectric constant by an applied field E directly proportional to the molecular weight according to (1).

In order to arrive at an estimate of the expected effect we have assumed a solvent like benzene with an index of refraction equal to 1.5. We have further assumed that in this medium a nonpolar polymer is dissolved to a concentration of 1%, which increases the refractive index by one unit in the third decimal. If finally it is assumed that this polymer has a molecular weight of 100,000 the characteristic field F becomes equal to 1.2×10^5 e.s. units or 36 million volts/cm. Since the field E, which can practically be applied, is certainly much smaller, it is evident that only a very small change in dielectric constant can be expected. So if we want to try and make use of this change for measuring molecular weights we will have to apply a method of observation which is highly sensitive to such a change.

(b) The beat method, in which the beats between the vibrations of a constant and a variable high frequency circuit are counted, is such a method. If the frequency is for instance 5 megacycles, measuring a change of the order 10^{-6} in frequency seems quite feasible with modern equipment.

The frequency of a circuit (number of vibrations per sec.) is $\frac{1}{2}\pi$ times the reciprocal of the square root of the product: self-induction L times capacity C. Therefore an arrangement is indicated which contains a capacitor filled with the solution and of such geometry that an inhomogeneous field is established. When a constant high potential is applied to this condensor a shift in frequency occurs. This shift is due to a capacity change, which in itself is a result of the fact that the medium which was homogeneous before the application of the static field becomes inhomogeneous with respect to concentration following the application.

As such a condensor we have adopted for the time being a cylinder-condensor consisting of a central wire with a diameter of about 0.02 cm. in an outer cylinder with a diameter of about 2 cm.

The theory of this arrangement shows that, when we denote the radius of the wire by a and the radius of the outer cylinder by A and apply a potential difference of V e.s. units, the quotient of the capacity C after application and the capacity C_0 before application is:

$$\frac{C}{C_0} = 1 + \frac{1}{2} \frac{1}{(\ln \frac{A}{a})^3} \frac{V^2/a^2}{F^2} \left(1 - \frac{a^2}{A^2}\right) \tag{4}$$

Considering the solution of our example under (a) for which F was 1.2×10^5 e.s. units and supposing that a potential V of 9000 Volts = 30 e.s. units is applied, relation (4) yields:

$$\frac{C}{C_0} = 1 + 3.2 \times 10^{-6}$$

If we had considered a solution containing the same kind of polymer at the same concentration of 1% but with a molecular weight 10 times larger, namely one million, the calculated change in capacity would also be 10 times larger. This change can not be expected to be instantaneous. If it is possible to follow the establishment of the equilibrium this will yield additional information concerning the diffusion constant of the polymer molecules and possibly lead to a direct method for determining molecular weight distributions.

(c) Disturbing effects, due to an original excentric position of the central wire, or to electrostriction of the solution, or to the Kerr effect, or to heating as a result of a very small conductivity of the solvent have been considered. They are or can be made too small to be of importance.

During the past six months P. P. Debye has assembled equipment which seems adequate for the purpose. A frequency of 5 megacycles is used, beats are counted by a so-called EPUT (Events per Unit Time) Meter, which automatically counts the number of beats either during 1 or during 10 seconds. The voltage has been 7500 volts.

Taking into account that due to deviations from ideality of the polymer solutions the effect at higher concentrations does not increase proportionally to the concentration, frequency changes of the correct order of magnitude and direction have been observed in solutions of polystyrene in benzene and in cyclohexane.